**Volume 2**

# GENETIC RESOURCES, CHROMOSOME ENGINEERING, AND CROP IMPROVEMENT

## Cereals

**GENETIC RESOURCES, CHROMOSOME ENGINEERING, AND CROP IMPROVEMENT SERIES**

Series Editor, Ram J. Singh

Volume 2

# GENETIC RESOURCES, CHROMOSOME ENGINEERING, AND CROP IMPROVEMENT

## Cereals

EDITED BY

RAM J. SINGH AND PREM P. JAUHAR

CRC Press
Taylor & Francis Group
Boca Raton London New York

CRC Press is an imprint of the
Taylor & Francis Group, an **informa** business

CRC Press
Taylor & Francis Group
6000 Broken Sound Parkway NW, Suite 300
Boca Raton, FL 33487-2742

First issued in paperback 2019

© 2006 by Taylor & Francis Group, LLC
CRC Press is an imprint of Taylor & Francis Group, an Informa business

No claim to original U.S. Government works

ISBN-13: 978-0-8493-1432-2 (hbk)
ISBN-13: 978-0-367-39125-6 (pbk)

---

**Library of Congress Cataloging-in-Publication Data**

---

Catalog record is available from the Library of Congress

---

**Visit the Taylor & Francis Web site at**
**http://www.taylorandfrancis.com**

**and the CRC Press Web site at**
**http://www.crcpress.com**

# Dedication

And, he gave it for his opinion that whoever could make two ears of corn or two blades of grass to grow upon a spot of land where only one grew before, would deserve better of mankind, and do more essential service to humanity, than the full race of politicians put together.

**—Swift**

Our efforts in the preparation of this volume are dedicated to our wives but for whose patience and sacrifice this volume would not have been completed.

**—The Editors**

# Preface

Cereal crops — chiefly wheat, rice, maize, sorghum, and pearl millet — are the main food source for more than two thirds of the world population. From time immemorial humans have relied heavily on cereals for their dietary carbohydrates. Thus, cereals have had a profound impact on the development of human societies and influenced civilization — perhaps in more ways than any other group of crops. Ancient Egyptian tomb paintings depict cultivation of wheat and harvesting and grinding of wheat grain to make bread. Today, cereals supply over 80% of the dietary protein for most Asian and African countries. Being devoid of cholesterol, cereal grains provide wholesome food for human consumption, and there is an inverse association between intake of whole grains and cardiovascular disease (simin.liu@channing.harvard.edu). Severe protein malnutrition among the poor masses in Asian and African countries, where cereals constitute the staple human diet, is a serious problem of alarming proportions. Some 842 million people worldwide are malnourished (fao.org/newsroom/en/news/2004), and this number is likely to increase with the projected increase in human population from 6.1 billion to 8.0 billion by 2030. To meet the ever-increasing demand for food, genetic improvement of grain yields and nutritive value of cereal crops cannot be adequately emphasized.

Most improvement in cereal crops has been achieved so far through conventional breeding, aided to some extent by knowledge from agronomy, cytogenetics, plant pathology, entomology, and related disciplines. The improved wheat and rice cultivars in the 1960s and 1970s launched the Green Revolution, averting starvation among the poor masses in Asia. Sustained improvement in grain yields and nutritional quality must remain the ultimate goal of plant scientists to ensure global food security. Continued crop improvement will necessitate the employment of all available tools: germplasm collection and conservation, conventional breeding, cytogenetics, biotechnology, and molecular genetics, among others. Improving yields and nutrition of cereal crops have been the primary goals of international centers like the International Maize and Wheat Improvement Center (CIMMYT) in Mexico (maize and wheat); the International Rice Research Institute (IRRI) in the Philippines (rice); the International Crops Research Institute for the Semi-Arid Tropics (ICRI-SAT) in India (sorghum and pearl millet); and the International Center for Agricultural Research in the Dry Areas (ICARDA) in Syria (barley and wheat). Because there is no consolidated account of cereal crop improvement using conventional and modern tools, we planned to assemble such a book that constitutes Volume 2 in the series "Genetic Resources, Chromosome Engineering, and Crop Improvement." This book is also an outgrowth of a symposium on "Alien Gene Transfer and Cereal Crop Improvement" that one of us (P.P.J.) organized and chaired at the Crop Science Society of America Meetings, Salt Lake City, Utah, in November 1999. We invited world-renowned scientists from several countries to contribute chapters on a cereal crop of their expertise. This volume consists of 13 chapters dealing with major cereal crops: wheat (durum wheat and bread wheat), rice, maize, oat, barley, pearl millet, sorghum, rye, and triticale.

The introductory chapter by Jauhar outlines the cytogenetic architecture of cereal crops, describes the principles and strategies of cytogenetic and breeding manipulations, and summarizes the landmarks of research done on various crops. Thus, the author has attempted to set the stage for the reader to comprehend the ensuing chapters. Each chapter generally provides a comprehensive account of the crop; its origin; wild relatives; exploitation of genetic resources in the primary, secondary, and tertiary gene pools through breeding and cytogenetic manipulation; and genetic enrichment using the tools of molecular genetics and biotechnology. Durum wheat, being the forerunner of bread wheat, is dealt with first by Ceoloni and Jauhar in Chapter 2. Chapter 3 by Mujeeb-Kazi provides details on the utilization of genetic resources for bread wheat improvement, while wheat genomics is covered by Lapitan and Jauhar in Chapter 4. In Chapter 5, Brar and Khush give a comprehensive account of genetic resources and chromosome engineering in genetic improvement of rice. Genetic enhancement of maize for yield and protein quality is dealt with in

Chapter 6 by Vasal, Riera-Lizarazu, and Jauhar. The subsequent chapters deal with oat, barley, pearl millet, sorghum, rye, and triticale.

Each chapter provides an authoritative account of the topic covered and was written by one or more experts in the field. We are privileged to have known the authors both professionally and personally and are very grateful to them for their invaluable contributions. Certain topics and research organisms are closely related, which has inevitably led to some overlap and duplication among chapters, although repetitions were minimized by giving cross-references. Each chapter can be read independently in this coherent volume on cereals.

We are also grateful to all the scientists who reviewed various chapters. Our communications with them were always cordial and friendly. We are particularly indebted to Charles Crane, Pierre Devaux, Sally Dillon, Pat Hayes, Eric Jellen, Daryl Klindworth, Mike McMullen, Richard Pickering, and Richard Cross for critically reviewing some of the chapters. Although every chapter has been appropriately reviewed by the editors and other experts in the field, the authors are ultimately responsible for the accuracy and completeness of their respective chapters. One of us (R.J.S.) expresses his gratitude to Dr. Steven G. Pueppke, Associate Dean and Research Director, University of Illinois, Urbana, for all his support and encouragement. Prem Jauhar is sincerely grateful to his wife, Raj Jauhar, for her help, patience, and understanding; she spent countless weekends and evenings at home alone when he was at work. But for her encourgement and unconditional support, this arduous journey would have been even more difficult.

This book is intended for scientists, professionals, and graduate students interested in genetic improvement of crops in general and cereals in particular. The book will be useful for plant breeders, agronomists, geneticists, cytogeneticists, taxonomists, evolutionists, molecular biologists, and biotechnologists. Graduate-level students in these disciplines with adequate background in genetics and a spectrum of other researchers interested in biology and agriculture will also find this volume a worthwhile reference. We sincerely hope that the information embodied in the book will help in the much-needed genetic amelioration of cereal crops to feed the ever-expanding human population. In addition, we hope that it helps to raise awareness of the importance of conserving wild genetic resources that may be exploited in improving their cultivated cereal relatives through cytogenetics and biotechnology.

**Ram J. Singh**
*Urbana-Champaign, Illinois*

**Prem P. Jauhar**
*Fargo, North Dakota*

# The Editors

**Ram Jag Singh, M.Sc., Ph.D.**, is an agronomist-plant cytogeneticist in the Department of Crop Sciences, University of Illinois at Urbana-Champaign. He received his Ph.D. degree in plant cytogenetics under the guidance of the late Professor Takumi Tsuchiya from Colorado State University, Fort Collins, Colorado. He benefited greatly from Dr. Tsuchiya's expertise in cytogenetics.

Dr. Singh conceived, planned, and conducted pioneering research related to cytogenetic problems in barley, rice, rye, wheat, and soybean. Thus, he isolated monotelotrisomics and acrotrisomics in barley, identified them by Giemsa C- and N-banding techniques, and determined chromosome arm-linkage group relationships. In soybean (*Glycine max*), he produced fertile plants with $2n = 40$, 41, or 42 chromosomes, from an intersubgeneric cross between soybean and a wild species, *Glycine tomentella* ($2n = 78$), and obtained certain lines with resistance to the soybean cyst nematode (SCN). Singh constructed, for the first time, a soybean chromosome map based on pachytene chromosome analysis and laid the foundation for creating a global soybean map. By using fluorescent genomic *in situ* hybridization he confirmed the tetraploid origin of the soybean.

Singh has published 67 research papers, mostly in reputable international journals, including the *American Journal of Botany, Chromosoma, Crop Science, Genetics, Genome, Journal of Heredity, Plant Breeding*, and *Theoretical and Applied Genetics*. In addition, he summarized his research results by writing nine book chapters. His book *Plant Cytogenetics* is widely used for teaching graduate students. Dr. Singh has presented research findings as an invited speaker at national and international meetings. He is a member of the Crop Science Society of America and the American Society of Agronomy. In 2000, he received the Academic Professional Award for Excellence: Innovation and Creativity from the University of Illinois at Urbana-Champaign.

**Prem Prakash Jauhar, M.Sc., Ph.D.**, is a senior research geneticist with the U.S. Department of Agriculture–Agricultural Research Service, Northern Crop Science Laboratory, Fargo, North Dakota. He also holds the position of professor of cytogenetics with North Dakota State University, Fargo. He is the principal investigator on the USDA project "Genomic Relationships in the Triticeae and Enhancement of Wheat Germplasm by Classical and Molecular Techniques."

Prem earned his Ph.D. from the Indian Agricultural Research Institute, New Delhi, in 1963 when he was appointed to the faculty of this institute, doing research and teaching cytogenetics to graduate students. From 1972 to 1975, he served as a senior scientific officer at the University College of Wales, Welsh Plant Breeding Station, Aberystwyth, Wales, U.K.

Prem Jauhar's research interests have centered on various facets of cytogenetics and biotechnology and their relevance to plant breeding. He has been particularly interested in chromosome pairing. In 1975, he discovered the regulatory mechanism that controls chromosome pairing in polyploid species of *Festuca* (*Nature* 254: 595–597, 1975) and originated the concept of hemizygous-ineffective genetic control of pairing — a phenomenon that has major implications in cytogenetics, plant breeding, and evolution. After establishing an efficient *in vitro* regeneration system for durum wheat, Jauhar's lab produced the first transgenic durum wheat and standardized the technology in 1996, paving the way for direct gene transfer into commercial durum cultivars. Jauhar is also involved in germplasm enhancement by genomic reconstruction through wide hybridization coupled with manipulation of homoeologous chromosome pairing. By transferring a part of a wild grass chromatin into the durum wheat genome, Jauhar produced durum germplasm with scab resistance.

Working on *Ph1*- and *ph1*-euhaploids in bread wheat ($2n = 3x = 21$; ABD genomes) and durum wheat ($2n = 2x = 14$; AB genomes) that he synthesized, Jauhar elucidated inter- and intragenomic relationships in these polyploid wheats. He demonstrated that the A and D genomes of bread wheat are more closely related to each other than either one is to B — a finding that contributed to the understanding of the phylogeny of wheat. Jauhar's haploidy research produced the first clear evidence of sexual polyploidization via 2n gamete formation in durum wheat haploids (*Crop Science* 40: 1742–1749, 2000), demonstrating how polyploids are produced in nature.

Jauhar has published in international journals, including *Nature, Chromosoma, Theoretical and Applied Genetics, Genome, Journal of Heredity, Genetica, Plant Breeding*, and *Mutation Research*. He has 120 publications, including 90 research papers, 3 books (two authored and one coauthored and edited), and 17 invited book chapters. His research papers and books are used in graduate teaching and research worldwide. Jauhar has a keen interest in disseminating science and serving the scientific community. He has given invited seminars in 15 countries, organized and chaired symposia and scientific sessions at national and international conferences, and served on international advisory committees. Most recently, he was a keynote speaker at the National Symposium on Classical Cytogenetics and Modern Biotechnology organized by the Centre for Advanced Study in Cell and Chromosome Research, Calcutta University, January 24–25, 2005. He also delivered the Panchanan Maheshwari Memorial Lecture at the centennial celebrations of the legendary scientist's birth, held at Delhi University, February 15, 2005. Since 1991, Prem Jauhar has served as an associate editor of the *Journal of Heredity*.

Prem P. Jauhar has received several awards and professional recognitions. Some recent awards include his election as Fellow of the Crop Science Society of America (1995), the American Society of Agronomy (1996), and the American Association for the Advancement of Science (2002).

# Contributors

**D.S. Brar**
Plant Breeding, Genetics and Biochemistry
  Division
International Rice Research Institute
Los Baños, Philippines

**Carla Ceoloni**
Department of Agrobiology and Agrochemistry
University of Tuscia
Viterbo, Italy

**Zhenbang Chen**
Department of Crop and Soil Sciences
Georgia Experiment Station
Griffith, Georgia

**Jose M. Costa**
Department of Natural Resource Sciences and
  Landscape Architecture
University of Maryland
College Park, Maryland

**Wayne W. Hanna**
Department of Crop and Soil Sciences
University of Georgia
Tifton, Georgia

**Prem P. Jauhar**
USDA–Agricultural Research Service
Northern Crop Science Laboratory
Fargo, North Dakota

**Eric N. Jellen**
Department of Plant and Animal Sciences
Brigham Young University
Provo, Utah

**G.S. Khush**
Plant Breeding
Genetics and Biochemistry Division
International Rice Research Institute
Los Baños, Philippines

**Nora Lapitan**
Department of Soil and Crop Sciences
Colorado State University
Fort Collins, Colorado

**J. Michael Leggett**
Institute for Grassland and Environmental
  Research
Welsh Plant Breeding Station
Aberystwyth, Dyred, Wales, U.K.

**Tamás Lelley**
University of Natural Resources and Applied
  Life Science
Vienna Department of Agrobiotechnology
Vienna, Austria

**A. Mujeeb-Kazi**
International Maize and Wheat Improvement
  Center (CIMMYT)
Mexico, D.F., Mexico

**Peggy Ozias-Akins**
Department of Horticulture
University of Georgia Tifton Campus
Tifton, Georgia

**Kedar N. Rai**
International Crops Research Institute for the
  Semi-Arid Tropics (ICRISAT)
Andhra Pradesh, India

**S. Ramesh**
International Crops Research Institute for the
  Semi-Arid Tropics (ICRISAT)
Andhra Pradesh, India

**Belum V.S. Reddy**
International Crops Research Institute for the
  Semi-Arid Tropics (ICRISAT)
Andhra Pradesh, India

**P. Sanjana Reddy**
International Crops Research Institute for the
  Semi-Arid Tropics (ICRISAT)
Andhra Pradesh, India

**Oscar Riera-Lizarazu**
Department of Crop and Soil Science
Oregon State University
Corvallis, Oregon

**Rolf Schlegel**
Gatersleben, Germany

**Ram J. Singh**
Department of Crop Sciences
University of Illinois
Urbana, Illinois

**Surinder K. Vasal**
Sasakawa Africa Association
International Maize and Wheat Improvement
  Center (CIMMYT)
Mexico, D.F., Mexico

# Contents

**Chapter 1**   Cytogenetic Architecture of Cereal Crops and Their Manipulation to Fit Human Needs: Opportunities and Challenges.................................................... 1
**Prem P. Jauhar**

**Chapter 2**   Chromosome Engineering of the Durum Wheat Genome: Strategies and Applications of Potential Breeding Value........................................27
**Carla Ceoloni and Prem P. Jauhar**

**Chapter 3**   Utilization of Genetic Resources for Bread Wheat Improvement........................61
**A. Mujeeb-Kazi**

**Chapter 4**   Molecular Markers, Genomics, and Genetic Engineering in Wheat.....................99
**Nora Lapitan and Prem P. Jauhar**

**Chapter 5**   Cytogenetic Manipulation and Germplasm Enhancement of Rice (*Oryza sativa* L.) ................................................................. 115
**D.S. Brar and G.S. Khush**

**Chapter 6**   Genetic Enhancement of Maize by Cytogenetic Manipulation, and Breeding for Yield, Stress Tolerance, and High Protein Quality........................ 159
**Surinder K. Vasal, Oscar Riera-Lizarazu, and Prem P. Jauhar**

**Chapter 7**   Cytogenetic Manipulation in Oat Improvement.................................................. 199
**Eric N. Jellen and J. Michael Leggett**

**Chapter 8**   Utilization of Genetic Resources for Barley Improvement ............................... 233
**Ram J. Singh**

**Chapter 9**   Chromosome Mapping in Barley (*Hordeum vulgare* L.)..................................... 257
**Jose M. Costa and Ram J. Singh**

**Chapter 10**   Genetic Improvement of Pearl Millet for Grain and Forage Production: Cytogenetic Manipulation and Heterosis Breeding ........................................... 281
**Prem P. Jauhar, Kedar N. Rai, Peggy Ozias-Akins, Zhenbang Chen, and Wayne W. Hanna**

**Chapter 11**   Sorghum Genetic Resources, Cytogenetics, and Improvement .......................... 309
**Belum V.S. Reddy, S. Ramesh, and P. Sanjana Reddy**

**Chapter 12**   Rye (*Secale cereale* L.): A Younger Crop Plant with a Bright Future............... 365
**Rolf Schlegel**

**Chapter 13**   Triticale: A Low-Input Cereal with Untapped Potential.................................... 395
**Tamás Lelley**

**Index** ......................................................................................................................... 431

# Cytogenetic Architecture of Cereal Crops and Their Manipulation to Fit Human Needs: Opportunities and Challenges

Prem P. Jauhar*

## CONTENTS

1.1   Introduction ................................................................................................................2
1.2   Cereal Crops: A Source of Sustenance to Humankind .........................................4
1.3   Polyploid Cereals: Their Cytogenetic Architecture .............................................5
    1.3.1   Polyploid Wheats: A Model for Evolution by Allopolyploidy .................5
        1.3.1.1   Durum Wheat: A Forerunner of Bread Wheat .........................6
        1.3.1.2   Establishment of Genetic Control of Chromosome Pairing ......6
        1.3.1.3   Origin of Bread Wheat: An Important Evolutionary Step .........7
    1.3.2   Cytogenetic Makeup of Hexaploid Oat .....................................................7
1.4   Genetic Control of Chromosome Pairing: Major Implications ...........................7
    1.4.1   Cytogenetic and Evolutionary Implications .............................................9
        1.4.1.1   Usefulness in Genome Analysis ...............................................9
        1.4.1.2   Gene Introgressions and Changes in Base Chromosome Numbers ..........10
    1.4.2   Breeding Implications: Homoeologous Pairing, the Key to Gene Transfer .............10
1.5   Cytogenetic Manipulation of Polyploid Cereal Crops .......................................11
    1.5.1   Wheat ..........................................................................................................11
        1.5.1.1   Specificity of Chromosome Pairing and Induction of Alien
                    Integration into Wheat ...............................................................12
    1.5.2   Oat ..............................................................................................................12
1.6   Diploid or Diploidized Cereals: Genomic Evolution .........................................12
    1.6.1   Cytogenetic Makeup and Ancient Polyploid Origin .............................12
    1.6.2   Diversity of Origin of Cereal Genomes: Genomic Diversity and Synteny .............14
    1.6.3   Cytogenetic Manipulation and Breeding Work in Diploid Cereals .........15
        1.6.3.1   Rice: A Model Cereal Crop .......................................................15
        1.6.3.2   Maize: A Cytogeneticist's Delight and a Breeder's Paradise .........15
        1.6.3.3   Pearl Millet: Poor Man's Bread: Heterosis Breeding ...............16
        1.6.3.4   Barley, Sorghum, Rye, and Triticale .......................................16

Mention of trade names or commercial products in this publication is solely to provide specific information and does not imply recommendation or endorsement by the U.S. Department of Agriculture.

* U.S. governement employee whose work is in the public domain.

1.7    Perspectives and Challenges ................................................................................................17
References ......................................................................................................................................19

## 1.1 INTRODUCTION

Cereals are members of the grass family Poaceae (Gramineae), whose seeds are used for food. The word *cereal* is derived from *Ceres*, the Greek goddess of agriculture. Cereal grains have been the staple human diet since prehistoric times. The cultivation of cereals for human consumption began around 10,000 B.C., ranking them as the earliest cultivated staple food plants of many human societies. Their cultivation signified the dawn of the era of stable civilization, which replaced the primitive nomadic way of life. Common cereals are wheat (bread wheat, *Triticum aestivum* L., and durum wheat, *Triticum turgidum* L.); rice (*Oryza sativa* L.); maize (*Zea mays* L. ssp. *mays*); oats (*Avena sativa* L.); barley (*Hordeum vulgare* L.); sorghum (*Sorghum bicolor* L. Moench); pearl millet (*Pennisetum glaucum* (L.) R. Brown = *Pennisetum typhoides* (Burm.) Stapf et Hubb.); rye (*Secale cereale* L.); and the man-made cereal, triticale (*Triticosecale* Wittmack). The total world production of cereal crops was 1835.2 million tonnes in 2002–2003, and is estimated at 1886.6 million tonnes in 2003–2004 (http://faostat.fao.org). Total world acreage of major cereal crops and their production are given in Figure 1.1 and Figure 1.2, respectively. Wheat, rice, and maize are undoubtedly the most important cereals worldwide.

Currently, the cereal crops are the main food source for more than two thirds of the world population. In most Asian and African countries, cereals supply over 80% of the dietary protein. Severe malnutrition among the poor masses poses a serious problem; some 842 million people worldwide are malnourished (fao.org/newsroom/2003). To meet the ever-growing demand for food, genetic improvement of cereal crops cannot be overemphasized. Conventional breeding practiced for over a century has resulted in cultivars with high yields and superior agronomic traits. Thus, largely through exploitation of hybrid vigor, grain yields of maize, pearl millet, and sorghum registered a substantial increase from 1965 to 1990 (Khush and Baenziger 1996; Jauhar and Hanna

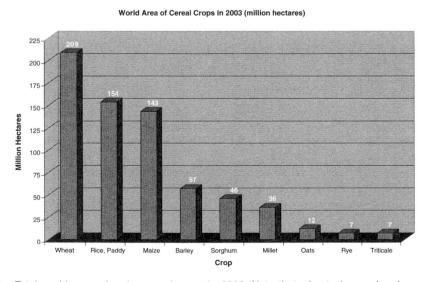

**Figure 1.1**  Total world area of major cereal crops in 2003. Note that wheat, rice, and maize are the most important cereals with the most acreage. Triticale, the new man-made cereal, occupies a small area. (*Source*: faostat.fao.org.)

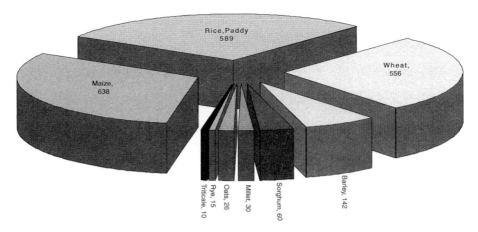

**Figure 1.2** Pie diagram showing the world production of the major cereals in 2003. (*Source*: faostat.fao.org.)

1998). More remarkably, the development of semidwarf improved wheat and rice cultivars in the 1960s launched the most welcome Green Revolution in Asia, averting mass-scale starvation (Khush 1999). Sustained improvement of grain yields and nutritional status of cereal crops should remain the principal goals of crop scientists and agriculturists.

A full understanding of a crop — its genomic constitution and nature of its polyploidy, if any — is very helpful in planning breeding strategies and refining various tools for its genetic improvement. Bread wheat (AABBDD) and durum wheat (AABB) are natural hybrids, having resulted from hybridization between related wild species and doubling of chromosome number (Figure 1.3). Although their genomes (chromosome sets) are genetically similar (homoeologous), a gene, *Ph1*, on the long arm of chromosome 5B (chromosome number 5 of the B genome) ensures diploid-like pairing, i.e., pairing between homologous partners only (Riley and Chapman 1958; Sears and Okamoto 1958), which confers disomic inheritance. Oat is also an allohexaploid, and its cytogenetic architecture is similar to that of bread wheat, with genetic control of chromosome pairing (Rajhathy and Thomas 1972; Jauhar 1977) similar to the *Ph1* system of wheat. It is interesting that maize, rice, pearl millet, and sorghum have been shown to be diploidized cereals, having resulted from ancestral rounds of polyploidy (see Section 1.6). Appropriate chromosome engineering and cytogenetic manipulations have been used for the improvement of these crops.

An important threat to global food security is the occurrence of numerous diseases and pests and the emergence or introduction of new ones. Landraces and wild relatives of cereal crops are rich reservoirs of genes for resistance to diseases, pests, and abiotic stresses; these genes can be incorporated into cereal crops through hybridization. Cytogenetic manipulations, including the suppression of *Ph1*-pairing regulation for recombining desirable alien chromatin or genes into hexaploid wheat, have been termed chromosome engineering (Sears 1972, 1981). Durum wheat is less genetically buffered than bread wheat, and hence the former is less amenable to cytogenetic or chromosomal manipulation. Although other techniques of transferring alien chromatin into a crop like wheat are known, e.g., through X-irradiation (Sears 1993), the promotion of homoeologous pairing offers a more precise means of "chromosome surgery" to recombine chromosome segments, and hence a more desirable method of alien gene transfer. Some of the cytogenetic manipulations used in wheat have been fruitfully employed in hexaploid oat (see Chapter 7 by Jellen and Leggett in this volume).

This volume on cereals covers the improvement of important cereal crops utilizing all available tools: conventional breeding, chromosome engineering by interspecific hybridization coupled with manipulation of chromosome pairing, and use of molecular tools, including markers and genetic transformation. Each cereal crop has been dealt with by an expert in the field. In this introductory chapter, I have given an overview of the cytogenetic architecture of cereal crops and discussed the

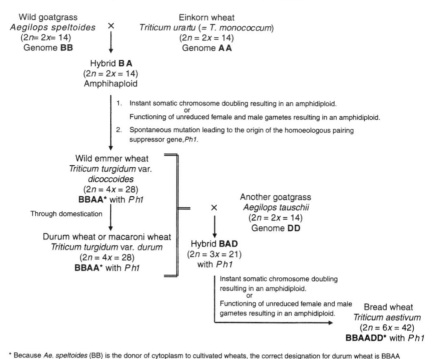

**Figure 1.3**    Diagram showing different steps in the evolution of bread wheat, with durum wheat as its forerunner. Note how allopolyploidy gave rise to our most important crop in a few steps of evolution.

attributes of their genomes, including size, gene density, and synteny with other cereal genomes. I have also summarized major landmark studies leading to the improvement of the major cereal crops. Subsequent chapters present the details.

## 1.2 CEREAL CROPS: A SOURCE OF SUSTENANCE TO HUMANKIND

Cereal crops occupy about two thirds of all cultivated land, and their importance lies in the fact that they have fair to good nutritive value and are relatively easy to grow, store, and transport. Although they are classified as carbohydrate-rich foods, they are also a major source of protein of fair to good quality in much of the world. They are in fact an important source of sustenance for humankind. Table 1.1 presents the food value (the energy, protein, and lipid content) of four important cereals compared to three major vegetable foods. Energy content depends mostly upon carbohydrate contents in these foods. Interestingly, in addition to energy figures, the protein content is higher in cereal grains than in potatoes or peas; these high values are, of course, partly due to the fact that cereal grains contain a much lower proportion of water than potatoes and peas. Even more importantly, cereals do not contain cholesterol, and there is an inverse relationship between intake of whole grains and coronary artery disease (Kushi et al. 1999). In a study over a period of 5.5 years, both total mortality and rate of occurrence of cardiovascular disease were found to be inversely associated with intakes of whole-grain but not refined-grain breakfast cereals (Liu et al. 2003).

Another important factor contributing to the worldwide importance of cereals is the large number of diverse species that can grow in different parts of the world, including temperate and tropical

**Table 1.1   Comparative Values of Energy Provided, and the Total Content of Protein and Lipid in 100 g of Cereal Grains and Other Common Foods**

| Food Crop | Energy (kJ) | Protein (g) | Lipid (g) |
|-----------|-------------|-------------|-----------|
| Wheat     | 1420        | 12.0        | 2.0       |
| Rice      | 1296        | 8.0         | 2.0       |
| Maize     | 1471        | 10.0        | 4.0       |
| Sorghum   | 1455        | 10.0        | 5.0       |
| Potatoes  | 347         | 2.0         | 0.1       |
| Peas      | 293         | 4.9         | 0.4       |
| Lettuce   | 63          | 1.2         | 0.2       |

*Source*: http://www.cix.co.uk/.

climates, and humid and arid or semiarid regions. Thus, bread wheat, durum wheat, barley, oats, and rye are cultivated in temperate regions throughout the world. Maize grows best in hotter regions and is important in tropical and subtropical areas. Rice is mainly a crop of the wet tropics, whereas sorghum and pearl millet can survive in hot, dry conditions, such as the drought-prone Sahel of Africa. As stated above, cereal crops are of diverse origins; some are diploid while others are allopolyploids that enjoy the benefits of polyploidy and hybridity. Thus, rice, maize, barley, sorghum, pearl millet, and rye are diploid or diploidized crops, whereas bread wheat, durum wheat, and oats are obvious polyploids (allopolyploids, to be precise) (see below).

## 1.3 POLYPLOID CEREALS: THEIR CYTOGENETIC ARCHITECTURE

Polyploidy is recognized as a dominant factor in plant speciation (Masterson 1994; Leitch and Bennett 1997; Soltis and Soltis 1999; Jauhar 2003a). Some plant groups seem to have undergone several cycles of chromosome doubling. The proportion of angiosperms that have had one or more events of polyploidy somewhere in their ancestry lies between 50 and 70% (Stebbins 1971; Lewis 1980; Masterson 1994). Allopolyploidy or amphidiploidy, resulting from interspecific or intergeneric hybridization coupled with chromosome doubling, has in fact produced the majority of our most important crop plants, including the cereal crops — bread wheat, durum wheat, and oats.

### 1.3.1   Polyploid Wheats: A Model for Evolution by Allopolyploidy

Allopolyploids enjoy the benefits of both polyploidy and stable hybridity and are highly adaptable to adverse environmental conditions. The enzyme diversity coded by related genes in different genomes seems to contribute to their selective advantage and fitness (Adams and Allard 1977; see also Jauhar 2003a). Sexual polyploidization, which results from functioning of meiotically unreduced gametes, is a significant source of allopolyploids in nature (Harlan and de Wet 1975; Jauhar et al. 2000; Jauhar 2003a). Since corresponding chromosomes of different genomes are genetically similar (or homoeologous) and hence capable of pairing with one another, a genetic control restricting pairing to homologous partners would be necessary for meiotic regularity and reproductive stability of allopolyploids. Thus, sexual polyploidization and genetically regulated chromosome pairing promote the founding of successful allopolyploid species (Jauhar 2003a). Wheat offers an excellent example of evolution by allopolyploidy.

That the successful establishment of a sexually reproducing allopolyploid depends upon the integration of constituent genomes into a meiotically and reproductively stable form is well exemplified by bread wheat with its three related genomes. Hexaploid bread wheat ($2n = 6x = 42$; AABBDD genomes) and its tetraploid forerunner, durum wheat ($2n = 4x = 28$; AABB genomes), are stabilized natural hybrids of wild diploid species. The steps in the evolution of polyploid wheats are outlined below and in Figure 1.3.

### 1.3.1.1  Durum Wheat: A Forerunner of Bread Wheat

Tetraploid durum wheat (macaroni wheat) is a predecessor of hexaploid bread wheat. Its two genomes, A and B, came from two diploid wild grasses. The donor of the A genome is *Triticum urartu* Tumanian (Nishikawa 1983; Dvořák et al. 1993), a species closely related to einkorn wheat (*Triticum monococcum* L.), which was domesticated in southeastern Turkey about 12,000 years ago (Heun et al. 1997). The B genome was probably derived from *Aegilops speltoides* Tausch (Sarkar and Stebbins 1956; Wang et al. 1997;  Dvořák 1998). The two progenitors, both native to the Middle East, hybridized in nature about half a million years ago (Huang et al. 2002) and gave rise to tetraploid wild emmer wheat (*T. turgidum* var. *dicoccoides* Körn), presumably in one step as a result of somatic chromosome doubling in the BA hybrid during premeiotic mitotic divisions or by meiotic nonreduction. Unreduced (2n) gametes could arguably have functioned in both progenitors, thereby producing an instant amphidiploid, which became emmer wheat. However, 2n gametes occur very rarely, if at all, in diploid species. Chromosome doubling most plausibly occurred via fusion of unreduced gametes in the BA hybrid (amphihaploid), since such gametes are known to occur in interspecific and intergeneric hybrids (see also Chapter 2 by Ceoloni and Jauhar and Chapter 3 by Mujeeb-Kazi) rather than in diploid parents. This step of evolution can in fact be recreated by inducing BA haploids (amphihaploids, to be precise) of durum wheat (see Figure 1.3). It has been shown that synthetic durum haploids produce unreduced gametes by first division restitution (FDR), resulting in a viable seed set and then tetraploid (disomic) durum plants (Jauhar et al. 2000). As noted by these workers, the *Ph1*-induced failure of homoeologous pairing may be a prerequisite for the formation of FDR nuclei (Jauhar 2003a), implying that *Ph1* was likely present in the hybrid that inherited it from its B- or the A-genome parent. If one of the diploid progenitors of emmer wheat did harbor *Ph1*, then what regulatory function, if any, it would have played there is a matter of speculation. But then, if *Ph1* had not been present in the BA hybrid, the derived wild emmer and durum wheat would be expected to show a number of B/A translocations at homoeologous breakpoints. However, we do not find such translocations, implying that the two genomes maintained their meiotic integrity perhaps through *Ph1* regulation. One might speculate, therefore, that the hybrid had somehow acquired *Ph1*, although its precise origin remains enigmatic.

If the BA hybrid did not have the benefit of FDR or some similar mechanism of forming 2n gametes, it would perhaps not have survived in nature. The establishment of *Ph1* regulation was also vital for the survival and success of the derived amphidiploid BBAA (Figure 1.3). Because *Ae. speltoides* (BB) functioned as the female parent (Wang et al. 1997), the correct genomic designation for emmer wheat would be BBAA, although it is generally given as AABB.

Wild emmer was domesticated in the Fertile Crescent, where it acquired the *Q* gene for free threshing (Muramatsu 1986) and gave rise to cultivated emmer or durum wheat, which is one of the earliest domesticated crops. The acquisition of *Q* gene marked the dawn of human civilization in the Near East. Emmer was the main crop during the spread of Neolithic agriculture from the Fertile Crescent to Eurasia and Africa.

### 1.3.1.2  Establishment of Genetic Control of Chromosome Pairing

The corresponding chromosomes of the A and B genomes are closely related genetically and are referred to as homoeologous chromosomes, their own partners being homologous. Because homoeologous chromosomes, e.g., 1A and 1B, are genetically similar and hence capable of pairing with one another, some sort of regulation of pairing would be necessary for meiotic regularity. Therefore, either *Ph1* was acquired at the BA hybrid level (see Section 1.1 above), or it can be hypothesized that at (or before) the origin of the tetraploid wild emmer, a spontaneous mutation gave rise to the homoeologous chromosome pairing suppressing gene, *Ph1*, located in the long arm of chromosome 5B. Thus, *Ph1* permitted pairing only between homologous partners, ensuring diploid-like pairing and disomic inheritance, and has helped to maintain the meiotic integrity of the A and B genomes. But for this rigid regulation of chromosome pairing, the two genomes would have converged during the period of about 500,000 years that they have been together. However, some intergenomic rearrangements, such as those involving chromosomes 4A, 5A, and 7B, have occurred both in bread wheat (Naranjo et al. 1987; Naranjo 1990) and durum wheat (Naranjo 1990; Doğramacı-Altuntepe and Jauhar 2001; see Figure 2.3). This cyclic translocation seems to confer some selective advantage and appears to be the evolutionary signature of polyploid wheats.

### 1.3.1.3 Origin of Bread Wheat: An Important Evolutionary Step

Another cycle of spontaneous hybridization (see Figure 1.3) took place between tetraploid wheat and diploid goatgrass (*Aegilops tauschii* Coss., 2n = 2x = 14, DD genome) (McFadden and Sears 1946) about 8000 years ago (Huang et al. 2002) and gave rise to hexaploid bread wheat that sustains humankind. It is likely that, as a result of hybridization between durum wheat and *Ae. tauschii*, a triploid hybrid ABD was first formed, which produced unreduced male and female gametes and a set hexaploid seed, giving rise to bread wheat (Matsuoka and Nasuda 2004). This parallels the events in durum haploids (amphihaploids BA) that set seed (Jauhar et al. 2000).

The *Ph1*-regulated diploid-like pairing that was established in its ancestral allotetraploid conferred meiotic regularity, and hence reproductive stability, to hexaploid wheat also. The three genomes, AA, BB, and DD, have since maintained their meiotic integrity. But for this stringent regulation of chromosome pairing, the wheat crop would not have evolved and human civilization would perhaps not have progressed the way it has.

### 1.3.2 Cytogenetic Makeup of Hexaploid Oat

Common oat (*Avena sativa* L.) and red oat (*Avena byzantina* C. Koch) are allohexaploid species with three genomes (2n = 6x = 42; AACCDD genomes), although only the allotetraploid progenitor has been identified so far. Unlike wheat, discrimination between two of the oat genomes, A and D, has been problematic (Linares et al. 1998; Drossou et al. 2004). It is clear, however, that diploid-like pairing in hexaploid oats is under genetic control (Rajhathy and Thomas 1972, 1974; Jauhar 1977), which is hemizygous effective, as evidenced by lack of pairing in oat polyhaploids with one dose of the diploidizing gene(s) (Nishiyama and Tabata 1964), and genetically repressible, as in hexaploid bread wheat (Table 1.2). Based on chromosome pairing in oat polyhaploids and several amphiploids (Thomas and Jones 1964; Rajhathy and Thomas 1974), Jauhar (1977) concluded the presence of a rigid and complex genetic control of chromosome pairing in oat. He further hypothesized that the diploidizing gene(s) was located in the A genome and that multiple copies of these genes in the AAAABBCCDD decaploids (Thomas and Jones 1964) drastically reduced the frequency of multivalents (Jauhar 1977). Although the pairing control mechanism in hexaploid oat has not been fully elucidated so far, genetic suppression of homoeologous pairing is an important consideration in alien gene transfers into this cereal.

## 1.4 GENETIC CONTROL OF CHROMOSOME PAIRING: MAJOR IMPLICATIONS

Allopolyploidy, resulting from interspecific or intergeneric hybridization coupled with chromosome doubling, and in conjunction with genetic regulation of chromosome pairing, has been instrumental in the production of many of our important grain, forage, and fiber crops. Thus, a

**Table 1.2    Comparison of the Chromosome Pairing Control Mechanisms in Polyploid Grasses**

| Features | 5B Control | Reference | A Control | Reference | C Control | Reference |
|---|---|---|---|---|---|---|
| Species in which chromosome pairing is regulated | Tetraploid and hexaploid wheats | Okamoto 1957; Sears and Okamoto 1958; Riley and Chapman 1958 | Tetraploid and hexaploid species of *Avena* | Rajhathy and Thomas 1972 | *Festuca arundinacea, Festuca rubra*, and other polyploid fescues | Jauhar 1975a,b |
| Location | 5BL | Sears and Okamoto 1958; Riley 1960 | A genome? | Jauhar 1975c | C genome? | Jauhar 1975c, 1977 |
| Effectiveness | Hemizygous effective | Riley and Chapman 1958; Riley 1960 | Hemizygous effective | Nishiyama and Tabata 1964; Rajhathy and Thomas 1972 | Hemizygous ineffective | Jauhar 1975a,b |
| Dosage effect | Yes | Feldman 1966, 1968; Martinez et al. 2001 | Yes | Jauhar 1975c; Ladizinsky 1973 | Yes | Jauhar 1975c |
| Genetically repressible | Yes | Riley 1960; Dvořák 1972 | Yes | Rajhathy and Thomas 1972, 1974 | Yes | Jauhar 1975b,c, 1977 |
| Species/genotype suppressing control | *Aegilops speltoides* | Riley et al. 1961 | *Avena longiglumis* Accession CW57 | Rajhathy and Thomas 1972 | Diverse ecotypes of tall fescue | Jauhar 1975c, 1991 |

*Source*: Paraphrased from Jauhar, P.P., *Theor. Appl. Genet.*, 49, 287–295, 1977.

combination of sexuality, polyploidy, and genetic control of chromosome pairing provides an ideal recipe for evolution of successful plant species (Jauhar 2003a). Polyploid wheats and the more recent man-made cereal, triticale, provide excellent examples of cataclysmic evolution or evolution by large quantum jumps. Although most allopolyploids in nature may have developed genetic control of chromosome pairing, the mechanism has been clearly elucidated in bread wheat and durum wheat (Okamoto 1957; Sears and Okamoto 1958; Riley and Chapman 1958), and to some extent in hexaploid oat (Rajhathy and Thomas 1972, 1974; Jauhar 1977) and hexaploid tall fescue, *Festuca arundinacea* Schreb. (2n = 6x = 42; AABBCC) (Jauhar 1975a–c, 1991). Such a control results in diploid-like chromosome pairing, which in turn ensures disomic inheritance. It is likely that disomic polyploidy and sexuality may not coexist in nature without such a regulatory mechanism (Jauhar 1975d, 2003a). However, polysomic polyploids like *Medicago sativa* and potato are successful species.

When one dose of the pairing control gene(s) is enough to enforce diploid-like chromosome pairing, such a gene(s) is hemizygous effective. As evidenced by the absence of homoeologous chromosome pairing in their polyhaploids, the genetic control of chromosome pairing is hemizygous effective in durum wheat (Figure 1.4), bread wheat (Figure 1.5), and hexaploid oat, whereas it is hemizygous ineffective in hexaploid tall fescue (Jauhar 1975a,b). Thus, one dose of *Ph1* effectively suppresses homoeologous pairing in durum haploids (Figure 1.4B; Jauhar et al. 1999) and bread wheat haploids (Figure 1.5A; Jauhar et al. 1991), whereas absence of *Ph1* results in extensive homoeologous pairing in both durum (Figure 1.4C–G) and bread wheat (Figure 1.5B) *ph1b* haploids. The similarities and differences among the three regulatory mechanisms are given in Table 1.2. Genetic pairing regulation has important implications in cytogenetics, evolution, and plant breeding (Jauhar 1975c).

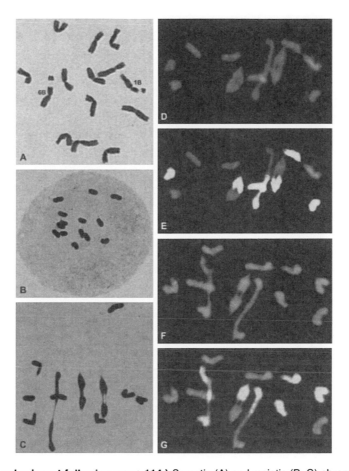

**Figure 1.4** **(See color insert following page 114.)** Somatic (A) and meiotic (B–G) chromosomes of durum haploids derived by hybridization with maize: a pollen mother cell (PMC) from a haploid with *Ph1* (B) and PMCs from haploids without *Ph1* showing high homoeologous pairing (C–G). A: 14 somatic chromosomes; note one dose each of the satellited chromosomes 1B and 6B. B: A PMC with 14 univalents; note the total absence of pairing in the presence of *Ph1*. C: Four bivalents (two ring II and two rod II) + six univalents; note high pairing in the absence of *Ph1*. D, E: Fluorescent genomic *in situ* hybridization (fl-GISH) analysis of chromosome pairing: two ring II + one rod II counterstained with propidium iodide (PI) (D) and the same cell as D probed with biotinylated A genome probe (E); the preparation was blocked with the genomic DNA of *Ae. speltoides* (B genome) and the probe was detected with FITC. The A genome chromosomes are brightly lit in green color. Note pairing between the A genome chromosomes and the B genome chromosomes. F, G: Fl-GISH analysis of chromosome pairing: one ring II + two rod II counterstained with PI (F) and same cell as F after probing with the A genome probe (G). Note the intergenomic ring bivalent involving an A- and a B-genome chromosome, an intergenomic rod bivalent, and an intragenomic (within the A genome) bivalent.

## 1.4.1   Cytogenetic and Evolutionary Implications

### 1.4.1.1   Usefulness in Genome Analysis

The *Ph1* gene of wheat suppresses pairing between less related (homoeologous) chromosomes. Therefore, pairing or lack of pairing between chromosomes of two genomes in the presence of *Ph1* in the wheat background would generally provide a stringent test of their relationship (Jauhar and Joppa 1996). Adopting this approach, it was found that the seven chromosomes of the J genome of diploid *Thinopyrum bessarabicum* and seven of the E genome of diploid *Lophopyrum elongatum*

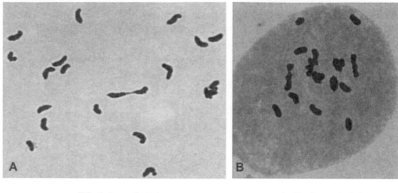

*Ph1*-haploid                    *ph1b*-haploid

**Figure 1.5**  Chromosome pairing in PMCs of bread wheat haploids with and without *Ph1*. Note that one gene can make such a difference. A: One rod II + 19 I in the presence of *Ph1* in a *Ph1* haploid. B: Six II (three ring II + three rod II) + nine I. Note extensive pairing in the haploid with the *ph1b* allele.

in the AABBDDJE amphiploids (Forster and Miller 1989) showed no pairing in the presence of *Ph1*, clearly indicating that the J and E genomes are not closely related (Jauhar 1990a,b).

### 1.4.1.2  Gene Introgressions and Changes in Base Chromosome Numbers

The hemizygous ineffectiveness of the genetic control of chromosome pairing in hexaploid tall fescue and other polyploid fescues is of evolutionary significance in that it allows gene flow from one species to another (Jauhar 1975d), which would explain the widespread introgression of characters among taxa. Such a regulatory mechanism could have played a role in bringing about changes in base chromosome numbers of the type present in the genus *Pennisetum* (Jauhar 1981b; see also Chapter 10 in this volume) and numerous other taxa (Jauhar 1975d). Hemizygous-ineffective control would produce irregular meiosis with multivalent formation, which would result in loss or gain of chromosomes in hybrids. Subsequent spontaneous chromosome doubling of such hybrids could have produced aneuploid taxa; such taxa may not compete their euploid relatives, and hence become apomictic (Jauhar 1975c, 2003a).

### 1.4.2  Breeding Implications: Homoeologous Pairing, the Key to Gene Transfer

As stated above, the regulator of chromosome pairing, e.g., the *Ph1* gene in bread wheat, suppresses homoeologous pairing, resulting in diploid-like pairing involving homologous partners only and ensuring meiotic regularity and reproductive stability of the polyploid species. The function of *Ph1* is to discipline chromosomes and prevent "adultery" among them, i.e., suppress recombination between less related chromosomes, and thereby ensure disomic inheritance. Because of its function as a regulator of chromosome pairing, I called *Ph1* a policeman. Although there are other genes with some regulatory effect on pairing, *Ph1* is the principal regulator.

*Ph1* suppresses pairing between unrelated or less related chromosomes and may inhibit pairing between wheat chromosomes and alien chromosomes in wheat × alien species hybrids, thereby impeding alien gene transfers into wheat. Because plant breeding depends on adultery among wheat and alien chromosomes to capture desirable segments or to trim off alien chromosome segments that bear unwanted genes, methods of promoting homoeologous pairing through cytogenetic manipulation must be adopted.

## 1.5 CYTOGENETIC MANIPULATION OF POLYPLOID CEREAL CROPS

### 1.5.1 Wheat

As stated above, the origin of a rigid regulatory mechanism in the form of *Ph1* was essential for suppressing homoeologous chromosome pairing, and hence for reproductive stability and survival of polyploid wheats. Because *Ph1* restricts pairing to identical (homologous) partners, it does not permit adultery among less related (homoeologous) chromosomes. However, since plant improvement is facilitated by adultery and recombination among related or even less related chromosomes, *Ph1* poses an obstacle to the incorporation of alien genes into bread wheat and durum wheat. Because chromosome pairing between related or less related chromosomes is the key to gene transfer across species, appropriate means of circumventing the *Ph1*-created barrier will need to be adopted. Elegant means of incorporating alien genes into wheat were devised by Sears and other wheat researchers. Cytogenetic manipulations, including those based on suppression or inactivation of the *Ph1* system, for engineering desirable alien chromatin into wheat were termed chromosome engineering (Sears 1972). A high-pairing mutation involving a small intercalary deficiency for *Ph1* was produced in bread wheat and designated *ph1b* (Sears 1977). Since this mutation raises the level of homoeologous pairing, it may be employed for alien gene transfers into wheat.

The use of the *ph1bph1b* mutant of wheat (as a female parent) provides one means of promoting homoeologous pairing. Use of 5B-deficient stocks such as the 5D(5B) substitution lines also promotes intergenomic chromosome pairing and intergeneric gene transfer. Such a strategy of alien gene introgression in wheat has been employed (Sears 1981; see Jauhar and Chibbar 1999).

The use of 5B-deficient stocks, such as the 5D(5B) disomic substitution line of durum wheat, has been successfully employed to promote intergeneric chromosome pairing, for example, in an intergeneric hybrid (Figure 1.6A) between durum wheat and a diploid wheatgrass, *Th. bessarabicum* (2n = 2x = 14; JJ). In the presence of *Ph1* there is no chromosome pairing or minimal pairing in the hybrid (Figure 1.6B), but in the absence of *Ph1* extensive homoeologous chromosome pairing occurs (Figure 1.6C), a welcome feature from the breeding standpoint. We are using the 5D(5B)

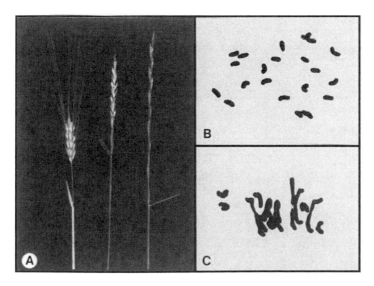

**Figure 1.6**    An intergeneric hybrid (2n = 3x = 21; ABJ) between durum wheat and a diploid wheatgrass, *Th. bessarabicum* (2n = 2x = 14; JJ). A: Spikes of durum parent, intergeneric hybrid, and the diploid wheatgrass. Note intermediate characteristics of the hybrid (the awns of durum parent missing). B: A PMC with 21 I. Note the absence of pairing in the presence of *Ph1*. C: A PMC with one III + one ring II + five rod II + six I. Note extensive homoeologous pairing in the absence of *Ph1*.

substitution line to promote homoeologous pairing and recombination between chromosomes of durum wheat and *Lophopyrum elongatum* (2n = 2x = 14; EE) (Jauhar and Chibbar 1999; Repellin et al. 2001).

Another and perhaps more suitable means of inducing wheat–alien species chromosome pairing is by crossing wheat with alien species that inactivate the homoeologous pairing suppressor *Ph1*. Certain genotypes of wild grasses, a potential donor of desirable genes, are known to inactivate *Ph1* at least partially (see, for example, Jauhar 1992; Jauhar and Almouslem 1998) and may be employed to promote homoeologous pairing and hence alien gene introgression. We are exploiting such genotypes in our wheat germplasm enhancement program (Repellin et al. 2001; Jauhar 2003b).

### 1.5.1.1 *Specificity of Chromosome Pairing and Induction of Alien Integration into Wheat*

Excellent means are now available for studying the nature and specificity of chromosome pairing in haploid complements and intergeneric hybrids (Jauhar et al. 1999, 2004). Thus, Jauhar (1992) combined both the E and J genomes with durum wheat in trigeneric hybrids with genomic constitution of ABJE and used fl-GISH to study the specificity of chromosome pairing: wheat–wheat, grass–grass, and wheat–grass (Jauhar et al. 2004). Wheat–grass pairing is, of course, essential for incorporation of alien segments into the wheat complement.

While introduction and integration of alien genetic material in the wheat genome is very important, its characterization (physical size and precise location) will also be very useful. Recently, we have witnessed a dramatic improvement in monitoring the alien transfer process in all its phases (Ceoloni et al. 1998). Thus, both molecular marker technology as well as molecular cytogenetic techniques, such as nonradioactive *in situ* hybridization, and such as fluorescent genomic *in situ* hybridization (fl-GISH), can effectively complement classical diagnostic tools for efficient and accurate characterization of introgression products (see Chapter 2 by Ceoloni and Jauhar in this volume). These procedures would also facilitate elimination of alien chromatin carrying undesirable traits of the wild donor. Details of the alien gene transfer work in polyploid wheats are given in Chapters 2 and 3 in this volume.

### 1.5.2  Oat

Hexaploid bread wheat (2n = 6x = 42; AABBDD) and hexaploid oat (2n = 6x = 42; AACCDD) have essentially similar genomic constitutions. (Note, however, that there is no correspondence between A genomes in the two Poaceae tribes to which wheat and oat belong.) As in bread wheat, the diploid-like pairing in oat is under genetic control. In both, the regulation is essentially similar (Table 1.2) and can be suppressed by appropriate genotypes of their wild relatives, leading to homoeologous pairing. In oat, a locus in the wild diploid *Avena longiglumis* CW 57 suppresses homoeologous pairing regulation. Through wide hybridization coupled with manipulation of chromosome pairing, some desirable genes have been transferred into cultivated oat. Thus, Thomas et al. (1980) transferred a mildew resistance gene from *Avena barbata* into oat. Jellen and Leggett present details of such transfers in Chapter 7 in this volume.

## 1.6 DIPLOID OR DIPLOIDIZED CEREALS: GENOMIC EVOLUTION

### 1.6.1  Cytogenetic Makeup and Ancient Polyploid Origin

Eukaryotic evolution is known to be accompanied by gene duplication. Duplications of genes, chromosomal segments, chromosomes, or whole genomes have played an important role in eukaryotic genome evolution (Koszul et al. 2004; Goffeau 2004). It has long been known that duplicated

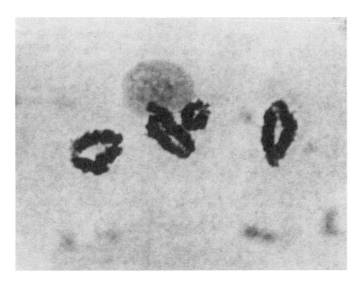

**Figure 1.7**    Meiotic pairing at a stage comparable to diakinesis in barley haploids (2n = x = 7). Note complete pairing: three large ring bivalents and fold-back pairing of the remaining univalent. (Courtesy of Sadasivaiah, R.S. and Kasha, K.J., *Chromosoma* 35, 247–263, 1971.)

genetic material confers adaptive advantage, and having extra gene copies is essential for an organism to evolve (Ohno 1970). Recent work has further shown that diversification of gene functions during evolution requires prior gene duplication (Dujon et al. 2004; Kellis et al. 2004).

Several of our crop plants are supposedly diploid. Genetic mapping studies have shown, however, that several crop species traditionally considered diploid are in fact diploidized polyploids derived from ancient polyploids through some sort of structural repatterning. Recent studies have shown that allopolyploidy-induced sequence elimination of low-copy DNA sequences or those in noncoding regions can occur in plant genomes (Feldman et al. 1997; Ozkan et al. 2001). Among these cereals, barley (2n = 2x = 14) is perhaps a typical diploid, although the formation of bivalents during meiosis in its haploids (2n = x = 7) (Sadasivaiah and Kasha 1971; Figure 1.7) would suggest chromosomal duplication. Haploids have half the normal somatic chromosome number and offer an excellent opportunity for studying inter- or intragenomic homologies, which are masked when every chromosome has an identical partner. Bivalent formation in haploid (monoploid, to be precise) barley suggests that the barley genome itself might have arisen from a lower basic chromosome number. Based on the presence of up to two bivalents in haploids (2n = x = 7) of pearl millet (Jauhar 1970), and on intergenomic and intragenomic chromosome pairing in its interspecific hybrids (Jauhar 1968, 1981a), it was inferred that the pearl millet complement was derived from an ancestral base chromosome number of x = 5 as a result of duplication during the course of evolution. Thus, Jauhar (1968) called pearl millet a secondarily balanced species (see Chapter 10 in this volume). From a study of restriction fragment length polymorphism (RFLP) linkage maps, Liu et al. (1994) provided corroborating evidence of the presence of duplicate loci in pearl millet.

Maize has long been considered to be an ancient polyploid having extensive chromosome duplications that were initially revealed from meiotic pairing in its haploids (Ting 1966), and by its ability to tolerate chromosome deficiencies. In diploid species, chromosome deficiency such as monosomy is not tolerated. In maize, however, 9 of the possible 10 monosomics (2n − 1 = 19) were produced along with some occasional double monosomics (2n − 1 − 1 = 18) and even triple monosomics (2n − 1 − 1 − 1 = 17) (Weber 1970, 1973, 1994), clearly suggesting maize to be "an ancient, secondarily balanced species with an extensive duplication (and probably redundance) of genetic information in the form of whole chromosomes" (Jauhar 1981b, p. 91). In this respect, surprisingly, there seems to be more genetic buffering in maize than in tetraploid durum wheat because in the latter, even simple monosomy for a chromosome is not well tolerated. The DNA sequence data of Gaut and Doebley (1997) supported the polyploid origin of maize. They studied 14 pairs of duplicated loci and noted two different groups of coalescence times, which they attributed to ancestral tetraploidization between two diploids whose genomes were partially differentiated from each other. Thus, maize has clearly resulted from an ancient polyploidization event (Gaut et al. 2000; Wendel 2000; Gaut 2001).

Sorghum (2n = 2x = 20) has also been considered to be an ancient polyploid; the genus *Sorghum* has several extant species with 2n = 10. Using bacterial artificial chromosomes (BACs) in conjunction with fluorescent *in situ* hybridization (FISH), Gómez et al. (1998) found that a 45-kb sorghum BAC preferentially hybridized to centromeric regions of 5 of the 10 chromosomes of *Sorghum bicolor* (2n = 20), supporting earlier evidence of the tetraploid nature of this crop and also revealing two genomic sets of 5 chromosomes each. Clearly, several of the supposedly diploid crop plants have been shown to result from ancient polyploidization events. It will not be surprising if rye (2n = 14) also turns out to be paleopolyploid, like maize, sorghum, and pearl millet.

The diploid nature of rice (2n = 24) has long been questioned. Based on chromosome morphology and secondary associations in units of five at meiosis, Nandi (1936) hypothesized that rice is a secondarily balanced species. That rice has undergone an ancestral round (or perhaps more than one round) of genome duplication is borne out by recent studies. Because its chromosome complement (2n = 24) is not a multiple of x = 5, 7, 9, or 10 (the common base numbers reported in the Poaceae), it is likely that the present-day rice is an ancient aneuploid, having undergone perhaps only partial genome duplication. Interestingly, a systematic sequence analysis indicated that 15% of the rice genome is in duplicated blocks (Vandepoele et al. 2003). It is not surprising that intragenomic duplications have been revealed in rice. Earlier data on genetic and physical mapping demonstrated the presence of duplicated segments between chromosomes 1 and 5 (Kurata et al. 1994) and between chromosomes 11 and 12 (Wu et al. 1998).

Recently, the genome of subspecies *japonica* of rice was sequenced using a whole-genome shotgun approach. The availability of the draft sequence and the comparison of 2000 mapped cDNAs suggested that large-scale duplication had occurred during the evolution of rice (Goff et al. 2002). The recent release by the Institute of Genome Research of 12 rice pseudochromosome sequences has allowed the investigation of intragenomic duplications, showing that the rice genome contains extensive chromosomal duplication accounting for 53% of the available sequences (Guyot and Keller 2004).

## 1.6.2  Diversity of Origin of Cereal Genomes: Genomic Diversity and Synteny

The grass family Poaceae is an assemblage of diverse, widely adapted species (including cereals), which have been classified into two major clades based on molecular phylogenetic studies (Soreng and Davis 1998). One clade contains the subfamily Panicoideae, which includes the cereal crops maize, sorghum, and pearl millet. The other clade contains the subfamily Pooideae, which includes the cereal crops wheat, barley, oat, and rye. Minor clades include the subfamily Oryzoideae, which contains the model cereal — rice that is recognized as an early diverging lineage. The cereal genomes vary considerably in size, ploidy, and taxonomic affinity. The haploid nuclear genomes of sorghum, maize, barley, and wheat are estimated to be 1000, 3000, 5000, and 16,000 mega base pairs (Mbp), respectively, while rice has a much smaller genome of about 420 Mbp (Goff et al. 2002).

Despite the phylogenetic diversity and the consequent evolutionary distance among these cereal crop species resulting from millions of years of evolution, their genomes show a high degree of conservation, gene similarity, and genome synteny. Molecular mapping of the cereal nuclear genomes using RFLP has allowed the development of comparative chromosome maps (Devos and Gale 1997; Paterson et al. 2000; Ilic et al. 2003; Varshney et al. 2005). Synteny and gene homology between rice and the other cereal genomes are extensive; homologs of 98%

of the known maize, wheat, and barley proteins exist in rice (Goff et al. 2002; Yu et al. 2002). The high genetic colinearity of the rice genome with the larger cereal genomes makes rice the model genome for studying genome evolution and facilitating gene isolation from related cereals.

### 1.6.3 Cytogenetic Manipulation and Breeding Work in Diploid Cereals

Maize, barley, sorghum, pearl millet, and rye are diploidized to the extent that they behave as true diploids, despite the ancestral duplication of genetic material at least in some of them. Thus, they show disomic inheritance. These cereals are relatively easy to manipulate by tools of cytogenetics or by traditional breeding.

#### 1.6.3.1 Rice: A Model Cereal Crop

Being the primary food source for more than a third of the world's population, rice is an extremely important cereal crop. It is the only cereal used almost exclusively for human consumption. Because of several desirable characteristics, including its diploidized nature with 2n = 24 chromosomes, availability of an extensive germplasm collection, availability of well-characterized cytogenetic stocks, and a large number of mutant markers, rice is a highly desirable organism for molecular cytogenetic and breeding research. However, with a symmetrical karyotype of small chromosomes, rice is not an attractive organism for traditional cytogenetic studies. Brar and Khush (Chapter 5 in this volume) provide a detailed account of its improvement via hybridization with wild species and by modern tools such as genetic transformation and functional genomics. Molecular markers have been employed to facilitate introgression of genes for resistance to diseases and pests (Gupta et al. 1999, 2002; Zhou et al. 2002; Somers et al. 2003; Liu and Anderson 2003; Adhikari et al. 2004; Dubcovsky 2004; Varshney et al. 2004; Zhou et al. 2004; Helguera et al. 2005).

As stated earlier, rice has a genome size of about 420 Mbp. Its small genome with high gene density, coupled with its synteny with other cereal crops, makes rice the standard model for cereal gene discovery.

#### 1.6.3.2 Maize: A Cytogeneticist's Delight and a Breeder's Paradise

Because of its amenability to pachytene analysis and the availability of translocation stocks involving supernumerary chromosomes, maize is an ideal organism for basic studies in cytogenetics. Several basic phenomena in genetics, such as linkage, have been elucidated using maize as an experimental organism (e.g., McClintock and Hill 1931). Such eminent geneticists as Barbara McClintock spent their lifetime working on cytogenetics of maize. Unlike many other crops, genetic gains obtained through breeding have been consistent with maize, and the methodologies developed for maize have been applied to other crops (Duvick 1992). Through hybridization with other crops, followed by chromosome elimination, maize offers an excellent system for inducing haploids in unrelated cereals (e.g., Jauhar 2003c), facilitating their cytogenetic manipulation. Thus, using oat × maize crosses, maize chromatin has been added to the oat genome (Riera-Lizarazu et al. 2000; Kynast et al. 2002) and oat–maize addition lines (Kynast et al. 2001) have been isolated to facilitate work on functional genomics.

As a cross-pollinated crop, maize has tremendous genetic diversity (Figure 1.8). It offers enormous possibilities for heterosis breeding, which has considerably improved its yield and nutritional quality. The development of quality protein (high-lysine) maize is a landmark achievement that has helped to alleviate malnourishment among the poor who depend on maize as a primary food source (Vasal 2002; Chapter 6 in this volume).

**Figure 1.8    (See color insert following page 114.)** A collection of maize cobs showing a wide range of diversity. (From www.tropag-fieldtrip.cornell.edu/tradag/Maize.jpg.)

### 1.6.3.3   Pearl Millet: Poor Man's Bread: Heterosis Breeding

Pearl millet, another diploidized cereal, is a dual-purpose crop providing both grain and fodder. As a poor man's source of dietary energy, it sustains a large proportion of the population in hot, arid regions in Africa and Asia. With 2n = 14, large chromosomes, and other desirable attributes, pearl millet is well suited for basic research in cytogenetics (Jauhar and Hanna 1998). Moreover, like maize, pearl millet is an open-pollinated crop that responds very well to heterosis breeding. Single-cross hybrids yield about 25 to 30% more than open-pollinated varieties, and in 2001 more than 70 hybrids were cultivated on about 6 million ha of the total 10-million-ha pearl millet area in India (see Chapter 10 in this volume). Research on nutritional enhancement of pearl millet is also in progress.

Interspecific hybridization followed by cytogenetic manipulation has produced several desirable, heterotic hybrids with high fodder yield and quality (Jauhar and Hanna 1998; Chapter 10 by Jauhar et al. in this volume). Exploitation of hybrid vigor will continue to be the most important means of increasing both grain and fodder yield.

### 1.6.3.4   Barley, Sorghum, Rye, and Triticale

While sorghum does seem to have undergone a cycle of tetraploidization, there is no clear evidence of ancient genomic duplication in barley and rye, even though their haploids show meiotic pairing (Sadasivaiah and Kasha 1971; Levan 1942) indicating chromosome duplication. In terms of total production, barley ranks fourth among the cereal crops (Figure 1.1 and Figure 1.2). Genomic constitution of *Hordeum* species, germplasm resources as donors of desirable traits, and germplasm enhancement of cultivated barley are covered by Singh in Chapter 8 in this volume. Thus, tolerance to biotic and abiotic stresses has been transferred from its wild relative *Hordeum bulbosum* into barley (Pickering 2000). Because the bulbosum method of producing barley haploids is very reliable, the doubled haploid technique has been fruitful in producing several barley cultivars (Pickering and Devaux 1992). Wheat-barley addition lines were produced by Islam and are being used in isolating wheat–barley recombinant chromosomes (Islam et al. 1981; Islam and Shepperd 1992, 2000). Barley is predominantly self-pollinated, and methods such as pedigree, backcrossing, and bulk breeding have been successfully employed for its improvement.

Sorghum is the world's fifth major cereal crop and is mostly grown in the semiarid tropics. It is primarily self-pollinated, but the discovery of cytoplasmic nuclear male sterility has facilitated the exploitation of hybrid vigor (Rooney and Klein 2000). Breeding for resistance to insect pests and diseases has produced excellent results. Marker-assisted selection has proved helpful in breeding for midge resistance (Henzell et al. 2002), although work in this area needs to be expanded. Development of high-lysine lines has improved the nutritional status of this important cereal for the benefit of the malnourished consumers. As in maize, the combination of high protein quality with high grain yield in sorghum will be highly welcome. Details of sorghum improvement using conventional and modern techniques, including transgenic technology, are given by Reddy et al. in Chapter 11 in this volume.

Rye is a relatively young cereal, with the world production amounting to about 30 million metric tonnes and the cool, temperate regions of Europe being the major growing areas. It is an outbreeder and shows inbreeding depression like pearl millet. However, suitable inbred lines with adequate vigor can be isolated for production of synthetics and exploitation of hybrid vigor. Hybrid varieties have been released in Europe and occupy 60% of the total rye acreage. The presence of cytoplasmic genic male sterility is necessary for making rye hybrids. Improvement in grain yield, protein content, insect pest resistance, and baking quality are among the goals of breeding. Varieties with large leaf mass would be suitable for green fodder.

Rye germplasm has also been used for wheat improvement. Hybridization between wheat and rye, coupled with intergeneric chromosome manipulation, led to the production of various combinations of genomes, wheat–rye addition lines involving individual rye chromosomes or chromosome segments (Schlegel 1990). Details of these cytogenetic manipulations are covered by Schlegel in Chapter 12 in this volume.

Most interestingly, wheat × rye hybridization also resulted in a new man-made cereal, called triticale or *Triticosecale* Wittmack (AABBRR). Endowed with improved protein content, disease and pest resistance, and cold tolerance, hexaploid triticale promises to enlarge the spectrum of cereal crops for the benefit of humankind. An extensive coverage of this new cereal is given by Tamás Lelley in Chapter 13 in this volume. He describes triticale as a typically human inspiration, a remarkable feat of evolution.

## 1.7 PERSPECTIVES AND CHALLENGES

Cereal crops constitute the most important food source for humans, sustaining about two thirds of the world population. The current world population of about 6.1 billion is projected to reach 8.0 billion by 2030, posing a great challenge for plant scientists and agriculturalists to help cope with the ever-growing demand for food. According to FAO estimates, 842 million people worldwide were malnourished in 1999 to 2001, the most recent years for which figures are available. This total also includes 10 million people in the industrialized countries, 34 million in countries in transition, and 798 million in developing countries (fao.org/newsroom/news/2003). These alarming numbers underscore the need for increasing yields and upgrading the nutritional status of these important food crops, utilizing all available tools of conventional breeding, cytogenetics, molecular genetics, and biotechnology. Despite recent successes in improving cereal crop yields, which brought about the Green Revolution and thankfully saved numerous lives, the momentum of crop improvement must be maintained to ensure future food security.

The art of plant breeding was developed long before the laws of genetics became known. The advent of the principles of genetics and cytogenetics at the turn of the last century catalyzed the growth of plant breeding, making it a science-based technology that was instrumental in substantial genetic improvement of crop yields. Grain yields of major crops, namely, wheat, rice, maize, sorghum, and pearl millet, have increased steadily since 1930, when principles of genetics were applied to plant breeding and heterosis breeding was adopted first in maize and later in pearl millet and sorghum. Although heterosis breeding has helped produce many superior hybrids in maize and pearl millet, the genetic base of such hybrids must be broadened to ensure protection against new pathogens (see Chapter 6 by Vasal et al. and Chapter 10 by Jauhar et al. in this volume). Apomixis could help maintain heterozygosity through seed production and perpetuate hybrid vigor. However, the potential of apomixis in harnessing hybrid vigor has not been exploited so far, although research on these lines has been under way for some time. If apomixis could be introduced in superior hybrids, they would clone themselves, thereby eliminating the need to produce commercial hybrids year after year. Exploitation of hybrid vigor will continue to be an important strategy in maize and pearl millet. Heterosis breeding could also prove very beneficial in other cereal crops, including the inbreeders, rice and wheat.

Alien genetic resources have been widely used to enrich cereal crops. Wide hybridization, coupled with manipulation of chromosome pairing, has been utilized to introgress desirable chromosome segments or genes from wild wheatgrasses into wheat. A novel technique of inducing crop–alien chromosome translocations involves the use of the chromosome-breaking action of gametocidal chromosomes (Endo 2003). Thus, certain *Aegilops* chromosomes become gametocidal when introduced into common wheat and induce chromosome breakage. By introducing these gametocidal genes into a wheat–alien addition or substitution line, random wheat–alien chromosome translocations have been recovered in the selfed progenies. Masoudi-Nejad et al. (2002) transferred rye chromosome segments into wheat by employing a gametocidal system. Structural changes in barley chromosomes added to wheat were induced by a gametocidal chromosome derived from *Aegilops cylindrica* (Shi and Endo 1999).

The availability of molecular markers has helped to map genes for superior agronomic traits, such as resistance to diseases and pests in wheat, and to identify quantitative trait loci. Both have accelerated breeding programs (Gupta et al. 1999; Buerstmayer et al. 2002; Anderson et al. 2003; Steiner et al. 2004). Discovery of more precise markers will aid genetic improvement of cereal crops. Recently, Hu and Vick (2003) developed a polymerase chain reaction (PCR)-based technique to target specific chromosome regions. The target region amplification polymorphism (TRAP) technique has been used on Langdon durum–*T. dicoccoides* substitution lines to generate chromosome-specific markers in wheat (Xu et al. 2003).

Since the mid-1980s, the availability of the tools of biotechnology, collectively termed genetic engineering, has helped to asexually incorporate into crop cultivars new traits that are otherwise very difficult to introduce by conventional breeding. Thus, this technology allows access to an unlimited gene pool for genetic enrichment of cereal crops. Most major crops have been genetically transformed by direct DNA delivery via microprojectile bombardment or other means (Jauhar and Khush 2002). The genetic transformation of cereal crops lagged behind other crops, mainly because *in vitro* regeneration techniques, a prerequisite for genetic transformation, were not available, and cereals also showed resistance to *Agrobacterium* infection. However, *in vitro* regeneration and transformation protocols have been developed for bread wheat (Vasil et al. 1992, 1993; see Repellin et al. 2001), durum wheat (Bommineni and Jauhar 1996; Bommineni et al 1997; He et al. 1999; Satyavathi and Jauhar 2003), and other cereals (see Chapters 5, 6, and 8 in this volume). Therefore, valuable genes can now be moved into elite varieties, further enhancing their quality, disease resistance, or productivity, to the limits of our understanding of the genetic basis of these traits. Using transgenic technology, it may be possible to asexually introduce value-added traits, including genes for disease resistance, into otherwise superior cereal cultivars. The transgenic production of golden rice, which is rich in vitamin A and iron, is a remarkable development (Ye et al. 2000) that has the potential of saving millions of lives and averting blindness among millions of children who subsist mainly on rice (see Chapter 5 by Brar and Khush). Thus, *in vitro* approaches to gene transfer can effectively supplement conventional breeding programs.

The advent of molecular tools has helped in the understanding of crop genomes. Cereal crops being the most important, their genomes have been subjected to extensive analyses, ushering in the era of functional genomics. Rich germplasm resources of cereal crops are being tapped to identify valuable genes for crop enrichment. High-throughput genomics strategies are providing new, precise methods for identifying genes for disease resistance, abiotic stress tolerance, and improved nutritional value. New genomics information is also providing molecular markers to accelerate breeding programs and mapped sequences of candidate genes and the traits they control. Thus, cereal genomics is already aiding breeding programs (see Chapter 4 by Lapitan and Jauhar). Gupta and Varshney (2004) further discuss structural, functional, and comparative genomics, and marker-assisted selection, in relation to crop improvement.

Although cereals have evolved independently, perhaps from a common ancestral species, for some 50 to 70 million years (Kellogg 2001; Gaut 2002), and have developed drastically different genome sizes (see Goff et al. 2002), they display a remarkable degree of synteny (Gale and Devos 1998; Gaut 2002; Ilic et al. 2003; Sorrells et al. 2003). Thus, synteny and gene homology between rice and other cereal genomes is very high, and homologs of 98% of the known maize, wheat, and barley proteins are present in rice (Goff et al. 2002). By virtue of the synteny of cereal genomes, it makes more sense to conduct gene search studies on a model crop with the smallest genome. And with a genome of only 430 Mbp and high gene density, rice merits to be such a model crop. Thus, synteny permits gene searches in rice to be applied to other cereals, which is more efficient and better leverages resources than do direct searches in the larger genomes, as long as the trait of interest is shared by rice and other cereals.

Modern biotechnology has great potential to accelerate crop improvement, and the results obtained so far are very encouraging (Borlaug 2000; Swaminathan 1999; Cook 2000; Jauhar and Khush 2002). However, the new technology will augment, but not replace, conventional plant breeding. The old and new technologies should go hand in hand to accelerate cereal crop improvement to sustain global food security. Adoption of a combination of crop improvement tools would help in the sustained improvement of cereal crops that play such an important role in the sustenance and welfare of humankind. With continued improvement in cereal grain yields, nutritional value, and other desirable attributes, it should be possible to effectively feed the future generations of humanity, even if the projected population explosion were to occur.

## REFERENCES

Adams, W.T. and Allard, R.W. 1977. Effect of polyploidy on phosphoglucose isomerase diversity in *Festuca microstachys*. *Proc. Natl. Acad. Sci. U.S.A.* 74: 1652–1656.

Adhikari, T.B., Cavaletto, J.R., Dubcovsky, J., Gieco, J.O., Schlatter, A.R., and Goodwin, S.B. 2004. Molecular mapping of the *Stb4* gene for resistance to *Septoria tritici* blotch in wheat. *Phytopathology* 94: 1198–1206.

Anderson, G.R., Papa, D., Peng, J., and Lapitan, N.L.V. 2003. Genetic mapping of *Dn7*, a rye gene conferring resistance to the Russian wheat aphid in wheat. *Theor. Appl. Genet.* 107: 1297–1303.

Bommineni, V.R. and Jauhar, P.P. 1996. Regeneration of plantlets through isolated scutellum culture of durum wheat. *Plant Sci.* 116: 197–203.

Bommineni, V.R., Jauhar, P.P., and Peterson, T.S. 1997. Transgenic durum wheat by microprojectile bombardment of isolated scutella. *J. Hered.* 88: 475–481.

Borlaug, N.E. 2000. Ending world hunger. The promise of biotechnology and the threat of antiscience zealotry. *Plant Physiol.* 124: 487–490.

Buerstmayer, H., Lemmens, M., Hartl, L., Doldi, L., Steiner, B., Stierschneider, M., and Ruckenbauer, P. 2002. Molecular mapping of QTLs for Fusarium head blight resistance in spring wheat. I. Resistance to fungal spread (type II resistance). *Theor. Appl. Genet.* 104: 84–91.

Ceoloni, C., Vitellozzi, F., Forte, P., Basili, F., Biagetti, M., Bitti, A., and Delre, V. 1998. Wheat chromosome engineering in the light of advanced genetic and cytogenetic marker-mediated approaches. In *Current Topics in Plant Cytogenetics Related to Plant Improvement*, T. Lelley, Ed. WUV-Universitätsverlag, Vienna, Austria, pp. 43–53.

Cook, R.J. 2000. Science based risk assessment for the approval and use of plants in agricultural and other environments. In *Agricultural Biotechnology and the Poor: Proceedings of an International Conference*, G.J. Persley and M.M. Lantin, Eds. Washington, DC, pp. 123–130.

Devos, K.M. and Gale, M.D. 1997. Comparative genetics in the grasses. *Plant Mol. Biol.* 35: 3–15.

Doğramaci-Altuntepe, M. and Jauhar, P.P. 2001. Production of durum wheat substitution haploids from durum × maize crosses and their cytological characterization. *Genome* 44: 137–142.

Drossou, A., Katsiotis, A., Leggett, J.M., Loukas, M., and Tsakas, S. 2004. Genome and species relationships in genus *Avena* based on RAPD and AFLP molecular markers. *Theor. Appl. Genet.* 109: 48–54.

Dubcovsky, J. 2004. Marker assisted selection in public breeding programs: the wheat experience. *Crop Sci.* 44: 1895–1898.

Dujon, B., Sherman, D., Fischer, G., Durrens, P., Casaregola, S., et al. 2004. Genome evolution in yeast. *Nature* 430: 35–44.

Duvick, D.N. 1992. Genetic contributions to advances in yield of US maize. *Maydica* 37: 69–79.

Dvořák, J. 1972. Genetic suppression of homoeologous chromosome pairing in hexaploid wheat. *Can. J. Genet. Cytol.* 14: 39–42.

Dvořák, J., 1998. Genome analysis in the *Triticum-Aegilops* alliance. In *Proceedings of the 9th International Wheat Genetics Symposium*, University of Saskatchewan, Saskatoon, Canada, 1: 8–11.

Dvořák, J., DiTerlizzi, P., Zhang, H.-B., and Resta, P. 1993. The evolution of polyploid wheats: identification of the A genome donor species. *Genome* 36: 21–31.

Endo, T.R. 2003. Wheat stocks carrying chromosomal segments induced by the gametocidal system. In *10th International Wheat Genetics Symposium*, Paestum, Italy, pp. 69–72.

Feldman, M. 1966. The effect of chromosomes 5B, 5D and 5A on chromosome pairing in *Triticum aestivum*. *Proc. Natl. Acad. Sci. U.S.A.* 55: 1447–1453.

Feldman, M. 1968. Regulation of somatic association and meiotic pairing in common wheat. In *Proceedings of the 3rd International Wheat Genetics Symposium*, K.W. Finlay and K.W. Shepherd, Eds., Australian Academy of Science, Canberra, pp. 169–178.

Feldman, M., Liu, B., Segal, G., Abbo, S., Levy, A.A., and Vega, J.M. 1997. Rapid elimination of low-copy DNA sequences in polyploid wheat: a possible mechanism for differentiation of homoeologous chromosomes. *Genetics* 147: 1381–1387.

Forster, B.P. and Miller, T.E. 1989. Genome relationship between *Thinopyrum bessarabicum* and *Th. elongatum*. *Genome* 32: 930–931.

Gale, M.D. and Devos, K.M. 1998. Comparative genetics in the grasses. *Proc. Natl. Acad. Sci. U.S.A.* 95: 1971–1974.

Gaut, B.S. 2001. Patterns of chromosomal duplication in maize and their implications for comparative maps of the grasses. *Genome Res.* 11: 55–66.

Gaut, B.S. 2002. Evolutionary dynamics of grass genomes. *New Phytol.* 154: 15–28.

Gaut, B.S., d'Ennequin, M.L.T., Peek, A.S., and Sawkins, M.C. 2000. Maize as a model for the evolution of plant nuclear genomes. *Proc. Natl. Acad. Sci. U.S.A.* 97: 7008–7015.

Gaut, B.S. and Doebley, J.F. 1997. DNA sequence evidence for the segmental allotetraploid origin of maize. *Proc. Natl. Acad. Sci. U.S.A.* 94: 6808–6814.

Goff, S.A., Ricke, D., Lan, T.-H., Presting, G., Wang, R., et al. 2002. A draft sequence of the rice genome (*Oryza sativa* L. ssp. *japonica*). *Science* 296: 92–100.

Goffeau, A. 2004. Evolutionary genomics: seeing double. *Nature* 430: 25–26.

Gómez, M.I., Islam-Faridi, M.N., Zwick, M.S., Czeschin, D.G., Hart, G.E., Wing, R.A., Stelly, D.M., and Price, H.J. 1998. Tetraploid nature of *Sorghum bicolor* (L.) Moench. *J. Hered.* 89: 188–190.

Gupta, P.K., Balyan, H.S., Edwards, K.J., Isaac, P., Korzun, V., et al. 2002. Genetic mapping of 66 new microsatellite (SSR) loci in bread wheat. *Theor. Appl. Genet.* 105: 413–422.

Gupta, P.K. and Varshney, R.K., Eds. 2004. *Cereal Genomics*. Kluwer Academic Publishers, Dordrecht, The Netherlands.

Gupta, P.K., Varshney, R.K., Sharma, P.C., and Ramesh, B. 1999. Molecular markers and their applications in wheat breeding. *Plant Breed.* 118: 369–390.

Guyot, R. and Keller, B. 2004. Ancestral genome duplication in rice. *Genome*, 47: 610–614.

Harlan, J.R. and de Wet, J.M.J. 1975. On Ö. Winge and a prayer: the origins of polyploidy. *Bot. Rev.* 41: 361–390.

He, G.Y., Rooke, L., Steele, S., Békés, F., Gras, P., Tatham, A.S., Fido, R., Barcelo, P., Shewry, P.R., and Lazzeri, P.A. 1999. Transformation of pasta wheat (*Triticum turgidum* L. var. *durum*) with high-molecular-weight glutenin subunit genes and modification of dough functionality. *Mol. Breed.* 5: 377–386.

Helguera, M., Vanzetti, L., Soria, M., Khan, I.A., Kolmer, J., and Dubcovsky, J. 2005. PCR markers for *Triticum speltoides* leaf rust resistance gene *Lr51* and their use to develop isogenic hard red spring wheat lines. *Crop Sci.* 45: 728–734.

Henzell, B., Jordan, D., Tao, Y., Hardy, A., Franzmann, B., Fletcher, D., MacCosker, T., and Bunker, G. 2002. Grain sorghum breeding for resistance to sorghum midge and drought, In *Plant Breeding for the 11th Millennium, Proceedings on the 12th Australian Plant Breeding Conference*, J.A. McComb, Ed., Australian Plant Breeding Association, Perth, W. Australia, September 15–20, pp. 81–86.

Heun, M., Schäfer-Pregl, R., Klawan, R.C., Accerbi, M., Borghi, B. and Salamini, F. 1997. Site of einkorn wheat domestication identified by DNA fingerprinting. *Science* 278: 1312–1314.

Hu, J. and Vick, B. 2003. Target region amplification polymorphism: a novel marker technique for plant genotyping. *Plant Mol. Bio. Rep.* 21: 289–294.

Huang, S., Sirikhachornkit, A., Su, X., Faris, J., Gill, B., Haselkorn, R., and Gornicki, P. 2002. Genes encoding plastid acetyl-CoA carboxylase and 3-phosphoglycerate kinase of the *Triticum/Aegilops* complex and the evolutionary history of polyploid wheat. *Proc. Natl. Acad. Sci. U.S.A.* 99: 8133–8138.

Ilic, K., SanMiguel, P.J., and Bennetzen, J.L. 2003. A complex history of rearrangement in an orthologous region of the maize, sorghum, and rice genomes. *Proc. Natl. Acad. Sci. U.S.A.* 100: 12265–12270.

Islam, A.K.M.R. and Shepherd, K.W. 1992. Production of wheat-barley recombinant chromosomes through induced homoeologous pairing. 1. Isolation of recombinants involving barley arms 3HL and 6HL. *Theor. Appl. Genet.* 83: 489–494.

Islam, A.K.M.R. and Shepherd, K.W. 2000. Isolation of a fertile wheat-barley addition line carrying the entire barley chromosome 1H. *Euphytica* 111: 145–149.

Islam, A.K.M.R., Shepherd, K.W., and Sparrow, D.H.B. 1981. Isolation and characterization of euplasmic wheat-barley chromosome addition lines. *Heredity* 46: 161–174.

Jauhar, P.P. 1968. Inter- and intra-genomal chromosome pairing in an interspecific hybrid and its bearing on basic chromosome number in *Pennisetum*. *Genetica* 39: 360–370.

Jauhar, P.P. 1970. Haploid meiosis and its bearing on phylogeny of pearl millet, *Pennisetum typhoides* Stapf et Hubb. *Genetica* 41: 532–540.

Jauhar, P.P. 1975a. Genetic control of diploid-like meiosis in hexaploid tall fescue. *Nature* 254: 595–597.

Jauhar, P.P. 1975b. Genetic regulation of diploid-like chromosome pairing in the hexaploid species, *Festuca arundinacea* Schreb. and *F. rubra* L. (Gramineae). *Chromosoma* 52: 363–382.

Jauhar, P.P. 1975c. Genetic control of chromosome pairing in polyploid fescues: its phylogenetic and breeding implications. In *Rep. Welsh Plant Breeding Station (Aberystwyth, Wales, U.K.) for 1974*, pp. 114–127.

Jauhar, P.P. 1975d. Polyploidy, genetic control of chromosome pairing and evolution in the *Festuca-Lolium* complex. Paper presented at *Proceedings of the 178th Meeting of the British Genetical Society*, John Innes Institute, Norwich, England, April.

Jauhar, P.P. 1977. Genetic regulation of diploid-like chromosome pairing in *Avena*. *Theor. Appl. Genet.* 49: 287–295.

Jauhar, P.P. 1981a. Cytogenetics of pearl millet. *Adv. Agron.* 34: 407–479.

Jauhar, P.P. 1981b. *Cytogenetics and Breeding of Pearl Millet and Related Species*. Alan R. Liss, New York, p. 91.

Jauhar, P.P. 1990a. Dilemma of genome relationship in the diploid species *Thinopyrum bessarabicum* and *Thinopyrum elongatum* (Triticeae: Poaceae). *Genome* 33: 944–946.

Jauhar, P.P. 1990b. Multidisciplinary approach to genome analysis in the diploid species, *Thinopyrum bessarabicum* and *Th. elongatum* (*Lophopyrum elongatum*), of the Triticeae. *Theor. Appl. Genet.* 80: 523–536.

Jauhar, P.P. 1991. Recent cytogenetic studies of the *Festuca-Lolium* complex. In *Chromosome Engineering in Plants: Genetics, Breeding, Evolution*, Vol. 2B, T. Tsuchiya and P.K. Gupta, Eds. Elsevier Science Publishers, Amsterdam, The Netherlands, pp. 325–362.

Jauhar, P.P. 1992. Chromosome pairing in hybrids between hexaploid bread wheat and tetraploid crested wheatgrass (*Agropyron cristatum*). *Hereditas* 116: 107–109.

Jauhar, P.P. 2003a. Formation of 2n gametes in durum wheat haploids: sexual polyploidization. *Euphytica* 133: 81–94.

Jauhar, P.P. 2003b. Genetics of crop improvement: chromosome engineering. In *Encyclopedia of Applied Plant Sciences*, Vol. 1, B. Thomas, D.J. Murphy, and B. Murray, Eds. Elsevier Academic Press, London, pp. 167–179.

Jauhar, P.P. 2003c. Haploid and doubled haploid production in durum wheat by wide hybridization. In *Manual on Haploid and Double Haploid Production in Crop Plants*, M. Maluszynski, K.J. Kasha, B.P. Forster, and I. Szarejko, Eds. Kluwer Academic Publishers, Dordrecht, The Netherlands, pp. 161–167.

Jauhar, P.P. and Almouslem, A.B. 1998. Production and meiotic analyses of intergeneric hybrids between durum wheat and *Thinopyrum* species. In *Proceedings of the Third International Triticeae Symposium*, A.A. Jaradat, Ed., Scientific Publishers, Enfield, NH, pp. 119–126.

Jauhar, P.P., Almouslem, A.B., Peterson, T.S., and Joppa, L.R. 1999. Inter- and intragenomic chromosome pairing relationships in synthetic haploids of durum wheat. *J. Hered.* 90: 437–445.

Jauhar, P.P. and Chibbar, R.N. 1999. Chromosome-mediated and direct gene transfers in wheat. *Genome* 42: 570–583.

Jauhar, P.P., Doğramaci, M., and Peterson, T.S. 2004. Synthesis and cytological characterization of trigeneric hybrids involving durum wheat with and without *Ph1*. *Genome* 47: 1173–1181.

Jauhar, P.P., Doğramaci-Altuntepe, M., Peterson, T.S., and Almouslem, A.B. 2000. Seedset on synthetic haploids of durum wheat: cytological and molecular investigations. *Crop Sci.* 40: 1742–1749.

Jauhar, P.P. and Hanna, W.W. 1998. Cytogenetics and genetics of pearl millet. *Adv. Agron.* 64: 1–26.

Jauhar, P.P. and Joppa, L.R., 1996. Chromosome pairing as a tool in genome analysis: merits and limitations. In *Methods of Genome Analysis in Plants*, P.P. Jauhar, Ed. CRC Press, Boca Raton, FL, pp. 9–37.

Jauhar P.P. and Khush, G.S. 2002. Importance of biotechnology in global food security. In *Food Security and Environmental Quality in the Developing World*, R. Lal, D.O. Hansen, N. Uphoff, and N. Slack, Eds. CRC Press, Boca Raton, FL, pp. 107–128.

Jauhar, P.P., Riera-Lizarazu, O., Dewey, W.G., Gill, B.S., Crane, C.F., and Bennett, J.H. 1991. Chromosome pairing relationships among the A, B, and D genomes of bread wheat. *Theor. Appl. Genet.* 82: 441–449.

Kellis, M., Birren, B.W., and Lander, E.S. 2004. Proof and evolutionary analysis of ancient genome duplication in the yeast *Saccharomyces cerevisiae*. *Nature* 428: 617–624.

Kellogg, E.A. 1998. Relationships of cereal crops and other grasses. *Proc. Natl. Acad. Sci. U.S.A.* 95: 2005–2010.

Kellogg, E.A. 2001. Evolutionary history of the grasses. *Plant Physiol.* 125: 1198–1205.

Khush, G.S. 1999. Green revolution: preparing for the 21st century. *Genome* 42: 570–583.

Khush, G.S. and Baenziger, P.S. 1996. Crop improvement: emerging trends in rice and wheat. In *Crop Productivity and Sustainability: Shaping the Future, Proceedings of the Second International Crop Science Congress*, V.L. Chopra, R.B. Singh, and A. Varma, Eds. Oxford and IBH Publishing Co. Pvt. Ltd., New Delhi, pp. 113–125.

Koszul, R., Caburet, S., Dujon, B., and Fischer, G. 2004. Eucaryotic genome evolution through the spontaneous duplication of large chromosomal segments. *EMBO J.* 23: 234–243.

Kurata, N., Moore, G., Nagamura, Y., et al. 1994. Conservation of genome structure between rice and wheat. *Bio/Technology* 12: 276–278.

Kushi, L.H., Meyer, K.A., and Jacobs, D.R. 1999. Cereals, legumes, and chronic disease risk reduction: evidence from epidemiologic studies. *Am. J. Clin. Nutr.* 70 (Suppl.): 451S–458S.

Kynast, R.G., Okagaki, R.J., Rines, H.W., and Phillips, R.L. 2002. Maize individualized chromosome and derived radiation hybrid lines and their use in functional genomics. *Funct. Integr. Genomics* 2: 60–69.

Kynast, R.G., Riera-Lizarazu, O., Isabel Vales, M., Okagaki, R.J., Maquieira, S.B., Chen, G., Ananiev, E.V., Odland, W.E., Russell, C.D., Stec, A.O., Livingston, S.M., Zaia, H.A., Rines, H.W., and Phillips, R.L. 2001. A complete set of maize individual chromosome additions to the oat genome. *Plant Physiol.* 125: 1216–1227.

Ladizinsky, G. 1973. Genetic control of bivalent pairing in the *Avena strigosa* polyploid complex. *Chromosoma* 42: 105–110.

Leitch, I.J. and Bennett, M.D. 1997. Polyploidy in angiosperms. *Trends Plant Sci.* 2: 470–476.

Levan, A. 1942. Studies on the meiotic mechanism of haploid rye. *Hereditas* 28: 6–11.

Lewis, W.H., Ed. 1980. *Polyploidy: Biological Relevance*. Plenum Press, New York.

Linares, C., Ferrer, E., and Fominaya, A. 1998. Discrimination of the closely related A and D genomes of the hexaploid oat, *Avena sativa* L. *Proc. Natl. Acad. Sci. U.S.A.* 95: 12450–12455.

Liu, C.J. and Anderson, J.A. 2003. Marker assisted evaluation of Fusarium head blight resistant wheat germplasm. *Crop Sci.* 43: 760–766.

Liu, C.J., Witcombe, J.K., Pittaway, T.S., Nash, M., Hash, C.T., Busso, C.S., and Gale, M.D. 1994. An RFLP-based genetic map of pearl millet (*Pennisetum glaucum*). *Theor. Appl. Genet.* 89: 481–487.

Liu, S., Sesso, H.D., Manson, J.E., Willett, W.C., and During, J.E. 2003. Is intake of breakfast cereals related to total and cause-specific mortality in men? *Am. J. Clin. Nutr.* 77: 594–599.

Martinez, M., Naranjo, T., Cuadrado, C., and Romero, C. 2001. The synaptic behaviour of *Triticum turgidum* with variable doses of the *Ph1* locus. *Theor. Appl. Genet.* 102: 751–759.

Masoudi-Nejad, A., Nsauda, S., McIntosh, R.A., and Endo, T.R. 2002. Transfer of rye chromosome segments to wheat by a gametocidal system. *Chromosome Research* 10: 349–357.

Masterson, J. 1994. Stomatal size in fossil plants: evidence for polyploidy in majority of angiosperms. *Science* 264: 421–424.

Matsuoka, Y. and Nasuda, S. 2004. Durum wheat as a candidate for the unknown female progenitor of bread wheat: an empirical study with a highly fertile $F_1$ hybrid with *Aegilops tauschii* Coss. *Theor. Appl. Genet.* 109: 1710–1717.

McClintock, B. and Hill, H.E. 1931. The cytological identification of the chromosome associated with the R-G linkage group in *Zea mays*. *Genetics* 16: 175–190.

McFadden, E.S. and Sears, E.R. 1946. The origin of *Triticum spelta* and its free threshing hexaploid relatives. *J. Hered.* 37: 81–89.

Muramatsu, M. 1986. The *vulgare* super gene, Q: its universality in *durum* wheat and its phenotypic effects in tetraploid and hexaploid wheats. *Can. J. Genet. Cytol.* 28: 30–41.

Nandi, H.K. 1936. The chromosome morphology, secondary association and origin of cultivated rice. *J. Genet.* 33: 315–336.

Naranjo, T. 1990. Chromosome structure of durum wheat. *Theor. Appl. Genet.* 79: 397–400.

Naranjo, T., Roca, A., Goicoechea, P.G., and Giraldez, R. 1987. Arm homoeology of wheat and rye chromosomes. *Genome* 29: 873–882.

Nishikawa, K. 1983. Species relationship of wheat and its putative ancestors as viewed from isozyme variation. In *Proceedings of the 6th International Wheat Genetics Symposium*, Kyoto, Japan, pp. 59–63.

Nishiyama, I. and Tabata, M. 1964. Cytogenetic studies in *Avena*. XII. Meiotic chromosome behaviour in a haploid cultivated oat. *Jap. J. Genet.* 38: 311–316.

Ohno, S. 1970. *Evolution by Gene Duplication*. Springer-Verlag, Berlin.

Okamoto, M. 1957. Asynaptic effect of chromosome V. *Wheat Inform. Serv.* 5: 6.

Ozkan, H., Levy, A.A., and Feldman, M. 2001. Allopolyploidy-induced rapid genome evolution in the wheat (*Aegilops-Triticum*) group. *Plant Cell* 13: 1735–1747.

Paterson, A.H., Bowers, J.E., Burow, M.D., Draye, X., Elsik, C.G., Jiang, C.-X., Katsar, C.S., Lan, T.-H., Lin, Y.-R., Ming, R., Wright, R.J. 2000. Comparative genomics of plant chromosomes. *Plant Cell* 12: 1523–1539.

Pickering, R.A. 2000. Do the wild relatives of cultivated barley have a place in barley improvement? In *Barley Genetics VIII. Proceedings of the 8th International Barley Genetics Symposium*, Vol. I, S. Logue, Ed., Department of Plant Science, Waite Campus, Adelaide University, Australia, 2000, pp. 223–230.

Pickering R.A. and Devaux, P. 1992. Haploid production: approaches and use in plant breeding. In *Biotechnology in Agriculture No. 5. Barley: Genetics, Biochemistry, Molecular Biology and Biotechnology*, P.R. Shewry, Ed. CAB International, Wallingford, U.K., pp. 519–547.

Rajhathy, T. and Thomas, H. 1972. Genetic control of chromosome pairing in hexaploid oats. *Nat. New Biol.* 239: 217–219.

Rajhathy, T. and Thomas, H. 1974. Cytogenetics of oats (*Avena* L.). *Miscellaneous Publications of the Genetics Society of Canada* — No. 2. 90 pp.

Repellin, A., Båga, M., Jauhar, P.P., and Chibbar, R.N. 2001. Genetic enrichment of cereal crops by alien gene transfers: new challenges. In *Annual Review of Plant Biotechnology and Applied Genetics*. Kluwer Academic Publishers, Dordrecht, The Netherlands, pp. 159–183.

Riera-Lizarazu, O., Vales, M.I., Ananiev, E.V., Rines, H.W., and Phillips, R.L. 2000. Production and characterization of maize chromosome 9 radiation hybrids derived from an oat-maize addition line. *Genetics* 156: 327–339.

Riley, R. 1960. The diploidisation of polyploid wheat. *Heredity* 15: 407–429.

Riley, R. and Chapman, V. 1958. Genetic control of the cytologically diploid behaviour of hexaploid wheat. *Nature* 182: 713–715.

Riley, R., Kimber, G., and Chapman, V. 1961. Origin of genetic control of diploid like behaviour of polyploid wheat. *J. Hered.* 52: 22–25.

Rooney, W.L. and Klein, R.R. 2000. Potential of marker-assisted selection for improving grain mold resistance in sorghum. In *Technical and Institutional Options for Sorghum Grain Mold Management: Proceedings of an International Consultation*, A. Chandrashekar, R. Bandyopadhyay, and A.J. Hall, Eds., Patancheru 502324, Andhra Pradesh, India, International Crops Research Institute for the Semi-Arid Tropics, May 18–19, pp. 183–194.

Sadasivaiah, R.S. and Kasha, K.J. 1971. Meiosis in haploid barley: an interpretation of non-homologous chromosome association. *Chromosoma* 35: 247–263.

Sarkar, P. and Stebbins, G.L. 1956. Morphological evidence concerning the origin of the B genome in wheat. *Am. J. Bot.* 43: 297–304.

Satyavathi, V.V. and Jauhar, P.P. 2003. *In vitro* regeneration of commercial durum cultivars and transformation with antifungal genes. In *Proceedings of the 2003 National Fusarium Head Blight Forum*, Minneapolis, December 12–15, pp. 32–35.

Schlegel, R. 1990. Efficiency and stability of interspecific chromosome and gene transfer in hexaploid wheat, *Triticum aestivum* L. *Kulturpflanze* 38: 67–78.

Sears, E.R. 1972. Chromosome engineering in wheat. *Stadler Genet. Symp.* 4: 23–38.

Sears, E.R. 1977. An induced mutant with homoeologous pairing in common wheat. *Can. J. Genet. Cytol.* 19: 585–593.

Sears, E.R. 1981. Transfer of alien genetic material to wheat. In *Wheat Science: Today and Tomorrow*, L.T. Evans and W.J. Peacock, Eds. Cambridge Univeristy Press, Cambridge, U.K., pp. 75–89.

Sears, E.R. 1993. Use of radiation to transfer alien segments to wheat. *Crop Sci.* 33: 897–901.

Sears, E.R. and Okamoto, M. 1958. Intergenomic chromosome relationships in hexaploid wheat. *Proc. X Intern. Cong. Genet.* 2: 258–259.

Shi, F. and Endo, T.R. 1999. Genetic induction of structural changes in barley chromosomes added to common wheat by a gametocidal chromosome derived from *Aegilops cylindrica*. *Genes Genet. Syst.* 74: 49–54.

Soltis, D.E. and Soltis, P.S. 1999. Polyploidy: origins of species and genome evolution. *Trends Ecol. Evol.* 9: 348–352.

Somers, D.J., Fedak, G., and Savard, M. 2003. Molecular mapping of novel genes controlling Fusarium head blight resistance and deoxynivalenol accumulation in spring wheat. *Genome* 49: 555–564.

Soreng, R.J. and Davis, J.I. 1998. Phylogenetics and character evolution in the grass family (Poaceae): simultaneous analysis of the morphological and chloroplast DNA restriction site character sets. *Bot. Rev.* 64: 1–85.

Sorrells, M.E., La Rota, M., Bermudez-Kandianis, C.E., Greene, R.A., Kantety, R., et al. 2003. Comparative DNA sequence analysis of wheat and rice genomes. *Genome Res.* 13: 1818–1827.

Stebbins, G.L. 1971. *Chromosomal Evolution in Higher Plants*. Edward Arnold, London.

Steiner, B., Lemmens, M., Griesser, M., Scholz, U., Schondelmaier, J., and Buerstmayr, H. 2004. Molecular mapping of resistance to Fusarium head blight in the spring wheat cultivar Frontana. *Theor. Appl. Genet.* 109: 215–224.

Swaminathan, M.S. 1999. Harness the Gene Revolution to Help Feed the World. *International Herald Tribune*, October 23.

Thomas, H., Powell, W., and Aung, T. 1980. Interfering with regular meiotic behaviour in *Avena sativa* as a method of incorporating the gene for mildew resistance from *A. barbata*. *Euphytica* 29: 635–640.

Thomas, J. and Jones, M.L. 1964. Cytological studies of pentaploid hybrids and a decaploid in *Avena*. *Chromosoma* 15: 132–139.

Ting, Y.C. 1966. Duplications and meiotic behavior of the chromosomes in haploid maize (*Zea mays* L.). *Cytologia* 31: 324–329.

Vandepoele, K., Simillion, C., and Van der Peer, Y. 2003. Evidence that rice and other cereals are ancient aneuploids. *Plant Cell* 15: 2192–2202.

Varshney, R.K., Korzun, V., and Börner, A. 2004. Molecular maps in cereals: methodology and progress. In *Cereal Genomics*, P.K. Gupta and R.K. Varshney, Eds. Kluwer Academic Publishers, Dordrecht, The Netherlands, pp. 35–82.

Varshney, R.K., Sigmund, R., Börner, A., Korzum, V., Stein, N., Sorrells, M.E., Langridge, P., and Graner, A. 2005. Interspecific transferability and comparative mapping of barley EST-SSR markers in wheat, rye and rice. *Plant Sci.* 168: 195–202.

Vasal, S.K. 2002. Quality protein maize: overcoming the hurdles. *J. Crop Prod.* 6: 193–227.

Vasil, V., Castillo, A.M., Fromm, M.E., and Vasil, I.K. 1992. Herbicide resistant fertile transgenic wheat plants obtained by microprojectile bombardment of regenerable embryogenic callus. *Bio/Technology* 10: 667–674.

Vasil, V., Srivastava, V., Castillo, A.M., Fromm, M.E., and Vasil, I.K. 1993. Rapid production of transgenic wheat plants by direct bombardment of cultured immature embryos. *Bio/Technology* 11: 1553–1558.

Wang, G.Z., Miyashita, N.T., and Tsunewaki, K. 1997. Plasmon analyses of *Triticum* (wheat) and *Aegilops*: PCR single-strand conformational polymorphism (PRC-SSCP) analyses of organellar DNAs. *Proc. Natl. Acad. Sci. U.S.A.* 94: 14570–14577.

Weber, D.F. 1970. Doubly and triply monosomic *Zea mays*. *Maize Genet. Coop. Newsl.* 44: 203.

Weber, D.F. 1973. A test of distributive pairing in *Zea mays* utilizing doubly monosomic plants. *Theor. Appl. Genet.* 43: 167–173.

Weber, D.F. 1994. Use of maize monosomics for gene localization and dosage studies. In *The Maize Handbook*, M. Freeling and V. Walbot, Eds. Springer-Verlag, New York, pp. 350–358.

Wendel, J.F. 2000. Genome evolution in polyploids. *Plant Mol. Biol.* 42: 225–249.

Wu, J., Kurata, N., Tanoue, H., Shimokawa, T., Umehara, Y., Yano, M., and Sasaki, T. 1998. Physical mapping of duplicated genomic regions of two chromosome ends in rice. *Genetics* 150: 1595–1603.

Xu, S.S., Hu, J., and Faris, J.D. 2003. Molecular characterization of the Langdon durum-*Triticum dicoccoides* chromosome substitution lines using TRAP (target region amplification polymorphism) markers. In *10th International Wheat Genetics Symposium*, Paestum, Italy, pp. 91–94.

Ye, X., Al-Babill, S., Klötl, A., Zhang, J., Lucca, P., Beyer, P., and Potrykus, I. 2000. Engineering the provitamin A (carotene) biosynthetic pathway into (carotenoid-free) rice endosperm. *Science* 287: 303–305.

Yu, J., Songian, H., Wang, J., Wong, G.K.-S., Li, S., et al. 2002. A draft sequence of the rice genome (*Oryza sativa* ssp. *indica*). *Science* 296: 79–92.

Zhou, W., Kolb, F.L., Bai, G., Shaner, G., Domier, L.L. 2002. Genetic analysis of scab resistance QTL in wheat with microsatellite and AFLP markers. *Genome* 45: 719–727.

Zhou, W., Kolb, F.L., Yu, J., Bai, G., Boze, L.K., and Domier, L.L. 2004. Molecular characterization of Fusarium head blight resistance in Wangshuibia with simple sequence repeat and amplified fragment length polymorphism markers. *Genome* 47: 1137–1143.

# CHAPTER 2

# Chromosome Engineering of the Durum Wheat Genome: Strategies and Applications of Potential Breeding Value

Carla Ceoloni and Prem P. Jauhar*

## CONTENTS

2.1   Introduction..................................................................................................28
2.2   The Evolutionary Pathways of Allopolyploid Wheats ...............................29
     2.2.1   Conservation of Intergenomic Relatedness.....................................30
     2.2.2   Mechanisms of Diploidization and Their Effects at Different Ploidy Levels ..........31
     2.2.3   Induced Haploidy: Its Use in Basic Studies and Genomic Reconstruction..............32
2.3   Wild Relatives as Sources of Desirable Genes............................................33
2.4   Transfer of Alien Genetic Material into Cultivated Wheats........................33
     2.4.1   Synthesis of Hybrids: The First Important and Informative Step...........................34
           2.4.1.1   Hybridization Involving Species with Genomes Homologous to Those of Durum Wheat ...................................35
           2.4.1.2   Hybridizing Durum Wheat with Donor Species with Homoeologous Genomes in the Presence of *Ph1*....................36
           2.4.1.3   Durum Wheat Mutants Lacking *Ph1* and Their Use in Complete Hybrid Combinations .........................37
     2.4.2   Engineering the Durum Wheat Genome with Targeted Introgressions of Limited Size..............................39
           2.4.2.1   Whole-Arm Translocations .................................................39
           2.4.2.2   Transfer of Chromosomal Segments ...................................41
           2.4.2.3   Multiple Combinations of Different Alien Segments ..........47
2.5   Direct Gene Transfer in Durum Wheat.......................................................48
2.6   Conclusions..................................................................................................49
References ............................................................................................................49

Mention of trade names or commercial products in this publication is solely to provide specific information and does not imply recommendation or endorsement by the U.S. Department of Agriculture.

* U.S. government employee whose work is in the public domain.

## 2.1 INTRODUCTION

The agricultural scene characterizing the onset of the third millennium appears to be profoundly different from that of a few decades ago, with concerns about natural resources increasingly acquiring a global dimension. So far, thanks to the widespread adoption of Green Revolution technology (Khush 1999), the demand for increased agricultural productivity has been met by combining genetic improvements with greater farming inputs and cultivation of more land. However, due to a progressive shortage of available farmland, water, and energy reserves, as well as to increased problems concerning the environment's capacity to assimilate the multiple forms of pollution generated by the economic growth, different food production methods need to be investigated. These will have to keep pace not only with the expanding human population's food needs (Braun et al. 1998; Khush 1999), but also with an array of newly arisen needs of environmental and socioeconomic relevance.

In this context, interventions aimed at enriching the seriously threatened genetic base of wheat cultivars with new variability from exotic sources may have great potential. The successful application of transgenic technology to bread wheat (*Triticum aestivum* L., 2n = 6x = 42; AABBDD) and durum wheat (*Triticum turgidum* L., 2n = 4x = 28; AABB) has recently opened up new and promising avenues by giving access to otherwise inaccessible gene pools (reviewed in Jauhar and Chibbar 1999; Jauhar 2001a). On the other hand, cytogenetic approaches, although unable to effect single-gene transfers, permit the engineering of the wheat genome with alien chromosomal introductions, thereby inspiring E.R. Sears to coin the term chromosome engineering (Sears 1972). Thus, wheat improvement can be brought about by these well-established methodologies (Sears 1972, 1981; Ceoloni 1987; Gale and Miller 1987) by exploiting diverse gene pools of Triticeae species (Feldman 1979, 1988; Feldman and Sears 1981). They represent an extremely rich reservoir of desirable genes that can significantly contribute to meet the present and future human needs to which the wheat crop is expected to respond.

The reason why we have been only modestly successful in taking advantage of alien genes for the development of improved wheats of commercial value does not reside in the lack of sufficient knowledge on the evolutionary and cytogenetic relationships between the wild and the cultivated wheat relatives or of proper transfer methods. Indeed, wheat cytogeneticists currently working on chromosome engineering are essentially following the footsteps of E.R. Sears, who highlighted and successfully exploited the main avenues that could lead to "transferring of segments of alien chromosomes carrying particular desired genes to wheat chromosomes" (Sears 1972, 1981, 1983).

Recently, we have witnessed a substantial improvement in monitoring the alien transfer process in all its phases (Ceoloni et al. 1998). To this end, both molecular marker technology and molecular cytogenetic techniques such as nonradioactive *in situ* hybridization — e.g., genomic *in situ* hybridization (GISH) and fluorescent *in situ* hybridization (FISH), or fluorescent GISH (fl-GISH) — can effectively complement classical diagnostic tools for efficient and accurate detection and characterization of introgression products. These procedures facilitate elimination of chromosomes carrying unfavorable traits of the wild donor and retention of only the most suitable ones for the target breeding goal(s).

Such a plentiful and diversified array of analytical methods is effectively assisting the work of wheat chromosome engineers, giving renewed and increased potential for meaningful practical achievements. This is particularly significant in the case of durum wheat, whose evolutionary history, as outlined below, is associated with an overall lower tolerance for genome alterations as compared to common wheat.

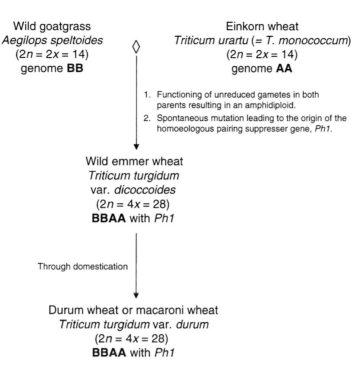

**Figure 2.1**    Steps in the evolution of durum wheat via functioning of unreduced gametes in both progenitors. (From Jauhar, P.P., *Euphytica*, 133, 81–94, 2003.)

## 2.2 THE EVOLUTIONARY PATHWAYS OF ALLOPOLYPLOID WHEATS

One of the most spectacular facets of the studies on plant and animal evolution has been the demonstration that it has not always proceeded by slow, even steps but that there have been "bursts of creative activity" (Anderson and Stebbins 1954). A salient factor in these evolutionary bursts is interspecific hybridization, followed or accompanied in some cases by the stabilizing force of chromosome doubling. These phenomena are well exemplified by the evolution of durum wheat (Figure 2.1 and Figure 2.2).

Durum wheat evolved in nature long before bread wheat. Its two genomes, A and B, were derived, respectively, from diploid wild grasses *Triticum urartu* Tum. (Nishikawa 1983; Dvořák et al. 1993) and *Aegilops speltoides* Tausch (Sarkar and Stebbins 1956; Wang et al. 1997). The two wild progenitors hybridized in nature about half a million years ago (Huang et al. 2002) and gave rise to tetraploid wild emmer wheat in one step because of functioning of unreduced gametes in both parents (Figure 2.1) or, alternatively, in the diploid AB hybrid (amphiploid), as illustrated in Figure 2.2 (Jauhar 2003a). The second route appears more plausible, because unreduced gametes are more likely to occur in the hybrid than in its diploid parents (see Chapter 1 by Jauhar in this volume). At the time of the origin of tetraploid emmer, a spontaneous mutation occurred to produce the homoeologous pairing suppressor gene *Ph1* that permitted pairing only among homologous partners, thereby ensuring diploid-like pairing and disomic inheritance (see Section 2.2.2).

It is through such an evolutionary pathway that cultivated polyploid wheats originated and became perhaps the most outstanding example of successful allopolyploids within the plant kingdom. Both the tetraploid and the later arisen hexaploid wheat are the outcome of hybridization–amphidiploidization events involving different diploid progenitors from the genera *Triticum* and *Aegilops* (Feldman et al. 1995; Jauhar 2003a).

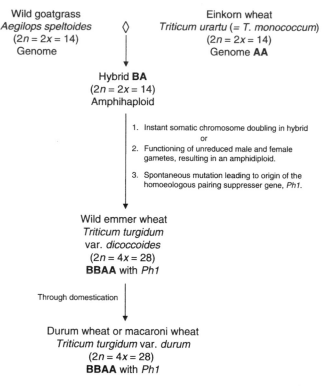

**Figure 2.2**  Steps in the evolution of durum wheat via instant somatic chromosome doubling or functioning of 2n male and female gametes in the BA hybrid (amphihaploid). (From Jauhar, P.P., *Euphytica*, 133, 81–94, 2003.)

## 2.2.1  Conservation of Intergenomic Relatedness

Polyploid wheats can be considered segmental rather than typical genomic allopolyploids (Feldman et al. 1995). Classical cytogenetic studies (Sears 1952, 1954, 1966) provided clear evidence for considerable genetic similarity shared by different genomes contributed by the diploid donors, with the partially homologous (homoeologous) chromosomes of the two (AB) or three (ABD) genomes falling into seven distinct groups of homoeology.

Comparative mapping analyses, extensively carried out in the last decade using a wide array of molecular markers and genes (Ahn et al. 1993; Devos and Gale 1997, 2000; Van Deynze et al. 1998; Keller and Feuillet 2000) have not only largely confirmed the significance of these first observations but also provided corroborating evidence for the concepts of intergenomic relatedness both within the tribe Triticeae and among diverse species of the grass family Poaceae (see also Chapter 1 in this volume). Recent phylogenetic (Kellogg 1998, 2001) and comparative genomic studies (Keller and Feuillet 2000; Feuillet et al. 2001; Feuillet and Keller 2002) have demonstrated the occurrence of many events of genome expansion, contraction, and rearrangements and, at the same time, the maintenance of a remarkable degree of overall conservation among the grass genomes. As for polyploid wheats, despite the existence of intergenomic rearrangements, such as those involving chromosomes 4A, 5A, and 7B of both *T. turgidum* (Naranjo, 1990; Jauhar et al. 2000; Doğramaci-Altuntepe and Jauhar 2001; see also Section 2.2.3, Figure 2.3) and *T. aestivum* (Naranjo et al. 1987; Anderson et al. 1992; Devos et al. 1995), a large body of evidence indicates that chromosomes belonging to each homoeologous group retained considerable gene orthology and colinearity during the course of evolution (e.g., Hart 1987; Anderson et al. 1992; Devos et al. 1993b; Van Deynze et al. 1995), with not only overall gene content but also physical location, structural organization, and gene density of gene-rich regions being similar among the component

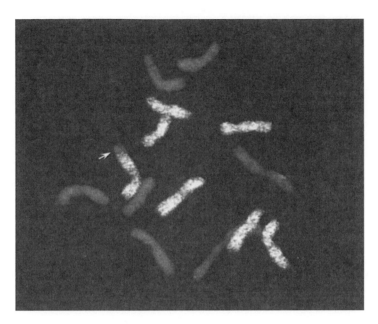

**Figure 2.3**  **(See color insert following page 114.)** Fourteen somatic chromosomes of a substitution haploid (derived from a D genome disomic substitution line 5D(5B) of Langdon durum wheat) after fluorescent genomic *in situ* hybridization. The brightly fluorescing (bright yellow) chromosomes belong to the A genome. (Chromosome 5D, partially hybridized with the A genome probe but masked by the propidium iodide counterstain, is not clearly seen in the photograph.) The six B genome chromosomes are in red. Note one dose of the 4A/7B translocation chromosome (arrow), which is a part of the evolutionary translocation 4A/5A/7B present both in durum and in common wheat; in this translocated chromosome, the distal segment of 7BS constitutes approximately 24% of the long arm of 4A. The 5A chromosome segment cannot be visualized because it stains the same as the 4A segment. (From DoğramacI-Altuntepe, M. and Jauhar, P.P., *Genome*, 44, 137–142, 2001.)

genomes (Keller and Feuillet 2000; Sandhu et al. 2001). As expected, the degree of such a structural and functional similarity turned out to be higher among Triticeae genomes than between these and the genomes of more distant grass species (Moore et al. 1995; Devos and Gale 2000).

## 2.2.2  Mechanisms of Diploidization and Their Effects at Different Ploidy Levels

Several different genetic and epigenetic mechanisms contributed to the successful establishment of allopolyploid wheats. Prevention of pairing between homoeologous chromosomes belonging to the constituent diploid genomes was an essential step, as exclusive bivalent pairing of homologous chromosomes ensured regular chromosome segregation and hence disomic inheritance and full fertility to the newly formed polyploid species. Since the early 1960s, several studies (e.g., Riley, 1960; Sears, 1976; Feldman, 1993) showed suppression of homoeologous pairing in wheat to be due to the action of a complex genetic system, with the *Ph1* gene, located in the long arm of chromosome 5 of genome B, exerting the strongest effect. Fixation of this gene in the primitive tetraploid wheat (and later of additional though less potent ones at the hexaploid level) represented an essential step in the cytological diploidization of the newly arisen polyploids. However, other mechanisms, being brought about concurrently with or immediately after the formation of the polyploids, probably provided the physical basis for the diploid-like meiotic behavior of polyploid wheats, later reinforced by the *Ph1* system (Feldman et al. 1997).

As inferred by recent studies on newly synthesized wheat allopolyploids (Feldman et al. 1997, Liu et al. 1998a,b; Ozkan et al. 2001; Shaked et al. 2001; Feldman et al. 2002; Kashkush et al. 2002), the mechanisms of polyploidization resulted in a variety of rapid genetic and epigenetic changes that affected, in a nonrandom and highly reproducible manner, both coding and noncoding

DNA sequences, with elimination of chromosome- and genome-specific sequences of the latter type being particularly relevant in accentuating the physical divergence among homoeologs (Feldman et al. 1997; Liu et al. 1998b, Ozkan et al. 2001; Shaked et al. 2001).

On the other hand, changes in the structure and expression of many coding DNA sequences contributed to the genetic diploidization and thus to the overall fitness of newly formed allopolyploids, avoiding negative effects due to gene redundancy. Processes such as selective gene silencing or even gene loss were recently shown to start at early stages postpolyploidization (Kashkush et al. 2002). Also, analyses at the tetraploid level clearly indicated the progression of loss of duplicate gene expression in the course of evolution and domestication. In contrast to the relatively little diploidization undergone by storage protein genes in the wild tetraploid wheat *T. dicoccoides*, a massive and nonrandom diploidization of these genes evidently occurred in the primitive cultivated type *T. turgidum* var. *dicoccum* and proceeded even further in modern cultivars of var. *durum* (Galili and Feldman 1983; Feldman et al. 1986).

Based on the available evidence, including recent results on synthetic allotetraploid (Kashkush et al. 2002) and allohexaploid (He et al. 2003) wheats, it may be assumed that the variety of regulatory processes mentioned above, including selective gene elimination, silencing, or activation, affecting one or both homoeoalleles of the natural tetraploid wheat forms, continued to take place at the hexaploid level, independent of the diploidization process at the tetraploid level (e.g., Garcia-Olmedo and Carbonero 1980; Galili and Feldman 1983; Feldman et al. 1986; Galili et al. 1986; He et al. 2003). However, their overall effect on cultivated tetraploid wheat, due to its lower ploidy level and probably also to its more ancient origin (Feldman et al. 1995; Salamini et al. 2002), was more extreme.

Several studies concerning mutational effects on cultivated *Triticum* and other Triticeae species at different ploidy levels (Mac Key 1959, 1981) clearly showed that resistance to the mutagenic treatments and tolerance to mutations increased as a whole from diploid to hexaploid types. However, the mutation spectrum for essential genes (exemplified by chlorophyll mutations) of the 4x *Triticum dicoccum* was in sharp contrast with that of 6x wheat, nearly coinciding with 2x *Triticum monococcum*. Such observations are in line with a large body of additional evidence indicating a definitely lower buffering ability of the *T. durum* genome toward genic and chromosomal alterations, compared to that of *T. aestivum*.

A clear proof of low buffering ability of the durum genome is provided by the production of aneuploids. Whereas a wide array of intra- as well as interspecific aneuploid types have been successfully produced in the background of common wheat (e.g., Sears 1954, 1966, 1975), considerable difficulty has been encountered in developing and maintaining aneuploid stocks of *T. durum* (reviewed in Joppa 1988, 1993; Ceoloni et al. 2005b). Reduced cytological stability, plant vigor, and fertility characterize not only intraspecific (particularly hypoploid) aneuploids but also additions and substitutions of alien chromosomes to tetraploid wheat, thus limiting their use in both basic and applied research. In this regard, a remarkable exception is presented by the D genome disomic substitution lines, of which a complete set was developed in the variety Langdon (Joppa and Williams 1988). In such lines, a pair of D genome chromosomes, derived from common wheat cv. Chinese Spring, replaces the corresponding A or B genome pairs. Because of a generally satisfactory compensation exerted by the D genome homoeologs, most of these lines are sufficiently stable and fertile to be used in determining the chromosomal location of genes and quantitative trait loci (QTL) in durum wheat (see, e.g., Joppa et al. 1983; Gorham et al. 1987), and in transferring desirable D genome characters in it (see Sections 2.4.2.1 and 2.4.2.2.3).

### 2.2.3  Induced Haploidy: Its Use in Basic Studies and Genomic Reconstruction

Haploid sporophytes with half the chromosome number afford opportunities for studying chromosome relationships within and between species. An easy technique of producing wheat haploids in large numbers, by hybridizing appropriate wheat material with maize, has been standardized (Almouslem et al. 1998; Jauhar 2003b). Study of chromosome pairing in synthetic wheat haploids, with and without *Phl*, allows assessment of intergenomic relationships. Thus, the closer affinity relating the A and D genomes compared to that of either one to B clearly emerged from the study of *Phl* and *phlb* haploids of wheat (Jauhar et al. 1991). The specificity of chromosome pairing in haploids with and without *Phl* can be even better highlighted by fl-GISH (Jauhar et al. 1999). By fl-GISH, then, physical demonstration of the 4AL/7BS translocation, which is part of the evolutionary translocation 4A/5A/7B present in durum wheat (Naranjo 1990), as well as in hexaploid wheat (Naranjo et al. 1987; Anderson et al. 1992; Devos et al. 1995), was also provided (Figure 2.3; see also Jauhar et al. 2000; Dorğamaci-Altuntepe and Jauhar 2001).

Induced haploidy offers an easy means of stabilizing heterozygous wheat material with desirable alien genes. Advanced hybrid derivatives may be haplodized by crossing with maize (Jauhar 2003b). Subsequent chromosome doubling would bring about homozygosity for the introduced alien genes and hence stabilize the reconstructed genome. Direct introduction of genes at the haploid level, followed by chromosome doubling, may also help in stable genetic transformation (Jauhar and Chibbar 1999).

## 2.3 WILD RELATIVES AS SOURCES OF DESIRABLE GENES

Wild relatives of wheat are useful sources of genes that can be utilized for durum improvement. Wild emmer (*Triticum turgidum* var. *dicoccoides* Körn) is a wild tetraploid wheat that shares the A and B genomes with durum wheat. It is in the primary gene pool of durum wheat and crosses readily with it. Wild emmer is known to have genes for resistance to several diseases, including stem rust (Rajaram et al. 2001), Fusarium head blight (FHB) (Miller et al. 1998; Stack et al. 2002), stripe rust (Sun et al. 1997), and powdery mildew (Liu et al. 2002; Xie et al. 2003), as well as for higher protein content (Joppa et al. 1991; Elias et al. 1996) and for novel glutenin subunits and gliadins (Vallega and Waines 1987; Levy et al. 1988; Pflüger et al. 2001; Xu et al. 2004). The diploid donors of the A and B genomes to durum wheat also hold considerable potential as sources of agronomically desirable genes.

Wild perennial relatives are also important sources of genes that could be used for genetic improvement of wheat via wide hybridization. Hybridization with perennial grasses in the genera *Agropyron*, *Thinopyrum*, and *Lophopyrum* has contributed to the genetic enrichment of bread wheat (Feldman and Sears 1981; McIntosh 1991; Jiang et al. 1994; Friebe et al. 1996; Jauhar 1993; Jauhar and Chibbar 1999; Repellin et al. 2001). These genera harbor genes for resistance to several diseases, including FHB, that hold promise for durum improvement. Two of the wheatgrasses —tetraploid wheatgrass (*Thinopyrum junceiforme* (Löve & Löve) Löve, $2n = 4x = 28$; $J_1J_1J_2J_2$ genomes) and diploid wheatgrass (*Lophopyrum elongatum* (Host) Á. Löve, $2n = 2x = 14$; EE) — are excellent sources of resistance to FHB (Jauhar 2001b; Jauhar and Peterson 1998, 2001). From the tertiary gene pool of wheat, Canadian workers have found one accession of *Elymus humidus* to be immune to FHB (Fedak 2000).

## 2.4 TRANSFER OF ALIEN GENETIC MATERIAL INTO CULTIVATED WHEATS

For cultivated wheats, with particular reference to durum wheat, the choice of the most suitable procedure for interspecific and intergeneric gene transfer depends primarily on whether the alien chromosome(s) carrying the desirable gene(s) is completely or only partially homologous (= homoeologous) to the corresponding chromosome(s) of the recipient species. Using this criterion, different methods for the exploitation of alien variability can be explored for incorporation of as much as the entire alien genome, particularly in the form of chromosome-doubled hybrids, i.e., amphidiploids, down to a single chromosome or chromosome arm pair (either added or substituted) or just a small chromosomal segment (Feldman 1983; Gale and Miller 1987; Jauhar 2003c). In practice, however, only rarely have sizable wheat–alien chromosome manipulations resulted in valuable materials for breeding, particularly when durum wheat is the recipient (see also Section 2.4.2).

Synthetic amphiploids have generally proved inferior to established crops. In this respect, the man-made cereal, triticale — the amphiploid derivative from the cross between tetraploid *T. durum* and rye, *Secale cereale* L. (Larter 1976) — probably represents the only exception (see Chapter 13 in this volume). In fact, in spite of a greatly stabilized pairing behavior and hence largely recovered fertility with respect to the corresponding $F_1$ hybrids, the excessive content of undesirable genes from the alien parent makes synthetic amphiploids unable to compete with well-adapted wheat cultivars. However, they have been usefully employed as bridging material for the transfer of specific alien genes into common and durum wheat (for the latter, see, e.g., Sections 2.4.1.1 and 2.4.1.3). They can also serve as valuable starting points for the production of alien chromosome addition and substitution lines in a wheat background, which in turn serve as ideal materials for the introduction of alien subchromosomal segments. Such a sequence of steps, leading to a progressive reduction of unwanted genetic contribution from the alien donor and allowing early verification of the alien gene expression in wheat, indeed represents the most elegant layout for targeted chromosome engineering.

However, as stated above (see Section 2.2.2), additions or substitutions of alien chromosomes or even chromosome arms (see also Section 2.4.2.1) to tetraploid wheat generally cause more detrimental effects, including lower cytological stability and plant fertility, than corresponding aneuploid conditions at the hexaploid level (reviewed in Ceoloni et al. 2005b). Consequently, except for the previously stated cases involving D genome disomic substitution lines, alien addition, substitution, or translocation lines of the hexaploid *T. aestivum* have been employed in several cases as donors of defined alien chromosomal segments to durum wheat (see Section 2.4.2).

## 2.4.1   Synthesis of Hybrids: The First Important and Informative Step

Whether followed by conventional breeding procedures involving crossing and backcrossing or by more sophisticated cytogenetic methodologies, the synthesis of hybrids between the recipient and the donor Triticeae species represents the first important step for alien gene transfer into wheat via sexual means. Indeed, a hybrid state, with all the wheat and alien genomes residing in a common nucleus, probably represents the best stage where all the new, exotic variation, together with even unpredictable, positive interactions or epistatic effects among the new gene combinations, can be fully revealed. Synthesis of hybrids and backcross derivatives remains the most direct route to follow in cases where the chromosomal location of the target gene(s) is unknown or when the desired trait is under a multigenic control (Jauhar and Almouslem 1998; Jauhar and Peterson 2001).

In all cases, apart from the possible accomplishment of practical targets, a remarkable achievement of wide hybridization is the production of fundamental knowledge on the intergenomic affinities between the donor and the recipient species, which of course is an important prerequisite for undertaking a proper transfer strategy. In this regard, much information has been gained from the analysis of the meiotic process, particularly from metaphase I pairing behavior of chromosomes of different genomes brought together in $F_1$ hybrids and derivatives. In fact, despite a number of limitations, assessment of chromosome pairing ability remains a valuable indicator of intergenomic affinity (Jauhar and Joppa 1996). This is particularly the case when this classical method of genome analysis is assisted by advanced cytogenetic approaches, such as the use of *in situ* hybridization onto meiotic metaphase I cells to discriminate the individual or genomic identity of the pairing partners (e.g., Cuadrado et al. 1997; Jauhar et al. 1999; Jauhar and Peterson 2001).

On the other hand, erroneous conclusions can be drawn if genomic affinities are estimated only on the basis of fertility/sterility of $F_1$ hybrids. In fact, observation of several hybrids between *T. durum* and *Aegilops* (Vardi 1973; Maan and Sasakuma 1977; Maan et al. 1980) as well as *Thi-*

*nopyrum* species (Jauhar and Peterson 2001) indicated pollen fertility and seed set to be positively correlated with reduced metaphase I pairing or even complete asynapsis, which eventually produced a higher number of unreduced functional gametes. Lack of pairing was in fact hypothesized to be the prerequisite for the occurrence of meiotic restitution and hence spontaneous chromosome doubling in haploids of durum wheat carrying the wild-type allele of the homoeologous pairing suppressor gene *Ph1* (Jauhar et al. 2000).

### 2.4.1.1 Hybridization Involving Species with Genomes Homologous to Those of Durum Wheat

At the beginning of the 20th century, even before the main principles of genetics and cytogenetics became known, a pioneering intuition that infusion of alien variation into the cultivated germplasm could be beneficial to enhance the wheat crop led a number of enlightened geneticists and breeders around the world to practice interspecific and intergeneric hybridization (e.g., Carleton 1901; Strampelli 1932; McFadden and Sears 1947). Because of durum wheat's more restricted area of cultivation and lower economic importance on a world scale, and also because of greater difficulties inherently associated with manipulations of its genome as compared to that of hexaploid wheat, durum wheat has been less frequently employed in interspecific crosses.

Initially, such crosses were almost exclusively performed with closely related species sharing one or both genomes with those of the durum parent. In a number of early, successful attempts, tetraploid wheat was crossed with common wheat, where it worked as a donor of desirable traits for the improvement of the latter. Thus in 1921, the Italian breeder Nazareno Strampelli made a cross between *T. durum* cv. Duro di Puglia and the Japanese *T. aestivum* cv. Akagomughi, from which the common wheat variety Balilla was developed (Strampelli 1932). At about the same time, Hayes et al. (1920) crossed Yumillo *durum* with common wheat cv. Marquis and obtained the leaf and stem rust-resistant line Marquillo, from which the successful cultivar Thatcher was derived. Similarly, McFadden (1930) crossed Yaroslav emmer with Marquis and produced the leaf and stem rust-resistant lines Hope and H44-24, both of which were widely used in breeding for rust resistance.

Later, the gene *Pm4a*, conferring resistance to powdery mildew, was transferred from *T. dicoccum* chromosome arm 2AL into that of durum and common wheat varieties (Zitelli 1973). Among other important genes for breeding, the *Rht-B1b* (formerly *Rht1*) dwarfing gene, located on 4BS, was transferred in the 1970s from the 6x variety Norin 10 into most Italian durum wheats (Vallega and Zitelli 1973). Similarly, *Rht9*, located on 7BS, was introduced from common wheat into Italian durum varieties (B. Giorgi, personal communication). In all such cases, transfer of the desired genes could be successfully achieved because, in the absence of cross-compatibility barriers, the complete homology between the A and B genomes of the tetraploid with those of the hexaploid parent posed no obstacle to gene flow between them.

This of course applies also to interspecific crosses involving wild *Triticum* species, such as *T. boeoticum* and *T. dicoccoides*, which share, respectively, one A or both A and B genomes with cultivated tetraploid wheat. Several examples are available of useful traits, including resistance to various diseases and other characters of agronomic value (e.g., high spikelet number, large grain size, high protein and high lysine content) being introduced from its immediate progenitors into durum wheat. Given the complete genomic correspondence and the consequent regular homologous pairing between the A and B genome chromosomes of wild tetraploid *T. dicoccoides* and those of the cultivated *T. durum*, exploitation of the rich gene pool of the former for both disease resistance and quality traits has been achieved with relative ease in several instances (e.g., Gerechter-Amitai and Grama 1974; Avivi et al. 1983; Pasquini et al. 1992; Singh et al. 1998a).

As stated above (see Section 2.3), *T. dicoccoides* is being exploited as a source of FHB resistance. Langdon–*dicoccoides* disomic substitution lines (LDN(Dic)) were evaluated for type II resistance, and the Langdon durum line with a pair of chromosomes 3A from a *T. dicoccoides* accession, i.e., LDN(Dic-3A), showed reduced levels of FHB infection compared to the controls

(Elias et al. 1996). More recently, Stack et al. (2002) studied LDN(Dic) disomic chromosome substitution lines for their hypersensitivity to FHB and found LDN(Dic-3A) to have the least infection (19.8%). This line is now being used in durum breeding programs; however, material showing stable FHB resistance has not been obtained so far.

In the case of diploid A genome sources (*T. boeoticum, T. urartu, T. monococcum*), the sterility problems associated with the parental ploidy differences were successfully overcome by using either the resulting triploid hybrids (Vardi and Zohary 1967; Gerechter-Amitai et al. 1971) or the corresponding amphiploids (Mujeeb-Kazi 1998) as bridges for gene transfer between the diploid and the tetraploid species. Cytologically stabilized products with a remarkable restoration of fertility could be recovered after a number of backcrosses with the recipient durum wheat.

### 2.4.1.2 Hybridizing Durum Wheat with Donor Species with Homoeologous Genomes in the Presence of Ph1

Hybridization has also been attempted between durum wheat and less closely related species than those mentioned above. In principle, the more distant the alien species from the cultivated crop, the greater the possibility of introducing totally new genetic material into the cultivated background. At the same time, however, exploitation of distant gene pools in genetic improvement of cultivated forms becomes more complicated when intergenomic relationships between the donor and recipient species are more homoeologous than homologous, with consequent difficulties in coping with possible unfavorable linkage drags from the donor genome(s).

This looks somewhat paradoxical when one considers that, in addition to the classical test of homoeology, i.e., the ability of an alien chromosome to adequately compensate for a wheat chromosome that it replaces (Dvořák 1980; Tuleen and Hart 1988), a variety of more rapid and accurate methods of assigning alien chromosomes to homoeologous groups have proven that even the chromosomes of the more distant relatives share with cultivated wheats substantial similarity in relative gene content and order (Hart and Tuleen 1983; Ono et al. 1983; Miller and Reader 1987; Urbano et al. 1988; Gill et al. 1988; Yang et al. 1996; see also Section 2.2.1 and references therein). Even in cases where complete colinearity between wheat chromosomes and alien chromosomes, which is clearly reminiscent of a common ancestry (see also Chapter 1), is interrupted by evolutionary translocations, as typically occurs in the rye genome relative to that of wheat (Devos et al. 1993a), a regional conservation of homoeology is disclosed when constraints to the inherent wheat–alien chromosome pairing ability are removed (Naranjo et al. 1987; Cuadrado et al. 1997).

Whereas the extremely low pairing frequency usually observed between wheat chromosomes and alien homoeologous chromosomes or chromosome segments can be partly attributed to their genetic differentiation, it largely reflects the activity of genes that suppress homoeologous pairing (see Section 2.2.2). Of these, the most potent is the *Ph1* gene, which exerts its effect in restricting pairing to homologous chromosomes both within the wheat complement and in interspecific and intergeneric hybrids with alien Triticeae.

Thus, pairing or lack of pairing of chromosomes of different genomes in the wheat background in the presence of *Ph1* provides a crucial test of intergenomic relationships (Jauhar and Crane 1989; Jauhar and Joppa 1996). For instance, the high pairing detected in hybrids between durum wheat and tetraploid *Dasypyrum hordeaceum*, showing no appreciable difference in the presence or absence of the *Ph1* gene, revealed the largely autotetraploid origin of the wild perennial species, and at the same time indicated a low degree of affinity between the wheat and the *Dasypyrum* genomes (Blanco and Simeone 1995). A similar situation was revealed by fl-GISH using total genomic DNA of the alien species as a probe on meiocytes of tetraploid hybrids ($ABJ_1J_2$) between durum wheat and the tetraploid wheatgrass *Thinopyrum junceiforme* (Jauhar and Peterson 2001). The considerable amount of pairing (64%) observed within the *Thinopyrum* complement indicated a close relationship between the $J_1$ and $J_2$ genomes. In the same hybrids, 37% pairing also occurred within the durum complement, a much higher value than that observed in *Ph1* haploids of durum wheat (Jauhar et al. 1999). This outcome, together with some low intercomplement pairing (0.06%) revealed by fl-GISH, suggested at least a partial inactivation of the wheat *Ph1* gene by the genotype of the grass parent, thereby increasing the possibility of transferring at least some of the genetic determinants of the Fusarium head blight resistance from the wild parent into durum wheat.

The E genome of diploid *L. elongatum* and the J genome of diploid *Thinopyrum bessarabicum* (Savul. & Rayss) Á. Löve (= *Agropyron bessarabicum* Savul. & Rayss) are particularly valuable sources of genes for several agronomically desirable traits (Dvořák and Ross 1986) and have been separately incorporated into durum wheat (Jenkins and Mochizuki 1957; Jauhar 1991). Jauhar (1992) combined both genomes with durum wheat in trigeneric hybrids with genomic constitution of ABJE and used fl-GISH to study the specificity of chromosome pairing: wheat–wheat, grass–grass, and wheat–grass (Jauhar et al. 2004). The seven chromosomes of the E genome show close genetic correspondence to the seven homoeologous groups of wheat (Dvořák 1980). More-over, the E genome is reported to have genes that promote or suppress homoeologous pairing (Dvořák 1987; Charpentier et al. 1988). Taking advantage of this promotion of pairing, Jauhar and Peterson (2000a,b) produced FHB-resistant germplasm of durum wheat.

The existence of *Ph1*-antagonistic genes was also hypothesized in other interspecific and intergeneric combinations involving *T. durum* and a number of wild relatives, including *Aegilops ovata* (Simeone et al. 1984), *D. villosum* (Blanco et al. 1988), and *Thinopyrum curvifolium* (Jauhar and Almouslem 1998). However, the ability of alien genes to counteract the action of the main homocologous pairing suppressor, *Ph1*, is generally limited to certain genotypes of the alien species, and their efficiency in inducing pairing between homoeologs, particularly of distant genomes, does not appear to be as high as that produced by the absence of chromosome 5B or by mutations at the *Ph1* locus (reviewed in Jauhar and Chibbar 1999). Therefore, as illustrated in the following sections, use of *ph1* mutants represents the most effective strategy to obtain wheat–alien transfers.

In cases where the wheat and alien genomes are distantly related, translocations induced by ionizing radiations, or even obtained during *in vitro* culture, have been suggested as an alternative to the genetically mediated homoeologous exchanges. However, in spite of a number of successful results (reviewed in Feldman 1983; Jauhar and Chibbar 1999), this essentially remains a random process, with rather laborious procedures and a low probability of yielding acceptable transfers.

### 2.4.1.3 Durum Wheat Mutants Lacking Ph1 and Their Use in Complete Hybrid Combinations

Under conditions in which the entire 5B chromosome or a subchromosomal portion of it comprising the *Ph1* locus is missing, the pairing potential inherent in homoeologous chromosomes becomes manifested, both within the wheat complement and in complete or partial wheat–alien combinations. Two types of durum wheat genotypes have been produced that determine such an effect: the Langdon 5D(5B) disomic substitution line (Joppa and Williams 1979, 1988) and the *ph1c* mutant lines (Giorgi 1978, 1983).

Although the presence of chromosome 5D provides a good compensation for the absence of 5B, considerable homoeologous pairing occurs in 5D(5B) disomic substitution plants (Joppa and Williams 1979). The resulting translocations and their accumulation during repeated selfings were apparently responsible for a drastic decline in vigor and fertility of such a line following its initial isolation. The line is therefore maintained as monosomic 5B or monotelosomic 5BL. In either case, the presence of even a single dose of the *Ph1* gene is sufficient to suppress homoeologous pairing, and thus circumvents the associated sterility problems (Joppa and Williams 1988). However, selection must then be carried out in the cross-progeny for plants lacking the extra 5B monosome or telosome to have the desired induction of homoeologous pairing.

Nonetheless, the use of the 5D(5B) disomic substitution line proved to be an efficient system to promote pairing in a number of hybrids involving durum wheat and *Thinopyrum* species carrying desirable genes for resistance to wheat rusts, barley yellow dwarf virus, and Fusarium head blight

(Jauhar and Almouslem 1998). Thus, in the ABJ triploid hybrids obtained by crossing diploid *Th. bessarabicum* and 5D(5B) disomic substitution line, chromosome pairing increased more than fourfold in the absence of 5B as compared to their counterparts with a normal *Ph1*. A considerable increase in pairing was also observed in hybrids with tetraploid *Th. curvifolium* when 5B was replaced by 5D. Indeed, in a comparison between durum wheat haploids carrying either the mutant *ph1c* allele or chromosome 5D in place of 5B, intergenomic pairing, as revealed by fl-GISH, seemed to be higher in the latter than in the former genotype (Jauhar et al. 1999). Such a result might be attributed to 5D, which carries a pairing promoter on each arm, with an overall effect, at least in *T. aestivum* (Sears 1976), estimated to be greater than that of the 5BS promoter, which is lacking in the substitution line while present in the *ph1c* mutant.

Although the possible promoting effect of 5D could be advantageous, the *ph1c* mutation has been perhaps more frequently used to induce homoeologous pairing in wheat–alien hybrid combinations and in targeted alien transfers into durum wheat via chromosome engineering (see Section 2.4.2). The *ph1c* mutation, determining a clearly *Ph1⁻* meiotic pairing phenotype, was obtained by seed treatment of the durum wheat cv. Cappelli (Giorgi 1978). The same mutation was later incorporated via hybridization and backcrosses into the background of the high-yielding, short-strawed variety Creso (Giorgi 1983). In the original line, the 5B pair appeared to be heteromorphic, with one member of the pair being shorter and the other longer than normal 5B. Upon selfing of this line, each of the modified 5Bs was isolated in homozygous condition, giving rise to two different lines. C-banding analysis of these lines (Dvořák et al. 1984) demonstrated the presence of a deletion and a tandem duplication, respectively, of the same chromosomal region containing the *Ph1* locus on 5BL, probably originating from a single event of unequal interchange between 5B homologs or sister chromatids.

Availability of zero, two, and four doses of the *Ph1* gene in Cappelli background represents an ideal tool to investigate the mechanism of action of this crucial pairing regulator. The results obtained from a recent study on the synaptic behavior of the above *T. durum* genotypes indicate that the effects on the *Ph1*-mediated diploidization mechanism of tetraploid wheat are exerted during meiotic prophase I (Martinez et al. 2001b), the same as in hexaploid common wheat (Martinez et al. 2001a). Correction of homoeologous synapsis and suppression of crossing-over between homoeologous regions would be very efficient in the genotypes with four and two doses of *Ph1*. On the contrary, in the absence of *Ph1*, the mechanism checking for homology would be less efficient, allowing crossing-over to take place between homoeologous chromosomes, with consequent formation of multivalent associations (Martinez et al. 2001b).

Physical and genetic map positions of the *Ph1* locus along the 5BL arm could be precisely determined by use of the *ph1c* mutant genotype (Jampates and Dvořák 1986; Gill et al. 1993). Within the deleted region, whose size was estimated to be less than 3 Mb (Gill et al. 1993), markers in addition to a relatively faint C-band have been located, which are very useful to screen for the presence or absence of the *Ph1* gene in materials of specific interest (e.g., Sections 2.4.2.2.2 and 2.4.2.2.3). They include molecular genetic markers of the restriction fragment length polymorphism (RFLP) type (Gill and Gill 1991; Gill et al. 1993), polymerase chain reaction (PCR) based (Gill and Gill 1996; Segal et al. 1997; Qu et al. 1998), and a physical marker consisting of an *in situ* hybridization site determined by the pSc119.2 highly repeated DNA sequence on normal 5BL and absent in *Ph1* deleted genotypes (Gill et al. 1993, see Figure 2.4). Use of this last probe is particularly advantageous, as it allows discrimination of all possible allelic conditions, including heterozygosity, at the critical locus.

Prior to the development of specific markers, detection of micronuclei in tetrads was taken as a good indicator of the presence of the mutated 5B in segregating populations (Giorgi 1983). In fact, dyads and tetrads with micronuclei, as well as laggards at anaphases I and II, can be observed as a result of aberrant segregation of univalents and multivalents, which can be present, albeit rarely, in meiocytes of the *ph1c* mutant.

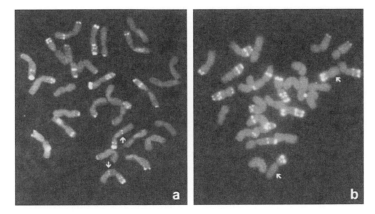

**Figure 2.4** **(See color insert following page 114.)** FISH with the highly repeated pSc119.2 sequence as probe of both normal *Ph1* (a) and *ph1c* mutant (b) of durum wheat cv. Creso. Arrows point to the pSc119.2 site, which is present on wheat 5BL arm (a) while lacking in the *ph1c* mutant (b).

In spite of a somewhat reduced seed set, both in the *ph1c* mutant lines and in their hybrids and amphiploids with alien Triticinae, crossability turned out to be fairly good. Several hybrids were developed involving different *Aegilops* species (*Ae. longissima, Ae. kotschyi, Ae. columnaris,* and *Ae. triuncialis;* Giorgi et al. 1981) as well as more distant relatives, such as *Secale cereale* (Giorgi et al. 1981) and *Th. bessarabicum* (King et al. 1993). A remarkable increase in the amount of pairing was observed in all these cases. The use of the fl-GISH technique on meiocytes of the latter hybrid allowed discrimination of the pairing partners and estimation of the relative frequency of wheat–wheat (over 82%) and wheat–*Th. bessarabicum* (about 13%) associations (King et al. 1993).

From a number of the $F_1$ hybrids reported above, amphiploids lacking the *Ph1* gene were developed by colchicine treatment, including a primary hexaploid triticale (Giorgi et al. 1981; Giorgi 1983; Giorgi and Ceoloni 1985). They showed fairly good fertility, though slightly lower than their controls bearing the wild-type allele *Ph1*. Potentially useful genes, such as those for resistance to wheat leaf rust, stem rust, and mildew, present in the alien species and proven to be expressed in their amphiploid derivatives (Giorgi, unpublished), could be transferred into the A and B genomes of durum wheat. Using the durum wheat recombinants as donors, or by directly crossing the *ph1c* amphiploids with a *ph1b* mutant of common wheat, the same traits could also be incorporated into the latter.

## 2.4.2 Engineering the Durum Wheat Genome with Targeted Introgressions of Limited Size

In one of his milestone papers on wheat chromosome engineering, E.R. Sears stated that, barring prior knowledge of the existence of other advantageous alien genes in addition to the target ones, "the cytogeneticist's goal must be to deliver to the breeder a transfer chromosome that includes the shortest possible alien segment" (Sears 1981). This is particularly true of alien transfers involving durum wheat, where incorporation of alien segments of minimal size becomes an essential requisite for overall stability and consequent practical usefulness. To this end, great progress can be achieved by using tools of molecular genetics and cytogenetics, which can make selection and characterization of desired genotypes far more efficient and accurate (Ceoloni et al. 1998).

### 2.4.2.1 Whole-Arm Translocations

One possibility of introducing into a cultivated background less than an entire alien chromosome is to substitute one arm of the latter for a corresponding wheat arm (Sears 1972). In wheat, such transfers are made possible by the strong tendency of univalent chromosomes to misdivide at meiosis and give rise to one-armed (telocentric) chromosomes. If both an alien chromosome and its wheat homoeolog are present as univalents and misdivide in the same sporocyte, the resulting telocentrics can occasionally fuse at the centromere and produce a bibrachial chromosome, with one alien and one wheat arm. One major limit of such a method for introducing alien variation resides in the low frequency with which the correct, balanced combination can be recovered, whether from spontaneous or from radiation-induced events. Moreover, a whole arm often possesses too much unwanted alien chromatin along with the target gene(s). However, before the extensive use of induced homoeologous pairing as the method of choice and, in some instances, to make possible use of desired introgressions involving A or B genome chromosomes already obtained at the 6x level, a number of alien transfers had been attempted into durum wheat involving entire chromosomal arms.

Among the few that met with relatively good success in a tetraploid background is the 1BL.1RS translocation. This translocation, together with the corresponding one involving wheat chromosome 1A (1AL.1RS), represents the most successful wheat–alien transfer effectively employed in common wheat breeding worldwide (Braun et al. 1998; Graybosch et al. 1999; Lukaszewski 2000). The short arm of rye chromosome 1R is known to carry many important genes for resistance to wheat pathogens, including the yellow rust resistance gene *Yr9*, the leaf rust resistance gene *Lr26*, the stem rust resistance gene *Sr31*, the powdery mildew resistance genes *Pm8* and *Pm17*, and the green bug resistance genes *Gb2* and *Gb6* (Friebe et al. 1996). Although resistance conferred by some of the 1RS genes has been overcome in a number of countries, a substantial interest remains in the use of such a translocation for improving several agronomic traits (Villareal et al. 1998). On the other hand, breeding lines and cultivars carrying 1RS have been found to produce flour with a pronounced dough-quality defect (Graybosch et al. 1993), which has so far prevented their use in the development of high-quality bread wheats. Poor-quality characteristics, attributed to the presence of secalins encoded by the rye *Sec-1* locus or to the absence of the wheat *Glu-B3* encoded glutenins (see also Graybosch et al. 1993), were also detected in tetraploid lines homozygous for the same 1BL.1RS whole-arm translocation (Boggini et al. 1998). However, further engineering of the 1RS arm present in hexaploid lines (Koebner and Shepherd 1986; Koebner et al. 1986; Lukaszewski 2000) and also in a tetraploid background (Pogna et al. 1993; Mazza et al. 1995) seems to hold good promise for overcoming the quality defects associated with the whole-arm translocation.

Transfer of such a translocation to tetraploid wheat was initially obtained by crossing the common wheat cv. Veery to the durum wheat cv. Cando (Friebe et al. 1987). Of the two plants with 28 chromosomes that were eventually isolated, in which C-banding revealed the presence of the 1RS arm pair, one survived and had good seed set. However, in $F_2$ progeny derived from crosses such as a 1BL.1RS tetraploid line, named CV256, and different Italian durum wheats, a deviation from the expected 1:2:1 ratio was observed for the presence of the 1RS and 1BL storage protein markers, indicating a reduced transmission of the 1BL.1RS chromosome through the male gametes (Mazza et al. 1995). The same competitive disadvantage of the translocated chromosome when in competition with a normal 1B was also observed at the hexaploid level (Koebner and Shepherd 1986; Lukaszewski 2000).

An even less satisfactory compensation was obtained in other durum wheat–whole alien arm translocations. This is exemplified by two translocations involving the short arm of Imperial rye (*Secale cereale*) chromosome 3R, harboring the *Sr27* stem rust resistance gene, and of one durum wheat–*Agropyron elongatum* translocation carrying the alien *Sr26* gene (Rao 1978). In the latter case, a translocation line of the hexaploid wheat variety Thatcher, incorporating a nearly whole 6AeL alien chromosome arm in place of the wheat 6AL (Friebe et al. 1996), was used as starting material for crosses and backcrosses to a durum-susceptible parent. A common wheat cv. Chinese Spring ditelosomic addition line with a pair of added rye telocentrics for the short arm of chromosome 3R was instead used as a donor to durum of the *Sr27* gene. After two backcrosses to a susceptible durum wheat, stem rust-resistant monotelosomic additions (14" + t') were isolated and then subjected to different radiation treatments. No homozygous lines could be derived from either of the two resistant translocations thus obtained, exhibiting a 2n = 28 and presumably containing an entire 3RS arm, because of absence of transmission of the translocated chromosome through the male germ line. Similarly, totally adverse selection was suffered by the *Ag. elongatum* translocation bearing *Sr26* when in a durum background. This same translocation was well tolerated by the hexaploid common wheat genome and, as such, has been extensively used in breeding in Australia (McIntosh et al. 1995).

Another case illustrating a differential impact of a given alien translocation at the hexaploid and tetraploid levels is represented by the 2BS.2RL centric-break fusion translocation carrying the rye *H21* gene for resistance to the Hessian fly, *Mayetiola destructor*. Such a whole-arm transfer, originally obtained as a result of tissue culturing of common wheat–rye hybrids, was shown not only to well compensate for the missing 2BL arm in terms of vigor and fertility but also to have some heterotic effects in hexaploid wheat germplasm (Friebe et al. 1990). In the course of the transfer of the same translocation into durum wheat, although male transmission was recovered in later stages, plant vigor and fertility remained considerably reduced in homozygous compared to heterozygous carriers (Friebe et al. 1999). This of course prevents the direct exploitation of the 2BS.2RL germplasm for durum wheat improvement, for which further shortening of the alien chromatin introgression appears a necessary additional step.

In contrast to the above translocations, good overall performance was apparently associated with the substitution of the long arm of common wheat chromosome 1D for durum's 1AL homoeologous arm (Joppa et al. 1998). The transferred 1DL carried the *GluD1d* allele coding for the high-molecular-weight glutenin (HMWG) subunits 5 + 10, with which high-quality bread-making properties are known to be associated. After crossing the donor bread wheat cultivar Len and the Langdon 1D(1A) disomic substitution line, followed by backcross of tetraploid or near-tetraploid $F_2$ derivatives to Langdon durum, double monosomic lines for chromosome 1A and a Len/Chinese Spring recombined 1D were isolated. Selected tetraploid plants homozygous for the 1DL translocation exhibited not only positive effects in terms of gluten quality but also good agronomic characteristics compared to common wheat Len and several durum wheat checks. This can probably be attributed to the considerable affinity and consequent high compensating ability relating the A and D genome chromosomes (Naranjo et al. 1987).

### 2.4.2.2 Transfer of Chromosomal Segments

The general outcome of the research described above demonstrates that a whole alien arm, particularly if it originates from relatively distant alien species, finds little acceptance by durum wheat. Among the possible strategies of chromosome engineering that enable controlled introductions of chromosome segments from related Triticeae into cultivated wheats (Sears 1972), the one based on manipulations of the wheat chromosome pairing control system, particularly the use of mutations of *Ph1*, is by far the most effective.

As mentioned earlier (see Section 2.4.1.2), radiation-induced transfers have the inherent drawback of producing essentially random translocations. Although many such translocations in wheat tend to involve homoeologous chromosomal segments, these are greatly outnumbered by those involving nonhomoeologs. Moreover, even for compensating transfers, attainment of intercalary translocations remains an extremely rare event. The only documented case of an intercalary wheat–alien transfer produced by radiation treatment appears to be that of a noncompensating translocation of a relatively short 6RL segment containing the *H25* gene for resistance to a Hessian fly proximally inserted into the wheat 4AL arm (Friebe et al. 1996). This introduction, originally obtained in common wheat, was recently incorporated via homologous recombination into durum. Gametic transmission of the translocated chromosome, showing disturbance on the male side in the $BC_1$ generation, appeared normal in $BC_2$ and $BC_3$ derivatives. Moreover, the resultant translocation stock was vigorous and had a seed set similar to the durum parent Cando (Friebe et al. 1999), thus demonstrating its usefulness in breeding.

Apart from this exceptional case, the past and more recent experience on wheat chromosome engineering clearly highlights that the *Ph1*-mediated approach offers the greatest promise for obtaining alien transfers. This is because disturbance of the recipient chromosome and genotype balance, if any, is minimal, and the benefits of the alien genetic contribution(s) are more readily available for practical utilization. One reason for this is the cytogenetic affinity of the recipient chromosome to the donor chromosome, because homoeologs are almost exclusively involved in *ph1*-promoted pairing. When the critical alien chromosome or chromosome arm is as an addition or substitution into the wheat genome, the potential for pairing between the alien and the wheat homoeologs is the highest if the two are present as univalents (Sears 1972, 1981; Ceoloni et al. 1988). Another important advantage of the *ph1*-mediated approach resides in the possibility to progressively shorten the amount of alien chromatin flanking the desired gene(s) during successive steps of the transfer procedure. This can be accomplished in different ways, depending on the materials available as well as on the position of the target gene(s) on the chromosome. If the alien gene is located toward the end of an arm, a single distal exchange can produce wheat recombinant products with short, terminal alien segments. If, however, the alien gene has a median or more proximal location, additional manipulations are needed to further shorten the alien segment due to the expected low frequency of double crossovers, especially between homoeologous chromosomes. One possibility consists of allowing the donor and recipient chromosomes to undergo repeated rounds of *ph1*-induced homoeologous recombination (see Section 2.4.2.2.3). This, however, can cause an excessive accumulation of unwanted background translocations, leading to considerable gametic and zygotic instability and eventually to loss of potentially desirable types (Ceoloni et al. 1996). An alternative and perhaps better strategy was originally suggested and successfully applied by Sears (1981, 1983). By combining two complementary transfer chromosomes, which resulted from single exchanges on the proximal and distal sides of the target gene, respectively, crossing-over can occur in the homologous region shared by them. This will give rise to a product equivalent to that of a double exchange, i.e., an alien insert containing the gene of interest. For such an elegant approach to be effective in bringing about a maximum reduction of the unwanted alien chromatin, sufficient numbers of the primary recombinant chromosomes are required, among which the ideal candidates can be selected (Lukaszewski 1998; Section 2.4.2.2.2).

To reach a high level of resolution in the course of selections of this kind, adequate tools are necessary and are currently available. A number of examples described below illustrate the progress attained in chromosome engineering of the durum wheat genome by means of refined cytogenetic approaches assisted by a variety of efficient selection procedures.

### 2.4.2.2.1 The Value of Genetic and Physical Mapping of Wheat–Alien Recombinant Chromosomes

Detection and precise analysis of wheat–alien recombination events can be achieved by both genetic and physical mapping of the breakpoint positions along the wheat–alien exchanged chromosomes. However, even though genetic maps can be useful to establish a relative ranking among a series of recombinant products (e.g., Donini et al. 1995; Cenci et al. 1999), they represent poor indicators of physical distances along chromosomes. One example is provided by the transfer of a distal segment of the short arm of *Ae. longissima* $3S^1$ chromosome. A recombinant $3BS/3S^1S$ chromosome arm, containing the alien powdery mildew resistance gene *Pm13*, had been homologously introduced into durum from a primary common wheat homoeologous transfer (Ceoloni et al. 1988, 1992, 1996). Normal transmission of the recombinant chromosome through both germ lines in its tetraploid derivatives (Ceoloni et al. 1996) indicated that the same primary recombination product obtained at the 6x level was also well tolerated at the 4x level. In fact, the size of this specific $3S^1S$ segment was proved to represent less than 20% of the physical arm length. This could be determined by applying FISH with a highly repeated (pSc119.2) and low-copy RFLP sequence (PSR907) as probes (Biagetti et al. 1999). These physical markers allowed the alien segment to be precisely located distal to the 3BS *Xpsr907* locus, in the adjacent subtelomeric interval separating the two most distal pSc119.2 sites of 3BS. The *Xpsr907* locus, thus shown to be physically located in the proximal border of the most distal quarter of 3BS, is known to be genetically positioned at less than 25 centimorgans (cM) from the 3B centromere (Devos et al. 1992).

The above example, together with additional ones given below, highlights how molecular cytogenetic techniques, such as FISH using specific DNA sequences or total genomic DNA (fl-GISH) of the alien species as probes, allow a precise assessment of the physical amount of exchanged material. This represents a critical parameter to estimate the potential impact of an alien transfer on the recipient genotype, particularly when operating at the less tolerant 4x ploidy level.

### 2.4.2.2.2 Reducing the Size of a Targeted Chromosomal Region Originating from Th. ponticum

The validity of physical mapping as a characterization and selection methodology in the course of transfer processes can be further exemplified by the work on incorporation into durum wheat of *Lr19* (resistance to leaf rust) and *Yp* (yellow pigmentation) genes, known to be closely associated on the long arm of chromosome 7Ag of *Th. ponticum* (syn. *A. elongatum* and *Lophopyrum ponticum*, $2n = 10x = 70$). Such a tight linkage, which is unfavorable for common wheat breeding (Sears 1978; Knott 1980), was instead considered to be of great interest for durum wheat improvement, for which both the leaf rust resistance and the yellow color of semolina and pasta products could prove beneficial. In a first step, a 7A/7Ag common wheat recombinant line (Sears 1973; Eizenga 1987) was employed to move the 7Ag segment from the hexaploid into a tetraploid background by homologous recombination. However, almost no male transmission of the carrier chromosome was observed in the progeny of tetraploid heterozygous resistant plants. A plausible explanation was provided by the results of a fl-GISH analysis using total *Th. ponticum* DNA as a probe (Ceoloni et al. 1996), which revealed the alien portion to span the whole long arm and about half of the short arm of the primary recombinant chromosome, thus probably being too large to be tolerated in a tetraploid background.

To develop suitable genotypes for homoeologous pairing and recombination to occur between portions of the critical chromosome pair 7A and 7Ag, and thus reduce the size of the alien segment, tetraploid plants bearing the primary 7A/7Ag chromosome were crossed and backcrossed to the *ph1c* mutant of the durum wheat cv. Creso. Among heterozygous 7A–7A/7Ag individuals selected by fl-GISH, those homozygous for the *ph1c* mutation were further selected by applying FISH with the pSc119.2 highly repeated DNA sequence as probe (see Section 2.4.1.3; see also Figure 2.4). GISH-based physical maps of 10 secondary recombinants with exchanges involving the 7AL and 7AgL critical arms showed recombination to have occurred in all cases in the distal half of the arms. Six such recombinants possessed a 7A chromosome with a distal 7AgL segment (spanning from 22 to 40% of the 7AL/7AgL arm length), and four had a 7A/7Ag chromosome containing terminal 7AL segments (from 10 to 44% of the recombinant arm length).

The correlation between the physical maps of the recombinant chromosomes and the *Lr19* and *Yp* phenotypes of the corresponding lines indicated that the alien genes, of which *Lr19* is more proximally located, are included in the most distal quarter of 7AgL. *Lr19*, in particular, can be allocated in the 1% fraction differentiating the two recombinants with the smallest distal alien segments, one resistant to leaf rust and possessing 23% of 7AgL (Figure 2.5), the other susceptible, with 22% of 7AgL. GISH screening of selfed heterozygous recombinants also revealed a clearly inverse correlation between transmission ability of the 7A/7Ag chromosomes and their relative alien chromatin content, with 7AgL distal segments spanning 28% of the arm length corresponding

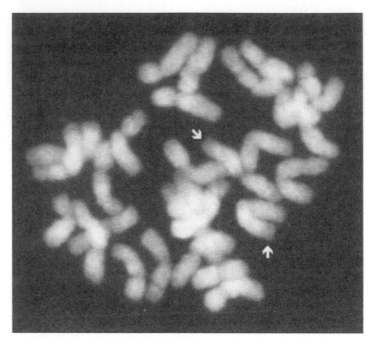

**Figure 2.5    (See color insert following page 114.)** GISH, preceded by preannealing of total genomic DNAs of the donor and recipient species, of a chromosomally engineered durum wheat line homozygous for a distal *Thinopyrum ponticum* segment. This segment, spanning 23% of the 7AL/7AgL recombinant arm, contains the *Lr19* + *Yp* alien genes (see Section 2.4.2.2.2).

to the uppermost limit for normal transmission of the recombinant chromosome through both germ lines.

Although some of the secondary recombinants already represent promising candidates for breeding, such as the line containing all the desired *Thinopyrum* genes in a distal 7AgL 23% long segment (Figure 2.5), the structure and genetic content of two of them provided the opportunity for further manipulation of the target alien region. The two selected lines, both expressing the desired *Lr19* and *Yp* phenotypes, have 7A/7Ag chromosomes with complementary patterns of 7AL and 7AgL chromatin (one having a 23% distal 7AgL and the other a 10% distal 7AL). From homologous recombination that occurred in their $F_1$ hybrid combination within the shared 7AgL region, plants with a 13% subterminal, intercalary 7AgL segment have been recently isolated (Ceoloni et al. 2000, 2005a). Such a tertiary recombinant chromosome, while equivalent to the best of the secondary types in terms of presence of the targeted genes, shows a considerable reduction of associated alien chromatin. This chromosome makeup might then have an even better impact on the recipient genotype than any of the ones previously recovered. Preliminary field tests have indicated a highly satisfactory performance of backcross derivatives obtained from crosses of durum cultivars with the secondary recombinant possessing a 23% 7AgL distal segment for both plant vigor and fertility, as well as for other parameters of agronomic relevance (Ceoloni et al. 2000; Gennaro et al. 2003).

Homozygous plants with the intercalary 7AgL segment, together with most of the 7AL/7AgL secondary recombinants, are being currently used to develop genetic maps of the various durum wheat–*Th. ponticum* chromosomes, which will be compared to their corresponding GISH-based maps. By means of such a translocation mapping approach the physical location is being determined for a number of molecular markers (RFLPs, simple sequence repeats (SSRs), etc.), which will provide good marker coverage of Triticeae group 7 long-arm distal regions.

## 2.4.2.2.3 Additional Useful Durum Wheat Transfers Containing D Genome Chromosomal Segments

The requisite size limit of the alien segment and consequent suitability for potential breeding use has been met by a number of transfers in which durum wheat was engineered with D genome chromosomal segments of various origins and gene content. One such work concerned the transfer of chromosome 1D segments derived from common wheat and containing the *Glu-D1* (1DL) and, separately, the *Gli-D1/Glu-D3* (1DS) storage protein genes (Ceoloni et al. 1995, 1996). The 1D-controlled alleles of *Glu-1* and *Glu-3* genes are known to have a major impact on quality attributes typical of *T. aestivum* (Gupta et al. 1994), which are essentially absent in or little expressed by *T. durum*. Moreover, results from a comparative analysis of technological properties of the complete set of Langdon D genome substitutions into durum had shown chromosome 1D to contribute significantly to dough quality improvement, particularly when it replaced chromosome 1A (Liu et al. 1995).

To induce recombination between chromosome 1D and its durum wheat homoeologs, a common wheat variety possessing the *Glu-D1d* allele (HMWG subunits 5 + 10) was initially crossed with Chinese Spring *ph1b* mutant. After crossing and backcrossing of their $F_1$ with the *ph1c* mutant of the durum cv. Cappelli, two lines were eventually isolated that separately integrated a 1D portion on the short and long arms of chromosome 1A, respectively (Ceoloni et al. 1995, 1996).

To ascertain the recombinant nature of the *Gli-D1/Glu-D3* carrier line, the sole use of endosperm protein markers was sufficient (Ceoloni et al. 1996). On the other hand, for the line expressing the HMWG subunits 5 + 10 coded by the *Glu-D1d* locus, evidence for a 1AL/1DL exchange was provided by FISH using the pAs1 highly repeated DNA sequence as a probe. In fact, pAs1 shows characteristic hybridization sites in the distal portions of 1DL and 1DS, with no site on 1AL and a minor, distal one on 1AS (Vitellozzi et al. 1997). When probed with this sequence, the 5 + 10 durum wheat line exhibited a normal 1AS and a clearly recombined 1AL/1DL chromosome, in which only the minor and more distal of the two pAs1 sites typical of 1DL, located in the distal third of the arm, was retained. On this basis, and assuming the likely occurrence of a single, distal crossover between 1AL and 1DL, the 1DL portion was first estimated to represent about 25 to 30% of the recombinant 1AL (Vitellozzi et al. 1997). However, a recent reinvestigation of the 1AS.1AL/1DL chromosome by means of a technical variant of the fl-GISH technique, where *in situ* hybridization is preceded by a preannealing step between the donor and recipient probe DNAs (Anamthawat-Jónsson and Reader 1995), indicates the 1DL segment to be intercalary and subterminally located (Figure 2.6), with a physical length corresponding to nearly half of the previous

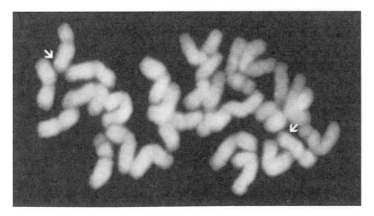

**Figure 2.6** **(See color insert following page 114.)** Preannealing GISH of D and AB total genomic DNAs highlights an interstitial 1DL common wheat segment (arrowed), containing the *Glu-D1* locus, transferred into 1AL of durum wheat by chromosome engineering (see Section 2.4.2.2.3).

estimate (Carozza et al. 2003). Whether the two crossing-over events (which evidently produced the intercalary segment) occurred in the course of the same meiosis or in subsequent generations has not been ascertained so far. Though rather exceptional between homoeologous chromosomes, the former possibility cannot be excluded, given the overall close affinity between the A and D genomes, and, specifically, the particularly high levels of 1AL–1DL homoeologous pairing under both *ph1* permissive and even normal *Ph1* conditions (see Naranjo et al. 1987; Jauhar et al. 1991).

As to the recombinant 1AS arm, both the pAs1 and the preannealing fl-GISH patterns suggested the occurrence of a single-exchange event, which left on the recipient arm less than 20% of the 1DS terminal segment. Similar to what is reported above for the durum wheat–*Th. ponticum* chromosomes, physical mapping is being combined with corresponding genetic mapping for the 1A/1D recombinants (Carozza et al. 2003, unpublished). Both the 1AL.1AS/1DS and 1DL/1AL.1AS recombinant chromosomes exhibited normal transmission through both germ lines (Ceoloni et al. 1996; Vitellozzi et al. 1997). Moreover, their presence was found to be associated with a highly positive enhancement of gluten quality of the carrier durum wheat lines (Ceoloni et al. 2000, 2003) but did not have any apparent detrimental effects on agronomic performance of derivatives of both 1A/1D recombinant types (Ceoloni et al. 2000).

A seemingly good acceptance was also shown by the durum wheat genome toward the introgression of a 1AL.1AS/1DS translocation similar to that described above but with the bread wheat cv. Perzivan-1 as the donor parent (Pogna et al. 1993). Two gliadin biotypes were identified in this variety (Metakovsky et al. 1990); biotype 1 expressed the protein phenotype of the *Gli-D1g* allele (Cheyenne type) associated with good bread-making quality, whereas biotype 2 was quite unusual in having both the *Gli-D1g* and *Gli-D1a* (Chinese Spring type) alleles, the latter associated with lower quality. Segregation in the progeny from the biotype 1 × biotype 2 cross suggested the presence in both Perzivan-1 genotypes of a spontaneously translocated *Gli-D1g* locus on the short arm of chromosome 1A, with the resident *Gli-D1* locus being different (*Gli-D1g* in biotype 1 and *Gli-D1a* in biotype 2). Genetic mapping of the translocated 1AS/1DS arm, whose origin is unknown, allowed allocation of the 1DS segment marked by *Gli-D1g* to a distal position relatively close to *Gli-A1* (Redaelli et al. 1992). The segment was found to contain the *Glu-D3* locus, coding for several low-molecular-weight (LMW) glutenin subunits, along with the *Gli-D1* locus (Pogna et al. 1993). Positive results from quality tests suggested the possibility of enhancing bread-making qualities of durum wheat by introducing the Perzivan-1 translocation. To this end, biotype 2 was crossed and backcrossed with the durum wheat cv. Rodeo. Tetraploid lines were identified in the resulting progeny, in which the Perzivan-1-derived translocation was stably incorporated (Mazza et al. 1995).

Another potentially beneficial trait for durum wheat is the $Na^+/K^+$ discrimination ability controlled by a gene(s) located on the D genome of common wheat. Examination of $Na^+/K^+$ ratios in Langdon D genome disomic substitution lines showed that at least 50 to 60% of the higher efficiency in excluding $Na^+$ and accumulating $K^+$ in the leaves (which makes common wheat more salt tolerant than durum) was accounted for by chromosome 4D (Dvořák and Gorham 1992). This chromosome was thus induced to recombine in a durum wheat background by crossing and backcrossing the Langdon 4D(4B) substitution with the *ph1c* Cappelli mutant. Genetic and cytogenetic evidence indicated that the $Na^+/K^+$ discrimination is controlled by a single gene, *Kna1* (on the long arm of chromosome 4D), which was incorporated on the long arm of 4B in all recombinant lines obtained (Dvořák and Gorham 1992). Molecular mapping showed *Kna1* to be located within the distal half of the 4DL genetic map, with all 4B recombinant chromosomes with *Kna1* harboring relatively long 4DL terminal segments (Dubcovsky et al. 1996), with associated potentially detrimental effects on yield (Dvořák et al. 1994). Because the spectrum of recombinants precluded the attainment of a *Kna1*-carrying intercalary 4DL segment by homologous recombination between two 4D/4B chromosomes with overlapping 4DL segments, as previously reported for durum wheat–*Th. ponticum* derivatives, an alternative strategy was adopted (Luo et al. 1996). Two of the recombinants isolated, possessing the shortest 4DL segments still including *Kna1*, were subjected to a second round of induced homoeologous recombination. As in the first cycle, heterozygosity for the 4B–4D/4B chromosome pair was selected on the basis of the C-banding pattern of the critical chromosomes, whereas homozygosity for *ph1c* was verified by the absence of an RFLP marker located within the *ph1c* deletion (see Section 2.4.1.3). A number of 4B chromosomes containing intercalary 4D segments with *Kna1* were thus recovered and characterized by RFLP mapping, with the shortest 4DL segment spanning only a few centimorgans along the 4BL arm (Luo et al. 1996).

Similar to the 4B/4D case, resorting to a second cycle of homoeologous recombination was considered a possibility to shorten a 7D$^v$L segment originating from *Aegilops ventricosa* (2n = 4x = 28; genomes D$^v$M$^v$) and containing the eyespot resistance gene *Pch1*, which was recently introduced into the 7AL arm of durum wheat (Huguet-Robert et al. 2001). Development of resistant durum lines was planned as a means of eventually increasing the copy number of the *Pch1* gene in common wheat genotypes, as well as of improving resistance of hexaploid triticale. In fact, common wheat lines carrying the *Ae. ventricosa* gene *Pch1* transferred into chromosome 7D are already available (Jahier et al. 1978). However, to reach the goal of incorporating an additional dose of the gene into either chromosome 7A or 7B, a strategy was adopted that did not rely on the use of the existing common wheat recombinants as donors. Instead, an ABD$^v$M$^v$ F$_1$ hybrid was produced with a different *Ae. ventricosa* accession (no. 11), which not only showed a highly resistant phenotype, but also had a different allele at the *Ep-D1* locus (tightly linked to *Pch1*) from that present in the line previously employed. This polymorphism enables marker-assisted selection of the double *Pch1* introgression (on 7D and on 7A or 7B) in hexaploid lines (Huguet-Robert et al. 2001).

In the above hybrid, the Creso *ph1c* mutant was used as the durum parent. The colchicine-induced amphiploid was then backcrossed either twice to Creso *ph1c* or, alternatively, once to the *ph1c* mutant and subsequently to the *ph1b* mutant of the hexaploid wheat cv. Courtot. From the latter backcross, selfing of one resistant BC$_2$F$_3$ plant with 2n = 32 chromosomes produced homozygous resistant individuals with 2n = 28, which were then backcrossed to normal Creso. Meiosis of heterozygous derivatives was regular, with 14 bivalents in most cells, indicating that the *Pch1* transfer into durum consisted of a chromosomal segment and not of an entire alien chromosome. Evidence from both the zymogram of endoptidase (*Ep-1*) markers and the PCR-amplified patterns of group 7-specific microsatellites suggested that in all of the five lines isolated, presumably derived from the same exchange event, recombination involved the distal portions of the 7D$^v$L and 7AL homoeologous arms. Although F$_2$ segregation of heterozygous recombinant plants showed no distortion with respect to a normal 3:1 transmission of *Pch1*, the presence of distal 7D$^v$L portions on both 7DL and 7AL might create problems in terms of meiotic stability of the double recombinant product in a hexaploid background. In fact, homologous 7D$^v$L pairing may occur between the otherwise homoeologous 7AL and 7DL arms, as it seems possible given the distal location of the alien segments, with consequent formation of quadrivalent associations. In that case, production of a further engineered product, resulting in an intercalary transfer, would be an effective strategy to circumvent the problem and make the hexaploid double transfers of practical usefulness.

### 2.4.2.3 *Multiple Combinations of Different Alien Segments*

It would be highly desirable to combine in the same wheat genotype different alien genetic contributions effecting an enhancement of a number of relevant traits of the recipient genome. For the gene pyramiding approach undertaken via chromosome engineering strategies, a reasonable expectation would be an overall suffering from limitations that are inherent even in the most refined cytogenetic manipulations, such as the inability to shorten the alien segment size beyond certain limits. Nonetheless, as exemplified above, in chromosome-mediated transfers assisted by high-resolution analyses, isolation of introgression products with the size of single alien segments conveniently minimized is a feasible objective. This opens, at least in principle, the possibility of their successful pyramiding. In common wheat, perhaps the only documented case is that of the combination of the 7DS.7AgL (from *Th. ponticum*) and the 1BS.1RS (from *S. cereale*), both being whole-arm translocations, which did not seem to have detrimental effects on the recipient wheat for the simultaneous alien introduction per se (Singh et al. 1998b). To our knowledge, no case history is so far available in durum wheat. However, in such a species, as adequately outlined above, the contemporary presence of two complete alien arms would certainly not be tolerated.

Recently, attempts have been made to create double and even triple alien segment combinations by intercrossing the best selections of different chromosomally engineered lines, separately carrying segments with the *Lr19 + Yp* genes from *Th. ponticum*, the *Pm13* gene from *Ae. longissima*, and the *Glu-D1* or *Gli-D1/Glu-D3* genes from common wheat (see Sections 2.4.2.2.1 to 2.4.2.2.3). For the genotyping of $F_2$ populations obtained from double heterozygotes, sequence-tagged site (STS) markers associated with *Glu-1* (D'Ovidio and Anderson 1994; D'Ovidio et al. 1996) and *Glu-3* (D'Ovidio et al. 1997) loci, as well as with the *Pm13* transfer (Cenci et al. 1999), have been used for PCR analysis of half-seed extracted DNAs. On the other hand, fl-GISH on root-tip chromosome preparations has been applied to detect the presence as well as the dose of the 7AL/7AgL chromosome containing a 23% distal *Th. ponticum* chromosomal segment. The recent development of a dominant STS marker, associated with *Lr19*-containing segments of *Th. ponticum* introgressed into common wheat lines (Prins et al. 2001), is currently facilitating selection of progeny segregating for the *Lr19 + Yp* transfer. Segregants exhibiting the above STS can then be subjected to further screening (by fl-GISH or use of 7AL markers, e.g., microsatellites) for an early discrimination of their homozygous or heterozygous condition.

Results from analysis of $F_2$ progeny of double heterozygotes for the *Th. ponticum* 7AgL segment and either the 1DS (*Gli-D1/Glu-D3*), 1DL (*Glu-D1*), or *Ae. longissima* 3S¹S (*Pm13*) transfer (Micali et al. 2003) indicate simultaneous transmission of the two different recombinant chromosomes to be normal through both germ lines. The same has been observed in the progeny of a triple combination involving the 7AL/7AgL, 1AS/1DS, and the 3BS/3S¹S recombinant chromosomes (Micali et al. 2003). Also, in terms of zygotic tolerance, no differential performance seems to characterize the various genotypes at seedling and later vegetative stages. Preliminary analysis of a number of agronomic traits has also revealed no significant differences associated with the presence of two or even three wheat–alien recombinant chromosome pairs (Micali et al. 2003, unpublished).

## 2.5 DIRECT GENE TRANSFER IN DURUM WHEAT

Wide hybridization coupled with manipulation of chromosome pairing has proved useful in producing superior wheat cultivars. However, these procedures are often time-consuming and tedious. *In vitro* tools of direct introduction of DNA into regenerable calli of wheat are developing rapidly and serve as a useful adjunct to conventional breeding programs. Thus, hexaploid bread wheat has been transformed by several groups (Vasil et al. 1992; Weeks et al. 1993; Becker et al. 1994; Nehra et al. 1994; Anand et al. 2003). But durum wheat was difficult to transform because an efficient *in vitro* regeneration system, a prerequisite for genetic transformation of any crop plant, was not developed for this species. Therefore, a suitable regeneration protocol for durum wheat was established (Bommineni and Jauhar 1996), and using this protocol, the first transgenic durum wheat was produced (Bommineni et al. 1997). Once durum was shown to be amenable to genetic transformation, more reports of production of transgenic durum followed (He et al. 1999; Pellegrineschi et al. 2002).

Having standardized the transgenic technology and the regeneration protocol for commercial durum cultivars, antifungal genes, including *Tri101*, *PDR5*, and thaumatin-like protein genes are being incorporated into them in an effort to produce scab-resistant materials (Satyavathi and Jauhar 2003).

## 2.6 CONCLUSIONS

The results reported above lead us to conclude that properly tailored alien chromosomal transfers, both single and multiple segments, that fit the requirements of the recipient genome for an overall positive acceptance could be an effective strategy to enhance the genetic variability and thus the breeding performance of an even reluctant species such as durum wheat toward chromosome manipulations. In fact, what pioneer scientists such as McFadden (McFadden 1928; McFadden and Sears 1947) had called radical wheat breeding, revealing their farsighted perception of the great and innovative potential of this approach, undoubtedly remained for long a little-exploited and also a little-rewarding strategy considering the overall amount of work invested in it.

However, the rediscovery of the principles of genetics at the beginning of the last century and the progressively acquired knowledge about the genomic makeup of allopolyploid wheats and their cytogenetic relationships with alien Triticeae species from the 1930s to the 1970s gave impetus to the wide hybridization approach, contributing to its transition from a fine art to a true science-based methodology. The development of molecular techniques in the following decades added new dimensions to alien gene transfer into wheat. Thanks to the recent availability of an extremely rich arsenal of tools, wheat chromosome engineering is currently experiencing a new age. In fact, assisted by the molecular diagnostics, identification of alien introgressions can now be much more easily achieved and efficiently exploited than in the past.

While awaiting further progress in the area of direct gene transfer via transformation methodologies, and when this becomes more readily accessible, it seems reasonable to conclude that both the chromosome and the genetic engineering approaches will contribute to wheat germplasm enhancement. In fact, the vast gene pool of closely related Triticeae species holds an enormous potential for enrichment of durum wheat. In view of the recent achievements, including those at international institutions such as CIMMYT (see Chapter 3 by Mujeeb-Kazi in this volume), the alien gene transfer approach can be complementary to and integrated with conventional breeding programs, as also with the direct gene transfer approach (Jauhar and Chibbar 1999; Jauhar and Khush 2002).

## REFERENCES

Ahn, S., Anderson, J.A., Sorrells, M.E., and Tanksley, S.D. 1993. Homoeologous relationships of rice, wheat and maize chromosomes. *Mol. Gen. Genet.* 241: 483–490.

Almouslem, A.B., Jauhar, P.P., Peterson, T.S., Bommineni, V.R., and Rao, M.B. 1998. Haploid durum wheat production via hybridization with maize. *Crop Sci.* 38: 1080–1087.

Anamthawat-Jónsson, K. and Reader, S.M. 1995. Pre-annealing of total genomic DNA probes for simultaneous genomic *in situ* hybridization. *Genome* 38: 814–816.

Anand, A., Zhou, T., Trick, H.N., Gill, B.S., Bockus, W.W., and Muthukrishnan, S. 2003. Greenhouse and field testing of transgenic wheat plants stably expressing genes for thaumatin-like protein, chitinase and glucanase against *Fusarium graminearum*. *J. Exp. Bot.* 54: 1101–1111.

Anderson, E. and Stebbins, G.L., Jr. 1954. Hybridization as an evolutionary stimulus. *Evolution* 8: 378–388.

Anderson, J.A., Ogihara, Y., Sorrells, M.E., and Tanksley, S.D. 1992. Development of a chromosomal arm map for wheat based on RFLP markers. *Theor. Appl. Genet.* 83: 1035–1043.

Avivi, L., Levy, A.A., and Feldman, M. 1983. Studies on high protein durum wheat derived from crosses with the wild tetraploid wheat, *Triticum turgidum* var. *dicoccoides*. In *Proceedings of the 6th Wheat International Genetics Symposium*, S. Sakamoto, Ed., Kyoto, Japan, pp. 199–204.

Becker, D., Brettschneider, R., and Lorz, H. 1994. Fertile transgenic wheat from microprojectile bombardment of scutellar tissue. *Plant J.* 5: 299–307.

Biagetti, M., Vitellozzi, F., and Ceoloni, C. 1999. Physical mapping of the wheat-*Aegilops longissima* breakpoints in mildew resistant recombinant lines using FISH with highly repeated and low-copy DNA probes. *Genome* 42: 1013–1019.

Blanco, A., Perrone, V., and Simeone, R. 1988. Chromosome pairing variation in *durum* wheat × *Dasypyrum villosum* (L.) Candargy hybrids and genome affinities. In *Proceedings of the 7th International Wheat Genetics Symposium*, T.E. Miller and R.M.D. Koebner, Eds., Institute of Plant Science Research, Cambridge, U.K., pp. 63–67.

Blanco, A. and Simeone, R. 1995. Chromosome pairing in hybrids and amphiploids between durum wheat and the tetraploid *Dasypyrum hordeaceum*. In *Proceedings of the 8th International Wheat Genetics Symposium*, Z.S. Li and Z.Y. Xin, Eds., Beijing, China, pp. 305–309.

Boggini, G., Tusa, P., Di Silvestro, S., and Pogna N.E. 1998. Agronomical and quality characteristics of durum wheat lines containing the 1BL/1RS translocation. *J. Genet. Breed.* 53: 167–172.

Bommineni, V.R. and Jauhar, P.P. 1996. Regeneration of plantlets through isolated scutellum culture of durum wheat. *Plant Sci.* 116: 197–203.

Bommineni, V.R., Jauhar, P.P., and Peterson, T.S. 1997. Transgenic durum wheat by microprojectile bombardment of isolated scutella. *J. Hered.* 88: 475–481.

Braun, H.J., Payne, T.S., Morgounov, A.I., van Ginkel, M., and Rajaram, S. 1998. The challenge: one billion tons of wheat by 2020. In *Proceedings of the 10th International Wheat Genetics Symposium*, Vol. 1, A.E. Slinkard, Ed., Saskatoon, Saskatchewan, Canada, pp. 33–40.

Carleton, M.A. 1901. Emmer: a grain for the semiarid regions. *U.S.D.A. Farmers' Bull.* 139.

Carozza, R., Pagnotta, M.A., and Ceoloni, C. 2003. Cytogenetic characterization of chromosomally engineered wheat chromosomes by FISH-based physical maps, molecular linkage maps and meiotic pairing analysis. *Ann. Génét.* 46: 301.

Cenci, A., D'Ovidio, R., Tanzarella, O.A., Ceoloni, C., and Porceddu, E. 1999. Identification of molecular markers linked to *Pm13*, an *Aegilops longissima* gene conferring resistance to powdery mildew of wheat. *Theor. Appl. Genet.* 98: 448–454.

Ceoloni, C. 1987. Current methods of chromosome engineering in wheat. In *EWAC (European Wheat Aneuploid Cooperative) Newsletter 1987*, J. Sutka and A.J. Worland, Eds., pp. 95–109.

Ceoloni, C., Biagetti, M., Ciaffi, M., Forte, P., and Pasquini M. 1996. Wheat chromosome engineering at the 4x level: the potential of different alien gene transfers into *durum* wheat. *Euphytica* 89: 87–97.

Ceoloni, C., Ciaffi, M., Lafiandra, D., and Giorgi, B. 1995. Chromosome engineering as a means of transferring 1D storage protein genes from common to *durum* wheat. In *Proceedings of the 8th International Wheat Genetics Symposium*, Z.S. Li and Z.Y. Xin, Eds., Beijing, China, pp. 159–163.

Ceoloni, C., Del Signore, G., Ercoli, L., and Donini, P. 1992. Locating the alien chromatin segment in common wheat-*Aegilops longissima* mildew resistant transfers. *Hereditas* 116: 239–245.

Ceoloni, C., Del Signore, G., Pasquini, M., and Testa, A. 1988. Transfer of mildew resistance from *Triticum longissimum* into wheat by induced homoeologous recombination. In *Proceedings of the 7th International Wheat Genetics Symposium*, T.E. Miller and R.M.D. Koebner, Eds., Institute of Plant Science Research, Cambridge, U.K., pp. 221–226.

Ceoloni, C., Forte, P., Ciaffi, M., Nenno, M., Bitti, A., De Vita, P., and D'Egidio, M.G. 2000. Chromosomally engineered durum wheat: the potential of alien gene introgressions affecting disease resistance and quality. *Options Méditerranéennes* A-40: 363–371.

Ceoloni, C., Forte, P., Gennaro, A., Micali, S., Carozza, R., and Bitti, A. 2005a. Recent developments in durum wheat chromosome engineering. *Cytogenet. Genome Res.* 109: 328–344.

Ceoloni, C., Margiotta, B., Colaprico, G., D'Egidio, M.G., Carozza, R., and Lafiandra, D. 2003. Introgression of D-genome associated gluten protein genes into durum wheat. In *Proceedings of the 10th International Wheat Genetics Symposium*, Vol. 2, N.E. Pogna, M. Romanò, E.A. Pogna, G. Galterio, Eds., S.I.M.I., Rome, Paestum, Italy, pp. 1320–1322.

Ceoloni, C., Pasquini, M., and Simeone, R. 2005b. The cytogenetic contribution to the analysis and manipulation of the durum wheat genome. In *Durum Wheat Breeding: Current Approaches and Future Strategies*, C. Royo, M.N. Nachit, N. Di Fonzo, J.L. Araus, W.H. Pfeiffer, and G.A. Slafer, Eds. The Haworth Press, New York, in press.

Ceoloni, C., Vitellozzi, F., Forte, P., Basili, F., Biagetti, M., Bitti, A., and Delre, V. 1998. Wheat chromosome engineering in the light of advanced genetic and cytogenetic marker-mediated approaches. In *Current Topics in Plant Cytogenetics Related to Plant Improvement*, T. Lelley, Ed. WUV-Universitätsverlag, Wien, pp. 43–53.

Charpentier, A., Cauderon, Y., and Feldman, M. 1988. Control of chromosome pairing in *Agropyron elongatum*. In *Proceedings of the 7th International Wheat Genetics Symposium*, Vol. 1, T.E. Miller and R.M.D. Koebner, Eds., Institute of Plant Science Research, Cambridge, U.K., pp. 231–236.

Cuadrado, A., Vitellozzi, F., Jouve, N., and Ceoloni, C. 1997. Fluorescence *in situ* hybridization with multiple repeated DNA probes applied to the analysis of wheat/rye chromosome pairing. *Theor. Appl. Genet.* 94: 347–355.

Devos, K.M., Atkinson, M.D., Chinoy, C.N., Francis, H.A., Harcourt, R.L., Koebner, R.M.D., Liu, C.J., Masojc, Xie, D.X., and Gale, M.D. 1993a. Chromosomal rearrangements in the rye genome relative to that of wheat. *Theor. Appl. Genet.* 85: 673–680.

Devos, K.M., Atkinson, M.D., Chinoy, C.N., Liu, C., and Gale, M.D. 1992. RFLP based genetic map of the homoeologous group 3 chromosomes of wheat and rye. *Theor. Appl. Genet.* 83: 931–939.

Devos, K.M, Dubcovsky, J., Dvořák, J., Chinoy, C.N., and Gale, M.D. 1995. Structural evolution of wheat chromosomes 4A, 5A and 7B and its impact on recombination. *Theor. Appl. Genet.* 91: 282–288.

Devos, K.M. and Gale, M.D. 1997. Comparative genetics in the grasses. *Plant Mol. Biol.* 35: 3–15.

Devos, K.M. and Gale, M.D. 2000. Genome relationships: the grass model in current research. *Plant Cell* 12: 637–646.

Devos, K.M., Millan, T., and Gale, M.D. 1993b. Comparative RFLP maps of homoeologous chromosomes of wheat, rye and barley. *Theor. Appl. Genet.* 85: 784–792.

Doğramaci-Altuntepe, M. and Jauhar, P.P. 2001. Production of durum wheat substitution haploids from durum × maize crosses and their cytological characterization. *Genome* 44: 137–142.

Donini, P., Koebner, R.M.D., and Ceoloni, C. 1995. Cytogenetic and molecular mapping of the wheat-*Aegilops longissima* chromatin breakpoints in powdery mildew resistant introgression lines. *Theor. Appl. Genet.* 91: 738–743.

D'Ovidio, R. and Anderson, O.D. 1994. PCR analysis to distinguish between a member of a multigene family correlated with wheat bread-making quality. *Theor. Appl. Genet.* 88: 759–763.

D'Ovidio, R., Masci, S., and Porceddu, E. 1996. Sequence analysis of the 5' non-coding regions of active and inactive *1Ay* HMW glutenin genes from wild and cultivated wheats. *Plant Sci.* 114: 61–69.

D'Ovidio, R., Simeone, M., Masci, S., and Porceddu, E. 1997. Molecular characterization of a LMW-GS gene located on chromosome 1B and the development of primers specific for the *Glu-B3* complex locus in durum wheat. *Theor. Appl. Genet.* 95: 1119–1126.

Dubcovsky, J., Santa Maria, G., Epstein, E., Luo, M.C., and Dvořák, J. 1996. Mapping of the K⁺/Na⁺ discrimination locus *Kna1* in wheat. *Theor. Appl. Genet.* 92: 448–454.

Dvořák, J. 1980. Homoeology between *Agropyron elongatum* chromosomes and *Triticum aestivum* chromosomes. *Can. J. Genet. Cytol.* 22: 237–259.

Dvořák, J. 1987. Chromosomal distribution of genes in diploid *Elytrigia elongata* that promote or suppress pairing of wheat homoeologous chromosomes. *Genome* 29: 34–40.

Dvořák, J., Chen, K.C., and Giorgi, B. 1984. The C-band pattern of a *Ph1⁻* mutant of durum wheat. *Can. J. Genet. Cytol.* 26: 360–363.

Dvořák, J., DiTerlizzi, P., Zhang, H.-B., and Resta, P. 1993. The evolution of polyploid wheats: the identification of the A genome donor species. *Genome* 36: 21–31.

Dvořák, J. and Gorham, J. 1992. Methodology of gene transfer by homoeologous recombination into *Triticum turgidum*: transfer of K⁺/Na⁺ discrimination from *T. aestivum*. *Genome* 35: 639–646.

Dvořák, J., Noaman, M.M., Goyal, S., and Gorham, J. 1994. Enhancement of the salt tolerance of *Triticum turgidum* L. by the *Kna1* locus transferred from the *Triticum aestivum* L. chromosome 4D by homoeologous recombination. *Theor. Appl. Genet.* 87: 872–877.

Dvořák, J. and Ross, K. 1986. Expression of tolerance of Na⁺, K⁺, Mg²⁺, Cl⁻, and SO²⁻₄ ions and sea water in the amphiploid of *Triticum aestivum* × *Elytrigia elongata*. *Crop Sci.* 26: 658–660.

Eizenga, G.C. 1987. Locating the *Agropyron* segment in wheat-*Agropyron* 'transfer no. 12.' *Genome* 29: 365–366.

Elias, E.M, Steiger, D.K., and Cantrell, R.G. 1996. Evaluation of lines derived from wild emmer chromosome substitutions. II. Agronomic traits. *Crop Sci.* 36: 228–233.

Fedak, G. 2000. Sources of resistance to Fusarium head blight. In *Proceedings of the International Symposium on Wheat Improvement for Scab Resistance*, Suzhou and Nanjing, China, May 5–7, p. 4.

Feldman, M. 1979. Genetic resources of wild wheats and their use in breeding. *Monogr. Genet. Agraria* IV: 9–26.

Feldman, M. 1983. Gene transfer from wild species into cultivated plants. *Genetika* 15: 145–161.

Feldman, M. 1988. Cytogenetic and molecular approaches to alien gene transfer in wheat. In *Proceedings of the 7th International Wheat Genetics Symposium*, T.E. Miller and R.M.D. Koebner, Eds., Institute of Plant Science Research, Cambridge, U.K., pp. 23–32.

Feldman, M. 1993. Cytogenetic activity and mode of action of the pairing homoeologous (*Ph1*) gene of wheat. *Crop Sci.* 33: 894–897.

Feldman, M., Galili, G., and Levy, A.A. 1986. Genetic and evolutionary aspects of allopolyploidy in wheat. In *The Origin and Domestication of Cultivated Plants*, C. Barigozzi, Ed. Elsevier, Amsterdam, pp. 83–100.

Feldman, M., Liu, B., Segal, G., Abbo, S., Levy, A.A., and Vega, J.M. 1997. Rapid elimination of low-copy DNA sequences in polyploid wheat: a possible mechanism for differentiation of homoeologous chromosomes. *Genetics* 147: 1381–1387.

Feldman, M., Lupton, F.G.H., and Miller, T.E. 1995. Wheats. In *Evolution of Crop Plants*, J. Smartt and N. W. Simmonds, Eds. Longman Group, London, pp. 184–192.

Feldman, M., Ozkan, H., Kashkush, K., Shaked, H., and Levy, A.A. 2002. Allopolyploidization: an inducer of rapid genome evolution in wheat. In *Proceedings of the Plant, Animal and Microbe Genome X Conference*, San Diego, CA, January 12–16 (abstract).

Feldman, M. and Sears, E.R. 1981. The wild gene resources of wheat. *Sci. Am.* 244: 98–109.

Feuillet, C. and Keller, B. 2002. Comparative genomics in the grass family: molecular characterization of grass genome structure and evolution. *Ann. Bot.* 89: 3–10.

Feuillet, C., Penger, A., Gellner, K., Mast, A., and Keller, B. 2001. Molecular evolution of receptor-like kinase genes in hexaploid wheat. Independent evolution of orthologs after polyploidization and mechanisms of local rearrangements of paralogous loci. *Plant Physiol.* 125: 1304–1313.

Friebe, B., Hatchett, J.H., Sears, R.G., and Gill, B.S. 1990. Transfer of Hessian fly resistance from 'Chaupon' rye to hexaploid wheat via a 2BS/2RL wheat-rye chromosome translocation. *Theor. Appl. Genet.* 79: 385–389.

Friebe, B., Jiang, J., Raupp, W.J., McIntosh, R.A., and Gill, B.S. 1996. Characterization of wheat-alien translocations conferring resistance to diseases and pests: current status. *Euphytica* 91: 59–87.

Friebe, B., Kynast, R.G., Hatchett, J.H., Sears, R.G., Wilson, D.L., and Gill, B.S. 1999. Transfer of wheat-rye translocation chromosomes conferring resistance to Hessian fly from bread wheat into durum wheat. *Crop Sci.* 39: 1692–1696.

Friebe, B., Zeller, F.J., and Kunzmann, R. 1987. Transfer of the 1BL/1RS wheat-rye translocation from hexaploid bread wheat to tetraploid durum wheat. *Theor. Appl. Genet.* 74: 423–425.

Gale, M.D. and Miller, T.E. 1987. The introduction of alien genetic variation into wheat. In *Wheat Breeding: Its Scientific Basis*, F.G.H. Lupton, Ed. Chapman & Hall, London, pp. 173–210.

Galili, G. and Feldman, M. 1983. Diploidization of endosperm protein genes in polyploid wheats. In *Proceedings of the 6th International Wheat Genetics Symposium*, S. Sakamoto, Ed., Kyoto, Japan, pp. 1119–1123.

Galili, G., Levy, A.A., and Feldman, M. 1986. Gene-dosage compensation of endosperm proteins in hexaploid wheat *Triticum aestivum*. *Proc. Natl. Acad. Sci. U.S.A.* 83: 6524–6528.

Garcia-Olmedo, F. and Carbonero, P. 1980. Chromosomes, genes, proteins and wheat endosperm. In *Proceedings of the 3rd International Wheat Conference*, Madrid, Spain, May 22–June 3, pp. 664–670.

Gennaro, A., Borrelli, G.M., D'Egidio, M.G., De Vita, P., Ravaglia, S., and Ceoloni, C. 2003. A chromosomally engineered durum wheat-*Thinopyrum ponticum* recombinant line with novel and promising attributes for varietal development. In *Proceedings of the 10th International Wheat Genetics Symposium*, Vol. 2, N.E. Pogna, M. Romanò, E.A. Pogna, and G. Galterio, Eds., Paestum, Italy, pp. 881–883.

Gerechter-Amitai, Z.K. and Grama, A. 1974. Inheritance of resistance to stripe rust (*Puccinia striiformis*) in crosses between wild emmer (*Triticum dicoccoides*) and cultivated tetraploid and hexaploid wheats. I. *Triticum durum*. *Euphytica* 23: 387–392.

Gerechter-Amitai, Z.K., Wahl, I., Vardi, A., and Zohary, D. 1971. Transfer of stem rust seedling resistance from wild diploid einkorn to tetraploid *durum* wheat by means of a triploid hybrid bridge. *Euphytica* 20: 281–285.

Gill, B.S., Morris, K.L.D., and Appels, R. 1988. Assignment of the genomic affinities of chromosomes from polyploid *Elymus* species added to wheat. *Genome* 30: 70–82.

Gill, K.S. and Gill, B.S. 1991. A DNA fragment mapped within the sub microscopic deletion of *Ph1*, a chromosome pairing regulator gene in polyploid wheat. *Genetics* 134: 1231–1236.

Gill, K.S. and Gill, B.S. 1996. A PCR-based screening assay of *Ph1*, the chromosome pairing regulator gene in polyploid wheat. *Crop Sci.* 36: 719–722.

Gill, K.S., Gill, B.S., Endo, T.R., and Mukai, Y. 1993. Fine physical mapping of *Ph1*, a chromosome pairing regulator gene in polyploid wheat. *Genetics* 129: 257–259.

Giorgi, B. 1978. A homoeologous pairing mutant isolated in *Triticum durum* cv. Cappelli. *Mutat. Breed. Newsl.* 11: 4–5.

Giorgi, B. 1983. Origin, behaviour and utilization of a *Ph1* mutant of durum wheat, *Triticum turgidum* (L.) var. *durum*. In *Proceedings of the 6th International Wheat Genetics Symposium*, S. Sakamoto, Ed., Kyoto, Japan, pp. 1033–1040.

Giorgi, B., Barbera, F., and Bitti, O. 1981. Intergeneric hybrids, meiotic behaviour and amphiploids involving a *Ph* mutant of *Triticum turgidum* ssp. *durum*. In *Proceedings of the Symposium on Induced Variability in Plant Breeding*, EUCARPIA, Wageningen, The Netherlands, pp. 96–100.

Giorgi, B. and Ceoloni, C. 1985. A ph1 hexaploid triticale: production, cytogenetic behaviour and use for intergeneric gene transfer. In *Proceedings of the EUCARPIA Meeting on Genetics and Breeding of Triticale*, Clermont-Ferrand, France, July 2–5, 1984, pp. 105–117.

Gorham, J., Hardy, C., Wyn Jones, R.G., Joppa, L.R., and Law, C.N. 1987. Chromosomal location of a K/Na discriminating character in the D-genome of wheat. *Theor. Appl. Genet.* 74: 584–588.

Graybosch, R.A., Lee, J.-H., Peterson, C.J., Porter, D.R., and Chung, O.K. 1999. Genetic, agronomic and quality comparisons of two 1AL.1RS wheat-rye chromosomal translocations. *Plant Breed.* 118: 125–130.

Graybosch, R.A., Peterson, C.J., Hansen, L.E., Worral, D., Shelton, D.R., and Lukaszewski, A.J. 1993. Comparative flour quality and protein characteristics of 1BL/1RS and 1AL/1RS wheat-rye translocations. *J. Cereal Sci.* 17: 95–106.

Gupta, R.B., Paul, J.G., Cornish, G.B., Palmer, G.A., Bekes, F., and Rathjen, A.J. 1994. Allelic variation at glutenin subunit and gliadin loci, *Glu-1*, *Glu-3* and *Gli-1*, of common wheats. I. Additive and interaction effects on dough properties. *J. Cereal Sci.* 19: 9–17.

Hart, G.E. 1987. Genetic and biochemical studies of enzymes. In *Wheat and Wheat Improvement*, Vol. 13, E.G. Heyne, Ed. American Society of Agronomy, Madison, WI, pp. 199–214.

Hart, G.E. and Tuleen, N.A. 1983. Introduction and characterization of alien genetic material. In *Isozymes in Genetics and Plant Breeding*, Part A, S.D. Tanksley and T.J. Orton, Eds. Elsevier Science Publishers, Amsterdam, pp. 339–362.

Hayes, H.K., Parker, J.H., and Kurtzweil, C. 1920. Genetics of rust resistance in crosses of varieties of *Triticum vulgare* with varieties of *T. durum* and *T. dicoccum*. *J. Agric. Res.* 19: 523–542.

He, G.Y., Rooke, L., Steele, S., Békés, F., Gras, P., Tatham, A.S., Fido, R., Barcelo, P., Shewry, P.R., and Lazzeri, P.A. 1999. Transformation of pasta wheat (*Triticum turgidum* L. var. *durum*) with high-molecular-weight glutenin subunit genes and modification of dough functionality. *Mol. Breed.* 5: 377–386.

He, P., Friebe, B.R., Gill, B.S., and Zhou, J.-M. 2003. Allopolyploidy alters gene expression in the highly stable hexaploid wheat. *Plant Mol. Biol.*, 52: 401–414.

Huang, S., Sirikhachornkit, A., Su, X., Faris, J., Gill, B., Haselkorn, R., and Gornicki, P. 2002. Genes encoding plastid acetyl-CoA-carboxylase and 3-phosphoglycerate kinase of the *Triticum/Aegilops* complex and the evolutionary history of polyploid wheat. *Proc. Natl. Acad. Sci. U.S.A.* 99: 8133–8138.

Huguet-Robert, V., Dedryver, F., Röder, M.S., Korzun, V., Abélard, P., Tanguy, A.M., Jaudeau, B., and Jahier, J. 2001. Isolation of a chromosomally engineered durum wheat line carrying the *Aegilops ventricosa* *Pch1* gene for resistance to eyespot. *Genome* 44: 345–349.

Jahier, J., Doussinault, G., Dosba, F., and Bourgeois, F. 1978. Monosomic analysis of resistance to eyespot in the variety "Roazon." In *Proceedings of the 5th International Wheat Genetics Symposium*, S. Ramanujam, Ed., New Delhi, pp. 336–347.

Jampates, R. and Dvořák, J. 1986. Location of the *Ph1* locus in the metaphase chromosome map and the linkage map of the 5Bq arm of wheat. *Can. J. Genet. Cytol.* 28: 511–519.

Jauhar, P.P. 1991. Hybrid between durum wheat and diploid *Thinopyrum bessarabicum*. *Genome* 34: 283–287.

Jauhar, P.P. 1992. Synthesis and cytological characterization of trigeneric hybrids involving durum wheat, *Thinopyrum bessarabicum*, and *Lophopyrum elongatum*. *Theor. Appl. Genet.* 84: 511–519.

Jauhar, P.P. 1993. Alien gene transfer and genetic enrichment of bread wheat. In *Biodiversity and Wheat Improvement*, A.B. Damania, Ed. John Wiley & Sons, Chichester, England, pp. 103–119.

Jauhar, P.P. 2001a. Genetic engineering and accelerated plant improvement: opportunities and challenges. *Plant Cell Tissue Organ Cult.* 64: 87–91.

Jauhar, P.P. 2001b. Problems encountered in transferring scab resistance from wild relatives into durum wheat. *Proceedings of the 2001 National Fusarium Head Blight Forum*, Cincinnati, December 8–10, pp. 188–191.

Jauhar, P.P. 2003a. Formation of 2*n* gametes in durum wheat haploids: sexual polyploidization. *Euphytica* 133: 81–94.

Jauhar, P.P. 2003b. Haploid and doubled haploid production in durum wheat by wide hybridization. In *Manual on Haploid and Double Haploid Production in Crop Plants*, M. Maluszynski, K.J. Kasha, B.P. Forster, and I. Szarejko, Eds. Kluwer Academic Publishers, Dordrecht, The Netherlands, pp. 161–166.

Jauhar, P.P. 2003c. Genetics of crop improvement: chromosome engineering. In *Encyclopedia of Applied Plant Sciences*, Vol. 1, B. Thomas, D.J. Murphy, and B. Murray, Eds. Elsevier-Academic Press, London, pp. 167–179.

Jauhar, P.P. and Almouslem, A.B. 1998. Production and meiotic analyses of intergeneric hybrids between durum wheat and *Thinopyrum* species. In *Proceedings of the 3rd International Triticeae Symposium*, A.A. Jaradat, Ed. Scientific Publishers, Enfield, NH, pp. 119–126.

Jauhar, P.P., Almouslem, A.B., Peterson, T.S., and Joppa, L.R. 1999. Inter- and intragenomic chromosome pairing in haploids of durum wheat. *J. Hered.* 90: 437–445.

Jauhar, P.P. and Chibbar, R.N. 1999. Chromosome-mediated and direct gene transfers in wheat. *Genome* 42: 570–583.

Jauhar, P.P. and Crane, C.F. 1989. An evaluation of Baum et al.'s assessment of the genomic system of classification in the Triticeae. *Am. J. Bot.* 76: 571–576.

Jauhar, P.P., Doǧramaci, M., and Peterson, T.S. 2004. Synthesis and cytological characterization of trigeneric hybrids involving durum wheat with and without *Ph1*. *Genome* 47: 1173–1181.

Jauhar, P.P., Doǧramaci-Altuntepe, M., Peterson, T.S., and Almouslem, A.B. 2000. Seedset on synthetic haploids of durum wheat: cytological and molecular investigations. *Crop Sci.* 40: 1742–1749.

Jauhar, P.P. and Joppa, L.R. 1996. Chromosome pairing as a tool in genome analysis: merits and limitations. In *Methods of Genome Analysis in Plants*, P.P. Jauhar, Ed. CRC Press, Boca Raton, FL, pp. 9–37.

Jauhar, P.P. and Khush, G.S. 2002. Importance of biotechnology in global food security. In *Food Security and Environment Quality: A Global Perspective*, R. Lal, D.O. Hansen, N. Uphoff, and N. Slack, Eds. CRC Press, Boca Raton, FL, pp. 107–128.

Jauhar, P.P. and Peterson, T.S. 1998. Wild relatives of wheat as sources of Fusarium head blight resistance. In *Proceedings of the 1998 National Head Blight (Scab) Forum*, Michigan State University, East Lansing, pp. 77–79.

Jauhar, P.P. and Peterson, T.S. 2000a. Progress in producing scab-resistant germplasm of durum wheat. In *Proceedings of the International Symposium on Wheat Improvement for Scab Resistance*, Suzhou and Nanjing, China, May 5–7, pp. 77–81.

Jauhar, P.P. and Peterson, T.S. 2000b. Toward transferring scab resistance from a diploid wild grass, *Lophopyrum elongatum*, into durum wheat. In *Proceedings of the 2000 National Fusarium Head Blight Forum*, Cincinnati, OH, December, pp. 201–204.

Jauhar, P.P. and Peterson, A.B. 2001. Hybrids between durum wheat and *Thinopyrum junceiforme*: prospects for breeding for scab resistance. *Euphytica* 118: 127–136.

Jauhar, P.P., Riera-Lizarazu, O., Dewey, W.G., Gill, B.S., Crane, C.F., and Bennett, J.H. 1991. Chromosome pairing relationships among the A, B, and D genomes of bread wheat. *Theor. Appl. Genet.* 82: 441–449.

Jenkins, B.C. and Mochizuki, A. 1957. A new amphiploid from a cross between *Triticum durum* and *Agropyron elongatum* (2n = 14). *Wheat Inf. Serv.* 5: 15.

Jiang, J., Friebe, B., and Gill, B.S. 1994. Recent advances in alien gene transfer in wheat. *Euphytica* 73: 199–212.

Joppa, L.R. 1988. Cytogenetics of tetraploid wheat. In *Proceedings of the 7th International Wheat Genetics Symposium*, T.E. Miller and R.M.D. Koebner, Eds. Institute of Plant Science Research, Cambridge, U.K., pp. 197–202.

Joppa, L.R. 1993. Chromosome engineering in tetraploid wheat. *Crop Sci.* 33: 908–913.

Joppa, L.R., Hareland, G.A., and Cantrell, R.G. 1991. Quality characteristics of the Langdon durum-*dicoccoides* chromosome substitution lines. *Crop Sci.* 31: 1513–1517.

Joppa, L.R., Khan, K., and Williams, N.D. 1983. Chromosomal location of genes for gliadin polypeptides in durum wheat *Triticum turgidum* L. *Theor. Appl. Genet.* 64: 289–293.

Joppa, L.R., Klindworth, D.L., and Hareland, G.A. 1998. Transfer of high molecular weight glutenins from spring wheat to durum wheat. In *Proceedings of the 9th International Wheat Genetics Symposium*, Vol. 1, A.E. Slinkard, Ed., Saskatoon, Saskatchewan, Canada, pp. 257–260.

Joppa, L.R. and Williams, N.D. 1979. A disomic 5D-nullisomic 5B substitution line of durum wheat. *Crop Sci.* 19: 509–511.

Joppa, L.R. and Williams, N.D. 1988. Langdon durum disomic substitution lines and aneuploid analysis in tetraploid wheat. *Genome* 30: 222–228.

Kashkush, K., Feldman, M., and Levy, A. 2002. Gene loss, silencing and activation in a newly synthesized allotetraploid. *Genetics* 160: 1651–1659.

Keller, B. and Feuillet, C. 2000. Colinearity and gene density in grass genomes. *Trends Plant Sci.* 5: 246–251.

Kellogg, E.A. 1998. Relationships of cereal crops and other grasses. *Proc. Natl. Acad. Sci. U.S.A.* 95: 2005–2010.

Kellogg, E.A. 2001. Evolutionary history of the grasses. *Plant Physiol.* 125: 1198–1205.

Khush, G.S. 1999. Green revolution: preparing for the 21st century. *Genome* 42: 646–655.

King, I.P., Purdue, K.A., Orford, S.E., Reader, S.M., and Miller, T.E. 1993. Detection of homoeologous chiasma formation in *Triticum durum* × *Thinopyrum bessarabicum* hybrids using genomic *in situ* hybridization. *Heredity* 71: 369–372.

Knott, D.R. 1980. Mutation of a gene for yellow pigment linked to *Lr19* in wheat. *Can. J. Genet. Cytol.* 22: 651–654.

Koebner, R.M.D. and Shepherd, K.W. 1986. Controlled introgression to wheat of genes from rye chromosome arm 1RS by induction of allosyndesis. 1. Isolation of recombinants. *Theor. Appl. Genet.* 73: 197–208.

Koebner, R.M.D., Shepherd, K.W., and Appels, R. 1986. Controlled introgression to wheat of genes from rye chromosome arm 1RS by induction of allosyndesis. 2. Characterization of recombinants. *Theor. Appl. Genet.* 73: 209–217.

Larter, E.N. 1976. Triticale. In *Evolution of Crop Plants*, N.W. Simmonds, Ed. Longman Group, London, pp. 117–120.

Levy, A.A., Galili, G., and Feldman, M. 1988. Polymorphism and genetic control of high molecular weight glutenin subunits in wild tetraploid wheat *Triticum turgidum* var. *dicoccoides*. *Heredity* 61: 63–72.

Liu, B., Vega, J.M., and Feldman, M. 1998a. Rapid genomic changes in newly synthesized amphiploids of *Triticum* and *Aegilops*. II. Changes in low-copy coding DNA sequences. *Genome* 41:535–542.

Liu, B., Vega, J.M., Segal, G., Abbo. S., Rodova, M., and Feldman, M. 1998b. Rapid genomic changes in newly synthesized amphiploids of *Triticum* and *Aegilops*. I. Changes in low-copy non-coding DNA sequences. *Genome* 41:272–277.

Liu, C.Y., Rathjen, A.J., Shepherd K.W., Gras, P.W., and Giles, L.C. 1995. Grain quality and yield characteristics of D-genome disomic substitution lines in 'Langdon' (*Triticum turgidum* var. *durum*). *Plant Breed.* 114: 34–39.

Liu, Z.Y., Sun, Q.X., Ni, Z.F., Nevo, E., and Yang, T.M. 2002. Molecular characterization of a novel powdery mildew resistance gene *Pm30* in wheat originating from wild emmer. *Euphytica* 123: 21–29.

Lukaszewski, A.J. 1998. Experimental designs in induced homoeologous recombination in wheat. In *Current Topics in Plant Cytogenetics Related to Plant Improvement*, T. Lelley, Ed. WUV-Universitätsverlag, Wien, pp. 54–62.

Lukaszewski, A.J. 2000. Manipulation of the 1RS.1BL translocation in wheat by induced homoeologous recombination. *Crop Sci.* 40: 216–225.

Luo, M.C., Dubcovsky, J., Goyal, S., and Dvořák, J. 1996. Engineering of interstitial foreign chromosome segments containing the K+/Na+ selectivity gene *Kna1* by sequential homoeologous recombination in durum wheat. *Theor. Appl. Genet.* 93: 1180–1184.

Maan, S.S. and Sasakuma, T. 1977. Fertility of amphihaploids in Triticinae. *J. Hered.* 94: 87–94.

Maan, S.S., Sasakuma, T., and Tsuji, S. 1980. Fertility of intergeneric hybrids in Triticinae. *Seiken Zihô* 29: 78–86.

Mac Key, J. 1959. Mutagenic response in *Triticum* at different levels of ploidy. In *Proceedings of the 1st International Wheat Genetics Symposium*, Winnipeg, Canada, 1958, pp. 88–109.

Mac Key, J. 1981. Mutations as a means for analysing polyploid systems. In *Induced Mutations: A Tool in Plant Research, FAO-IAEA International Symposium*, Session II, Vienna, Austria, March 9–13, pp. 143–151.

Martinez, M., Cuñado, N., Carcelén, N., and Romero, C. 2001a. The *Ph1* and *Ph2* loci play different roles in the synaptic behaviour of hexaploid wheat *Triticum aestivum. Theor. Appl. Genet.* 103: 398–405.

Martinez, M., Naranjo, T., Cuadrado, C., and Romero, C. 2001b. The synaptic behaviour of *Triticum turgidum* with variable doses of the *Ph1* locus. *Theor. Appl. Genet.* 102: 751–758.

Mazza, M., Pagliaricci, S., Pasquini, M., Iori, A., Cacciatori, P., Castagna, R., and Pogna, N.E. 1995. Analisi genetica dei bracci cromosomici 1BS, 1RS e 1AS e loro utilizzazione nel miglioramento genetico del grano duro. In *Proceedings of the V Giornate Internazionali sul Grano Duro*, N. Di Fonzo and G. Ronga, Eds., Foggia, Italy, May 2–4, pp. 113–122 (in Italian, with abstract in English).

McFadden, E.S. 1928. Possibilities and difficulties in the field of radical wheat crossing. In *Proceedings of the 1st Annual Hard Spring Wheat Conference*, Fargo, ND, pp. 58–61.

McFadden, E.S. 1930. A successful transfer of emmer characters to *vulgare* wheat. *J. Am. Soc. Agron.* 22: 1020–1034.

McFadden, E.S. and Sears, E.R. 1947. The genomic approach in radical wheat breeding. *J. Am. Soc. Agron.* 39: 1011–1026.

McIntosh, R.A. 1991 Alien sources of disease resistance in bread wheats. In *Nuclear and Organellar Genomes of Wheat Species, Proceedings of the Dr. H. Kihara Memorial International Symposium on Cytoplasmic Engineering in Wheat*, T. Sasakuma and T. Kinoshita, Eds., Hokkaido University, Japan, July, pp. 321–332.

McIntosh, R.A., Wellings, C.R., and Park, R.F. 1995. *Wheat Rusts: An Atlas of Resistance Genes.* CSIRO, Sydney, Australia.

Metakovsky, E.V., Akhmedov, M.G., and Payne, P.I. 1990. A case of spontaneous intergenomic transfer of genetic material containing gliadin genes in bread wheat. *J. Genet. Breed.* 44: 127–132.

Micali, S., Forte, P., Bitti, A., D'Ovidio, R., and Ceoloni, C. 2003. Chromosome engineering as a tool for effectively introgressing multiple useful genes from alien Triticeae into durum wheat. In *Proceedings of the 10th International Wheat Genetics Symposium*, Vol. 2, N.E. Pogna, M. Romanò, E.A. Pogna, G. Galterio, Eds., Paestum, Italy, pp. 896–898.

Miller, J.D., Stack, R.W., and Joppa, L.R. 1998. Evaluation of *Triticum turgidum* L. var. *dicoccoides* for resistance to Fusarium head blight and stem rust. In *Proceedings of the IX International Wheat Genetics Symposium*, Vol. 13, A.E. Slinkard, Ed., University of Saskatoon, Saskatoon, Canada, pp. 292–293.

Miller, T.E. and Reader, S.M. 1987. A guide to the homoeology of chromosomes within the Triticeae. *Theor. Appl. Genet.* 74: 214–217.

Moore, G., Devos, K.M., Wang, Z., and Gale, M.D. 1995. Cereal genome evolution: grasses, line up and form a circle. *Curr. Biol.* 5: 737–739.

Mujeeb-Kazi, A. 1998. Evolutionary relationships and gene transfer in the Triticeae. In *Proceedings of the 3rd International Triticeae Symposium*, A.A. Jaradat, Ed. Scientific Publishers, Enfield, NH, pp. 59–65.

Naranjo, T. 1990. Chromosome structure of durum wheat. *Theor. Appl. Genet.* 79: 397–400.

Naranjo, T., Roca, A., Goicoechea, P.G., and Giraldez, R. 1987. Arm homoeology of wheat and rye chromosomes. *Genome* 29: 873–882.

Nehra, N.S., Chibbar, R.N., Leung, N., Caswell, K., Mallard, C., Steinhauer, L., Båga, M., and Kartha, K.K. 1994. Self-fertile transgenic wheat plants regenerated from isolated scutellar tissues following microprojectile bombardment with two distinct gene constructs. *Plant J.* 5: 285–297.

Nishikawa, K. 1983. Species relationship of wheat and its putative ancestors as viewed from isozyme variation. In *The 6th International Wheat Genetics Symposium*, Kyoto, Japan, 1983, pp. 59–63.

Ono, H., Nagayoshi, T., Nakamura, C., and Shinotani, K. 1983. *Triticum durum–Elytrigia elongata* addition line: cytological, morphological and biochemical characteristics. In *Proceedings of the 6th International Wheat Genetics Symposium*, S. Sakamoto, Ed., Kyoto, Japan, pp. 1049–1053.

Ozkan, H., Levy, A.A., and Feldman, M. 2001. Allopolyploidy-induced rapid genome evolution in the wheat (*Aegilops-Triticum*) group. *Plant Cell* 13: 1735–1747.

Pasquini, M., Biancolatte, E., and Galterio, G. 1992. Wild emmer (*Triticum dicoccoides*) as a valuable source of powdery mildew resistance and high protein content. *J. Genet. Breed.* 46: 173–178.

Pellegrineschi, A., Brito, R.M., Velazquez, L., Noguera, L., Pfeiffer, M.W., McLean, S., and Hoisington, D. 2002. The effect of pretreatment with mild heat and drought stresses on the explant and biolistic transformation frequency of three durum wheat cultivars. *Plant Cell Rep. 20: 955–960.*

Pfluger L.A., Martin, L.M., and Alvarez, J.B. 2001. Variation in the HMW and LMW glutenin subunits from Spanish accessions of emmer wheat (*Triticum turgidum* ssp. *dicoccum* Schrank). *Theor. Appl. Genet.* 102: 767–772.

Pogna, N.E., Redaelli, R., Pasquini, M., Curioni, A., Dal Belin Peruffo, A., and Castagna, R. 1993. Inheritance studies of two chromosome translocations in bread and durum wheat. In *5th Gluten Proteins*, Detmold, Germany, pp. 308–318.

Prins, R., Groenewald, J.Z., Marais, G.F., Snape, J.W., and Koebner, R.M.D. 2001. AFLP and STS tagging of *Lr19*, a gene conferring resistance to leaf rust in wheat. *Theor. Appl. Genet.* 103: 618–624.

Qu, L.J., Foote, T.N., Roberts, M.A., Money, T.A., Aragón-Alcaide, L., Snape, J.W., and Moore, G. 1998. A simple PCR-based method for scoring the *ph1b* deletion in wheat. *Theor. Appl. Genet.* 96: 371–375.

Rajaram, S., Peña, R.J., Villareal, R.L., Mujeeb-Kazi, A., Singh, R., and Gilchrest, L. 2001. Utilization of wild and cultivated emmer and of diploid wheat relatives in breeding. *Israel J. Plant Sci.* 49:S-93–S-104.

Rao, M.V.B. 1978. The transfer of alien genes for stem rust resistance to durum wheat. In *Proceedings of the 5th International Wheat Genetics Symposium*, S. Ramanujam, Ed., New Delhi, India, pp. 338–341.

Redaelli, R., Pogna, N.E., Dachkevitch, T., Cacciatori, P., Biancardi, A.M., and Metakovsky, E.V. 1992. Inheritance studies of the 1AS/1DS chromosome translocation in the bread wheat variety Perzivan-1. *J. Genet. Breed.* 46: 253–262.

Repellin, A., Båga, M., Jauhar, P.P., and Chibbar, R.N. 2001. Genetic enrichment of cereal crops by alien gene transfers: new challenges. In *Reviews of Plant Biotechnology and Applied Genetics*. Kluwer Academic Publishers, Dordrecht, The Netherlands, pp. 159–183.

Riley, R. 1960. The diploidization of polyploid wheat. *Heredity* 15: 407–429.

Salamini, F., Ozkan, H., Brandolini, A., Schafer-Progl, R., and Martin, W. 2002. Genetics and geography of wild cereal domestication in the Near East. *Nat. Rev.* 3: 429–441.

Sandhu, D., Champoux, J.A., Bondareva, S.N., and Gill, K.S. 2001. Identification and physical localization of useful genes and markers to a major gene-rich region on wheat group 1S chromosomes. *Genetics* 157: 1735–1747.

Sarkar, S. and Stebbins, G.L. 1956. Morphological evidence concerning the origin of the B genome in wheat. *Am. J. Bot.* 43: 297–304.

Satyavathi, V.V. and Jauhar, P.P. 2003. *In vitro* regeneration of commercial durum cultivars and transformation with antifungal genes. In *Proceedings of the 2003 National Fusarium Head Blight Forum*, Minneapolis, MN, December 12–15, pp. 32–35.

Sears, E.R. 1952. Homoeologous chromosomes in *Triticum aestivum*. *Genetics* 37: 624.

Sears, E.R. 1954. The aneuploids of common wheat. *Univ. Mo. Agric. Exp. Stn. Res. Bull.* 572: 1–59.

Sears, E.R. 1966. Nullisomic-tetrasomic combinations in hexaploid wheat. In *Chromosome Manipulations and Plant Genetics*, Vol. 20, R. Riley and K.R. Lewis, Eds. Oliver and Boyd, Edinburgh, pp. 29–45.

Sears, E.R. 1972. Chromosome engineering in wheat. In *Proceedings of the Stadler Genetics Symposium* 4: 23–38.

Sears, E.R. 1973. *Agropyron*-wheat transfers obtained by homoeologous pairing. In *Proceedings of the 4th International Wheat Genetics Symposium*, E.R. Sears and L.M.S. Sears, Eds., Columbia, MO, pp. 191–199.

Sears, E.R. 1975. The wheats and their relatives. In *Handbook of Genetics*, Vol. 2, C. King, Ed. Plenum Press, New York, pp. 59–91.

Sears, E.R. 1976. Genetic control of chromosome pairing in wheat. *Ann. Rev. Genet.* 10: 31–51.

Sears, E.R. 1978. Analysis of wheat-*Agropyron* recombinant chromosomes. In *Proceedings of the 8th EUCARPIA Congress on Interspecific Hybridization in Plant Breeding*, Madrid, Spain, 1977, pp. 63–72.

Sears, E.R. 1981. Transfer of alien genetic material to wheat. In *Wheat Science: Today and Tomorrow*, L.T. Evans and W.J. Peacock, Eds. Cambridge University Press, Cambridge, U.K., pp. 75–89.

Sears, E.R. 1983. The transfer to wheat of interstitial segments of alien chromosomes. In *Proceedings of the 6th International Wheat Genetics Symposium*, S. Sakamoto, Ed., Kyoto, Japan, pp. 5–12.

Segal, G., Liu, B., Vega, J.M., Abbo, S., Rodova, M., and Feldman, M. 1997. Identification of a chromosome-specific probe that maps within the *Ph1* deletions in common and durum wheat. *Theor. Appl. Genet.* 94: 968–970.

Shaked, H., Kashkush, K., Ozkan, H., Feldman, M., and Levy, A.A. 2001. Sequence elimination and cytosine methylation are rapid reproducible responses of genome to wide hybridization and allopolyploidy in wheat. *Plant Cell* 13: 1749–1759.

Simeone, R., Blanco, A., and Bisignano, V. 1984. Production, morphology and cytology of the hybrids between *T. durum* Desf. and *Aegilops ovata* L. In *Proceedings of the Workshop on "Breeding Methodologies on Durum Wheat and Triticale,"* Viterbo, Italy, 1983, pp. 41–47.

Singh, H., Phutela, S., Kaur, P.P., Harinder, K., and Dhaliwal, H.S. 1998a. Transfer of novel HMW glutenin subunits from wild *Triticum* species into *Triticum durum*. In *Proceedings of the 9th International Wheat Genetics Symposium*, Vol. 4, A.E. Slinkard, Comp., Saskatoon, Saskatchewan, Canada, pp. 268–270.

Singh, R.P., Huerta-Espino, J., Rajaram, S., and Crossa, J. 1998b. Agronomic effects from chromosome translocations 7DL.7Ag and 1BL.1RS in spring wheat. *Crop Sci.* 38: 27–33.

Stack, R.W., Elias, E.M., Mitchell, J.F., Miller, J.D., and Joppa, L.R. 2002. Fusarium head blight reaction of Langdon Durum–*Triticum dicoccoides* chromosome substitution lines. *Crop Sci.* 42: 637–642.

Strampelli, N. 1932. *Origine, sviluppi, lavori e risultati*. Istituto Nazionale di Genetica per la Cerealicoltura in Roma: S.A. Stab, Arti Grafiche Alfieri & Lacroix, Milano, Italy.

Sun, G.L., Fahima, T., Korol, A.B., Turpeinen, T., Grama, A., Ronin, Y.I., and Nevo, E. 1997. Identification of molecular markers linked to the *Yr15* stripe rust resistance gene of wheat originated in wild emmer wheat, *Triticum dicoccoides*. *Theor. Appl. Genet.* 95: 622–628.

Tuleen, N.A. and Hart, G.E. 1988. Isolation and identification of wheat-*Elytrigia elongata* chromosome 3E and 5E addition and substitution lines. *Genome* 30: 519–524.

Urbano, M., Resta, P., Benedettelli, S., and Blanco, A. 1988. A *Dasypyrum villosum* (L.) Candargy chromosome related to homeologous group 3 of wheat. In *Proceedings of the 7th International Wheat Genetics Symposium*, T.E. Miller and R.M.D. Koebner, Eds., Institute of Plant Science Research, Cambridge, U.K., pp. 169–173.

Vallega, J. and Zitelli, G. 1973 New high yielding Italian durum wheat varieties. In *Proceedings of the Symposium on Genetics and Breeding of Durum Wheat*, Bari, Italy, pp. 373–399.

Vallega, V, and Waines, J.G. 1987. High molecular weight glutenin subunit variation in *Triticum turgidum* var. *dicoccum*. *Theor. Appl. Genet.* 74: 706–710.

Van Deynze, A.E., Dubcovski, J., Gill, K.S., Nelson, J.C., Sorrells, M.E., Dvořák, J., Gill, B.S., Lagudah, E.S., McCouch, S.R., and Appells, R. 1995. Molecular-genetic maps for group 1 chromosomes of Triticeae species and their relation to chromosomes in rice and oat. *Genome* 38: 45–59.

Van Deynze, A.E., Sorrells, M.E., Park, W.D., Ayres, N.M., Fu, H., Cartinhour, S.W., Paul, E., and McCouch, S.R. 1998. Anchor probes for comparative mapping of grass genera. *Theor. Appl. Genet.* 97: 356–369.

Vardi, A. 1973. Introgression between different ploidy levels in the wheat group. In *Proceedings of the 4th International Wheat Genetics Symposium*, E.R. Sears and L.M.S. Sears, Eds., Columbia, MO, pp. 131–141.

Vardi, A. and Zohary, D. 1967. Introgression in wheat via triploid hybrids. *Heredity* 22: 541–560.

Vasil, V., Castillo, A.M., Fromm, M.E., and Vasil, I.K. 1992. Herbicide resistant fertile transgenic wheat plants obtained by microprojectile bombardment of regenerable embryogenic callus. *Bio/Technology* 10: 667–674.

Villareal, R.L., Banuelos, O., Mujeeb-Kazi, A., and Rajaram, S. 1998. Agronomic performance of chromosomes 1B and T1BL.1RS near-isolines in the spring bread wheat Seri M82. *Euphytica* 103: 195–202.

Vitellozzi, F., Ciaffi, M., Dominici, L., and Ceoloni, C. 1997. Isolation of a chromosomally engineered durum wheat line carrying the common wheat *Glu-D1d* allele. *Agronomie* 17: 413–417.

Wang, G.Z., Miyashita, N.T., and Tsunewaki, K. 1997. Plasmon analyses of *Triticum* (wheat) and *Aegilops*: PCR single-strand conformational polymorphism (PRC-SSCP) analyses of organellar DNAs. *Proc. Natl. Acad. Sci. U.S.A.* 94: 14570–14577.

Weeks, J.T., Anderson, O.D., and Blechl, A.E. 1993. Rapid production of multiple independent lines of fertile transgenic wheat (*Triticum aestivum*). *Plant Physiol.* 102: 1077–1084.

Xie, C., Sun, Q., Ni, Z., Yang, T., Nevo, E., and Fahima, T. 2003. Chromosomal location of a *Triticum dicoccoides*-derived powdery mildew resistance gene in common wheat by using microsatellite markers. *Theor. Appl. Genet.* 106: 341–345.

Xu, S.S., Khan, K., Klindworth, D.L., Faris, J.D., and Nygard, G. 2004. Chromosomal locations of genes for novel glutenin subunits and gliadins in wild emmer wheat (*Triticum tugidum* L. var. *dicoccoides*). *Theor. Appl. Genet.* 108: 1221–1228.

Yang, Y.C., Tuleen, N.A., and Hart, G.E. 1996. Isolation and identification of *Triticum aestivum* L. em. Thell. cv Chinese Spring-*T. peregrinum* Hackel disomic chromosome addition lines. *Theor. Appl. Genet.* 92: 591–598.

Zitelli, G. 1973. Genetic improvement of durum wheat for diseases resistance. In *Proceedings of the Symposium on Genetics and Breeding of Durum Wheat*, Bari, Italy, pp. 473–487.

CHAPTER **3**

# Utilization of Genetic Resources for Bread Wheat Improvement

A. Mujeeb-Kazi

## CONTENTS

3.1   Introduction...................................................................................................61
3.2   Genetic Diversity and Its Distribution............................................................62
    3.2.1   Primary Gene Pool ..............................................................................63
    3.2.2   Secondary Gene Pool ..........................................................................63
    3.2.3   Tertiary Gene Pool ..............................................................................63
3.3   Production of Intergeneric Hybrids................................................................63
    3.3.1   Phenology ............................................................................................65
    3.3.2   Cytology ..............................................................................................65
    3.3.3   Maintenance.........................................................................................66
    3.3.4   Asymmetric Synthetic Genomes..........................................................67
3.4   Interspecific Hybridization .............................................................................68
3.5   Intergeneric Hybridization .............................................................................74
3.6   Cytogenetic Manipulation and Alien Gene Transfer.....................................74
3.7   Utilization and Practicality of Wide Crosses Germplasm.............................79
3.8   Impact from Documented Transfer.................................................................80
3.9   Impact through Undocumented Transfer ........................................................81
3.10  Futuristic Anticipation ...................................................................................82
3.11  Conclusions.....................................................................................................86
References ...............................................................................................................88

## 3.1 INTRODUCTION

Tetraploid durum wheat (*Triticum turgidum* L., 2n = 4x = 28, AABB) and hexaploid bread wheat (*Triticum aestivum* L., 2n = 6x = 42, AABBDD) are central to world agriculture. Although bread wheat came on the world scene much later than its predecessor durum wheat, it has become the most important cereal crop for feeding humankind (see Chapter 1 by Jauhar in this volume). Genetic diversity and wide hybridization in relation to bread wheat improvement have been studied for almost two centuries. According to Ciferri (1955), the first interspecific cross was made in 1806 by Barelle. Subsequently, the first wheat–rye hybrid was reported by Wilson (1876). The first fertile Triticale came on the scene 5 years later (Rimpau 1881), followed by a mention of the complex wheat–barley hybrid (Farrer 1904), although of questionable authenticity (Shepherd and Islam 1981). Greater attention was paid to such divergent crosses for several reasons (McFadden and Sears 1946; Gupta and Priyadarshan 1982), primarily to enhance the genetic diversity in wheat that had a narrow genetic base because of adherence to the requirements of uniformity (Kronstad 1998), even though extensive genetic variation is present and stored in *ex situ* gene banks (Valkoun 2001). The holdings were estimated some years ago to be around 8,000,000 accessions (FAO 1996). Only 3% of these gene bank holdings are comprised of wild wheat relatives distributed within three gene pools (Harlan and de Wet 1971) possessing annual and perennial species (Dewey 1984). These are a valuable source of genes for biotic and abiotic stresses (Kimber and Feldman 1987). It is well recognized (Mujeeb-Kazi and Rajaram 2002) that genetic diversity is a prerequisite for ensuring durability of resistance and a major factor for sustainable agricultural systems that are challenged with the responsibility of producing a target of 1 billion tons of wheat over the next two decades (Braun et al. 1998). This output figure means a shift from the current wheat global average yield of 2.5 K/ha to roughly 4 K/ha (Rajaram 2001) to adequately address the projected world population needs of about 8.2 billion people.

Extensively utilized methods of wheat improvement are various conventional plant breeding protocols that have significantly contributed to global wheat production and will continue to do so. However, greater emphasis is required for charting the course of future wheat research around integrated multidisciplinary activity, visionary/functional astute policy setting, and use of novel technologies to ensure crop production outputs that are efficient. The latter forms the principal theme in this documentation, categorized under wide crosses or wide hybridization, which during the 1920s and 1930s were the *Triticum × Aegilops* combinations (Kihara 1937).

The *Agropyron* combinations with wheat were reviewed and summarized by White (1940) and Smith (1943). Then followed cross listings (Knobloch 1968), hybrid cytological data compilation (Kimber and Abu-Bakar 1979) on some 1104 hybrids and 270 bibliographic references, and a review on transfer of useful alien genes in various field crops (Stalker 1980). Over these past decades emerged at least two significant events that changed the course of wide hybridization research. These were the discovery of colchicine in the late 1930s (Eigsti and Dustin 1955) and the integration of embryo culture technology in the 1960s (Murashige and Skoog 1962), which set the stage for active wide cross research with wheat (Kruse 1967, 1969, 1973, 1974a,b) to be pursued rigorously by several groups. Some of these experimentations have led to the contributions of Islam et al. (1981), Sharma and Gill (1983a,b,c), Mujeeb-Kazi and Kimber (1985), Gill and Raupp (1987), Mujeeb-Kazi et al. (1987, 1989), Gill (1989), Mujeeb-Kazi and Asiedu (1989, 1990), Jauhar (1993), Friebe et al. (1993, 1996a), Jiang et al. (1994), Sharma (1995), Mujeeb-Kazi et al. (1996a), Jauhar and Chibbar (1999), and Mujeeb-Kazi (2001a,b, 2003a, 2004). This chapter presents a broad scenario of crucial aspects governing wide cross research across basic parameters and their extension into applied outputs. The major focus shall be on bread wheat with a minor coverage of complementing technologies that promote functional efficiency via short-term or rapid outputs for fostering durability of resistance and agricultural sustainability.

## 3.2 GENETIC DIVERSITY AND ITS DISTRIBUTION

The tribe Triticeae is comprised of about 75 annual and 250 perennial species (Dewey 1984). This is the range of natural diversity available for wheat improvement as located in the conventional wheat germplasm and in closely or distantly related alien species sources. Genetic diversity can also be induced via mutagenic agents (ionizing, nonionizing irradiation or chemical mutagens; see review in Mujeeb (1970)). Preference is, however, generally for controlled, well-directed compensating homologous chromosomal exchanges, tilting the emphasis toward a preferred use of naturally present diversity available in various gene pools of the Triticeae (Harlan and de Wet 1971): the primary, secondary, and tertiary. The gene pool species categorization depends upon the genomic constitution of the species, thus forming a structure that dictates practical agricultural gains spread over short- or long-term time frames. The gene pool descriptions, with selective examples of each, are described below.

### 3.2.1 Primary Gene Pool

The species include hexaploid wheat landraces, cultivated tetraploid durum ($2n = 4x = 28$, AABB), wild *T. dicoccoides*, and the diploid ($2n = 2x = 14$) donors of the A and D genomes to durum and bread wheats. Genetic transfer in primary gene pool results through direct hybridization, homologous recombination, and relatively simple breeding strategies. Some combinations require assistance of embryo rescue and are of greater interest for enhancing diversity in wheat. *Aegilops tauschii* ($2n = 2x = 14$, DD) (syn. *Ae. squarrosa*, *T. tauschii*) (goatgrass) currently occupies a high priority (Mujeeb-Kazi 2003a) in wheat breeding.

### 3.2.2 Secondary Gene Pool

The polyploid *Aegilops* and *Triticum* species sharing one genome with wheat are included in this gene pool. Also included are the diploid species of the Sitopsis section. Genetic transfers are routine within homologous genomes but require manipulative protocols between nonhomologous types. Embryo rescue is a complementary aid for obtaining hybrids. Very limited usage exists for wheat improvement, but priority has been suggested for exploiting the Sitopsis diploid accessions of *Aegilops speltoides* ($2n = 2x = 14$, BB or $B^sB^s$) for durum and bread wheat improvement (Mujeeb-Kazi 2003b, 2004). Breeding protocols are more complex since manipulation strategies associated with alien gene transfer often incorporate undesirable traits together with the target gene of interest.

### 3.2.3 Tertiary Gene Pool

Diploid and polyploid species with nonhomologous genomes to those of wheat are included in this gene pool. Alien genetic transfer requires complex cytogenetic manipulation methods that facilitate homoeologous exchanges. Irradiation or tissue culture is another option when homoeologous chromosomal exchanges are not possible. Protocols promoting homoeologous exchanges are preferred since the alien introduced segments would be placed in the best location in the recipient chromosomes. Wheat–alien chromosome translocations are the general outputs of irradiation or tissue or callus culture, but these protocols are less favored since compensating exchange products are rarely obtained. Hybridization requires embryo rescue, success is of low frequency, and practical agricultural outputs are time-consuming unless modified approaches get incorporated (Mujeeb-Kazi 2003a). Mujeeb-Kazi and Wang (1995) elucidated the exploitation of the tertiary gene pool.

## 3.3 PRODUCTION OF INTERGENERIC HYBRIDS

Intergeneric hybrids in the Triticeae have been studied for considerable time. The objectives have focused on combining divergent species in order to obtain basic cytological, evolutionary, or phylogenetic information about the parental species and, in many cases, to introduce desirable alien genes into elite cultivars lacking that particular trait. The production of the earliest hybrids was accomplished by the aid of conventional simple techniques of emasculation and pollination. This progress, up to 1974, was reviewed (Armstrong 1936; Kihara 1937; Smith 1943; Knobloch 1968; Kruse 1974a,b).

It is felt in many research circles that the contributions of Kruse provided a stimulus for intensive research with wide crosses, of which the priority was given to barley × wheat hybridization (Bates et al. 1976; Islam et al. 1978; Mujeeb et al. 1978). The outputs since then formed another phase via the diversified contributions of Islam et al. (1981), Mujeeb-Kazi and Rodriguez (1983a,b), Sharma and Gill (1983a), Mujeeb-Kazi and Kimber (1985), Wang (1989), Mujeeb-Kazi and Asiedu (1990), Pienaar (1990), Jauhar (1993), and Jauhar and Chibbar (1999). Over this decade and a half it became clear that hybrid production and growth had overridden the previous complexity stage, and numerous complex combinations were made rather simplistically (Sharma 1995). Researchers, however, still enjoyed making new hybrids, but the desire to find workable means to introgress alien genes required efficient strategies, and from doing this emerged novel means, of which the *ph* locus got greater attention (Sharma and Gill 1986), as did interspecific hybridization around the D genome (Alonso and Kimber 1984; Gill and Raupp 1987; Mujeeb-Kazi and Hettel 1995). This focus shift may have been a consequence of the agriculturists seeing the practical impacts of wide crossing, which has now formed the research thrust over the last decade (Ma et al. 1995a,b; Villareal et al. 1995b, 1996a, 1997; Ayala et al. 2001; Mujeeb-Kazi 2001a,b, 2003a, 2004). The last decade of outputs has addressed production constraints for various biotic and abiotic stresses, of which globally the most significant are for *Neovossia indica* (Karnal bunt), *Helminthosporium sativum* (spot blotch), *Septoria tritici* (leaf blotch), *Fusarium graminearum* (head scab), rust, salinity (Pritchard et al. 2002), and drought (Trethowan et al. 2003).

It is appropriate to recognize here that about three decades ago more attention was given to the barriers of production of wide hybrids that existed at various stages in the ontogeny of the hybrids. They were divided into six parts: (1) choice of parents, (2) emasculation procedures, (3) prepollination treatment, (4) pollination, (5) postpollination treatment, and (6) embryo rescue and culture. The intricacies of each aspect are reviewed by Mujeeb-Kazi and Kimber (1985). It is safe to conclude that researchers can make complex hybrids at will, fully endorsing the conclusion made a decade ago by Sharma (1995). Briefly, the key steps that facilitate production of wide hybrids relate to generally having the higher ploidy species as the female parent (for exceptions, see Delgado and Mujeeb-Kazi 2002), a pre- or postpollination hormone treatment, adequate pollen load with bud or late pollination timings, using wheat genotypes with recessive crossability genes (Riley and Chapman 1967; Snape et al. 1979; Falk and Kasha 1981; Yen et al. 1986; Luo et al. 1992; Zheng et al. 1992; Jauhar 1995a; Fedak 1998), and appropriate embryo culture media and regeneration strategies, not discounting the all-important environmental input. The literature cited above with several additional cross-references address most of these facets, prompting this author to conclude that one can hybridize any species within the Triticeae with each other.

Inherent variations in hybrid production outputs do occur between and within the intergeneric and interspecific categories (Mujeeb-Kazi and Rajaram 2002; Mujeeb-Kazi 2003a), with results also differing for similar combinations across various laboratories. In essence, one viable hybrid of a combination is enough to exploit for practical programs objectives associated with desired trait transfer. Important factors that also contribute to hybrid outputs are associated with the involved germplasms that are comprised of durum or bread wheats, their spring, winter, or facultative habit, the nature/polyploidy of the alien parent, i.e., annual/perennial nature, and diploidy to polyploidy.

Wide hybrids of present focus in this article are those that capture unique genetic diversity from divergent alien species being generally self-sterile, require embryo rescue, and possess a mitotic cytological complement that combines the basic genomic set of each parent. When both parents have the same ploidy level with no karyotypic difference, then differential staining procedures become necessary. Often hybrids exhibit a co-dominant phenotype, but this may not be essential (Kimber 1983), in which case unequivocal validation of a hybrid emerges from meiotic data — information that also sets the hybrid exploitation strategy.

### 3.3.1 Phenology

$F_1$ hybrids expressing a co-dominant phenotype have been considered ideal candidates for wheat improvement since the alien genetic expression is readily observed. Mujeeb-Kazi et al. (1987, 1989) felt that a modified $F_1$ phenotype should be used as a preselection parameter in practical studies. Delgado and Mujeeb-Kazi (2002), upon analyzing over 185 intergeneric combinations, found this co-dominant trend prevalent. Parameters studied were for increased spike length and lax internodal distance. Other features studied in wide hybrids have been spikelet/spike, glume body length, awn length, lemma body length, and anther length. Parental phenotypic dominance is a rare event, of which barley × wheat (or wheat × barley) hybrids are good examples. Although morphologically the wheat phenology dominates, subtle biochemical variations may exist. These variations, however, have yet to contribute toward agricultural practicality, for which the expressed $Yd_2$ gene in a barley disomic chromosome addition line has yet to exhibit its BYDV resistance transfer effect in wheat germplasm. The extensive array of intergeneric hybrids also observed by other groups show a similar $F_1$ modified phenotype trend (Sharma and Gill 1983a; Mujeeb-Kazi et al. 1987, 1989; Jauhar 1993). The logic of using co-dominance in $F_1$ as a selection sieve is that if the entire alien chromosome contribution cannot phenotypically express in a wheat background, there would be negligible or no chance for a single alien chromosome (added, substituted, or introgressed) to express the trait of interest.

In interspecific $F_1$ combinations where A, B, and D genome species are utilized, phenotypic co-dominance is the norm for over 1000 combinations observed so far (Mujeeb-Kazi 2001b, 2003a). It must be recognized, however, that often co-dominant $F_1$ hybrids may also mask (suppress) or dilute the alien species trait of interest (Ma et al. 1995b). Such observations influence alien species utilization protocols. For targeted programs, the alien species are first screened for the stress constraints. Those found to be positive contributors for the traits are then involved in crossing programs yielding hybrids that may or may not express the trait based upon genetic expressivity of the alien genes. Alternatively, as done for the D genome synthetics in CIMMYT, the diploid Ae. tauschii accessions are randomly hybridized with durum wheat to yield $F_1$ hybrids and eventually fertile hexaploids ($2n = 6x = 42$, AABBDD) which become the source of providing resistant products upon screening for use in breeding. Phenotypic co-dominance exists and trait expression validates alien genetic contribution since the maternal durum cultivars are trait susceptible. This untargeted strategy may not capture precise data of Ae. tauschii accessional resistance/suppression, but it is a preferred approach for delivering usable applied agricultural materials, as well as being very rapid and efficient.

In classical wide crossing/cytogenetics programs, during the production of alien disomic chromosome addition or substitution lines, protocols used typical backcrossing, where the same wheat cultivar was involved as present in the $F_1$ combination — generally being Chinese Spring. This facilitated convenient detection of the alien chromosomes, phenotypic expression, and unique characterization. The use of such Chinese Spring-based genetic products did not find extensive utility in global wheat breeding programs since the Chinese Spring agronomic type could not blend with normal field-oriented growth cycles due to lateness and the various susceptibilities that prevailed. Consequently, our applied usage strategy employed backcrossing or top-crossing the Chinese Spring-based $F_1$ hybrids with a high yielding wheat cultivar or using two such cultivars that gave genetic stocks in a quality wheat background more amenable to breeders' usage criteria (Mujeeb-Kazi and Asiedu 1990).

### 3.3.2 Cytology

Genuine wheat–alien species hybrids possess half the chromosome number of each parent involved in the cross combination, the initial validation being based upon root-tip mitotic compositions (Mujeeb-Kazi and Miranda 1985). Where both parents have different ploidy levels a mere number count is ample evidence of hybridity. Combinations with similar ploidy levels (e.g., *T. aestivum* × *Thinopyrum intermedium*; $2n = 6x = 42$) may complicate validation unless superior mitotic cytology can discern the single-satellited wheat chromosomes. Recourse to Giemsa C-banding helps confirmation (Jahan et al. 1990), and more depth can also be achieved by fluorescent *in situ* hybridization (FISH) (Islam-Faridi and Mujeeb-Kazi 1995). Validating the hybridity status, though essential to convey data concerning a cross combinations status, does not provide complete details for the utility of the hybrid in wheat breeding where alien chromatin transfer is crucial. Meiotic analyses provide this backbone information as well as complement the hybrid status. It is quite rare to find $F_1$ hybrids where meiocytes exhibit a high recombination frequency that involves wheat and alien species chromosomes. Meiotic data enable genomic analyses and provide a practical basis for advancing $F_1$ hybrids, further strengthened through the development of numerical methods of assessing genomic affinity (Alonso and Kimber 1981; Kimber and Alonso 1981; Jauhar and Joppa 1996). In general, chromosome association complexities are a part of combinations where diverse tertiary gene pool species are involved, classified as intergeneric hybrids. For combinations where primary gene pool species are involved, gene transfers are routine and such interspecific programs currently receive more attention. Researchers are cognizant of these two distinctions and maintain a blend of the total diversity prevalent that could be incorporated, thereby enhancing chances of delivering a final product with durable traits and pyramided genes.

Cytological evaluations, apart from validating crosses or permitting the development of a wide cross breeding program, also provide novel outlets as a consequence of abnormalities in chromosome behavior, generally involving alien species chromosomes. Reports of these aneuploid hybrid progenies are extensive (Fedak 1977, 1980; Islam et al. 1981; Mujeeb-Kazi and Rodriguez 1983a,b), and if these variations are put to stringent diagnostic tests, they do prove beneficial in promoting wheat–alien recombination events as reported by Jewell and Mujeeb-Kazi (1982) for a bread wheat × *Ae. variabilis* cross where absence of chromosome 5B is advantageous. A wide array of such aneuploid anomalies has since been reported (Mujeeb-Kazi et al. 1996c) and occur in backcross I selfed derivatives and in amphiploid maintenance.

Aneuploidy has provided outputs that otherwise remained elusive to researchers. Intergeneric hybrids between *T. aestivum* and *Th. repens–A. desertorum* gave several $F_1$ hybrids with stable mitotic counts of $21 + 21 + 14 = 56$ chromosomes. Recovery of two $F_1$ events (10565 and 10585) possessed 35 chromosomes comprised of 21 wheat + 14 *A. desertorum*; a cross combination that had not been produced by hybridizing wheat directly with *A. desertorum* but was realized via the complete elimination of *Th. repens* genomes — a fortuitous happening. Of greater significance has been total paternal genome elimination in wheat × maize intergeneric hybrids, which has led to wheat haploids, a boon to wheat breeding programs and contributing to several other research areas (Mujeeb-Kazi 2000, 2004).

### 3.3.3  Maintenance

Wheat × alien species hybrids are the primary material sources from which basic, strategic, and applied programs emanate. Valid $F_1$ hybrids are either annual or perennial, the growth habits that influence the nature of their exploitation and maintenance. Such combinations provide a rich source of unique information that, over the last two decades after the initial observations of Kruse (1967, 1969), has enabled numerous researchers to produce more hybrids and describe similar complex undertakings. Several projections advocating applied agricultural potentials of wide hybrids have been regularly made, but in order to go beyond the $F_1$ and deliver a usable final agricultural product, we must recognize that this is a tall order. Not much success has been achieved, for which a few obvious long-standing or unrealized projections have been for transfer of salt tolerance (*Lophopyrum elongatum* to wheat; Dvořák et al. 1988), barley yellow dwarf (barley to wheat; Qualset, personal communication, 2003), and head scab resistance (*Th. elongatum* to durum wheat; Jauhar and Peterson 2000, 2001). Nevertheless, having a promising trait attributed to an alien species source is a good basis to justify making a hybrid that gives a practical dimension to wide hybridization programs where this criterion is essential. Consequently, while it is paramount to exploit the $F_1$ hybrid for applied goals, it is also essential to maintain hybrid integrity for basic studies, i.e., keep the $F_1$ viable for continued research, and for achieving these ends, some maintenance options are available and practiced. Annual $F_1$ hybrids have a limited growth span, and being self-sterile, they tiller, grow luxuriously up to a certain stage, produce a lot of spikes, and eventually die. These growth stages permit solid data to be generated, e.g., on cytology and phenology, and biochemical, molecular, submicroscopic characteristics without any provision on hand to allow the $F_1$ to survive, as is possible for perennial $F_1$ combinations. There are two options that have been effective in maintaining $F_1$ germplasm: (1) physically cloning the annual $F_1$ hybrid at an early seedling stage and then using the clonal progeny to produce amphiploids (Mujeeb-Kazi et al. 1987), or (2) pollinating the self-sterile but female-fertile $F_1$ annual hybrid spikes with wheat (same cultivar as in the hybrid or any other elite line ($F_1$ top-cross)) to give backcross I ($BC_1$) seed. The advantages of amphiploids are greater (Gill 1989), while the $BC_1$ route is an added security if the amphiploid cannot be produced.

In the case of perennial $F_1$ hybrids, inducing amphiploidy or advancing $F_1$ combinations by back- or top-crossing remain two options. However, the perenniality gives more operational flexibility in that more attempts are possible to try and obtain amphiploids; that are a definite preference; and also generate more $BC_1$ seed. The majority of the perennial $F_1$s are outputs of utilizing perennial alien tetraploid and hexaploid species where the tertiary gene pool sources have been abundantly exploited. In most cases, such $BC_1$ derivatives can also be seed increased by selfing because of their high self-fertility. Generally, in $BC_1$ derivatives, when alien tetraploid sources are utilized, selfed products produce good seed, have stable meiosis, and possess 56 chromosomes (42 wheat + 14 of the two closely related genomes of the alien species, e.g., $E_1$ and $E_2$ of *Thinopyrum curvifolium*). Repeated selfings of these BC progenies enable rapid seed increase and facilitate biotic and abiotic stress screening but are subject to subtle chromosomal changes due to recombinational events likely to happen when the 14 $E_1$ and $E_2$ chromosomes associate in meiosis. This product output is also known as complete synthetic genome progeny (Mujeeb-Kazi and Miranda 1984). Applied cytogeneticists prefer such changes since they may be a source of genomic restructuring, and if individual addition lines are developed after several selfing cycles, these addition stocks could be a lot different in their structure than ones obtained from the straight $BC_1$ advance to $BC_2$, $BC_3$, etc., where selfing is avoided (Figure 3.1). Commonly preferred tetraploids that could be so incorporated are *Thinopyrum scirpeum*, *Thinopyrum junceiforme*, *Thinopyrum distichum*, *Thinopyrum rechingeri*, and *Thinopyrum scythicum*.

### 3.3.4 Asymmetric Synthetic Genomes

Perennial tertiary gene pool species that are segmental allohexaploids (*Th. junceum*, 2n = 6x = 42) or segmental autoallohexaploids (*Th. intermedium* ssp. *acutum, glaucum, pulcherrimum, trichophorum, varnense*) that possess two related genomes (E1 and E2) and a distinctly different third genome (X or Z) yield an $F_1$ hybrid product that for *Th. intermedium* would be 2n = 6x = 42, $ABDE_1E_2X$ or Z. The resulting amphiploids possess 84 chromosomes but are beset with aneuploidy during their subsequent selfings when making seed increase, suggesting that the threshold for stable and normal amphiploid maintenance seems to be at an octoploid level rather than decaploid or higher ploidy. The derived $BC_1$ progeny ($F_1$ × wheat, or amphiploid × wheat) possesses 63 chromosomes, genomically $AABBDDE_1E_2X$ or Z. Like the complete synthetic genome germplasm, these $BC_1$s are self-fertile and can also undergo genomic restructuring around the E1 and E2 genomes. Upon repeated selfings of the 63-chromosome $BC_1$s, it is fairly common to encounter derivatives that are mixed with an octoploid structure of 56 chromosomes as a consequence of preferential alien genome loss. This phenomenon has been described as genome splitting or asymmetric genome reduction (Gottschalk 1971; Cauderon 1977; Ladinsky and Fainstein 1978;

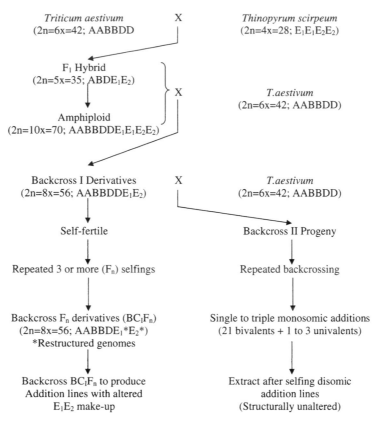

**Figure 3.1**   Synthetic genome development in a *Triticum aestivum* × *Th. scirpeum* hybrid combination, also showing production of alien disomic chromosome addition lines (modified vs. normal).

Dewey 1980). The $F_1$ schematic will also be essentially applicable to this category of materials, maintaining a similar promise of promoting gene pyramiding, thus agglomerating polygenes in new restructured packages.

Perennial species-based $F_1$ hybrids, apart from leading to generation of amphiploids and $BC_I$ derivatives, can also be maintained as a living herbarium via a biannual physical cloning coupled with a cytological check for continued stability. Close to 185 unique $F_1$ hybrids are maintained in the wide hybridization program at CIMMYT, Mexico, which are a perpetual source to service global needs for preferred $BC_I$s around specific regional and national cultivars. This service assists researchers internationally to swiftly exploit alien genetic diversity without having to struggle at making complex $F_1$s.

## 3.4 INTERSPECIFIC HYBRIDIZATION

Conventional wheat breeding programs are built around diverse cross combinations of germplasms residing in the same gene pool that undergo genetic recombination followed by trait segregation, evaluation, and ultimate varietal release. In order to amplify the genetic diversity of the crop, novel genetic resources become a focus for which the close progenitors of wheat are preferred. These are the numerous accessions of the A, B, and D genomes. Donors within this spectrum, A and D genomes, have greater advantage than the B genomes, essentially because of their proximity to the A and D sets present in bread wheat and also based upon cytogenetic test

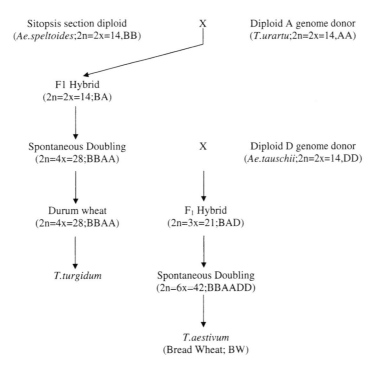

**Figure 3.2** The origin and evolution scheme of durum and bread wheat involving the three diploid progenitors *Ae. speltoides, T. urartu,* and *Ae. tauschii.*

analyses that indicate greater closeness of the seven chromosomes of the D genome wild diploid than the A genome chromosomes with the respective D and A genomes of cultivated bread wheat. Accessions of these two diversity sources reside in the primary gene pool, can be hybridized with ease, allow for swift gene transfer via homologous recombination, and have extensive diversity for global biotic and abiotic stress constraints that limit wheat production. A look at the evolution scheme of wheat (Figure 3.2) elucidates several options for making parental choices to enrich bread wheat. Logic and scientific reasoning, however, tilt the optimum choice toward the exploitation of the D genome diploid *Ae. tauschii* $(2n = 2x = 14)$, because it is doubtful if numerous accessions got involved in the natural hybridization/amphiploidization event, thus giving rise to a crop with an extremely narrow genetic base (Metakovsky et al. 1984; Zohary and Hopf 1988). Additionally, the observations of Kihara (1944) and McFadden and Sears (1946) associated with the *Ae. tauschii* role have enabled current investigators to focus their wheat improvement efforts around this wild diploid via various protocols, three of which are:

1. Producing synthetic hexaploid wheats $(2n = 6x = 42, AABBDD)$ by crossing *T. turgidum* cultivars with *Ae. tauschii* accessions (Figure 3.3A)
2. Crossing elite but specific trait-susceptible *T. aestivum* cultivars with resistant *Ae. tauschii* accessions, backcrossing the ABDD $F_1$ hybrid with the elite cultivar, and selecting euploid $(2n = 6x = 42, AABBDD)$ derivatives (Figure 3.4)
3. Extracting the AABB genomes from commercial *T. aestivum* cultivars and then developing synthetic hexaploids by crossing with desired *Ae. tauschii* accessions (Figure 3.5)

Greater preference has been for (1) bridge crosses and (2) direct crosses (Gill and Raupp 1987; Mujeeb-Kazi 2001b). Practical proponents of using the strategy involving bridge crossing are indicated in Mujeeb-Kazi (2003a). The direct crossing described by Alonso and Kimber (1984) and Gill and Raupp (1987) gained practical importance via several outputs during the 1990s (Cox

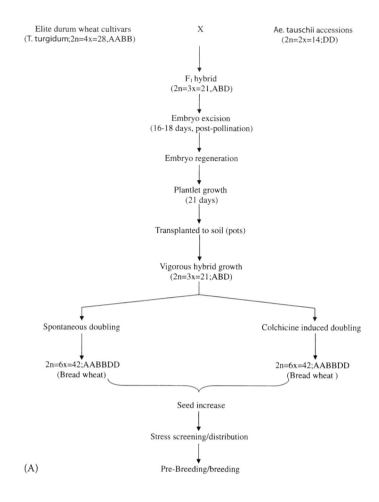

(A)

(B)

**Figure 3.3**    Production of synthetic hexaploid wheats (SH) from crosses between durum cultivars (*Triticum turgidum*) and *Ae. tauschii*.

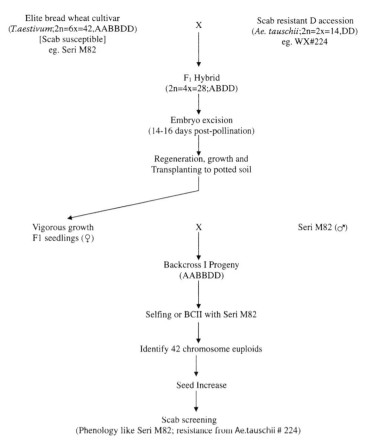

**Figure 3.4**  Schematic showing the direct crossing scheme derived from Alonso and Kimber (1984) and Gill and Raupp (1987), with subtle modifications (Mujeeb-Kazi and Hettel 1995).

and Hatchett 1994; Cox et al. 1994, 1995a,b). For a successful trait transfer, *Ae. tauschii* must possess the stress resistance/tolerance, and it should express in the recipient wheat background whether that is a durum (Figure 3.3B) or an elite bread wheat (Figure 3.4). Thus, in a targeted program the diploid accessions need to be screened and desirable ones identified, which can then be incorporated in synthetic combinations or in susceptible bread wheats. If the resistance is genetically expressed, then both these products can enter a breeding program. Often a resistant accession, when used to produce a synthetic haploid (SH) wheat, does not express its resistance or this is diluted as a consequence of the durum genomic effect (Ma et al. 1995b); hence, an option is to use trait-susceptible durum wheats for being randomly hybridized with all possible *Ae. tauschii* accessions yielding SHs. Upon screening the SH products for various stresses and their durum parents as susceptible controls, if an SH is identified as resistant/tolerant, then it can be safely concluded that the attribute was contributed by the D genome parent. However, for a susceptible SH there is no positive assurance whether the grass parent was resistant and its resistance was masked in the SH.

This untargeted strategy has been adopted by Mujeeb-Kazi (2001a, 2003a) due to its efficiency and the necessity for addressing multiple stress factors globally with swiftness. Around a decade and a half ago it was well accepted that *Ae. tauschii* accessions embodied unparalleled wealth of genetic diversity that could resolve numerous production constraints for bread wheat — an acceptance strengthened through the distribution expanse of the accessions from Turkey into Pakistan, inclusive of Afghanistan and China. The untargeted use of the accessions to randomly produce SH

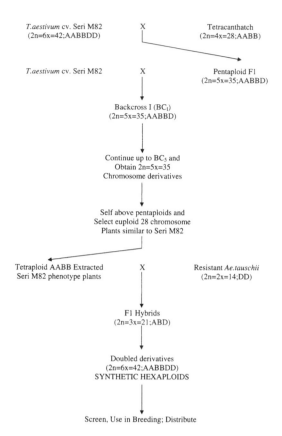

**Figure 3.5**   Extraction scheme of the tetraploid genomes (AABB) from bread wheat and their usage in production of synthetic hexaploids via crosses with *Ae. tauschii* accessions.

wheats received its share of criticism, but those that were pessimistic about the approach had failed to realize that for a global alien gene transfer program to be functional across all country boundaries and mega- or microenvironments, a different and radical approach was required for which some basic scientific data could be sacrificed. From the close to 1000 SHs produced so far, the global impact of their usage has been very promising. Diversity has been identified for numerous biotic and abiotic stresses, varietal releases have occurred, and usage has also extended into molecular areas, of which the International Triticeae Mapping Initiative (ITMI) mapping population is a significant output.

Additionally, the DNA polymorphisms that abound in *Ae. tauschii* accessions have paved the way for development and exploitation of D genome/chromosome-specific microsatellites (Roder et al. 1998a,b; Pestova et al. 2000, 2001; Plaschke et al. 1995). Production of SH wheats does require meticulous procedural steps of careful emasculation, bud pollinations, hormonal post-pollination treatments, embryo, culture, regeneration, and chromosome doubling similar to what are the norm for all interspecific and intergeneric hybridization protocols. These state-of-the-art procedures have now become so routine that the earlier descriptions of crossability are no longer deemed essential. It may suffice to mention that all cross combinations are highly successful (Mujeeb-Kazi and Rajaram 2002). Worthwhile to express is the fact that when durum wheats are generally the female parents of the SH cross, the embryos excised are of a definitive shape and easy to handle, giving a high regeneration rate, healthy seedlings, and almost 100% success across all genotypes and accession sources. The SH output range has been from 1.0 to 60.0% across diverse crossing locations in Mexico spread over 18 years. Reciprocal combinations have

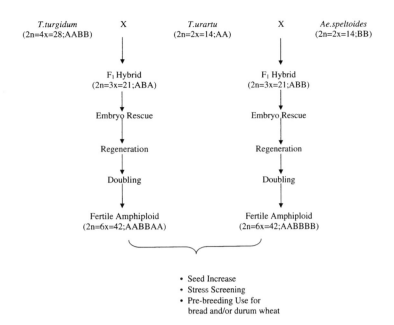

**Figure 3.6** The strategy for exploiting A and B genome diploids for wheat improvement akin to the D genome showing the bridge cross options only.

not been extensively practiced, i.e., *Ae. tauschii* (♀)–*T. turgidum* (♂). From the limited crosses made and SHs produced, observations indicate that hybrid seed set is high (60 to 85%), but embryo excision yields ill-differentiated minute embryos that may require special culture media to capture the full benefits of the high embryo recovery frequencies — an aspect that may be worthwhile to explore in the future.

Stress screening for applied goals using SH wheats is the main emphasis. Nevertheless, three categories of SHs have also been formulated to encourage basic/strategic research investigations (Mujeeb-Kazi 2003a). The categories are:

1. Same durum cultivar as the female parent combined with different *Ae. tauschii* accessions in order to evaluate accessional diversity and its expression in a uniform maternal background
2. Different durum cultivars as the female parent combined with the same *Ae. tauschii* accessions so as to assess the expressivity of the accession in diverse durum backgrounds
3. Durum (♀)–*Ae. tauschii* (♂) combinations and their reciprocal cross (*Ae. tauschii* (♀)–Durum (♂)) to study cytoplasmic influences, if any

Currently available are about 750 *Ae. tauschii* accessions that may have several duplications. Kawahara et al. (2003) placed 450 accessions in a core set that has been almost totally combined in the SH germplasm produced at CIMMYT (Mujeeb-Kazi 2003a) comprising 1000 entries. For convenience in application various SH subgroups have been made known as elite I of 95 SHs, elite II of 34 SHs with multiple disease resistances, and various subsets that address biotic and abiotic stress constraints. Molecular fingerprinting of these SHs would appear meaningful for starters instead of covering the entire SH materials at this stage.

The above D genome exploitation strategy has also been extended to cover the A and B genomes (Figure 3.6). Though the outputs would be more amenable for durum wheat improvement, usage of the AAAABB and AABBBB hexaploid stocks are also a potent means to introduce A and B genomic diversity into recipient bread wheats. The direct cross options also exist for both these genomes with bread wheat. The hexaploid A and B genome stocks have been produced and show promise but have not entered volatile usage in wheat breeding so far, since the D genome diversity has maintained its greater interest (Mujeeb-Kazi 2003a; Villareal and Mujeeb-Kazi 2003). Based upon the positive traits that have been identified in the D genome-based SH wheats, where durums are susceptible, we are currently developing a wheat prebreeding strategy where specific *Ae. tauschii* accessions are being targeted for improving location-specific wheats and important stress constraints via direct crossing. Similar ramifications could be expected for the two other genomes.

## 3.5 INTERGENERIC HYBRIDIZATION

In contrast to interspecific hybrid production discussed in the previous section, intergeneric crosses involve alien species that are genomically extremely diverse and complex to hybridize with wheat. When hybridized, the combinations exhibit little to no intergenomic chromosome associations. Despite these constraints, research interest has been high and goes back to the late 1980s. Gauging the successes and advancements made over the last three decades, one can safely conclude that hybridization per se is no longer a major obstacle (Sharma 1995). Some intergeneric hybrids are quite easy to produce, e.g., *T. aestivum–Th. intermedium* (Mujeeb-Kazi et al. 1987), while others are more difficult, e.g., *T. aestivum–Hordeum* (Islam et al. 1978). Other complex combinations have been with all *Agropyron* species, *Psathyrostachys juncea*, and the small-anther *Elymus* species. Subtle manipulations and careful choice of parents are a solace and have widened the hybridization categories. One variation utilized alien species as female parents in crosses with wheat in order to overcome failure that prevailed for the normal reverse cross, e.g., *A. ciliare* and *A. yezoense*–wheat (Sharma and Gill 1983a,b,c), *A. trachycaulum*–wheat (Mujeeb-Kazi and Bernard 1982), *A. fibrosum*–wheat (Mujeeb-Kazi and Bernard 1982), *E. canadensis–T. aestivum* (Mujeeb-Kazi and Bernard 1982). It thus appears that other hybrids that have not been possible to date may be recovered if reciprocal combinations are attempted.

Further, reciprocal crossing may also allow higher frequencies of embryo recovery as observed in *A. smithii* combinations with wheat (0.8 vs. 6.7%) (Sharma and Gill 1981). Another strategy that facilitates the crossing success involves pollinating wheat with an induced autotetraploid of an alien species. The hybrid possesses the ABD genomes of wheat and the normal diploid status of the alien source. Backcrossing this $F_1$ leads to an amphiploid very simplistically, as demonstrated for the *T. aestivum–Ps. juncea* intergeneric hybrid (Mujeeb-Kazi and Asiedu 1990). Not much utilized is the alien accessional effect, particularly in intergeneric hybrids where varied success rates prevail among a species accession, as shown for *Ps. juncea* (Plourde et al. 1990) and *Elymus scabrus* (Ahmad and Comeau 1991).

Circumventing the low $F_1$ meiotic pairing constraint of most intergeneric hybrids is a necessity in order to deliver a practically useful commercial product. Such outcomes are rare and particularly so for polygenic characters, thus necessitating the use of special protocols, which are considered below.

## 3.6 CYTOGENETIC MANIPULATION AND ALIEN GENE TRANSFER

There seems to be no parallel to the chromosome 5B manipulation approach, where its genetic stocks are used as maternal parents in crosses with alien species/genera. In common usage have been stocks of monosomic 5B, *ph ph*, and nullisomic 5B–tetrasomic 5A or 5D combinations. $F_1$ hybrids produced with these stocks exhibit enhanced wheat–alien chromosome pairing and recombinations due to the absence of *Ph1* (Sharma and Gill 1983a,b,c; Darvey 1984; Mujeeb-Kazi et al. 1984; Sharma and Baenziger 1986; Jauhar and Almouslem 1998). Manipulating low meiotic chromosome pairing associations to high is paramount to achieving alien genetic transfers, and the majority of the wide hybrids have the low pairing characteristics that would reduce any chance of desirable alien gene transfer. In rare situations, it is fortuitous that high pairing is present, even in the presence of the *Ph1* locus, indicative of the alien genomic influence, e.g., wheat × *Thinopyrum bessarabicum* and *Th. curvifolium* (Almouslem and Jauhar 1998). A systematic treatment (Jauhar and Chibbar 1999) briefly addressed here shall elucidate the progress with cytogenetic manipulation options that wide cross researchers more often than not seriously contemplate when attempting to make alien gene transfers. It is paramount that wheat and alien species chromosomes associate (pair) in order to realize the alien exchange. The earliest protocols for promoting chromosome pairing are attributed to Sears (1956, 1972, 1981, 1984); they aided in making gene transfers. The phase where it was demonstrated that some alien species inactivated the homoeologous pairing suppressor *Ph1* (Sears 1976) also emerged. Using *Ae. speltoides* to suppress *Ph1* regulatory activity enabled Riley et al. (1968) to transfer stripe rust resistance from *Ae. comosa* to wheat. Additional findings led to the conclusion that perennial or annual grasses are rich in materials that influence the *Ph1* locus (Jauhar 1975, 1977, 1992a,b, 1993, 1995a,b; Knott and Dvořák 1981; Mujeeb-Kazi et al. 1987; Jauhar and Joppa 1996; Farooq et al. 1996; Motsny and Simonenko 1996; Jauhar and Almouslem 1998; Simonenko et al. 1998) and are potential sources to contend with. More recently, Chen et al. (1994) reported the development of wheat germplasm that possessed the *Ph^I* gene contributed by *Ae. speltoides*. This gene has the capacity to inhibit the homoeologous pairing suppressor gene, and since it is dominant, a single dosage is adequate to permit cytogenetic changes. It has not gained too extensive an application so far, but some limited outputs have been achieved (Aghaee-Sarbarzeh et al. 2000; Wang et al. 2003).

Researcher preference has also been directed toward exploiting 5B-deficient stocks mentioned at the start of this section: by using 5B-deficient stocks where the *Ph* gene is absent, ensuing intergeneric hybrids exhibit increased intergenomic chromosome pairing (Jauhar 1991; Jauhar and Almouslem 1998; Murai et al. 1997). Achieving high intergenomic pairing, however, is no guarantee that successful, practically useful agricultural products will emerge. Three scenarios must be recognized:

1. High pairing is restricted to events at the $F_1$ stage and is extremely encouraging. The subsequent step is to advance the $F_1$ to an amphiploid or produce a $BC_1$ derivative; earlier both these events remained unsuccessful. The constraint could subsequently be overcome with new complex techniques that are currently prevalent (Mujeeb-Kazi 2004) and used earlier with success (Ter-Kuile et al. 1987; Rosas et al. 1988). Forster and Miller (1985) obtained normal (2n = 4x = 28 ABDJ) and 5B-deficient (2n = 4x = 27, ABDJ — chromosome 5B) hybrids from wheat–*Th. bessarabicum* crosses. One 5B-deficient hybrid exhibited significantly high pairing (8.46I + 1.71 II rings + 2.39 II rods + 2.54 III + 0.58 IV + 0.07 V).

2. The second hybrid had a similar enhancement of pairing extended to hexa- and heptavalent associations. These $F_1$ hybrids could not be advanced to realize the full advantage of the demonstrated $F_1$ recombinations. Consequently, the potential of how good *Th. bessarabicum* was for addressing wheat production in saline soils (a topmost global priority) could not be fully exploited.

3. Often genomes of the alien species may not be homoeologous (genetically related) to the wheat genomes; thus, manipulations cannot be made via the standard chromosome pairing regulator systems. Under such remote relationship scenarios it is essential to resort to methods causing the breakage and reunion of chromosomes, as happens in irradiation and tissue culture (Sears 1956; Knott 1968; Friebe et al. 1991, 1992, 1993; Mukai et al. 1993; Lapitan et al. 1984; Larkin et al. 1990). The outcome events from both these strategies would yield random recombinants for which a selective sieve would be required in order to capture those that are compensating and have greater agronomic advantage. A new technique that can also cause random chromosome breaks leading to aberrant compositions utilizes gametocidal genes from *Aegilops* species (Endo 1988, 1990 Tsujimoto and Noda 1988) and also has an application in wide crosses. It is well recognized that the introduction of genetic material from alien species with close evolutionary ties to wheat would be expected to have the most potential. Further, processes that can induce recombination between homoeologous chromosomes would provide the means to place introduced segments in the best

location in recipient chromosomes. Consequently, greater exploitation of the *phph* genetic stock (Sears 1977, 1984) in wide crosses has been advocated (Darvey 1984), attempted (Sharma and Gill 1986), and innovatively modified for enhanced efficiency (Mujeeb-Kazi 2000, 2001b, 2003b). The level of homoeologous pairing in this *ph* genetic stock is the highest captured via the various manipulative options available. Sharma and Gill (1986) put the *ph* stock to use for several *T. aestivum–Aegilops* species hybrid combinations and did indeed observe high intergenomic pairing. All $F_1$ hybrids were self-sterile. Those with normal *Ph* produced $BC_1$ seed but no seed set on any of the *ph*-based $F_1$ combinations. Possibly the high homoeologous pairing in such $F_1$s interfered with the production of the rare restitution events, also augmented by sterility/poor seed set that may be a cause of translocation events present during the maintenance of the *ph* Chinese spring (CS) genetic stock. Though the advance of *ph*-based $F_1$s to BCI was subsequently overcome (Ter-Kuile et al. 1987; Rosas et al. 1988), the protocol for achieving this success was very taxing, and any alternative route to utilize the *ph* locus would be a better option.

In conventional intergeneric programs the standard protocols (classical) undergo progressive steps of $F_1$ hybrid production, its advance to $BC_1$ via restitution nucleus ($F_1 \times$ wheat) formation or via amphiploid × wheat, subsequent backcrossing plus cytology to advance $BC_1$ progeny to $BC_n$, and ending with alien monosomic to alien double/triple monosomic chromosome addition ($2n = 6x = 42 + 1$ or $2 + 3$), which upon selfing will yield the alien disomic addition progeny, hopefully representing the complete alien species. If the alien species is a diploid ($2n = 2x = 14$, $E^bE^b$), as is *Th. bessarabicum*, then seven addition lines are possible (William and Mujeeb-Kazi 1995; Zhang et al. 2002; Cortes et al. 2002). The next step is to establish homoeology of the addition lines with groups 1 to 7 of bread wheat so that alien $1E^b$ is identified as homoeologous to 1A, 1B, or 1D of wheat. The homoeology in turn leads to substitution of $1E^b$ for 1A, 1B, and 1D, yield testing and identifying the most suitable substitution that does not have any yield penalty effects. Based upon stress screening, it is possible to identify the alien added chromosome that is preferred for a research objective, e.g., $5E^b$ for salinity (Forster et al. 1988). Thus, the strategy would be to substitute $5E^b$ for 5A, 5B, and 5D, detect the ideal substitution, and then engineer a recombination event between $5E^b$ and the preferred corresponding wheat chromosome pairs (5A or 5B or 5D).

For greater depth on the addition line or substitution line and eventual manipulation, details can be harnessed through current texts (e.g., Singh 2003) and also several papers emanating from several research groups who have presented at or reported in the *International Wheat Genetic Symposia* (Volumes 1 to 10). It will suffice to mention here an early paper by O'Mara (1940) where rye addition line production was described. The developments of state of the art giving similar and more stringent protocol involvements that now prevail are apparent in some recent papers — a few being on addition line productions (Islam et al. 1981; Friebe and Larter 1988; William and Mujeeb-Kazi 1995; Friebe et al. 1996b, 2000; Zhang et al. 2002; Cortes et al. 2002).

The research effort of producing a complete alien addition line set is phenomenal, and achieving that ultimate beneficial alien transfer for enhancing wheat production is even more demanding. Hence, in order to hasten the process, alternate options have been explored that could (1) utilize existing *phph*-based genetic stocks found to be positive for stress resistances/tolerances, (2) utilize addition line stocks available that have resistances/tolerances identified, or (3) use one's own program germplasm differently from the classical route.

Described is the third option, and this can readily address the other two categories. The case study is with a *T. aestivum* (CS) *phph* × *Th. bessarabicum* perennial intergeneric combination: $2n = 4x = 28$, $ABDE^b$ (*ph*). The alien species is an excellent source of salt tolerance and head scab resistance. Being a diploid, it becomes the top choice for wide crossing if one is looking for rapid product outputs since complexities would be minimal for a diploid compared to using higher ploidy alien sources with similar attributes; thus, choice of the alien parent is paramount for a successful practical output.

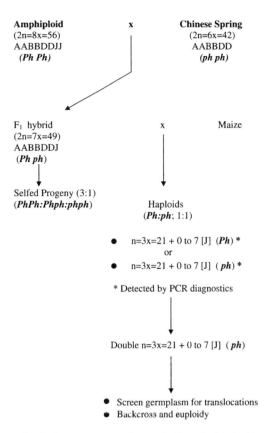

**Figure 3.7** Intergeneric genetic stocks as a source for production of wheat–alien chromosome translocations: the protocol and stages of obtaining stable euploid-resistant products.

An integrated cytogenetic manipulation scheme was theoretically put together by Mujeeb-Kazi (2000) that was then tested (Mujeeb-Kazi 2001b) and found to be useful (Mujeeb-Kazi 2003a,b, 2004). Details are given in Figure 3.7. The strategy is aimed at providing maximum recombination between wheat and alien species chromosomes in the early hybrid generation stages or at the amphiploid level. These initial germplasms have the *Ph* dominant gene, and crossing either with CS *phph* will give $BC_1$ progeny that is *Phph*. This would be selfed to yield 1 *PhPh*:2 *Phph*:1 *phph* derivatives. If crossed with maize, it would generate haploids that have *Ph* or *ph* (1:1). The materials have all the wheat chromosomes (42 or 21) plus several of *Th. bessarabicum* because in this alien source the univalent transmission frequency is very high. The *phph* or *ph* plants are readily identified by polymerase chain reaction (PCR) technology (Qu et al. 1998) and advanced to maximize recombination events. For the *phph* derivatives a few selfings facilitate in enhancement of recombination. For the *ph* haploid plants, doubled haploids are required (*phph*), which then could be selfed to increase wheat–alien chromosomal recombination frequency. Meiotic analyses (conventional and fluorescent *in situ* hybridization (FISH)) demonstrate the influence of the *ph* presence in excessive multivalent formation (Figure 3.8b). FISH analysis of such high-pairing derivatives enables identification of wheat–alien chromosome translocations (Figure 3.9). A second root-tip sampling of the growing plants targeted for C-banding then characterizes the translocated chromosomes. The chromosome number is usually more than 42 in a *phph* structure. At this stage, after translocations are detected, backcrossing to an elite bread wheat cultivar is done to eventually restore the *PhPh* system and drop to a euploid state around the presence of a disomic translocated chromosome (Figure 3.10). Translocations occur within homoeologous and nonhomoeologous

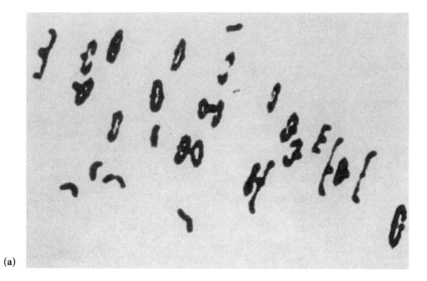

(a)

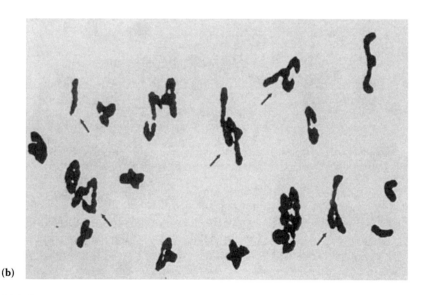

(b)

**Figure 3.8**   Meiotic analyses of backcross derivatives with *Ph* or *ph* loci showing (a) normal bivalent pairing with alien chromosomes either paired or as univalents and (b) multivalent associations due to the *ph* locus between wheat–alien chromosomes. The cross combination is bread wheat–*Th. bessarabicum*.

groups; the former are desirable (Mujeeb-Kazi 2003a; Mujeeb-Kazi and Van-Ginkel 2004). The strategy is also easily applicable to a targeted alien disomic addition line where crossing the *Ph*-based addition (42 wheat + 1E$^b$1E$^b$) by CS (*phph*) will give a BC$_1$ product that is *Phph* (42 wheat + 1E$^b$). This, when crossed with maize, will give haploids from which those with 21 wheat + 1E$^b$ (*ph*) can be selected, becoming the potential source for homoeologous translocations between 1E$^b$ and 1A, 1B, or 1D.

With adequate data support obtained from using the above *ph* manipulation protocol, a step was taken to generate various bread wheat *Phph* heterozygote stocks from the wheat intergeneric germplasm available. The heterozygote stocks were derived from the living F$_1$ herbarium of 185 hybrids that are in CIMMYT's wide crosses program (Delgado and Mujeeb-Kazi 2002) and also

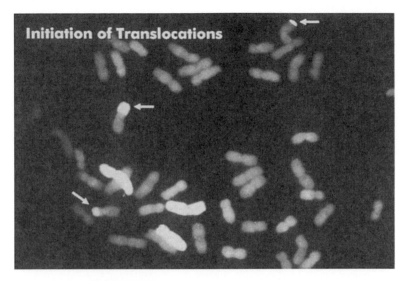

**Figure 3.9** **(See color insert following page 114.)** FISH of a *phph* backcross derivative showing wheat–alien chromosome translocations being initiated (the combination is *T. aestivum–Th. bessarabicum*).

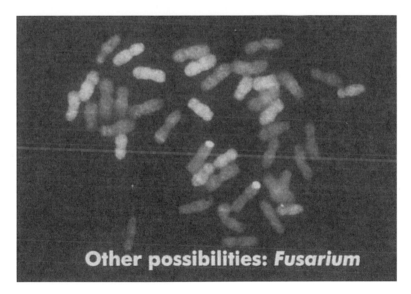

**Figure 3.10** **(See color insert following page 114.)** FISH of a translocation homozygote derived from the *phph* backcross plants with a hypoploid count (Figure 3.9) to yield a 2n = 6x = 42 translocation derivative that is *PhPh*.

from the amphiploids (Mujeeb-Kazi 2003a).This could be extended to include all disomic addition lines of alien sources available in wheat, from which specific preferred ones can be swiftly exploited for achieving particular recombinational targets.

## 3.7 UTILIZATION AND PRACTICALITY OF WIDE CROSSES GERMPLASM

Producing $F_1$ hybrids within the Triticeae species that are genomically distant, or close but require special tools to salvage the event, is a definite art that leads to scientific knowledge of enormous magnitude. The most common inferences in literature that one comes across are the basic/logical interpretations that alien species were screened and found desirable for such traits. The next step commonly reported is that hybrids have been made, cytological validation done, and projections made to improve wheat, thus alleviating the production constraints. Realizing this final output substantiated with concrete measures, however, eludes many wide cross researchers, essentially because a single trait improvement promise rarely gives the balance of productivity that a cultivar requires; there is no magic tons per hectare output since it necessitates multiple inputs. We will address this aspect of utilization and practicality of the wide cross germplasm, going beyond what has been superbly compiled in evidence of alien gene transfers, of which a few (Sharma and Gill 1983a; Wang 1989; Pienaar 1990; Jiang et al. 1994; Friebe et al. 1996a; Jauhar and Chibbar 1999) can be cited.

In general, for a wheat cultivar to be grown for human consumption, it must be high yielding, have good quality, and be resistant/tolerant to the growing environment's biotic and abiotic stresses, implying that a genetic component managing multiple stresses would be essential.

## 3.8 IMPACT FROM DOCUMENTED TRANSFER

The one alien contribution that has had the greatest impact on practical agriculture is the renowned spontaneous Robertsonian translocation product called the 1B1R translocation (T1BL1RS). Over 5 million ha wheats possessing this translocation are grown globally because of their biotic and abiotic resistances, wide adaptability, and yield advantage (Rajaram et al. 1983). Currently, about 58% of CIMMYT germplasm accessions possess the T1BL1RS chromosome. Stringent experimentations have followed over the last two decades to unravel the 1RS contribution (Villareal et al. 1998; Mujeeb-Kazi et al. 2001a,b,c) to such wheats. The weakest aspect that has emerged is the bread-making quality (Carver and Rayburn 1995; McKendry et al. 1996; Graybosch et al. 1993), which subsequently has received favorable research attention (Baenziger et al. 2001). Presumably next in value for wheat is the T1AL1RS translocation (Villareal et al. 1996b), which was the product of an x-ray-mediated event, but may also be the outcome of a centric-break fusion univalency event. These two translocations in wheat improvement have contributed to research interest in using such materials or direct focus on those stocks that possess subtle alien chromatin introgressions where the insert magnitudes could be from intercalary (Friebe et al. 1991; Mukai et al. 1993), as shown for the rye/wheat transfer to various sizes of terminal arm exchanges as present for bread wheats that have alien genes for $Lr19$, $Lr25$, $Bdv_2$, and various others (Friebe et al. 1996a). In essence, any introgression that is homoeologous (compensatory) and minute would be ideal for practical usage. The whole-arm contributions of T1BL1RS and T1AL1RS are rare events of practicality that have been very impacting — a fortuitous event.

Based upon some studies on T1BL1RS wheats, it was reported that the rye segment (1RS) may enhance grain yield (Villareal et al. 1995a; Schlegel and Meinel 1994; Carver and Rayburn 1994; Moreno-Sevilla et al. 1995a). However, no such positive effects were seen when studied by Villareal et al. (1991, 1994), Moreno-Sevilla et al. (1995b), and McKendry et al. (1996). This suggested that the importance of the genetic background may be a significant factor and needs due attention when greater emphasis is given to the role of translocations in wheat improvement.

Projecting ahead is the potential impacting role of the $Lr19$-related segment not linked with yellow pigment. Its contribution to wheat production through significantly high-yield outputs on limited trials has been identified (Singh et al. 2001). Also in contention is the TC14 Australian germplasm conditioning barley yellow dwarf virus resistance ($Bdv_2$) marked by gwm37 (Ayala et al. 2001). More recently the contribution from $Th.\ junceiforme$ for salt tolerance to bread wheat via $Ph^l$ manipulation has emerged (Wang et al. 2003). Another possibility for scab resistance comes from the translocation stocks that originate from $Th.\ bessarabicum$ (Mujeeb-Kazi 2003a; Mujeeb-Kazi and Van-Ginkel 2004). None of these translocations cover more than 25% of a wheat chromosome arm and are of top choice in wheat breeding. Several other examples can be tabulated (McIntosh et al. 1995; Sharma 1995; Jiang et al. 1994; Friebe et al. 1996a, 2001), from which germplasm candidacy for wheat improvement is justified. In order to accomplish the positive goals of such introgressions, the measure would be a varietal release that must be high yielding (tons per hectare), have multiple/durable stress resistance, have quality, and cover substantial hectarage. Researchers are cognizant of the fact that use of cultivars with single-gene resistances permits the selection of mutations at a single pathogen locus to render resistances ineffective in a relatively short time. The breakdown of the stripe rust gene from *Ae. comosa* is just one example to elucidate this. Hence, use of gene combinations has been recognized as the best strategy for biotic stress control (Singh et al. 1998). The situation is different for abiotic stresses, e.g., drought, salinity, etc., since in the absence of a pathogen exists a static system and single-gene transfers could be long lasting, as there is no pathogen present that is prone to any mutational change.

The focus so far has been on tertiary gene pool species yielding documented transfers. Even though these are complex to harness, this author believes that it is a genetic resource that must be tapped and exploited via gene pyramiding in order to achieve a sustainable output productivity globally.

## 3.9 IMPACT THROUGH UNDOCUMENTED TRANSFER

To strengthen this aspect, the production limit caused by spot blotch (*Cochliobolus sativus* syn. *H. sativum*) is elucidated; spot blotch is the leading production constraint in Bangladesh and also a limiting biotic stress in another 14 countries spread between the west (Argentina on the southern tip of Latin America) and the east (South Vietnam on the eastern tip of Southeast Asia). The tertiary gene pool resource *Th. curvifolium* (2n = 4x = 28, $E_1E_2$) combined with bread wheat, prebred and advanced using a novel breeding methodology, led to derivatives (CIGM295), with its superior resistance transferred to wheat (Mujeeb-Kazi et al. 1996b). This germplasm led to lines Chirya and Mayoor, which have expressed multilocational resistance for a decade and a half, culminating in a varietal release in Bolivia called Azubiciat. The alien transfer has yet to be identified and has eluded conventional cytology diagnostics plus initial molecular assessments. The level of resistance in this germplasm surpasses all that have been seen in conventional wheat cultivars (Figure 3.11) and exemplifies why complex species should receive due attention.

Other releases from undocumented alien transfers were the cultivars Pasban 90 and Rohtas 90 in Pakistan derived from a *Th. distichum* hybrid combination, which have now succumbed to leaf rust. A related sister line combination also gave cultivar Luan, a Mexican release for karnal bunt resistance.

Shifting impact emphasis to interspecific germplasm, it is well recognized that primary gene pool species, e.g., *Ae. tauschii*, are indeed easy to utilize and have a high practicality index across numerous stress constraints, essentially because of trait diversity and possibility of successful gene introgressions by homologous exchanges. From the synthetic hexaploids derived by crossing durums/*Ae. tauschii* accessions, entries were identified with stress resistance/tolerance that completed the basic phase of the wide cross strategic activity around which anticipations were made for improved wheat productivity. This utilization phase involved the crossing of useful synthetic hexaploids (SHs) with target quality bread wheats missing the trait needed for ideal productivity to get $F_1$ combinations. A subsequent $F_1$ advance was based upon appropriate conventional breeding strategies (limited backcrossing, top-crossing, etc.), selections, and stability via doubled haploidy (wheat × maize; Mujeeb-Kazi 2000), generating products of practical value. Such outputs have been realized for biotic stresses and abiotic tolerances. Notable are resistances/tolerances to head scab (Mujeeb-Kazi 2003a), *Septoria* leaf blotch (Mujeeb-Kazi et al. 2000a), spot blotch (Mujeeb-Kazi et al. 1996b), Karnal bunt (Mujeeb Kazi et al. 2001d; Villareal et al. 1996a), salinity tolerance

**Figure 3.11** **(See color insert following page 114.)** A spot blotch-resistant advanced derivative from an intergeneric wheat–*Th. curvifolium* combination (right) compared to a bread wheat-susceptible cultivar (left) based upon field screening in a disease hot site in Mexico. The line Chirya was released as the variety Azubiciat in Bolivia.

(Pritchard et al. 2002), drought tolerance (Trethowan et al. 2003), waterlogging tolerance (Villareal et al. 2001), leaf/stripe rust (Mujeeb-Kazi 2003a), and yield (Mujeeb-Kazi 2003a).

The use of SH germplasms has been emerging in numerous global breeding programs. Some essential considerations to be wary of are:

1. $F_1$ hybrid necrosis if *ne1ne2* are in a recessive combination
2. The dominance of the tenacious glumes necessitating large segregating populations in order to select free threshing individuals
3. Genetic control of stress traits due to major genes, hence making it imperative to use gene pyramiding across different accessions or bread wheats in the crossing that themselves possess major plus minor genes

In order to facilitate wide use of SH germplasm, the wide cross program in CIMMYT has assembled an elite I set of 95 SH entries that have been made available globally in the last 7 years. Initial impacts of this novel germplasm are now surfacing through varietal releases that recently occurred in China and Spain. These cultivars are Chaun 38 and Chaun 42 in China and Carmora THA453 in Spain.

## 3.10 FUTURISTIC ANTICIPATION

More promise exists in the near future from wide cross outputs around objectives of biotic stresses, high yield (irrigated), and drought and salinity tolerance. In CIMMYT's base program in Mexico, taking scab and spot blotch as two stresses, one can see the potency of SH compositions. For advanced lines like Mayoor// TKSN1081/*Ae. tauschii*(222) that possess all four categories of scab resistance with additional stress resistances, the wide usage of this line as a donor source is phenomenal. This line's performance for spot blotch resistance is also superior (Figure 3.12), and when 15 countries are beneficiaries, outputs are bound to occur swiftly, thus taking a promising genetic stock to wide-scale application. For spot blotch where *Th. curvifolium*-derived Mayoor and Chirya have dominated the scene, additional SH inputs add more leverage for even greater resistance

**Figure 3.12 (See color insert following page 114.)** Spot blotch resistance of a wheat/synthetic hexaploid advanced derivative compared to a susceptible entry.

performance. Disease scores around 9-5 for CIGM295 give a low coefficient of infection in Mexican hot disease sites (Mujeeb-Kazi et al. 1996b, 1998). SH wheats have disease scores ranging from 2-2 to 3-3 (Mujeeb-Kazi et al. 2001e), and their prebred free-threshing lines average 2-2 to 3-2, distinctly superior to CIGM295's disease score of 9-5. Hence, one can envision that when breeders combine such improved germplasm, not only will the products have low infection levels, but also the pyramided diverse genes will offer great security to farmers who grow wheat in such stressed conditions.

Going back in time to 1980, one can set some facts in place to show that genetic diversity in conventional bread wheats was not available or very limited for Karnal bunt, *H. sativum*, *S. tritici*, *Fusarium*, barley yellow dwarf, drought, and salinity. The use of alien sources has alleviated all of these limitations, in some cases delivering the finished end cultivar also, or has provided tremendous impetus to reach that goal in the very near future. It is not uncommon to find breeding programs that possess from 12% to around 50% of their materials based upon alien germplasm addressing various growing environments. In addition, when a gigantic breeding program, such as CIMMYT, can create an international bread wheat screening nursery (IBWSN) with about 28% lines having SH influence, the global impact appears optimistic and imminent. Locally (in Mexico), in the irrigated environment yield trials, SH- and (double haploid) DH-based lines were three of the top five entries in very recent studies (M. Van-Ginkel, personal communication). SH influence in drought breeding has increased from 8 to 40% over the past 6 years in CIMMYT (Trethowan et al. 2003).

Over the last decade and a half SH wheats have shown promise in exhibiting genetic diversity for numerous production constraints. It is therefore not surprising to envision SH usage for supporting traits like resistance to *S. nodorum*, powdery mildew, tan spot, cold tolerance, sprouting tolerance, bread-making quality, and, more recently, micronutrients like high iron and zinc. If the potential of tertiary gene pool species is also accessed, then the coverage is broadened and additional traits like beta-carotene and copper tolerance can also be addressed. Thus, applications of wide cross germplasm in the future have the capacity to produce almost at will wheat productivity impacts around the diversity that these Triticeae species possess. Futuristic trends will lead researchers to several novel activities around the basic structure of wide crosses that is presented here. Some of these are briefly alluded to and encompass:

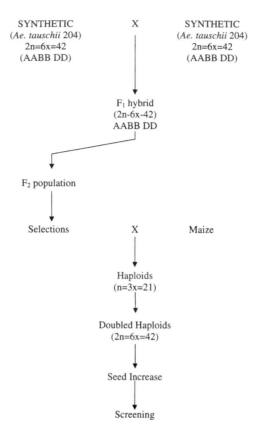

**Figure 3.13**  Schematic showing gene pyramiding across two *Ae. tauschii*-based synthetic wheats resistant to *Helminthosporium sativum*. Both *Ae. tauschii* accessions were of diverse origin and allelic structure.

1. New tetraploids. Direct crosses that involve targeted *Ae. tauschii* accessions for bread wheat improvement are an efficient way to achieve rapid practical goals. Bridge crosses where SHs are used in the improvement protocol offer the advantage of exploiting the *Ae. tauschii* diversity as well as the genetic potential of the durum genomes (AB) present in the SH wheats. Thus, gene pyramiding occurs across all three (ABD) genomes. To exemplify the benefits of this approach, karnal bunt resistant wheats provide necessary evidence where *Ae. tauschii* contributes resistance and a 1000-kernel weight between 55 to 58 g is a durum cultivar output (Mujeeb-Kazi et al. 1998): based upon the observations that relate with gene pyramiding, production of new tetraploids was envisioned and taken up via the involvement of A plus D genome diploids where accessions of both grasses showed good resistance to a practical trait, e.g., *S. tritici* (Mujeeb-Kazi 2004). Involved resistant accessions of *Triticum urartu* ssp. *boeoticum* or *monococcum* (2n = 2x = 14, AA) combined with those of *Ae. tauschii* (2n = 2x = 14, DD) yielded $F_1$ hybrids (2n = 2x = 14, AD) and led to doubled tetraploid (2n = 4x = 28, AADD) products with superior *S. tritici* resistance — a co-dominant expressivity of both genomes. Using these stocks in breeding wheats is a preferred means to harness many genes with expediency.

2. Gene pyramiding within the same alien group is another strategy where accessional diversity can be exploited. *Ae. tauschii* is a superior source for *H. sativum* resistance, a biotic stress spread across the globe and country boundaries. One approach has addressed pyramiding-resistant genes from various SHs using the scheme illustrated in Figure 3.13. Based upon data that validated *H. sativum* resistance of SH wheats, those with *Ae. tauschii* accessions belonging to diverse regions and with DNA polymorphic profiles became candidates for gene pyramiding. Superior resistant $F_2$ plants could be selected (parental comparison) and stabilized via the pedigree breeding protocol. However, use of doubled haploidy has greater advantage, particularly when bread wheat is the crop, since there is no genotypic specificity and outputs emerge with almost perfect (100%) success

(Mujeeb-Kazi 2003b). By selecting highly resistant $F_2$ individual plants from diverse SH/SH populations, using the detached tiller protocol (Mujeeb-Kazi et al. 2003b), genetically stable $F_2$ SH/SH DH derivatives resulted, exhibiting disease scores of 1-1 or 2-1 vs. 9-9 (susceptible), according to the double-digit scoring scale (Saari and Prescott 1975). This disease score was an indicator of genetic co-dominance since the disease symptomology was better (i.e., less infection) than either parent used in the crossing. Hence, additive gene action seemed to be in operation, an aspect that deserves greater assessment in the future. Such genetic stocks provide a better resistance diversity source to breeding programs. This is apparent in our program, where advanced segregates reach maturity with a 2-1 or 2-2 score — a tremendous source for resistance security even around major genes, which have become more potent via pyramiding.

3. Doubled haploidy based on the classical findings of Laurie and Bennett (1986, 1988), Suenaga and Nakajima (1989), and Riera-Lizarazu and Mujeeb-Kazi (1990) apparently will become an important tool in bread wheat research. The current status indicates frequencies of embryo excision at 20 to 25%, regeneration at 80 to 90%, and doubling at 80 to 95% (Mujeeb-Kazi and Van-Ginkel 2004). The protocol is 100% effective across all wheat genotypes with spring, winter, or facultative habit, plus economically efficient with good integration in research functions. A few DH research involvements are in cytogenetics, enhancing breeding efficiency, producing molecular mapping populations, stability of transgenes, genetic studies, and complete/partial monosomic analyses (Mujeeb-Kazi 2000). The author feels that a significant future trend allowing wheat breeders to cope with the food productivity challenge of the next two decades (Braun et al. 1998) will be integrating DH activity with wheat breeding, where selections in segregating generations can be readily stabilized. This sexual DH application strategy will apparently only be offset if the attractive microspore culture methodology (Liu et al. 2002) gets to a level where it can deliver similar outputs expediently.

4. Wide cross-based germplasms are a paradise for molecular diagnostics and applications essentially because these distant species of the various gene pools have high degrees of DNA polymorphisms that facilitate development of maps and markers. The interspecific germplasms built around the A, B, and D genome accessions are rich sources that permit microsatellite usage and offer a high level of discrimination at the genomic and chromosomal levels. For bread wheat, since *Ae. tauschii* has demonstrated tremendous practical utility focusing alone on the D genome, microsatellites will generate extensive molecular insights. Where chromosomal saturation is less than optimum, strategies like comparative genome mapping can be incorporated to strengthen diagnostics around chromosome 4D for salinity tolerance. The well-recognized example of DNA diversity tied with molecular aspects was the use of a synthetic hexaploid (Altar84–*Ae. tauschii*) crossed onto a bread wheat cultivar (Opata) to yield the mapping population. It gave a very dense map, and the genotyping and phenotyping of the population's entries led to the generation of crucially essential molecular data. It also became the basis for Roder et al. (1998a,b) to place several hundred loci on the map. Since the integration of molecular output data (current plus futuristic) is associated with enhancing breeding efficiency, development of molecular mapping populations via DH or SSD routes will continue to receive due attention, adding to the strength of marker-assisted selection via user-friendly techniques.

5. Cytogenetic manipulations in intergenerics with outputs involving intercalary alien introgressions and translocations (terminal or Robertsonian) will form an investigation area around wheat breeding strategies based upon the huge impact from limited genetic materials like T1BL.1RS (Mujeeb-Kazi et al. 2000b) or T1AL.1RS. Generating new translocations (targeted or random) would focus more on the *ph* locus, since this strategy supports homoeologous and compensatory outputs. Germplasms with translocations are also ideal candidates for being associated with specific molecular markers, of which gwm37 for *Bvd2* in TC14 germplasm is one example (Ayala et al. 2001).

6. Briefly it must be mentioned that though the main focus of this article is on bread wheat, wide crossing for durum wheat improvement follows a similar interspecific and intergeneric route in order to address biotic and abiotic stress production constraints. Envisioned by this author for the future in durum wide crosses is greater emphasis on realizing intergenomic D-to-A transfers via cytogenetic manipulation based upon the usable diversity identified in *Ae. tauschii*-based SH wheats. Homoelogous A and D genome chromosome pairing has been reported earlier (Okamoto and Sears 1962; Alonso and Kimber 1983). Jauhar et al. (1991) during an analysis of chromosome

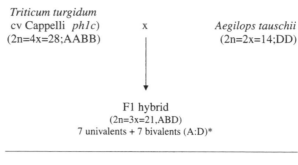

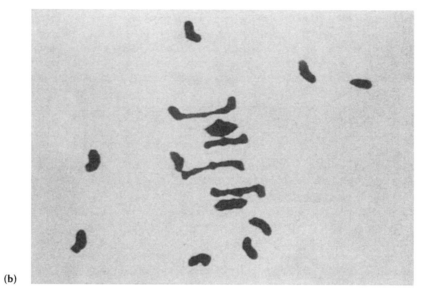

**Figure 3.14** (a) Schematic showing the protocol of a homoeologous A and D genome relationship for affecting transfers into durums from *Ae. tauschii* (2n = 2x = 14, DD). (b) An F₁ meiocyte from a 2n = 3x = 21 ABD plant involving a durum (*ph1c*) × *Ae. tauschii* cross. The bivalents are the preferential pairs between several A and D genome chromosomes (FISH not shown).

pairing in *ph1b* haploids (2n = 3x = 21, ABD) demonstrated that the A and D genome chromosomes paired preferentially, suggesting important breeding implications. This author has addressed durum improvement for scab resistance using the Figure 3.14 protocol, a study that is in progress. Using chromosome engineering strategies, Ceoloni et al. (2003) successfully transferred gluten protein genes into durum wheat, providing optimism that more traits can be transferred similarly into durums from the promising D genome accessions of *Ae. tauschii*.

## 3.11 CONCLUSIONS

Experts predict that today's worldwide population of just over 6.0 billion people will grow to 8.2 billion over the next two decades. Then by 2050, 12 billion people will crowd the planet, with more than 90% of the growth occurring in developing nations. Estimates are that the world would require 1 billion metric tons of wheat over the next two decades, as compared to the current production of slightly over 600 million metric tons. Hence, one can extrapolate and conclude that the mean global yield of wheat would have to shift from 2.5 t/ha to over 4 t/ha. These ominous circumstances are placing a formidable task before agricultural scientists and the food management sector. This productivity can only happen around an invigorated research program with superintegration of multidisciplinary activities in which germplasm and genetic resources would remain paramount. Plant breeders involved in crop improvement efforts in order to meet the ever-increasing demand for food are finding less and less appropriate germplasm with desired traits among cultivated crops themselves with which to make the needed improvements. Fortunately, new and useful genetic resources are being found in wild, uncultivated plants. The challenge is to incorporate this germplasm into existing food crops, which forms the crux of this chapter.

The main focus addressed here has been on utilizing novel genetic diversity that resides in the wheat progenitor species, and access to such materials; their exploitation has spanned over several decades (*Triticale*, barley/wheat), and tremendous progress has been made to allow us to answer several basic questions — answers that are now leading investigators to generate finished cultivar product outputs. Such germplasms have paved the way for sophisticated technologies to promise a scientific revolution as a consequence of the molecular advancements, which are ongoing and continuously being refined as we advance in time. This chapter has not explored the intricacies of the molecular potentials that surround wheat improvement, but one must remain cognizant of the strength that is available for addressing crop production in an efficient manner. The focus of this compilation stays with plant-level manipulation, covering the whole spectrum of activities from the crucial hybrid production to its successful culmination via varietal outputs. It is well recognized that many factors impinge upon generating finished cultivars, and presumably this is an assemblage of multiple stress attributes that are to be interknit in such a manner that durable resistance is ensured. Such limits are often overlooked by wide cross researchers, for whom making a hybrid combination with the potential of countering a biotic or abiotic stress is a major accomplishment. The reality, however, is entirely different and goes beyond the stage when the cultivar is developed, since its production on the farm is beset with numerous policy input constraints that scientists often cannot comprehend. The diversity of alien species is a boon to plant improvement, and this has been addressed in the chapter via several strategies that basically address swift practical outputs for the benefit of humankind.

Because cereal crops provide the structural base for world food production, it is fortuitous that most alien genetic transfers, to date, have involved the *Triticum* species within the tribe Triticeae — where the greatest emphasis has been placed on using these introgressions to improve bread wheat and to some extent durum wheat. Wheat has received the most attention because of its global importance and because genetic manipulation techniques have become well established for wheat and its relatives. Using wide crosses to improve bread wheat is an area that CIMMYT has pursued vigorously over the past three decades. Then research efforts and findings of other researchers are detailed in this chapter, encompassing a wide area of basic to applied aspects. There are different methodologies for transferring desired resistances or tolerances from the alien species to wheat. Irrespective of the procedure adopted, production of hybrids (intergeneric or interspecific) is the key to accomplishing useful genetic transfers. The art and technology of doing this have been presented. Alien transfers can diversify variability for both dynamic biotic situations and static abiotic circumstances. Alien sources of variability are normally inaccessible to breeders who work in conventional crop improvement programs, and hence wide hybridization is considered additive to traditional plant breeding efforts.

Actual successes of incorporating usable alien genetic variation have been relatively few and, for the most part, have involved simply inherited genetic traits. So, in order to ensure faster practical returns to agriculture, it is felt that simply inherited traits should be the major emphasis when making intergeneric crosses, while traits with complex heritability (involving the introgression of several genes simultaneously) and ill-defined genetic information should be limited to the less complex interspecific crosses. Some of the complex traits receiving attention are associated with resistances or tolerances to *H. sativum*, *F. graminearum*, *N. indica* (syn. *Tilletia indica*), drought, and salinity.

Major problems that earlier limited the use of wide hybrids, i.e., crossability and embryo development, have been largely resolved, thus encouraging researchers to examine the possibilities of yet even wider hybrids for complementing wheat improvement strategies.

Some major prerequisites for success in wide cross research are a long-term commitment, collaboration among specialists and institutions, and free sharing of germplasm and technologies. The long-term commitment can be appreciated when some output facts are addressed, like:

1. Nearly a century of accomplishments with wheat × rye (Triticale) and wheat × barley crosses
2. Nearly 50 years of astounding successes with wheat cytogenetic stocks (late Dr. E.R. Sears)
3. Some 20 to 25 years of work involving alien genetic stock developments involving *Dasypyrum villosum, Th. elongatum, Th. bessarabicum, H. vulgare*, and revitalized exploitation of *Ae. tauschii* over the last two decades — in each case pursued by independent researchers handling major wheat improvement and cytogenetic programs

A lesson from the few examples mentioned above endorses the opinion that to counter the known complexities of wide crosses explicitly identified in the scientific literature, researchers and administrators must recognize that an inescapable long-term commitment should prevail in order to achieve practical success. In essence, a successful wide cross program requires, apart from the commitment, integration and cooperation of literally hundreds of specialists located across country boundaries in elite research centers across diversified disciplines. The spread may be across areas like genetics, cytogenetics, pathology, entomology, physiology, prebreeding/breeding, nutrition, molecular biology/genetics, genomics, genetic engineering, and agronomy. For a who's who listing of selected specialists involved in such a program targeted at harnessing novel genetic diversity through nature's rich resources, one can scan the references cited in this chapter and cross-reference for added depth.

## REFERENCES

Aghaee-Sarbarzeh, M., Singh, H., and Dhaliwal, H.S. 2000. *Ph^l* gene derived from *Aegilops speltoides* induces homoeologous chromosome pairing in wide crosses of *Triticum aestivum. J. Hered.* 91: 417–421.

Ahmad, F. and Comeau, A. 1991. Production, morphology, and cytogenetics of *Triticum aestivum* (L.) Thell. × *Elymus scabrus* (R. Br.) Löve intergeneric hybrids obtained by *in ovulo* embryo culture. *Theor. Appl. Genet.* 81: 833–839.

Almouslem, A.B. and Jauhar, P.P. 1998. Chromosome pairing in synthetic hybrids between bread wheat and *Thinopyrum* species: breeding implications. In *Proceedings of the Third International Triticeae Symposium*, A.A. Jaradat, Ed. Scientific Publishers, Enfield, NH, pp. 127–133.

Alonso, L.C. and G. Kimber, 1981. The analysis of meiosis in hybrids. II. Triploid hybrids. *Can. J. Genet. Cytol.* 23: 221–234.

Alonso, L.C. and Kimber, G. 1983. A study of genome relationships in wheat based on telocentric chromosome pairing II. *Z. Planzenzucht.* 90: 273–284.

Alonso, L.C. and Kimber, G. 1984. Use of restitution nuclei to introduce alien genetic variation into hexaploid wheat. *Z. Pflanzenzucht.* 92: 185–189.

Armstrong, J.M. 1936. Hybridization of *Triticum* and *Agropyron*. I. Crossing results and description of the first generation hybrids. *Can. J. Res.* C14: 190–202.

Ayala, L., Khairallah, M., González de León, D., Van Ginkel, M., Mujeeb-Kazi, A., Keller, B., and Henry, M. 2001. Identification and use of molecular markers to detect barley yellow dwarf virus resistance derived from *Th. intermedium* in bread wheat. *Theor. Appl. Genet.* 102: 942–949.

Baenziger, P.S., Shelton, D.R., Shipman, M.J., and Graybosch, R.A. 2001. Breeding for end-use quality: reflections on the Nebraska experience. In *Wheat in a Global Environment*, Z. Bedo and L. Lang, Eds. Kluwer Academic Publishers, The Netherlands, pp. 255–262.

Bates, L.S., Mujeeb-Kazi., A., and Waters, R.F. 1976. Wheat × barley hybrids: problems and potentials. *Cereal Res. Commun.* 4: 377–386.

Braun, H.J., Payne, T.S., Morgounov, A.I., van Ginkel, M., and Rajaram, S. 1998. The challenge: one billion tons by year 2020. In *9th International Wheat Genetics Symposium*, Saskatoon, Canada, August 2–7, pp. 33–40.

Carver, B.F. and Rayburn, A.L. 1994. Comparison of related wheat stocks possessing 1B or 1RS.1BL chromosomes: agronomic performance. *Crop Sci.* 35: 1505–1510.

Carver, B.F. and Rayburn, A.L. 1995. Comparison of related wheat stocks possessing 1B or T1BL.1RS chromosomes: grain and flour quality. *Crop Sci.* 35: 1316–1321.

Cauderon, Y. 1977. Allopolyploidy. In *Interspecific Hybridization in Plant Breeding, Proceedings of the 8th Eucarpia Congress*, Madrid, Spain, pp. 131–143.

Ceoloni, C., Margiotta, B., Colaprico, G., D'Egidio, M.G., Carozza, R., and Lafiandra, D. 2003. Introgression of D-genome associated gluten protein genes into durum wheat. In *Tenth International Wheat Genetics Symposium*, Italy, pp. 1320–1322.

Chen, P.D., Tsujimoto, H., and Gill, B.S. 1994. Transfer of *Ph1* genes promoting homoeologous pairing from *Triticum speltoides* to common wheat. *Theor. Appl. Genet.* 88: 97–101.

Ciferri, R. 1955. The first interspecific wheat hybrids. *J. Hered.* 46: 81–83.

Cortes, A., Delgado, R., Vahidy, A.A., Razzaki, T., Diaz-de-Leon, J.L., and Mujeeb-Kazi, A. 2002. Disomic chromosome addition lines of *Thinopyrum bessarabicum* in bread wheat. In *Agronomy Abstracts*. American Society of Agronomy, Madison, WI (CD-ROM).

Cox, T.S. and Hatchett, J.H. 1994. Hessian fly-resistance gene *H26* transferred from *Triticum tauschii* to common wheat. *Crop Sci.* 34: 958–960.

Cox, T.S., Raupp, W.J., and Gill, B.S. 1994. Leaf rust-resistance genes *Lr41, Lr42*, and *Lr43* transferred from *Triticum tauschii* to common wheat. *Crop Sci.* 34: 339–349.

Cox, T.S., Sears, R.G., and Bequette, R.K. 1995a. Use of winter wheat × *Triticum tauschii* backcross populations for germplasm evaluation. *Theor. Appl Genet* 90: 571–577.

Cox, T.S., Sears, R.G., Bequette, R.K., and Martin, T.J. 1995b. Germplasm enhancement in winter wheat × *Triticum tauschii* backcross populations. *Crop Sci.* 35. 913–919.

Darvey, N.L. 1984. Alien wheat bank. *Genetics* 107: 24.

Delgado, R. and Mujeeb-Kazi, A. 2002. A living herbarium of intergeneric hybrids in CIMMYT: characterization, cytogenetics and maintenance. In *Agronomy Abstracts*. American Society of Agronomy, Madison, WI (CD-ROM).

Dewey, D.R. 1980. Some applications and misapplications of induced polyploidy to plant breeding. In *Polyploidy: Biological Relevance*, W.H. Lewis, Ed. Plenum Publishing Corp., New York, pp. 445–470.

Dewey, D.R. 1984. The genomic system of classification as a guide to intergeneric hybridisation with the perennial Triticeae. In *Gene Manipulation in Plant Improvement*, J.P. Gustafson, Ed. Plenum Press, New York, pp. 209–279.

Dvořák, J., Edge, M., and Ross, K. 1988. On the evolution of the adaptation of *Lophopyrum elongatum* to growth in saline environments. *Proc. Natl. Acad. Sci. U.S.A.* 85: 3805–3809.

Eigsti, O.J. and Dustin, A.P. 1955. *Colchicine in Agriculture, Medicine, Biology and Chemistry*. Iowa State College Press, Ames.

Endo, T.R. 1988. Chromosome mutation induced by gametocidal chromosomes in common wheat. In *Proceedings of the 7th International Wheat Genetics Symposium*, T.E. Miller and R.M.D. Koebner, Eds. Cambridge, England, pp. 259–265.

Endo, T.R. 1990. Gametocidal chromosomes and their induction of chromosome mutation in wheat. *Jpn. J. Genet.* 65: 135–152.

Falk, D.E. and Kasha, K.J. 1981. Comparison of the crossability of rye (*Secale cereale*) and *Hordeum bulbosum* onto wheat (*Triticum aestivum*). *Can. J. Genet. Cytol.* 23: 81–88.

FAO. 1996. *The State of the World's Plant Genetic Resources for Food and Agriculture*. FAO, Rome, Italy.

Farooq, S., Shah, T.M., and Asghar, M. 1996. Intergeneric hybridization for wheat improvement: production of and metaphase I analysis in $F_1$ hybrids of wheat (*Triticum aestivum*) with *Aegilops ovata* L. *Cereal Res. Commun.* 24: 155–161.

Farrer, W. 1904. Some notes on the wheat ìBobsî: its peculiarities, economic value and origin. *Agric. Gaz. N.S.W.* 15: 849–854.

Fedak, G. 1977. Increased homoeologous chromosome pairing in *Hordeum vulgare* × *Triticum aestivum* hybrids. *Nature* 266: 529–530.

Fedak, G. 1980. Production, morphology, and meiosis of reciprocal barley wheat hybrids. *Can. J. Genet. Cytol.* 22: 117–123.

Fedak, G. 1998. Procedures for transferring agronomic traits from alien species to crop plants. In *Cytogenetics and Evolution, Proceedings of the 9th International Wheat Genetics Symposium*, A.E. Slinkard, Ed., Saskatoon, Saskatchewan, Canada, August 2–7, pp. 1–7. Saskatoon, Canada. University of Saskatchewan, University Extension Press.

Forster, B.P. and Miller, T.E., 1985. 5B-deficient hybrid between *Triticum aestivum* and *Agropyron junceum*. *Cereal Res. Comm.* 13: 93–95.

Forster, B.P., Miller, T.E., and Law, C.N. 1988. Salt tolerance of two wheat-*Agropyron junceum* disomic addition lines. *Genome* 30: 559–564.

Friebe, B., Hatchett, J.H., Sears, R.G., and Gill, B.S. 1991. Transfer of Hessian fly resistance from rye to wheat via radiation-induced terminal and intercalary chromosome translocations. *Theor. Appl. Genet.* 83: 33–40.

Friebe, B., Jiang, J., Gill, B.S., and Dyck, P.L. 1993. Radiation induced nonhomoeologous wheat-*Agropyron intermedium* chromosomal translocations conferring resistance to leaf rust. *Theor. Appl. Genet.* 86: 141–149.

Friebe, B., Jiang, J., Raupp, W.J., McIntosh, R.A., and Gill, B.S. 1996a. Characterization of wheat alien translocations conferring resistance to diseases and pests: current status. *Euphytica* 91: 59–87.

Friebe, B. and Larter, E.N. 1988. Identification of a complete set of isogenic wheat/rye D-genome substitution lines by means of Giemsa C-banding. *Theor. Appl. Genet.* 76: 473–479.

Friebe, B., Qi, L.L., Nasuda, S., Zhang, P., Tuleen, N.A., and Gill, B.S. 2000. Development of a complete set of *Triticum aestivum-Aegilops speltoides* chromosome addition lines. *Theor. Appl. Genet.* 101: 51–58.

Friebe, B., Raupp, W.J., and Gill, B.S. 2001. Alien genes in wheat improvement. In *Wheat in a Global Environment*, Z. Bedo and L. Lang, Eds. Kluwer Academic Publishers, The Netherlands, pp. 709–720.

Friebe, B., Tuleen, N.A., Badaeva, E.D., and Gill, B.S. 1996b. Cytogenetic identification of *Triticum peregrinum* chromosomes added to common wheat. *Genome* 39: 272–276.

Friebe, B., Zeller, F.J., Mukai, Y., Forster, B.P., Bartos, P., and McIntosh, R.A. 1992. Characterization of rust-resistant wheat-*Agropyron intermedium* derivatives by C-banding, *in situ* hybridization and isozyme analysis. *Theor. Appl. Genet.* 83: 775–782.

Gill, B.S. 1989. The use of chromosome banding and *in situ* hybridization for the study of alien introgression in plant breeding. In *Review of Advances in Plant Biotechnology, 1985–88*, A. Mujeeb-Kazi and L.A. Sitch, Eds. CIMMYT, Mexico, D.F., pp. 157–163.

Gill, B.S. and Raupp, W.J. 1987. Direct gene transfers from *Aegilops squarrosa* L. to hexaploid wheat. *Crop Sci.* 27: 445–450.

Gottschalk, W. 1971. The phenomenon of "asymmetric genomic reduction." *J. Indian Bot. Soc. Golden Jubilee* 50A: 308–317.

Graybosch, R.A., Peterson, C.J., Hansen, L.E., Worrall, D., Shelton, D.R., and Lukaszewski, A. 1993. Comparative flour quality and protein characteristics of 1BL/1RS and 1AL/1RS wheat/rye translocation lines. *J. Cer. Sci.* 17: 95–106.

Gupta, P.K. and Priyadarshan, P. M. 1982. Triticale: present status and future prospects. *Adv. Genet.* 21: 255–345.

Harlan, J. and de Wet, J.M.J. 1971. Towards a rational classification of cultivated plants. *Taxon* 20: 509–517.

Islam, A.K.M.R., Shepherd, K.W., and Sparrow, D.H.B. 1978. Production and characterization of wheat-barley addition lines. In *Proceedings of the 5th IWG Symposium*, New Delhi, pp. 365–371.

Islam, A.K.M.R., Shepherd, K.W., and Sparrow, D.H.B. 1981. Isolation and characterization of euplasmic wheat-barley chromosome addition lines. *Heredity* 46: 161–174.

Islam-Faridi, M.N. and Mujeeb-Kazi, A. 1995. Visualization of *Secale cereale* DNA in wheat germplasm by fluorescent *in situ* hybridization. *Theor. Appl. Genet.* 90: 595–600.

Jahan, Q.N., Ter-Kuile, N., Hashmi, N., Aslam, M., Vahidy, A. and Mujeeb-Kazi, A. 1990. The status of 1B/1R translocation chromosome in some released wheat varieties of Pakistan. *Pakistan J. Bot.* 22: 1–10.

Jauhar, P.P. 1975. Genetic regulation of diploid-like chromosome pairing in the hexaploid species, *Festuca arundinacea* Schreb. and *F. rubra* L. (Gramineae). *Chromosoma* 52: 363–382.

Jauhar, P.P. 1977. Genetic regulation of diploid-like chromosome pairing in *Avena*. *Theor. Appl. Genet.* 49: 287–295.

Jauhar, P.P. 1991. Hybrid between durum wheat and diploid *Thinopyrum bessarabicum*. *Genome* 34: 283–287.

Jauhar, P.P. 1992a. High chromosome pairing in hybrids between hexaploid bread wheat and tetraploid crested wheat grass (*Agropyron cristatum*). *Hereditas* 116: 107–109.

Jauhar, P.P. 1992b. Synthesis and cytological characterization of trigeneric hybrids involving durum wheat, *Thinopyrum bessarabicum*, and *Lophopyrum elongatum*. *Theor. Appl. Genet.* 84: 511–519.

Jauhar, P.P. 1993. Alien gene transfer and genetic enrichment bread wheat. In *Biodiversity and Wheat Improvement*, A.B. Damania, Ed. John Wiley & Sons, New York, pp. 103–119.

Jauhar, P.P. 1995a. Morphological and cytological characteristics of some wheat × barley hybrids. *Theor. Appl. Genet.* 90: 872–877.

Jauhar, P.P. 1995b. Meiosis and fertility of F$_1$ hybrids between hexaploid bread wheat and decaploid tall wheat grass (*Thinopyrum ponticum*). *Theor. Appl. Genet.* 90: 865–871.

Jauhar, P.P. and Almouslem, A.B. 1998. Production and meiotic analyses of intergeneric hybrids between durum wheat and *Thinopyrum* species. In *Proceedings of the Third International Triticeae Symposium*, May 4–8, 1997, A.A. Jaradat, Ed. Scientific Publishers, Enfield, NH, pp. 119–126.

Jauhar, P.P. and Chibbar R.N. 1999. Chromosome-mediated and direct gene transfers in wheat. *Genome* 42: 570–583.

Jauhar, P.P. and Joppa, L.R. 1996. Chromosome pairing as a tool in genome analysis: merits and limitations. In *Methods of Genome Analysis in Plants*, P.P. Jauhar, Ed. CRC Press, Boca Raton, FL, pp. 9–37.

Jauhar, P. and Peterson, T.S. 2000. Toward transferring scab resistance from a diploid wild grass, *Lophopyrum elongatum*, into durum wheat. In *Proceedings of the 2000 U.S. Wheat and Barley Scab Initiative*, Michigan State Univ., pp. 201–204.

Jauhar, P. and Peterson, T.S. 2001. Hybrids between durum wheat and *Thinopyrum junceiforme*: prospects for breeding for scab resistance. *Euphytica* 118: 127–136.

Jauhar, P.P., Riera-Lizarazu, O., Dewy, W.G., Gill, B.S., Crane, C.F., and Bennett, J.H. 1991. Chromosome pairing relationships among the A, B, and D genomes of bread wheat. *Theor. Appl. Genet.* 82: 441–449.

Jewell, D. and Mujeeb-Kazi, A. 1982. Unexpected chromosome numbers in backcross 1 generations of F$_1$ hybrids between *Triticum aestivum* and related alien genera. *Wheat Inf. Serv.* 55: 5–9.

Jiang, J., Friebe, B., and Gill, B.S. 1994. Recent advances in alien gene transfer in wheat. *Euphytica* 73: 199–212.

Kawahara, T., Takumi, S., Matsuoka, Y., Mori, N., and Yasui, Y. 2003. *Tauschii* core collection, a powerful tool for utilize wheat D genome genetic resources. In *Tenth International Wheat Genetics Symposium*, Paestum, Italy, pp. 584–586.

Kihara, H. 1937. Genomanalyse bei *Triticum* und *Aegilops*. VII. Kurze Uebersicht ber die Ergebnisse der Jahre 1934–36. *Men. Coll. Agr. Kyoto Imp. Univ. Mo.* 41: 1–61.

Kihara, H. 1944. Discovery of the DD-analyser, one of the ancestors of vulgare wheats. *Agric. Horticult.* 19: 889–890.

Kimber, G. 1983. The suppression of characters of generic importance in hybrids and amphiploids in the Triticeae. *Cereal Res. Commun.* 11: 9–14.

Kimber, G. and Abu-Bakar, M. 1979. A wheat hybrid information system. *Cereal Res. Commun.* 7: 237–260.

Kimber, G. and Alonso, L.C. 1981. The analysis of meiosis in hybrids. II. Tetraploid hybrids. *Can. J. Genet. Cytol.* 23: 235–254.

Kimber, G. and Feldman, M. 1987. Wild Wheat: An Introduction, Special Report 353. College of Agriculture, University of Missouri–Colombia, pp. 129–131.

Knobloch, I.W. 1968. *A Check List of Crosses in Gramineae*. University of Michigan, pp. 1–170.

Knott, D.R. 1968. Translocations involving *Triticum* chromosomes carrying rust resistance. *Can. J. Genet. Cytol.* 10: 695–696.

Knott, D.R. and Dvořák, J. 1981. Agronomic and quality characteristics of wheat lines with leaf rust resistance derived from *Triticum speltoides*. *Can. J. Genet. Cytol.* 23: 475–480.

Kronstad, W.E. 1998. Agricultural development and wheat breeding in the 20th century. In *Wheat: Prospect for Global Improvement*, H.J. Braun et al., Eds. Kluwer Academic Publishers, Dordrech, The Netherlands, pp. 1–10.

Kruse, A. 1967. Intergeneric hybrids between *Hordeum vulgare* L ssp. *distichum* (v Pallas 2n = 14) and *Secale cereale* L (v Petkus 2n = 14). In *Royal Veterinary and Agricultural College Yearbook 1967*. Copenhagen, pp. 82–92.

Kruse, A. 1969. Intergeneric hybrids between *Triticum aestivum* L (v Koga II 2n = 42) and *Avena sativa* L(v Stal 2n = 42) with pseudogamous seed formation. In *Royal Veterinary and Agricultural Yearbook.* Copenhagen, pp. 188–200.

Kruse, A. 1973. *Hordeum × Triticum* hybrids. *Hereditas* 73: 157–161.

Kruse, A. 1974a. A 2,4-D treatment prior to pollination eliminates the haplontic (gametic) sterility in wide intergeneric crosses with 2-rowed barley. *Hordeum vulgare* spp. *distichum*, as maternal species. *Hereditas* 78: 319.

Kruse, A. 1974b. *Hordeum vulgare* ssp *distichum* (var Bomi) × *Triticum aestivum* (var Koga). An $F_1$ hybrid with generative seed formation. *Hereditas* 78: 319.

Ladinsky, G. and Fainstein, R. 1978. A case of genome partition in polyploid oats. *Theor. Appl. Genet.* 51: 159–160.

Lapitan, N.L.V., Sears, R.G., and Gill, B.S. 1984. Translocations and other karyotypic structural changes in wheat × rye hybrids regenerated from tissue culture. *Theor. Appl. Genet.* 68: 547–554.

Larkin, P.J., Spindler, L.H., and Banks, P.M. 1990. The use of cell culture to restructure plant genomes for introgressive breeding. In *Proceedings of the 2nd International Symposium of Chromosome Engineering in Plants*, G. Kimber, Ed., Columbia, MO, pp. 80–89.

Laurie, D.S. and Bennett, M.D. 1986. Wheat × maize hybridization. *Can. J. Genet. Cytol.* 28: 313–316.

Laurie, D.A. and Bennett, M.D. 1988. The production of haploid wheat plants from wheat × maize crosses. *Theor. Appl. Genet.* 76: 393–397.

Liu, W., Zheng, M.Y., Polle, E.A., and Konzak, C.F. 2002. Highly efficient doubled-haploid production in wheat (*Triticum aestivum* L) via induced microspore embryogenesis. *Crop Sci.* 42: 686–692.

Luo, M.C., Yen, C., and Yang, J.L. 1992. Crossability percentages of bread wheat landraces from Shaanxi and Henan provinces, China with rye. *Euphytica* 67: 1–8.

Ma, H., Singh, R.P., and Mujeeb-Kazi, A. 1995a. Resistance to stripe rust in *Triticum turgidum*, *T. tauschii*, and their synthetic hexaploids. *Euphytica* 82: 117–124.

Ma, H., Singh, R.P., and Mujeeb-Kazi, A. 1995b. Suppression/expression of resistance to stripe rust in synthetic hexaploid wheat (*Triticum turgidum* × *T. tauschii*). *Euphytica* 83: 87–93.

McFadden, E.S. and Sears, E.R. 1946. The origin of *Triticum spelta* and its free-threshing relatives. *J. Hered.* 37: 81–89, 107–116.

McIntosh, R.A., Wellings, C.R., and Park, R.F. 1995. *Wheat Rusts: An Atlas of Resistance Genes*. CSIRO Publications, Australia.

McKendry, A.L., Tague, D.N., Finney, P.L., and Miskin, K.E. 1996. Effect of 1BL.1RS on milling and baking quality of soft red winter wheat. *Crop Sci.* 36: 848–851.

Metakovsky, E.V., Yu. Novoselskaya, A., Kopus, M.M., Sobko, T.A., and Sozinov, A.A. 1984. Blocks of gliadin components in winter wheat detected by one-dimensional polyacrylamide gel electrophoresis. *Theor. Appl. Genet.* 67: 559–568.

Moreno-Sevilla, B., Baenziger, P.S., Peterson, C.J., Graybosch, R.A., and McVey, D.V. 1995a. The 1BL/1RS translocation: agronomic performance of F3-derived lines from a winter wheat cross. *Crop Sci.* 35: 1051–1055.

Moreno-Sevilla, B., Baenziger, P.S., Shelton, D.R., Graybosch, R.A., and Peterson, C.J. 1995b. Agronomic performance and end-use quality of 1B vs. 1BL/1RS genotypes derived from winter wheat ëRawhide.íCrop Sci. 35: 1607–1612.

Motsny, I.I. and Simonenko, V.K. 1996. The influence of *Elymus sibiricus* L. genome on the diploidization system of wheat. *Euphytica* 91: 189–193.

Mujeeb, K.A. 1970. *Gamma Radiation Effects on Phaseolus vulgaris L. and Seed Radiosensitivity Determinations for Other Species*. Kansas State University, Manhattan, pp. 1–130.

Mujeeb, K.A., Thomas, J.B., Rodriguez, R., Waters, R.F., and Bates, L.S. 1978. Chromosome instability in hybrids of *Hordeum vulgare* L. with *Triticum turgidum* and *T. aestivum*. *J. Hered.* 69: 179–182.

Mujeeb-Kazi, A. 2000. An analysis of the use of haploidy in wheat improvement. In *Application of Biotechnologies to Wheat Breeding*, M.M. Kohli and M. Francis, Eds., La Estanzuela, Uruguay, November 19–20, 1998, pp. 33–48.

Mujeeb-Kazi, A. 2001a. Synthetic hexaploids for bread wheat improvement. In *International Triticeae IV Symposium*, P. Hernandez, M.T. Moreno, J.I. Cubero, and A. Martin, Eds., Cordoba, Spain, September 10–12, pp. 193–199.

Mujeeb-Kazi, A. 2001b. Intergeneric hybrids in wheat: current status. In *International Triticeae IV Symposium*, P. Hernandez, M.T. Moreno, J.I. Cubero, and A. Martin, Eds., Cordoba, Spain, September 10–12, pp. 261–264.

Mujeeb-Kazi, A. 2003a. New genetic stocks for durum and bread wheat improvement. In *Tenth International Wheat Genetics Symposium*, Paestum, Italy, pp. 772–774.

Mujeeb-Kazi, A. 2003b. Wheat improvement facilitated by novel genetic diversity and *in vitro* technology. *Plant Tissue Cult.* 13: 179–210.

Mujeeb-Kazi, A. 2004. Wide crosses for durum wheat improvement. In *Durum Wheat Breeding: Current Approaches and Future Strategies*, C. Royo, M.N. Nachit, N. DiFonzo, J.L. Araus, W.H. Pfeiffer, G.A. Slafer, Eds. The Haworth Press, in press.

Mujeeb-Kazi, A. and Asiedu, R. 1989. Alien germplasm for wheat (*Triticum aestivum* L.) improvement facilitated by cytogenetic manipulation and use of novel techniques. In *Strengthening Collaboration in Biotechnology: International Agriculture Research and Private Sector*. USAID Conference, VA, pp. 211–231.

Mujeeb-Kazi, A. and Asiedu, R. 1990. Wide hybridization: potential of alien genetic transfers for *Triticum aestivum* improvement. In *Biotech in Agriculture and Forestry*, Vol. 13, *Wheat*, Y.P.S. Bajaj, Ed. pp. 111–127.

Mujeeb-Kazi, A. and Bernard, M. 1982. Somatic chromosome variations in backcross 1 progenies from intergeneric hybrids involving some Triticeae. *Cereal Res. Commun.* 10: 41–45.

Mujeeb Kazi, A., Cano, S., Rosas, V., Cortes, A., and Delgado, R. 2001e. Registration of five synthetic hexaploid wheat and seven bread wheat germplasm lines resistant to wheat spot blotch. *Crop Sci.* 41: 1653–1654.

Mujeeb-Kazi, A., Cortes, A., Rosas, V., Cano, S., and Delgado, R. 2001b. Registration of 17 isogenic chromosome 1B and 17 T1BL.1RS chromosome translocation bread wheat germplasms. *Crop Sci.* 41: 596–597.

Mujeeb-Kazi, A., Cortes, A., Rosas, V., Cano, S., and Delgado, R. 2001c. Registration of six isogenic T1BL.1RS chromosome translocation and six chromosome 1B durum germplasms. *Crop Sci.* 41: 595–596.

Mujeeb-Kazi, A., Fuentes-Davila, G., Villareal, R.L., Cortes, A., Rosas, V., and Delgado, R. 2001d. Registration of 10 synthetic hexaploid wheat and six bread wheat germplasms resistant to karnal bunt. *Crop Sci.* 41: 1652–1653.

Mujeeb-Kazi, A., Gilchrist, L.I., Villareal, R.L., and Delgado, R. 1998. Production and utilization of D-genome synthetic hexaploids in wheat improvement. In *Triticeae III*, A.A. Jaradat, Ed. Scientific Publishers, Enfield, NH, pp. 369–374.

Mujeeb-Kazi, A., Gilchrist, L.I., Villareal, R.L., and Delgado, R. 2000a. Registration of ten wheat germplasm lines resistant to *Septoria tritici* leaf blotch. *Crop Sci.* 40: 590–591.

Mujeeb-Kazi, A. and Hettel, G.P. 1995. Utilising Wild Grass Biodiversity in Wheat Improvement: 15 Years of Wide Cross Research at CIMMYT, CIMMYT Research Report 2. CIMMYT, Mexico, D.F.

Mujeeb-Kazi, A. and Kimber, G. 1985. The production, cytology, and practicality of wide hybrids in the Triticeae. *Cereal Res. Commun.* 13: 111–124.

Mujeeb-Kazi, A. and Miranda, J.L. 1984. High frequency synthetic genome formation potentialities in backcross 1 selfed derivatives from some intergeneric hybrids involving *Triticum aestivum* and *Agropyron*. In *Agronomy Abstracts*. American Society of Agronomy, Madison, WI, p. 80.

Mujeeb-Kazi, A. and Miranda, J.L. 1985. Enhanced resolution of somatic chromosome constrictions as an aid to identifying intergeneric hybrids among some Triticeae. *Cytologia* 50: 701–709.

Mujeeb-Kazi, A. and Rajaram, S. 2002. Transferring alien genes from related species and genera for wheat improvement. In *Bread Wheat Improvement and Production*. FAO, pp. 199–215.

Mujeeb-Kazi, A. and Rodriguez, R. 1983a. Cytogenetics of a *Hordeum vulgare* × *Triticum turgidum* hybrid and its backcross progeny with *T. turgidum*. *J. Hered.* 74: 109–113.

Mujeeb-Kazi, A. and Rodriguez, R. 1983b. Meiotic instability in *Hordeum vulgare* × *Triticum aestivum* hybrids. *J. Hered.* 74: 292–296.

Mujeeb-Kazi, A., Roldan, S., and Miranda, J.L. 1984. Intergeneric hybrids of *Triticum aestivum* with *Agropyron* and *Elymus* species. *Cereal Res. Commun.* 12: 75–79.

Mujeeb-Kazi, A., Roldan, S., Suh, D.Y., Sitch, L.A., and Farooq, S. 1987. Production and cytogenetic analysis of hybrids between *Triticum aestivum* and some caespitose *Agropyron* species. *Genome* 29: 537–553.

Mujeeb-Kazi, A., Roldan, S., Suh, D.Y., Ter-Kuile, N., and Farooq, S. 1989. Production and cytogenetics of *Triticum aestivum* L hybrids with some rhizomatous species. *Theor. Appl. Genet.* 77: 162–168.

Mujeeb-Kazi, A., Rosas, V., and Roldán, S. 1996a. Conservation of the genetic variation of *Triticum tauschii* (Coss.) Schmalh. (*Aegilops squarrosa* auct. non L.) in synthetic hexaploid wheats (*T. turgidum* L. s. lat. × *T. tauschii*; 2n = 6x = 42, AABBDD) and its potential utilization for wheat improvement. *Genet. Resour. Crop Evol.* 43: 129–134.

Mujeeb-Kazi, A., Sitch, L.A., and Fedak, G. 1996c. The range of chromosomal variations in intergeneric hybrids involving some Triticeae. *Cytologia* 61: 125–140.

Mujeeb Kazi, A. and van-Ginkel, M. 2004. Genetic resources: extent of diversity, its usage, and purpose for wheat improvement. *Plant Anim. Genome* XII: 51.

Mujeeb-Kazi, A., Villareal, R.L., Gilchrist, L.I., and Rajaram, S. 1996b. Registration of five wheat germplasm lines resistant to *Helminthosporium* leaf blight. *Crop Sci.* 36: 216–217.

Mujeeb-Kazi, A. and Wang, R.R.C. 1995. Perennial and annual wheat relatives in the Triticeae. In *Utilizing Wild Grass Biodiversity in Wheat Improvement: 15 Years of Wide Cross Research at CIMMYT*, CIMMYT Research Report No. 2, A. Mujeeb-Kazi and G.P. Hettel, Eds. CIMMYT, Mexico, D.F., Mexico, pp. 5–13.

Mujeeb-Kazi, A., William, M.D.H.M., Villareal, R.L., Cortés, A., Rosas, V., and Delgado, R. 2000b. Registration of 11 new isogenic T1BL.1RS chromosome translocation and 11 extracted chromosome 1B lines in *Triticum turgidum* L. cv. 'Altar 84.' *Crop Sci.* 40: 588–589.

Mujeeb-Kazi, A., William, M.D.H.M., Villareal, R.L., Cortes, A., Rosas, V., and Delgado, R. 2001a. Registration of 10 isogenic chromosome 1B and 10 T1BL.1RS chromosome translocation bread wheat germplasms. *Crop Sci.* 41: 280–281.

Mukai, Y., Friebe, B., Hatchett, J.H., Yamamoto, M., and Gill, B.S. 1993. Molecular cytogenetic analysis of radiation-induced wheat-rye terminal and intercalary chromosomal translocations and the detection of rye chromatin specifying resistance to Hessian fly. *Chromosoma* 102: 88–95.

Murai, K., Taketa, S., Islam, A.K.M.R., and Shepherd, K.W. 1997. A simple procedure for the production of wheat-barley 5H chromosome recombinant lines utilizing 5B nullisomy and 5H-specific molecular markers. *Wheat Inf. Serv.* 84: 53–55.

Murashige, T. and Skoog, F. 1962. A revised medium for rapid growth and bioassays with tobacco tissue cultures. *Physiologia Plantarum* 15: 473–497.

Okamoto, M. and Sears, E.R. 1962. Chromosomes involved in translocations obtained from haploids of common wheat. *Can. J. Genet. Cytol.* 4: 24–30.

O'Mara, J.G. 1940. Cytogenetic studies on Triticale. 1. A method for determining the effects of individual *Secale* chromosomes on *Triticum*. *Genetics* 25: 401–408.

Pestova, E.G., Borner, A., and Roder, M.S. 2001. Development of wheat D-genome introgression lines assisted by microsatellite markers. In *International Triticeae IV Symposium*, P. Hernandez, M.T. Moreno, J.I. Cubero, and A. Martin, Eds., Cordoba, Spain, pp. 207–210.

Pestova, E.G., Ganal, M.W., and Roder, M.S. 2000. Isolation and mapping of microsatellite markers specific for the D genome of bread wheat. *Genome* 43: 689–697.

Pienaar, R. de V. 1990. Wheat × *Thinopyrum* hybrids. In *Biotechnology in Agriculture and Forestry 13*, Y.P.S. Bajaj, Ed. Springer-Verlag, Berlin, pp. 167–217.

Plaschke, J., Ganal, M.W., and Roder, M.S. 1995. Detection of genetic diversity in closely related bread wheat using microsatellite markers. *Theor. Appl. Genet.* 91: 1001–1007.

Plourde, A., Fedak, G., St.-Pierre, C.A., and Comeau, A. 1990. A novel intergeneric hybrid in the Triticeae: *Triticum aestivum* × *Psathyrostachys juncea*. *Theor. Appl. Genet.* 79: 45–48.

Pritchard, D.J., Hollington, P.A., Davies, W.P., Gorham, J., Diaz de Leon, J.L., and Mujeeb-Kazi, A. 2002. K+/Na+ discrimination in synthetic hexaploid wheat lines: transfer of the trait for K+/Na+ discrimination from *Aegilops tauschii* to *Triticum turgidum*. *Cereal Res. Commun.* 30: 261–267.

Qu, L.-J., Foote, T.N., Roberts, M.A., Money, T.A., Aragon-Alcaide, L., Snape, J.W., and Moore, G. 1998. A simple PCR-based method for scoring the *ph1b* deletion in wheat. *Theor. Appl. Genet.* 96: 371–375.

Rajaram, S. 2001. Prospects and promise of wheat breeding in the 21st century. *Euphytica* 119: 3–15.

Rajaram, S., Mann, C.E., Ortiz-Ferrara, G., and Mujeeb-Kazi, A. 1983. Adaptation, stability and high yield potential of certain 1B/1R CIMMYT wheats. In *Proceedings of the Sixth International Wheat Genetics Symposium 1983*, Kyoto, Japan, pp. 613–621.

Riera-Lizarazu, O. and Mujeeb-Kazi, A. 1990. Maize (*Zea mays* L) mediated wheat (*Triticum aestivum* L) polyhaploid production using various crossing methods. *Cereal Res. Commun.* 18: 339–345.

Riley, R. and Chapman, V. 1967. The inheritance in wheat of crossability with rye. *Genet. Res.* 9: 259–267.

Riley, R., Chapman, V., and Johnson, R. 1968. Introduction of yellow rust resistance of *Aegilops comosa* into wheat by genetically induced homoleologous recombination. *Nature* 217: 383–384.

Rimpau, V. 1881. Kreuzungsprodukte landwirtschaftlicher Kulturpflanzen. *Landwirtschaftl. Jahrb.* 20: 335–371.

Roder, M.S., Korzun, V., Gill, B.S., and Ganal, M.W. 1998a. The physical mapping of microsatellite markers in wheat. *Genome* 41: 278–283.

Roder, M.S., Korzun, V., Wendehake, K., Plaschke, J., Tixier, M.-H., Leroy, P., and Ganal, M.W. 1998b. A microsatellite map of wheat. *Genetics* 149: 1–17.

Rosas, V., Asiedu, R., and Mujeeb-Kazi, A. 1988. Callus culture induced amphiploids of *Triticum aestivum* chromosome 5B hybrids and their backcross 1 derivatives. In *Agronomy Abstracts*. American Society of Agronomy, Madison, WI, p. 94.

Saari, E.E. and Prescott, J.M. 1975. A scale of appraising the foliar intensity of wheat diseases. *Plant Dis. Rep.* 59: 377–380.

Schlegel, R. and Meinel, A. 1994. A quantitative trait locus (QTL) on chromosome arm 1RS of rue and its effect on yield performance of hexaploid wheat. *Cereal Res. Commun.* 2: 7–13.

Sears, E.R. 1956. The transfer of leaf-rust resistance from *Aegilops umbellulatum* to wheat. *Brookhaven Symp. Biol.* 9: 1–22.

Sears, E.R. 1972. Chromosome engineering in wheat. *Stadler Genet. Symp.* 4: 23–28.

Sears, E.R. 1976. Genetic control of chromosome pairing in wheat. *Ann. Rev. Genet.* 10: 31–51.

Sears, E.R. 1977. An induced mutant with homoeologous pairing in wheat. *Can. J. Genet. Cytol.* 19: 585–593.

Sears, E.R. 1981. Transfer of alien genetic material to wheat. In *Wheat Science: Today and Tomorrow*, L.T. Evans and W.J. Peacock, Eds. Cambridge University Press, Cambridge, U.K., pp. 75–89.

Sears, E.R. 1984. Mutations in wheat that raise the level of meiotic chromosome pairing. *Stadler Genet. Symp.* 16: 295–300.

Sharma, H.C. 1995. How wide can a wide cross be? *Euphytica* 82: 43–64.

Sharma, H.C. and Baenziger, P.S. 1986. Production, morphology, and cytogenctic analysis of *Elymus caninus* (*Agropyron caninum*) × *Triticum aestivum* F₁ hybrids and backcross-1 derivatives. *Theor. Appl. Genet.* 71: 750–756.

Sharma, H.C. and Gill, B.S. 1981. New hybrids between *Agropyron* and wheat. 1. *A. ciliare* × wheat and *A. smithii* × wheat. *Wheat Inf. Serv.* 52: 19–22.

Sharma, H.C. and Gill, B.S. 1983a. Current status of wide hybridization in wheat. *Euphytica* 32: 17–31.

Sharma, H.C. and Gill, B.S. 1983b. New hybrids between *Agropyron* and wheat. 2. Production, morphology and cytogenetic analysis of F₁ hybrids and backcross derivatives. *Theor. Appl. Genet.* 66: 111–121.

Sharma, H.C. and Gill, B.S. 1983c. New hybrids between *Agropyron* and wheat. III. Backcross derivatives, effect of *Agropyron* cytoplasm, and production of addition lines. In *Proceedings of the 6th International Whea Genetics Symposium*, Kyoto, Japan, pp. 213–221.

Sharma, H.C. and Gill, B.S. 1986. The use of *Phl* gene in direct transfer and search for *Ph*-like genes in polyploid *Aegilops* species. *Z. Planzenzücht.* 96: 1–7.

Shepherd, K.W. and Islam, A.K.M.R. 1981. Wheat:barley hybrids: the first eighty years. In *Wheat Science: Today and Tomorrow*, L.T. Evans and W.J. Peacock, Eds. Cambridge University Press, Cambridge, U.K., pp. 107–128.

Simonenko, V.K., Motsny, I.I., Sechnyak, A.L., and Kulbida, M.P. 1998. The use of wheat-alien amphiploids for introgression of genetic material to wheat. *Euphytica* 100: 313–321.

Singh, R.J. 2003. *Plant Cytogenetics*, 2nd ed. CRC Press, Boca Raton, FL.

Singh, R.P., Henry, M., Huerta-Espino, J., Mujeeb-Kazi, A., Pena, R.J., and Khairallah, M. 2001. Recombined *Thinopyrum* chromosome segments in wheat carrying genes *Lr19* and *Bdv2*. In *Warren E. Kronstad Symposium*, Cd. Obregon, Sonora, Mexico, March 15, pp. 142–144.

Singh, R.P., Huerta-Espino, J., Rajaram, S., and Crossa, J. 1998. Agronomic effects from chromosome translocations 7DL.7Ag and 1BL.1RS in spring wheat. *Crop Sci.* 38: 27–33.

Smith, D.C. 1943. Intergeneric hybridization of *Triticum* and other grasses, principally *Agropyron*. *J. Hered.* 34: 219–224.

Snape, J.W., Chapman, V., Moss, J., Blanchard, C.E., and Miller, T.E. 1979. The crossabilities of wheat varieties with *Hordeum bulbosum*. *Heredity* 42: 291–298.

Stalker, H.T. 1980. Utilization of wild species for crop improvement. *Adv. Agron.* 33: 111–147.

Suenaga, K. and Nakajima, K. 1989. Efficient production of haploid wheat (*Triticum aestivum*) through crosses between Japanese wheat and maize (*Zea mays*). *Pl. Cell Rep.* 8: 263–266.

Ter-Kuile, N., Rosas, V., Asiedu, R., and Mujeeb-Kazi, A. 1987. The role of some cytogenetic systems in affecting alien genetic transfers for *Triticum aestivum* improvement. In *Agronomy Abstracts*. American Society of Agronomy, Madison, WI, p. 82.

Trethowan, R.M., Borja, J., and Mujeeb-Kazi, A. 2003. Contribution of synthetic wheats to drought tolerance. In *Agronomy Abstracts*. American Society of Agronomy, Madison, WI (CD-ROM).

Tsujimoto, H. and Noda, K. 1988. Chromosome breakage in wheat induced by the gametocidal gene of *Aegilops triuncialis* L: its utilization for wheat genetics and breeding. In *Proceedings of the 7th International Wheat Genetics Symposium*, T.E. Miller and R.M.D. Koebner, Eds., Cambridge, England, pp. 455–460.

Valkoun, J.J. 2001. Wheat pre-breeding using wild progenitors. *Euphytica* 119: 17–23.

Villareal, R.L., BaÒuelos, O., and Mujeeb-Kazi, A. 1997. Agronomic performance of related durum wheat (*Triticum turgidum* L.) stocks possessing the chromosome substitution T1BL.1RS. *Crop Sci.* 37: 1735–1740.

Villareal, R.L., BaÒuelos, O., Mujeeb-Kazi, A., and Rajaram, S. 1998. Agronomic performance of chromosome 1B and T1BL.1RS genetic stocks in spring bread wheat (*Triticum aestivum* L.) cultivar Seri M82. *Euphytica* 103: 195–202.

Villareal, R.L., del-Toro, E., Mujeeb-Kazi, A., and Rajaram, S. 1995a. The 1BL/1RS chromosome translocation effect on yield characteristics in a *Triticum aestivum* L. cross. *Plant Breed.* 114: 497–500.

Villareal, R.L., del-Toro, E., Rajaram, S., and Mujeeb-Kazi, A. 1996a. The effect of chromosome 1AL/1RS translocation on agronomic performance of 85 F2-derived F6 lines from three *Triticum aestivum* L. crosses. *Euphytica* 89: 363–369.

Villareal, R.L. and Mujeeb-Kazi, A. 2003. Genetic diversity and wheat improvement for sustainable agriculture. In *Agronomy Abstracts*. American Society of Agronomy, Madison, WI (CD-ROM).

Villareal, R.L., Mujeeb-Kazi, A., Fuentes-Davilla, G., and Rajaram, S. 1996b. Registration of four synthetic hexaploid wheat germplasm lines derived from *Triticum turgidum* × *T. tauschii* crosses and resistant to karnal bunt. *Crop Sci.* 36: 218.

Villareal, R.L., Mujeeb-Kazi, A., Fuentes-Davilla, G., Rajaram, S., and Del-Toro, E. 1994. Resistance to karnal bunt (*Tilletia indica* Mitra) in synthetic hexaploids derived from *Tritcum turgidum* × *Triticum tauschii*. *Plant Breed.* 112: 63–69.

Villareal, R.L., Mujeeb-Kazi, A., Gilchrist, L., and del-Toro, E. 1995b. Yield loss to spot blotch in spring bread wheat in warm non-traditional wheat production areas. *Plant Dis.* 79: 893–897.

Villareal, R.L., Rajaram, S., Mujeeb-Kazi, A., and del-Toro, E. 1991. The effect of chromosome 1B/1R translocation on the yield potential of certain spring wheats (*Triticum aestivum* L) *Plant Breed.* 106: 77–81.

Villareal, R.L., Sayre, K., BaÒuelos, O., and Mujeeb-Kazi, A. 2001. Registration of four synthetic hexaploid wheat (*Triticum turgidum/Aegilops tauschii*) germplasm lines tolerant to waterlogging. *Crop Sci.* 41: 274.

Wang, R.R.-C. 1989. Intergeneric hybrids involving perennial Triticeae. *Genet. (Life Sci. Adv.)* 8: 57–64.

Wang, R.R.-C., Li, X.-M., Hu, Z.-M., Zhang, J.-Y., Larson, S.R., Zhang, X.-Y., Grieve, C.M., and Shannon, M.C. 2003. Development of salinity-tolerant wheat recombinant lines from a wheat disomic addition line carrying a *Thinopyrum junceum* chromosome. *Int. J. Plant Sci.* 164: 25–33.

White, W.J. 1940. Intergeneric crosses between *Triticum* and *Agropyron*. *Sci. Agri.* 21: 198–232.

William, M.D.H.M. and Mujeeb-Kazi, A. 1995. Biochemical and molecular diagnostics of *Thinopyrum bessarabicum* chromosomes in *Triticum aestivum* germplasm. *Theor. Appl. Genet.* 90: 952–956.

Wilson, A.S. 1876. On wheat and rye hybrids. *Trans. Proc. Bot. Soc. Edinburgh* 12: 286–288.

Yen, C., Dai, D.Q., and Luo, M.C. 1986. The high compatibility resources of wheat for generic hybridization among *Secale* and *Aegilops*. In *Proceedings of the International Triticale Symposium*, pp. 42–52.

Zhang, J.-Y., Li, X.-M., Wang, R.R.-C., Cortes, A., Rosas, V., and Mujeeb-Kazi, A. 2002. Molecular cytogenetic characterization of E$^b$-genome chromosomes in *Thinopyrum bessarabicum* disomic addition lines of bread wheat. *Int. J. Plant Sci.* 163: 167–174.

Zheng, Y., Luo, M., Yen, C., and Yang, J. 1992. Chromosome location of a new crossability gene in common wheat. *Wheat Inf. Serv.* 75: 36–40.
Zohary, D. and Hopf, M. 1988. *Domestication of Plants in the Old World.* Clarendon Press, Oxford.

# CHAPTER 4

# Molecular Markers, Genomics, and Genetic Engineering in Wheat

Nora Lapitan and Prem P. Jauhar[*]

## CONTENTS

4.1   Introduction.................................................................................................99
4.2   Genomic Makeup of Wheat.......................................................................100
4.3   Molecular Markers in Wheat Breeding .....................................................100
4.4   Wheat Genomics ........................................................................................102
      4.4.1   Wheat ESTs, Gene Organization, and Comparative Mapping.............102
      4.4.2   Map-Based Cloning in Wheat ...........................................................106
4.5   Wheat Transformation and Application in Wheat Breeding ......................107
      4.5.1   Production of Transgenic Wheat.......................................................107
      4.5.2   Engineering Insect Pests and Disease Resistance .................................108
      4.5.3   Improvement of Grain Quality.........................................................108
      4.5.4   Tolerance to Abiotic Stresses ..........................................................108
4.6   Conclusion.................................................................................................109
Acknowledgment.................................................................................................109
References ...........................................................................................................109

## 4.1 INTRODUCTION

Bread wheat, *Triticum aestivum* L. (2n = 6x = 42; AABBDD genomes), is the most widely grown cereal crop, occupying 17% of the cultivated area worldwide, and is a staple for 15% of the world's population (http://www.CIMMYT.org/). With an annual production of over 570 million metric tons, wheat leads all other crops in terms of seed planted and acreage harvested (FAO 2001). By 2020, it is expected that the demand for wheat will be 40% greater than its current level; however, the resources for wheat production are likely to be significantly lower (Pingali and Rajaram 1999). Durum wheat or macaroni wheat (*Triticum turgidum*, 2n = 4x = 28; AABB genomes) is

Mention of trademark or proprietary product does not constitute a guarantee or warranty of the product by the USDA or imply approval to the exclusion of other products that may also be suitable. This is a contribution from the Colorado Agricultural Experiment Station.
[*] U.S. government employee whose work is in the public domain.

also an important cereal used for human consumption worldwide. Its high protein content and gluten strength make it the choice wheat for preparing pasta products.

Genetic improvement of cereal crops, which account for two thirds of the world food supply (Borlaug 1998), has generally been accomplished by conventional breeding and, in some cases, by interspecific and intergeneric hybridizations. Since the 1950s, hybridization between cultivated and wild relatives (also known as wide hybridization) has been effectively used to develop wheat cultivars with improved agronomic performance, disease and pest tolerance, and high yields (Jauhar and Chibbar 1999). Wide hybridization with perennial grasses of the tribe Triticeae has contributed to the genetic improvement of bread wheat (Jauhar and Chibbar 1999; Repellin et al. 2001), although it has been a slow process. Tools of molecular genetics and biotechnology offer possibilities for significantly increasing the efficiency of wheat breeding. These tools can provide insights into the genetic control of key traits, markers for selection, genes for manipulation, and methods for introducing novel sources of genetic variation, thereby speeding up the breeding process (Jauhar and Chibbar 1999; Jauhar and Khush 2002; Janakiraman et al. 2002). This review presents a summary of three areas of research that offer great potential in wheat improvement: (1) application of molecular makers in wheat breeding, (2) development of genomics tools, and (3) wheat transformation.

## 4.2 GENOMIC MAKEUP OF WHEAT

Hexaploid bread wheat and tetraploid durum wheat are true breeding natural hybrids with the genomic constitutions of AABBDD and AABB, respectively. Durum wheat evolved long before bread wheat. Its A genome was derived from the diploid species *Triticum urartu* Tum., and the B genome was derived from *Aegilops speltoides* Tausch or a species closely related to it. The two diploid progenitors hybridized in nature, some half million years ago (Huang et al. 2002), and produced tetraploid wild emmer wheat in one step as a result of functioning of unreduced gametes in both parents or in the AB hybrid, as shown in Figure 2.1 and Figure 2.2 in this volume. The wild emmer, through domestication, gave rise to cultivated durum wheat. Hexaploid wheat (AABBDD) evolved later via natural hybridization of tetraploid emmer wheat, wild or cultivated, with a third diploid grass, *Aegilops tauschii* Coss., that contributed the D genome. Again, unreduced gametes in both parents or in the triploid hybrid ABD would have given rise to bread wheat (Jauhar 2003a; see Figure 1.3 in this volume).

Because homoeologous (genetically similar) chromosomes are closely related, and hence capable of pairing with one another, some control of pairing would be necessary to enforce regular, diploid-like pairing. At the time of origin of tetraploid emmer, a spontaneous mutation occurred to give rise to the homoeologous pairing suppressor gene, *Ph1*, which permitted pairing only among homologous partners, thereby ensuring diploid-like pairing and disomic inheritance. If not for this precise regulation of chromosome pairing, the A and B genomes might have converged during the period of the half million years they have been together (Jauhar 2003b). However, *Ph1* also prevented pairing between wheat chromosomes and alien chromosomes in interspecific and intergeneric hybrids, and thus offered an obstacle to the incorporation of alien genes into wheat. Different methods of circumventing the influence of *Ph1* are now known and have proved useful in genetic enrichment of wheat (see Chapter 1 for details).

## 4.3 MOLECULAR MARKERS IN WHEAT BREEDING

Molecular markers are typically DNA-based chromosome landmarks used to genetically map genes for important traits in sexually reproducing organisms. During the last 10 years, wheat researchers constructed detailed genetic maps containing more than 3000 molecular markers (Gill et al. 1991; Nelson et al. 1995; Roder et al. 1998; Somers et al. 2004) and physical maps with more than 16,000 loci (Delaney et al. 1995; Qi et al. 2004) (http://wheat.pw.usda.gov/NSF/). The markers include restriction fragment length polymorphisms (RFLPs) (Gill et al. 1991; Nelson et al. 1995), simple sequence repeats (SSRs) (Roder et al. 1998; Somers et al. 2004), amplified fragment length polymorphisms (AFLPs) (Peng et al. 2000), and expressed sequence tags (ESTs) (Lazo et al. 2004; Qi et al. 2004) (http://wheat.pw.usda.gov/NSF/). Wheat SSRs developed from ESTs provide another source of markers (Nicot et al. 2004; Peng and Lapitan 2005; Yu et al. 2004), although the average polymorphism information content of EST-derived SSRs (~45%) is lower than that of genomic SSRs (~54%) (Peng and Lapitan 2005). Single-nucleotide polymorphisms (SNPs) are the most abundant form of DNA variation and mutations in many genes (Useche et al. 2001). In wheat genes, a SNP is estimated to occur every 540 bp (Somers et al. 2003). A polymerase chain reaction (PCR)-based technique to target specific chromosome regions was recently developed (Hu and Vick 2003). The target region amplification polymorphism (TRAP) technique is based on the use of one fixed primer that is designed from a targeted EST sequence database and a second, arbitrary primer with either an AT- or GC-rich core to anneal with an intron or exon, respectively. This technique was used on Langdon durum–*Triticum dicoccoides* substitution lines to generate chromosome-specific markers in wheat (Xu et al. 2003).

Molecular markers have been successfully used in wheat to map genes for agronomic traits. A large proportion of mapped genes are for those conferring resistance to diseases (Ling et al. 2003; Liu and Anderson 2003; Steiner et al. 2004) and insects (Anderson et al. 2003; Dweikat et al. 2002; Ma et al., 1998; Zhu et al. 2004). Other mapped traits include quality parameters or characteristics such as hardness and grain protein content (Dubcovsky et al. 1998), developmental genes such as those controlling vernalization requirement and preharvest sprouting (Li et al. 2004; Yan et al. 2003), and quantitative trait loci for a variety of agronomic traits (Borner et al. 2002; Peng et al. 2003).

The value of markers in breeding programs is for selection of plants containing desired genes linked to them during breeding, a process called marker-assisted selection (MAS). The benefits of MAS in breeding have been extensively discussed in the literature (Gupta et al. 1999; Tanksley 1983). These include the ability to select for genes that are highly affected by the environment, incorporation of resistance to diseases and pests that cannot be easily screened for or where the pest or pathogen has not been introduced into a specific geographic region, and the ability to combine multiple genes for a given trait within the same cultivar in a process called gene pyramiding.

Despite the large number of wheat genes that have been genetically mapped, the deployment of markers in wheat breeding is not yet a common practice. Some of the reasons for this include cost in terms of time, labor, and equipment, lack of technical expertise, low throughput compared to many conventional screening methods, difficulty in recovering informative polymorphisms in breeding germplasm, and changing magnitude of effects of quantitative trait loci (QTL) across environments and in different genetic backgrounds (Anderson 1998). Many of the early genetic mapping studies were based on RFLPs, which involve tedious and laborious procedures for detection and oftentimes require the use of radioisotopes. This problem has been overcome with the use of markers based on PCR. PCR-based markers are simpler to use, involve less cost and time, and can often be detected with an inexpensive gel apparatus. SSRs and SNPs may be the most promising among PCR-based markers for application in MAS. SSR markers have a higher polymorphism index than other PCR-based markers, such as RAPDs (Nagaoka and Ogihara 1997; Roder et al. 1998). At present, there are over 1000 SSRs in wheat (Bryan et al. 1997; Kantety et al. 2002; Roder et al. 1998; Peng and Lapitan 2005; Somers et al. 2004; Yu et al. 2004); SNPs, on the other hand, are just beginning to be exploited for their potential importance in wheat MAS (Somers et al. 2003) (http://wheat.pw.usda.gov/ITMI/2002/WheatSNP.html). One of the challenges in applying SNPs in wheat is being able to distinguish genome-specific SNPs. Cytogenetic stocks have been used for this purpose (Mochida et al. 2003). ESTs were first grouped into contigs and were assigned to wheat chromosomes using nullisomic-tetrasomic lines in conjunction with pyrosequencing. The expression profiles of 90 homoeologous genes showed no preference for gene silencing in particular genomes or chromosomes of wheat.

Another reason for the lack of implementation of markers in breeding has been the lack of public funding for this type of work (Dubcovsky 2002). This scenario is already beginning to change. Public and private investments are now being made in this arena. Over the last 6 years, Australia has made significant investments in its national winter cereal breeding programs. In the U.S., a consortium (called MASwheat) consisting of wheat geneticists and breeders was funded by the U.S. Department of Agriculture to implement markers in public wheat breeding programs. The project's goal was to introgress, by backcrossing, more than 23 disease resistance genes and 21 quality-related gene variants in approximately 100 adapted parents from different market classes. The project has completed the introgression of valuable genes into 78 wheat lines, and 178 lines will be completed by late 2005, when the project finishes (Soria and Dubcovsky, personal communication) (http://maswheat.ucdavis.edu/). This project has a similar structure to the Australian National Wheat Molecular Marker Program (NWMMP) established earlier (1996) (Harker et al. 2001) (http://www.scu.edu.au/research/cpcg/wheat/index.php). The main objective of both projects is to support breeders by implementing MAS capacities within existing public breeding programs. MAS as a strategy for wheat breeding is also receiving support in Canada (D. Somers, personal communication). Expected outcomes from studies such as these include not only new varieties developed with the aid of MAS but also an assessment of the usefulness and practicality of MAS in wheat breeding. Implementation of MAS in Australian wheat breeding programs showed the usefulness of MAS for traits of substantial economic importance and where the nonmarker assay is expensive and unreliable (Eagles et al. 2001).

## 4.4 WHEAT GENOMICS

During the last 10 years, there has been an explosion of information in genomics, owing to the application of high-throughput technologies to the study of whole genomes. The complete genomes of model species, including *Escherichia coli* (Blattner et al. 1997), *Caenorhabitis elegans* (The *Caenorhabitis elegans* Consortium 1998), *Drosophila malanogaster* (Kornberg and Krasnow 2000), and human (The Human Genome Sequencing Consortium 2001; Venter et al. 2001), have been determined. The genomes of two plant species, *Arabidopsis thaliana* and *Oryza sativa*, have been sequenced recently (Goff et al. 2002; The Arabidopsis Genome Sequencing Initiative 2000; Yu et al. 2002). The small genome sizes of *Arabidopsis* (120 Mbp) and rice (430 Mbp) have made them the choice for sequencing of representative dicot and monocot species. The availability of genome sequences for these two plant species and other organisms can revolutionize research on plant genomes. Map-based cloning of important genes for which gene products are not known will require less investment, time, and effort. Comparative sequence analysis among species, genera, and even families enables the identification of conserved sequences as well as sequences that are unique to a species. Earlier comparative mapping studies among cereals revealed a remarkable conservation of a backbone of linkage groups among related species in the Poaceae, such as rice, maize, barley, wheat, sorghum, and sugarcane (Gale and Devos 1998). Based on this conservation, the basic assumption was that rice sequences could be used to find corresponding sequences in other grasses. However, recent studies using wheat ESTs (described below) reveal exceptions to this assumption.

### 4.4.1  Wheat ESTs, Gene Organization, and Comparative Mapping

Cognizant that the wheat genome was not going to be sequenced immediately because of its large size and complexity, wheat researchers initiated efforts to sequence expressed portions of the wheat genome. ESTs, which are produced by single-pass sequencing of cDNA clones, were first produced by individual researchers as part of a consortium called the International Triticeae EST

Cooperative (http://wheat.pw.usda.gov/genome/). Federal agencies in the U.S. and Canada later funded collaborative efforts to produce wheat ESTs (http://wheat.pw.usda.gov/NSF/) (Cloutier et al. 2002). Currently, wheat has the largest number of ESTs (almost 700,000 in http://www.ncbi.nlm.nih.gov/dbEST/index.html as of September 14, 2004) among plants. Wheat ESTs were generated from cDNA libraries representing all developmental stages and tissues, with emphasis on reproductive development subjected to environmental factors that ultimately affect grain quality and yield (Zhang et al. 2004). A total of 8241 wheat ESTs were mapped to chromosome bins (http://wheat.pw.usda.gov/NSF/), which are chromosome regions defined by breakpoints of the selected deletion lines. The strategy used for mapping took advantage of existing wheat cytogenetic stocks that are missing chromosomes, chromosome arms, or terminal regions in deletion lines (Qi et al. 2004). Mapping relied on hybridization of EST probes to Southern blots containing DNA from these lines and observing the presence or absence of bands. The example in Figure 4.1 shows mapping of three EST loci to group 2 chromosomes. Band 1 is absent in N2BT2D (nullisomic for 2B and tetrasomic for 2D), indicating the locus is present in 2B. Bands 2 and 3 mapped to 2A and 2D, respectively. The same principle was applied to further map the loci to chromosome bins 2BL6-0.89-1.00, C-2AL-0.85, and 2DL9-0.76-1.00 for bands 1, 2, and 3, respectively.

Each EST probe detected an average of 4.8 restriction fragments, and 2.8 loci were mapped per EST (Qi et al. 2004). The number of loci mapped in the three genomes were 5173, 5774, and 5146 for the A, B, and D genomes, respectively. The large number of wheat ESTs is opening up areas of investigation not previously possible. The resulting map of ESTs, also known as the transcriptome map, was used to study wheat gene organization in detail (Akhunov et al. 2003a,b; Qi et al. 2004). ESTs were not uniformly distributed among the three genomes, with the B genome having the largest number (Qi et al. 2004). This difference is not due to the difference in genome sizes, since the A genome is bigger (89.0 μm) than the B genome (87.2 μm). While the D genome (68.6 μm) is much smaller than the A genome, the two genomes contained roughly similar numbers of ESTs. In fact, the relative EST density as measured by the ratio of observed to expected loci based on chromosome length was significantly higher for the D genome than for the A or B genome. Qi et al. (2004) suggested that the difference in distribution of EST loci among the three genomes was likely explained by the evolutionary history of each of the hexaploid wheat genomes. The two parameters may be the breeding system and age of diploid donors. The A and D genome donors, *T. urartu* and *Ae. tauschii*, are self-pollinating, while the likely donor of the B genome donor (*Ae. speltoides*) is cross-pollinating. *Ae. speltoides* has been shown to exhibit greater RFLP variation than *Ae. tauschii* (Dvořák 1998). Based on DNA sequence analysis of three genes, greater poly- morphism in *Ae. spletoides* was attributed to the presence of duplicate copies of the genes compared to single copies in the self-pollinating A and D genomes (Huang et al. 2002; Li and Gill 2002). In terms of age of the lineages, the *Ae. speltoides* lineage is older (ca. 4 million years ago) than *Ae. tauschii* (ca. 2.5 million years ago), and the *T. urartu* lineage is the youngest (1 million years ago). Older lineages are expected to accumulate more polymorphisms and gene duplications. Indeed, when Akhunov et al. (2003a) analyzed a subset of ESTs, they observed twice as many unique loci in the B genome than in the A and D genomes. Within chromosomes, relative gene density and recombination frequency were not uniformly distributed either. These values increased with the relative distance of a bin from the centromere (Akhunov et al. 2003b; Qi et al. 2004), which is consistent with previous observations (Boyko et al. 1999; Sandhu et al. 2001). Akhunov et al. (2003a) showed that the degree of synteny (the number of loci shared) is more between the A and D genomes than between A and B, or B and D. This agrees with previous conclusions based on chromosome pairing affinities (Jauhar et al. 1991).

Single-gene loci, with one fragment in each of the three wheat genomes, were found to occur predominantly in proximal regions, while multigene loci (with more than one copy per genome) tended to be more frequent in distal, high-recombination regions (Akhunov et al. 2003b). One quarter of the wheat genes were found to be duplicated within a genome. Comparison of duplicated wheat ESTs with the rice genome sequence was used to distinguish ancestral loci from those derived

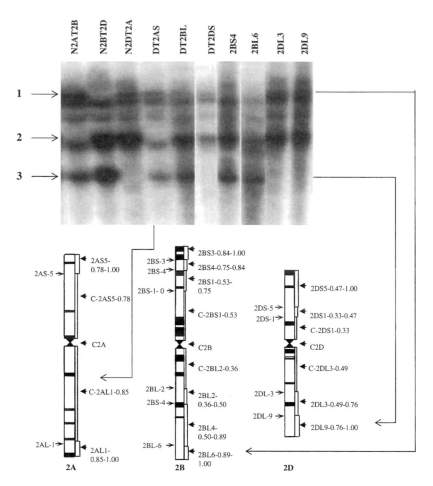

**Figure 4.1** Strategy for mapping ESTs on wheat chromosome bins using deletion lines of wheat. Top panel: Autoradiogram following hybridization of an EST probe to genomic DNA of cytogenetic lines cut with *EcoR*I. Bottom panel: Ideograms of chromosomes 2A, 2B, and 2D, showing the location of bins (shown to the right of each chromosome) and breakpoints (shown to the left of each chromosome). Lanes in top panel are N2AT2B (no 2A and four copies of 2B), N2BT2D (no 2B and four copies of 2D), N2DT2A (no 2D and four copies of 2A), DT2AS (no 2AL arm), DT2BL (no 1BS arm), and DT2DS (no 2DL arm); 2BS, 2BL6, 2DL3, and 2SDL contain terminal deletions (location of deletion shown in ideograms). Three bands were mapped. Band 1 is missing in N2BT2D and present in DT2BL, indicating it is on 2BL. Among deletion lines, only 2BL6 is missing band 1; thus it was assigned to 2BL6-0.89-1.00. Band 2 is absent in N2AT2B and DT2AS, and present in deletion line 2AL1-0.85-1.00 (not shown). Thus, it was assigned to the most proximal bin, C-2AL1-0.85. Band 3 was absent in N2DT2A, DT2DS, 2DL3, and 2DL9; thus, it was assigned to the most terminal bin, 2DL9-0.76-1.00.

from it by duplication. If a homologous gene was present on a rice homoeologous chromosome, the wheat and rice genes were assumed to be orthologous and the wheat locus was considered to be the ancestral locus. The locus (or loci) of the set present on a nonhomoeologous wheat chromosome was assumed to have originated by duplication of this ancestral locus. This analysis showed that duplications were most frequently located in distal, high-recombination chromosome regions, whereas ancestral loci were most frequently located proximally. These results suggest that recombination has played a central role in the evolution of wheat genome structure and that gradients of recombination rates along chromosome arms promote more rapid rates of genome evolution in distal, high-recombination regions than in proximal, low-recombination regions.

**Figure 4.2** **(See color insert following page 114.)** Rice genome view showing the wheat chromosome arm location for the most similar wheat gene sequences. Each colored box represents a rice–wheat gene sequence match at $\geq$80% identity. When the wheat EST mapped to more than one wheat chromosome, the other color coded locations adjacent to the first. Homoeologous wheat chromosome locations are grouped together. Rice BAC/PAC sequences that did not match any wheat sequences, as well as redundant sequences, are omitted. The rice centromere location is indicated by C. (From http://wheat.pw.usda.gov/pubs/2003/Sorrells/. Based on Sorrells, M.E., La Rota, M., Bermudez-Kandlanis, C.E., et al., *Genome Res.*, 13, 1818–1827, 2003. With permission from Dr. Sorrells.)

The transcriptome map was also used to investigate the degree of homology between wheat and rice. To use the rice genome sequence for gene discovery in wheat, it is important to have information on the degree of similarity at the DNA sequence level between the two genomes. Earlier comparative RFLP mapping analyses revealed a remarkable conservation of linkage groups between wheat and rice (Ahn et al. 1993; Gale and Devos 1998). Conserved linkage groups were found between wheat chromosome group 1 (W1) and rice chromosomes 5 and 10 (R5 and R10); W2 and R4 and R7; W3 and R1; W4 and R3; W5 and R9, R12, R11, and R3; W6 and R2 and R11; and W7 and R6, R8, and R2. Sequence-based analyses using wheat ESTs confirmed these relationships, but at the same time, the finer resolution achieved with this method revealed the complexity of relationships between the two genomes not previously detected with RFLPs (Sorrells et al. 2003). The degree of homology between wheat and rice varied among and within chromosomes. Although genes contained in each rice chromosome are predominantly from one or two wheat chromosomes, there are also many genes that are located in many different wheat chromosomes (Figure 4.2). These recent results support the view that grass genomes are labile, rapidly evolving entities and that structural and functional relationships are complex (Gaut 2002). These findings suggest that the use of rice as a model for cross-species transfer of information is complicated by the existence of many nonconserved regions.

## 4.4.2   Map-Based Cloning in Wheat

The ultimate goal of genomics is to identify genes for important traits. Map-based cloning is a strategy that has proven to be useful for genes with unknown gene products. The technique involves cloning of a chromosome fragment containing a gene of interest flanked by molecular markers. Isolation of the chromosome fragment defined by flanking markers is performed by screening a genomic library containing large-insert clones with the marker probes. Once a positive clone is identified, the isolated region is extended by identifying a second clone containing fragments that overlap with the first clone. This procedure, called chromosome walking, is repeated until the target chromosome region has been isolated. Map-based cloning has become a commonplace technique for cloning genes in model species, including humans and *Arabidopsis* (Harris 2000; Lukowitz et al. 2000). However, the large genome size (16,000 Mbp), high amounts of repetitive sequences, and polyploidy in the wheat genome all contribute to the difficulty of map-based cloning in this species. The large wheat genome requires that a wheat genomic library contain a large number of clones. For example, a wheat BAC library with an average 100-kb insert size requires ~470,000 clones to ensure a 95% chance of finding any single-copy sequence. High amounts of repetitive sequences and duplication of loci in the three genomes further complicate the process of identifying clones in a specific chromosome region. Until recently, map-based cloning was considered to be impossible in wheat. Today the picture is different. Five wheat genes have been cloned by this approach. These include *Lr10*, *Lr21*, *Vrn1*, *Vrn2*, and *Pm3b* (Feuillet et al. 2003; Huang et al. 2003; Yahiaoui et al. 2004; Yan et al. 2003, 2004). *Lr10*, *Lr21*, and *Pm3b* are genes conferring resistance to *Puccinia triticina* and *Blumeria graminis* f. sp. *tritici*, respectively (Feuillet et al. 2003; Huang et al. 2003; Yahiaoui et al. 2004). *Vrn1* and *Vrn2* control the vernalization response in wheat but have different modes of action. *Vrn1* is dominant for the spring growth habit (i.e., long exposure to low temperatures is not required in order to flower), while *Vrn2* is dominant for winter growth habit (Yan et al. 2003, 2004). Candidate genes have also been identified for other genes, including *Cre3* for nematode resistance (Lagudah et al. 1997), *Gpc-6B1* locus, which is responsible for differences in grain protein content among wheat lines (Distelfeld et al. 2004), and *Q*, which confers the free-threshing character to domesticated wheat (Faris et al. 2003).

What has made the cloning of these genes possible can be attributed to the availability of resources that were nonexistent only 5 years ago. Construction of large-insert libraries in bacterial artificial chromosomes (BACs) was critical in the success of these projects. BAC libraries were first made for diploid and tetraploid relatives of wheat, *Triticum monococcum* (A genome) (Lijavetzky et al. 1999), *Ae. tauschii* (D genome) (Huang et al. 2003; Moullet et al. 1999), and *T. turgidum* ssp. *durum* (A and B genomes) (Cenci et al. 2003). BAC libraries for hexaploid wheat are now also available (Ma et al. 2000; Nilmalgoda et al. 2003). Recently, flow sorting technology was used to isolate specific wheat chromosomes and arms, from which BAC libraries were created. Subgenomic libraries have been made for wheat chromosomes 1D, 4D, 6D, and 3B (Janda et al. 2004; Safar et al. 2004). Chromosome-specific BAC libraries provide solutions to many of the problems mentioned above and offer the possibility of studying genomic differences. Rice genomic sequences are another resource that have proven to be useful for saturating specific wheat regions in some cases (Yan et al. 2003, 2004).

Comparison of strategies used for the successful positional cloning of wheat genes shows that there is not one cut-and-dried method to accomplish this. Cloning of *Lr21*, *Vrn1*, and *Vrn2* was accomplished using diploid BAC or other large-insert libraries (Huang et al. 2003; Yan et al. 2003, 2004). The use of diploid species in these studies reduced the required number of clones to screen in order to detect positive clones containing the target sequences. However, while cloning of *Lr21* required a relatively small number of segregating progeny (520 F2s), cloning of *Vrn1* and *Vrn2* was accomplished by using 6000 gametes to obtain a high-resolution map. The difference in these studies is explained by the recombination frequencies in the chromosomal locations of these genes. *Lr21* is located in the distal gene-rich region of chromosome 1DS, which was shown to be highly recombinogenic (Huang et al. 2003). In comparison, *Vrn1* is at a proximal location in chromosome 5A (Yan et al. 2003). This region has a gene density of 6.2 Mbp/cM, which is two times higher than the average genome-wide gene density in wheat (3 Mbp/cM). Rice genomic and sorghum sequences were used to find markers tightly linked to the gene.

Cloning of *Lr10* and *Pm3b* provides the first examples of map-based cloning from hexaploid wheat (Feuillet et al. 2003; Yahiaoui et al. 2004). Both studies made use of combined analyses from wheat species with lower ploidy levels. To clone *Lr10*, which is located on chromosome 1AS, a subgenome walking strategy was used where genetic mapping was performed in hexaploid wheat and chromosome walking was performed by using BAC clones from *T. monococcum* (Feuillet et al. 2003; Stein et al. 2000). Sequencing of a 211-kb region from a BAC clone spanning the *Lr10* region identified two resistance gene analogs (RGAs) in a region that showed complete linkage to *Lr10*. The *T. monococcum* RGAs provided starting points to amplify the corresponding sequences from hexaploid wheat. Similarly, *Pm3b* was mapped in hexaploid wheat. *T. monococcum* and *T. turgidum* ssp. *durum* were used to establish a physical contig spanning the *Pm3* locus (Yahiaoui et al. 2004). A candidate gene for *Pm3b* was identified using partial sequence conservation between a hexaploid cultivar containing the resistance gene and *T. monococcum* cv. DV92.

## 4.5 WHEAT TRANSFORMATION AND APPLICATION IN WHEAT BREEDING

Conventional breeding alone or in conjunction with intervarietal, interspecific, and intergeneric hybridization has been mainly responsible for the production of superior wheat cultivars. However, these procedures are time-consuming and often tedious (see, for example, Jauhar and Chibbar 1999). Since the early 1990s, the development of novel techniques has allowed direct introduction of DNA into regenerable callus of wheat. These *in vitro* tools have made it possible to introduce novel traits from unrelated organisms into otherwise superior cultivars of wheat. Wheat genetic engineering involves the insertion of a characterized gene isolated from any organism into wheat cells and the subsequent recovery of fully fertile plants with the inserted gene(s) integrated into their genome. This technology gives access to an unlimited gene pool, without the constraint of sexual compatibility among plant species. However, there is yet to be a commercially released cultivar with a novel trait inserted by transformation.

### 4.5.1 Production of Transgenic Wheat

The availability of an efficient *in vitro* regeneration system is a prerequisite for production of transgenic plants. Initial success in the genetic transformation of wheat, incorporating glufosinate ammonium resistance, was made possible by the availability of a high-efficiency regeneration protocol in conjunction with a biolistics-mediated gene delivery system (Becker et al. 1994; Nehra et al. 1994; Vasil et al. 1992; Weeks et al. 1993). Soon afterward, several value-added traits were incorporated into various wheat cultivars.

Although bread wheat was genetically transformed by several groups (Becker et al. 1994; Nehra et al. 1994; Vasil et al. 1992; Weeks et al. 1993), durum wheat was difficult to transform because its *in vitro* regeneration protocol was not established. After standardizing a suitable regeneration protocol (Bommineni and Jauhar 1996), Jauhar's lab produced the first transgenic durum wheat (Bommineni et al. 1997). Since this first report, there have been several reports of production of transgenic durum (He et al. 1999; Pellegrineschi et al. 2002), facilitating incorporation of value-added traits in this important cereal.

## 4.5.2    Engineering Insect Pests and Disease Resistance

Insect pests and pathogens cause huge losses to wheat yields. Grain moth is one of the serious pests of wheat. By engineering the wheat cultivar Bobwhite with the trypsin inhibitor gene, Altpeter et al. (1999) produced enhanced resistance to Angoumois grain moth (*Sitotroga cerealella*). Wheat is also attacked by several fungal pathogens. For example, scab or Fusarium head blight (FHB), caused by *Fusarium graminearum* Schwabe, is a serious disease of both bread wheat and durum wheat. In view of the absence of a reliable source of FHB resistance in wheat, Jauhar's group produced FHB-resistant durum germplasm by crossing durum cultivars with wild wheatgrasses (Jauhar and Peterson 2000, 2001). However, in view of the ravaging nature of FHB, it would be advisable to also employ transgenic approaches to combat it in both bread wheat (Anand et al. 2003; Chen et al. 1999; Dahleen et al. 2001) and durum wheat. Having standardized the transgenic technology for durum wheat, Jauhar's group is introducing antifungal genes, including *Tri101* and *PDR5*, into commercial durum cultivars for protection against scab. In addition, genetic engineering has been employed to produce powdery mildew resistance in wheat (Bliffeld et al. 1999). FHB resistance in bread wheat was achieved by expressing pathogenesis-related proteins using particle bombardment (Anand et al. 2003). However, this technique of genetic transformation is hampered by the occurrence of multiple gene insertions and gene silencing. On the other hand, *Agrobacterium*-mediated transformation offers a number of advantages, including integration of defined transgene and, in low copy number, preferential integration into transcriptionally active regions of the chromosome (Hu et al., 2003). Recent advances in the *Agrobacterium*-mediated transformation of cereal crops, including bread wheat (Gelvin 2003), will help introduce several genes of interest into wheat cultivars.

## 4.5.3    Improvement of Grain Quality

Wheat flour quality is determined by the quantity and quality of the gluten that imparts viscoelastic properties to dough, an important factor in bread making. Gluten is comprised of gliadins and glutenins, the storage prolamins (Shewry et al. 1989). Among the glutenins, high-molecular-weight (HMW) subunits are of special interest with regard to bread-making properties of wheat flour. Thus, a higher quantity of HMW glutenin subunits (GSs) would result in more elastic gluten polymers, and hence an improved bread-making quality of the dough (Shewry et al. 1995; Vasil and Anderson 1997). Blechl and Anderson (1996) obtained expression of a hybrid GS under the control of a native HMW-GS regulatory sequence in Bobwhite. This work suggested that the addition of HMW-GS genes raised the HMW-GS accumulation in relation to other storage proteins. Following this first report of the genetic engineering of seed storage proteins in wheat grains, expression of the *HMW-GS-1Ax1* gene was reported to increase the HMW-GS proteins in the grain (Altpeter et al. 1996). The benefit of flours with such strong dough is that they can be blended with weaker dough found in poor-quality wheat to achieve desirable bread-making quality. Transgenic durum wheat that produced one additional HMW glutenin, 1Ax1 or 1Dx5, imparted enhanced dough strength and stability, thereby changing the properties of durum flour for both bread and pasta making (He et al. 1999).

## 4.5.4    Tolerance to Abiotic Stresses

Several abiotic stresses pose a limitation on wheat yields, drought being one of the major causes of yield loss. Although some progress has been made in improving the level of drought tolerance in wheat through conventional breeding (Trethowan et al. 2001), this procedure is lengthy and tedious. Several candidate genes are known to improve a plant's response under water-limited conditions (Soderman et al. 2000). *DREB* genes cloned from *Arabidopsis*, which code for proteins that bind to dehydration-responsive elements (DRE), have been reported to improve the level of tolerance to abiotic stresses. When the DREB1A gene was introduced in wheat, the transformed $T_1$ plants survived a short and intensive water stress, while the control plants dried up (Pellegrineschi et al. 2002).

Overexpression of *E. coli* trehalose biosynthetic genes (*otsA* and *otsB*) as a fusion gene was found to result in increased tolerance of abiotic stress in rice, resulting in elevated capacity for photosynthesis under drought and low-temperature stress conditions (Garg et al. 2002). This technique holds the possibility of engineering abiotic stress tolerance into other cereal crops, including wheat. Abebe et al. (2003) have demonstrated that wheat transformed with the *mtlD* gene from *E. coli* conferred on wheat improved tolerance to water stress and salinity. It appears that transgenic technology holds promise for engineering tolerance to abiotic stresses in wheat.

## 4.6 CONCLUSION

Conventional wheat breeding, sometimes combined with interspecific and intergeneric hybridizations, has resulted in substantial improvement of grain yields and quality traits. Over the years, several supplementary tools have become available to aid breeding programs. The availability of molecular markers helps to identify QTLs for complex traits and is of great value. The discovery of gene synteny even among remotely related crop plants facilitates isolation of gene homologs from various cereal crops. Thus, progress in molecular marker and mapping technologies combined with advances in genomics and proteomics technologies holds great promise for discovering and characterizing genes, thereby speeding up wheat improvement.

Conventional breeding is very slow in transferring a desirable trait into crop cultivars. The advent of genetic engineering has provided novel techniques of rapid gene transfer into crop plants. Thus, the use of transgenic approaches to combat pests and diseases, improve tolerance to abiotic stresses, and upgrade grain quality of wheat is opening up exciting possibilities for accelerating wheat breeding programs. Hopefully, the enormous potential of the new technologies will be harnessed to the best advantage of humankind.

## ACKNOWLEDGMENT

We thank Pat Byrne, Evans Lagudah, and Steve Stack for reviewing the manuscript and for their comments.

## REFERENCES

Abebe, T., Guenzi, A.C., Martin, B., et al. 2003. Tolerance of mannitol-accumulating transgenic wheat to water stress and salinity. *Plant Physiol.* 131: 1748–1755.

Ahn, S., Anderson, J.A., Sorrells, M.E., et al. 1993. Homoeologous relationships of rice, wheat and maize chromosomes. *Mol. Gen. Genet.* 241: 483.

Akhunov, E.D., Akhunova, A.R., Linkiewicz, A.M., et al. 2003a. Synteny perturbations between wheat homoeologous chromosomes caused by locus duplications and deletions correlate with recombination rates. *Proc. Natl. Acad. Sci. U.S.A.* 100: 10836–10841.

Akhunov, E.D., Goodyear, A.W., Geng, S., et al. 2003b. The organization and rate of evolution of wheat genomes are correlated with recombination rates along chromosome arms. *Genome Res.* 13: 753–763.

Altpeter, F., Diaz, I., McAuslane, H., et al. 1999. Increased insect resistance in transgenic wheat stably expressing trypsin inhibitor CME. *Mol. Breed.* 5: 53–63.

Altpeter, F., Vasil, V., Srivastava, V., et al. 1996. Expression of a novel high-molecular-weight glutenin subunit *1Ax1* gene into wheat. *Nat. Biotechnol.* 14: 1155–1159.

Anand, A., Zhou, T., Trick, H.N., et al. 2003. Greenhouse and field testing of transgenic wheat plants stably expressing genes for thaumatin-like protein, chitinase and glucanase against *Fusarium graminearum*. *J. Exp. Bot.* 54: 1101–1111.

Anderson, G.A., Papa, D., Peng, J.H., et al. 2003. Genetic mapping of *Dn7*, a rye gene conferring resistance to the Russian wheat aphid in wheat. *Theor. Appl. Genet.* 107: 1297–1303.

Anderson, J.A. 1998. Marker-assisted selection of disease resistance genes in wheat. In *Application of Biotechnologies to Wheat Breeding*, M.M. Kohli and M. Francis, Eds. CIMMYT, La Esuanzuela, Uruguay, pp. 71–84.

The Arabidopsis Genome Sequencing Initiative. 2000. Analysis of the genome sequence of the flowering plant *Arabidopsis thaliana*. *Nature* 408: 796–815.

Becker, D., Brettschneider, R., and Lorz, H. 1994. Fertile transgenic wheat from microprojectile bombardment of scutellar tissue. *Plant. J.* 5: 299–307.

Blattner, F.R., Plunkett, G., Bloch, C.A., et al. 1997. The complete genome sequence of *Escherichia coli* K-12. *Science* 277: 1453–1462.

Blechl, A.E. and Anderson, O.D. 1996. Expression of a novel high-molecular-weight glutenin subunit gene in transgenic wheat. *Nat. Biotechnol.* 14: 875–879.

Bliffeld, M., Mundy, J., Potrykus, I., et al. 1999. Genetic engineering of wheat for increased resistance to powdery mildew disease. *Theor. Appl. Genet.* 98: 1079–1086.

Bommineni, V.R. and Jauhar, P.P. 1996. Regeneration of plantlets through isolated scutellum culture of durum wheat. *Plant Sci.* 116: 197–203.

Bommineni, V.R., Jauhar, P.P., and Peterson, T.S. 1997. Transgenic durum wheat by microprojectile bombardment of isolated scutella. *J. Hered.* 88: 475–481.

Borlaug, N.E. 1998. Feeding a world of 10 billion people: the miracle ahead. *Plant Tiss. Cult. Biotechnol.* 3: 119–127.

Borner, A., Schumann, E., Furste, A., et al. 2002. Mapping of quantitative trait loci determining agronomically important characters in hexaploid wheat (*Triticum aestivum* L.). *Theor. Appl. Genet.* 105: 921–936.

Boyko, E.V., Gill, K.S., Mickelson-Young, L., et al. 1999. A high-density genetic linkage map of *Aegilops tauschii*, the D-genome progenitor of bread wheat. *Theor. Appl. Genet.* 99: 16–26.

Bryan, G.J., Collins, A.J., Stephenson, P., Orry, A., Smith, J.B., and Gale, M.D. 1997. Isolation and characterisation of microsatellites from hexaploid bread wheat. *Theor. Appl. Genet.* 94: 557–563.

The *Caenorhabitis elegans* Consortium. 1998. Genome sequence of the nematode *C. elegans*: a platform for investigating biology. *Science* 282: 2012–2018.

Cenci, A., Chantret, N., Kong, X., et al. 2003. Construction and characterization of a half million BAC library of durum wheat (*Triticum turgidum* ssp. *durum*). *Theor. Appl. Genet.* 107: 931–939.

Chen, W.P., Chen, P.D., Liu, D.J., et al. 1999. Development of wheat scab symptoms is delayed in transgenic wheat plants that constitutively express a rice thaumatin-like protein gene. *Theor. Appl. Genet.* 99: 755–760.

Cheng, M., Fry, J.E., Pang, S., Zhou, H., Hironaka, C.M., Duncan, D.R., Conner, T.W., and Wan, Y. 1997. Genetic transformation of wheat mediated by *Agrobacterium tumefaciens*. *Plant Physiol.* 115: 971–980.

Cloutier, S., Beimcik, E., Gusdal, D., et al. 2002. A Canadian wheat EST initiative. In *Plant and Animal Genome X*. Scherago International, San Diego, p. 90.

Dahleen, L.S., Okubara, P.A., and Blechl, A.E. 2001. Transgenic approaches to combat Fusarium head blight in wheat and barley. *Crop. Sci.* 41: 628–637.

Delaney, D.E., Nasuda, S., Endo, T.R., et al. 1995. Cytologically based physical maps of the group-2 chromosomes of wheat. *Theor. Appl. Genet.* 91: 568–573.

Distelfeld, A., Uauy, C., Olmos, S., et al. 2004. Microcolinearity between a 2-cM region encompassing the grain protein content locus *Gpc-6B1* on wheat chromosome 6B and a 350-kb region on rice chromosome 2. *Funct. Integr. Genomics* 4: 59–66.

Dubcovsky, J. 2002. A Critical Look at Genomics: Will It Really Be Useful to Crop Breeding? Crop Science Society of America, USDA/CSREES, CSSA Div. C-1, Div. C-7, Indianapolis, Indiana.

Dubcovsky, J., Tranquilli, G., Lijavetzky, D., et al. 1998. Advances in molecular markers for bread making quality. In *Application of Biotechnologies to Wheat Breeding*, M.M. Kohli and M. Francis, Eds. CIMMYT, La Esuanzuela, Uruguay, pp. 57–70.

Dvořák, J. 1998. Genome analysis in the *Triticum-Aegilops* alliance. In *The 9th International Wheat Genetics Symposium*, University of Saskatchewan, Saskatoon, Canada, pp. 8–11.

Dweikat, I., Zhang, W., and Ohm, H. 2002. Development of STS markers linked to Hessian fly resistance gene H6 in wheat. *Theor. Appl. Genet.* 105: 766–770.

Eagles, H.A., Bariana, H.S., Ogbonnaya, F.C., et al. 2001. Implementation of markers in Australian wheat breeding. *Austr. J. Agric. Res.* 52: 1349–1356.

FAO 2001. FAOSTAT agricultural database. Food and Agriculture Organization of the United Nations (FAO). Available at http://fao.org/.

Faris, J.D., Fellers, J.P., Brooks, S.A., et al. 2003. A bacterial artificial chromosome contig spanning the major domestication locus *Q* in wheat and identification of a candidate gene. *Genetics* 164: 311–321.

Feuillet, C., Travella, S., Stein, N., et al. 2003. Map-based isolation of the leaf rust disease resistance gene *Lr10* from the hexaploid wheat (*Triticum aestivum* L.) genome. *Proc. Natl. Acad. Sci. U.S.A.* 100: 15253–15258.

Gale, M.D. and Devos, K.M. 1998. Plant comparative genetics after 10 years. *Science* 282: 656–659.

Garg, A.K., Kim, J.K., Owens, T.G., et al. 2002. Trehalose accumulation in rice plants confers high tolerance levels to different abiotic stresses. *Proc. Natl. Acad. Sci. U.S.A.* 99: 15898–15903.

Gaut, B.S. 2002. Evolutionary dynamics of grass genomes. *New Phytol.* 154: 15–28.

Gelvin, S.B. 2003. *Agrobacteroium*-mediated plant transformation: the biology behind the "gene-jockeying" tool. *Microbiol. Mol. Biol. Rev.* 67: 16–37.

Gill, K.S., Lubbers, E.L., Gill, B.S., et al. 1991. A genetic linkage map of *Triticum tauschii* (DD) and its relationship to the D genome of bread wheat (AABBDD). *Genome* 34: 362–374.

Goff, S.A., Ricke, D., Lan, T.-H., et al. 2002. A draft sequence of the rice genome (*Oryza sativa* L. ssp. *japonica*). *Science* 296: 92–100.

Gupta, P.K., Varshney, R.K., Sharma, P.C., et al. 1999. Molecular markers and their applications in wheat breeding. *Plant Breed.* 118: 369–390.

Harker, N., Rampling, L.R., Shariflou, M.R., et al. 2001. Microsatellites as markers for Australian wheat improvement. *Austr. J. Agric. Res.* 52: 1121–1130.

Harris, T. 2000. Genetics, genomics, and drug discovery. *Med. Res. Rev.* 20: 203–211.

He, G.Y., Rooke, L., Steele, S., et al. 1999. Transformation of pasta wheat (*Triticum turgidum* L. var. *durum*) with high-molecular-weight glutenin subunit genes and modification of dough functionality. *Mol. Breed.* 5: 377–386.

Hu, T., Metz, S., Chay, C., Zhou, H.P., Biest, N., Chen, G., Cheng, M., Feng, X., Radipnenko, M., Lu, F., and Fry, J. 2003. *Agrobacterium*-mediated large-scale transformation of wheat (*Triticum aestivum* L.) using glyphosate selection. *Plant Cell Rep.* 21: 1010–1019.

Hu, J. and Vick, B. 2003. Target region amplification polymorphism: a novel marker technique for plant genotyping. *Plant Mol. Bio. Rep.* 21: 289–294.

Huang, L., Brooks, S.A., Li, W., et al. 2003. Map-based cloning of leaf rust resistance gene *Lr21* from the large and polyploid genome of bread wheat. *Genetics* 164: 655–664.

Huang, S., Sirikhachornkit, A., Su, X., et al. 2002. Genes encoding plastid acetyl-CoA-carboxylase and 3-phosphoglycerate kinase of the *Triticum/Aegilops* complex and the evolutionary history of polyploid wheat. *Proc. Natl. Acad. Sci. U.S.A.* 99: 8133–8138.

Janakiraman, V., Steinau, M., McCoy, S.B., et al. 2002. Recent advances in wheat transformation. *In Vitro Cell. Dev. Biol. Plant* 38: 404–414.

Janda, J., Bartos, J., Safar, J., et al. 2004. Construction of a subgenomic BAC library specific for chromosomes 1D, 4D, and 6D of hexaploid wheat. *Theor. Appl. Genet.,* 109: 1337–1345.

Jauhar, P.P. 2003a. Formation of 2n gametes in durum wheat haploids: sexual polyploidization. *Euphytica* 133: 81–94.

Jauhar, P.P. 2003b. Genetics of crop improvement: chromosome engineering. In *Encyclopedia of Applied Plant Sciences*, B. Thomas, D.J. Murphy, and B. Murray, Eds. Academic Press, London, U.K. pp. 167–179.

Jauhar, P.P. and Chibbar, R.N. 1999. Chromosome-mediated and direct gene transfers in wheat. *Genome* 42: 570–583.

Jauhar, P.P. and Khush, G.S. 2002. Importance of biotechnology in global food security. In *Food Security and Environment Quality: A Global Perspective*, R. Lal, D.O. Hansen, N. Uphoff, and S. Slack, Eds. CRC Press, Boca Raton, FL, pp. 107–128.

Jauhar, P.P. and Peterson, T.S. 2000. Progress in producing scab-resistant germplasm of durum wheat. In *Proceedings of the International Symposium on Wheat Improvement for Scab Resistance*, Suzhou and Nanjing, China, pp. 77–81.

Jauhar, P.P. and Peterson, T.S. 2001. Hybrids between durum wheat and *Thinopyrum junceiforme*: prospects for breeding for scab resistance. *Euphytica* 118: 127–136.

Jauhar, P.P., Riera-Lizarazu, O., Dewey, W.G., et al. 1991. Chromosome pairing relationships among the A, B, and D genomes of bread wheat. *Theor. Appl. Genet.* 82: 441–449.

Kantety, R.V., La Rota, M., Matthews, D.E., and Sorrells, M. 2002. Data mining for simple sequence repeats in expressed sequence tags from barley, maize, rice, sorghum and wheat. *Plant Mol. Biol.* 48: 501–510.

Kornberg, T.B. and Krasnow, M.A. 2000. The *Drosophila* genome sequence: implications for biology and medicine. *Science* 287: 2218–2220.

Lagudah, E.S., Moullet, O., and Appels, R. 1997. Map-based cloning of a gene sequence encoding a nucleotide-binding domain and a leucine-rich region at the *Cre3* nematode resistance locus of wheat. *Genome* 40: 659–665.

Lazo, G.R., Chao, S., Hummel, D., et al. 2004. Development of an expressed sequence tag (EST) resource for wheat (*Triticum aestivum* L.): EST generation, unigene analysis, probe selection, and bioinformatics for a 16,000-locus bin-delineated map. *Genetics*, 168: 585–593.

Li, C., Ni, P., Francki, M., et al. 2004. Genes controlling seed dormancy and pre-harvest sprouting in a rice-wheat-barley comparison. *Funct. Integr. Genomics* 4: 84–93.

Li, W.L. and Gill, B.S. 2002. The colinearity of the *Sh2/A1* orthologous region in rice, sorghum, and maize is interrupted and accompanied by genome expansion in the *Triticeae*. *Genetics* 160: 1153–1162.

Lijavetzky, D., Muzzi, G., Wicker, T., et al. 1999. Construction and characterization of a bacterial artificial chromosome (BAC) library for the A genome of wheat. *Genome* 42: 1176–1182.

Ling, H.Q., Zhu, Y., and Keller, B. 2003. High-resolution mapping of the leaf rust disease resistance gene *Lr1* in wheat and characterization of BAC clones from the *Lr1* locus. *Theor. Appl. Genet.* 106: 875–882.

Liu, S. and Anderson, J.A. 2003. Targeted molecular mapping of a major wheat QTL for Fusarium head blight resistance using wheat ESTs and synteny with rice. *Genome* 46: 817–823.

Lukowitz, W., Gillmor, C.S., and Scheible, W.R. 2000. Positional cloning in *Arabidopsis*. Why it feels good to have a genome initiative working for you. *Plant Physiol.* 123: 795–805.

Ma, Z., Weining, S., Sharp, P.J., et al. 2000. Non-gridded library: a new approach for BAC (bacterial artificial chromosome) exploitation in hexaploid wheat (*Triticum aestivum*). *Nucleic Acids Res.* 28: E106.

Ma, Z.Q., Saidi, A., Quick, J.S., et al. 1998. Genetic mapping of Russian wheat aphid resistance genes *Dn2* and *Dn4* in wheat. *Genome* 41: 303–306.

Mochida, K., Yamazaki, Y., and Ogihara, Y. 2003. Discrimination of homoeologous gene expression in the hexaploid wheat by SNPs analysis of contigs grouped from a large number of expressed sequence tags. *Mol. Genet. Genomics* 270: 371–377.

Moullet, O., Zhang, H.-B., and Lagudah, E.S. 1999. Construction and characterization of a large DNA insert library from the D genome of wheat. *Theor. Appl. Genet.* 99: 303–313.

Nagaoka, T. and Ogihara, Y. 1997. Applicability of inter-simple sequence repeat polymorphisms in wheat for use as DNA markers in comparison to RFLP and RAPD markers. *Theor. Appl. Genet.* 94: 597–602.

Nehra, N., Chibbar, R., Leung, N., et al. 1994. Self-fertile transgenic wheat plants regenerated from isolated scutellar tissues following microprojectile bombardment with two distinct gene constructs. *Plant J.* 5: 285–297.

Nelson, J.C., Sorrells, M.E., Van Deynze, A.E., et al. 1995. Molecular mapping of wheat: major genes and rearrangements in homoeologous groups 4, 5, and 7. *Genetics* 141: 721–731.

Nicot, N., Chiquet, V., Gandon, B., et al. 2004. Study of simple sequence repeat (SSR) markers from wheat expressed sequence tags (ESTs). *Theor. Appl. Genet.* 109: 800–805.

Nilmalgoda, S.D., Cloutier, S., and Walichnowski, A.Z. 2003. Construction and characterization of a bacterial artificial chromosome (BAC) library of hexaploid wheat (*Triticum aestivum* L.) and validation of genome coverage using locus-specific primers. *Genome* 46: 870–878.

Pellegrineschi, A., Brito, R.M., Velazquez, L., et al. 2002. The effect of pretreatment with mild heat and drought stresses on the explant and biolistic transformation frequency of three durum wheat cultivars. *Plant Cell. Rep.* 20: 955–960.

Peng, J., Korol, A.B., Fahima, T., et al. 2000. Molecular genetic maps in wild emmer wheat, *Triticum dicoccoides*: genome-wide coverage, massive negative interference, and putative quasi-linkage. *Genome Res.* 10: 1509–1531.

Peng, J. and Lapitan, N.L.V. 2005. Characterization of EST-derived microsatellites in the wheat genome and development of eSSR markers. *Funct. Integr. Genomics*, 5: 80–96.

Peng, J., Ronin, Y., Fahima, T., et al. 2003. Domestication quantitative trait loci in *Triticum dicoccoides*, the progenitor of wheat. *Proc. Natl. Acad. Sci. U.S.A.* 100: 2489–2494.

Pingali, P.L. and Rajaram, S. 1999. Global wheat research in a changing world: options for sustaining growth in wheat productivity. In *Global Wheat Research in a Changing World: Challenges and Achievements*, Pingali, P.L., Ed. CIMMYT, Mexico, pp. 1–18.

Qi, L., Echalier, B., Chao, S., et al. 2004. A chromosome bin map of 16,000 expressed sequence tag loci and distribution of genes among the three genomes of polyploid wheat. *Genetics* 168: 701–712.

Repellin, A., Båga, M., Jauhar, P.P., et al. 2001. Genetic enrichment of cereal crops via alien gene transfer: new challenges. *Plant Cell. Tiss. Org. Cult.* 64: 159–183.

Roder, M.S., Korzun, V., Wendehake, K., et al. 1998. A microsatellite map of wheat. *Genetics* 149: 2007–2023.

Safar, J., Bartos, J., Janda, J., et al. 2004. Dissecting large and complex genomes: flow sorting and BAC cloning of individual chromosomes from bread wheat. *Plant J.* 2004: 960–968.

Sandhu, D., Champoux, J.A., Bondareva, S.N., et al. 2001. Identification and physical localization of useful genes and markers to a major gene-rich region on wheat group 1S chromosomes. *Genetics* 157: 1735–1747.

Shewry, P.R., Halford, N.G., and Tatham, A.S. 1989. The high molecular weight subunits of wheat, barley and rye: genetics, molecular biology, chemistry and role in wheat gluten structure and functionality. In *Oxford Surveys of Plant Molecular and Cell Biology*, B.J. Miflin, Ed. Oxford University Press, Oxford, pp. 163–219.

Shewry, P.R., Tantham, A.S., Barro, F, et al. 1995. Biotechnology of breadmaking: unraveling and manipulating the multi-protein gluten complex. *Biotechnology* 13: 1185–1190.

Söderman, E.M., Brocard, I.M., Lynch, T.J., et al. 2000. Regulation and function of the Arabidopsis *ABA-insensitive4* gene in seed and abscisic acid response signaling networks. *Plant Physiol.* 124: 1752–1765.

Somers, D.J., Isaac, P., and Edwards, K. 2004. A high-density microsatellite consensus map for bread wheat (*Triticum aestivum* L.). *Theor. Appl. Genet.* 109: 1105–1114.

Somers, D.J., Kirkpatrick, K.R., Moniwa, M., et al. 2003. Mining single-nucleotide polymorphisms from hexaploid wheat ESTs. *Genome* 46: 431–437.

Sorrells, M.E., La Rota, M., Bermudez-Kandianis, C.E., et al. 2003. Comparative DNA sequence analysis of wheat and rice genomes. *Genome Res.* 13: 1818–1827.

Stein, N., Feuillet, C., Wicker, T., et al. 2000. Subgenome chromosome walking in wheat: a 450-kb physical contig in *Triticum monococcum* L. spans the *Lr10* resistance locus in hexaploid wheat (*Triticum aestivum* L.). *Proc. Natl. Acad. Sci. U.S.A.* 97: 13436–13441.

Steiner, B., Lemmens, M., Griesser, M., et al. 2004. Molecular mapping of resistance to Fusarium head blight in the spring wheat cultivar Frontana. *Theor. Appl. Genet.* 109: 215–224.

Tanksley, S.D. 1983. Molecular markers in plant breeding. *Plant Mol. Biol. Rep.* 1: 3–8.

The Human Genome Sequencing Consortium. 2001. Initial sequencing and analysis of the human genome. *Nature* 409: 860–921.

Trethowan, R.M., Crossa, J., van Ginkel, M., et al. 2001. Relationships among bread wheat international yield testing locations in dry areas. *Crop Sci.* 41: 1461–1469.

Useche, F.J., Gao, G., Hanafrey, M., et al. 2001. High-throughput identification, database storage and analysis of SNPs in EST sequences. *Genome Informatics* 12: 194–203.

Vasil, I.K. and Anderson, O.D. 1997. Genetic engineering of wheat gluten. *Trends Plant Sci.* 2: 292–297.

Vasil, V., Castillo, A.M., Fromm, M.E., et al. 1992. Herbicide resistant fertile transgenic wheat plants obtained by microprojectile bombardment of regenerable embryogenic callus. *Bio/Technology* 10: 667–674.

Venter, J.C., Adams, M.D., Myers, E.W., et al. 2001. The sequence of the human genome. *Science* 291: 1304–1351.

Weeks, J.T., Anderson, O.D., and Blechl, A.E. 1993. Rapid production of multiple independent lines of fertile transgenic wheat (*Triticum aestivum*). *Plant Physiol.* 102: 1077–1084.

Xu, S.S., Hu, J., and Faris, J.D. 2003. Molecular characterization of the Langdon durum-*Triticum dicoccoides* chromosome substitution lines using TRAP (target region amplification polymorphism) markers. In *10th International Wheat Genetics Symposium*, Paestum, Italy, pp. 91–94.

Yahiaoui, N., Srichumpa, P., Dudler, R., et al. 2004. Genome analysis at different ploidy levels allows cloning of the powdery mildew resistance gene *Pm3b* from hexaploid wheat. *Plant J.* 37: 528–538.

Yan, L., Loukoianov, A., Blechl, A., et al. 2004. The wheat *Vrn2* gene is a flowering repressor down-regulated by vernalization. *Science* 303: 1607–1644.

Yan, L., Loukoianov, A., Tranquilli, G., et al. 2003. Positional cloning of the wheat vernalization gene *Vrn1*. *Proc. Natl. Acad. Sci. U.S.A.* 100: 6263–6268.

Yu, J., Hu, S., Wang, J., et al. 2002. A draft sequence of the rice genome (*Oryza sativa* L. ssp. *indica*). *Science* 296: 79–92.

Yu, J.K., La Rota, M., Kantety, R.V., et al. 2004. EST derived SSR markers for comparative mapping in wheat and rice. *Mol. Genet. Genomics* 271: 742–751.

Zhang, D., Choi, D.W., Wanamaker, S., et al. 2004. Construction and evaluation of cDNA libraries for large-scale EST sequencing in wheat (*Triticum aestivum* L.). *Genetics*, 168: 595–608.

Zhu, L.C., Smith, C.M., Fritz, A., et al. 2004. Genetic analysis and molecular mapping of a wheat gene conferring tolerance to the greenbug (*Schizapis graminum* Rondani). *Theor. Appl. Genet.* 109: 289–293.

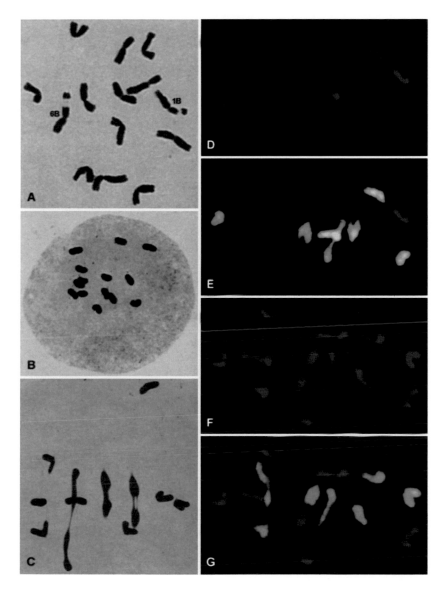

**Figure 1.4** Somatic (A) and meiotic (B–G) chromosomes of durum haploids derived by hybridization with maize: a pollen mother cell (PMC) from a haploid with *Ph1* (B) and PMCs from haploids without *Ph1* showing high homoeologous pairing (C–G). A: 14 somatic chromosomes; note one dose each of the satellited chromosomes 1B and 6B. B: A PMC with 14 univalents; note the total absence of pairing in the presence of *Ph1*. C: Four bivalents (two ring II and two rod II) + six univalents; note high pairing in the absence of *Ph1*. D, E: Fluorescent genomic *in situ* hybridization (fl-GISH) analysis of chromosome pairing: two ring II + one rod II counterstained with propidium iodide (PI) (D) and the same cell as D probed with biotinylated A genome probe (E); the preparation was blocked with the genomic DNA of *Ae. speltoides* (B genome) and the probe was detected with FITC. The A genome chromosomes are brightly lit in green color. Note pairing between the A genome chromosomes and the B genome chromosomes. F, G: Fl-GISH analysis of chromosome pairing: one ring II + two rod II counterstained with PI (F) and same cell as F after probing with the A genome probe (G). Note the intergenomic ring bivalent involving an A- and a B-genome chromosome, an intergenomic rod bivalent, and an intragenomic (within the A genome) bivalent.

**Figure 1.8** A collection of maize cobs showing a wide range of diversity. (From www.tropag-fieldtrip.cornell.edu/tradag/Maize.jpg.)

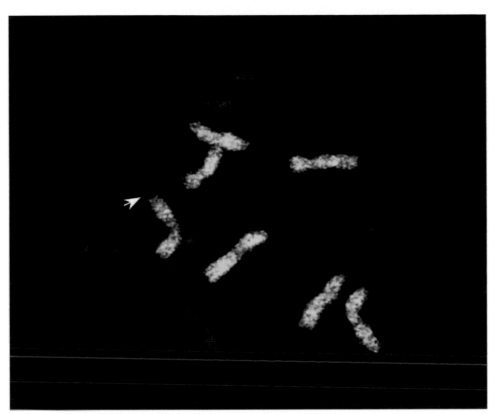

**Figure 2.3**   Fourteen somatic chromosomes of a substitution haploid (derived from a D genome disomic substitution line 5D(5B) of Langdon durum wheat) after fluorescent genomic *in situ* hybridization. The brightly fluorescing (bright yellow) chromosomes belong to the A genome. (Chromosome 5D, partially hybridized with the A genome probe but masked by the propidium iodide counterstain, is not clearly seen in the photograph.) The six B genome chromosomes are in red. Note one dose of the 4A/7B translocation chromosome (arrow), which is a part of the evolutionary translocation 4A/5A/7B present both in durum and in common wheat; in this translocated chromosome, the distal segment of 7BS constitutes approximately 24% of the long arm of 4A. The 5A chromosome segment cannot be visualized because it stains the same as the 4A segment. (From Doğramacı-Altuntepe, M. and Jauhar, P.P., *Genome*, 44, 137–142, 2001.)

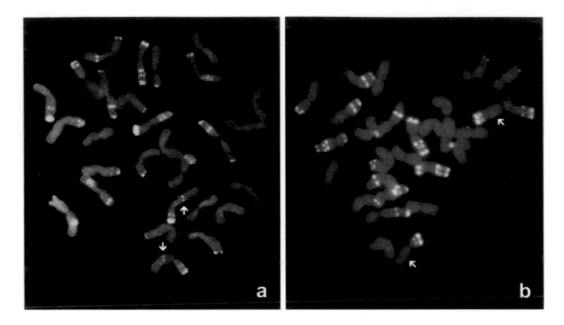

**Figure 2.4** FISH with the highly repeated pSc119.2 sequence as probe of both normal *Ph1* (a) and *ph1c* mutant (b) of durum wheat cv. Creso. Arrows point to the pSc119.2 site, which is present on wheat 5BL arm (a) while lacking in the *ph1c* mutant (b).

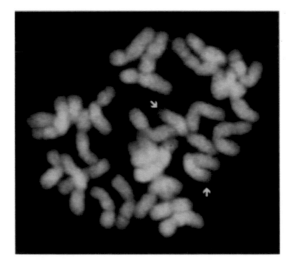

**Figure 2.5** GISH, preceded by preannealing of total genomic DNAs of the donor and recipient species, of a chromosomally engineered durum wheat line homozygous for a distal *Thinopyrum ponticum* segment. This segment, spanning 23% of the 7AL/7AgL recombinant arm, contains the *Lr19* + *Yp* alien genes (see Section 2.4.2.2.2).

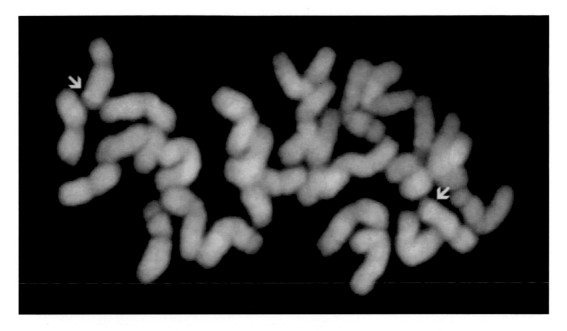

**Figure 2.6** Preannealing GISH of D and AB total genomic DNAs highlights an interstitial 1DL common wheat segment (arrowed), containing the *Glu-D1* locus, transferred into 1AL of durum wheat by chromosome engineering (see Section 2.4.2.2.3).

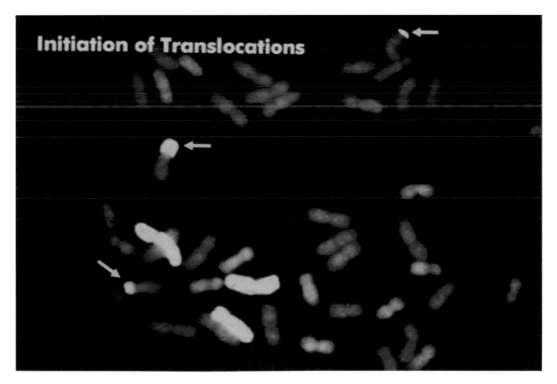

**Figure 3.9** FISH of a *phph* backcross derivative showing wheat–alien chromosome translocations being initiated (the combination is *T. aestivum–Th. bessarabicum*).

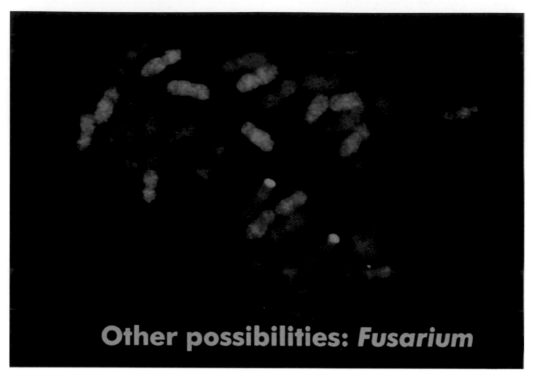

**Figure 3.10** FISH of a translocation homozygote derived from the *phph* backcross plants with a hypoploid count (Figure 3.9) to yield a 2n = 6x = 42 translocation derivative that is *PhPh*.

**Figure 3.11** A spot blotch-resistant advanced derivative from an intergeneric wheat–*Th. curvifolium* combination (right) compared to a bread wheat-susceptible cultivar (left) based upon field screening in a disease hot site in Mexico. The line Chirya was released as the variety Azubiciat in Bolivia.

**Figure 3.12** Spot blotch resistance of a wheat/synthetic hexaploid advanced derivative compared to a susceptible entry.

**Figure 4.2** Rice genome view showing the wheat chromosome arm location for the most similar wheat gene sequences. Each colored box represents a rice–wheat gene sequence match at ≥80% identity. When the wheat EST mapped to more than one wheat chromosome, the other color coded locations adjacent to the first. Homoeologous wheat chromosome locations are grouped together. Rice BAC/PAC sequences that did not match any wheat sequences, as well as redundant sequences, are omitted. The rice centromere location is indicated by C. (From http://wheat.pw.usda.gov/pubs/2003/Sorrells/. Based on Sorrells, M.E., La Rota, M., Bermudez-Kandianis, C.E., et al., *Genome Res.*, 13, 1818–1827, 2003. With permission from Dr. Sorrells.)

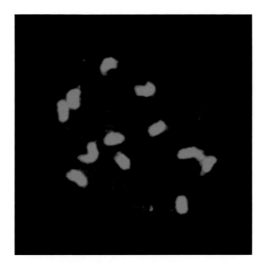

**Figure 5.2** Characterization of parental genomes in wide hybrids of *Oryza* through GISH. Metaphase cell showing 36 chromosomes in F1 of *O. australiensis* (2n = 24,EE) × *O. ridleyi* (2n = 48; HHJJ). Genomic DNA of *O. australiensis* labeled with biotin14–dATP was used as a probe. The chromosomes of two parental species are clearly differentiated: the 12 labeled chromosomes of *O. australiensis* flouresced yellow due to FITC, while 24 unlabeled chromosomes of *O. ridleyi* appear red due to counterstaining with PI.

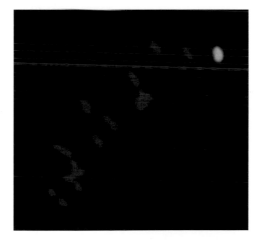

**Figure 5.3** Characterization of wild species chromosome through GISH in MAAL of rice. Metaphase I showing 12II of *O. sativa* and one extra chromosome of *O. ridleyi*. Genomic DNA of *O. ridleyi* labeled with biotin14–dATP was used as a probe. The single chromosome of *O. ridleyi* flouresced yellow due to FITC, while all chromosomes (12II) of *O. sativa* appeared red due to counterstaining with PI.

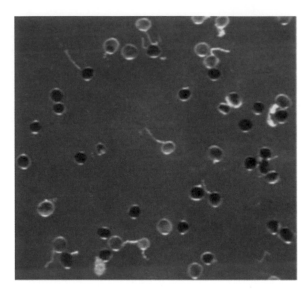

**Figure 6.2**  Pollen germination assay in plants heterozygous for a mutation and a *waxy1*-tagged chromosome translocation (*Wx1;rop2*/T4-9b *wx1;rop2+*). The waxy phenotype was observed by staining pollen with an iodine solution after pollen grains were permitted to germinate. Pollen grains carrying the *wx1* and mutant *rop2* alleles stained brown (no amylose), while pollen grains carrying the *Wx1* and wild-type *rop2+* alleles stained black (amylose present). *Wx1* pollen grains (black) showed reduction in germination linked to mutant *rop2* compared to *wx1* grains (brown). (This photograph was graciously provided by John E. Fowler and Kirstin M. Arthur, Oregon State University.)

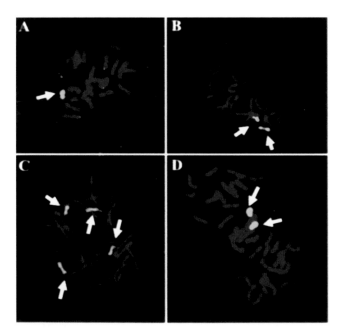

**Figure 6.3**  Genomic *in situ* hybridization (GISH) of partial oat × maize hybrids and a disomic maize chromosome addition line of oat. Unlabeled red oat chromosomes were counterstained with propidium iodide, while fluorescein-labeled maize genomic DNA was used as the probe. Yellow-green fluorescence marks maize chromatin (white arrows). (A) GISH of a chromosome spread of a 22-chromosome oat × maize hybrid with 21 oat (red) and 1 maize chromosome (yellow). (B) GISH of a chromosome spread of a 23-chromosome individual showing 21 oat (red) and 2 maize (yellow) chromosomes. (C) GISH of a chromosome spread of a 25-chromosome individual showing 21 oat (red) and 4 maize (yellow) chromosomes. (D) GISH of a chromosome spread of a 44-chromosome individual with 42 oat chromosomes (red) and a pair of maize chromosomes (yellow).

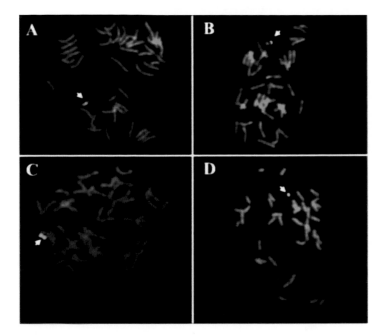

**Figure 6.4** Fluorescence *in situ* hybridization (FISH) of four maize chromosome 9 radiation hybrid lines. Unlabeled blue oat chromosomes were counterstained with DAPI. Fluorescein-labeled maize genomic DNA (yellow-green) was used as the probe for genomic *in situ* hybridization in C and D. Yellow-green fluorescence marks maize chromatin. Multicolor fluorescence *in situ* hybridization using two maize-specific repetitive probes (180-bp knob-specific sequence and the maize-specific multiprobe) was used in A and B. Red fluorescence marks the 180-bp knob homologous sequences, and yellow fluorescence marks sequences homologous to the maize-specific multiprobe. (A) A line with an intergenomic translocation (arrowhead). (B) A line with a complex rearrangement that involved at least one deletion, an inversion, plus an intergenomic translocation (arrowhead). (C) A line with an isochromosome (arrowhead). (D) A line with a diminutive maize chromosome 9 that resulted from multiple radiation-induced deletions (arrowhead).

**Figure 7.1** Photographs of wild-weedy oat species. (Top) *A. murphyi* growing in southern Spain. (Bottom) *A. barbata* and *A. sterilis* growing with wild *Hordeum* in Izmir Province, Turkey.

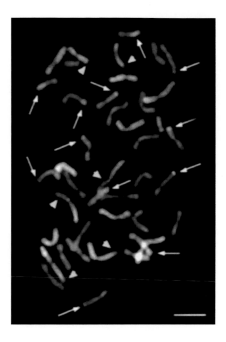

**Figure 7.5** Mitotic metaphase chromosome spread of *A. sativa* variety Sun II probed with rhodamine-labeled total genomic DNA derived from the C genome diploid *A. eriantha* and counterstained with DAPI. The pink fluorochrome indicates the C genome chromatin, while blue counterstain indicates chromatin from the A and D genomes. White arrows point to C genome translocations on A or D genome chromosomes, and yellow arrowheads point to A/D genome translocations on C genome chromosomes. Bar = 10 microns.

**Figure 8.1** Spikes of wheat–barley disomics addition lines. From left to right: Chinese Spring wheat, Betzes barley, 2H (E), 3H (F), 4H (A), 5H (B), 6H (C), and 7H (D). (Homoeologous designation is from R. Islam, personal communication, 2003. Letters in parentheses are from Islam, A.K.M.R. et al., *Heredity*, 46, 161–174, 1981. With permission.)

CS Wheat     1H/6H     1H/1HS

**Figure 8.2**  Mature spike of Chinese Spring wheat (left), double monosomic wheat–barley addition line having barley 1H and 6H (center), and disomic-monotelodisomic wheat–barley addition line having a pair of barley 6H and a heteromorphic 1H/1HS pair. (From Islam, A.K.M.R. and Shepherd, K.W., *Euphytica*, 111, 145–149, 2000. With permission.)

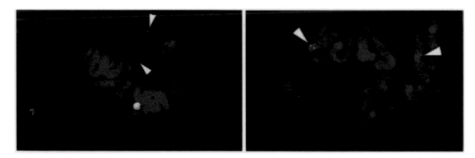

**Figure 8.3**  FISH of inserted foreign genes (*sgfp* (S65T) or *uidA*)) in metaphase chromosomes of transgenic barley plants. (A) Two *sgfp* (S65T) signals in a homozygous $T_1$ diploid plant. (B) Two *uidA* signals in a homozygous $T_1$ diploid plant. Arrowheads show signals. (Choi, H.W. et al., *Theor. Appl. Genet.*, 106, 92–100, 2002. With permission.)

**Figure 11.1**  Genetic variability in sorghum is shown by a wide range of panicle colors.

**Figure 11.2**  (Left) Technique to screen sorghum lines resistant to midge. (Right) An improved sorghum cultivar resistant to midge produced by ICRISAT.

**Figure 11.3**  (Left) Screening of grain mold in sorghum by sprinkler irrigation method. (Right) An ICRISAT public sector-bred grain mold-resistant sorghum variety.

# CHAPTER 5

# Cytogenetic Manipulation and Germplasm Enhancement of Rice (*Oryza sativa* L.)

D.S. Brar and G.S. Khush

## CONTENTS

5.1   Introduction ........................................................................................................117
5.2   Origin of Cultivated Rice ...................................................................................118
        5.2.1   Wild Progenitors of Cultivated Rice .......................................................118
        5.2.2   Polyphyletic Origin of *O. sativa* ...........................................................121
5.3   Rice Genetic Resources: Exploration and Conservation .....................................123
        5.3.1   Exploration and Collection of Germplasm ...............................................123
        5.3.2   Genetic Erosion ......................................................................................123
        5.3.3   Conservation of Germplasm ...................................................................124
                   5.3.3.1   *Ex Situ* Conservation ...............................................................124
                   5.3.3.2   *In Situ* Conservation .................................................................125
5.4   Taxonomy ...........................................................................................................125
        5.4.1   *O. sativa* Complex ..................................................................................125
        5.4.2   *Oryza officinalis* Complex .....................................................................126
        5.4.3   *Oryza ridleyi* Complex ..........................................................................126
        5.4.4   *Oryza meyeriana* Complex .....................................................................126
        5.4.5   Unknown Complex ..................................................................................127
                   5.4.5.1   *Oryza brachyantha* ...................................................................127
                   5.4.5.2   *Oryza schlechteri* .....................................................................127
        5.4.6   Related Genera .......................................................................................127
        5.4.7   Genomic Relationships ...........................................................................127
                   5.4.7.1   Molecular Approaches .................................................................128
                   5.4.7.2   Seed Protein Analysis .................................................................128
5.5   Cytogenetics .......................................................................................................128
        5.5.1   Somatic Karyotype ..................................................................................129
        5.5.2   Pachytene Karyotype ..............................................................................129
        5.5.3   Translocations .........................................................................................129
        5.5.4   Haploids, Triploids, and Aneuploids ......................................................130
        5.5.5   Trisomics ................................................................................................130
                   5.5.5.1   Primary Trisomics .......................................................................130

5.5.5.2    Secondary and Telotrisomics ................................................................130
5.5.5.3    Monosomic Alien Addition Lines................................................................130
5.5.5.4    Comparative Genetic Maps................................................................131
5.6    Germplasm Enhancement through Interspecific Hybridization ................................132
   5.6.1    Strategies for Alien Gene Transfer ................................................................132
      5.6.1.1    Search for Useful Genetic Variability for Target Traits ....................132
      5.6.1.2    Production of Hybrids and Introgression Lines ................................133
      5.6.1.3    Evaluation of Introgression Lines for Transfer of Target Traits............133
      5.6.1.4    Molecular Characterization of Introgression and Construction of
              Alien Chromosome Segment Substitution Lines ....................................133
      5.6.1.5    Mapping of Introgressed Alien Genes/QTL................................................133
      5.6.1.6    Characterization of Wide Cross Progenies Using GISH........................133
   5.6.2    Production of Interspecific Hybrids, MAALs, and Advanced-Backcross
      Progenies................................................................133
   5.6.3    Introgression from AA Genome Wild Species ................................................134
      5.6.3.1    Introgression of Gene(s) for Resistance to Grassy Stunt Virus..............134
      5.6.3.2    Introgression of Gene(s) for Resistance to Tungro Disease ....................135
      5.6.3.3    Introgression of Gene(s) for Resistance to Bacterial Blight....................136
      5.6.3.4    Incorporation of CMS Sources from Wild Species................................136
      5.6.3.5    Introgression for Tolerance to Abiotic Stresses................................136
      5.6.3.6    Introgression from African rice (*O. glaberrima*) into Asian Rice
              (*O. sativa*)................................................................137
   5.6.4    Development of Doubled Haploids from *O. sativa* × *O. glaberrima*....................137
   5.6.5    Construction of Chromosome Segment Substitution Lines of *O. glaberrima*
      and *O. rufipogon* in the Background of *O. sativa*................................137
   5.6.6    Identification and Introgression of Yield-Enhancing Loci/QTL Wild
      Species Alleles from AA Genome Species................................................138
   5.6.7    Introgression of Genes from Distantly Related Genomes................................138
      5.6.7.1    Introgression from the CC Genome Species................................139
      5.6.7.2    Introgression from the BBCC Genome Species ................................139
      5.6.7.3    Introgression from the CCDD Genome Species ................................139
      5.6.7.4    Introgression from the EE Genome Species ................................139
      5.6.7.5    Introgression from the FF Genome Species................................140
      5.6.7.6    Introgression from the GG Genome Species ................................140
      5.6.7.7    Introgression from the HHJJ Genome Species ................................140
      5.6.7.8    Introgression from the HHKK Genome Species................................140
   5.6.8    Molecular Mapping of Introgressed Alien Genes ................................140
      5.6.8.1    Mapping of the *Xa21* Gene for Bacterial Blight Resistance ..................141
      5.6.8.2    Mapping of Genes for Tungro Tolerance ................................141
      5.6.8.3    Mapping of Genes for BPH Resistance ................................141
      5.6.8.4    Mapping of *Pi9t* for Blast Resistance ................................141
      5.6.8.5    Mapping QTL for Tolerance to Aluminum Toxicity................................141
   5.6.9    Molecular Characterization of Alien Introgression ................................142
      5.6.9.1    Introgression from *O. glaberrima* ................................142
      5.6.9.2    Introgression from Distant Genomes of *Oryza* ................................142
   5.6.10  Characterization of Parental Genomes, MAALs, and Homoeologous Pairing
      in *Oryza* through GISH................................................................142
5.7    Germplasm Enhancement through Cell and Tissue Culture ................................144
   5.7.1    Anther Culture................................................................144
   5.7.2    Somaclonal Variation................................................................145
      5.7.2.1    Enhancing Alien Introgression through Somaclonal Variation..............145

|  | 5.7.3 | Somatic Cell Hybridization | 146 |
|  |  | 5.7.3.1 Production of Cybrids and Organelle Recombinants | 147 |
|  |  | 5.7.3.2 Transfer of Cytoplasmic Male Sterility | 147 |
| 5.8 | Germplasm Enhancement through Induced Mutations | | 147 |
| 5.9 | Apomixis for Germplasm Enhancement | | 148 |
|  | 5.9.1 | Screening Wild *Oryza* Germplasm for Apomixis | 148 |
|  | 5.9.2 | Mutagenesis-Induced Apomixis | 149 |
|  | 5.9.3 | Genetic Engineering to Induce Apomixis | 150 |
| 5.10 | Genetic Enhancement through Transformation | | 150 |
| 5.11 | Functional Genomics | | 152 |
| References | | | 152 |

## 5.1 INTRODUCTION

Rice (*Oryza sativa* L.) is the primary food source for more than a third of the world's population. More than 90% of the world's rice is grown and consumed in Asia alone, where 60% of the Earth's people live. Rice accounts for 35 to 60% of the calories consumed by 2.9 billion Asians. Rice is planted on almost 150 million ha, with annual production of 600 million tonnes. It is the only major cereal crop that is consumed almost exclusively by humans. Rice production was 589 million tons in 2003. China, the largest producer, produced 166 million tons, followed by India (132 million tons), Indonesia (52 million tons), Vietnam (35 million tons), Bangladesh (38 million tons), Thailand (27 million tons), and Myanmar (25 million tons) (Table 5.1).

Major advances have been made in increasing rice productivity. World rice production more than doubled from 257 million tonnes in 1966 to 599 million tonnes in 2000. This was mainly

**Table 5.1   Rice Production, Area, and Productivity in 2003**

| Region/Country | Area (ha) | Production (tonnes) | Yield (t/ha) |
|---|---|---|---|
| Asia | 135,666 | 534,821 | 3.94 |
| Bangladesh | 11,100 | 38,060 | 3.43 |
| Cambodia | 2,000 | 4,300 | 2.15 |
| China | 27,398 | 166,417 | 6.07 |
| India | 44,000 | 132,013 | 3.00 |
| Indonesia | 11,477 | 52,079 | 4.54 |
| Iran | 560 | 3,300 | 5.89 |
| Japan | 1,665 | 9,740 | 5.85 |
| Korea, DPR | 593 | 2,284 | 3.85 |
| Korea, Republic of | 1,013 | 6,068 | 5.99 |
| Laos | 754 | 2,500 | 3.32 |
| Malaysia | 675 | 2,145 | 3.18 |
| Myanmar | 6,650 | 24,640 | 3.71 |
| Nepal | 1,550 | 4,155 | 2.68 |
| Pakistan | 2,210 | 6,751 | 3.05 |
| Philippines | 4,094 | 14,031 | 3.43 |
| Sri Lanka | 911 | 3,071 | 3.37 |
| Thailand | 11,000 | 27,000 | 2.45 |
| Vietnam | 7,449 | 34,519 | 4.63 |
| South America | 5,129 | 20,133 | 3.93 |
| Africa | 10,228 | 19,080 | 1.87 |
| Australia | 38 | 391 | 10.29 |
| U.S. | 1,213 | 9,034 | 7.45 |
| World | 153,522 | 589,126 | 3.84 |

*Source*: FAO 2004.

achieved through the application of principles of classical Mendelian genetics and conventional plant breeding. The current world population of 6.1 billion is expected to reach 8.0 billion by 2030, and rice production must increase by 50% to meet the growing demand. Further, several biotic (diseases, insects) and abiotic (drought, salinity, iron and aluminum toxicity) stresses continue to lower rice productivity. To overcome these constraints, there is an urgent need to develop rice varieties with higher yield potential and durable resistance to diseases, insects, and abiotic stresses. Recent advances in cellular and molecular biology, particularly in genomics, have provided new opportunities to develop improved germplasm with new genetic properties and to understand the function of rice genes. Rice has become a model plant for genetic and breeding research. Several factors have contributed to this: diploid species with 2n = 24 chromosomes, self-pollination, a short growth duration, the most extensive germplasm collection comprising more than 100,000 accessions of rice and wild species, a large number of mutant markers, well-characterized cytogenetic stocks (primary trisomics, secondary trisomics, monosomic alien addition lines), the smallest genome (450 Mbp) among the cereals, a dense molecular map consisting of more than 4000 DNA markers, extensive synteny with genomes of other cereals, ease in genetic transformation, availability of new genetic resources for functional genomics (expressed sequence tag (EST), bacterial artificial chromosome (BAC) libraries, T-DNA insertion mutants, retrotransposons, deletion mutants), and, above all, a huge database on genome sequences for both indica and japonica rices.

The genus *Oryza*, to which cultivated rice belongs, probably originated at least 130 million years ago and spread as a wild grass in Gondwanaland, the super continent that eventually broke up and drifted apart to become Asia, Africa, the Americas, Australia, and Antartica (Chang 1976). Today's species of the genus *Oryza* are distributed in all of these continents except Antartica (Table 5.2).

Of the two cultivated species, *O. sativa* (2n = 24 AA), the Asian rice, is grown worldwide, but *Oryza glaberrima* (2n = 24 AA), the African cultivated rice, is limited to some parts of West Africa. The genus has 22 wild species (2n = 24, 48), with genomic constitution AA, BB, CC, BBCC, CCDD, EE, FF, GG, HHJJ, and HHKK. Cultivated varieties of *O. sativa* are grouped into three types: indica, japonica, and javanica (tropical japonica). Rice cultivation is spread from the latitude from 40° south to 44° north and grows from sea level to altitudes of 2500 m or even higher. It is grown under a wide range of agroclimatic conditions, including irrigated or rain-fed puddle lands, low-land soils in river deltas, with water as deep as 4 m, and as an upland crop without any standing water.

## 5.2 ORIGIN OF CULTIVATED RICE

The common rice, *O. sativa*, and the African rice, *O. glaberrima*, are thought to be examples of parallel evolution in crop plants. The wild progenitor of *O. sativa* is the Asian common wild rice, *Oryza rufipogon*, which shows a range of variation from perennial to annual types. In a parallel evolutionary path, *O. glaberrima* was domesticated from the annual *Oryza breviligulata*, which in turn evolved from perennial *Oryza longistaminata*. There is a general agreement that *O. rufipogon* is the immediate ancestor of *O. sativa* and that cultivated rice originated polyphyletically.

### 5.2.1  Wild Progenitors of Cultivated Rice

*Oryza rufipogon* is distributed from Pakistan to China and Indonesia, and its populations vary between perennial and annual types, which differ markedly in life history traits (Oka 1988). The perennial types grow in deep swamps, which retain moisture throughout the year, while the annual types occur in temporary marshes, which are parched in the dry season. All these wild rices cross with cultivated rice under natural conditions, producing hybrid swarms in the field.

Domestication of wild rices probably started about 9000 years ago. Linguistic evidence points to the early origin of cultivated rice in Asian arc. In several regional languages, the general terms

Table 5.2 Chromosome Number, Genomic Composition, and Distribution of *Oryza* Species and Their Useful Traits

| Species | 2n | Genome | Number of Accessions[a] | Distribution | Useful Traits |
|---|---|---|---|---|---|
| **O. sativa Complex** | | | | | |
| O. sativa L. | 24 | AA | 96,564 | Worldwide | Cultigen |
| O. glaberrima Steud. | 24 | $A^gA^g$ | 1543 | West Africa | Cultigen; tolerance to drought, acidity, iron toxicity; resistance to RYMV, African gall midge, nematodes; weed competitiveness |
| O. nivara Sharma et Shastry | 24 | AA | 1260 | Tropical and subtropical Asia | Resistance to grassy stunt virus |
| O. rufipogon Griff. | 24 | AA | 1005 | Tropical and subtropical Asia, tropical Australia | Resistance to BB, tungro virus; tolerance to aluminum and soil acidity; source of CMS |
| O. breviligulata A. Chev. et Roehr. (O. barthii) | 24 | $A^gA^g$ | 218 | Africa | Resistance to GLH, BB; drought avoidance |
| O. longistaminata A. Chev. et Roehr. | 24 | $A^lA^l$ | 207 | Africa | Resistance to BB, nematodes; drought avoidance |
| O. meridionalis Ng | 24 | $A^mA^m$ | 56 | Tropical Australia | Elongation ability; drought avoidance |
| O. glumaepatula Steud. | 24 | $A^{gp}A^{gp}$ | 54 | South and Central America | Elongation ability; source of CMS |
| **O. officinalis Complex** | | | | | |
| O. punctata Kotschy ex Steud. | 24, 48 | BB, BBCC | 71 | Africa | Resistance to BPH, zigzag leafhopper |
| O. minuta J.S. Presl. ex C.B. Presl. | 48 | BBCC | 64 | Philippines and Papua New Guinea | Resistance to BB, blast BPH, GLH; tolerant to Shb |
| O. officinalis Wall ex Watt | 24 | CC | 279 | Tropical and subtropical Asia, tropical Australia | Resistance to thrips, BPH, GLH, WBPH, BB, stem rot |
| O. rhizomatis Vaughan | 24 | CC | 19 | Sri Lanka | Drought avoidance |
| O. eichingeri A. Peter | 24 | CC | 30 | South Asia and East Africa | Resistance to BPH, WBPH, GLH |
| O. latifolia Desv. | 48 | CCDD | 58 | South and Central America | Resistance to BPH; high biomass production |
| O. alta Swallen | 48 | CCDD | 6 | South and Central America | Resistance to striped stemborer; high biomass production |
| O. grandiglumis(Doell) Prod. | 48 | CCDD | 10 | South and Central America | High biomass production |
| O. australiensis Domin. | 24 | EE | 36 | Tropical Australia | Resistance to BPH, BB; drought avoidance |

*(Continued)*

Table 5.2 Chromosome Number, Genomic Composition, and Distribution of *Oryza* Species and Their Useful Traits (continued)

| Species | 2n | Genome | Number of Accessions[a] | Distribution | Useful Traits |
|---|---|---|---|---|---|
| **O. meyeriana Complex** | | | | | |
| *O. granulata* Nees et Arn. ex Watt | 24 | GG | 24 | South and Southeast Asia | Shade tolerance; adaptation to aerobic soil |
| *O. meyeriana* (Zoll. et (Mor. ex Steud.) Baill. | 24 | GG | 11 | Southeast Asia | Shade tolerance; adaptation to aerobic soil |
| **O. ridleyi Complex** | | | | | |
| *O. longiglumis* Jansen | 48 | HHJJ | 6 | Irian Jaya, Indonesia, and Papua New Guinea | Resistance to blast, BB |
| *O. ridleyi* Hook. F. | 48 | HHJJ | 14 | South Asia | Resistance to BB, blast, stemborer, whorl maggot |
| **Unclassified** | | | | | |
| *O. brachyantha* A. Chev. et Roehr. | 24 | FF | 19 | Africa | Resistance to BB, yellow stemborer, leaf folder, whorl maggot; tolerance to laterite soil |
| *O. schlechteri* Pilger | 48 | HHKK | 1 | Papua New Guinea | Stoloniferous |
| Hybrids or unidentified species | – | – | 941 | – | – |
| Related genera | – | – | 15 | – | – |
| *Chikusiochloa* Koidz. | 24 | – | 1 | China, Japan, Indonesia | – |
| *Hygroryza* Nees | 24 | – | 4 | Asia | – |
| *Leersia* Soland | 24, 48, 60, 96 | – | 8 | Worldwide | – |
| *Luziola* Juss. | 24 | – | 1 | North and South America | – |
| *Prosphytochloa* Schweickerdt | Unknown | – | 0 | Southern Africa | – |
| *Rhynchoryza* Baill. | 24 | – | 1 | South America | – |
| *Zizania* Gronov. Ex Linn | 30, 34 | – | 0 | Europe, Asia, North America | – |
| *Zizaniopsis* Doell. & Aschers | 24 | – | 1 | North and South America | – |
| *Porteresia* Tateoka | 48 | – | 1 | South Asia | – |
| *Potamorphila* R. Br. | 24 | – | 1 | Australia | – |

BPH = brown planthopper; GLH = green leafhopper; WBPH = white-backed planthopper; BB = bacterial blight; Shb = sheath blight; CMS = cytoplasmic male sterility; RYMV = rice yellow mottle virus.

[a]Accessions maintained at IRRI, Philippines. Modified from Vaughan (1994) and Brar and Khush (2002).
Modified from Vaughan (1994) and Brar and Khush (2002).

for rice and food or for rice and agriculture are synonymous. The earliest and most convincing archeological evidence for domestication of rice in Southeast Asia was discovered by Welhelm G. Solheim II in 1966 (Solheim 1972). Pottery shards bearing the imprints of grain and husks of *O. sativa* were discovered at Non Nok Tha in the Korat area of Thailand. These remains have been confirmed by $^{14}$C and thermoluminescence testing as dating to at least 4000 B.C.

Ancient India is undoubtedly one of the oldest regions where cultivation of *O. sativa* began. The oldest grain samples excavated at Mohenjodaro, now in Pakistan, date back to about 2500 B.C. The oldest carbonized grains found in India date back to about 6750 B.C. The antiquity of rice cultivation in China has long been a subject of debate. The oldest remains of cultivated rice date back to five centuries before Christ. Carbonized rice grains from Tongxieng County of Zhejiang Province were identified as 7040 years old. The second oldest, 6960 years old, is from Hemdu relic in Yuyao County of Zhejiang Province. Suh et al. (2000) analyzed carbonized rice aged 13,010 and 17,310 years. Carbonized rice and quasi-rice hull were excavated from the peat soil layer in Sorori village located in the middle part of South Korea. The upper peat–soil layer, where the carbonized rice hull was excavated, was evaluated to be 13,010 years by carbon isotope dating. Based on randomly amplified polymorphic DNA (RAPD) variation, the ancient rices (carbonized and quasi-rice) were classified into one group and the present rice into another group. Genetic similarity between the ancient and present rice groups was approximately 57%.

The African cultivar, *O. glaberrima*, originated in the Niger River Delta. The primary center of diversity for *O. glaberrima* is the swampy basin of the upper Niger River and two secondary centers to the southwest near the Guinean coast. The primary center was probably formed around 1500 B.C., while the secondary centers were formed 500 years later (Porteres 1956).

## 5.2.2  Polyphyletic Origin of *O. sativa*

*Oryza sativa* is a highly variable species and is distributed worldwide. The Chinese have recognized two rice varietal groups, Hsien and Keng. They correspond to indica and japonica classification, introduced by Kato et al. (1928). Indica and japonica cultivars differ in many characters when typical varieties are compared but show overlapping variations. The indica and japonica types are each characterized by an association of certain diagnostic characters, such as $KClO_3$ resistance, cold tolerance, apiculus hair length, and phenol reaction (Oka 1958).

Morinaga (1954) proposed a third group to include bulu and gundil varieties of Indonesia under the name javanica. Several authors have ranked javanicas at the same taxonomic level as indicas and japonicas. As shown by Glaszmann (1987), on the basis of genetic affinity using isozyme analysis, javanica varieties fall within the japonica group and are now referred to as tropical japonicas, and the so-called typical japonicas are referred to as temperate japonicas.

Most of the earlier studies on rice varietal classification were confined to varieties from East and Southeast Asia, and the majority of the varieties from this region belong to the two (indica and japonica) groups. However, Glaszmann (1987) examined 1688 varieties from a wider geographical distribution for allelic frequencies at 15 isozyme loci and analyzed the data by multivariate analysis. On the basis of this classification, 95% of the cultivars fell into six distinct groups (I to VI), the remaining 5% being scattered over intermediate positions. This classification involved no morphological criteria. Group I corresponded to the indica and group VI to the japonica, including the bulu and gundil varieties, or the so-called javanicas. Groups II, III, and IV were atypical but also classified as indicas in the conventional classification. Group V includes aromatic rices of the Indian subcontinent. Khush et al. (2003) analyzed 25,519 varieties originating from different Asian countries using 20 loci for 11 enzymes. The vast majority of varieties belonged to group I (73.4%) and group VI (23%). Less than 1% of the varieties could be assigned to each of groups II, III, IV, and V.

Various hypotheses have been proposed about the origin of indica and japonica rices. Kato et al. (1928) expressed the opinion that indica and japonica rices originated independently from a wild ancestor. Ting (1957), on the other hand, proposed that japonicas were derived from the indicas. Earlier studies primarily focused on indica–japonica differentiation. However, the so-called indicas are such a diverse group that several morphological types can be recognized that correspond to Glaszmann's classification based on isozymes. Thus, information from genetic affinity (isozyme analysis), isolation barriers ($F_1$ sterility), and morphological classification suggests that the six groups may have been domesticated from different populations of *Oryza nivara* at different locations and on different timescales. Rayada rices (group IV) of Bangladesh adapted to deepwater conditions, for example, may have been domesticated in only recent times as some of the deepwater areas were brought under cultivation.

Recently, molecular techniques have been used to understand the possible pathways for the origin of cultivated rice. Ishii et al. (1995) compared results of restriction fragment length polymorphism (RFLP) analysis with previous isozyme grouping. Classification on isozyme analysis matched well with that on nuclear genome, indicating synchronous differentiation of isozyme constitutions and nuclear genome in Asian rice varieties.

Wang et al. (1992) carried out RFLP analysis of 93 accessions belonging to 21 species. Classification of *Oryza* species based on RFLPs matched remarkably well with the classifications based on morphological and hybridization studies. Four species complexes were identified, which corresponded to those proposed by Vaughan (1989). Within the *O. sativa* complex, accessions of *O. rufipogon* from Asia and Australia clustered together with *O. sativa*. Two groups of cultivated rices, indica and japonica, showed close affinity with different accessions of *O. rufipogon*, supporting the hypothesis of independent domestication from these two types of rices. CCDD tetraploid species (*Oryza latifolia*, *Oryza alta*, and *Oryza grandiglumis*) are of ancient origin and show closer affinity to each other than to any known diploid species. Their C genome donor may be *Oryza eichingeri*, and the D genome donor may have been related to the E genome of *Oryza australiensis*.

Aggarwal et al. (1999) used amplified fragment length polymorphism (AFLP) markers to study phylogenetic relationships in *Oryza* species. Seventy-seven accessions of 23 *Oryza* species, 5 related genera, and 3 outgroup taxa were fingerprinted using AFLP. A total of 1191 polymorphic markers were obtained using five AFLP primer combinations. Species relationships were studied using different clustering algorithms. The results showed common ancestry to the genus *Oryza*. Further evolution in *Oryza* has followed a polyphyletic path wherein multiple lineages underwent independent divergence after separation from a common ancestor/pool of related taxa.

Representational differential analysis (RDA) was applied to characterize genomic differentiation in rice (Panaud et al. 2002). Rice was used as the tester and millet as the driver. The RDA clones were used as probes in Southern hybridization experiments with genomic DNAs of several species from the family Poaceae. The results suggest that the genomic differentiations associated with the activity of transposable elements are of relatively recent origin. Comparison of the hybridization patterns obtained for several *Oryza* species suggests that several independent amplifications of these transposable elements might have occurred within the genus.

Cheng et al. (2003) determined polyphyletic origin of cultivated rice based on the interspersion pattern of *SINEs*. The retrotransposon, *p-SINE1*, which shows insertion polymorphism in the *O. sativa–O. rufipogon* population, was identified and used to bar code each of the 101 cultivated and wild species based on the presence or absence of the *p-SINE1* members at the respective loci. The phylogenetic tree showed that *O. sativa* strains fall into two groups, corresponding to japonica and indica, whereas *O. rufipogon* strains were in four groups, in which annual *O. rufipogon* strains formed a single group, differing from the perennial *O. rufipogon* strains of the other three groups. Japonica strains were closely related to the *O. rufipogon* perennial strains of one group, and the indica strains were closely related to the *O. rufipogon* annual strains, indicating that *O. sativa* has been derived polyphyletically from *O. rufipogon*.

In another study, Cheng et al. (2002) identified new *p-SINE1* members showing interspecific insertion polymorphisms from representative strains of four species (*Oryza barthii*, *Oryza glumae-patula*, *O. longistaminata*, and *Oryza meridionalis*) with the AA genome. Some of these members were present only in strains of one species, whereas the others were present in strains of two or more species. Phylogenetic analysis based on the *p-SINE1* insertion patterns showed that the strains of each of the five wild rice species formed a cluster. The strains of *O. longistaminata* appear to be distantly related to those of *O. meridionalis*. The strains of these two species appear to be distantly related to those of three other species, *O. rufipogon*, *O. barthii*, and *O. glumaepatula*. The latter three species are closely related to one another, with *O. barthii* and *O. glumaepatula* being most closely related. A phylogenetic tree including a hypothetical ancestor with all loci empty for *p-SINE1* insertion showed that the strains of *O. longistaminata* are related most closely to the hypothetical ancestor. This indicates that *O. longistaminata* and *O. meridionalis* diverged early on, whereas the other species diverged relatively recently.

## 5.3 RICE GENETIC RESOURCES: EXPLORATION AND CONSERVATION

Rice genetic resources comprise landrace varieties, modern and obsolete varieties, genetic stocks, and the wild rices (Table 5.2). The International Rice Gene Bank at the International Rice Research Institute (IRRI) in the Philippines conserves the largest (>100,000) and most diverse collection of rice germplasm. The facilities of the gene bank ensure the long-term conservation of this valuable gene pool.

### 5.3.1 Exploration and Collection of Germplasm

Records of early occasional exploration and collection for wild rice samples in different countries can be found in various literatures. Specimens of wild *Oryza* species can be found in many herbaria, particularly in India, China, Indonesia, Thailand, Singapore, and Malaysia.

Efforts for collection and conservation of wild *Oryza* species were initiated in the late 1950s by many National Agricultural Research Systems (NARS) on small scales, along with conservation programs for cultivated rice varieties and landraces. India and China are the world's two largest rice-producing and -consuming countries, and wild *Oryza* species are found abundantly in both countries. Lu and Sharma (2003) have reviewed the exploration and collection of wild *Oryza* species.

The workshops organized by IRRI and the International Board for Plant Genetic Resources on genetic conservation of rice, held at IRRI in 1977 and 1983, and the "Third International Workshop on Rice Germplasm: Collection, Preservation, and Use," held at IRRI in 1990, were three workshops important to international cooperation and conservation activities of wild *Oryza* species. Since then collection of wild *Oryza* species has gradually received more attention by the NARS, particularly in Asian countries, and more intensive and systematic collecting activities have been conducted in different countries. Based on reports by several NARS programs, a nominal number of seed samples of wild *Oryza* species was collected in the early 1980s, except in Thailand, where more than 100 accessions of wild *Oryza* species were collected.

Much germplasm exploration for rice was completed by the early 1990s. By the end of 1962, the IRRI varietal collection contained 6867 accessions from 73 countries. By 1972, the collection had grown to 14,600 accessions (Chang 1972). By the early 1980s, the number of accessions in the IRRI gene bank reached 49,027. More than 200 accessions of rice were collected during the second half of 1995 from Lao People's Democratic Republic (PDR). The IRRI gene bank now contains 3696 accessions of 21 wild species of *Oryza*, besides 1543 accessions of African cultivated rice (*O. glaberrima*).

### 5.3.2 Genetic Erosion

Genetic erosion or loss of biodiversity of rice varieties has been recognized as a problem since the 1960s. Factors such as the adoption of high-yielding rice varieties, farmer's increased integration into the markets, change of farming systems, industrialization, human population increases, and cultural change have significantly accelerated continual erosion of the rice gene pool (Bellon et al. 1998). A similar situation has also been observed for wild *Oryza* species. In many places of Asia, populations of wild *Oryza* species are becoming extinct or are threatened because their natural habitats are seriously damaged by extension of cultivation areas, expansion of communication systems such as road construction, and urban pressures. According to unpublished data collected by the Chinese Academy of Sciences in 1994, nearly 80% of the common wild rice (*O. rufipogon*) sites recorded during the 1970s have already disappeared (cited by Lu and Sharma 2003). The size of some surviving *O. rufipogon* populations was also found to be significantly reduced. A similar situation has been observed in other countries such as Vietnam, Thailand, Nepal, Indonesia, Malaysia, India, and Bangladesh.

The problems of genetic erosion are severe, but international efforts to conserve rice genetic resources, in which IRRI has taken a leading role, have led to the establishment of several gene banks in Asia. These joint efforts between national, regional, and international organizations ensure the long-term conservation of the biodiversity of the rice gene pool.

### 5.3.3 Conservation of Germplasm

For many plant species, *ex situ* conservation of seeds is safe and cost efficient, provided proper attention is paid to seed drying and storage conditions. Fortunately, rice seeds exhibit orthodox storage behavior and can be dried to a low moisture content of ca. 6% and stored at –20°C, retaining their viability for decades, if not longer. Jackson (1997) has summarized the strategies on conservation of genetic resources. There are two basic approaches to germplasm conservation (*ex situ* and *in situ* conservation).

#### 5.3.3.1  *Ex Situ Conservation*

In this approach, genetic resources are actually removed from their original habitat or natural environment. *Ex situ* conservation provides efficient means for germplasm preservation, utilization, exchange, and information generation through effective management and value-added research of the conserved wild rice species. However, seed samples placed under *ex situ* conservation in a gene bank become isolated from the *Oryza* ecosystem where they originated and grow. The expected microevolution of these *Oryza* species in their original environment is stopped, particularly the adaptive variations that could occur during change in environmental conditions. Therefore, in evolutionary terms, *ex situ* conservation is static (Bellon et al. 1998). Concerns have been raised following the observation that static conservation may reduce the adaptive potential of wild *Oryza* species and their populations in the future. Thus, *ex situ* conservation cannot be considered the only approach for conserving biodiversity of wild *Oryza* species. Complementary dynamic approaches such as *in situ* conservation are also necessary.

The long-term conservation of rice genetic resources is the principal aim of the International Rice Gene Bank (IRG). The gene bank has operated since 1977, although genetic conservation activities started in the early 1960s. For several countries, including Sri Lanka, Cambodia, Lao PDR, and the Philippines, the germplasm conserved in the IRG represents a more or less complete duplicate of their national collections. For other countries, such as India and the People's Republic of China, only part of their national collections are duplicated at IRRI. Nevertheless, the IRG has provided an important safety net for national conservation efforts. On several occasions, it has been possible to restore rice germplasm that had been lost in national gene banks with accessions already conserved at IRRI. IRRI maintains an active collection for medium-term storage and distribution of rice germplasm, at +2°C in sealed laminated aluminum foil packets, and long-term (50 to 100 years) conservation, at –20°C, each with two vacuum-sealed aluminum cans.

The germplasm collection is held in trust by IRRI under the auspices of FAO in an International Network of *Ex Situ* Collections. Duplicate storage of the IRG collection is carried out at the National

Seed Storage Laboratory (NSSL), Fort Collins, CO, and about 75% of the collection is currently stored under black-box conditions. Duplicate storage of African rices is shared between IRRI, the International Institute of Tropical Agriculture (IITA) in Nigeria, and the West Africa Rice Development Association (WARDA) in Cote d'Ivoire.

### 5.3.3.2 In Situ Conservation

This method attempts to preserve the integrity of genetic resources by conserving them within the evolutionary dynamic ecosystems of their original habitat or natural environment. Under *in situ* conservation, local control of traditional rice varieties will ensure that benefits accrue to farmers and communities that have developed them. For long-term and dynamic conservation, the *in situ* approach has great value. However, for some reason, *in situ* conservation has, in general, received the least attention and even been rejected. Limited scientific and financial inputs are constraints in *in situ* conservation and its design and management for wild *Oryza* species.

## 5.4 TAXONOMY

Cultivated rice (*O. sativa* L.) belongs to the family Poaceae (Gramineae), subfamily Bambuosoideae, and tribe Oryzeae. This tribe has 11 genera, of which genus *Oryza* is the only one with cultivated species. Roschevicz (1931) published a comprehensive study of 19 species that provided a basis for later taxonomic investigations of the genus. Chatterjee (1948) and Sampath (1962) listed 23 species, and Tateoka (1962a,b) listed 21 species. Two cultivated and 22 wild species are recognized in the genus *Oryza*. The species of the genus have been characterized into several groups on the basis of morphological characteristics. The genus *Oryza* has been divided into four species complexes: (1) *sativa* complex, (2) *officinalis* complex, (3) *meyeriana* complex, and (4) *ridleyi* complex. Two species, *Oryza brachyantha* and *Oryza schlechteri* cannot be placed in any of these groups (Vaughan 1989, 1994).

### 5.4.1 O. sativa Complex

This complex consists of two cultivated species, *O. sativa* and *O. glaberrima*, and six wild taxa (Table 5.2). All of them have AA genomes and form the primary gene pool for rice improvement. Wild species closely related to *O. sativa* have been variously named. The species name *Oryza perennis* has been widely used for the perennial wild relative of rice found in Asia, Africa, and Latin America. Tateoka (1963) considers *O. rufipogon* a taxonomically valid name for this species.

Sharma and Shastry (1965a) divided *O. rufipogon* of Asia into two categories, recognizing a new species, *O. nivara*, as an annual form of wild rice and retaining *O. rufipogon* for perennial populations. The three taxa — *O. sativa*, *O. nivara*, and *O. rufipogon* — together with the weedy race (*O. sativa* f. *spontanea*) form a large species complex. Recent literature seems to agree on giving the perennial wild relatives of rice separate species names, *O. rufipogon* and *O. nivara*, to the annual forms.

The weedy types of rice have been given various names such as fatua and spontanea in Asia and *Oryza stapfii* in Africa. Depending on the location, these weedy rices may be more closely related to *O. rufipogon* and *O. nivara* in Asia and *O. longistaminata* or *O. breviligulata* in Africa. The *O. sativa* complex includes two other species. The one distributed across tropical Australia is called *O. meridionalis* (Ng et al. 1981). This species has many characteristics similar to those of *O. nivara* described by Sharma and Shastry (1965b). However, it has longer awns, narrower spikelets, and a more compact panicle. This species is often sympatric with *O. australiensis* in Australia.

The species closely related to the West African cultivated rice, *O. glaberrima*, are somewhat easier to distinguish from each other. The perennial and annual relatives of *O. glaberrima* are *O.*

*longistaminata* and *O. breviligulata*, respectively. Previously, *O. longistaminata* was called *O. barthii*. This species is easily distinguished from the annual wild species of this complex by its strong rhizomes and long anthers. *Oryza longistaminata* differs morphologically from its Asian counterpart primarily in its panicle branching and short ligule. *Oryza glaberrima* is distinguished from *O. sativa* by its short, rounded ligule, panicle-lacking secondary branches, and almost glabrous lemma and palea. *Oryza glaberrima* is not as variable as *O. sativa*. However, in some areas of West Africa, local people prefer the taste of *O. glaberrima*, and the species is better adapted to upland habitats than the introduced common rice from Asia. *Oryza longistaminata* is more widely distributed in Africa than *O. glaberrima*.

### 5.4.2 *Oryza officinalis* Complex

The largest complex in the genus is *O. officinalis* complex. It consists of nine species, also called the *O. latifolia* complex by Tateoka (1962a). This complex has related species groups in Asia, Africa, and Latin America. In Asia, the most common species is *O. officinalis*, widely distributed in South and Southeast Asia and South and Southwest China. *Oryza officinalis* thrives in partial shade or full sun. In the Philippines it is called bird rice.

The tetraploid species *Oryza minuta* grows in the Philippines. *Oryza minuta* is sympatric with *O. officinalis* in the central islands of Bohol and Leyte. It grows in shade or partial shade along stream edges as a minor member of the flora. Only a few populations of *O. minuta*, also called *Oryza malampuzhaensis*, have been found localized in neighboring parts near the town of Malampuzha of Kerala and Tamil Nadu in South India.

A new species from Sri Lanka belonging to this complex was described by Vaughan (1989) as *Oryza rhizomatis*. A form of the *O. officinalis* complex from Sri Lanka was named *Oryza collina* (Sharma and Shastry 1965b). Another species of this complex, *O. eichingeri*, grows in the forest shade in Uganda (Tateoka 1965). It was recently found distributed in Sri Lanka (Vaughan 1989).

In Africa, two species of the *O. officinalis* complex are *Oryza punctata* and *O. eichingeri*. Both have been reported as having diploid and tetraploids forms (Tateoka 1965). The American species of this complex (*O. latifolia*, *O. alta*, and *O. grandiglumis*) are allotetraploid with a CCDD genome. *Oryza latifolia* is widely distributed, growing in Central and South America as well as in the Caribbean islands; *O. alta* and *O. grandiglumis* grow only in South America, primarily in the Amazon basin, except for one population of *O. alta* reported from Belize (Oka 1961). These species are all allotetraploid. A diploid species of this complex, *O. australiensis*, occurs in northern Australia in isolated populations.

### 5.4.3 *Oryza ridleyi* Complex

This complex comprises two tetraploid species, *O. ridleyi* and *Oryza longiglumis*, which usually grow in shaded habitats beside rivers, streams, or pools. The *O. ridleyi* complex is primarily found in the lowland forests of insular Southeast Asia and New Guinea. *O. longiglumis* is known only from a few sites along the Koembe River, Irian Jaya, Indonesia, and in Papua New Guinea. *Oryza longiglumis* is related to *O. ridleyi* but has much longer sterile lemmas in relation to the palea and lemma length. *Oryza ridleyi* grows across Southeast Asia and as far east as Papua New Guinea.

### 5.4.4 *Oryza meyeriana* Complex

This complex has two species (*O. meyeriana* and *Oryza granulata*). The most common and widespread species of the complex, *O. granulata*, grows in South Asia, Southeast Asia, and Southwest China. *O. meyeriana* is found in Southeast Asia. Another species, *Oryza indandamanica*, from Andaman Islands, India, was described recently (Ellis 1985). However, Khush and Jena (1989) consider it to be a subspecies of *O. granulata*. This is the only group of species in the genus *Oryza* that is not found in or near permanently or seasonally standing or flooded water. All members of this species complex grow in the shade or partial shade of degraded primary or well-established secondary forests, often on sloping terrain. *Oryza granulata* is called forest rice by tribal people in Kerala, South India, peacock rice in parts of Vietnam, and bamboo rice in the Philippines. Members of the *O. meyeriana* complex grow at higher elevation than other wild species, as high as 1000 m. All species of this complex have unbranched panicles with small spikelets (Tateoka 1962b).

### 5.4.5 Unknown Complex

#### 5.4.5.1 *Oryza brachyantha*

*Oryza brachyantha* is distributed in the African continent. Of all the species, it is the most closely related to the genus *Leersia*. This species has a small, narrow spikelet with long awns (6 to 17 cm). Features of the awn, such as its coriaceous, rigid structure served with a single vascular bundle, ally this species with *Oryza* rather than with *Leersia* (Launert 1965). This species grows in the Sahel zone and in East Africa in small, temporary pools, often in laterite soils. It is often sympatric with *O. longistaminata*.

#### 5.4.5.2 *Oryza schlechteri*

*Oryza schlechteri* is the least studied species in the genus. Richard Schlechter collected it in 1907 from Northeast New Guinea. Vaughan and Sitch (1991) recollected it as living material from the same location. Naredo et al. (1993) found that the presence of a sterile lemma and a striated spikelet epidermal (abaxial) surface lacking siliceous triads in *O. schlechteri* allies this species with other *Oryza* species rather than with *Leersia*. It is tufted perennial, less than 1 m tall, with an erect, 4- to 5-cm panicle and small, unawned spikelets. It is tetraploid, but its relationship to other species is unknown. It is a stoloniferous species of unstable stony soil, such as riverbanks, and grows in full or semishade.

### 5.4.6 Related Genera

Besides *Oryza*, the tribe Oryzeae contains 10 other genera (Table 5.2). Vaughan (1989, 1994) has given brief descriptions of these genera. These genera include *Chikusichloa*, *Hygrooryza*, *Leersia*, *Luziola*, *Prosphytochloa*, *Rhynchoryza*, *Zizania*, *Zizaniopsis*, *Potaomophila*, and *Porteresia*. Among these 10 genera, *Porteresia* has been studied by many workers. This genus has only one species, *P. coarctata*, which is commonly found in coastal areas of South Asia. It has unusual leaf anatomy, including glands to secrete salts. It has rough, erect leaves and occurs in brackish water. The species is characterized by large caryopses with a somewhat bent apex, a large embryo relative to the endosperm, and a short petiole attachment at the base. The leaf blade is coriaceous with prickly tuberculate margins and has a peculiar arrangement of vascular bundles; each rib contains one smaller vascular bundle and below it a larger one. Earlier, many workers classified *P. coarctata* as *O. coarctata*. Later it was included under *Porteresia*, but now the species has again been classified as *O. coarctata* (Ge et al. 1999).

### 5.4.7 Genomic Relationships

Two common approaches, such as chromosome pairing in $F_1$ hybrids and molecular divergence, are used to determine genomic relationships in *Oryza* species. On the basis of chromosome pairing in $F_1$ hybrids, various authors have assigned genome symbols AA for the *O. sativa* complex, BB, CC, BBCC, CCDD, and EE for the *O. officinalis* complex, and FF for *O. brachyantha* (see Nayar 1973; Khush and Brar 2002; Vaughan 2003). Genome symbols could not be assigned for *O.*

*meyeriana* and *O. ridleyi* complexes as well as for *O. schlechteri* due to lack of data on chromosome pairing in $F_1$ hybrids of these species with other species of the genus *Oryza*. However, molecular approaches have revealed genomic composition of these two complexes (Aggarwal et al. 1997).

### 5.4.7.1  Molecular Approaches

In several cases, due to strong incompatibility barriers, hybrids are difficult to produce between divergent species. For example, the hybrids between species of the *O. ridleyi* complex and *O. meyeriana* complex with other species could not be produced, and genome analysis based on meiotic pairing in $F_1$ hybrids could not be carried out. Under such situations, molecular approaches have been used to determine genomic relationships in *Oryza* species. Aggarwal et al. (1997) used molecular divergence analysis based on total genomic DNA hybridization. Genomic DNA (after restriction digestion) of 79 accessions of 23 *Oryza* species, 6 related genera, and 5 outgroup taxa (2 monocots, 3 dicots) were hybridized individually with $^{32}$P-labeled total genomic DNA from 12 *Oryza* species: *O. ridleyi*, *O. longiglumis*, *O. granulata*, *O. meyeriana*, *O. brachyantha*, *O. punctata*, *O. officinalis*, *O. eichingeri*, *O. alta*, *O. latifolia*, *O. australiensis*, and *O. sativa*. The labeled genomic DNAs representing the *ridleyi* and *meyeriana* complexes cross-hybridized best to all the accessions of their respective species, less to those representing other genomes of *Oryza* and related genera, and least to outgroup taxa. In general, the hybridization differential measured in terms of signal intensities was more than 50-fold under conditions that permit detection of 70 to 75% homologous sequences, both in the presence and in the absence of *O. sativa* DNA as competitor. In contrast, when total DNA representing other *Oryza* genomes was used as a probe, species of the *O. ridleyi* and *O. meyeriana* complexes did not show any significant cross-hybridization (<5%). The results revealed that the genome(s) of both of these complexes are highly diverged and distinct from all other known genomes of *Oryza*. Based on molecular divergence analysis, new genomic designations GG for the diploids of the *O. meyeriana* complex and HHJJ for the allotetraploids of the *O. ridleyi* complex were proposed.

Ge et al. (1999) used two nuclear genes, *Adh1* and *Adh2*, and one chloroplast gene (*matK*). Based on sequences of these genes and clades, it was concluded that *O. schlechteri* and *O. coarctata* share the same genome HHKK. The *Adh2* phylogeny suggested a close relationship between the HH genome and the diploid FF genome species. The JJ and KK genomes are not grouped strongly with any diploid species on either *Adh* phylogeny. These results suggest that the diploid species with the HH, JJ, or KK genomes are either extinct or undiscovered.

### 5.4.7.2  Seed Protein Analysis

Sarkar and Raina (1992) studied profiles of soluble protein of cultivated rice and wild species belonging to different groups. Protein analysis confirmed the previous classification based on morphological and cytological criteria. The information from diverse sources confirms the genome designations of various *Oryza* species as shown in Table 5.2. Four species complexes are clearly known. *Oryza brachyantha* and *O. schlechteri* do not belong to any of the four complexes and are not related to each other.

## 5.5 CYTOGENETICS

Kuwada (1910) was the first to report chromosome number of rice as 2n = 2x = 24. Since then, various workers have confirmed this chromosome number for rice and wild species with 2n = 24, 48 chromosomes (Table 5.2). The karyotype of rice has been studied at the somatic prometaphase as well as at the pachytene stage of meiosis, and individual members of the chromosome complement have been identified. A number of marker genes affecting various morphological and physiological traits have been assigned to respective linkage groups. All of the 12 linkage groups are known and have been associated with respective chromosomes.

### 5.5.1   Somatic Karyotype

Somatic chromosomes of rice are in general very small, 0.7 to 2.8 μm length (Nandi 1936) to 2.0 to 5.0 μm (Yasui 1941). On the other hand, pachytene chromosomes are fairly large (12 to 79 μm) (Shastry et al. 1960). Rau (1929) reported that of the 12 pairs of the complement, five were large, four were medium, and the remaining three were small. Nandi (1936) hypothesized that *O. sativa* is a secondarily balanced allotetraploid. Recent evidence has shown that rice may indeed have undergone genomic duplication; intragenomic duplications have been revealed (see Chapter 1 by Jauhar in this volume). Sen (1963) reported that of the 12 chromosome pairs, 2 were metacentric, 9 submetacentric, and 1 subtelocentric. Kurata and Omura (1978) reported chromosome length at prometaphase from 1.5 to 4.1 μm and from 1.0 to 1.9 μm at metaphase. The karyotype consisted of five metacentric, five submetacentric, and two subtelocentric pairs.

### 5.5.2   Pachytene Karyotype

The pachytene chromosome complement of rice was first studied by Shastry et al. (1960). Each chromosome was identified on the basis of length, arm ratio, and presence or absence of dark staining knobs. The chromosomes were numbered in decreasing order of length — the longest being chromosome 1 and the smallest chromosome 12. The karyotype consisted of eight submetacentric, two metacentric, and two subtelocentric chromosomes. The centromeres of all chromosomes were flanked by darkly stained heterochromatic chromomeres, although the centromere position was not unequivocally clear in all cases. Kurata et al. (1981) also analyzed the pachytene chromosome complement of japonica cultivar Nipponbare. The chromosome designation agreed in these two studies except for chromosomes 11 and 12, which were interchanged. Now the pachytene numbering system of Shastry et al. (1960) is adopted by the rice community. Khush et al. (1984) identified an extra chromosome in each of the 12 primary trisomics of rice at the pachytene stage of meiosis following the chromosome designation of Shastry et al. (1960). Thus, the trisomic for chromosome 1 (longest chromosome) was designated as triplo 1, and so on. Trisomic segregations were observed for 18 genes, and each trisomic showed trisomic segregation for at least 1 gene. Thus, the associations were determined between the 12 linkage groups and 12 cytologically identifiable pachytene chromosomes.

Cheng et al. (2001a) constructed a karyotype of pachytene chromosomes in japonica cultivar Nipponbare. A set of 24 chromosomal arm-specific BACs was established. An idiogram depicting the distribution of heterochromatin in the rice genome was developed based on the staining patterns of 4,6-diamidono-2-phenylindole (DAPI) of pachytene chromosomes. The majority of heterochromatin is distributed in the pericentric regions with some rice chromosomes containing a significantly higher proportion of heterochromatin than other chromosomes. Yu et al. (2003) reported the sequence of chromosome 10, which contains considerable heterochromatin with an enrichment of repetitive elements on 10s and an enrichment of expressed genes on 10L. Multiple insertions from the organelle genome were also detected.

### 5.5.3   Translocations

Radiation-induced chromosome translocations have been studied by many workers. Nishimura (1961) identified individual chromosomes involved in translocations from intercrosses of translocation stocks. The 12 chromosomes were arbitrarily numbered (I to XII) but not related to the number assigned to individual chromosomes on the basis of cytological observations. These translocation stocks have been used to locate marker genes on chromosomes.

### 5.5.4  Haploids, Triploids, and Aneuploids

Haploids of spontaneous origin and through anther culture have been reported. Haploids are characterized by complete sterility and show reduced height and smaller leaves. Tetraploids have been induced through colchine treatment and also occur spontaneously. Tetraploids in general have reduced seed fertility and are inferior to disomics in agronomic performance; no commercial release has been made. Triploids also occur spontaneously but have also been produced through crosses of $4X \times 2X$. Triploids have proved to be the best source of primary trisomics.

### 5.5.5  Trisomics

#### 5.5.5.1  Primary Trisomics

Primary trisomics have been obtained by many workers. These trisomics are morphologically distinct from the disomic sibs as well as from each other. A complete series of trisomics representing each of the 12 chromosomes has been established (Khush et al. 1984). In general, trisomics have short stature, delayed heading, and vary in size of panicles. Seed setting is generally low. These trisomics have been used to locate several genes on rice chromosomes.

#### 5.5.5.2  Secondary and Telotrisomics

Singh et al. (1996a) isolated secondary trisomics and telotrisomics representing the 12 chromosomes from the progenies of primary trisomics. Plants showing variation in gross morphology compared to the primary trisomics and disomic sibs were selected and analyzed cytologically at diakinesis and pachytene. Secondary trisomics for both arms of chromosomes 1, 2, 6, 7, and 11 and for one arm of chromosomes 4, 5, 8, 9, and 12 were identified. Telotrisomics for short arms of chromosomes 1, 8, 9, and 10 and for long arms of chromosomes 2, 3, and 5 were isolated. Genetic segregation of 43 marker genes was studied in the $F_2$ or backcross progenies. On the basis of segregation data, these genes were delimited to specific chromosome arms. Correct orientation of 10 linkage groups was determined. In another study, Singh et al. (1996b) used secondary and telotrisomics to assign restriction fragment length polymorphism (RFLP) markers to specific chromosome arms and thereby to map the positions of centromeres. More than 170 RFLPs were assigned to specific chromosome arms through gene dosage analysis using the secondary and telotrisomics, and the centromere positions were mapped on all 12 linkage groups. The orientations of seven linkage groups were reversed and the corrected map was presented.

Recently, Cheng et al. (2001b) developed a complete set of telotrisomics covering all 24 chromosome arms. The telotrisomics were obtained from the progeny of 30,000 plants of primary trisomics and aneuploids grown from 1994 to 1999. The identities of the extra chromosomes were further confirmed by dosage analysis of the RFLP markers on extra chromosome arms. The telocentric nature of the extra chromosomes in these stocks was verified by fluorescence *in situ* hybridization (FISH) using a rice centromeric BAC clone as a marker probe. In general, the shorter the extra chromosome arm of a telotrisomic, the stronger the resemblance it bears to the diploid; the longer the extra chromosome arm, the stronger the resemblance to the corresponding primary trisomic.

#### 5.5.5.3  Monosomic Alien Addition Lines

Monosomic alien addition lines (MAALs) $(2n = 25)$ representing extra chromosomes from wild species in the genetic background of *O. sativa* have been reported. MAALs are produced during backcrossing of wide cross progenies ($BC_1F_1$, $BC_2F_1$, $BC_3F_1$, etc.) with the recurrent parents (Figure 5.1). MAALs are characterized based on plant morphology and through isozyme and molecular marker analysis. MAALs representing 6 to 11 chromosomes have been obtained

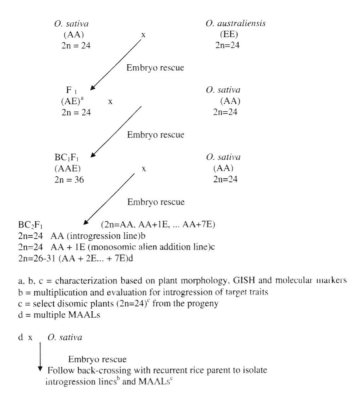

**Figure 5.1**    Production and characterization of monosomic alien addition lines (2n = 25) from crosses of rice and distantly related wild species of *Oryza*.

in *O. officinalis* (2n = 24 CC), *O. minuta* (2n = 48 BBCC), *O. latifolia* (2n = 48 CCDD), *O. australiensis* (2n = 24 EE), *O. brachyantha* (2n = 24 FF), *O. granulata* (2n = 24 GG), and *O. ridleyi* (2n = 48 HHJJ) (see Brar and Khush 2002). The MAALs show characteristic features of changes in morphology like primary trisomics of rice. In general, MAALs show reduced plant height, and seed setting is also low. The female transmission rates of the alien chromosome could vary from 4.4 to 35.5% in MAALs of *O. latifolia* (Multani et al. 2003). MAALs are useful cytogenetic stocks in mapping genes introgressed from wild species and also serve as an important source of alien genetic variation.

### 5.5.5.4  Comparative Genetic Maps

The development of molecular genetic maps has been of great value in understanding the homoeologous relationships between the genomes of various crop plants. Ahn et al. (1993) found extensive homoeologies in several regions of the genomes of wheat, rice, and maize. Later studies in rice and wheat have shown many wheat chromosomes contain homoeologous genes and genomic DNA fragments in an order similar to that found in rice. Comparative genome mapping in rice, maize, wheat, barley, sorghum, foxtail millet, and sugarcane into a single synthesis demonstrates that gene content and orders are highly conserved at both the map and megabase level between different species within the grass family, but the amount and organization of repetitive sequences have diverged considerably (Devos and Gale 1997). Comparative genomics reveals that cereal genomes are composed of similar genomic building stocks (linkage blocks).

The genomes of major cereals have been aligned by dissecting the individual chromosome into segments and rearranging these blocks into highly similar structures. Based on comparative mapping, gene location in one species can be used to predict presence and location of orthologous loci in other species. Comparative mapping is accelerating map-based cloning of orthologous genes. The synteny relationships among cereals have resulted in the discovery of common genes such as the dwarf phenotype in maize (*D8*) and wheat (*Rht1*) based on genomic information derived from rice.

## 5.6 GERMPLASM ENHANCEMENT THROUGH INTERSPECIFIC HYBRIDIZATION

Recent advances in tissue culture, genetic engineering, molecular cytogenetics, comparative genetics, and genomics, particularly in rice genome sequencing, have opened new opportunities to develop improved rice germplasm with novel genetic properties and in understanding the function of rice genes. Breeders have successfully used conventional breeding methods and exploited the rice (*O. sativa*) gene pool to develop high-yielding improved rice varieties resistant to pests with improved quality characteristics. The major emphasis has been on utilizing indica, japonica, and javanica germplasm through intraspecific hybridization (indica × indica, japonica × japonica, indica × japonica). More recently, indica × tropical japonica (javanica) crosses have been used to develop new plant-type rices and the commercial high-yielding hybrids. In several cases, genetic variability for target agronomic traits is limited in the cultivated rice gene pool. Under such situations, interspecific hybridization is an important plant breeding approach to introduce novel genes for different agronomic traits from wild species into rice.

The genus *Oryza* has 22 wild species (2n = 24, 48) representing 10 genomic types (AA, BB, CC, BBCC, CCDD, EE, FF, GG, HHJJ, and HHKK). We have more than 3000 accessions of wild species and 1500 accessions of cultivated African rice (*O. glaberrima*) in the rice gene bank. These wild species are reservoirs of many useful genes, particularly for resistance to major biotic and abiotic stresses. However, these wild species are associated with several weedy traits, such as grain shattering, poor plant type, poor grain characteristics, and low seed yield. Besides, several incompatibility barriers limit the transfer of useful genes from wild species into cultivated species (Brar and Khush 1986, 1997). The major consideration in alien gene transfer is to selectively transfer agronomically important genes from wild species, avoiding linkage drag. To achieve precise transfer of genes from wild to cultivated species, strategies involving a combination of conventional plant breeding methods with tissue culture and molecular approaches have become important (Brar and Khush 2002). Advances in tissue culture, molecular marker technology, genomics, and fluorescence *in situ* hybridization have opened new opportunities to tap alien genetic variability from distant genomes of *Oryza* through interspecific hybridization.

At IRRI, useful genes for resistance to BPH, BB, blast, and tungro, and tolerance to acid sulfate conditions and cytoplasmic male sterility have been transferred from wild species into rice. Some of the IRRI breeding lines resistant to BPH have been released as varieties (MTL98, MTL103, MTL110) in Vietnam. During 2002, three varieties (Matatag 9, AS996, NSIC RC112) have been released through wide hybridization.

### 5.6.1  Strategies for Alien Gene Transfer

#### 5.6.1.1  Search for Useful Genetic Variability for Target Traits

This involves screening wild species germplasm to identify particular accession(s) possessing useful variability for specific target traits. In a number of cases, the desired genetic variability is present only in selected accessions of a particular species. Emphasis should be placed on identifying variability in the closely related (AA genome) species, followed by CC genome species, and later screening distantly related species such as *O. brachyantha* (FF), *O. granulata* (GG), and *O. ridleyi* (HHJJ), which show limited homoeologous pairing with the AA genome.

### 5.6.1.2 Production of Hybrids and Introgression Lines

Interspecific hybrids are produced between elite breeding lines with the wild species accessions carrying useful genes for target traits of immediate interest to the breeder. Such hybrids are produced through direct crosses between rice and AA genome wild species. However, embryo rescue is required to produce hybrids and backcross progenies (introgression lines) between rice and all the wild species of *Oryza* except AA genome species.

### 5.6.1.3 Evaluation of Introgression Lines for Transfer of Target Traits

Advanced introgression lines generated through backcrossing are evaluated for the transfer of target traits. This involves extensive laboratory, screen house, and field testing using various screening and inoculation protocols and testing in hot-spot nurseries for major biotic and abiotic stresses. Evaluation is carried out in different rice-growing countries in collaboration with National Agricultural Research and Extension Systems (NARES).

### 5.6.1.4 Molecular Characterization of Introgression and Construction of Alien Chromosome Segment Substitution Lines

Molecular markers are used to characterize introgression from wild species during backcross breeding. The availability of a dense molecular map of rice comprising simple sequence repeat (SSR) markers has facilitated large-scale analysis to determine the extent and process of introgression and characterize the alien chromosome segments' substitution into the rice genome.

### 5.6.1.5 Mapping of Introgressed Alien Genes/QTL

Monosomic alien addition lines (MAALs) are used to locate the introgressed alien gene(s). Different types of mapping populations are generated through wide crosses such as recombinant inbred lines (RIL), doubled haploids (DH), and near-isogenic alien introgression lines, including segregating populations $F_2$, $F_3$, and BC (alien introgression lines × recurrent parent). Introgressed alien genes and QTL are mapped and tagged with molecular markers for use in marker-assisted selection.

### 5.6.1.6 Characterization of Wide Cross Progenies Using GISH

Genomic *in situ* hybridization (GISH) has become popular in characterization of parental genomes in interspecific progenies and for locating introgressed segments on rice chromosomes. Fluorescence *in situ* hybridization (FISH) techniques are used on both mitotic and meiotic chromosomes. Total genomic DNA of wild species is used as a probe in FISH experiments. Centromere-specific probes and BAC clones are also used to map the introgressed alien segments/genes on rice chromosomes.

### 5.6.2 Production of Interspecific Hybrids, MAALs, and Advanced-Backcross Progenies

Hybrids have been produced between rice (AA genome) and wild species of *Oryza*, representing all 10 genomes (AA, BB, CC, BBCC, CCDD, EE, FF, GG, HHJJ, HHKK) through direct crosses or through embryo rescue (Brar et al. 1991; Brar and Khush 2002). MAALs are isolated from the wide cross progenies (Figure 5.1). A large number of alien introgression lines have been produced

**Table 5.3   Wide Cross Progenies between Rice and Wild Species of *Oryza* Produced at IRRI through Direct Crosses and Embryo Rescue**

| Cross Combination | F₁ | Method | Advanced Introgression Lines (2n = 24) | MAALs (2n = 25) |
|---|---|---|---|---|
| *O. sativa* (AA) × *O. rufipogon* (AA) | + | Direct cross | | − |
| *O. sativa* (AA) × *O. glumaepatula* (AA) | + | Direct cross | + | − |
| *O. sativa* (AA) × *O. longistaminata* (AA) | + | Direct cross | + | − |
| *O. sativa* (AA) × *O. glaberrima* (AA) | + | Direct cross | + | − |
| *O. sativa* (AA) × *O. punctata* (BB) | + | Embryo rescue | n/a | − |
| *O. sativa* (AA) × *O. officinalis* (CC) | + | Embryo rescue | + | + |
| *O. sativa* (AA) × *O. rhizomatis* (CC) | + | Embryo rescue | n/a | n/a |
| *O. sativa* (AA) × *O. eichingeri* (CC) | + | Embryo rescue | n/a | n/a |
| *O. sativa* (AA) × *O. minuta* (BBCC) | + | Embryo rescue | + | + |
| *O. sativa* (AA) × *O. latifolia* (CCDD) | + | Embryo rescue | + | + |
| *O. sativa* (AA) × *O. alta* (CCDD) | + | Embryo rescue | n/a | n/a |
| *O. sativa* (AA) × *O. glandiglumis* (CCDD) | + | Embryo rescue | n/a | n/a |
| *O. sativa* (AA) × *O. australiensis* (EE) | + | Embryo rescue | + | + |
| *O. sativa* (AA) × *O. brachyantha* (FF) | + | Embryo rescue | + | + |
| *O. sativa* (AA) × *O. granulata* (GG) | + | Embryo rescue | + | + |
| *O. sativa* (AA) × *O. ridleyi* (HHJJ) | + | Embryo rescue | + | n/a |
| *O. sativa* (AA) × *Porteresia coarctata* (HHKK) | + | Embryo rescue | n/a | n/a |
| *O. sativa* (AA) × *Zizania palustris*[a] | − | Embryo rescue | n/a | n/a |

*Note*: Minus sign (−) indicates that because of homologous genomes, MAALs are not recovered. +, available; n/a, not available; MAALs, monosomic alien addition lines.

[a]Asymmetrical hybrid.

through backcrossing with the recurrent rice parents. At IRRI, a series of hybrids, MAALs representing 7 to 12 chromosomes of 7 wild species and introgression lines, have been produced through direct crosses, as well as through embryo rescue (Table 5.3). Most of these wide crosses have been used to transfer useful genes into rice (Table 5.4). Some examples include transfer of resistance to BPH, BB, blast, and tungro, and tolerance to acid sulfate conditions, including introgression of genes for cytoplasmic male sterility from different wild species into rice.

## 5.6.3   Introgression from AA Genome Wild Species

Crosses between cultivated rice (*O. sativa*, 2n = 24 AA) and the AA genome wild species can be easily made and the genes transferred through conventional crossing and backcrossing procedures. Hybrids between *O. sativa* and *O. rufipogon* are partially fertile; however, *O. sativa* × *O. glaberrima* and *O. sativa* × *O. longistaminata* F₁s are highly sterile. Among the classical examples are the introgression of a gene for grassy stunt virus resistance from *O. nivara* to cultivated rice varieties (Khush 1977), and the transfer of a cytoplasmic male sterile (CMS) source from wild rice, *O. sativa* f. *spontanea*, to develop CMS lines for commercial hybrid rice production (Lin and Yuan 1980). Other useful genes, such as *Xa21* for bacterial blight resistance, were transferred into rice from *O. longistaminata*, and new CMS sources from *O. perennis* and *O. glumaepatula*. More recently, genes for tungro tolerance and tolerance to acid sulfate soil conditions have been transferred from *O. rufipogon* into indica rice cultivar. Three varieties (Matatag 9, AS996, NSIC RC112) have been released from crosses of *O. sativa* × *O. rufipogon* and *O. sativa* × *O. longistaminata*. A summary of genes transferred from wild species to cultivated rice is given in Table 5.4.

### 5.6.3.1   Introgression of Gene(s) for Resistance to Grassy Stunt Virus

The grassy stunt virus is a serious disease transmitted by the brown planthopper (BPH). Of the 6000 accessions of cultivated rice and several wild species screened, only one accession of *O.*

**Table 5.4 Introgression of Genes of Wild *Oryza* Species into Rice**

| Trait | Donor *Oryza* Species | | | |
| | Wild Species | Gene | Genome | Accession Number |
|---|---|---|---|---|
| **A. Transferred to *Oryza sativa*** | | | | |
| Grassy stunt resistance | *O. nivara* | ae | AA | 101508 |
| Bacterial blight resistance | *O. rufipogon* | *Xa23*(t) | AA | RBB16 |
| | *O. longistaminata* | *Xa21* | AA | 110404 |
| | *O. officinalis* | ae | CC | 100896 |
| | *O. minuta* | ae | BBCC | 101141 |
| | *O. latifolia* | ae | CCDD | 100914 |
| | *O. australiensis* | ae | EE | 100882 |
| | *O. brachyantha* | ae | FF | 101232 |
| Blast resistance | *O. minuta* | *Pi9*(t) | BBCC | 101141 |
| Brown planthopper resistance | *O. officinalis* | *bph11*(t), *bph12*(t) | CC | 100896 |
| | *O. minuta* | ae | BBCC | 101141 |
| | *O. latifolia* | ae | CCDD | 100914 |
| | *O. australiensis* | *Bph10* | EE | 100882 |
| Whitebacked planthopper resistance | *O. officinalis* | ae | CC | 100896 |
| | *O. latifolia* | ae | CCDD | 100914 |
| Cytoplasmic male sterility | *O. sativa f. spontanea* | ae | AA | |
| | *O. perennis* | ae | AA | 104823 |
| | *O. glumaepatula* | ae | AA | 100069 |
| | *O. rufipogon* | ae | AA | |
| Tungro tolerance | *O. rufipogon* | ae | AA | 105908 |
| Tolerance to iron toxicity | *O. glaberrima* | ae | AA | TOG5675 |
| Tolerance to aluminum toxicity | *O. rufipogon* | QTL | AA | 106424 |
| Yield-enhancing loci | *O. rufipogon* | *yld1, yld2* | AA | |
| **B. (Progenies) Introgression Lines under Evaluation** | | | | |
| Yellow stemborer | *O. longistaminata* | ae | AA | 110404 |
| Sheath blight resistance | *O. minuta* | ae | BBCC | 101141 |
| Increased elongation ability | *O. rufipogon* | ae | AA | CB751 |
| Tolerance to acidity, aluminum toxicity | *O. glaberrima* | ae | AA | Many |
| | *O. rufipogon* | ae | AA | 106412 |
| Weed competitive ability | *O. glaberrima* | ae | AA | TOG6216 |

*Source*: Modified from Brar and Khush (2002).

*nivara* (accession 101508) was found to be resistant (Ling et al. 1970). Following backcrosses with improved rice varieties, the gene for grassy stunt resistance was transferred into cultivated germplasm (Khush 1977). The first grassy stunt-resistant varieties, IR28, IR29, and IR30, were released for cultivation in 1974. Subsequently, many such varieties, e.g., IR34, IR36, IR38, IR40, IR48, IR50, IR56, and IR58, have been released, some developed at IRRI, others by NARES.

### 5.6.3.2 Introgression of Gene(s) for Resistance to Tungro Disease

Rice tungro disease (RTD) is the most serious viral disease in southern and Southeast Asia. The disease is caused either by a single infection or by a double infection with two viral particles, the rice tungro bacilliform virus (RTBV), a double-stranded DNA virus, and the rice tungro spherical virus (RTSV), a single-stranded RNA virus. There is limited variability in cultivated rice germplasm for resistance to RTBV, the main cause of tungro symptoms. Kobayashi et al. (1993) found 15 accessions of eight wild species resistant to RTBV. Three accessions of *O. rufipogon* (IRGC accessions 105908, 105909, and 105910) showed a low or moderate level of antibiosis to the vector green leafhopper (GLH). From the crosses of IR64 with *O. rufipogon*, many tungro-tolerant lines have been developed. One of the elite breeding lines (IR73885-1-4-3-2-1-6) resistant to tungro has been released as a stopgap variety (Matatag 9) for cultivation in tungro-prone areas of the Philippines. The recombinant inbred lines (RILs) and SSR markers are being used to map the introgressed genes for tungro tolerance from *O. rufipogon* into IR64.

### 5.6.3.3  Introgression of Gene(s) for Resistance to Bacterial Blight

The bacterial blight caused by *Xanthomonas oryzae* pv. *oryzae* is one of the most destructive diseases of rice in Asia. Many wild species such as *O. longistaminata*, *O. officinalis*, *O. latifolia*, *O. brachyantha,* and *O. ridleyi* have been found to be resistant to bacterial blight. Through backcrossing with the recurrent rice parent, a gene for resistance to bacterial blight was transferred from *O. longistaminata* and designated *Xa21* (Khush et al. 1990). This gene has a wide spectrum of resistance to bacterial blight. The *Xa21* has been transferred through marker-assisted selection in several other indica lines, such as IR64 and PR106, Pusa Basmati, and Samba Mahsuri, including elite breeding lines of new plant-type (NPT) rice (Sanchez et al. 2000; Singh et al. 2001, 2004; Sonti et al. 2004).

Zhang et al. (1998) transferred bacterial blight resistance from *O. rufipogon* (RBB16) into the rice cultivar Jiagang 30. The resistance gene was mapped to chromosome 11 and designated as *Xa23*(t). This gene also conferred a very wide spectrum of resistance and showed a highly resistant reaction to all nine races of bacterial blight of the Philippines.

### 5.6.3.4  Incorporation of CMS Sources from Wild Species

The A genome wild species have been an important source of CMS, the major tool to breed commercial rice hybrids. A number of CMS sources have been developed in rice. However, the most commonly used CMS source in hybrid-rice breeding is derived from the wild species *O. sativa* f. *spontanea* (Lin and Yuan 1980). The cytoplasmic source has been designated as wild abortive (WA), which refers to a male sterile wild rice plant having abortive pollen. About 95% of the male sterile lines used in commercial rice hybrids grown in China and other countries have the WA type of cytoplasm.

A new CMS source from *O. perennis* (accession 104823) was transferred into indica rice (Dalmacio et al. 1995). The male sterility source of the new line (IR66707A) is different from that of WA. Southern hybridization using mtDNA-specific probes showed an identical banding pattern between IR66707A (recipient) and *O. perennis* (donor), indicating that CMS may not be caused by any major rearrangement or modification of mtDNA. Another CMS line (IR69700A) having the cytoplasm of *O. glumaepatula* (A genome species) and the nuclear genome of IR64 has been developed (Dalmacio et al. 1996). No good restorer could be identified for these two CMS lines. Many laboratories have transferred CMS from other AA genome wild species; however, due to lack of good restorers, none of these sources except WA are used in commercial hybrid rice breeding.

### 5.6.3.5  Introgression for Tolerance to Abiotic Stresses

Little or no work has been done on the transfer of genes for tolerance to abiotic stresses from wild species into rice. We have evaluated several introgression lines derived from the crosses of *O. sativa* × *O. rufipogon* and *O. sativa* × *O. glaberrima* at hot spots under field conditions for tolerance to abiotic stresses at Iloilo, Philippines. Elite breeding lines with good agronomic traits and moderately tolerant to iron toxicity, aluminum toxicity, and acid sulfate conditions have been identified. One of the wild species (*O. rufipogon*) that grows under natural conditions in acid sulfate soils of Vietnam was used in crosses with IR64. Three promising lines were selected and tested through the yield-testing network of Cuu Long Delta Rice Research Institute (CLRRI), Vietnam.

Of the three breeding lines, IR73678-6-9-B has been released as a variety (AS996) for commercial cultivation in Mekong Delta, Vietnam. This variety has become popular and occupies 100,000 ha (Bui Chi Buu, personal communication). It is a short-duration (95 to 100 days) semidwarf variety with good plant type suitable for moderately acid sulfate soils and is tolerant to BPH and blast.

### 5.6.3.6   Introgression from African rice (O. glaberrima) into Asian Rice (O. sativa)

Cultivars of Asian rice *O. sativa* are high yielding, whereas African rice, *O. glaberrima*, is low yielding. However, *O. glaberrima* has several desirable traits, such as resistance to rice yellow mottle virus (RYMV), African gall midge, and nematodes, and tolerance to drought, acidity, and iron toxicity. Another important feature of *O. glaberrima* is its strong weed competitive ability. Thus, the interspecific hybridization among Asian and African species offers tremendous potential for combining the high productivity of *O. sativa* with tolerance to biotic and abiotic stresses of *O. glaberrima*. The $F_1$ hybrids between *O. sativa* and *O. glaberrima*, in spite of complete chromosome pairing, are highly sterile. Backcrossing is used to restore fertility and derive agronomically desirable lines. Molecular analysis has revealed frequent exchange of segments between *O. sativa* and *O. glaberrima*.

Efforts have been made at the West Africa Rice Development Association (WARDA) to introgress genes for weed competitive ability from *O. glaberrima* into elite breeding lines of *O. sativa* (Jones et al. 1997). At IRRI, a large number of advanced introgression lines have been produced from crosses of *O. sativa* and several accessions of *O. glaberrima*. These progenies are being evaluated in collaborative projects with WARDA and NARES for introgression for tolerance to RYMV, African gall midge, and abiotic stresses. Some lines for tolerance to iron toxicity have been identified. Major efforts are under way to introgress weed competitive ability into high-yielding indica cultivars.

### 5.6.4   Development of Doubled Haploids from O. sativa × O. glaberrima

Hybrids between *O. sativa* and *O. glaberrima* can be produced easily through direct crosses; however, the $F_1$ in spite of complete chromosome pairing is highly sterile. Several sterility genes differentiate these two species (Oka 1988). Anther culture is an important technique to overcome sterility and to produce homozygous lines, fix recombinants, and use such doubled haploid (DH) lines as mapping populations to locate genes governing agronomic traits.

We cultured 45,400 anthers from 75 $F_1$s; no calli were produced from 34 crosses (Enriquez et al. 2000). The other 41 $F_1$s showed, on average, 1.3% callus formation from 144,160 cultured anthers. The anther-derived calluses from 16 $F_1$s showed no plant regeneration. Strong genotypic differences for anther culturability for both callus induction and plant regeneration were observed. Furthermore, callus induction and plant regeneration from anther culture were found to be independent of each other. Although 562 DH lines could be produced from different crosses, the majority showed high seed sterility (56.2 to 100%). Such high sterility of DH lines is indicative of the presence of several loci for sterility that differentiate between the Asian and African rice species. Genotyping of DH lines using microsatellite markers indicated frequent exchange of chromosome segments between *O. sativa* and *O. glaberrima*. Some of the markers located on chromosomes 1, 6, 9, 11, and 12 showed distorted segregation favoring alleles of either parents (Enriquez 2001).

### 5.6.5   Construction of Chromosome Segment Substitution Lines of O. glaberrima and O. rufipogon in the Background of O. sativa

Chromosome segment substitution lines (CSSLs) are valuable cytogenetic stocks for mapping of genes governing agronomic traits. Molecular markers have made it possible to develop well-defined CSSLs. Doi et al. (2003) constructed a series of *O. glaberrima*, *O. glumaepatula*, and

*O. meridionalis* CSSLs in the background of japonica rice Taichung 65. These lines cover most parts of the genome of donor species. We are using microsatellite markers to identify and develop CSSLs of *O. glaberrima* and *O. rufipogon* in the background of the high-yielding indica cv. IR64 and elite breeding line of NPT rice. Such CSSLs are an important genetic resource for functional genomics of rice.

### 5.6.6    Identification and Introgression of Yield-Enhancing Loci/QTL Wild Species Alleles from AA Genome Species

Wild species are phenotypically inferior to the cultivated species. Transgressive segregation for yield in crosses of cultivated and wild species suggests that, despite inferior phenotypes, wild species contain genes that can improve quantitative traits, such as yield. Molecular markers have made it possible to identify and introgress desirable QTL from wild species into elite breeding lines. Tanksley and Nelson (1996) proposed advanced-backcross (AB) QTL analysis to discover and transfer valuable QTL alleles from unadapted germplasm, such as wild species into elite breeding lines.

QTL from AA genome wild species of rice for increased yield have been identified. Xiao et al. (1996) analyzed 300 $BC_2$ test-cross families produced from the cross of *O. sativa* × *O. rufipogon*. *Oryza rufipogon* alleles at two marker loci, RM5 (*yld1-1*) on chromosome 1 and RG256 on chromosome 2 (*yld2-1*), however, were associated with enhanced yield. Both alleles *yld1-1* and *yld2-1* were associated with a significant increase in grains per plant. In another experiment, Xiao et al. (1998) identified 68 QTL. Of these, 35 (51%) had trait-improving alleles derived from the phenotypically inferior wild species. Nineteen of these beneficial QTL alleles had no deleterious effects on other characters.

Moncada et al. (2001) followed an advanced-backcross breeding strategy and analyzed $BC_2F_2$ populations derived from the cross involving an upland japonica rice cultivar, Caiapo, from Brazil and an accession of *O. rufipogon* from Malaysia. The populations were tested under drought-prone acid soil conditions. Based on analyses of 125 SSR and RFLP markers, 2 putative *O. rufipogon*-derived QTL were detected for yield, 13 for yield components, 4 for maturity, and 6 for plant height. Recently, Septiningsih et al. (2003) used advanced-backcross QTL analysis to identify and introduce agronomically useful genes from *O. rufipogon* into the cultivated gene pool. A total of 165 markers consisting of 131 SSRs and 34 RFLPs were used to construct the genetic linkage map. Despite its inferior performance, 33% of the QTL alleles originating from *O. rufipogon* had a beneficial effect for yield and yield components in the IR64 background. Twenty-two QTL (53.4%) were located in similar regions as previously reported rice QTL, suggesting the existence of stable QTL across genetic backgrounds and environments. Twenty QTL (47.6%) were exclusively detected in this study, uncovering potentially novel alleles from the wild, some of which might improve the performance of the tropical indica variety IR64. Additionally, several QTL for plant height, grain weight, and flowering time detected in this study corresponded to homeologous regions in maize containing previously detected maize QTL for these traits.

Our preliminary results at IRRI of advanced-backcross progenies derived from the crosses of an elite breeding line of NPT rice, with *O. longistaminata* and IR64 × *O. rufipogon*, also support transgressive segregation for yield and yield components. These findings show that genes from wild species can increase the yield of elite rice lines, even though wild species are phenotypically inferior to cultivated rice.

### 5.6.7    Introgression of Genes from Distantly Related Genomes

Introgression lines have been produced from crosses of *O. sativa* with distantly related species with CC, BBCC, CCDD, EE, FF, GG, and HHJJ genomes. However, gene transfer has been achieved only from CC, BBCC, CCDD, EE, and FF genomes. So far, no introgression could be achieved from GG and HHJJ genomes.

### 5.6.7.1 Introgression from the CC Genome Species

Jena and Khush (1990) produced several introgression lines from the cross of *O. sativa* × *O. officinalis*. One of the most successful examples on transfer of genes from C genome wild species to rice is that of brown planthopper (BPH). Three genes, *Bph10*, *bph11*, and *bph12*, have been transferred from wild species to rice. Four breeding lines have been released as varieties (MTL95, MTL98, MTL103, MTL110) for commercial cultivation in the Mekong Delta, Vietnam.

Hirabayashi et al. (2003) analyzed recombinant inbred lines (RILs) from the cross between Hinohikari (susceptible japonica) and the indica introgression line derived from *O. sativa* × *O. officinalis*. Two genes for BPH resistance, *bph11*(t) and *bph12*(t), were identified and mapped to chromosomes 3 and 4 of rice. Huang et al. (2001) also transferred BPH resistance from *O. officinalis* into Zhensheng 97B.

We produced advanced-backcross progenies from crosses of an elite breeding line of NPT rice (IR65600-81-5-3-2) with two accessions of *O. officinalis* and another with tetraploid species *O. minuta*. The NPT line is highly susceptible to bacterial blight. More than 1053 $BC_2F_3$ progenies derived from NPT × *O. officinalis* (accession 101399) were screened after inoculation with race 1. Several progenies resistant to bacterial blight have been identified.

### 5.6.7.2 Introgression from the BBCC Genome Species

Interspecific hybrids have been produced between *O. sativa* and the tetraploid wild species *O. minuta* (BBCC). Following backcrossing and embryo rescue, advanced introgression lines were evaluated and resistance to bacterial blight and blast was transferred (Amante-Bordeos et al. 1992). The introgressed blast gene has been designated as *Pi9*(t). It has resistance to several isolates of blast. Introgression lines were produced from the cross of NPT × *O. minuta* and evaluated for resistance to 10 Philippine races of bacterial blight. The NPT parent is susceptible to each of the 10 races. One of the family, WHDIS 1958-19, derived from *O. sativa* × *O. minuta* was found to have a broad spectrum of resistance to all 10 races tested. The genes introgressed from *O. minuta* seem to have a wide spectrum of resistance, and also the number of genes introgressed could be more than one. Similarly, BPH resistance from *O. minuta* has been transferred to rice (Brar et al. 1996). These lines have shown a wide spectrum of resistance to BPH in the Philippines and Korea.

### 5.6.7.3 Introgression from the CCDD Genome Species

A number of workers have produced hybrids between rice and CCDD genome species (Sitch 1990; Brar et al. 1991). Several introgression lines derived from *O. sativa* × *O. latifolia* have been evaluated for introgression of useful traits (Multani et al. 2003). Of the 2295 disomic $BC_3F_3$ progenies, 309 showed introgression for resistance to BPH and 188 each for WBPH and BB resistance. Four plant progenies that were resistant to both BPH and WBPH were also resistant to BB race 2 of the Philippines. Introgression for 10 allozymes of *O. latifolia*, such as *Est5*, *Amp1*, *Pgi1*, *Mdh3*, *Pgi2*, *Amp3*, *Pgd2*, *Est9*, *Amp2*, and *Sdh1*, located on 8 of the 12 chromosomes was observed. Alien introgression was also detected for morphological traits such as long awns, earliness, black hull, purple stigma, and apiculus. Abnormal plants with many wild species traits suddenly appeared in normal disomic progenies. These plants showing instability and abnormal segregation behavior are being investigated for the activation of transposons.

### 5.6.7.4 Introgression from the EE Genome Species

Multani et al. (1994) produced hybrids between colchicine-induced autotetraploids of rice and *O. australiensis* (2n = 24 EE). Introgression was detected for morphological traits, such as long awns and earliness, and for *Amp-3* and *Est-2* allozymes. Of the 600 $BC_2F_4$ progenies, four were resistant to BPH. BPH resistance was found to be controlled by a recessive gene in two of the four lines, but was controlled by a dominant gene in the other two lines. Two of these lines (IR65482-4-136, IR65482-7-216) have proven resistant to the Korean BPH population (K.K. Jena, personal communication). One of the lines (IR65782-4-136-2-2) carried the *Bph10* gene. Of the eight MAALs, only MAAL 12 segregated for resistance to BPH. The data on BPH segregation in 2n progenies and MAAL 12 plants (2n = 25) suggested that the gene for BPH resistance is located on chromosome 12.

### 5.6.7.5 Introgression from the FF Genome Species

A series of introgression lines has been derived from the cross of *O. sativa* cv. IR56 and the wild species *O. brachyantha* (2n = 24 FF). IR56 is susceptible to bacterial blight races 1, 4, and 6 from the Philippines, whereas *O. brachyantha* is resistant. Of the 149 backcross progenies analyzed, 27 showed introgression for resistance to bacterial blight races 1, 4, and 6 (Brar et al. 1996). Further, introgression for awning and growth duration has also been obtained.

### 5.6.7.6 Introgression from the GG Genome Species

Hybrids have been produced from the cross of *O. sativa* and *O. granulata* (Brar et al. 1991). Advanced progenies have also been produced; however, none of the lines tested has shown introgression of traits from *O. granulata* into rice.

### 5.6.7.7 Introgression from the HHJJ Genome Species

The tetraploid *ridleyi* complex comprises two species: *O. ridleyi* and *O. longiglumis*. *Oryza ridleyi* shows strong resistance to all 10 Philippine races of BB. Hybrids between rice cv. IR56 and *O. ridleyi* (accession 100821) have been produced. However, the cross shows strong necrosis. Thus, only a few introgression lines (BC$_3$F$_3$) from this cross have been produced, but no introgression could be detected.

### 5.6.7.8 Introgression from the HHKK Genome Species

Intergeneric hybrids between *O. sativa* and *P. coarctata* have been produced through both sexual crosses following embryo rescue (Brar et al. 1997) and protoplast fusion (Jelodar et al. 1999). The hybrid (2n = 36) is sterile and shows no chromosome elimination of either parent. Due to strong incompatibility barriers, no backcross progenies could be obtained.

BC$_2$ progenies derived from crosses of *O. sativa* with *O. officinalis* (CC), *O. australiensis* (EE), *O. brachyantha* (FF), and *O. granulata* (GG) resembled the recurrent rice parent in most morphological traits. This suggested that only a limited amount of recombination between the A genome of *O. sativa* and C, E, F, and G genomes of wild species occurred. Progenies recovered in BC$_2$ of *O. sativa* × *O. officinalis* were so similar in morphology to *O. sativa* that they were evaluated in field trials and released as varieties for commercial cultivation in Vietnam. Most introgressed segments were detected via single RFLP and SSR markers; the flanking markers were negative for introgression. This also supports the conclusion regarding limited recombination and the possible cause for the rapid recovery of the recurrent parent phenotype.

## 5.6.8 Molecular Mapping of Introgressed Alien Genes

Traits introgressed from different wild species into rice are listed in Table 5.4. Some of the introgressed genes have been mapped via linkage to molecular markers.

### 5.6.8.1 Mapping of the Xa21 Gene for Bacterial Blight Resistance

The *Xa21* gene, introgressed from *O. longistaminata*, has been mapped to chromosome 11, close to the RG103 marker (Ronald et al. 1992). This gene has been transferred through marker-assisted breeding into many rice cultivars — IR64, PR106, Pusa Basmati, Samba Mahsuri — and NPT rice (Sanchez et al. 2000; Singh et al. 2001, 2004; Sonti et al. 2004). Wang et al. (1995) used a bacterial artificial chromosome (BAC) library and isolated 12 BAC clones that hybridized with the three DNA markers linked to the *Xa21* locus. Jiang et al. (1995) used BAC clones and FISH and physically mapped the *Xa21* locus to chromosome 11 of rice. The *Xa21* gene has been isolated (Song et al. 1995) via positional cloning. The transgenic plants carrying the cloned *Xa21* show a high level of resistance to bacterial blight pathogen. Field testing of transgenic rice carrying the *Xa21* gene at the Philippine Rice Research Institute (PhilRice), Maligaya, Philippines, showed resistance to bacterial blight.

### 5.6.8.2 Mapping of Genes for Tungro Tolerance

We have produced a number of tungro-tolerant elite breeding lines with genes introgressed from *O. rufipogon*. One tungro-tolerant introgression line (IR73385-1-4-3-2-1-6) derived from IR64 × *O. rufipogon* was used in molecular analysis. Of the 181 SSR markers analyzed, 11 showed introgression of *O. rufipogon* alleles. On average, 6% of the *O. rufipogon* genome was introgressed into *O. sativa*. No marker could be associated with tungro resistance. We are now analyzing backcross progenies derived from the cross of tungro-tolerant introgression lines × IR64 using 11 putative SSR markers to map the gene(s) for tungro tolerance introgressed from *O. rufipogon* into rice.

### 5.6.8.3 Mapping of Genes for BPH Resistance

A gene conferring resistance to three BPH biotypes from the Philippines was introgressed from *O. australiensis* into rice (Multani et al. 1994). MAAL analyses showed that the gene for BPH resistance is located on chromosome 12 of *O. australiensis*. All 14 probes were polymorphic in the recurrent parent and wild species; however, only RG457 detected introgression from *O. australiensis* into the introgression line. Cosegregation for BPH reaction and molecular markers showed a gene for BPH resistance linked to RG457, with a distance of 3.68 cM (Ishii et al. 1994). Hirabayashi et al. (2003) located *bph11*(t), *bph12*(t) introgressed from *O. officinalis* into rice on chromosomes 3 and 4, respectively. In another study, we have identified putative markers for BPH resistance genes introgressed from *O. minuta* into rice.

### 5.6.8.4 Mapping of Pi9t for Blast Resistance

A gene for blast resistance (*Pi9t*) was introgressed from *O. minuta* (BBCC) into rice (Amante-Bordeos et al. 1992). A backcross population produced by crossing the introgression line and the susceptible parent IR31917-45-3-2 was analyzed. Three RAPD markers were found to be linked to *Pi9t* (Reimer and Nelson, unpublished). The *Pi9* shows resistance to several blast isolates.

### 5.6.8.5 Mapping QTL for Tolerance to Aluminum Toxicity

Nguyen et al. (2003) mapped QTL for tolerance to aluminum toxicity introgressed from *O. rufipogon* (accession 106424) into IR64. *Oryza rufipogon* contributed favorable alleles for each of the five QTL for relative root length (RRL), an important parameter for tolerance to aluminum toxicity, and individually explained 9.0 to 24.9% of the phenotyped variation. A major QTL (24.9% phenotypic variation) for RRL was found on chromosome 3 of rice, which is conserved across cereal species. Fine mapping of the QTL is being attempted using BAC clones from rye. Positional cloning of such QTL introgressed from *O. rufipogon* could lead to development of rice cultivars tolerant to aluminum toxicity for cultivation under different ecosystems.

### 5.6.9  Molecular Characterization of Alien Introgression

Both RFLP and SSR markers have been used to detect introgression from different wild species into the rice.

#### 5.6.9.1  *Introgression from O. glaberrima*

We have produced a large number of advanced-backcross progenies from the crosses of elite breeding lines of *O. sativa* with different accessions of *O. glaberrima*. Microsatellite markers were used to detect polymorphisms among different accessions of *O. glaberrima* and *O. sativa*. Introgression was detected from all 12 chromosomes of *O. sativa*, suggesting occurrence of recombination among AA genomes of *O. sativa* and *O. glaberrima*. Analysis of DH populations derived through anther cultures of *O. sativa* × *O. glaberrima* showed frequent exchange of chromosome segments between *O. sativa* and *O. glaberrima*. Distortion of segregation was observed in crosses of *O. sativa* × *O. glaberrima*. SSR analysis of introgression lines has revealed introgression from centromeric regions of *O. glaberrima* chromosomes. Chromosome regions (hot spots) for exchange of segments have been identified.

#### 5.6.9.2  *Introgression from Distant Genomes of Oryza*

Jena et al. (1992) analyzed 52 introgression lines ($BC_2F_8$) derived from crosses of *O. sativa* × *O. officinalis*. Of the 174 informative RFLP markers, 28 (16.1%) identified putative *O. officinalis* introgressed segments in one or more of the introgression lines. Individual introgression lines contained 1.1 to 6.8% introgressed *O. officinalis* segments. Introgressed segments were found on 11 of the 12 rice chromosomes. In most cases, *O. sativa* alleles were replaced by *O. officinalis* alleles. Introgressed segments were smaller in size and similar in plants derived from early and later generations. Single RFLP markers detected most introgressed segments, and the flanking markers were negative for introgression. Brar et al. (1996) analyzed 29 derivatives of *O. sativa* × *O. brachyantha* and 40 derivatives of *O. sativa* × *O. granulata*. Extensive polymorphism between rice and wild species was observed. Of the six chromosomes surveyed, no introgression was detected from chromosomes 7, 9, 10, and 12 of *O. granulata* and chromosomes 10 and 12 of *O. brachyantha*. For each of the remaining chromosomes, one to two RFLP markers showed introgression in some of the derived lines. Although the level of introgression was low, the results showed possibilities of introgressing chromosome segments even from distantly related genomes into cultivated rice, and thus the feasibility of transferring useful genes from distant *Oryza* species.

### 5.6.10  Characterization of Parental Genomes, MAALs, and Homoeologous Pairing in *Oryza* through GISH

*In situ* hybridization is a powerful technique to characterize parental genomes in interspecific hybrids, identify alien chromosome(s) in MAALs, locate introgressed segments, and detect homoeologous pairing, thus leading to the precise understanding of alien introgression into the rice genome. Total genomic DNA was used as a probe in *in situ* hybridization to detect parental genomes in interspecific hybrids (Figure 5.2) and to characterize homoeologous pairing in *Oryza*.

We have used GISH and characterized parental chromosomes in wide hybrids ($F_1$, $BC_1$) involving *O. sativa* (AA) × *O. officinalis* (CC), *O. sativa* (AA) × *O. brachyantha* (FF), *O. sativa* (AA) × *O. australiensis* (EE), *O. sativa* (AA) × *O. granulata* (GG), and *O. sativa* (AA) × *O. ridleyi*

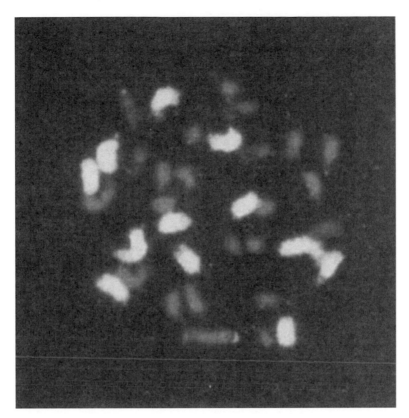

**Figure 5.2    (See color insert following page 114.)** Characterization of parental genomes in wide hybrids of *Oryza* through GISH. Metaphase cell showing 36 chromosomes in F1 of *O. australiensis* (2n = 24,EE) × *O. ridleyi* (2n = 48; HHJJ). Genomic DNA of *O. australiensis* labeled with biotin14–dATP was used as a probe. The chromosomes of two parental species are clearly differentiated: the 12 labeled chromosomes of *O. australiensis* flouresced yellow due to FITC, while 24 unlabeled chromosomes of *O. ridleyi* appear red due to counterstaining with PI.

(HHJJ). The alien extra chromosome in MAALs and introgressed segments could also be identified (Figure 5.3). Asghar et al. (1998) applied FISH for characterizing the chromosomes of *O. sativa* and *O. officinalis* and located rDNA loci on somatic chromosomes of both *O. sativa* and *O. officinalis*. Yan et al. (1999) used FISH to characterize A and C genome chromosomes in $F_1$ and $BC_1$ of *O. sativa* × *O. eichingeri*.

Abbasi et al. (1999) used total genomic DNA of *O. australiensis* as a probe and hybridized it with the meiotic chromosomes of the $F_1$ hybrid (*O. sativa* × *O. australiensis*). Both autosyndetic and allosyndetic pairing among A and E genomes could be detected. Meiotic analysis using GISH showed three types of pairing: (1) between A and E genome chromosomes, (2) within A genome chromosomes, and (3) within E genome chromosomes. Of the paired chromosomes, 78.8% involved A and E genomes, 16.8% showed pairing within A genome chromosomes, and 4.3% within E genome chromosomes. Similarly, autosyndetic and allosyndetic pairing have been characterized among chromosomes involving different genomic combinations such as AA × CC, AA × BBCC, AA × EE, AA × FF, AA × GG, and AA × HHJJ (Hue 2004). The results demonstrate the usefulness of FISH for precisely characterizing homoeologous pairing and highlighting overestimation resulting from autosyndetic pairing, which is otherwise difficult to detect through conventional cytogenetic analysis. FISH revealed one or two bivalents resulting from pairing among A genome chromosomes, indicating duplication of chromosome segments, and thus the possibility of rice being a secondary polyploid. Similar intragenomic duplications have been shown in pearl millet (see Chapter 1 by Jauhar in this volume). We are extending the FISH technique and centromere-specific probe to detect pairing among

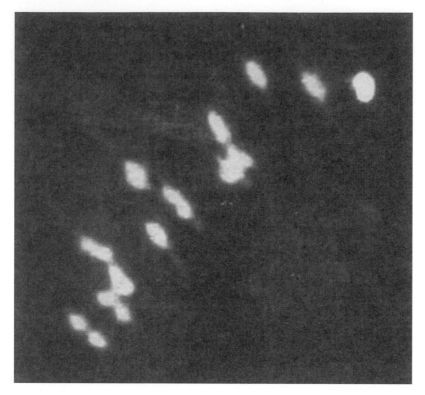

**Figure 5.3**   **(See color insert following page 114.)** Characterization of wild species chromosome through GISH in MAAL of rice. Metaphase I showing 12II of *O. sativa* and one extra chromosome of *O. ridleyi*. Genomic DNA of *O. ridleyi* labeled with biotin14–dATP was used as a probe. The single chromosome of *O. ridleyi* flouresced yellow due to FITC, while all chromosomes (12II) of *O. sativa* appeared red due to counterstaining with PI.

A genome and other distantly related genomes of *Oryza* at pachytene to precisely understand the process of alien introgression, particularly of smaller chromosome segments.

Variability in rDNA loci has been detected through FISH in *Oryza* species (Fukui et al. 1994). Jiang et al. (1995) mapped the *Xa21* gene derived from *O. longistaminata* to rice chromosomes, using FISH and BAC clones. Shishido et al. (1998) used multicolor FISH to characterize A, B, and C genomes in somatic hybrids of rice.

## 5.7 GERMPLASM ENHANCEMENT THROUGH CELL AND TISSUE CULTURE

During the last few decades, major advances have been made in refining the procedures for somatic cell culture, anther culture, and protoplast culture. It has become possible to regenerate from cultured cells using various explants from diverse genotypes. This has led to the use of cell and tissue culture techniques in rice improvement. Useful somaclonal variants and anther culture-derived lines have become available. Plants can be regenerated from protoplasts of both indica and japonica rices.

### 5.7.1   Anther Culture

As early as 1968, Niizeki and Oono reported the production of haploids from anther culture of rice. Since then, the anther culture technique has been greatly refined. It is now possible to produce haploids from the anther culture of many japonica and indica rice varieties, although the frequency of regenerated plants is relatively lower in indicas. More than 100 breeding lines and varieties have been developed through anther culture in China; notable examples include Huayu-1, Huayu-2, Tanfong 1, and Xin Xiu. Huayu-1 is a high-yielding variety resistant to bacterial blight and has wide adaptability. One anther culture-derived line, IR51500-AC-11-1, has been released as a variety (PSBRc50) in the Philippines. The Republic of Korea has also released several rice varieties (Hwacheongbyeo, Joryeongbyeo, Hwajinbyeo) through anther culture. Hwacheongbyeo is resistant to brown planthopper, blast, and bacterial blight and is recommended for southwestern plain and southwestern coastal areas of Korea. Hwajinbyeo was developed from the cross of Milyang 64 × Iri53 and is resistant to stripe virus and bacterial blight. Another cultivar, Texmont, has been developed by anther culture in the U.S. Most of the anther culture-derived varieties are japonicas; however, indica varieties are generally recalcitrant. The double-haploid (DH) lines produced from indica × japonica (IR64 × Azucena, CT9993 × IR62226) are being used in molecular mapping of genes and quantitative trait loci (QTL) governing agronomic traits. Many DH lines have been obtained from the crosses of *O. sativa* × *O. glaberrima* (Enriquez et al. 2000).

### 5.7.2   Somaclonal Variation

Somaclonal variation refers to the variation arising through cell and tissue culture in regenerated plants and their progenies. The technique consists of growing callus or cell suspension cultures for several cycles and regenerating plants from these long-term cultures. The regenerated plants and their progenies are evaluated in order to identify individuals with a new phenotype. Somaclonal variation has been reported in large numbers of plant species: potato, sugarcane, tobacco, tomato, wheat, rice, *Brassica*, and others for various agronomic traits, such as disease resistance, plant height, tiller number, and maturity, and for various physiological and biochemical traits. Several factors, such as genetic background, explant source, medium composition, and age of culture, affect somaclonal variation. Some useful somaclonal variants have been isolated and released as varieties in crop plants.

In rice, somaclonal variants for disease resistance and male sterility have been isolated. Heszky and Simon-Kiss (1992) tested several somaclonal variants of anther culture origin, one of which was released as variety Dama: it is resistant to *Pyricularia* and has good cooking qualities. Similarly, Ogura and Shimamoto (1991) identified useful somaclonal variants from protoplast-regenerated progenies of Koshihikari, and a new variety, Hatsuyeme, was released. This variety is late by 1 week, shorter in height, lodging resistant, and has 10% higher grain yield than the mother variety, Koshihikari. The exact mechanism of somaclonal variation is not understood. The probable causes include changes in karyotype (chromosome number and structure), cryptic changes associated with chromosome rearrangements, transposable elements, somatic gene arrangements, DNA methylation, deletion, etc. (Brar and Jain 1998). Epigenetic changes also occur frequently. It is possible that different processes cause such variation in different species, or one or several factors operate simultaneously during *in vitro* culture, resulting in somaclonal variation.

### 5.7.2.1   Enhancing Alien Introgression through Somaclonal Variation

The technique of somaclonal variation appears to be particularly important in enhancing variation in interspecific crosses, particularly where the parental genomes of the two species show little or no homoeology. Under such situations, chromosome breakage and reunion could result in new combinations and in the transfer of alien chromosome segments into the cultivated species. In hybrids of *Hordeum vulgare* × *H. jubatum*, enhanced variation in isozyme pattern and chromosome pairing was observed in contrast to the original hybrid, which was asynaptic.

Larkin et al. (1989) and Banks et al. (1995) reviewed the usefulness of cell culture to enhance alien introgression in wide crosses. Tissue culture of wheat–rye monosomic addition lines showed introgression of cereal cyst nematode resistance from rye to wheat. Similarly, tissue culture of monosomic alien addition lines of wheat–*Thinopyrum intermedium* showed introgression of barley yellow dwarf virus (BYDV) resistance into wheat. The technique appears equally promising to obtain chromosomal exchanges and derive progenies with introgression of useful genes from somatic hybrids produced through protoplast fusion among widely divergent species.

A tissue culture cycle of the wide hybrids of rice ($F_1$, monosomic alien chromosome addition lines, somatic hybrids) could enhance the frequency of genetic exchange, particularly among distant genomes (AA × FF, AA × GG, AA × HHJJ, AA × HHKK) of *Oryza* species, which otherwise lack or show very limited homoeologous pairing, and where gene transfer through conventional methods is difficult. However, so far the technique has not been employed for enhancing exchanges among cultivated rice and distant genomes of *Oryza*.

### 5.7.3  Somatic Cell Hybridization

Somatic cell hybridization involves isolation, culture, and fusion of protoplasts from different species and regeneration of somatic hybrid plants. It is an alternative to sex parasexual hybridization. The technology provides potential to develop cybrids and organelle recombinants and to (1) produce somatic hybrid novel plants between otherwise sexually incompatible species, (2) produce cybrids and exploit the cytoplasmic variability (mitochondrial recombination), (3) transfer cytoplasmic male sterility (CMS) into adapted cultivars through fusion of X-irradiated protoplasts of the CMS parent, (4) study compatibility between nuclei and cytoplasms of different species, and (5) in the uptake of chloroplasts and isolated chromosomes into protoplasts. Since the first production of somatic hybrids in tobacco in 1972, many laboratories worldwide have produced somatic hybrids in many species through protoplast fusion, but with limited success in applied breeding.

Somatic hybrids have been produced between rice and a few wild species (Table 5.5). Somatic cell hybrids were produced through protoplast fusion between rice and carrot (Sala et al. 1985) and between rice and soybean (Niizeki et al. 1985); however, most of the hybrid cells failed to regenerate plants. Terada et al. (1987) attempted protoplast fusion of rice and barnyard grass (*Echinochloa oryzicola* Vasing). A total of 166 calli were identified as hybrid by isozyme and chromosome analyses. Of the 44 regenerated shoots, 9 grew to plantlets.

Hayashi et al. (1988) produced 250 somatic hybrid plants through electrofusion of protoplasts of rice with four wild species. Somatic hybrids have also been produced through protoplast fusion of *O. sativa* and *Porteresia coarctata* (Jelodar et al. 1999). Liu et al. (1999) produced a highly asymmetrical and fertile somatic hybrid through protoplast fusion of *O. sativa* and *Zizania latifolia*. Southern analysis of hybrid plants (2n = 24) showed intergenomic exchange between rice and *Zizania*.

**Table 5.5  Intergeneric Somatic Hybrids Produced in Rice through Protoplast Fusion**

| Combination | Method | Reference |
|---|---|---|
| *O. sativa* + *Daucus carota*[a] | Polyethylene glycol | Sala et al. 1985 |
| *O. sativa* + *Glycine max*[a] | Polyethylene glycol | Niizeki et al. 1985 |
| *O. sativa* + *Echinochloa oryzicola* | Electrofusion | Terada et al. 1987 |
| *O. sativa* + *O. officinalis* | Electrofusion and nurse culture | Hayashi et al. 1988 |
| *O. sativa* + *O. eichingeri* | Electrofusion and nurse culture | Hayashi et al. 1988 |
| *O. sativa* + *O. brachyantha* | Electrofusion and nurse culture | Hayashi et al. 1988 |
| *O. sativa* + *O. perrieri* | Electrofusion and nurse culture | Hayashi et al. 1988 |
| *O. sativa* + *Daucus carota* | Electrofusion | Kisaka et al. 1994 |
| *O. sativa* + *Zizania latifolia* | Electrofusion | Liu et al. 1999 |
| *O. sativa* + *Porteresia coarctata* | Electrofusion | Jelodar et al. 1999 |
| *O. sativa* + *Zizania palustris*[b] | Bombardment with total genomic DNA of *Zizania* | Abedinia et al. 2000 |
| *O. sativa* + *Hordeum vulgare* | Electrofusion | Kisaka et al. 2001 |

[a]Hybrid cells only.
[b]Transgenic plants.

Kisaka et al. (2001) produced a hybrid plant through electrofusion of rice and barley protoplast. The selection of somatic hybrids was based on the low rate of cell division of barley protoplasts and lack of regenerative ability of rice protoplasts. Only one plant regenerated which, upon Southern hybridization with the *trpB* gene of rice, showed bands specific to rice (9.0 kb) and barley (3.0 kbb). It showed novel bands of both mtDNA and ctDNA, which were not detected in either of the parents.

### 5.7.3.1  Production of Cybrids and Organelle Recombinants

The production of cytoplasmic hybrids (cybrids) following protoplast fusion is one of the most exciting applications of protoplast fusion. Protoplasts provide a unique opportunity to produce cybrids and recombine cytoplasmically inherited traits. In cybridization, the nuclear genome of one parent is combined with the organelles of a second parent. In effect, organelles are transferred from one parent to the other in a single step. Cybrids have been produced in several species, including rice. The cybridity is confirmed on the basis of mtDNA restriction patterns, morphological traits, and isozyme and cytological tests. Cybrid plants of rice have been created by donor–recipient protoplast fusion between CMS lines and fertile cultivars (Akagi et al. 1989; Kyozuka et al. 1989). Yang et al. (1989) produced cybrid plants in rice by electrofusing gamma-irradiated protoplasts of A-58 CMS and iodoacetamide-treated protoplasts of the fertile cultivar Fujiminori. Cybrids had peroxidase isozyme of the fertile parent but had four plasmid-like DNAs (B1, B2, B3, and B4) from the sterile A-58 CMS parent in their mitochondrial genomes.

### 5.7.3.2  Transfer of Cytoplasmic Male Sterility

Transfer of CMS to elite breeding lines requires five to seven repeated backcrosses. The success in protoplast fusion has made it possible to transfer CMS within several months. Donor–recipient protoplant fusion is an efficient procedure for the transfer of CMS. Protoplasts of the donor CMS line are exposed to high doses of irradiation and fused with the iodoacetamide-treated protoplasts of the recipient line. Irradiation inactivates the nucleus, while chemical treatment with iodoacetamide inhibits protoplast division. The donor–recipient method has been used in rice to transfer CMS into elite lines. The method has been used to transfer CMS from indica rice Chinsurah Boro into 35 japonica rice varieties (Kyozuka et al. 1989). Nakamura et al. (1995) developed the CMS line Bio-Mother 1 using this method. Akagi (2001) reported production of CMS hybrid plants using WA cytoplasm. The cybrid plants between Sasanishiki and IR58024A were successfully converted to CMS. The system enables transfer of CMS traits encoded in the mitochondrial genome to the fertile line in a single step. However, the method has been used to a very limited extent.

## 5.8 GERMPLASM ENHANCEMENT THROUGH INDUCED MUTATIONS

Since the first demonstration in 1927 by Muller of induced genetic variation by irradiation in *Drosophila*, and in 1928 by Stadler in barley, numerous reports have become available on mutagen-induced variability for several agronomic traits. Many kinds of mutants have become available, e.g., chlorophyll mutation, pigmentation, height, male sterility, disease resistance, and other agronomic traits. According to the FAO/IAEA database, more than 2359 varieties have been released through mutation breeding in 169 crop species. Among cereals, in rice alone more than 400 varieties have been released through mutation breeding.

However, only a few varieties released through induced mutagenesis have become popular. Rutgers (1992a) reviewed the status of induced mutation in rice improvement. Major improvements through mutation breeding have been earlier maturity, short stature, and grain character modifications. Reimei, a short-statured mutant cultivar in Japan, was the success story. Reimei's subsequent use in crossbreeding has led to the release of 21 additional cultivars. Reimei carries a semidwarf allele that is allelic to *sd1*. Cultivars that apparently have become important include Miyana-Nishiki, Miyuki-mochi, Shirankabanishiki, Shinano-Sakigake, and Iwate 21 (Kawai and Amano 1991). Varieties such as Yuanfangzao and Zhefu 802, each of which has been grown on over 1 million ha annually in China, and the numerous short-stature upland rice cultivars developed by IRAT and its cooperators in Africa and South America. Calrose 76 was the first U.S. semidwarf cultivar developed through induced mutation with acceptable grain quality. Although Calrose 76 itself was grown on only 7% of the total rice area in the U.S., it was quickly followed by nine additional semidwarf cultivars.

The most widely grown mutant rice cultivars have been Yuanfengzao and Zhefu 802, both in China. Yuanfengzao, a gamma-ray-induced early-maturity mutant was grown on over 1,100,000 ha in the lower Yangtze River region during the early 1980s. Zhefu 802 was also early maturing, had broader disease resistance and a higher yield than its parent, and was grown on more than 1,400,000 ha in China in 1989 (Wang 1991). Seven other mutant cultivars have been grown on 100,000 or more hectares in China. Some other important varieties include Qinghuaai 6, Nanjing 34, Reimei, Akihikari, Dongting 3, Wanhua, Aifu 9, Wangeng 257, and Hongnam.

A number of induced mutant marker genes have been used in genetic analysis. More recently, deletion mutants, T-DNA insertion lines, and retrotransposon-induced variation have become important genetic resources for functional genomics.

## 5.9 APOMIXIS FOR GERMPLASM ENHANCEMENT

Apomixis refers to asexual reproduction through seed. It is a method of reproduction in which the embryo (seed) develops without the union of the egg and sperm. Apomictic seed is genetically identical to the maternal parent. Apomixis is being explored as a new frontier project to exploit hybrid vigor and develop true breeding hybrid rice varieties. It will enhance the efficiency of hybrid rice breeding by producing many new true breeding hybrids, compared with those produced by using the three-line or two-line hybrid breeding system. The availability of a large number of hybrids will help increase genetic diversity and reduce genetic vulnerability. Vulnerability caused by cytoplasm would be virtually eliminated because a specific cytoplasmic-nuclear male sterility-inducing cytoplasm would not be needed to release a hybrid commercially.

Apomixis is widespread and more than 300 plant species are apomictic. It occurs mainly in polyploid species. Apospory, diplospory, and adventitious embryony are the most common types of apomixis. Various cytological, histological, and genetic tests are available to screen germplasm for apomixis. Genetic studies indicate that the switch between sexual reproduction and apomixis is controlled by one or two major genes. DNA marker analysis has shown restricted recombination around some apomixis loci. Khush et al. (1994) have reviewed different aspects of apomixis, such as occurrence, types, inheritance, screening techniques, and potential for exploiting hybrid vigor and cultivar development. Three approaches have been proposed to obtain apomictic rice: (1) searching for apomixis in the wild *Oryza* germplasm, (2) mutagenesis to induce apomixis, and (3) molecular approaches to engineer apomixis.

### 5.9.1 Screening Wild *Oryza* Germplasm for Apomixis

As discussed earlier, apomixis has been reported in the wild relatives of several crop species. Besides two cultivated species, the genus *Oryza* has 22 wild species. IRRI has a collection of more than 3000 accessions of wild species. Rutgers (1992b) screened 547 accessions of closely related wild species of rice with AA genomes through the pistil-clearing technique. Results were negative. Apomixis is rare in diploid species but is common in polyploid relatives of crop plants. Hence, tetraploid wild species of *Oryza* and other related genera were screened using two techniques: (1) pistil clearing and (2) callose detection using fluorescence microscopy.

**Table 5.6   Wild Species of *Oryza* and Related Genera Screened at IRRI for Apospory (Multiple Embryo Sac Development) and Diplospory (Callose Fluorescence) Types of Apomixis**

| Species | | Genome | Number of Accessions Analyzed | |
|---|---|---|---|---|
| | | | Apospory | Diplospory |
| **Diploid *Oryza* Species** | | | | |
| *O. sativa* (cultivated) | | AA | 10 | 10 |
| *O. nivara* | | AA | 4 | 4 |
| *O. longistaminata* | | AA | 1 | 1 |
| *O. barthii* | | AA | 6 | 0 |
| *O. glumaepatula* | | AA | 1 | 1 |
| *O. punctata* | | BB | 2 | 2 |
| *O. officinalis* | | CC | 5 | 3 |
| *O. australiensis* | | EE | 6 | 5 |
| *O. brachyantha* | | FF | 3 | 3 |
| *O. granulata* | | GG | 4 | 2 |
| **Tetraploid *Oryza* Species** | | | | |
| *O. punctata* | | BBCC | 16 | 15 |
| *O. minuta* | | BBCC | 32 | 28 |
| *O. malampuzhaensis* | | BBCC | 3 | 2 |
| *O. alta* | | CCDD | 9 | 8 |
| *O. latifolia* | | CCDD | 30 | 21 |
| *O. grandiglumis* | | CCDD | 8 | 4 |
| *O. ridleyi* | | HHJJ | 8 | 5 |
| *O. longiglumis* | | HHJJ | 2 | 3 |
| **Related Genera** | | | | |
| *Porteresia coarctata* | (2n = 48) | HHKK | 1 | 1 |
| *Hygroryza aristata* | (2n = 24) | Unknown | 1 | 1 |
| *Leersia perrieri* | (2n = 24) | Unknown | 1 | 1 |
| *Rhynchoryza subulata* | (2n = 24) | Unknown | 1 | 1 |
| *Chikusichloa aquatica* | (2n = 24) | Unknown | 1 | 1 |

About 200 accessions of tetraploid species are available in the IRRI Genetic Resources Center. We used the pistil-clearing technique to detect diplospory. We examined several accessions of diploid and tetraploid wild species of rice, including related genera (Table 5.6), for the possible occurrence of apospory based on multiple embryo sac development. None of the accessions, however, showed any evidence of apospory (Brar et al. 1995; Khush et al. 1998).

We have also examined several accessions of diploid and tetraploid wild species of *Oryza* and five accessions of related genera for diplospory through callose detection using fluorescence microscopy (Table 5.6). Callose is deposited in the cell walls of megaspore mother cells during megasporogenesis in sexual species, but this deposition is nearly absent in the cell walls of apomictic embryo sacs. None of the accessions of *Oryza* species and related genera examined so far, however, showed a diplospory type of apomixis.

## 5.9.2   Mutagenesis-Induced Apomixis

We used physical and chemical mutagens to explore the possibility of inducing mutations for apomixis in rice. Identifying mutants with the asexual mode of reproduction in a large population of mutagenized individuals is problematic. We employed a genetic male sterile line (msms) of rice variety IR36. Male sterile plants are pollinated with pollen from fertile (MsMs) plants. Fertilized egg cells (Msms), 8 to 16 h after pollination, are treated with 1.5 mM N-methyl-N-nitrosourea (MNU) for 1 h. We used a dominant marker — purple leaf mutant of rice — for identifying apomictic mutants following mutagenesis. Two approaches were followed: (1) emasculated panicles of IR36 were irradiated with gamma rays and immediately pollinated with the pollen of purple leaf mutant, and 2) fertilized eggs from the cross of IR36 × purple leaf mutant were mutagenized with MNU. $M_2$ was screened to identify any true breeding purple leaf progeny for dominant mutations for apomixis. None of the mutagenized population showed any evidence of apomixis.

### 5.9.3   Genetic Engineering to Induce Apomixis

Apomixis and sexual reproduction follow the same fundamental pathway from floral induction to seed maturation. They differ principally in the route by which a single nucellar cell in the ovule gives rise to an embryo. Furthermore, genetic studies indicate that only a few genes control apomixis. Moreover, it is now possible to isolate, identify, and characterize genes for the switch from sexual to apomixis. However, it is not clear whether diplospory and apospory are intrinsically incompatible with diploidy or have become associated with polyploidy for ecological or evolutionary reasons. One approach is to develop adventitious embryony, in which the embryo arises directly from a diploid cell of the maternal nucellus, whereas the endosperm arises from the fusion of a sperm cell from the pollen with the nuclei of the central cell of the sexual embryo sac. The zygotic embryo should either fail to form or should abort at a very early stage. The second route is to develop the *Panicum* type of apospory that is common in grasses. The embryo should arise by parthenogenesis from the diploid egg cell of an unreduced four-nucleate aposporous embryo sac, whereas the endosperm should arise by fusion of a sperm cell with the diploid polar nucleus in the same aposporous embryo sac.

Major advances have been made in isolation of important genes for components of apomixis, such as FIS and FIE genes for seed and endosperm development, respectively, without fertilization in *Arabidopsis* (Ohad et al. 1996; Luo et al. 1999). *LEC1* (leafy cotyledon I) from *Arabidopsis* and *DMC1* (disrupted meiosis cDNA) from yeast could be used to ablate sexual embryos. Homologs of *LEC1* and *DMC1* have been isolated from rice (Bennett et al. 2001). The molecular approach to achieve synthetic apomixis in rice is emphasized, in which a sexual embryo would be replaced by an asexual embryo induced in the nucellus. The key steps would be induction of a nucellar embryo and ablation of a zygotic embryo. It is important to identify embryo-inducing genes, placing them under the control of nucellus-specific promoters.

## 5.10 GENETIC ENHANCEMENT THROUGH TRANSFORMATION

The introduction of alien genes from bacteria, viruses, fungi, animals, and, of course, unrelated plants into crop species allows plant breeders to achieve breeding objectives that were once not considered possible. Several techniques are now available for the transformation of rice, e.g., electroporation, polyethylene-glycol-induced uptake of DNA, microprojectile bombardment, and, more recently, *Agrobacterium*-mediated transformation. Earlier protoplasts were used as the source material for transformation. Transgenic rices carrying multiple transgenes have been produced (Chen et al. 1998). A series of transgenic rices carrying agronomically important genes for resistance to major pests, herbicide tolerance, and with improved nutritional-quality golden rice have been produced in both japonica and indica rices (Table 5.7).

**Table 5.7 Some Examples of Transgenic Rice Plants with Agronomically Important Genes**

| Transgene | Transformation Method | Useful Trait | Reference |
|---|---|---|---|
| *bar* | Microprojectile bombardment | Tolerance to herbicide | Cao et al. 1992 |
| *bar* | PEG mediated | Tolerance to herbicide | Datta et al. 1992 |
| Coat protein | Electroporation | Tolerance to stripe virus | Hayakawa et al. 1992 |
| Coat protein | Particle bombardment | Tolerance to rice tungro spherical virus | Sivamani et al. 1999 |
| Chitinase | PEG mediated | Sheath blight resistance | Lin et al. 1995; Datta et al. 2000; Baisakh et al. 2001 |
| *cryIA(b)* | Electroporation | Tolerance to striped stemborer | Fujimoto et al. 1993 |
| *cryIA(b)* | Particle bombardment | Tolerance to yellow stemborer and striped stemborer | Wuhn et al. 1996; Ghareyazie et al. 1997 |
| *cryIA(c)* | Particle bombardment | Tolerance to yellow stemborer | Nayak et al. 1997 |
| *cry1A(b), cryIA(c)* | Agrobacterium mediated | Tolerance to striped stemborer and yellow stemborer | Cheng et al. 1998 |
| *cry1A(b), cryIA(c)* | Agrobacterium mediated | Tolerance to striped stemborer and yellow stemborer | Tu et al. 2000 |
| *cry1A(c), cry2A, gna* | Particle bombardment | Tolerance to stemborer, leaf folder, and brown planthopper | Maqbool et al. 1998, 2001 |
| *CpTi* | PEG mediated | Tolerance to striped stemborer and pink stemborer | Xu et al. 1996 |
| *gna* | Particle bombardment | Insecticidal activity for brown planthopper | Rao et al. 1998 |
| Corn cystatin (CC) | Electroporation | Insecticidal activity for Sitophilus zeamais | Irie et al. 1996 |
| *Xa 21* | Particle bombardment | Resistance to bacterial blight | Tu et al. 1998 |
| *Afp* | Agrobacterium mediated | Enhanced resistance to blast fungus | Coca et al. 2004 |
| *codA* | Electroporation | Increased tolerance to salt | Sakamoto and Murata 1998 |
| ferritin | Agrobacterium mediated | Increased iron content in seed | Goto et al. 1999; Lucca et al. 2001 |
| *psy, crt1, lcy* | Agrobacterium mediated | Provitamin A (Golden rice) | Ye et al. 2000 |
| *psy, crtl* | Agrobacterium mediated | Provitamin A (Golden rice) | Datta et al. 2003 |
| PEPC (C4 enzyme from maize) | Agrobacterium mediated | Enhanced photosynthesis | Ku et al. 1999 |

*Source*: Modified from Brar and Khush (2002).

## 5.11 FUNCTIONAL GENOMICS

Major progress has been made in structural genomics of rice. The availability of sequence data in indica and japonica rice has ushered in the era of functional genomics. The sequencing efforts under the International Rice Genome Sequencing Project (IRGSP) and by other public and private organizations will provide sequences for the estimated 30,000 rice genes and for the intergenic DNA that plays an important but poorly understood role in gene expression, DNA replication, chromosome organization, recombination, specialization, and evolution. The activator-dissociator (AC-DS) maize transposable elements, retrotransposons, miniature inverted repeat transposable elements (MITEs), and T-DNA insertions have provided a wealth of genetic resources for functional genomics. Some notable examples include T-DNA-tagged insertation mutants with 30,000 lines carrying 42,000 T-DNA inserts, Tos17 retrotransposon insertional mutants with about 30,000 lines carrying more than 250,000 Tos17s, and more than 40,000 deletion mutants produced by fast neutron, gamma radiation, and chemical mutagenesis. Maize transposon constructs have been used in the transformation of japonica and indica cultivars for knockout and gene detection insertion.

With the availability of high-throughput technologies using microarrays or gene chips, it has become possible to identify candidate genes for biotic and abiotic stress pathways and other agronomic traits and to assess gene functions at a genomic level. The combination of forward genetics and reverse genetics offers new potential to apply modern tools of genomics and cell biology for precise understanding of the complexity of the genetic architecture of the rice genome. Understanding the biological functions encoded by a sequence through genetic and phenotypic analysis is the major goal of functional genomics. Identification of genes through functional genomics and their manipulation would be another major breakthrough in rice genetics and breeding.

## REFERENCES

Abbasi, F.M. et al. 1999. Detection of autosyndetic and allosyndetic pairing among A and E genomes of *Oryza* through genomic *in situ* hybridization. *Rice Genet. Newsl.* 16: 24–29

Abedinia, M. et al. 2000. Accessing genes in the tertiary gene pool of rice by direct introduction of total DNA from *Zizania palustris* (wild rice). *Plant Mol. Biol. Rep.* 18: 133–138.

Aggarwal, R.K. et al. 1999. Phylogenetic relationships among *Oryza* species revealed by AFLP markers. *Theor. Appl. Genet.* 98: 1320–1328.

Aggarwal, R.K., Brar, D.S. and Khush, G.S. 1997. Two new genomes in the *Oryza* complex identified on the basis of molecular divergence analysis using total genomic DNA hybridization. *Mol. Gen. Genet.* 254: 1–12.

Ahn, S. et al., 1993. Homoeologous relationships of rice, wheat and maize chromosomes. *Mol. Gen. Genet.* 241: 483–490.

Akagi, H. 2001. Cybridization in *Oryza sativa* L. (rice). In *Biotechnology in Agriculture and Forestry 49. Somatic Hybridization in Crop Improvement II*, T. Nagata and Y.P.S. Bajaj, Eds. Spring-Verlag, Berlin, pp. 17–36.

Akagi, H. et al. 1989. Construction of rice cybrid plants. *Mol. Gen. Genet.* 215: 501–505.

Amante-Bordeos, A. et al. 1992. Transfer of bacterial blight and blast resistance from the tetraploid wild rice *Oryza minuta* to cultivated rice. *Theor. Appl. Genet.* 84: 345–354.

Asghar, M. et al. 1998. Characterization of parental genomes in a hybrid between *Oryza sativa* L. and *O. officinalis* Wall ex Watt. through fluorescence *in situ* hybridization. *Rice Genet. Newsl.* 15: 83–84.

Baisakh, N. et al. 2001. Rapid development of homozygous transgenic rice using anther culture harboring rice chitinase gene for enhanced sheath blight resistance. *Plant Biotechnol.* 18: 101–105.

Banks, P.M. et al. 1995. The use of cell culture for subchromosomal introgressions of barley yellow dwarf virus resistance from *Thinopyrum intermedium* to wheat. *Genome* 38: 395–405.

Bellon, M.R. et al. 1998. Rice genetic resources. In *Sustainability of Rice in the Global Food System*, N.G. Dowling, S.M. Greenfield, K.S. Fischer, Eds. Pacific Basin Study Center and International Rice Research Institute, Manila, Philippines, pp. 251–282.

Bennett, J. et al. 2001. Molecular tools for achieving synthetic apomixes in hybrid rice. In *Rice Genetics IV*, G.S. Khush, D.S. Brar, and B. Hardy, Eds. International Rice Research Institute, Manila, Philippines, pp. 377–401.

Brar, D.S. et al. 1995. Search for apomixis in wild species of rice. In *Proceedings of the International Conference on Harnessing Apomixis*, College Station, TX, September 25–27, p. 39.

Brar, D.S. et al. 1996. Gene transfer and molecular characterization of introgression from wild *Oryza* species into rice. In *Rice Genetics III*. International Rice Research Institute, Manila, Philippines, pp. 477–486.

Brar, D.S. et al. 1997. Cytogenetic and molecular characterization of an intergeneric hybrid between *Oryza sativa* L. and *Porteresia coarctata* (Roxb.) Tateoka. *Rice Genet. Newsl.* 14: 43–44.

Brar, D.S., Elloran, R., and Khush, G.S. 1991. Interspecific hybrids produced through embryo rescue between cultivated and eight wild species of rice. *Rice Genet. Newsl.* 8: 91–93.

Brar, D.S. and Jain, S.M. 1998. Somaclonal variation: mechanism and applications in crop improvement. In *Somaclonal Variation and Induced Mutation in Crop Improvement*, S.M. Jain, D.S. Brar, B.S. Ahloowalia, Eds. Kluwer Academic Publishers, Dordrect, The Netherlands, pp. 15–37.

Brar, D.S. and Khush, G.S. 1986. Wide hybridization and chromosome manipulation in cereals. In *Handbook of Plant Cell Culture*, Vol. 4, *Techniques and Applications*, D.H. Evans, W.R. Sharp, and P.V. Ammirato, Eds. MacMillan Publ. Co., New York, pp. 221–263.

Brar, D.S. and Khush, G.S. 1997. Alien introgression in rice. *Plant Mol. Biol.* 35: 35–47.

Brar, D.S. and Khush, G.S. 2002. Transferring genes from wild species into rice. In *Quantitative Genetics, Genomics and Plant Breeding*, M.S. Kang, Ed. CABI Publ., Wallingford, U.K., pp. 197–217.

Cao, J. et al. 1992. Regeneration of herbicide resistant transgenic rice plants following microprojectile-mediated transformation of suspension culture cells. *Plant Cell Rep.* 11: 586–591.

Chang, T.T. 1972. International cooperation in conserving and evaluation of rice germplasm resources. In *Rice Breeding*. International Rice Research Institute, Manila, Philippines, pp. 177–185.

Chang, T.T. 1976. The origin, evolution, cultivation, dissemination and diversification of the Asian and African rices. *Euphytica* 25: 425–444.

Chatterjee, D. 1948. A modified key and enumeration of the species of *Oryza* Linn. *Indian J. Agric. Sci.* 18: 185–192.

Chen, L. et al. 1998. Expression and inheritance of multiple transgenes in rice plants. *Nat. Biotechnol.* 16: 1060–1064.

Cheng, C. et al. 2002. Evolutionary relationships among rice species with AA genome based on SINE insertion analysis. *Gen. Genet. Syst.* 77: 323–334.

Cheng, C. et al. 2003. Polyphyletic origin of cultivated rice: based on interspersion pattern of SINEs. *Mol. Biol. Evol.* 20: 67–75.

Cheng, X. et al. 1998. *Agrobacterium*-transformed rice plants expressing synthetic *cry1A(b)* and *cry1A(c)* genes are highly toxic to striped stemborer and yellow stemborer. *Proc. Natl. Acad. Sci. U.S.A.* 95: 2767–2772.

Cheng, Z. et al. 2001a. Towards a cytological characterization of the rice genome. *Genome Res* II: 2133–2141.

Cheng, Z. et al. 2001b. Development and applications of a complete set of rice telotrisomics. *Genetics* 157: 361–368.

Coca, M. et al. 2004. Transgenic rice plants expressing the antifungal AFP protein from *Aspergillus giganteus* show enhanced resistance to the rice blast fungus *Magnaporthe grisea*. *Plant Mol. Biol.* 54: 245–259.

Dalmacio, R. et al. 1995. Identification and transfer of a new cytoplasmic male sterility source from *Oryza perennis* into indica rice (*O. sativa*). *Euphytica* 82: 221–225.

Dalmacio, R. et al. 1996. Male sterile line in rice (*Oryza sativa*) developed with *O. glumaepatula* cytoplasm. *Intern. Rice Res. Notes* 21: 22–23.

Datta, K. et al. 2000. *Agrobacterium*-mediated engineering for sheath blight resistance of indica rice cultivars from different ecosystems. *Theor. Appl. Genet.* 100: 832–839.

Datta, K. et al. 2003. Bioengineered 'golden' indica rice cultivars with β-carotene metabolism in the endosperm with hygromycin and mannose selection systems. *Plant Biotechnol. J.* 1: 81–93.

Datta, S.K. et al. 1992. Herbicide-resistant indica rice plants from IRRI breeding line IR72 after PEG-mediated transformation of protoplasts. *Plant Mol. Biol.* 20: 619–629.

Devos, K.M. and Gale, M.D. 1997. Comparative genetics in the grasses. *Plant Mol. Biol.* 35:3–15.

Doi, K. et al. 2003. Developing and evaluating rice chromosome segment substitution lines. In *Rice Science: Innovations and Impact Livelihood*, T.W. Mew, D.S. Brar, S. Peng, D. Dawe, H. Hardy, Eds. International Rice Research Institute and CASE and CAAS, Beijing, pp. 289–296.

Ellis, J.L. 1985. *Oryza indandamanica*. Ellis, a new rice plant from islands of Andamans. *Bot. Bull. Surv. India* 27: 225–227.

Enriquez, E.C. 2001. Production of Doubled Haploids from *Oryza sativa* L. × *O. glaberrima* Steud. and Their Characterization Using Microsatellite Markers. Ph.D. thesis, UPLB, Los Baños, Philippines, p. 116.

Enriquez, E.C. et al. 2000. Production and characterization of doubled haploids from anther culture of the $F_1$s of *Oryza sativa* L. × *O. glaberrima* Steud. *Rice Genet. Newsl.* 17: 67–69.

Fujimoto, H. et al. 1993. Insect resistant rice generated by introduction of a modified δ-endotoxin gene of *Bacillus thuringiensis*. *Bio/Technology* 11: 1151–1155.

Fukui, K. et al. 1994. Variability in rDNA loci in the genus *Oryza* as detected through fluorescence *in-situ* hybridization. *Theor. Appl. Genet.* 81: 589–596.

Ge, S. et al. 1999. Phylogeny of rice genomes with emphasis on origins of allotetraploid species. *Proc. Natl. Acad. Sci. U.S.A.* 96: 14400–14405.

Ghareyazie, B. et al. 1997. Enhanced resistance to two stemborers in an aromatic rice containing a synthetic *cry1A(b)* gene. *Mol. Breed.* 3: 401–414.

Glaszmann, J.C. 1987. Isozymes and classification of Asian rice varieties. *Theor. Appl. Genet.* 74: 21–30.

Goto, F. et al. 1999. Iron fortification of rice seed by the soybean ferritin gene. *Nat. Biotechnol.* 17: 282–286.

Hayakawa, T. et al. 1992. Genetically engineered rice resistant to rice stripe virus, an insect-transmitted virus. *Proc. Natl. Acad. Sci. U.S.A.* 89: 9865–9869.

Hayashi, Y., Kozuka, J., and Shimamoto, K. 1988. Hybrids of rice (*Oryza sativa* L.) and wild *Oryza* species obtained by cell fusion. *Mol. Gen. Genet.* 214: 6–10.

Heszky, L.E. and Simon-Kiss, I. 1992. DAMA, the first plant variety of biotechnology origin in Hungary registered in 1992. *Hung. Agric. Res.* 1: 30–32.

Hirabayashi, H. et al. 2003. Mapping QTLs for brown planthopper (BPH) resistance introgressed from *O. officinalis* in rice. In *Advances in Rice Genetics*, G.S. Khush, D.S. Brar, and B. Hardy, Eds. International Rice Research Institute, Manila, Philippines, pp. 268–270.

Huang, Z. et al. 2001. Identification and mapping of two brown planthopper resistance genes in rice. *Theor. Appl. Genet.* 102: 929–934.

Hue, N.T.N. 2004. Homoeologous Chromosome Pairing and Alien Introgression Analyses in Wide-Cross Derivatives of *Oryza* through Fluorescence *In Situ* Hybridization. Ph.D. thesis, UPLB, Los Baños, Philippines, p. 160.

Irie, K. et al. 1996. Transgenic rice established to express corn cystatin exhibits strong inhibitory activity against insect gut proteinases. *Plant Mol. Biol.* 30: 149–157.

Ishii, T. et al. 1994. Molecular tagging of genes for brown planthopper resistance and earliness introgressed from *Oryza australiensis* into cultivated rice, *O. sativa*. *Genome* 37: 217–221.

Ishii, T. et al. 1995. Nuclear genome differentiation in Asian cultivated rice as revealed by RFLP analysis. *Jpn. J. Genet.* 70: 643–652.

Jackson, M.T. 1997. Conservation of rice genetic resources: the role of the International Rice Gene Bank at IRRI. *Plant Mol. Biol.* 35: 61–67.

Jelodar, N.B. et al. 1999. Intergeneric somatic hybrids of rice [*Oryza sativa* L. (+) *Porteresia coarctata* (Roxb.) Tateoka]. *Theor. Appl. Genet.* 99: 570–577.

Jena, K.K. and Khush, G.S. 1990. Introgression of genes from *Oryza officinalis* Well ex Watt to cultivated rice, *O. sativa* L. *Theor. Appl. Genet.* 80: 737–745.

Jena, K.K, Khush, G.S., and Kochert, G. 1992. RFLP analysis of rice (*Oryza sativa* L.) introgression lines. *Theor. Appl. Genet.* 84: 608–616.

Jiang, J. et al. 1995. Metaphase and interphase fluorescence *in situ* hybridization mapping of the rice genome with bacterial artificial chromosomes. *Proc. Natl. Acad. Sci. U.S.A.* 92: 4487–4491.

Jones, M.P. et al. 1997. Interspecific *Oryza sativa* L. × *O. glaberrima* Steud. progenies in upland rice improvement. *Euphytica* 92: 237–246.

Kato, S., Kosaka, H., and Hara, S. 1928. On the affinity of rice varieties as shown by fertility of hybrid plants. *Bull. Sci. Fac. Agric. Kyushu Univ.* 3: 132–147.

Kawai, T. and Amano, E. 1991. Mutation breeding in Japan. In *Plant Mutation Breeding for Crop Improvement*, Vol. 1. IAEA, Vienna, pp. 47–66.

Khush, G.S. 1977. Disease and insect resistance in rice. *Adv. Agron.* 29: 265–341.

Khush, G.S. et al. 1984. Primary trisomics of rice: origin, morphology, cytology and use in linkage mapping. *Genetics* 107: 141–163.

Khush, G.S. et al. 1994. Apomixis for rice improvement. In *Apomixis: Exploiting Hybrid Vigor in Rice*, G.S. Khush, Ed. International Rice Research Institute, Manila, Philippines, pp. 1–21.

Khush, G.S. et al. 2003. Classifying Rice Germplasm by Isozyme Polymorphism and Origin of Cultivated Rice, IRRI Discussion Paper 46. International Rice Research Institute, Manila, Philippines, 279 pp.

Khush, G.S., Bacalangco, E., and Ogawa, T. 1990. A new gene for resistance to bacterial blight from *O. longistaminata*. *Rice Genet. Newsl.* 7: 121–122.

Khush, G.S. and Brar, D.S. 2002. Rice. In *Evolution and Adaptation of Cereal Crops*, V.L. Chopra and S. Prakash, Eds. Science Publishers, Enfield, NH, pp. 1–41.

Khush, G.S., Brar, D.S., and Bennett, J. 1998. Apomixis in rice and prospects for its use in heterosis breeding. In *Advances in Hybrid Rice Technology*, S.S. Virmani, E.A. Siddiq, and K. Muralidharan, Eds. International Rice Research Institute, Manila, Philippines, pp. 297–309.

Khush, G.S. and Jena, K.K. 1989. Biosystematic status of *Oryza indandamanica* Ellis. In *Proceedings of the 6th International Congress*, SABRAO, Tsukuba, Japan, pp. 179–182.

Kisaka, H. et al. 2001. Somatic hybridization between *Oryza sativa* L. (Rice) and *Hordeum vulgare* L. (Barley) in biotechnology. In *Agriculture and Forestry 49. Somatic Hybridization in Crop Improvement II*, T. Nagata and Y.P.S. Bajaj, Eds. Spring-Verlag, Berlin, pp. 37–47.

Kisaka, J. et al. 1994. Production and analysis of asymmetric hybrid plants between monocotyledon (*Oryza sativa* L.) and dicotyledon (*Daucus carota* L.). *Theor. Appl. Genet.* 89: 365–371.

Kobayashi, N. et al. 1993. Resistance to infection of rice tungro viruses and vector resistance in wild species of rice (*Oryza* spp.). *Jpn. J. Breed.* 43: 377–387.

Ku, S.B. et al. 1999. High level expression of maize phosphoenolpyruvate carboxylase in transgenic rice plants. *Nat. Biotechnol.* 17: 76–80.

Kurata, N., Iwata, N., and Omura, T. 1981. Karyotype analysis in rice. II. Identification of extra chromosomes in trisomic plants and banding structure on some chromosomes. *Jpn. J. Genet.* 56: 41–50.

Kurata, N. and Omura, T. 1978. Karyotype analysis in rice. I. A new method for identifying all chromosome pairs. *Jpn. J. Genet.* 53: 251–255.

Kuwada, Y. 1910. A cytological study of *Oryza sativa* L. *Shokubutsugaksu Zasshi* 24: 267–281.

Kyozuka, J., Kaneda, T., and Shimamoto, K. 1989. Production of cytoplasmic male sterile rice (*Oryza sativa* L.) by cell fusion. *Bio/Technology* 7: 1171–1174.

Larkin, P.J. et al. 1989. From somatic variation to variant plants: mechanisms and applications. *Genome* 31: 705–711.

Launert, E. 1965. A survey of the genus *Leersia* in Africa (Garmineae, Oryzoideae, Oryzeae). *Sneck Biol.* 46: 129–153.

Lin, S.C. and Yuan, L.P. 1980. Hybrid rice breeding in China. In *Innovative Approaches to Rice Breeding*. International Rice Research Institute, Manila, Philippines, pp. 35–51.

Lin, W. et al. 1995. Genetic engineering of rice for resistance to sheath blight. *Bio/Technology* 13: 686–691.

Ling, K.C., Aguiero, V.M., and Lee, S.H. 1970. A mass screening method for testing resistance to grassy stunt disease of rice. *Plant Dis. Rep.* 56: 565–569.

Liu, B., Liu, Z., and Li, X.W. 1999. Production of a highly asymmetric somatic hybrid between rice and *Zizania latifolia* (Griseb): evidence for intergenome exchange. *Theor. Appl. Genet.* 98: 1099–1103.

Lu, B. and Sharma, S.D. 2003. Exploration, collection and conservation of wild *Oryza* species. In *Monograph on Genus Oryza*, J.S. Nanda, S.D. Sharma, Eds. Science Publishers, Enfield, NH, pp. 263–283.

Lucca, P, Hurrell, R., and Potrykus, I. 2001. Genetic engineering approaches to improve the bioavailability and the level of iron in rice grains. *Theor. Appl. Genet.* 102: 392–397.

Luo, M. et al. 1999. Genes controlling fertilization-independent seed development in *Arabidopsis thaliana*. *Proc. Natl. Acad. Sci. U.S.A.* 96: 296–301.

Maqbool, S.B. et al. 1998. Effective control of yellow stemborer and leaf folder in transgenic rice indica varieties Basmati 370 and M7 using the novel δ-endotoxin *cryzA Bacillus thuringiensis* gene. *Mol. Breed.* 4: 501–507.

Maqbool, S.B. et al. 2001. Expression of multiple insecticidal genes confers broad resistance against a range of different rice pests. *Mol. Breed.* 7: 85–93.

Moncada, P. et al. 2001. Quantitative trait loci for yield and yield components in an *Oryza sativa* × *Oryza rufipogon* BC$_2$F$_2$ population evaluated in an upland environment. *Theor. Appl. Genet.* 102: 41–52.

Morinaga, T. 1954. Classification of rice varieties on the basis of affinity. In *Reports for the 5th Meeting of International Rice Commission's Working Party on Rice Breeding*. Ministry of Agriculture and Forestry, Tokyo, pp. 1–14.

Muller, H.J. 1927. Artificial transmutation of the gene. *Science* 66: 84–87.

Multani, D.S. et al. 1994. Development of monosomic alien addition lines and introgression of genes from *Oryza australiensis* Domin. to cultivated rice *O. sativa* L. *Theor. Appl. Genet.* 88: 102–109.

Multani, D.S. et al. 2003. Alien genes introgression and development of monosomic alien additional lines from *Oryza latifolia* Desv. to rice, *Oryza sativa* L. *Theor. Appl. Genet.* 107: 395–405.

Nakamura, A. et al. 1995. Breeding of cytoplasmic male sterile rice cultivar. *Biomother. Breed. Sci.*, Suppl. 1: 212 (in Japanese).

Nandi, H.K. 1936. The chromosome morphology, secondary association and origin of cultivated rice. *J. Genet.* 33: 315–336.

Naredo, E.B., Vaughan, D.A., and Cruz, F.S. 1993. Comparative spikelet morphology of *Oryza schlechteri* Pilger and related species of *Leersia* and *Oryza* (Poaceae). *J. Plant Res.* 106: 109–112.

Nayak, P. et al. 1997. Transgenic elite indica rice expressing Cry1Ac δ-endotoxin of *Bacillus thuringiensis* are resistant against yellow stemborer (*Scirophaga incertulas*). *Proc. Natl. Acad. Sci. U.S.A.* 94: 2111–2116.

Nayar, N.M. 1973. Origin and cytogenetics of rice. *Adv. Genet.* 17: 153–292.

Ng, N.Q. et al. 1981. The recognition of a new species of rice (*Oryza*) from Australia. *Bot. J. Linn. Soc.* 82: 327–330.

Nguyen, B.D. et al. 2003. Identification and mapping of the QTL for aluminum tolerance introgressed from new source, *Oryza rufipogon* Griff. into indica rice, *Oryza sativa* L. *Theor. Appl. Genet.* 106: 583–593.

Niizeki, H. et al. 1985. Callus formation of somatic hybrids of rice and soybean and characteristics of the hybrid callus. *Jap. J. Genet.* 60: 81–92.

Niizeki, H. and Oono, K. 1968. Induction of haploid rice plant from anther culture. *Proc. Jpn. Acad.* 44: 544–557.

Nishimura, Y. 1961. Studies on the reciprocal translocations in rice and barley. *Bull. Natl. Inst. Agric. Sci. Jpn. Ser.* D9: 171–235.

Ogura, H. and Shimamoto, K. 1991. Field performance of protoplasts derived rice plants and the release of a new variety. In *Biotechnology in Agriculture and Forestry*, Vol. 14, Y.P.S. Bajaj, Ed. Springer-Verlag, Berlin, pp. 269–282.

Ohad, N. et al. 1996. A mutation that allows endosperm development without fertilization. *Proc. Natl. Acad. Sci. U.S.A.* 93: 5319–5324.

Oka, H.I. 1958. Varietal variation and classification of cultivated rice. *Indian J. Genet.* 18: 78–89.

Oka, H.I. 1961. Report of Trip for Investigation of Rice in Latin American Countries. International Rice Research Institute Library, Manila, Philippines (unpublished).

Oka, H.I. 1988. *Origin of Cultivated Rice. Developments in Crop Science*, Vol. 14. Japan Scientific Society Press, Tokyo.

Panaud, O. et al. 2002. Characterization of transposable elements in the genome of rice (*Oryza sativa* L.) using representational difference analysis (RDA). *Mol Genet. Genomics* 268: 113–120.

Porteres, R. 1956. Taxonomie agrobotanique des Riz. Cultives, *O. sativa* L. et *glaberrima*. Steud. *J. Agr. Trop. Bot. Appl.* 3: 341–356.

Rao, K.V. et al. 1998. Expression of snowdrop lectin (GNA) in transgenic rice plants confers resistance to rice brown planthopper. *Plant J.* 15: 469–477.

Rau, N.S. 1929. Further contribution to the cytology of some crop plants of South India. *Bot. Soc.* 8: 201–206.

Ronald, P.C. et al. 1992. Genetic and physical analysis of rice bacterial blight resistance locus, *Xa21*. *Mol. Gen. Genet.* 236: 113–120.

Roschevicz, R. 1931. A contribution to the knowledge of rice. *Bull. Appl. Bot. Genet. Plant Breed.* 27: 1–33.

Rutgers, J.N. 1992a. Impact of mutation breeding in rice: a review. In *Mutation Breeding Review 8*. FAO/IAEA, Vienna, pp. 1–23.

Rutgers, J.N. 1992b. Searching for apomixis in rice. In *Proceedings of the Apomixis Workshop*. USDA-ARS, Atlanta, GA, pp. 36–39.

Sakamoto, A. and Murata, N. 1998. Metabolic engineering of rice leading to biosynthesis of glycine betaine and tolerance to environmental stress. In *International Workshop on Breeding and Biotechnology for Environmental Stress in Rice*, Sapporo, Japan, p. 164.

Sala C. et al. 1985. Selection and nuclear DNA analysis of cell hybrids between *Daucus carota* and *Oryza sativa*. *J. Plant Physiol.* 118: 409–419.

Sampath, S. 1962. The genus *Oryza*: its taxonomy and species relationship. *Oryza* 1: 1–29.

Sanchez, A.C. et al. 2000. Sequence tagged site marker-assisted selection for three bacterial blight resistance genes in rice. *Crop Sci.* 40: 792–797.

Sarkar, R. and Raina, S.N. 1992. Assessment of genomic relationships in the genus *Oryza sativa* L. based on seed-protein profile analysis. *Theor. Appl. Genet.* 85: 127–131.

Sen, S.K. 1963. Analysis of rice pachytene chromosomes. *Nucleus* 6: 107–120.

Septiningsih, E.M. et al. 2003. Identification of quantitative trait loci for yield and yield components in an advanced backcross population derived from the *Oryza sativa* variety IR64 and the wild relative *O. rufipogon*. *Theor. Appl. Genet.* 107: 1419–1432.

Sharma, S.D. and Shastry, S.V.S. 1965a. Taxonomic studies in genus *Oryza*. III. *O. rufipogon* Griff, *sensu stricto* and *O. nivara* Sharma et Shastry *nom nov*. *Indian J. Genet.* 25: 157–167.

Sharma, S.D. and Shastry, S.V.S. 1965b. Taxonomic studies in genus *Oryza*. IV. The Ceylones *Oryza* spp. affin. *O. officinalis* Wall ex Watt. *Indian J. Genet.* 25: 168–172.

Shastry, S.V.S., Rangao Rao, D.R., and Misra, R.N. 1960. Pachytene analysis in *Oryza*. I. Chromosome morphology in *Oryza sativa* L. *Indian. J. Genet. Plant Breed.* 20: 15–21.

Shishido, R. et al. 1998. Detection of specific chromosome reduction in rice somatic hybrids with the A, B, and C genomes by multi-color genomic *in situ* hybridization. *Theor. Appl. Genet.* 97: 1013–1018.

Singh, K. et al. 1996b. Centromere mapping and orientation of the molecular linkage map of rice (*Oryza sativa* L.) *Proc. Natl. Acad. Sci. U.S.A.* 93: 6163–6168.

Singh, K. et al. 2004. Utilization of molecular marker technology for rice improvement at PAU, Ludhiana. In *Ninth National Rice Biotechnology Meeting*, New Delhi, April 15–17, p. 7 (abstract).

Singh, K., Multani, D.S., and Khush, G.S. 1996a. Secondary trisomics and telotrisomics of rice: origin, characterization, and use in determining the orientation of chromosome map. *Genetics* 143: 517–529.

Singh, S. et al. 2001. Pyramiding three bacterial blight resistance genes (*xa5*, *xa13* and *Xa21*) using marker assisted selection into indica rice cultivar PR106. *Theor. Appl. Genet.* 102: 1011–1015.

Sitch, L.A. 1990. Incompatibility barriers operating in crosses of *Oryza sativa* with related species and genera. In *Genetic Manipulation in Plant Improvement II*, J.P. Gustafson, Ed. Plenum Press, New York, pp. 77–94.

Sivamani, E. et al. 1999. Rice plant (*Oryza sativa* L.) containing rice tungro spherical virus (RTSV) coat protein transgenes are resistant to virus infection. *Mol. Breed.* 5: 177.

Solheim, W.G., II. 1972. An earlier agricultural revolution. *Sci. Ann.* 266: 34–41.

Song, W.Y. et al. 1995. A receptor kinase like protein encoded by the rice disease resistance gene, *Xa21*. *Science* 270: 1804–1806.

Sonti, R.V. et al. 2004. Introduction of bacterial leaf blight resistance genes into Samba Mahsuri background. In *Ninth National Rice Biotechnology Meeting*, New Delhi, April 15–17, pp. 12–13 (abstract).

Stadler, L.J. 1928. Mutations in barley induced by x-rays and radium. *Science* 68: 186–187.

Suh, H.S. et al. 2000. RAPD variation of the carbonized rice aged 13,010 and 17,310 years. In *Fourth International Genetics Symposium*. International Rice Research Institute, Manila, Philippines, pp. 49 (abstract).

Tanksley, S.D. and Nelson, J.C. 1996. Advanced backcross QTL analysis: a method for the simultaneous discovery and transfer of valuable QTLs from unadapted germplasm into elite breeding lines. *Theor. Appl. Genet.* 92: 191–203.

Tateoka, T. 1962a. Taxonomic studies of *Oryza*. I. *O. latifolia* complex. *Bot. Mag.* 75: 418–427.

Tateoka, T. 1962b. Taxonomic studies of *Oryza*. II. Several species complexes. *Bot. Mag.* 75: 455–461.

Tateoka, T. 1963. Notes on some grasses. VIII. Relationship between *Oryzeae* and *Ehrharteae*, with special reference to leaf antomology and histology. *Bot. Gaz.* 124: 264–270.

Tateoka, T. 1965. A taxonomic study of *Oryza eichingeri* and *O. punctata*. *Bot. Mag.* 78: 156–163.

Terada, R. et al. 1987. Plantlet regeneration from somatic hybrids of rice (*Oryza sativa* L.) and barnyard grass (*Echinochloa oryzicola* Vasing). *Mol. Gen. Genet.* 210: 39–43.

Ting, Y. 1957. The origin and evolution of cultivated rice in China. *Acta Agron. Sinica* 8: 243–260 (in Chinese).

Tu, J. et al. 1998. Transgenic rice variety, IR72, with *Xa21* resistant to bacterial blight. *Theor. Appl. Genet.* 97: 31–36.

Tu, J. et al. 2000. Field performance of transgenic elite commercial hybrid rice expressing *Bacillus thuringiensis* δ-endotoxin. *Nat. Biotechnol.* 18: 1101–1104.

Vaughan, D.A. 1989. The Genus *Oryza* L.: Current Status of Taxonomy, IRRI Research Paper Series 138. International Rice Research Institute, Manila, Philippines, 21 pp.

Vaughan, D.A. 1994. *The Wild Relatives of Rice: A Genetic Resources Handbook.* International Rice Research Institute, Manila, Philippines, 137 pp.

Vaughan, D.A. 2003. Genepools of genus *Oryza.* In *Monograph on Genus Oryza*, J.S. Nanda and S.D. Sharma, Eds. Science Publishers, Enfield, NH, p. 113.

Vaughan, D.A. and Sitch, L.A. 1991. Gene flow from the jungle to farmers: wild-rice genetic resources and their uses. *Bioscience* 44: 22–28.

Wang, G.L. et al. 1995. Construction of a rice bacterial artificial chromosome library and identification of clones linked to the *Xa21* disease resistant locus. *Plant J.* 7: 525–533.

Wang, L.Q. 1991. Induced mutation for crop improvement in China. In *Plant Mutation Breeding for Crop Improvement*, Vol. 1. IAEA, Vienna, pp. 9–32.

Wang, Z.Y., Second, G., and Tanksley, S.D. 1992. Polymorphism and phylogenetic relationships among species in the genus *Oryza* as determined by analysis of nuclear RFLPs. *Theor. Appl. Genet.* 83: 565–581.

Wuhn, J. et al. 1996. Transgenic indica rice breeding line IR58 expressing a synthetic *cry1A(b)* gene from *Bacillus thuringiensis* provides effective insect pest control. *Bio/Technology* 14: 171–176.

Xiao, J. et al. 1996. Genes from wild rice improve yield. *Nature* 384: 223–224.

Xiao, J. et al. 1998. Identification of trait-improving quantitative trait loci alleles from a wild rice relative, *Oryza rufipogon. Genetics* 150: 899–909.

Xu, D. et al. 1996. Constitutive expression of a cowpea trypsin inhibitor gene, *CpTi*, in transgenic rice plants confers resistance to two major rice insect pests. *Mol. Breed.* 2: 167–173.

Yan, H. et al. 1999. Visualization of *Oryza eichingeri* chromosomes in intergeneric hybrid plants from *O. sativa* × *O. eichingeri* via fluorescence *in-situ* hybridization. *Genome* 42: 48–51.

Yang, Z-Q. et al. 1989. Plant regeneration from cytoplasmic hybrids of rice (*Oryza sativa* L.). *Theor. Appl. Genet.* 77: 305–310.

Yasui, K. 1941. Diploid bud formation in a haploid *Oryza* with some remarks on the behaviour of nucleolus in mitosis. *Cytologia* 11: 515–525.

Ye, X. et al. 2000. Engineering the provitamin A (β-carotene) biosynthetic pathway into (carotenoid-free) rice endosperm. *Science* 287: 303–305.

Yu, Y. et al. 2003. In-depth view of structure, activity and evolution of rice chromosome 10. *Science* 300: 1566–1569.

Zhang, Q. et al. 1998. Identification and tagging of a new gene for resistance to bacterial blight (*Xanthomonas oryzae* pv. *Oryzae*) from *O. rufipogon. Rice Genet. Newsl.* 15: 138–142.

# CHAPTER 6

# Genetic Enhancement of Maize by Cytogenetic Manipulation, and Breeding for Yield, Stress Tolerance, and High Protein Quality

Surinder K. Vasal, Oscar Riera-Lizarazu, and Prem P. Jauhar*

## CONTENTS

6.1   Introduction .......................................................................................................160
6.2   Maize: Its Cytogenetic Architecture ...............................................................161
6.3   Cytogenetic Manipulation of the Maize Genome ...........................................162
     6.3.1   Manipulations of the Ploidy Level.......................................................162
     6.3.2   Manipulations of Chromosome Number: Aneuploidy .........................163
     6.3.3   Manipulations of Chromosomal Rearrangements .................................163
          6.3.3.1   Reciprocal Translocations .....................................................164
          6.3.3.2   B-A Translocations ................................................................165
          6.3.3.3   Inversions...............................................................................166
     6.3.4   Manipulation of Maize Chromosomes in an Oat Background .............167
          6.3.4.1   Oat × Maize Crosses..............................................................167
          6.3.4.2   Oat–Maize Addition Lines and Their Uses............................167
          6.3.4.3   Maize Subchromosome Fragment Stocks and Their Use ......169
6.4   Genetic Transformation: Adding Value-Added Traits .....................................170
6.5   Germplasm Conservation and Early Breeding Work .......................................171
     6.5.1   Conservation and Utilization of Maize Genetic Resources ..................171
     6.5.2   Early Breeding Work: Exploring New Options and Methodologies......171
          6.5.2.1   Reducing Plant Height............................................................172
          6.5.2.2   Developing Early-Maturity Germplasm .................................172
          6.5.2.3   Breeding for Morphological Traits ........................................172
          6.5.2.4   Improved Nutritional Quality ................................................173
          6.5.2.5   Developing Potentially Useful Germplasm and Breeding Methodologies..173
6.6   Hybrid-Oriented Source Germplasm and Hybrid Development ........................173
     6.6.1   Characterizing Germplasm for Combining Ability and Heterotic Pattern(s) .........173
     6.6.2   Shifts in Recurrent Selection Procedures ............................................173

Mention of trademark or proprietary product does not constitute a guarantee or warranty of the product by the USDA or imply approval to the exclusion of other products that may also be suitable.

* U.S. government employee whose work is in the public domain.

6.6.3    Development of Inbred Progenitors ...................................................................174
6.6.4    Inbreeding Tolerance and Crossbred Performance ..............................................174
6.6.5    Hybrid Options and Their Relevance ................................................................174
6.6.6    Inbred Line Evaluation Nurseries .....................................................................175
6.6.7    Characterizing and Using Maize Inbred Progenitors .........................................175
6.6.8    Hybrid Development and Testing ......................................................................175
6.6.9    Research on Maize Testers.................................................................................176
6.6.10   Release of Maize Inbreds and Other Materials .................................................176
6.7    Breeding for Stress Tolerance .....................................................................................177
6.7.1    Enhancing Disease Resistance ..........................................................................177
6.7.2    Development of Insect-Resistant Germplasm....................................................179
6.7.2.1    Resistance to Crop Pests...................................................................179
6.7.2.2    Resistance to Stored Grain Pests .....................................................179
6.7.3    Germplasm Development for Abiotic Stresses ...................................................180
6.7.3.1    Drought-Tolerant Germplasm ..........................................................181
6.7.3.2    Waterlogging-Tolerant Germplasm...................................................181
6.7.3.3    Acid Soil Tolerant Germplasm ........................................................182
6.8    Breeding for Improved Nutritional Quality..................................................................182
6.8.1    Search for Useful Genetic Variation and Early Work .......................................182
6.8.2    Development and Improvement of Soft Opaques ...............................................183
6.8.3    Correcting the First-Generation Problems and Exploring New Alternatives..........183
6.8.4    Genetic Modifiers and Their Use as a Successful Strategy ................................184
6.8.5    Development of QPM Donor Stocks..................................................................184
6.8.6    Expanded QPM Germplasm Development Efforts..............................................185
6.8.7    QPM Hybrid Development and Testing .............................................................186
6.9    Apomixis as a Possible Means of Perpetuating Hybrid Vigor.......................................189
6.10    Conclusions and Perspectives .....................................................................................189
References .............................................................................................................................190

## 6.1 INTRODUCTION

Maize (*Zea mays* L.) is a cross-pollinated crop of worldwide economic importance (see Figure 1.1 and Figure 1.2 in Chapter 1 by Jauhar in this volume). It was domesticated from its wild progenitor teosinte some 6250 years ago (Piperno and Flannery 2001). Maize occupies an important position as a food, feed, and industrial grain crop, contributing to the global grain pool 1800 million metric tons annually. It is grown on 140 million ha with a production of about 600 million metric tons valued at U.S. $65 billion (James 2003). The maize plant has several unique features not encountered in other major cereal crops, such as wheat and rice. Being a C4 plant, it exhibits a great yield potential per unit of land compared to other crops, has wide adaptation, and certainly surpasses all other crops in genetic variability and diversity either naturally or created through conscious breeding efforts over the last 70 years. Maize is also a unique organism for cytogenetic research. Genetic, cytogenetic, and phylogenetic evidence suggests that maize (2n = 20) is an ancient segmental allotetraploid (Gaut et al. 2000), although it is a functional diploid. More importantly, it lends itself to cytogenetic manipulation.

Although a larger volume of maize produced annually is used as livestock feed, it constitutes a staple of the human diet consumed in a variety of ways in at least 22 countries, mostly in Africa and Latin America. With the projected annual population growth of 80 million people and doubling of demand for meat and other animal products by 2020, it is expected that demand for maize will increase dramatically in the next two decades. In developing countries alone, it is estimated that demand for maize will rise from 241 million tons in 1993 to 442 million tons by 2020. The maize crop has, in

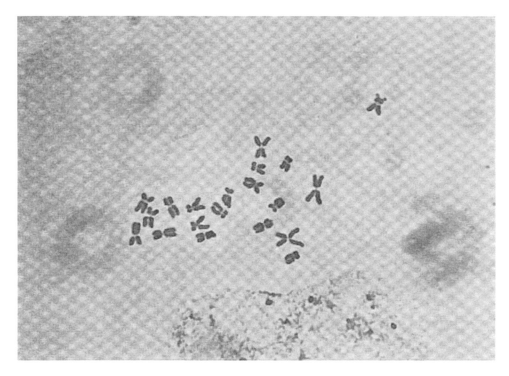

**Figure 6.1** Mitotic metaphase chromosome spread of maize (*Zea mays* l ; 2n = 20). This root-tip chromosome spread shows 10 chromosome pairs of the maize inbred line B73.

fact, a potential to meet this challenge. Genetic improvement of maize with regard to its yield and nutritional quality cannot be overemphasized. Heterosis breeding may be the most appropriate means to gain yield advantage and nutritional superiority relatively rapidly (Vasal et al. 1999a).

In this chapter we have attempted to provide an overview of basic cytogenetics of maize, its cytogenetic manipulation, and germplasm enhancement by incorporating chromatin from other cereals, and its improvement by traditional breeding, at the International Maize and Wheat Improvement Center (Centro Internacional de Mejoramiento de Maiz y Trigo, CIMMYT) and elsewhere. An in-depth treatment will be given to hybrid development and breeding for stress tolerance and improved nutritional quality.

## 6.2 MAIZE: ITS CYTOGENETIC ARCHITECTURE

Maize is an ancient tetraploid with 10 chromosome pairs (2n = 20; Figure 6.1). The first indication that the maize genome contained extensive regions of homology reflecting chromosomal duplications came from chromosome pairing analysis in haploids (2n = 10), where bivalents and multivalents were observed during the prophase of meiosis (Ting 1966; Snope 1967). The duplicated nature of the maize genome was reaffirmed with the discovery of duplicate genes (Rhoades 1951), isozymes (Goodman et al. 1980), and DNA-based markers (Helentjaris et al. 1988). The extent of gene and chromosome segment duplication in maize has also been demonstrated by various genome mapping studies (Ahn and Tanksley 1993; Davis et al. 1999; Wilson et al. 1999).

Even though maize has a purported polyploid origin, its diploidized nature is reflected in its sensitivity to chromosome duplications and deficiencies (see also Chapter 1 by Jauhar in this volume). Deficiencies of whole chromosomes or chromosome segments are deleterious during male and female gametophyte development, resulting in pollen and ovule abortion. Duplications, on the other hand, usually do not cause pollen or ovule abortion, but their transmission through the male is low due to a lack of pollen germination or a lack of competitive ability. Duplications can be transmitted through the female. This gametophytic screen is a natural mechanism to restrict the transmission of chromosome aberrations. Similarly, the gametophytic screen also limits the generation and maintenance of some chromosome aberrations that are useful for applied purposes or to study chromosome biology. Nevertheless, maize has been the focus of intense genetic and cytogenetic research due to chromosome manipulation strategies and mutations that have evaded the gametophytic screen (for review, see Sprague and Dudley 1988; Freeling and Walbot 1994).

## 6.3 CYTOGENETIC MANIPULATION OF THE MAIZE GENOME

### 6.3.1 Manipulations of the Ploidy Level

A euploid is an individual having a chromosome complement that is an exact multiple of the basic haploid chromosome number (x = 10). A normal diploid maize (2x) is part of a euploid series that also includes haploid (x), triploid (3x), tetraploid (4x), and other polyploids (5x, 6x, etc.). Haploids have been generated in various ways. Coe (1959) found a particular genetic stock (*stock 6*) that, when selfed, produces maternal haploids at a rate of ~3%. Kermicle (1969) identified a mutant gene *ig* (*indeterminate gametophyte*) that has pleiotropic effects on embryo sac development. One consequence of the *ig* mutation is a high frequency of haploid progeny (3%), of which the majority is paternal in origin. Thus, the *ig* mutation may be used to combine the nucleus from a male parent with cytoplasm from the female parent (Goodsell 1961). The possibility to instantly produce homozygous lines is arguably the most important application of haploids in plant breeding (Nitzsche and Wenzel 1977; Jain et al. 1996; Maluszynski et al. 2003).

The spontaneous production of triploids (3x) is rare. However, mutations that affect meiosis (*asynaptic* and *elongate*) can increase their frequency by increasing the production of unreduced gametes (Beadle 1930; Rhoades and Dempsey 1966). Alternatively, triploids may be produced by crossing diploid and tetraploid lines. The segregation of three chromosome sets in triploids results in spores that are mostly aneuploid and unbalanced. Thus, triploids have low levels of fertility, but they have been used to generate various aneuploids. McClintock (1929) identified a number of primary trisomics among progenies of diploid and triploid crosses that were subsequently used to associate genetic linkage groups with specific chromosomes (McClintock and Hill 1931). The association of a genetic linkage group with a particular chromosome is believed to mark the beginning of maize cytogenetics. Thus, the assignment of genes to chromosomes has been a major application of aneuploids.

Tetraploids can be produced by inducing chromosome doubling in diploids. In maize, the availability of the *asynaptic* and *elongate* mutants that produce unreduced gametes has allowed the production of tetraploids (Alexander 1957; Rhoades and Dempsey 1966). Unlike triploids, tetraploids are more fertile since a 2-2 disjunction of homologs is possible. Thus, tetraploid maize lines can be easily produced and maintained. In addition to tetraploids, plants with higher levels of polyploidy can be produced. For example, Rhoades and Dempsey (1966) utilized the *elongate* mutation to construct a ploidy series, including 3x, 4x, 5x, 6x, and 7x.

The ability to produce a ploidy series in maize has allowed researchers to assess genome dosage effects on various phenomena. For example, Guo et al. (1996) analyzed the expression of 18 genes located on all 10 maize chromosomes in leaf tissues of haploid (1x), diploid (2x), triploid (3x), and tetraploid (4x) plants. For most genes, the absolute level of gene expression per cell increased as the gene dosage increased in multiples of the monoploid complement. The manipulation of ploidy has also been used to address genome dosage effects on endosperm development (Leblanc et al. 2002).

### 6.3.2 Manipulations of Chromosome Number: Aneuploidy

Aneuploids are individuals with a chromosome complement that is not an exact multiple of the basic number (x). Although the number of possible aneuploids is limitless, the most commonly studied and used aneuploids include monosomics (2x − 1) and trisomics (2x + 1).

Plants with an additional chromosome are known as trisomics (2n = 2x + 1). There are several classes of trisomics, including primary, secondary, tertiary, telocentric, and translocation types. Primary trisomics are those in which one of the standard chromosomes is present in triplicate, while the rest are in duplicate. All 10 possible primary trisomics have been isolated from x + 1 gametes produced by triploids (McClintock 1929). Trisomics have also been selected in diploid populations (Ghidoni et al. 1982) and as products of the *r-x1*-mediated chromosome nondisjunction (Weber 1973).

As discussed earlier, primary trisomics were instrumental in the establishment of maize cytogenetics research. Primary trisomic crosses were first used by McClintock and Hill (1931) to place a gene (the *R* locus) on maize chromosome 10. The use of trisomic crosses to place genetic loci relies on the aberrant genetic ratios that are produced. Test-cross ratios for the primary trisomics are variable but always deviate significantly from the 1 dominant:1 recessive disomic ratio. Assuming that 50% of the ovules of a trisomic plant are x + 1, a 5 dominant:1 recessive ratio would be expected from a cross between a trisomic plant of the *AAa* constitution as a female and an *aa* male. When trisomics are used as pollen parents, the expected ratio for *aa* × *AAa* crosses is 2 dominant:1 recessive, since x + 1 pollen does not survive or is outcompeted by x pollen. The trisomic ratio can be variable because it is dependent upon the frequency of x + 1 gametes and the amount of crossing-over between the locus in question and the centromere. Still, the disparity between female and male ratios is a hallmark of primary trisomics. In addition to assigning genes to chromosomes, primary trisomics have also been used to study preferential chromosome pairing in maize (Doyle 1967).

Plants with one copy of a chromosome while the rest of the complement is disomic (2n = 2x − 1) are known as monosomics. Monosomics generate x and x − 1 spores, but because of abortion of x − 1 gametophytes, only balanced (x) gametophytes are produced by 2x − 1 individuals. Thus, the monosomic condition is not transmitted sexually. However, a system facilitates the production of monosomics. This system is based on the *r-x1* deficiency, which is a submicroscopic deletion on chromosome 10 that is transmitted through the female but not through the male (Weber 1983). More importantly, the *r-x1* deficiency causes nondisjunction in the second postmeiotic mitosis during embryo sac development and at the first postmeiotic division during microgametophyte development. Thus, stocks that have the *r-x1* deficiency produce monosomic progeny. Monosomics produced by the *r-x1* deficiency have been used to assign genes to specific chromosomes. The assignment is made by crossing a maize plant carrying a recessive allele of an unplaced gene as a male with plants homozygous for the dominant allele of this gene and that carry the *r-x1* deficiency. $F_1$ plants that are monosomic for the critical chromosome will express the recessive phenotype (Simcox and Weber 1985). This strategy was extended to the assignment of chromosomal location of biochemical and DNA-based markers (Helentjaris et al. 1986). In this regard, monosomics have been instrumental in associating genetic linkage maps to specific chromosomes. A detailed review of the *r-x1* system and uses of monosomics can be found in Weber (1991).

### 6.3.3 Manipulations of Chromosomal Rearrangements

Chromosome breakage induced by chemical treatment, irradiation, or chromosome manipulations can result in the production of a wide range of chromosomal aberrations. Commonly studied and used rearrangements include reciprocal translocations, B-A translocations, and to a lesser degree inversions.

### 6.3.3.1 *Reciprocal Translocations*

Reciprocal translocations are chromosomal rearrangements where the reciprocal exchange of segments between nonhomologous chromosomes has occurred. Investigations on the nature of semisterility (50% pollen and ovule abortion) in maize showed that this trait was not due to a gene mutation but instead was due to the production of balanced and unbalanced spores typical of reciprocal translocations (Brink 1927; Burnham 1930). In plants that are heterozygous for a reciprocal translocation, homologous segments of the normal and the interchange chromosomes pair during the meiotic prophase. During anaphase I, alternate disjunction of chromosomes results in viable spores containing either both normal or both interchange chromosomes. On the other hand, adjacent forms of disjunction produce nonviable spores with deficiencies and duplications, since they contain combinations of one normal and one interchange chromosome. Thus, plants that are heterozygous for a reciprocal translocation typically show 50% pollen and ovule abortion; they are semisterile. A detailed discussion on the behavior of translocations is given in Burnham (1962).

Although translocations of spontaneous origin have been described and studied, the majority of translocations (>1000) were induced by ionizing radiation (Longley 1961). The availability of this large collection of cytologically characterized translocations facilitates the integration of cytological and genetic linkage maps. In linkage tests, the inheritance of the translocation-based semisterility can be treated as being caused by a locus. The semisterility locus will map to the exchange point of the interchange. Loci on the chromosomes involved in the translocation will be linked with semisterility and with each other. If a trait or gene shows linkage to semisterility in crosses with a known translocation stock, it is located on the chromosomes involved in the translocation. Thus, translocations can be used to place a trait, marker, or genetic locus to a particular chromosome (Burnham 1966, 1982) or chromosome segment since most translocation breakpoints have been well defined physically (see Neuffer et al. 1997 for the comprehensive integration of genetic and cytogenetic maps).

A series of chromosome 9 translocations where the translocation site is linked to the *Waxy1* locus (Anderson 1956) has been very useful in the localization of new mutant loci. In crosses between a line with a mutation and all of the *waxy1*-tagged translocations, linkage to the *waxy1* phenotype suggests that the mutant locus must be located on chromosome 9. On the other hand, linkage to only one of these translocations suggests that the mutant locus is on the chromosome involved in the translocation with chromosome 9. Arthur et al. (2003) provided a recent illustration of how these translocations can also be used to characterize mutations in a gene (*rop2*) that affects male gametophyte function. In crosses between the mutant line (*Wx1;rop2*) and the pertinent translocation stock (wx13A, T4-9b *wx1;rop2+*), the normal allele *rop2+* is present in a chromosome translocation involving chromosomes 9 and 4 tagged with the *waxy1* allele, while the mutant allele *rop2* is present in a normal chromosome 4 (*Wx1;rop2*/T4-9b *wx1;rop2+*). Alternate disjunction results in viable gametophytes where half contain normal chromosomes (the majority having the *Wx1;rop2* genotype) and half contain the translocated chromosomes (the majority having the *wx1;rop2+* genotype). Since the waxy phenotype can be assayed in pollen grains (potassium iodide staining; Figure 6.2), the *wx1* marker allows the classification of individual pollen grains in a heterozygous population. Thus, the role of *rop2* in germination and pollen tube growth can be quickly assayed. These assays suggested that *rop2* plays an important role in germination and pollen tube growth. Because this series of *waxy1*-tagged translocations (~30) provides good coverage of the genome, other mutations that affect pollen function can be similarly analyzed.

Translocations that produce viable deficient-duplicate gametes (from adjacent disjunction) have been identified. These are instances where the deficiencies are small and genes that are essential for gametophyte development are not involved. Translocations that produce viable deficient-duplicate gametes can be used to place genes on chromosomes. When a deficient-duplicate egg is fertilized by a sperm cell carrying the recessive allele of a character in question, the recessive phenotype will be expressed if the locus lies within the deficient segment. Several loci have been placed on the cytological

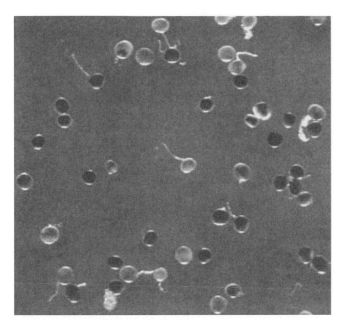

**Figure 6.2** **(See color insert following page 114.)** Pollen germination assay in plants heterozygous for a mutation and a *waxy1*-tagged chromosome translocation (*Wx1;rop2*/T4-9b *wx1;rop2+*). The waxy phenotype was observed by staining pollen with an iodine solution after pollen grains were permitted to germinate. Pollen grains carrying the *wx1* and mutant *rop2* alleles stained brown (no amylose), while pollen grains carrying the *Wx1* and wild-type *rop2+* alleles stained black (amylose present). *Wx1* pollen grains (black) showed reduction in germination linked to mutant *rop2* compared to *wx1* grains (brown). (This photograph was graciously provided by John E. Fowler and Kirstin M. Arthur, Oregon State University.)

maps of chromosomes 2 and 9 using this approach (Patterson 1952). Phillips et al. (1971) identified over 50 translocations yielding viable deficient-duplicate gametes covering the entire maize genome. Other applications for deficient-duplicate chromosomal complements include studies of chromosome synapsis, mutagenesis, and gene dosage effects (Patterson 1978).

The use of duplicate-deficient chromosomes from translocation heterozygotes to establish a nuclear male sterility system for hybrid maize production has been proposed (Patterson 1973). Crosses of male sterile (*ms ms*) females with duplicate-deficient *Ms/ms* males will give all male sterile progeny if the *Ms* locus is completely linked to a breakpoint and the duplicate-deficient chromosome is not male transmissible. Female transmissibility is needed to maintain the stock. Using the duplicate-deficient gamete from the T5-6b translocation opposite a normal chromosome 6 carrying the *polymitotic* (*po*) male sterility mutation (Beadle 1929, 1932) and crossing that heterozygote as the male to a female homozygote *po po* will produce progeny that are 98.3% male sterile (Phillips 1993).

### 6.3.3.2 B-A Translocations

Under experimental conditions, supernumerary B chromosomes have been found to be potentially useful for manipulations of dosage of chromosome segments via B-A translocations. A detailed review of the behavior and use of B-A translocations can be found in Beckett (1991). The first B-A translocation, analyzed by Roman (1947, 1948), involved the B chromosome and chromosome 4. The two new chromosomes formed by the translocation were designated 4B and B4. The 4B chromosome contained the terminal end of the long arm of the B chromosome. The B4 chromosome had the B centromere with its adjacent proximal region and a terminal segment of the short arm of chromosome 4. When plants that were homozygous for the translocation were crossed as males to a female tester, kernels with no B4 chromosomes in the endosperm and two B4 chromosomes in the embryo were found. Also, kernels with no $B^4$ chromosomes in the embryo and two $B^4$ chromosomes in the endosperm were identified. The dissimilarity in $B^4$ chromosome dosage between the embryo and endosperm was attributed to nondisjunction of the $B^4$ chromosome at the second pollen mitosis. Thus, the dosage of other A chromosome segments can be varied in the same fashion if the pertinent B-A translocations are available. Due to the inertness and dispensability of B chromosomes, another approach to vary dosage is to take advantage of adjacent disjunction in meiosis of B-A translocation heterozygotes. In a cross between a normal plant, as the female, and one plant heterozygous for the translocation, as the male, $4\ 4 \times 4\ 4^B\ B^4$, the male parent will produce microspores with varying constitutions, including the balanced translocation $4^B B^4$. Pollen formation by the $4^B B^4$ class will yield sperm with the $4^B B^4 B^4$ and $4^B$ constitutions due to nondisjunction of the $B^4$ chromosome. Individuals derived from a normal egg and sperm nuclei with the products of nondisjunction will either have the $4\ 4^B\ B^4\ B^4$ or $4\ 4^B$ constitution, depending on the pattern of fertilization. Individuals derived from a normal egg and normal sperm nuclei will have the $4\ 4^B\ B^4$ constitution. The $4\ 4^B\ B^4\ B^4$ individual has three copies of the chromosome 4 segment that was translocated to the B chromosome. This individual is referred to as a hyperploid or partial trisome. The $4\ 4^B$ individual has only one copy of the chromosome 4 segment and is referred to as a hypoploid or partial monosome. The $4\ 4^B\ B^4$ individual has two copies of the chromosome 4 segment. Altogether, these represent a segmental aneuploid series. A detailed review of other permutations and manipulations of the B-A translocation system to produce other segmental aneuploids can be found in Birchler (1991).

B-A translocations can also be used to localize genes to specific chromosomes (Roman and Ullstrup, 1951). Normally, a cross $aa \times AA$ produces $F_1$ progeny with the dominant $Aa$ phenotype only. If $A$ is carried on a $B^A$ chromosome, some $F_1$ progeny will have the recessive phenotype uncovered due to nondisjunction. Thus, the gene is located in the chromosome arm that is carried on the $B^A$ chromosome. The use of B-A translocations for each chromosome arm in the maize genome would permit the localization of any mutation (Beckett 1978). This approach has been used to localize a large number of induced mutations (Neuffer and Sheridan 1980). Like with reciprocal translocations, cytologically characterized B-A translocations can be used to integrate cytological and genetic linkage maps.

Another major use of B-A translocations in maize is the study of gene dosage effects. The phenotypic consequences of aneuploidy have been recognized in a variety of organisms. The prevailing theory has been that abnormal phenotypes are due to the imbalance of gene products encoded in the varied chromosome relative to those of the remainder of the genome. Guo and Birchler (1994) used B-A translocations to produce a dosage series (1, 2, and 3 in embryos and 2, 3, and 4 in endosperm) of 14 chromosomal regions in maize to study the effect of dosage on gene expression. Transcript amounts for *Adh1*, *Adh2*, *Glb1*, *Sh1*, *Sus1*, and *Zein* were measured on the different stocks. Surprisingly, genes whose dosage was altered showed compensation — their level of expression was the same regardless of dosage. On the other hand, genes whose dosage was not altered were affected by dosage of unlinked chromosomal segments. Thus, Guo and Birchler (1994) concluded that phenotypic effects of aneuploidy are probably due to consequences of altered dosage-sensitive regulatory systems. Similar dosage sensitivities were also observed on mitochondrial genes (Auger et al. 2001).

### 6.3.3.3 Inversions

Inversions in maize were first described by McClintock (1931). There are two types of inversions — paracentric and pericentric. In paracentric inversions the affected segment is confined to one chromosome arm, whereas in pericentric inversions the centromere is included in the inverted segment. In both types of inversions, a loop-shaped configuration is produced at pachytene by homologous synapsis of a normal and inversion chromosome. The subsequent behavior of the two types of inversions is different. If single- or three-strand double crossovers occur in the loop of a paracentric inversion heterozygote, a dicentric bridge will unravel in anaphase I. The dicentric bridge will also be accompanied by an acentric fragment. McClintock's (1938, 1939, 1941) work on the fate of dicentric bridges generated by paracentric inversion heterozygotes and other complex rearrangements led to the characterization of breakage-fusion-bridge (BFB) cycles. A study of BFB cycles caused by the chromosome-breaking activity of an unknown element led to McClintock's (1951) discovery of transposons or jumping genes.

Inversions have been used for a variety of genetic studies, including the localization of genes (Dobzhansky and Rhoades 1938; Sprague 1941; Chao 1959) and the study of synapsis and crossing-over (Bellini and Bianchi 1963, McKinley and Goldman 1979). A correlation between homologous pairing and chiasma formation and the phenomenon of synaptic adjustment were established by studying chromosome pairing in inversion heterozygotes (Maguire 1972, 1981).

### 6.3.4 Manipulation of Maize Chromosomes in an Oat Background

#### 6.3.4.1 Oat × Maize Crosses

In cultivated grasses, interspecific and intergeneric hybridization has been useful to widen the genetic base of a crop, for haploid production, and to construct stocks for genetic analysis. Hybridization between oat (*Avena sativa* L.) and maize is an example of an extreme wide cross (Riera-Lizarazu et al. 1996) since oat and maize are estimated to have diverged about 70 million years ago (Buckler and Holtsford 1996) and they belong to two different subfamilies of the Poaceae (Gramineae), namely, Pooideae and Panicoideae, respectively.

The reasons for attempting crosses between oat and maize stemmed from reports that wheat × maize crosses were feasible (Laurie and Bennett 1986) and uniparental elimination of maize chromosomes resulted in the production of wheat haploid plants with reasonably good efficiency (Laurie et al. 1990). Rines and Dahleen (1990) produced oat haploids from oat × maize crosses and found that oat haploid plants were partially self-fertile (~40% of the primary and secondary florets produced seed), probably due to a meiotic restitution process (Davis 1992). Riera-Lizarazu et al. (1996) later discovered that in addition to oat haploids, oat × maize crosses could result in the production of partial hybrids. About 69% of the plants produced from oat × maize crosses were oat haploids (2n = 3x = 21), and 31% were partial hybrids with one to four maize chromosomes (2n = 21 + 1 to 2n = 21 + 4; Figure 6.3). Maize chromosomes 2 to 9 were detected in various partial hybrids. Partial self-fertility of oat × maize hybrids, as observed in haploid oat, resulted in the production of maize chromosome addition lines of oat. Oat–maize addition lines have the disomic chromosome complement of 42 oat chromosomes plus a pair of maize chromosomes (2n = 6x = 42 + 2). Once stabilized as a disomic addition, the maize chromosome pair is, in most cases, transmitted consistently to its progeny. Riera-Lizarazu et al. (1996) reported the production of fertile addition lines for maize chromosomes 2, 3, 4, 7, and 9. In addition to disomic addition lines, a monosomic addition for maize chromosome 8 and a double disomic addition for chromosomes 4 and 7 were also isolated. Later studies showed that partial hybrids with up to eight maize chromosomes (2n = 21 + 8) could be produced from oat × maize crosses (Maquieira 1997). Among other addition lines, Maquieira (1997) produced and identified partial hybrids that yielded fertile disomic addition lines for maize chromosomes 6 and 8. Recently, disomic addition lines for maize chromosomes 1 and 5 and double disomic addition lines with maize chromosomes 1 and 9 and 4 and 6 have been isolated (Kynast et al. 2001, 2002). Thus, fertile disomic addition lines for each maize chromosome have been established, except for maize chromosome 10, where only short-arm ditelosomic lines are available (Kynast et al. 2004)

#### 6.3.4.2 Oat–Maize Addition Lines and Their Uses

Oat–maize addition lines are important genetic stocks for maize genome analysis because they constitute a heterologous system to study, analyze, and manipulate a given maize chromosome

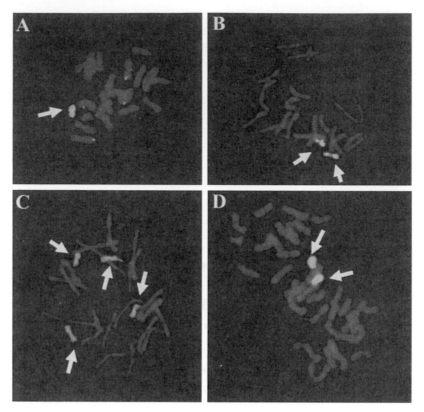

**Figure 6.3**    **(See color insert following page 114.)** Genomic *in situ* hybridization (GISH) of partial oat × maize
hybrids and a disomic maize chromosome addition line of oat. Unlabeled red oat chromosomes
were counterstained with propidium iodide, while fluorescein-labeled maize genomic DNA was
used as the probe. Yellow-green fluorescence marks maize chromatin (white arrows). (A) GISH of
a chromosome spread of a 22-chromosome oat × maize hybrid with 21 oat (red) and 1 maize
chromosome (yellow). (B) GISH of a chromosome spread of a 23-chromosome individual showing
21 oat (red) and 2 maize (yellow) chromosomes. (C) GISH of a chromosome spread of a 25-
chromosome individual showing 21 oat (red) and 4 maize (yellow) chromosomes. (D) GISH of a
chromosome spread of a 44-chromosome individual with 42 oat chromosomes (red) and a pair of
maize chromosomes (yellow).

separate from other chromosomes of its native genome. Thus, manipulation and analysis of a single
maize chromosome constitutes a 10-fold reduction of complexity when compared to working with
all 10 maize chromosomes simultaneously. A good example includes the use of oat–maize addition
lines for the physical isolation of maize chromosomes by flow cytometric sorting, since maize
chromosomes are significantly smaller than oat chromosomes. The feasibility of this approach was
demonstrated by the isolation of about $6 \times 10^3$ maize chromosomes from 30 root tips of a maize
chromosome 9 addition line (Li et al. 2000). The isolation of highly pure (90%) fractions of maize
chromosome 9 was a major accomplishment because maize chromosome 9 cannot be separated
from maize chromosomes 6, 7, and 8 when normal maize is used. Thus, oat–maize addition lines
may allow the development of single maize chromosome genomic DNA libraries.

Oat–maize addition lines can also be used to isolate and clone DNA from specific maize
chromosomes because oat and maize genomic DNA differ significantly for their repetitive DNA
composition (Ananiev et al. 1997). This is achieved by using a collection of highly repetitive maize-
specific sequence probes to selectively isolate DNA clones containing maize genomic DNA from
an oat–maize addition line genomic DNA library. This targeted cloning approach was demonstrated
by the efficient (95%) isolation of cosmid clones originating from maize chromosome 9 from an

oat–maize addition line genomic library (Ananiev et al. 1997). Targeted cloning has also been used to study the structure and composition of maize heterochromatic knobs (Ananiev et al. 1998a,c) and centromeric regions (Ananiev et al. 1998b).

Another example of the usefulness of oat–maize addition lines is their demonstrated value as chromosomal mapping stocks for a variety of markers, including those that are monomorphic. The ease with which markers can be assigned to maize chromosomes using oat–maize addition lines was demonstrated by Okagaki et al. (2001), who localized about 350 maize expressed sequence tag (EST) and sequence-tagged site (STS) markers to specific maize chromosomes by simple polymerase chain reaction (PCR)-based assays. Similarly, this approach can be extended to the localization of large-insert DNA (cosmid, BAC, or YAC) clones to specific maize chromosomes, thus helping with contig construction at a chromosomal scale. The targeted integration of cytogenetic and genetic linkage maps of maize chromosomes is also possible by using maize chromosome addition lines of oat. Koumbaris and Bass (2003) used a maize chromosome 9 addition line and an improved single-locus fluorescence *in situ* hybridization (FISH) protocol to localize loci on pachytene chromosomes. A comparison of markers mapped cytogenetically and by linkage analysis revealed the discrepancy between physical and genetic distance for maize chromosome 9. A system to develop cytogenetic maps of other maize chromosomes with the help of oat–maize addition lines is now possible. From this discussion it is clear that oat–maize addition lines represent unique materials that complement other genetic and physical mapping efforts that are currently ongoing.

Besides mapping and genomics applications, oat–maize addition lines have been used for a variety of other studies. For example, Bass et al. (2000) used a maize chromosome 9 addition line to show that maize homologs lack premeiotic pairing. Instead, maize homologs paired and synapsed during the telomere bouquet stage of the meiotic prophase. Muehlbauer et al. (2000) used an oat–maize addition line to study the expression of a maize gene in an oat background. A maize chromosome 3 addition line exhibiting several morphological abnormalities, including blade-to-sheath transformations resulting in liguleless leaves, was studied. RNA expression analysis showed that the maize *lg3* gene was ectopically expressed in sheath tissues of oat plants carrying maize chromosome 3. In maize, ectopic expression of a dominant maize *Lg3* allele is also observed in leaf sheath tissues, and this ectopic expression is believed to cause the liguleless phenotype. It is conceivable that oat–maize addition lines will be useful heterologous systems to study other properties of individual maize chromosomes in the future.

### 6.3.4.3 Maize Subchromosome Fragment Stocks and Their Use

Maize chromosome addition lines are also suitable materials for the dissection of single maize chromosomes using radiation. Riera-Lizarazu et al. (2000) demonstrated that gamma-ray treatments (30 to 50 krad) of monosomic maize chromosome 9 addition line seed, followed by self-pollination of surviving plants, resulted in the production of oat lines possessing different fragments of maize chromosome 9. These maize subchromosome fragment stocks are referred to as radiation hybrids since they are equivalent to human–rodent somatic cell hybrids used for radiation hybrid mapping (Goss and Harris 1975; Cox et al. 1990). The maize chromosome 9 radiation hybrids (M9RHs) (Figure 6.4) included plants with apparently normal maize chromosome 9 as well as plants with various maize–chromosome 9 rearrangements (intergenomic translocations, deletions, and a combination of both). The number of radiation-induced breaks per chromosome ranged from 0 to 10. The usefulness of these radiation hybrid derivatives for the dissection of a maize chromosome was demonstrated by the identification of M9RHs with unique chromosome breakage patterns that allowed the dissection of maize chromosome 9 into 27 distinct regions. This panel of M9RHs would allow the placement of a marker to 1 of 27 regions on maize chromosome 9. Because mapping with radiation hybrid derivatives involves assays for the presence or absence of a given marker, monomorphic markers such as STS and EST can be quickly and efficiently mapped, making this system particularly amenable to automation and high-throughput formats. Besides mapping,

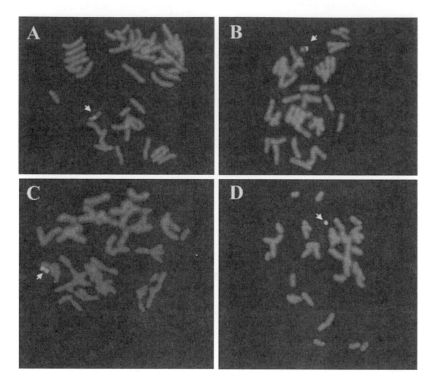

**Figure 6.4**   **(See color insert following page 114.)** Fluorescence *in situ* hybridization (FISH) of four maize chromosome 9 radiation hybrid lines. Unlabeled blue oat chromosomes were counterstained with DAPI. Fluorescein-labeled maize genomic DNA (yellow-green) was used as the probe for genomic *in situ* hybridization in C and D. Yellow-green fluorescence marks maize chromatin. Multicolor fluorescence *in situ* hybridization using two maize-specific repetitive probes (180-bp knob-specific sequence and the maize-specific multiprobe) was used in A and B. Red fluorescence marks the 180-bp knob homologous sequences, and yellow fluorescence marks sequences homologous to the maize-specific multiprobe. (A) A line with an intergenomic translocation (arrowhead). (B) A line with a complex rearrangement that involved at least one deletion, an inversion, plus an intergenomic translocation (arrowhead). (C) A line with an isochromosome (arrowhead). (D) A line with a diminutive maize chromosome 9 that resulted from multiple radiation-induced deletions (arrowhead).

radiation hybrid derivatives also represent sources of chromosome region-specific DNA for cloning applications. The development of radiation hybrid derivatives for other maize chromosomes present in various maize chromosome addition lines of oat is currently under way (Kynast et al. 2002, 2004).

## 6.4 GENETIC TRANSFORMATION: ADDING VALUE-ADDED TRAITS

Genetic transformation is becoming a useful technique to incorporate certain useful traits into crop plants by asexual means. Ohta (1986) brought about transformation by self-pollination of the recipient maize plant along with DNA of a donor plant, which was applied onto silks in a pollen DNA pasty mixture. The exogenous DNA transferred into endosperm expressed itself in endosperm formation, resulting in 9.3% frequency of transformed endosperm.

Stable transformation of maize cells was achieved through direct uptake of DNA by protoplasts that had been permeablized by electroporation (Fromm et al. 1986), but no plants were obtained. Rhodes et al. (1988) were able to obtain genetically transformed plants from protoplasts treated with recombinant DNA. Thus, electroporation proved to be a convenient method for maize transformation. Recent work has shown that use of electroporation for gene transfer into seed-derived maize embryos facilitates uptake, expression, and stable integration of exogenous DNA. Using this technique, Sawahel (2002) produced herbicide-resistant transgenic maize plants in which foreign DNA was stably integrated and inherited in a Mendelian fashion. Other methods of maize transformation include microprojectile bombardment (Gordon-Kamm et al. 1990; Fromm et al. 1990) and that mediated by *Agrobacterium* (Escudero et al. 1996).

Genetic engineering is emerging as a useful adjunct to conventional plant breeding and has helped solve several agricultural problems, such as pest control. Postharvest loss due to insects, mostly in the developing countries, is estimated at 15% of the world production. European corn borer (ECB), for example, causes a loss amounting to $1 to $2 billion in the U.S. alone (Hyde et al. 1999). Incorporating ECB resistance by conventional means is tedious and slow, even if a suitable source of pest resistance can be found. Through 12 years of conventional breeding, Syngenta, a seed and agrochemical company, produced a corn cultivar with only 10% resistance to ECB. A bacterium, *Bacillus thuringiensis* (Bt), has a gene that confers resistance to ECB. By engineering the Bt gene into the corn genome, ECB-resistant corn varieties have been produced (Ostlie et al. 1997) and are under commercial cultivation in the U.S. ECB-resistant corn inbreds have been used to produce superior Bt–corn hybrids with 4 to 8% higher grain yields than standard hybrids when infested with ECB (Lauer and Wedberg 1999).

## 6.5 GERMPLASM CONSERVATION AND EARLY BREEDING WORK

### 6.5.1 Conservation and Utilization of Maize Genetic Resources

Conserving maize genetic resources, supplying to users, and effectively utilizing them in breeding programs have been pursued vigorously ever since the inception of CIMMYT. The CIMMYT maize gene bank is perhaps the world's largest gene bank, the others being in Russia, Yugoslavia, and INIFAP in Mexico. The bank is regarded as a working and functional bank, as opposed to merely a conserving house for freezing evolution. The majority of the accessions held in the CIMMYT bank are of Latin American and Caribbean origin, while some are the result of CIMMYT's breeding activities in Mexico and its regional breeding activities in Asia, South America, and Africa. Like other gene banks, the maize gene bank also performs routine activities of regenerating accessions for keeping good seed viability and preserving in long-term (–15°C) and medium- to short-term (–2°C) storage.

At least 26,000 accessions are documented from Latin America and the Caribbean region. Of these, 17,000 are conserved in the CIMMYT gene bank and the remaining 9000 in other gene banks of Latin America (Taba and Krakowsky 2002). Database information on bank accessions is available on CD-ROM, facilitating users to pick and request what they need, and thus encouraging the use of extensive collections of maize genetic resources for the benefit of humankind. The bank in recent years has also attempted to identify core subsets of major and potentially useful maize landraces, such as the Tuxpeno race and the Caribbean race (Taba and Krakowsky 2002). Accessions numbering 50 to 100 are identified based on some sort of evaluation and are regarded as breeder-targeted core subsets.

### 6.5.2 Early Breeding Work: Exploring New Options and Methodologies

Early breeding work at CIMMYT paved the way to sort out options and methodologies that would be relevant and appropriate in breeding different traits in maize. Prior to CIMMYT's establishment in the 1960s, important work had been done to collect and preserve maize samples from Mexico and other countries — the work that led to identification of potentially useful races and, in some instances, to specific accessions within a particular race for undertaking breeding and improvement work. With the inception of CIMMYT, germplasm activities were strengthened, and breeding methodologies suited to conducting improvement work were employed. A functional gene bank was also established at CIMMYT to maintain, multiply, and preserve bank accessions and to characterize for phenotypic diversity.

Some of the important early accomplishments (e.g., Vasal et al. 1982, 1997a, 1999a,b; Pandey and Gardner 1992) are summarized below.

### 6.5.2.1 Reducing Plant Height

Several height-reducing mutants, Brachytic-2 (*br2*), *pigmy*, and others, were investigated together with recurrent selection procedures to reduce plant height in tropical maize genotypes, which happened to be too tall and prone to lodging. Both approaches were deployed and researched for several years, but by about 1976, the recurrent selection approach appeared to be promising and led to steady progress (Johnson et al. 1986; Fischer et al. 1989). Work was done on Tuxpeno type germplasm, which is virtually described as a gold mine of high productivity and of enormous genetic diversity for further improvement of maize in the tropics. Two potentially useful populations, 21 and 49, resulted from this work and were used for further improvement as part of CIMMYT's recurrent selection program. Both materials have been used extensively for reducing height and developing different versions of these populations for maize streak virus (MSV) resistance, drought tolerance, and improved nutritional value. Changes in combining ability of Tuxpeno-derived versions have been studied using testers from the same and opposite heterotic groups (Srinivasan et al. 1992). Both populations have contributed significantly to the development of new sources of germplasm and are important components of several populations and gene pools developed at CIMMYT and in the national programs. This germplasm has also been improved for photoperiod insensitivity as an important source of tropical germplasm for maize improvement efforts in temperate conditions (Hallauer and Miranda 1988). In addition to excellent donors for reduced height, these materials have provided the largest number of good combining lines, as well as important sources of genetic resistance to biotic and abiotic stresses prevalent in the tropical and subtropical regions (CIMMYT 2004; Vasal et al. 1999a).

### 6.5.2.2 Developing Early-Maturity Germplasm

Developing early-maturing maize genotypes that combine good yield and earliness is considered difficult, challenging, and a perpetual concern of maize scientists worldwide. Several approaches were tried, including recurrent selection for earliness in normal genotypes, compositing and genetic recombination of known well-performing early genotypes, and the third approach of crossing early × late genotypes, followed by selection of early well-performing genotypes in every cycle of selection. Of the three approaches listed, the first and third approach appeared promising and have been successfully exploited in maize programs to develop early-maturing maize gene pools and populations. The most important germplasm resulting from the recurrent selection approach was Seleccion Precoz, which was later numbered and named population 31 (Amarillo Cristalino-2). Considering the importance of early germplasm, disease-resistant versions of downy mildew and MSV have been developed in regional programs of Asia and Africa. Well-combining, agronomically desirable inbred lines have been derived and released that are being used as parent(s) of early-maturing hybrids in the Asian region (Vasal and Mclean 1995; Pandey and Gardner 1992; CIMMYT 2004).

### 6.5.2.3 Breeding for Morphological Traits

In addition to reducing plant height, recurrent selection was also practiced for reduced tassel size and reduced foliage above the ear in Tuxpeno-1, and Antigua Republica Domonicana. In all populations studied, there were significant reductions in tassel branch number and leaf area density above the ear, accompanied by significant increase in grain yield (Fischer et al. 1987).

### 6.5.2.4  Improved Nutritional Quality

Several aspects of nutritional quality of maize endosperm protein were investigated. A few of the most important ones include introduction of high-quality protein mutants, development of soft, nutritionally enhanced maize populations, screening of normal maize genotypes for high lysine and tryptophane content, and introduction of gametophytic factor(s) to prevent contamination of soft opaques by normal pollen and selection of modified opaques (Paez et al. 1969a,b). A detailed discussion on this topic that emerged as an important scientific breakthrough will be presented later in this chapter.

### 6.5.2.5  Developing Potentially Useful Germplasm and Breeding Methodologies

Of the various breeding methodologies studied, full-sib recurrent selection and modified ear-to-row selections and their modifications proved useful for future maize improvement efforts (Pandey et al. 1984; Vasal and Mclean 1995). A few unique germplasm development efforts were undertaken to form the so-called world composite involving a huge number of bank accessions and a Tuxpeno-based composite involving practically all known and available accessions of this race in the gene bank. Screening and evaluation of Caribbean accessions for insects and diseases assisted in the formation of insect disease resistant nurseries (IDRN) and Cogollero populations with low levels of tolerance to diseases and insects.

## 6.6 HYBRID-ORIENTED SOURCE GERMPLASM AND HYBRID DEVELOPMENT

Hybrid development activities in CIMMYT's maize program were introduced in 1984. For almost two decades during the prehybrid era, intrapopulation recurrent selection methodologies were emphasized with a strong focus on the development and promotion of open-pollinated varieties (OPVs). Growing interest in hybrids and the strong presence of the private sector in some countries prompted this change in product development at CIMMYT. This indeed was a major shift in strategy to initiate research on maize hybrids, but still continuing a major emphasis and focus on OPVs. In the last 15 years, the hybrid development activities have witnessed many positive changes and accomplishments in both generating scientific information and product development as it relates to inbred parents and hybrids.

### 6.6.1  Characterizing Germplasm for Combining Ability and Heterotic Pattern(s)

Potentially important germplasm types were characterized for their combining ability and heterotic behavior. All available materials, including pools and populations, plus a few others, were grouped by adaptation, maturity, and in some instances grain color. Quality protein maize (QPM) germplasm, both tropical and subtropical, was studied separately in two groups. Overall, eight separate groups were constituted, and $F_1$ crosses were developed within each group using either a diallel or design-2 mating system. In all cases, parents and the $F_1$ crosses resulting within each group were evaluated in multilocation trials involving several interested countries, and important heterotic relationships emerged from this study (Beck et al. 1991; Vasal et al. 1992a). The results from QPM trials are discussed in later sections.

### 6.6.2  Shifts in Recurrent Selection Procedures

The initiation of a hybrid development program prompted a shift in recurrent selection strategy. Two populations, 21 and 32, already known to be heterotic, were subjected to an interpopulation improvement program using modified reciprocal recurrent selection (MRRS). A generation of inbreeding and recombination was necessary to get rid of deleterious alleles and develop heterotic populations. Several heterotic populations were developed through conscious efforts using selfed progenies × testers approach. In interpopulation improvement, the two schemes used are modified reciprocal recurrent selection using half-sib test-cross progenies (MRRS-HS) or modified reciprocal recurrent selection with full-sib progenies (MRRS-FS). The latter scheme was used at CIMMYT headquarters in some populations and is still in use in the Asian Regional Maize Program (ARMP). The preliminary results have been encouraging with respect to per se improvement and crossbred performance (Cordova and Trifunovic 2002). The gains in yield accompanied gains in other traits. Plant height was reduced, populations became earlier in maturity, and the ear rot incidence was also reduced.

### 6.6.3  Development of Inbred Progenitors

With the start of the hybrid program, inbred line development was initiated. To begin with, the early-generation lines $S_1$ and $S_2$, resulting from ongoing population improvement activities, were secured and advanced to the next inbreeding generation by selfing. Inbreeding was continued every season or cycle to attain a higher degree of fixation of alleles to result in homozygosity and uniformity. Between- and within-line selection was practiced during inbreeding generations, and more importantly, evaluation per se and advancing inbreeding generation were consciously practiced at every step. This was accomplished by pollinating the second half of the row and leaving the first half for observation and visual evaluation. Differential densities were also practiced in the same row to select for some traits, such as husk cover and standability. As inbreeding progressed, densities were increased to 80,000 plants per hectare.

### 6.6.4  Inbreeding Tolerance and Crossbred Performance

Extraction of good inbreds from any given maize population is a difficult and challenging task. To improve tolerance to inbreeding depression in some of the potentially important tropical and subtropical maize populations, a simple selfed progeny $S_3$ recurrent selection was followed in some materials for at least two cycles, which led to encouraging results (Vasal et al. 1995a). The selection process improved the grain yield of the selfed progeny from the advanced cycle, compared to the original or $C_0$ cycle. The linear gain per cycle ranged from 4.21 to 11.76% in four tropical late white populations: 21, 25, 29, and 32. The inbreeding depression for grain yield across populations decreased from 39.4% in $C_0$ to 34.9% in $C_2$. The selection was thus effective in reducing inbreeding depression and improving the yield of the selfed progeny. The populations improved for this trait served as better germplasm sources for the extraction of vigorous and better-performing inbred progenitors. Other traits remained unchanged, but the plant height increased slightly. From this work, 12 hybrid-oriented maize germplasm tolerant to inbreeding depression were registered (Vasal et al. 1995b).

### 6.6.5  Hybrid Options and Their Relevance

In the mid-1980s, CIMMYT embarked upon a hybrid initiative, and all types of conventional and nonconventional hybrids were produced in the developing countries (Vasal et al. 1999a). The best hybrid showed a yield superiority of 26% or more over the best OPV included as a check in the trial. Two-parent hybrids, either top crosses or single crosses, were the best, with a yield superiority of 33 to 40% and 40 to 53%, respectively, over the best OPV check entry (Vasal et al. 1994). These results were quite conclusive in suggesting that it is better to strive for two-parent single-cross hybrids to achieve maximum yield potential.

A new hybrid development strategy of forming hybrids between tropical and subtropical inbreds has added a new dimension to enhancing the yield potential of hybrids (Cordova and Trifunoovic 2002). A whole array of hybrid trials was constituted from tropical, subtropical, and highland subprograms in Mexico, and others from the regional programs in Asia, South America, and eastern and southern Africa, as part of a global international testing network. The response for such trials has been exceedingly good from public and private institutions, as CIMMYT provides a complete and open pedigree of all hybrids tested and makes freely available the parents of the superior hybrids.

### 6.6.6 Inbred Line Evaluation Nurseries

Evaluation per se of maize inbreds is not a common practice or strategy in most hybrid development programs. However, there are good examples where inbred lines have been distributed widely for disease evaluation, as is the case with maize inbred resistance (MIR) nurseries (Brewbaker et al. 1991). The MIR project has studied tolerance/resistance to virus and virus-like diseases and to insects, parasitic weeds, and other diseases. As a result of this effort, a number of lines were identified as good/excellent sources of resistance to viral diseases. Inbred line evaluation nurseries have become an important component of hybrid research and development efforts in all subprograms at CIMMYT, Mexico, and in the regional programs. Selfed progenies are evaluated as part of ongoing population improvement activities, but in addition, inbreds resulting from the inbred development programs are evaluated routinely for agronomic performance as well as tolerance/resistance to different biotic or abiotic stresses. The success of this strategy is evident from the varying number of lines that have been identified with tolerance/resistance to a whole spectrum of diseases, insects, and abiotic stresses (Vasal et al. 1999b). Several of the lines found resistant to tarspot disease complex, fusarium ear and stalk rots, gray leaf spot, rust, maize streak virus (MSV), corn stunt, downy mildew, fall armyworm, borers, and acid soil tolerance had no previous history of selection against these stresses. Such inbred line evaluation nurseries also serve as an important mechanism for germplasm distribution and accelerating hybrid development efforts in the developing countries.

### 6.6.7 Characterizing and Using Maize Inbred Progenitors

Both combining ability and testers are important aspects of hybrid research and development. The information resulting from such studies was also used in forming inbred-based populations (IBP-1, IBP-2, IBP-3, and IBP-4). In addition, new heterotic populations were formed and designated as heterotic group-1 (THG-A) and heterotic group-2 (THG-B). A few other populations were also developed, demonstrating the power of inbreeding as an important tool. The inbreeding process exposes deleterious alleles but also exposes rare useful alleles that are difficult to encounter at the population and family levels. Several special-trait inbreds can be and have been identified during the inbreeding process and later recombined to form special-trait populations such as a lodging-resistant population, long-ear synthetic, stay-green population, tarspot-resistant synthetics, and rust-resistant synthetics. These conscious efforts were considered important in identifying tester lines and in permitting follow-up line × tester studies (Vasal et al. 1992b). These studies provided useful information on heterotic behavior of lines with different testers.

### 6.6.8 Hybrid Development and Testing

A set of diallels and the sequential ones with the superior-performing lines, as well as the line × tester studies, provided useful information on intra- and interpopulation interline hybrids. The studies demonstrated clearly that interpopulation interline hybrids were on the average superior to the intrapopulation interline hybrids (Han et al. 1991). Averaged over several hybrids, the hybrids between lines from different populations showed a slight to substantial edge over the intrapopulation hybrids, although a few hybrids in the latter category were as good or even better than the former category (Vasal and Srinivasan 1991). The year 1994 marked a new beginning in the testing of hybrids in the international hybrid trials.

### 6.6.9    Research on Maize Testers

Testers, whether inbreds or noninbred progenitors, should to be available for use in a whole array of hybrid-related activities. As to the number of testers, it could vary with the size and diversity of the program. Institutions like CIMMYT, which handle a huge and diverse germplasm volume adapted to different ecologies, will certainly need many testers for different classes of germplasm. It is preferable to have testers that are appropriate in terms of ecological adaptation, maturity, seed color, and sometimes in possessing stress resistances that are relevant to maize production.

Research on finding new and better successor testers with high inbreeding levels has continued (Vasal et al. 1995a, 2001). Conscious efforts have also been under way to find some white and yellow single-cross testers. A series of new diallels with highly inbred materials and line × single-cross testers representing three-way hybrids were developed and tested in a few locations in Mexico and Central America. These studies turned out to be exceedingly important in identifying hybrids for the first set of international hybrid trials commenced in 1994. Some hybrids from the international hybrid trial (CHHT-Y) showed 3% superiority over the best check across locations.

Success in identifying testers is evident from the large number of lines that have been identified and some released as CIMMYT maize lines (CMLs) (Table 6.1). Practically all programs have identified two or more tester lines to meet their hybrid development needs. However, testers in early and extraearly categories are still not available in some programs, but they may become available with concerted and conscious efforts in the future.

### 6.6.10    Release of Maize Inbreds and Other Materials

The maize program at CIMMYT makes periodic announcements of maize inbreds to public and private institutions to accelerate hybrid development efforts and to strengthen the inbred base germplasm of these programs. Complete information on pedigree is provided to permit recycling between CIMMYT inbreds or in crosses with their own developed inbreds. The first announcement, of 139 tropical and subtropical inbreds, was made in 1991. Since then, additional announcements have been made practically every year, with a total number of released inbreds touching a new high of 497. The announcement of maize inbreds in different years is given in Table 6.1. As can be seen, the CMLs released so far have different adaptations, maturities, grain colors, and special attributes. At least 59 inbreds are QPM with improved nutritional quality. Several of the released CMLs have special traits and possess tolerance/resistance to drought, acid soils, low N, downy mildew, corn stunt, maize streak virus, rust, stalk and ear rots, and other diseases. A few of the lines carry tolerance to borers and fall armyworm. Samples of released lines are made freely available on request (Vasal et al. 1997a,b). In addition to released inbreds, selfed progeny $S_3$ bulks

**Table 6.1    Maize Inbreds Released by CIMMYT Maize Program 1991–2003**

| Year of Release | No. of Lines | CML Numbers | Description |
|---|---|---|---|
| 1991 | 139 | 1–139 | Tropical, subtropical, stress tolerant |
| 1992 | 99 | 140–238 | QPM, maize streak virus (MSV) resistant |
| 1993 | 8 | 239–246 | Highland |
| 1994 | 64 | 247–310 | Tropical lines |
| 1995 | 19 | 311–329 | Subtropical lines |
| 1997 | 37 | 330–336 | Tropical, subtropical, and acid soil tolerance |
| 1998 | 58 | 367–424 | Tropical, subtropical, and mid-altitude |
| 2000 | 30 | 425–454 | Stress-tolerant lines, downy mildew, acid soil tolerance, drought, low N, and MSV |
| 2002 | 21 | 455–475 | Highland, transition zone, downy mildew tolerant |
| 2002 | 14 | 476–489 | Tropical, subtropical, and mid-altitude |
| 2003 | 8 | 490–497 | Tropical |

from pedigree populations have also been made available at least from Asian regional maize programs. Maize inbreds are also available from IITA in Nigeria, and several of them are resistant to maize streak virus (Kim et al. 1989).

## 6.7 BREEDING FOR STRESS TOLERANCE

This section deals with research initiatives related to biotic and abiotic stresses. In biotic stresses, the emphasis will be on diseases and insect pests that affect maize crop during various stages of crop growth from planting to harvest, as well as losses suffered in storage due to stored grain pests. More progress has been made in the development of maize germplasm that is tolerant to different diseases, compared to insect pests-tolerant or abiotic stress-tolerant germplasm. Most national programs in the developing countries lack physical and human resources to conduct systematic and sustained research in stress breeding. Even when resources and research projects are approved in some countries, lack of cooperation and integration in interdisciplinary research efforts fail to produce concrete results. In the past decade or so, there has been increasing awareness of the need to do more research and develop stress-tolerant maize germplasm. Concerted efforts of researchers involving different disciplines are the key to achieving desired success in stress-related objectives.

### 6.7.1    Enhancing Disease Resistance

Disease resistance in source populations is important to reduce and prevent losses. A whole spectrum of diseases affects maize crop during different stages, from planting until harvest. The nature, severity, and extent of damage by different diseases vary in different agroclimatic regions around the world. Environmental factors coupled with crop rotations, cultural practices, and occasionally release of susceptible material(s) help build up inoculum and incidence of what earlier was not an important disease. The foliar diseases regarded as important are maydis leaf blight (*Bipolaris maydis*), turcicum leaf blight (*Exserohilum turcicum*), leaf rusts (*Puccinia polysora* and *Puccinia sorghi*), several species of downy mildew (genera *Peronosclerospora* and *Sclerophthora*), corn stunt complex, and the maize streak virus. In recent years, gray leaf spot caused by *Cercospora maydis* has become an important disease in some countries. Two of the above-mentioned diseases, downy mildew and maize streak virus, are of importance in Asia and Africa, respectively. The stunt virus complex is limited to the Latin American region. Another disease, known as tarspot complex, caused by *Phyllacora maydis* and associated fungi *Monographella maydis* and *Coniothorium phyllachorae,* is considered a station disease in Mexico but is also known to occur in Costa Rica and Honduras. Banded leaf and sheath blight caused by *Rhizoctonia solani* is a newly emerging disease, and its incidence is increasing in several countries of Asia; it is also reported from Venezuela. In stalk and ear rots, several different organisms may be involved, but the ones considered most important are *Fusarium moniliforme*, *Gibberella zeae*, and *Diplodia maydis*. Ear rots may also be the result of *Aspergillus flavus*, particularly in hot, humid environments.

Selection and improvement against maize diseases have been practiced and have evolved over the years in National Agricultural Research Systems (NARs) and at CIMMYT (Jeffers et al. 2000). In CIMMYT gene pools and populations described earlier, disease resistance has been handled as an integral part of the overall improvement process. Selection is practiced against diseases occurring naturally in all populations in different stages of improvement, including progeny regeneration, multilocation testing, intrafamily improvement, and in companion disease nurseries if necessary. During intrafamily improvement, artificial disease pressure may also be exercised for a specific leaf disease and/or for stalk and ear rots. Different populations have been subjected to different disease pressure, depending upon their susceptibility to a particular disease or to their use in areas where they will be grown. Artificial disease screening pressure has also been applied to the pools using the above criteria. With the introduction of inbreeding as an important tool, disease screening was practiced in both noninbred and selfed progenies as part of ongoing population improvement programs. Inbreeding also helped in exposing deleterious and susceptible allele(s) and their genetic resistance to some diseases, such as tarspot complex (Vasal et al. 1992c). De León and Pandey (1989) conducted studies to determine progress from selection for ear and stalk rot resistance. They reported that a modified ear-to-row scheme was effective in increasing grain yield and resistance to ear rot and stalk rots in two early and two intermediate–late pools.

Recurrent selection programs have been conducted to improve resistance to turcicum leaf blight in eight subtropical pools using a selection procedure involving alternate generations of full-sib and $S_1$ families. Four cycles of selection were completed, resulting in 19% improvement in resistance to turcicum and 6% improvement in resistance to *P. sorghi* per cycle across all pools (Ceballos et al. 1991). Simultaneous improvements in grain yield and in the levels of downy mildew resistance were also achieved in four populations improved for these traits in the CIMMYT–Asian Regional Maize Program in Thailand (De León et al. 1993, 1997).

As pointed out earlier, downy mildew is an important disease in Asia, though lately it has also been reported in Venezuela and in a few countries of Africa. Similarly, maize streak virus (MSV) is distributed in Africa and is considered an important constraint to maize production. Corn stunt is prevalent in Latin America but is much more important in Central America. The work on these three diseases was taken up as a collaborative effort between CIMMYT and the selected national programs in Central America, Asia, and Africa. Three maize populations, namely, tropical late white dent (TLWD), tropical intermediate white flint (TIWF), and tropical late yellow flint-dent (TLYFD), were selected for this purpose and were improved for four cycles. The incidence of downy mildew was reduced to less than 2% in two populations, and improvement to corn stunt was observed in the TIWF population (Pandey and Gardner 1992). As a follow-up to this work, three corn stunt-resistant populations, 73, 76, and 79, were developed for future improvement efforts. Installation of greenhouse and artificial screening facilities in recent years at Poza Rica, and now at the new station Agua Fria, has greatly facilitated expanding and systematizing work on corn stunt. A number of white lines (CMLs 16, 42, 254, and 266) and yellow lines (CMLs 31, 282, 284, and 305) have been found to carry resistance to this disease.

The establishment of the CIMMYT–Asian Regional Maize Program (CIMMYT-ARMP) in the early 1970s helped expanding downy mildew-resistant (DMR) work in Thailand. Over the years, more DMR populations have been developed, such as populations 102, 145, 147, 300, 345, AMATL, 351, 352, and P49(Y). Other DMR sources, Suwan-1, Suwan-2, Suwan-3, and NS-1, were developed by Kasetsart University and the Department of Agriculture, Thailand. At least 20 DMR lines have also been released as CMLs in 2000 and 2001 (CMLs 425 to 433 and 465 to 475).

In the case of maize streak virus (MSV), a large volume of MSV-resistant germplasm has been developed by IITA as well as a joint initiative with CIMMYT. A late white dent population, 43 (La Posta), a few QPM populations, 62 and 63, and several experimental varieties (EVs) were improved for MSV resistance at IITA. Later efforts by CIMMYT breeders in the Ivory Coast helped to develop more MSV-resistant populations singly and in combination with DMR. MSV-resistant versions of populations 32 and 49 were also developed (Diallo and Dosso 1994). A huge volume of MSV-resistant germplasm has been developed and is available from the CIMMYT outreach program based in Harare, Zimbabwe (CIMMYT 1998). Several of the useful sources worth mentioning are INT-A, INT-B, LAT-A, LAT-B, DR-A, DR-B, P501 MSV, P502 MSV, ZM421, ZM521, ZM621, ZM601, and ZM609. Several released CMLs, 195 to 216, 386 to 395, 440 to 445, 488, and 489, are fairly tolerant to MSV. CMLs 202, 390, and 395 are excellent in MSV resistance and have flint kernel texture. An early QPM material, Pool 15QPM SR, has also been developed and is being further improved.

## 6.7.2    Development of Insect-Resistant Germplasm

Breeding for insect resistance is a difficult task, and a few programs in the developing countries are engaged in this work. There is a lack of systematic effort, slow progress, and traditional methodologies often deployed are inappropriate and lack continuity. The CIMMYT maize program has a well-established insect-rearing laboratory.

### 6.7.2.1  Resistance to Crop Pests

Four insect species, fall armyworm (FAW), sugarcane borer (SCB), southwest corn borer (SWCB), and corn earworm, are mass reared for artificial infestation using bazooka (Mihm et al. 1994). Several pools and populations in the mid-1970s were designated for insect improvement work to reduce damage from fall armyworm and borers.

A shift in insect resistance breeding strategy took place in 1984–1985, which coincided with the initiation of hybrid development efforts in the maize program. It was decided to emphasize multiple resistances to several species of *Lepidopterous* pests. Two breeding populations were formed to serve tropical lowland and subtropical temperate ecologies. A multiple-borer-resistant population, 590, was formed with subtropical and mid-altitude adaptation in mind. The components involved in its formation included insect-resistant lines from population 47, Antigua-based lines from Mississippi, known resistant lines to ECB from New York and Missouri, and some plant introductions reported resistant in Iowa. Full sibs from this population were sent out for international screening in 1986. Resistance to one or more insect species was encountered in varying frequencies. Approximately 39% of the families were resistant to one species, 24% to two species, 9% to three species, 2.5% to four species, and 0.5% to five species (Smith et al. 1989). Based on information obtained, at least four species-specific and one across-species experimental varieties (EVs) were formed, and multilocation tests were continued. As of 2002, P590 has completed seven cycles and P590 B five cycles of selfed progeny $S_3$ recurrent selection (Bergvinson 2002).

The second population, 390, was formed and named multiple-insect-resistant tropical (MIRT). This population is also mixed in kernel color and texture and is comprised of several CIMMYT populations and pools improved earlier for insect resistance, Antigua-based germplasm, lines from CIMMYT and IITA hybrid programs, MSV-resistant EVs, population 10 from Tanzania, and DMR sources Suwan-1 and Suwan-2. It is undergoing improvement and being tested in insect-resistant progeny trials (IRPT) internationally. This population has good resistance to tropical diseases, and the collaborators have been able to find resistance to species that are important under their conditions. This population has already completed six cycles of $S_3$ recurrent selection.

Cycles of selection trials have also been conducted with populations 390 and 590. The results demonstrate progress in yield and an increase in the level of resistance to FAW and stem borers under artificial infestation conditions (Mihm et al. 1997; Bergvinson 2002).

### 6.7.2.2  Resistance to Stored Grain Pests

Work on stored grain pests is becoming increasingly important in recent years. The two pests being emphasized are maize weevils (*Sitophilus zeamais*) and larger grain borer (LGB) (*Prostephanus truncates*). A few bank accessions (Sinaloa 35, Yucatan 7, and Sinaloa 2) have been found to be resistant to maize weevils by Canadian researchers, and it appears that resistance resides in pericarp or aleurone layers. A released CML 268 also shows moderate levels of resistance to LGB. Population 84, resistant to LGB, has also been formed and has completed three cycles of $S_3$ recurrent selection. The population is based on bank accessions largely from the Caribbean islands.

Inbred and hybrid development is progressing well in both populations. Lines and hybrids have been identified that possess good levels of resistance (Bergvinson 2002). Insect-resistant lines combining resistance to corn stunt, MSV, and downy mildew are also being developed in collaboration with the regional programs in India, Colombia, Kenya, and Zimbabwe using a shuttle breeding strategy. Program released lines are also being characterized for FAW, SCB, and SWCB. At a later date, CMLs will also be characterized for weevils and larger grain borer.

### 6.7.3  Germplasm Development for Abiotic Stresses

The importance of abiotic stresses and their ever-increasing global concern cannot be underestimated. Losses occurring every year due to one reason or another are massive and invariably result in fluctuating production and market price. The environmental stresses include drought, heat, high or low temperatures, air pollutants, high-velocity winds, hail, and frost. The soil-related stresses may be due to mineral element deficiencies, toxicities, extremes in soil pH, low N, and excess soil moisture. The discussion in this chapter, however, will be limited to drought, low N, waterlogging, and acid soil tolerance, as CIMMYT and some NARs are addressing these problems. Environmental stresses generally are unpredictable in both time and severity. In contrast, soil-related stresses are known with some scientific facts, and their effect on production potential or losses can be estimated reasonably well. These stresses are important globally, and their awareness in recent years has increased as being the primary or major constraints to limiting maize production in the developing world.

The quest for abiotic stress-tolerant germplasm can make a massive impact in reducing and preventing losses worth billions of U.S dollars. As we stretch and move to more marginal lands, the need and demand for this germplasm will grow as well. The success in research and breeding efforts will depend to a large extent on genetic variation that can be encountered for these traits in either existing germplasm under improvement or what is conserved in several maize gene banks around the world, including CIMMYT's bank. Differences among and within species have been reported for these traits (Clark and Duncan 1993). Fortunately, genetic variation is present in maize for stress traits and can be exploited in superior-performing genetic backgrounds (Fischer et al. 1983; Edmeades and Lafitte 1987; Vasal et al. 1999a,b; Banziger et al. 2000). The genetics of resistance may be simple or complex and, in some instances, may be influenced by modifying a gene complex. Only consciously planned and well-executed studies will help us in this resolve.

In the quest for tolerant germplasm for each stress trait, it would be desirable to identify a few key traits or secondary traits that can facilitate the selection process. It will be preferable if these traits can be visually observed, are highly heritable, and are less prone to mistakes in data recording and in either selection or rejection of the material in the field. Excellent examples already exist in this regard. In maize, anthesis-silking interval (ASI) is considered a good secondary trait to facilitate selection for drought tolerance (Bolaños and Edmeades 1996; Edmeades et al. 1993, 2000). It is also an important trait for other stresses.

In stress breeding, the choice of germplasm is a key to rapid success in providing superior germplasm in the hands of users as quickly as possible. Generally, high-performing maize populations with sufficient genetic variation for the trait in question are selected. Preference may also be given to those populations that have some level of tolerance for the stress trait. Other aspects or features worth considering will be inbreeding tolerance behavior, heterosis expression, and good combining ability if such populations are to serve as dual-purpose materials for extracting OPVs and hybrids. In the absence of such materials, new populations and synthetics can be formed using a selected fraction of prescreened lines for this trait.

Breeding methodologies for stress work should be relevant and appropriate for improvement in stress trait and to the needs of the germplasm products. Both direct and indirect breeding options and alternatives can be deployed considering the robustness of population and hybrid research activities (Edmeades et al. 1998; Vasal et al. 1998). For drought, a rain-free season is a must to control and manage stress levels. Aside from creating stress field conditions, the management of stress levels in the field is equally important. If materials are to perform well under stress and nonstress conditions, the success of the selection program will depend very much on evaluating genotypes/families under no stress, intermediate stress, and severe stress conditions.

### 6.7.3.1 Drought-Tolerant Germplasm

Drought-tolerant source Michoacan 21, commonly referred to as Latente, has been known for several decades. Because of its poor agronomic performance and other undesirable attributes, it has been difficult to exploit and transfer this trait to other genetic backgrounds. In the absence of specific genes controlling this trait, the research initiatives for improving it have been limited worldwide. An exploratory and modest effort was undertaken at CIMMYT in the early 1970s to improve drought tolerance of the most productive and widely grown Tuxpeno population adapted to lowland conditions (Fischer et al. 1983).

Recurrent full-sib family procedure was used to improve this material for drought tolerance using the winter season at the CIMMYT experimental station Tlaltizapan to evaluate families for drought at different water regimes or stress levels. Progress from eight cycles of selection was encouraging, as it demonstrated in a convincing manner that the performance of maize populations experiencing water deficits at flowering and during grain filling can be improved by recurrent selection at no cost to performance in well-watered conditions (Bolaños and Edmeades 1993a,b). The selection resulted in a significant increase in grain yield of 108 kg/ha/cycle at yield levels ranging from 1 to 8 tons/ha. At a yield level of 2 tons/ha, this represents a gain of 6.3% per cycle. In a later study, the gains averaged 90 kg/ha/cycle across all sites at a mean yield level of 5.6 tons/ha (Byrne et al. 1995). A more recent finding is exciting in that it shows that drought selections also perform well under low-N conditions (Banziger et al. 1999, 2002). Today several improved populations with tolerance to drought are available with additional cycles of selection completed (TS6C3, La Posta sequia C3, Pool 18 sequia C4, Pool 26 sequia C3, Pool 16 C2, DTP-1C7, DTP-2 C5). Both white- and yellow-grained versions are now available in DTP-1 and DTP-2.

The recurrent selection procedures described above were quite effective in developing source drought-tolerant populations. Intensive inbreeding efforts have also been under way in these populations to develop inbred lines. Recently, a few drought-tolerant lines have been released as CMLs (CMLs 339 to 344, 347, 348, 488, 489). Inbred line evaluation trials under drought of lines with no previous history of selection for drought have helped to identify a number of good drought-tolerant lines (Vasal et al. 1999b, 2001).

### 6.7.3.2 Waterlogging-Tolerant Germplasm

As part of cooperative efforts between CIMMYT-ARMP and the Department of Agriculture (DOA), Thailand, inbred lines and hybrids not previously selected for this trait were tested in waterlogged conditions. Surprisingly, some of the early and late lines and hybrids did very well (Table 6.2). It was also interesting to observe that some of the lines and hybrids did very well

**Table 6.2  Top Performing Early Hybrids Grown under Different Stresses for Grain Yield (tons/ha)**

| Entry no. | Hybrid | Normal | High Density | Low N | Drought | Water-logging | Average |
|---|---|---|---|---|---|---|---|
| 2 | NSX991002 | 8.08 | 7.88 | 3.74 | 4.72 | 6.55 | 6.20 |
| 12 | CA03130/CML429 | 7.67 | 7.62 | 3.06 | 5.83 | 6.39 | 6.12 |
| 13 | CA14512/CML421 | 7.92 | 8.51 | 3.89 | 5.93 | 5.48 | 6.35 |
| 17 | CA14514/CML421 | 7.84 | 8.11 | 3.11 | 5.08 | 7.51 | 6.33 |
| 18 | CA1415/CML421 | 8.12 | 8.44 | 3.18 | 5.29 | 6.40 | 6.29 |
| 19 | CP999 (Check) Late | 9.19 | 9.37 | 4.01 | 8.08 | 7.10 | 7.55 |
| 20 | NSX 982029 (Check) Late | 8.74 | 8.96 | 4.59 | 5.78 | 7.45 | 7.10 |

across other stresses as well. At source population level, materials developed and available are limited. Two national programs, EMBRAPA in Brazil and DMR in India, are working on improving tolerance to waterlogging conditions.

### 6.7.3.3 Acid Soil Tolerance Germplasm

Two programs, EMBRAPA in Brazil and CIMMYT's South American Regional Maize Program at Cali, Colombia, are deeply engaged in developing this kind of germplasm. CIMMYT maize researchers at Cali have developed a number of acid soil-tolerant maize populations (Pandey et al. 1997, Narro et al. 1997). In recent years, the emphasis has been placed on development of inbreds and hybrids tolerant to acid soils and aluminum toxicity. In two separate announcements, the program has released 14 yellow inbreds (CMLs 357 to 364 and 434 to 439) and two white inbreds (CMLS 365 and 366) that have tolerance to acid soil conditions. Several countries, including Colombia, Argentina, Bolivia, Peru, Venezuela, Indonesia, and the Philippines, have released acid soil-tolerant varieties and hybrids (De León and Narro 2002).

## 6.8 BREEDING FOR IMPROVED NUTRITIONAL QUALITY

Improving nutritional quality of maize is an important goal. It was demonstrated several decades ago that maize protein is nutritionally deficient because of the limiting quantities of two essential amino acids, lysine and tryptophane (Osborne and Mendel 1914).

### 6.8.1 Search for Useful Genetic Variation and Early Work

In the absence of specific genes that could elevate the protein profile with respect to the two amino acids, serious research efforts by and large were limited to screening elite and bank maize accessions in the hope of finding useful genetic variation for these traits (Aguirre et al. 1953; Paez et al. 1969a; Zuber and Helm 1975). As expected, variation was encountered, but breeding efforts were not pursued to make use of these materials in developing high-protein-quality strains. Research initiatives had to wait almost 50 years until three distinguished Purdue researchers discovered high lysine mutants such as opaque-2 (Mertz et al. 1964) and floury-2 (Nelson et al. 1965). The discovery of the biochemical effects of these two mutants paved the way for genetic manipulation in breeding to improve the nutritional quality of maize endosperm protein. Introduction of mutant genes into normal maize brings a twofold increase in the levels of lysine and trytophane. Lysine-deficient zein or prolamine fraction is reduced dramatically by about 50%, while other fractions, albumins, globulins, and glutelins, rich in lysine show a marked increase.

Most maize scientists envisioned this report with great optimism and high hopes. Worldwide conversion programs were initiated producing a wide array of opaque-2 varieties and hybrids. These high lysine conversions were tested extensively in the developed and developing world. However, the yield and agronomic performance of these materials did not match the normal counterparts (Alexander 1966; Lambert et al. 1969; Glover and Mertz 1987; Vasal 1975; Vasal et al. 1979, 1980, 1984a, Bjarnason and Vasal 1992). In general, these materials had 10 to 15% less grain yield, a dull and chalky kernel appearance, slower drying, greater kernel rots, and greater vulnerability to stored grain pests. The reasoning and causes behind these problems have been outlined earlier (Bjarnason and Vasal 1992; Vasal 2000a,b, 2002; Prasanna et al. 2001). Thus, in the very first decade this maize did not pass the test of its strength. Frustration and declining interest were a consequence of its lacking competitive performance.

Following the mid-1970s, only a few institutes and programs showed continuing interest and enthusiasm, CIMMYT being one of them. The other institutions and programs that sustained and persisted in their breeding efforts were Purdue University, Crow's hybrid seed company in Milford,

IL, and the South African program at the University of Natal. CIMMYT's work on quality protein maize (QPM) is considered significant, as it has achieved scientific breakthrough in developing germplasm products that are competitive and meet acceptance of farmers and consumers (Vasal 2000a,b). Several elements that have contributed to this success include sustained funding, readiness to deploy alternate sound options in breeding when the rest of the world was witnessing declining interest, continuous evolving breeding methodologies and strategies, and making constant changes and adjustments as needed during different phases of germplasm development.

### 6.8.2 Development and Improvement of Soft Opaques

In the period from 1966 to 1973, the emphasis was on developing high-lysine, soft-endosperm versions of normal materials. Initially both opaque-2 (*o2o2*) and floury-2 (*fl2fl2*) mutants were used extensively, but very soon the use of floury-2 was discontinued in the breeding program. Being semidominant, the kernel expression and quality depended on dosage of *fl2* allele in the triploid endosperm. The floury-2 conversion programs were thus difficult to handle, and deployment of any double mutant combination strategy was even more difficult. In contrast, the opaque-2 mutant had some obvious advantages. It was inherited in a simple recessive manner, the conversion process was easy, and no dosage effects were encountered in the endosperm. The soft chalky phenotype associated with the opaque-2 kernels was an additional help, as it served as a marker and greatly facilitated the conversion process. All potentially important maize materials in the maize program were converted to opaque-2 with varying numbers of backcrosses. A few opaque-2 materials were also introduced from other countries, such as Thai opaque-2 composite, Venzuela-1 opaque-2, and Ver.-Antigua opaque-2. In addition, at least three broad-based opaque-2 composites, K, J, and I, were formed to serve lowland, mid-altitude, and highland agroecologies.

Population improvement efforts were undertaken in some populations, and population crosses were also developed. A larger share of the resources was invested in developing and improving soft opaques. The materials so developed were tested nationally and internationally in OMPT-11 trials. In the absence of a hybrid program, no efforts were made to develop and convert maize lines. This was, however, attempted in many countries with emphasis on hybrid maize, as was the case with Brazil and Colombia. It may be interesting to point out that partially modified kernels (Figure 6.5) were observed during the conversion process, but their significance and importance were not realized and appreciated until the first published report (Paez et al. 1969b). In addition to large-scale research effort on soft opaques, the program during this period included screening bank accessions for lysine and trytophane content as well as developing genetic isolation mechanisms to prevent contamination in opaques. Double mutant combinations of *o-2* and *fl-2* with *waxy* were also developed. A beginning in modified opaques had already begun due to the wisdom and vision of Dr. Lonnquist and Dr. Asnani. By the early to mid-1970s, the problems plaguing soft opaques were well recognized and documented in the developing and developed world.

### 6.8.3 Correcting the First-Generation Problems and Exploring New Alternatives

The search continued for new and better mutants, developing double mutant combinations and using a combined *o-2* and genetic modifier approach. Unfortunately, these mutants did not offer any advantage over the *opaque-2* gene. The double mutant combination *su2o2* appeared promising, as it had vitreous kernels and protein quality marginally better than the *opaque-2* gene alone (Paez 1973). The kernels, however, are smaller in size, reflecting reduced grain yield of about 15 to 25%, depending on the genetic background. The yield gap can be narrowed down through recurrent selection, but it is highly improbable that the gap can be eliminated completely (Vasal et al. 1980, 1984b; Vasal 2000a,b, 2002). The double mutant *o2fl2* has resulted in vitreous kernels, but unfortunately, it happens in only rare backgrounds. In addition, the double mutant shows effects on lysine, protein, and light transmission (Paez et al. 1970).

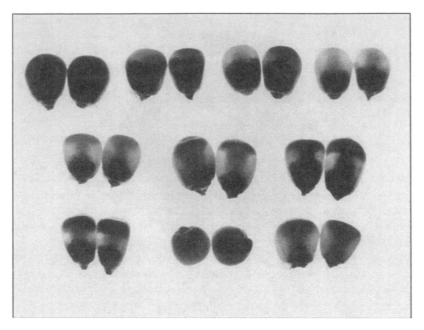

**Figure 6.5**   Genetic variation for kernel modification in maize opaque-2.

From the soft opaque materials we had to look desperately for variation in modified kernel behavior and received initial hints as to the types of genetic backgrounds, which will enhance our chances of success in developing QPM (Vasal 2000b, 2002).

### 6.8.4   Genetic Modifiers and Their Use as a Successful Strategy

The breeding options described above had shortcomings of one kind or the other. The approach that seemed to have promise involved exploitation of genetic modifiers of the *opaque-2* locus. It offered hope in improving phenotype of soft opaques to hard and in remedying other drawbacks in these materials. CIMMYT scientists viewed this approach with great optimism and felt that it would have great application in developing high-quality protein materials and in meeting producer, consumer, and industrial acceptance. The genetic modifiers are complex in their inheritance, the additive genetic effects being more important than the dominance effects. Also, their expression varies with the genetic background of the material in regards to kernel modification, density, and biochemical composition of the grain. However, in some genotypes protein quality may suffer a decline. Monitoring of protein quality thus becomes an important component in accumulating modifiers not affecting quality. The new strategy brought not only successes but also new challenges previously unknown. The progress was slow and frustrating in the beginning, as the starting point was zero, or so-called soft opaque-2 materials. Modifiers also exert maternal influence, thus complicating the selection process. Negative influence on protein quality in some genotypes was disturbing as protein quality had to be monitored continuously, requiring laboratory analyses. Thus, the conversion program became lengthy and cumbersome. The segregating generations needed to be handled much more carefully to avoid mistakes in selection.

### 6.8.5   Development of QPM Donor Stocks

Moving from soft opaques to hard endosperm was highly critical if such materials were to be accepted in developing countries. Any approach under consideration should therefore involve a change in kernel phenotype. Of the three options at hand at that time, a combination of the *opaque-2* gene and genetic modifiers was considered most appropriate to rectify phenotypic appearance and other agronomic problems affecting this maize. CIMMYT's choice of this approach proved to be a successful and viable strategy. Deploying this strategy on a large scale required development of white and yellow seeded QPM donor stocks. Genetic variation for kernel modification was encountered in several opaque-2 conversions and opaque-2 populations under improvement (Figure 6.5). Frequencies of modified kernels, however, varied considerably among different genotypes. Higher frequencies were observed in flint Cuban and Caribbean germplasm as well as in Thai opaque-2 composite and composite K. Modified opaque-2 ears were selected independently from several genetic backgrounds during the conversion process and during maintenance and multiplication of opaque-2 materials. Selected kernels were planted each generation separately from each ear for several generations. This effort led to the development of a large number of white and yellow hard endosperm opaque-2 selfed families. Later, white and yellow families were recombined separately to give rise to white and yellow hard endosperm opaque-2 composites, respectively. This constituted the first approach in developing QPM donor stocks. Genetic mixing coupled with selection for modified ears and modified kernels of good protein quality was practiced for three or four cycles.

The second approach involved intrapopulation selection for modifiers in some populations exhibiting a higher frequency of modified kernels. Four tropical and one highland population that met these criteria were chosen. These populations were Composite K, Ver.181-Ant.gpo.2 $x_1$ Venezuela-1 opaco-2, Thai opaque-2 Composite, PD (MS6) H.E.o2, and Composite I. The breeding procedure used was controlled by full-sib pollinations initially followed by the modified ear-to-row system suggested by Lonnquist (1964). Between- and within-ear selection for modification was practiced at all stages.

Some unexpected chance events inspired confidence and aided in a big way in deploying a two-genetic system strategy on a grand scale. This was particularly true in a population cross of two opaque-2 versions originating independently in Thailand and Venezuela, which exhibited a very high frequency of mosaic modified kernels. QPM donor stocks developed using the above-mentioned approaches improved considerably in the frequency of modifiers by the mid-1970s. At this point it was decided to use these materials as donor stocks as well as QPM populations for further improvement using appropriate schemes.

## 6.8.6  Expanded QPM Germplasm Development Efforts

Through the combined use of the *opaque-2* gene and genetic modifiers, QPM germplasm development efforts were applied to several genetic backgrounds. Different approaches and strategies were put into action to develop a wide array of germplasm products to meet requirements of different agroecologies.

In developing a large volume of QPM germplasm, the thrust was on two principal approaches. The first was to convert most of the advanced unit populations representing tropical, subtropical, and highland adaptation involving different maturities, grain color, and texture. A backcross-cum-recurrent selection procedure was used for this purpose (Vasal et al. 1984b). During the conversion process emphasis was placed on yield, kernel modification and appearance, reduced ear rot incidence, rapid drying, and other agronomic attributes. In evaluation trials, QPM versions performed quite similar to normal counterparts, in yield and agronomic traits (Vasal et al. 1984a,b, 1994).

The second approach involved formation and improvement of QPM gene pools. These pools were developed using different procedures but had a homozygous opaque-2 background to facilitate selection and accumulation of genetic modifiers. A modified half-sib system was used for the improvement of these pools. A total of seven tropical and six subtropical QPM gene pools were developed and improved continuously for varying numbers of cycles. There was clear evidence of improvement in germplasm (Bjarnason and Vasal 1992; Vasal 2000a,b, 2002). Improvements were clearly evident in grain yield (Table 6.3 and Table 6.4) and kernel modification in several different genetic backgrounds.

**Table 6.3   Performance of Late Maturing Tropical QPM Germplasm Compared with Similar Normal Materials**

| Pool/Population | Grain Yield (kg/ha) | Days to Silk | Plant Height (cm) | % Tryptophan in Endosperm Protein |
|---|---|---|---|---|
| **White Dent** | | | | |
| Population 68 Q | 6101[a] | 55.8 | 230 | 0.82[a] |
| Population 42 N | 5709 | 55.8[a] | 230 | 0.47 |
| | | | | |
| Pool 24 Q | 5457 | 56.3 | 226 | 0.82[a] |
| Pool 24 N | 5680 | 59.1[a] | 225 | 0.47 |
| | | | | |
| High oil selection Q | 5240 | 55.8 | 223 | 0.86[a] |
| Population 21 N | 5742[a] | 58.4[a] | 224 | 0.44 |
| **White Flint** | | | | |
| Population 62 Q | 5571 | 55.8[a] | 220[a] | 0.85[a] |
| Population 25 N | 5420 | 53.9 | 210 | 0.42 |
| | | | | |
| Pool 23 Q | 5334 | 55.4 | 220 | 0.81[a] |
| Pool 23 N | 5496 | 58.1[a] | 220 | 0.43 |
| **Yellow Flint** | | | | |
| Population 65 Q | 5300 | 55.5 | 227 | 0.81[a] |
| Population 27 N | 5305 | 57.6[a] | 225 | 0.41 |
| C.V. | 7.5 | 2.0 | 3.7 | 4.9 |

[a]Highly significant ($p = 0.01$).

*Source*: CIMMYT (Bjarnason, Short, Vasal, and Villegas).

**Table 6.4   Performance of Normal and Corresponding Tropical QPM Germplasm 1987**

| Material | Grain Yield | | QPM as Percentage of Normal |
|---|---|---|---|
| | Normal | QPM | |
| | Kg ha$^{-1}$ | | |
| Pool 23 | 5405 | 5330 | 98.6 |
| Pool 24 | 5706 | 5457 | 95.6 |
| Tropical high oil | 5733 | 5170 | 90.2 |
| Population 62 | 5347 | 5484 | 102.6 |
| Population 65 | 5255 | 5369 | 102.2 |
| Population 63 | 5705 | 6236 | 109.3 |

*Source*: Bjarnason and Short, 1988.

Merging, consolidation, and reorganization of QPM germplasm were done to reduce the number of QPM gene pools and populations to a level that could be handled in a more systematic way and also had some correspondence between pools and populations. From the available QPM germplasm, seven tropical and six subtropical gene pools were formed. In addition, six tropical and four subtropical populations were developed (Table 6.5). The handling of QPM gene pools and QPM populations has been discussed in earlier publications (Vasal et al. 1979, 1984a,b, 1997b).

## 6.8.7   QPM Hybrid Development and Testing

There were several turning points in QPM research at CIMMYT, which led to changing the emphasis in the program more than once in different years and at different stages of germplasm development. These changes related to shifts in the use of mutants or combined use of more than

**Table 6.5   QPM Gene Pools and Corresponding Populations**

| Pool No. | Name | Population | Name |
|----------|------|------------|------|
| 15 QPM | Tropical early white flint QPM | — | — |
| 17 QPM | Tropical early yellow flint QPM | 61 | Early yellow flint QPM |
| 18 QPM | Tropical early yellow dent QPM | — | |
| 23 QPM | Tropical late white flint QPM | 62[a] | White flint QPM |
| 24 QPM | Tropical late white dent QPM | 63, 64 | Blanco dentado-1 and Blanco dentado-2 QPM |
| 25 QPM | Tropical late yellow flint QPM | 65[a] | Yellow flint QPM |
| 26 QPM | Tropical late yellow dent QPM | 66 | Yellow dent QPM |
| 27 QPM | — | — | |
| 29 QPM | — | — | |
| 31 QPM | Subtropical intermediate white flint QPM | 67 | Templado blanco cristalino QPM |
| 32 QPM | Subtropical intermediate white dent QPM | 68 | Templado blanco dentado QPM |
| 33 QPM | Subtropical intermediate yellow flint QPM | 69[a] | Templado amarillo |
| 34 QPM | Subtropical intermediate yellow dent QPM | 70 | Templado amarillo dentado QPM |

**Table 6.6   Performance of Superior Tropical QPM Hybrids across 41 Locations in Latin America and Asia, 1999–2000**

| Pedigree | GrainYield (tons/ha) | Ear Rot (%) | Silking (%) | Ear Modification[a] | Tryptophan |
|----------|----------------------|-------------|-------------|---------------------|------------|
| (CML-141 × CML-144) | 6.40 | 5.5 | 55 | 1.6 | 0.088 |
| (CML-141 × CML-144) CML-142 | 6.29 | 6.2 | 55 | 1.7 | 0.081 |
| (CML-142 × CML-146) | 6.28 | 6.3 | 55 | 2.2 | 0.100 |
| (CML-142 × CML-150) | 6.20 | 7.8 | 55 | 2.0 | 0.089 |
| (CML-142 × CML-150) CML-176 | 6.08 | 7.5 | 55 | 2.0 | 0.086 |
| (CLQ-6203 × CML-150) | 5.80 | 7.2 | 55 | 2.3 | 0.090 |
| (CML-144 × CML-159) CML-176 | 5.64 | 6.0 | 56 | 1.7 | 0.094 |
| (CML-144 × CML-159 (RE) | 5.93 | 5.9 | 56 | 1.9 | 0.093 |
| Local check-1 | 5.95 | 7.6 | 55 | 1.9 | 0.050 |

[a]Rating scale, 1–5: 1 = completely hard, 5 – completely soft.

one genetic system, changes in emphasis from soft to hard endosperm opaques, breeding approaches and strategies, huge germplasm development efforts initially, later consolidated into a more definite number to satisfy germplasm needs and permit handling in a more systematic manner, and, more importantly, work in homozygous *opaque-2* genetic backgrounds. These constant changes and adjustments were made to achieve rapid progress in overcoming problems affecting QPM materials and to emphasize competitive agronomic performance of these materials to similar or comparable normal counterparts.

Perhaps the next turning point was the most important and significant in QPM research efforts. This was a new initiative related to introduction of QPM hybrid research and development at CIMMYT (Vasal 1986, 1987, 2002). This decision was particularly important for QPM research as the use of the inbred-hybrid approach offered several advantages, including seed purity; enhanced kernel modification and added uniformity and stability; ease in monitoring and maintaining protein quality, realizing enhanced yield potential, and addressing and circumventing problems more effectively and efficiently; greater market acceptability; and, more importantly, spurring seed industry growth. With the initiation of hybrid development activities, combining ability information was generated and heterotic patterns of important classes of QPM germplasm were established (Vasal et al. 1993a,b).

The hybrid initiative turned out to be exceedingly important for QPM efforts and success (Table 6.6 and Table 6.7). Without hybrid development efforts, the QPM OPVs would not have made the

**Table 6.7   Some Prominent QPM Varieties and Hybrids Released in Different Countries**

| | | | |
|---|---|---|---|
| 1 | (CML144 × 159) × CML176 | — | Nicargua, Guatemala, El Salvador, Honduras, Mexico, Venezuela, Colombia |
| 2 | CML176 × CML186 | — | India, Mexico |
| 3 | CML161 × CML165 | — | Peru, Vietnam, Guangzi (China) |
| 4 | (CML142 × CML150) × CML176 | — | India, Mexico |
| 5 | Across 83635R | — | Mozambigue, Mali, Uganda, Benia, Burkina Faso, Guinea |
| 6 | Poza Rica 8763 | — | Nicaragua, Mexico |

**Table 6.8   Recent QPM Releases in Some Countries**

| Name | Type | Pedigree | Country |
|---|---|---|---|
| HQ INTA-993 | Hybrid | (CML144 × CML159) CML176 | Nicaragua |
| NB-Nutrinta | Open pollinated | Poza Rica 8763 | Nicaragua |
| HB-PROTICTA | Hybrid | (CML144 × CML159) CML176 | Guatemala |
| HQ-61 | Hybrid | (CML144 × CML159) CML176 | El Salvador |
| HQ-31 | Hybrid | (CML144 × CML159) CML176 | Honduras |
| Zhongdan 9409 | Hybrid | Pool 33 × Temp QPM | China |
| Zhongdan 3850 | Hybrid | | China |
| QUIAN2609 | Hybrid | Tai 19/02 × CML171 | China |
| ICA- | Hybrid | (CML144 × CML159) CML176 | Colombia |
| Susuma[a] | Open pollinated | Across 8363SR | Mozambique |
| Obatampa[a] | Open pollinated | Across 8363SR | Mali |
| Obangaina[a] | Open pollinated | Across 8663SR | Uganda |
| Obatampa[a] | Open pollinated | Across 8363SR | Benin |
| BR-473 | Open pollinated | | Brazil |
| BR-451 | Open pollinated | | Brazil |
| Assum Preto | Open pollinated | | Brazil |
| Obatampa[a] | Open pollinated | Across 8363SR | Burkina Faso |
| Obatampa[a] | Open pollinated | Across 8363SR | Guinea |
| QS-7705[a] | Hybrid | | South Africa |
| GH-132-28[a] | Hybrid | P62, P63 | China |
| INIA- | Hybrid | CML161 × CML165 | Peru |
| FONAIAP | Hybrid | (CML144 × CML159) CML176 | Venezuela |
| HQ-2000 | Hybrid | CML161 × CML165 | Vietnam |
| SHAKTIMAN-1 | Hybrid | (CML142 × CML150) CML176 | India |
| SHAKTIMAN-1 | Hybrid | CML176 × CML186 | India |

**In Mexico, 21 hybrids and 5 open pollinated varieties, including ...**

| Name | Type | Pedigree | Country |
|---|---|---|---|
| 44IC | Hybrid | CML142 × CML116 | Mexico |
| H-551C | Hybrid | CML142 × CML150 | Mexico |
| H-553C | Hybrid | (CML142 × CML150) CML176 | Mexico |
| H-519C | Hybrid | (CML144 × CML159) CML170 | Mexico |
| H-368EC | Hybrid | CML186 × CML149 | Mexico |
| H-369EC | Hybrid | CML176 × CML186 | Mexico |
| VS-537 C | Open pollinated | POZA RICA 8763 | Mexico |
| VS-538 C | Open pollinated | ACROSS 8762 | Mexico |

[a] Sasakawa-Global 2000, a nongovernmental organization dedicated to ending malnutrition and poverty in Africa and a leading promoter of QPM in the region, cooperated with national programs and CIMMYT for the release of these varieties.

same impact. It is encouraging to state that we are already harnessing the fruits of this important decision. Many countries in recent years have released directly CIMMYT-developed hybrids and, in a few cases, are using CIMMYT lines in combination with their own lines (Table 6.8). The outlook of QPM appears promising, and many countries in the developing world are becoming increasingly interested in QPM efforts.

QPM development required continuous monitoring of protein quality and thus needed a strong support from the biochemical laboratory (Villegas et al. 1992). Simple analytical techniques were developed and used to analyze a large number of samples in a rapid and efficient manner, to provide results to the breeders in a timely fashion to make the right decision at the right time. The roles of other disciplines also cannot be underestimated, especially pathology, entomology, and physiology, which rendered services and support in evaluating QPM germplasm under different stresses. Thus, interdisciplinary cooperation was crucial to the development of QPM. Deploying the right strategy of *o2* and genetic modifiers, all problems plaguing these materials were addressed, and good progress was made on all fronts. The development of quality protein maize is a scientific breakthrough and a great success story, whose fruits are being conveyed to several developing countries.

## 6.9 APOMIXIS AS A POSSIBLE MEANS OF PERPETUATING HYBRID VIGOR

As described in Sections 6.6 and 6.7, the introduction of hybrid maize has led to phenomenal increases in yields and in improvement of other agronomic traits. In the U.S., hybrid maize was introduced in the mid-1930s, and within about 10 years, almost all maize fields were planted with hybrids. However, hybrid seed needs to be produced year after year for distribution to farmers or growers. Apomixis is an interesting reproductive mechanism that enables a plant to produce true copies of itself. Therefore, the potential value of apomixis in perpetuating hybrids is high. It could provide the ability to maintain heterozygosity through seed production on the same hybrid and help perpetuate hybrid vigor. Thus, if apomixis could be introduced in otherwise desirable hybrids, they would clone themselves, thereby eliminating the need to produce commercial hybrids year after year.

Several researchers have attempted to transfer the apomictic mode of reproduction from eastern gamagrass (*Tripsacum dactyloides*) to maize, but they have been only partially successful so far (Kindinger, personal communication, 2004). Modern molecular tools may facilitate this process.

## 6.10 CONCLUSIONS AND PERSPECTIVES

Maize is a unique, cross-pollinated crop of great economic importance worldwide. Maize plant is also a highly suitable organism for basic studies in genetics and cytogenetics. Being a C4 plant, it has great yield potential. Heterosis breeding has proven to be the best means of increasing yields and nutritional superiority relatively rapidly. In addition, exploitation of hybrid vigor will continue to be an important strategy for increasing tolerance to biotic and abiotic stresses. If apomixis could be introduced into desirable hybrids with the right gene combination, it would enable the hybrids to clone themselves and perpetuate hybrid vigor over extended periods, thereby eliminating the need to produce hybrid seed year after year.

Genetic improvement of maize to feed an ever-increasing human population cannot be over-emphasized. The development of quality protein maize is a major breakthrough that would alleviate malnourishment among hundreds of millions of underprivileged people in developing countries dependent on maize as a primary food source. Modern biotechnology could also help introduce value-added traits, including perhaps apomixis, into otherwise superior maize cultivars. Therefore, all available tools should be used for genetic enhancement of maize so that it may better feed masses of poor people worldwide.

## REFERENCES

Aguirre, F., Robles, C.E., and Scrimshaw, N.S. 1953. The nutritive value of Central American corns. 11. Lysine and methionine content of twenty-three varieties in Guatemala. *Food Res.* 18: 268–272.

Ahn, S. and Tanksley, S.D. 1993. Comparative linkage maps of the rice and maize genomes. *Proc. Natl. Acad. Sci. U.S.A.* 90: 7980–7984.

Alexander, D.E. 1957. The genetic induction of autotetraploidy: a proposal for its use in corn breeding. *Agron. J.* 49: 40–43.

Alexander, D.E. 1966. Problems associated with breeding opaque-2 corn and some proposed solutions. In *Proceedings of the High Lysine Corn Conference Refiners Association*, Washington, DC, pp. 156–160.

Ananiev, E.V., Phillips, R.L., and Rines, H.W. 1998a. Complex structure of knob DNA on maize chromosome 9: retrotransposon invasion into heterochromatin. *Genetics* 149: 2025–2037.

Ananiev, E.V., Phillips, R.L., and Rines, H.W. 1998b. Chromosome-specific molecular organization of maize (*Zea mays* L.) centromeric regions. *Proc. Natl. Acad. Sci. U.S.A.* 95: 13073–13078.

Ananiev, E.V., Phillips, R.L., and Rines, H.W. 1998c. A knob-associated tandem repeat in maize capable of forming fold-back DNA segments: are chromosome knobs megatransposons? *Proc. Natl. Acad. Sci. U.S.A.* 95: 10785–10790.

Ananiev, E.V., Riera-Lizarazu, O., Rines, H.W., and Phillips, R.L. 1997. Oat-maize chromosome addition lines: a new system for mapping the maize genome. *Proc. Natl. Acad. Sci. U.S.A.* 94: 3524–3529.

Anderson, E.G. 1956. The application of chromosomal techniques to maize improvement. In *Brookhaven Symposium in Biology: Genetics in Plant Breeding*, No. 9. Brookhaven National Lab, Upton, NY, pp. 23–36.

Arthur, K.M., Vejlupkova, Z., Meeley, R.B., and Fowler, J.E. 2003. Maize ROP2 GTPase provides a competitive advantage to the male gametophyte. *Genetics* 165: 2137–2151.

Auger, D.L., Newton, K.J., and Birchler, J.A. 2001. Nuclear gene dosage effects upon the expression of maize mitochondrial genes. *Genetics* 157: 1711–1721.

Banziger, M., Edmeades, G.O., Beck, D.L., and Bellón, M.R. 2000. *Breeding for Drought and Nitrogen Stress Tolerance in Maize: From Theory to Practice.* CIMMYT, Mexico, D.F., Mexico, 68 pp.

Banziger, M., Edmeades, G.O., and Lafitte, H.R. 1999. Selection for drought tolerance increases maize yields across a range of nitrogen levels. *Crop Sci.* 39: 1035–1040.

Banziger, M., Edmeades, G.O., and Lafitte, H.R. 2002. Physiological mechanisms contributing to the increased N stress tolerance of tropical maize selected for drought tolerance. *Field Crops Res.* 75: 223–233.

Bass, H.W., Riera-Lizarazu, O., Ananiev, E.V., Bordoli, S.J., Rines, H.W., Phillips, R.L., Sedat, J.W. Agard, D.A., and Cande, W.Z. 2000. Evidence for the coincident initiation of homolog pairing and synapsis during the telomere-clustering (bouquet) stage of meiotic prophase. *J. Cell Sci.* 113: 1033–1042.

Beadle, G.W. 1929. A gene for supernumerary mitoses during spore development in *Zea mays. Science* 70: 406–407.

Beadle, G.W. 1930. *Genetical and Cytological Studies of Mendelian Asynapsis in Zea mays.* Ph.D. Thesis, Cornell Univ., Ithaca, NY.

Beadle, G.W. 1932. Genes in maize for pollen sterility. *Genetics* 17: 413–431.

Beck, D., Vasal, L., and Crossa, J. 1991. Heterosis and combining ability among subtropical and temperate intermediate-maturity maize germplasm. *Crop Sci.* 31: 68–73.

Beckett, J.B. 1978. B-A translocations in maize. *J. Hered.* 69: 27–36.

Beckett, J.B. 1991. Cytogenetics, genetic and plant breeding applications of B-A translocations in maize. In *Chromosome Engineering in Plants: Genetics, Breeding, Evolution, Part A*, P.K. Gupta and T. Tsuchiya, Eds. Elsevier, Amsterdam, pp. 493–529.

Bellini, G. and Bianchi, A. 1963. Interchromsomal effects of inversions on crossover rate in maize. *Z. Vererbungsl.* 94: 126–132.

Bergvinson, D. 2002. Maize Entomology. In *A Briefing Book, External Review of CIMMYT Maize Program*, M. Listman, Ed., CIMMYT, Mexico, September 23–27.

Birchler, J.A. 1991. Chromosome manipulations in maize. In *Chromosome Engineering in Plants: Genetics, Breeding, Evolution, Part A*, P.K. Gupta and T. Tsuchiya, Eds. Elsevier, Amsterdam, pp. 531–559.

Bjarnason, M. and Vasal, S.K. 1992. Breeding of quality protein maize (QPM). *Plant Breed. Rev.* 9: 181–216.

Bolaños, J. and Edmeades, G.O. 1993a. Eight cycles of selection for drought tolerance in lowland tropical maize. I. Responses in grain yield, biomass, and radiation utilization. *Field Crops Res.* 31: 233–252.

Bolaños, J. and Edmeades, G.O. 1993b. Eight cycles of selection for drought tolerance in lowland tropical maize. II. Responses in reproductive behavior. *Field Crops Res.* 31: 253–268.

Bolaños, J. and Edmeades, G.O. 1996. The importance of the anthesis-silking interval in breeding for drought tolerance in tropical maize. *Field Crops Res.* 48: 65–80.

Brink, R.A. 1927. The occurrence of semi-sterility in maize. *J. Hered.* 18: 266–270.

Brewbaker, J.L., Kim, S.K., and Logrono, M.L. 1991. Resistance of tropical maize inbreds to major virus like diseases. *Maydica* 36: 257–265.

Buckler and Holtsford. 1996. *Zea* Systematics: ribosomal evidence. *Mol. Biol. Evol.* 13: 612–622.

Burnham, C.R. 1930. Genetical and cytological studies of semisterility and related phenomena in maize. *Proc. Natl. Acad. Sci. U.S.A.* 16: 269–277.

Burnham, C.R. 1962. *Discussions in Cytogenetics.* Burgess Publ., Minneapolis.

Burnham, C.R. 1966. Cytogenetics in plant improvement. In *Plant Breeding* (a symposium held at Iowa State University). Iowa State University Press, Ames.

Burnham, C.R. 1982. The locating of genes to chromosomes by the use of chromosomal interchanges. In *Maize for Biological Research*, W.F. Sheridan, Ed. University of North Dakota Press, Grand Forks.

Byrne, P.F., Bolaños, J., Edmeades, G.O., and Eaton, D.L. 1995. Gains from selection under drought versus multilocation testing in related tropical maize population. *Crop Sci.* 35: 63–69.

Ceballos, H., Deutsch, J.A., and Gutierrez, H. 1991. Recurrent selection for resistance to *Exserohilum turcicum* in eight subtropical maize populations. *Crop Sci.* 31: 964–971.

Chao, C.Y. 1959. Heterotic effects of a chromosomal segment in maize. *Genetics* 44: 657–677.

CIMMYT. 1998. A complete listing of improved maize germplasm from CIMMYT. Maize Program Special Report, Mexico, D.F.

CIMMYT. 2004. *Maize Inbred Lines Released by CIMMYT*, a compilation of 497 CIMMYT maize lines (CMLS), CML1 to CML497. CIMMYT, Mexico, D.F., pp. 1–59.

Clark, R.B. and Duncan, B.R. 1993. Selection of plants to tolerate soil salinity, acidity, and mineral deficiencies. In *International Crop Science Congress 1*, Buxton, D.R. et al., Eds. CSSA, Madison, WI, pp. 371–379.

Coe, E.H. 1959. A line of maize with high haploid frequency. *Am. Nat.* 93: 381–382.

Cordova, H.S. and Trifunovic, S. 2002. Lowland tropical maize. In *A Briefing Book, External Review of CIMMYT Maize Program*, M. Listman, Ed. CIMMYT, Mexico, D.F., September 23–27.

Cox, D.R., Burmeister, M., Price, E.R., Kim, S., and Myers, R.M. 1990. Radiation hybrid mapping: a somatic cell genetic method for constructing high-resolution maps of mammalian chromosomes. *Science* 250: 245–250.

Davis, D.W. 1992. Characterization of Oat Haploids and Their Progeny. M.S. thesis, University of Minnesota, St. Paul.

Davis, G.L., McMullen, M.D., Baysdorfer, C., Musket, T., Grant, D., Staebell, M., Xu, G., Polacco, M., Koster, L., Melia-Hancock, S., Houchins, K., Chao, S., and Coe, E.H., Jr. 1999. A maize map standard with sequenced core markers, grass genome reference points and 932 expressed sequence tagged sites (ESTs) in a 1736-locus map. *Genetics* 152: 1137–1172.

De León, C., Granados, G., Wedderburn, R.N., and Pandey, S. 1993. Simultaneous improvement of downy mildew resistance and agronomic traits in tropical maize. *Crop Sci.* 33: 100–102.

De León, C. and Narro, L. 2002. The South American Regional Maize Program. In A *Briefing Book, External Review of CIMMYT Maize Program*, M. Listman, Ed., CIMMYT, Mexico, September 23–27.

De León, C. and Pandey, S. 1989. Improvement of resistance to ear and stalk rots and agronomic traits in tropical maize gene pools. *Crop Sci.* 29: 12–17.

De León, C., Vasal, S.K., and Lothrop, J.E. 1997. The Asian Regional Maize Program. In *Maize Research in 1995-96*, CIMMYT Maize Program Special Report. CIMMYT, Mexico, D.F., Mexico, pp. 58–60.

Diallo, A.O. and Dosso, Y. 1994. Germplasm development in sub-Saharan Africa with emphasis on streak resistance. In *The Lowland Tropical Maize Subprogram*, Maize Program Special Report, S.K.Vasal and S. Mclean, Eds. CIMMYT, Mexico, D.F., Mexico, pp. 47–58.

Dobzhansky, T. and Rhoades, M.M. 1938. A possible method for locating favorable genes in maize. *J. Am. Soc. Agron.* 30: 668–675.

Doyle, G.G. 1967. Preferential pairing in trisomics of *Zea mays*. In *Chromosomes Today*, Vol. 2. Plenum Press, New York.

Edmeades, G.O. and Lafitte, H.R. 1987. Crop physiology and maize improvement. In *CIMMYT 1987, CIMMYT Research Highlights 1986*. CIMMYT, Mexico, D.F., Mexico.

Edmeades, G.O., Bolaños, J., Banziger, M., Ribaut, J.M., White, J.W., Reynolds, M.P., and Lafitte, H.R. 1998. Improving crop yield under water deficits in the tropics. In *Crop Productivity and Sustainability Shaping the Future, Proceedings of the 2nd International Crop Science Congress*, V.L. Chopra et al., Eds., Oxford, pp. 437–451.

Edmeades, G.O., Bolaños, J., Eilings, A., Ribaut, J.M., Banziger, M., and Westgate, M.E. 2000. The role and regulation of anthesis silking interval in maize. In *Physiology and Modeling Kernel Set in Maize*, CSSA Special Publication 29. CSSA, Madison, WI, pp. 43–73.

Edmeades, G.O., Bolaños, J., Hernandez, M., and Bello, S. 1993. Causes for silk delay in a lowland tropical maize population. *Crop Sci.* 33: 1029–1035.

Escudero, J., Neuhaus, G., Sclappi, M., and Hohn, B. 1996. T-DNA transfer in meristematic cells of maize provided with intracellular *Agrobacterium*. *Plant J.* 10: 355–360.

Fischer, K.S., Edmeades, G.O., and Johnson, E.C. 1987. Recurrent selection for reduced tassel branch number and reduced leaf area density above the ear in tropical maize populations. *Crop Sci.* 27: 1150–1156.

Fischer, K.S., Edmeades, G.O., and Johnson, E.C. 1989. Selection for the improvement of maize yield under water deficits. *Field Crops Res.* 22: 227–243.

Fischer, K.S., Johnson, E.C., and Edmeades, G.O. 1983. *Breeding and Selection for Drought Resistance in Tropical Maize*. CIMMYT, El Batán.

Freeling, M. and Walbot, V., Eds. 1994. *The Maize Handbook*. Springer-Verlag, New York.

Fromm, M., Morrish, F., Armstrong, C., Williams, R., Thomas, J., and Klein, T. 1990. Inheritance and expression of chimeric genes in the progeny of transgenic maize plants. *Bio/Technology* 8: 833–839.

Fromm, M.F., Taylor, L.P., and Walbot, V. 1986. Stable transformation of maize after gene transfer by electroporation. *Nature* 319: 791–793.

Gaut, B.S., d'Ennequin, M.L.T., Peek, A.S., and Sawkins, M.C. 2000. Maize as a model for the evolution of plant nuclear genomes. *Proc. Natl. Acad. Sci. U.S.A.* 97: 7008–7015.

Ghidoni, A., Pogna, N.E., and Villa, N. 1982. Spontaneous aneuploids of maize (*Zea mays*) in a selected sample. *Can. J. Genet. Cytol.* 24: 705–713.

Glover, D.V. and Mertz, E.T. 1987. Corn. In *Nutritional Quality of Cereal Grains: Genetic and Agronomic Improvement*, ASA Monograph 28, A. Olson and K.J. Frey, Eds. American Society of Agronomy, Madison, WI, pp. 183–336.

Goodman, M.M., Stuber, C.W., Newton, K., and Weissinger, H.H. 1980. Linkage relationships of 19 enzyme loci in maize. *Genetics* 96: 697–710

Goodsell, S.F. 1961. Male sterility in corn by androgenesis. *Crop. Sci.* 1: 227–228.

Gordon-Kamm, W., Spencer, T., Mangano, M., Adams, T., Daines, R., Statr, W., O'Brian, J., Chambers, S., Adams, W., Willetts, J., Rice, N., Mackey, T., Krueger, R., Kausch, A., and Lemaux, P. 1990. Transformation of maize cells and regeneration of fertile transgenic plants. *Plant Cell* 2: 603–618.

Goss, S.J. and Harris, H. 1975. New method for mapping genes in human chromosomes. *Nature* 255: 680–684.

Guo, M. and Birchler. J.A. 1994. Trans-acting dosage effects on the expression of model gene systems in maize aneuploids. *Science* 266: 1999–2002.

Guo, M., Davis, D., and Birchler, J.A. 1996. Dosage effects on gene expression in a maize ploidy series. *Genetics* 142: 1349–1355.

Hallauer, A.R. and Miranda, J.B. 1988. *Quantitative Genetics in Maize Breeding*, 2nd ed. Iowa State University Press, Ames.

Han, G.C., Vasal, S.K., Beck, D.L., and Elias, E. 1991. Combining ability analysis of inbred lines derived from CIMMYT maize (*Zea mays* L.) germplasm. *Maydica* 36: 57–64.

Helentjaris, T., Weber, D., and Wright, S. 1986. Use of monosomics to map cloned DNA fragments in maize. *Proc. Natl. Acad. Sci. U.S.A.* 83: 6035–6039.

Helentjaris, T., Weber, D., and Wright, S. 1988. Identification of the genomic locations of duplicate nucleotide-sequences in maize by analysis of restriction fragment length polymorphisms. *Genetics* 118: 353–363.

Hyde, J., Martin, M.A., Preckel, P.V., and Edwards, C.R. 1999. The economics of Bt corn: valuing protection from the European corn borer. *Rev. Agri. Econ.* 21: 442–454.

Jain, S.M, Sopory, S.K., and Veilleux, R.E., Eds. 1996. *In Vitro Haploid Production in Higher Plants*, Vol. 2, *Applications*. Kluwer Academic Publishers, Dordrecht, The Netherlands.

James, C. 2003. *Global Review of Commercialized Transgenic Crops. 2002 Feature: Bt. Maize*, ISAAC Brief 29. ISAAC, Ithaca, NY.

Jeffers, D., Cordova, H., Vasal, S.K., Srinivasan, G., Beck, D., and Barandiaran, M. 2000. Status in breeding for resistance to maize diseases at CIMMYT. In *Proceedings of the Seventh Asian Regional Maize Workshop*, S.K. Vasal, F. Gonzalez, and Fan Xing Ming, Eds., PCARRD, Los Baños, Philippines, pp. 257–266.

Johnson, E.C., Fischer, K.S., Edmeades, G.O., and Palmer, A.F.E. 1986. Recurrent selection for reduced plant height in lowland tropical maize. *Crop Sci.* 26: 253–260.

Kermicle, J.L. 1969. Androgenesis conditioned by a mutation in maize. *Science* 166: 1422–1424.

Kim, S.K., Efron, Y., Fijemisin, J.M., and Buddenhagen, I.W. 1989. Mode of gene action for resistance to maize streak virus. *Crop Sci.* 29: 890–894.

Koumbaris, G.L. and Bass, H.W. 2003. A new single-locus cutogenetic mapping system for maize (*Zea mays* L.): overcoming FISH detection limits with marker-selected sorghum (*S. propinquum* L.) BAC clones. *Plant J.* 35: 647–659.

Kynast, R.G., Okagaki, R.J., Galatowitsch, M.W., Granath, S.R., Jacobs, M.S., Stec, A.O., Rines, H.W., and Phillips, R.L. 2004. Dissecting the maize genome by using chromosome addition and radiation hybrid lines. *Proc. Natl. Acad. Sci. U.S.A.* 101: 9921–9926.

Kynast, R.G., Okagaki, R.J., Rines, H.W., and Phillips, R.L. 2002. Maize individualized chromosome and derived radiation hybrid lines and their use in functional genomics. *Funct. Integr. Genomics* 2: 60–69.

Kynast, R.G., Riera-Lizarazu, O., Vales, M.I., Okagaki, R.J., Maquieira, S.B., Chen, G., Ananiev, E.V., Odland, W.E., Russel, C.D., Stec, A.O., Livingston, S.M., Zaia, H.A., Rines, H.W., and Phillips, R.L. 2001. A complete set of maize individual chromosome additions to the oat genome. *Plant Physiol.* 125: 1216–1227.

Lambert, R.J., Alexander, D.E., and Dudley, J.W. 1969. Relative performance of normal and modified protein (opaque-2) maize hybrids. *Crop Sci.* 9: 242–243.

Lauer, J. and Wedberg, J. 1999. Grain yields of initial Bt corn hybrid introductions to farms in the northern corn belt. *J. Prod. Agric.* 12: 373–376.

Laurie, D.A. and Bennett, M.D. 1986. Wheat × maize hybridization. *Can. J. Genet. Cytol.* 28: 313–316.

Laurie, D.A., O'Donoughue, L.S., and Bennett, M.D. 1990. Wheat × maize and other wide sexual hybrids: their potential for genetic manipulation and crop improvement. In *Gene Manipulation in Plant Improvement II*, J.P. Gustafson, Ed. Plenum Press, New York, pp. 95–126.

Leblanc, O., Pointe, C., and Hernandez, M. 2002. Cell cycle progression during endosperm development in *Zea mays* depends on parental dosage effects. *Plant J.* 32: 1057–1066.

Li, L.J., Arumuganathan, K., Rines, H.W., Phillips, R.L., Riera-Lizarazu, O., Sandhu, D., Zhou, Y., and Gill, K.S. 2000. Flow cytometric sorting of maize chromosome 9 from an oat-maize chromosome addition line. *Theor. Appl. Genet.* 102: 658–663.

Longley, A.E. 1961. *Breakage Points for Four Corn Translocation Series and Other Corn Chromosome Aberrations Maintained at the California Institute of Technology*, Pamphlet 34-16. USDA–ARS, Washington, DC.

Lonnquist, J.H. 1964. A modification of the eat-to-row procedure for the improvement of maize populations. *Crop Sci.* 4: 227–228.

Maguire, M.P. 1972. The temporal sequence of synaptic initiation, crossing-over and synaptic completion. *Genetics* 70: 353–370.

Maguire, M.P. 1981. A search for the synaptic adjustment phenomenon in maize. *Chromosoma* 81: 717–725.

Maluszynski, M., Kasha, K.J., Forster, B.P., and Szarejko, I., Eds. 2003. *Doubled Haploid Production in Crop Plants: A Manual*. Kluwer Academic Publishers, Dordrecht, The Netherlands.

Maquieira, S.B. 1997. Production and Characterization of Plants from Oat × Maize and Oat × Pearl Millet. M.S. thesis, University of Minnesota, St. Paul.

McClintock, B. 1929. A cytological and genetical study of triploid maize. *Genetics* 14: 180–222.

McClintock, B. 1931. Cytological observations of deficiencies involving known genes, translocations and an inversion in *Zea mays*. *Res. Bull. Mo. Agric. Exp. Stn.* 163: 1–30.

McClintock, B. 1938. The fusion of broken ends of sister half-chromatids following chromatid breakage at meiotic anaphases. *Res. Bull. Mo. Agric. Exp. Stn.* 290: 1–48.

McClintock, B. 1939. The behavior in successive nuclear divisions of a chromosome broken at meiosis. *Proc. Natl. Acad. Sci. U.S.A.* 26: 405–416.

McClintock, B. 1941. The stability of broken ends of chromosomes in *Zea mays*. *Genetics* 26: 234–282.

McClintock, B. 1951. Chromosome organization and genic expression. *Cold Spring Harbor Symp. Quant. Biol.* 16: 13–47.

McClintock, B. and Hill, H.E. 1931. The cytological identification of the chromosome associated with the *R-G* linkage group in *Zea mays. Genetics* 16: 175–190.

McKinley, C.M. and Goldman, S.L. 1979. The interchromosomal effect of inversions in maize. *Mol. Gen. Genet.* 172: 119–125.

Mertz, E.T., Bates, L.S., and Nelson, O.E. 1964. Mutant gene that changes protein composition and increases lysine content of maize endosperm. *Science* 145: 279–280.

Mihm, J.A., Bergvinson, D., and Kumar, H. 1997. Maize entomology. In *Maize Research in 1995–96*, CIMMYT Maize Program Special Report. CIMMYT, Mexico, D.F., Mexico, pp. 42–46.

Mihm, J.A., Deutch, J., Jewell, D., Hoissington, D., and Gonzalez-de-Leon, D. 1994. Improving maize with resistance to major insect pests. In *Stress Tolerance Breeding: Maize That Resists Insects, Drought, Low-N, and Acid Soils*, G.O. Edmeades et al., Eds. CIMMYT, Mexico, D.F., Mexico, pp. 1–20.

Muehlbauer, G.J., Riera-Lizarazu, O., Kynast, R.G., Martin, D., Phillips, R.L., and Rines, H.W. 2000. A maize-chromosome 3 addition line of oat exhibits expression of the maize homeobox gene *liguleless3* and alterations of cell fates. *Genome* 43: 1055–1064.

Narro, L., Pandey, S., De León, C., Pérez, J.C., Salazar, F., and Arias, M.P. 1997. Heterosis in acid soil-tolerant maize germplasm. In *The Genetics and Exploitation of Heterosis in Crops. An International Symposium*, Mexico City, Mexico, August 17–22, pp. 290–291.

Nelson, O.E., Mertz, E.T., and Bates, L.S. 1965. Second mutant gene affecting the amino acid pattern of maize endosperm proteins. *Science* 150: 1469–1470.

Neuffer, M.G., Coe, E.H., and Wessler, S.R. 1997. *Mutants of Maize*. Cold Spring Harbor Laboratory Press, Plainview, NY.

Neuffer, M.G. and Sheridan, W.F. 1980. Defective kernel mutants of maize. I. Genetic and lethality studies. *Genetics* 95: 929–944.

Nitzsche, W. and Wenzel, G. 1977. *Haploids in Plant Breeding*. Verlag Paul Parey, Berlin.

Ohta, Y. 1986. High-efficiency genetic transformation of maize by a mixture of pollen and exogenous DNA. *Proc. Natl. Acad. Sci. U.S.A.* 83: 715–719.

Okagaki, R.J., Kynast, R.G., Livingston, S.M., Russel, C.D., Rines, H.W., and Phillips, R.L. 2001. Mapping maize sequences to chromosomes using oat-maize chromosome addition materials. *Plant Physiol.* 125: 1228–1235.

Osborne, T.B. and Mendel, L.B. 1914. Nutritive properties of the maize kernel. *J. Biol. Chem.* 18: 1–16.

Ostlie, K.R., Hutchinson, W.D., and Hellmich, R.L. 1997. Bt corn and European corn borer; long term success through resistance management. *Univ. Minn. Ext. Serv. Bull.* 7055-GO: 17.

Paez, A.V. 1973. Protein quality and kernel properties of modified opaque-2 endosperm corn involving a recessive allele at the sugary-2 locus. *Crop Sci.* 13: 633–636.

Paez, A.V., Helm, J.L., and Zuber, M.S. 1969a. Lysine content of opaque-2 maize kernels having different phenotypes. *Crop Sci.* 29: 251–252.

Paez, A.V., Helm, J.L., and Zuber, M.S. 1970. Dosage effects of opaque-2 and floury-2 on lysine, protein and light transmission of maize endosperm. *Z. Pflanzenzuecht.* 63: 119–123.

Paez, A.V., Ussary, J.P., Helm, J.L., and Zuber, M.S. 1969b. Survey of maize strains for lysine content. *Agron. J.* 61: 886–889.

Pandey, S., De León, C., and Narro, L. 1997. The South American Regional Maize Program. In *Maize Research in 1995–96*, CIMMYT Maize Program Special Report. CIMMYT, Mexico, D.F., Mexico, pp. 61–63.

Pandey, S. and Gardner, C.O. 1992. Recurrent selection for population, variety, and hybrid improvement in tropical maize. *Adv. Agron.* 48: 1–87.

Pandey, S., Vasal, S.K., De León, C., Ortega, A., Granados, G., and Villegas, E. 1984. Development and improvement of maize populations. *Genética* 16: 23–42.

Patterson, E.B. 1952. The use of functional duplicate-deficient gametes in locating genes in maize. *Genetics* 37: 612–613 (abstract).

Patterson, E.B. 1973. Genic male sterility and hybrid maize production. Part 1. In *Proceedings of the 7th Meeting of the Maize and Sorghum Section*. Eur. Assoc. Res. Plant Breeding (EUCARPIA), Zagreb, Yugoslavia.

Patterson, E.B. 1978. Properties and uses of duplicate-deficient chromosome complements in maize. In *Maize Breeding and Genetics*, D.B. Walden, Ed. John Wiley & Sons, New York, pp. 693–710.

Phillips, R.L. 1993. Cytogenetic manipulation of polymitotic (*po*). *Maydica* 38: 85–92.

Phillips, R.L., Burnham, C.R., and Patterson, E.B. 1971. Advantages of chromosomal interchanges that generate haplo-viable deficiency-duplications. *Crop Sci.* 11: 525–528.

Piperno, D.R. and Flannery, K.V. 2001. The earliest archaeological maize (*Zea mays* L.) from highland Mexico: new accelerator mass spectrometry dates and their implications. *Proc. Natl. Acad. Sci. U.S.A.* 98: 2101–2103.

Prasanna, B.M., Vasal, S.K., Kassalun, B., and Singh, N.N. 2001. Quality protein maize. *Curr. Sci.* 81: 1308–1319.

Rhoades, M.M. 1951. Duplicate genes in maize. *Am. Naturalist* 85: 105–110.

Rhoades, M.M. and Dempsey, E. 1966. Induction of chromosome doubling at meiosis by the *elongate* gene in maize. *Genetics* 54: 505–522.

Rhodes, C.A., Pierce, D.A., Mettler, I.J., Mascarenhas, D., and Detmer, J.J. 1988. Genetically transformed maize plants from protoplasts. *Science* 240: 204–206.

Rines, H.W. and Dahleen, L.S. 1990. Haploid oat plants produced by application of maize pollen to emasculated oat florets. *Crop Sci.* 30: 1073–1078.

Riera-Lizarazu, O., Rines, H.W., and Phillips, R.L. 1996. Cytological and molecular characterization of oat × maize partial hybrids. *Theor. Appl. Genet.* 93: 123–135.

Riera-Lizarazu, O., Vales, M.I., Ananiev, E.V., Rines, H.W., and Phillips, R.L. 2000. Production and characterization of maize chromosome 9 radiation hybrids derived from an oat-maize addition line. *Genetics* 156: 327–339.

Roman, H. 1947. Mitotic nondisjunction in the case of interchanges involving the B-type chromosome in maize. *Genetics* 32: 391–409.

Roman, H. 1948. Directed fertilization in maize. *Proc. Natl. Acad. Sci. U.S.A.* 34: 36–42.

Roman, H. and Ullstrup, A.J. 1951. The use of A-B translocations to locate genes in maize. *Agron. J.* 43: 450–454.

Sawahel, W.A. 2002. Production of herbicide-resistant transgenic maize plants using electroporation of seed-derived embryos. *Plant Mol. Biol. Rep.* 20: 303a–303h.

Simcox, K.D. and Weber, D.F. 1985. Localization of the *benzoxazinless* (*bx*) locus in the maize by monosomic and B-A translocational analyses. *Crop Sci.* 25: 827–830.

Smith, M.E., Mihm, J.A., and Jewell, D.C. 1989. Breeding for multiple resistance to temperate, subtropical, and tropical maize insect pests at CIMMYT. In *Toward Insect Resistant Corn for the Third World: Proceedings of the International Symposium on the Methodologies for Developing Host Plant Resistance to Corn Insects*, CIMMYT, Mexico, D.F., pp. 109–121.

Snope, A.J. 1967. The relationship of abnormal chromosome 10 to B-chromosomes in maize. *Chromosoma* 21: 243–249.

Sprague, G.F. 1941. The location of dominant favorable genes in maize by means of an inversion. *Genetics* 26: 170 (abstract).

Sprague, G.F. and Dudley J.W., Eds. 1988. *Corn and Corn Improvement. Agronomy 18*, 3d ed. ASA, CSSA, SSSA Publishers, Madison, WI.

Srinivasan, G., Vasal, S.K., and Gonzalez, F.C. 1992. Combining ability of different versions of Tuxpeño-based maize germplasm developed at CIMMYT. *Maize Genet. Coop. Newsl.* 66: 72–73.

Taba, S. and Krakowsky, M. 2002. Maize genetic resources and pre-breeding. In *A Briefing Book, External Review of CIMMYT Maize Program*, M. Listman, Ed., CIMMYT, Mexico, D.F., Mexico, September 23–27.

Ting, Y.C. 1966. Duplications and meiotic behavior of the chromosomes in haploid maize (*Zea mays* L.). *Cytologia* 31: 324–329.

Vasal, S.K. 1975. Use of genetic modifiers to obtain normal-type kernels with the opaque-2 gene In *High Quality Protein Maize*. Hutchinson Ross Publishing Co., Stroudsburg, PA, pp. 197–216.

Vasal, S.K. 1986. Approaches and methodology in the development of QMP hybrids. In *Anais do XV Congresso National de Milho e Sorgo, Brasilia*, EMBRAPA-CNPMS, Documentos 5, pp. 419–430.

Vasal, S.K. 1987. *Development of Quality Protein Maize Hybrids*, K.S. Gill et al., Eds. A keynote paper presented at the First Symposium on Crop Improvement, Ludhiana, India, February 23–27, pp. 57–75.

Vasal, S.K. 2000a. High quality protein corn. In *Specialty Corns*, A.R. Hallauer, Ed. CRC Press, Boca Raton, FL, pp. 85–129.

Vasal, S.K. 2000b. The quality protein maize story. *Food Nutr. Bull.* 21: 445–450.

Vasal, S.K. 2002. Quality protein maize: overcoming the hurdles. *J. Crop Prod.* 6: 193–227.

Vasal, S.K., Córdova, H.S., Beck, D.L., and Edmeades, G.O. 1997a. Choices among breeding procedures and strategies for developing stress tolerant maize germplasm. In *Proceedings of Developing Drought- and Low Nitrogen Maize*, G.O. Edmeades et al., Eds., CIMMYT, El Batan, México, March 25–29, 1996.

Vasal, S.K., Córdova, H.S., Pandey, S., and Srinivasan, G. 1999a. Tropical maize and heterosis. In *The Genetics and Exploitation of Heterosis in Crops*, J.G. Coors and S. Pandey, Eds. ASA/CSSA/SSSA, Madison, WI, pp. 363–373.

Vasal, S.K., Dhillon, B.S., and Pandey, S. 1994. Recurrent selection methods based on evaluation-cum-recombination block. *Plant Breed. Rev.* 14: 139–163.

Vasal, S.K., Dhillon, B.S., Srinivasan, G., McLean, S.D., Crossa, J., and Zhang, S.H. 1995a. Improvement in selfed and random-mated generations of four subtropical maize populations through S3 recurrent selection. *Euphytica* 83: 1–8.

Vasal, S.K., Gonzalez Ceniceros, F., and Balla, O. 2001. Research activities and some achievements of CIMMYT-ARMP. Paper presented at the 30th National Corn and Sorghum Research Conference, Bangkok, Thailand, June 19–23.

Vasal, S.K., Gonzalez Ceniceros, F., and Srinivansa, G. 1992c. Genetic variation and inheritance of resistance to the "tar spot" disease complex. *Maize Genet. Coop. Newsl.* 66: 74.

Vasal, S.K. and Mclean, S. 1995. Past and Future Uses of Recurrent Selection Schemes. A Briefing Document Prepared by CIMMYT Maize Program for Internally Managed External Review on Breeding Strategies and Methodologies, August–September.

Vasal, S.K., Ortega, A., and Pandey, S. 1982. *CIMMYT's Maize Germplasm Management, Improvement and Utilization Program.* CIMMYT, El Batan, México, D.F., México.

Vasal, S.K., Pandey, S., Briandiaran, M., and Cordova, H. 1998. A Critique of Breeding Options for the Development of Abiotic Stress-Tolerant Maize Germplasm. Paper presented at the 7th Asian Regional Maize Workshop, Los Banos, Philippines.

Vasal, S.K. and Srinivasan, G. 1991. Performance of intra- and inter-population interline maize hybrids. In *Agronomy Abstracts.* ASA, Madison, WI, p. 119.

Vasal, S.K., Srinavasan, G.G., Córdova, H.S., Pandey, S., Jeffers, D., Bergvinson, D.J., and Beck, D.L. 1999b. Inbred line evaluation nurseries and their role in maize breeding at CIMMYT. *Maydica* 44: 341–351.

Vasal, S.K., Srinivasan, G., Crossa, J., and Beck, D.L. 1992b. Heterosis and combining ability of CIMMYT's subtropical and temperate early-maturity maize germplasm. *Crop Sci.* 32: 884–890.

Vasal, S.K., Srinivasan, G., Dhillon, B.S. Malean, S.D., Vergara, N., and Zhang, S.H. 1997b. Registration of 21 tropical yellow parental lines of maize. *Crop Sci.* 37: 1402–1403.

Vasal, S.K., Srinivasan, G., Gonzalez, F.C., Beck, D.L., and Crossa, J. 1993b. Heterosis and combining ability of CIMMYT's QPM maize germplasm. II. Subtropical. *Crop Sci.* 33: 51–58.

Vasal, S.K., Srinivasan, G., Gonzalez, F.C., Han, G.C., Pandey, S., Beck, D.L., and Crossa, J. 1992a. Heterosis and combining ability of CIMMYT's tropical × subtropical maize germplasm. *Crop Sci.* 32: 1483–1489.

Vasal, S.K., Srinivasan, G., Pandey, S., Gonzalez, F.C., Beck, D.L., and Crossa, J. 1993a. Heterosis and combining ability of CIMMYT's QPM maize germplasm. I. Lowland Tropical. *Crop Sci.* 33: 46–51.

Vasal, S.K., Srinivasan, G., and Vergara, N. 1995b. Registration of 12 hybrid-oriented maize germplasm tolerant to inbreeding depression. *Crop Sci.* 35: 1233–1234.

Vasal, S.K., Villegas, E., and Bauer, R. 1979. Present status of breeding quality protein maize. In *Proceedings of the Symposium on Seed Protein Improvement in Cereals and Grain Legumes*, Vol. 11. IAEA, Vienna, pp. 127–150.

Vasal, S.K., Villegas, E., Bjarnason, M., Gelaw, B., and Goertz, P. 1980. Genetic modifiers and breeding strategies in developing hard endosperm opaque-2 materials. In *Improvement of Quality Traits of Maize for Grain and Silage Use*, W.G. Pollmer and R.H. Phillips, Eds. Nijhoff, The Hague, pp. 37–73.

Vasal, S.K., Villegas, E., and Tang, C.Y. 1984b. Recent advances in the development of quality protein maize germplasm at CIMMYT. In *Panel Proceedings Series: Cereal Grain Protein Improvement*, December 6–10, 1982. IEAE, Vienna, pp. 167–189.

Vasal, S.K., Villegas, E., Tang, C.Y., Werder, J., and Read, M. 1984a. Combined use of two genetic systems in development and improvement of quality protein maize. *Kulturpflanza* 32: 171–185.

Villegas, E., Vasal, S.K., and Bjarnason, M. 1992. Quality protein maize: what is and how was it developed? In *Proceedings of the AAAC 75th Meeting*, Dallas, Texas, October 14–80, p. 90.

Weber, D.F. 1973. A test for distributive pairing in *Zea mays* utilizing doubly monosomic plants. *Theor. Appl. Genet.* 43: 167–173.

Weber, D.F. 1983. Monosomic analysis in diploid crop plants. In *Cytogenetics of Crop Plants*, M.S. Swaminathan et al., Eds. Macmillan, New Delhi, pp. 352–378.

Weber, D.F. 1991. Monosomic analysis in maize and other diploid crop plants. In *Chromosome Engineering in Plants: Genetics, Breeding, Evolution, Part A*, P.K. Gupta and T. Tsuchiya, Eds. Elsevier, Amsterdam, pp. 181–209.

Wilson, W.A., Harrington, S.E., Woodman, W.L., Lee, M., Sorrells, M.E., and McCouch, S.R. 1999. Inferences on the genome structure of progenitor maize through comparative analysis of rice, maize and the domesticated panicoids. *Genetics* 153: 453–473.

Zuber, M.S. and Helm, J.L. 1975. Approaches to improving protein quality in maize without the use of specific mutants. In *High Quality Protein Maize*. Hutchinson Ross Publishing Co., Stroudburg, PA, pp. 241–252.

# CHAPTER 7

# Cytogenetic Manipulation in Oat Improvement

**Eric N. Jellen and J. Michael Leggett**

## CONTENTS

7.1   Overview and History of Oats: Introduction and History ................................................200
7.2   Speciation in *Avena* ........................................................................................................201
    7.2.1   Introduction to the Genus ....................................................................................201
    7.2.2   Diploids ................................................................................................................202
    7.2.3   Tetraploids ...........................................................................................................204
    7.2.4   Hexaploids ...........................................................................................................205
7.3   Interspecific Hybridization ..............................................................................................206
    7.3.1   Introduction ..........................................................................................................206
    7.3.2   Diploid × Diploid Hybrids ...................................................................................206
        7.3.2.1   Interspecific Hybridization in the A Genome Diploid Group .................206
        7.3.2.2   Interspecific Hybridization in the C Genome Diploid Group .................207
    7.3.3   Tetraploid × Tetraploid Hybrids ..........................................................................207
        7.3.3.1   Interspecific Hybridization in the Section Pachycarpa Tetraploid
                     Group ........................................................................................................207
        7.3.3.2   Interspecific Hybridization in the *A. barbata* Tetraploid Group .............207
        7.3.3.3   Interspecific Hybridization Involving Tetraploids in Different
                     Groups .......................................................................................................208
    7.3.4   Triploid Hybrids ...................................................................................................208
        7.3.4.1   Hybrids Involving Diploids and the *A. barbata* Tetraploid Group ..........208
        7.3.4.2   Hybrids Involving Diploids and the Section Pachycarpa
                     Tetraploid Group ......................................................................................208
    7.3.5   Interspecies Hexaploid Hybrids ...........................................................................209
    7.3.6   Diploid × Hexaploid Hybrids ..............................................................................209
    7.3.7   Pentaploid Hybrids ...............................................................................................209
        7.3.7.1   Hybrids Involving Hexaploids and the *A. barbata* Tetraploid
                     Group ........................................................................................................209
        7.3.7.2   Hybrids Involving Hexaploids and the Section Pachycarpa
                     Tetraploids ................................................................................................210
        7.3.7.3   Hybrids Involving Hexaploids and the Remaining Tetraploids ...............210
7.4   Genetic Control of Chromosome Pairing .........................................................................210
7.5   Chromosome Rearrangements ..........................................................................................211

|       | 7.5.1 | Background | 211 |
|       | 7.5.2 | Detection of Translocations | 213 |
|       | 7.5.3 | Ancient Chromosome Structural Changes | 213 |
|       | 7.5.4 | Modern Chromosome Structural Changes | 214 |
| 7.6   | Gene Pools | | 215 |
|       | 7.6.1 | Introgression | 215 |
|       | 7.6.2 | Introgressions Utilizing the Primary Gene Pool | 215 |
|       | 7.6.3 | Introgressions Utilizing the Secondary Gene Pool | 216 |
|       | 7.6.4 | Introgressions Utilizing the Tertiary Gene Pool | 216 |
|       | 7.6.5 | The Effect of Direction of a Cross and the Cytoplasm Donor | 220 |
|       | 7.6.6 | Oat–Maize Hybridization | 220 |
| 7.7   | Development of Subarm Aneuploids | | 221 |
|       | 7.7.1 | Introduction | 221 |
|       | 7.7.2 | Duplicate-Deficient Segments and Crown Rust Resistance | 221 |
|       | 7.7.3 | Duplicate-Deficient Lines from Sun II × N770-165-2-1 | 221 |
|       | 7.7.4 | Segregation of 7C and 17 in Populations Derived from Translocation Heterozygotes | 222 |
| 7.8   | Induced Variation | | 222 |
|       | 7.8.1 | Ionizing Radiation | 222 |
|       | 7.8.2 | Other Methods of Inducing Variation | 223 |
| 7.9   | Future Uses of *Avena* Genetic Resources | | 223 |
|       | 7.9.1 | Adding Value to the Oat Crop | 223 |
|       | 7.9.2 | Expanding the Oat Crop into New Production Regions | 224 |
| References | | | 224 |

## 7.1 OVERVIEW AND HISTORY OF OATS: INTRODUCTION AND HISTORY

The primitive domesticated and wild gene pools of *Avena* are relatively underutilized resources for improving the four cultivated oat species: the common oat (*A. sativa* L., 2n = 6x = 42), the red oat (*A. byzantina* C. Koch, 2n = 6x = 42), the Ethiopian oat (*A. abyssinica* Hochst., 2n = 4x = 28), and the gray oat (*A. strigosa* Schreb., 2n = 2x = 14). This can be attributed to three main factors. First, oats are declining in economic importance. Second, oat research has attracted relatively little funding. Last, *Avena* has proven more recalcitrant to interspecies gene exchange than the cereals of the Triticeae (including wheat, durum, barley, and rye), primarily due to postzygotic sterility barriers. We will demonstrate in this chapter that although these introgression barriers are considerable, the potential payoffs are great.

Archaeological records provide tangible evidence for a long association between man and oats. The earliest archaeological record of *A. sterilis* dates to approximately 10,500 B.C. in Greece (Hansen and Renfrew 1978). In many of the early archaeological records, oats occurred only as a weed in mixtures of wheat and barley, and the wild-weedy forerunners of these three crop species are still found growing together in primary habitats in temperate regions of the world (Figure 7.1). The fact that emmer and some bread wheats were under cultivation around 5000 B.C. in Spain and 4000 B.C. in The Netherlands (Zohary and Hopf 1993) indicates that wild-weedy oats likely grew alongside these crops in Europe previous to oats' acquisition of the domestication syndrome. The earliest reference to nonshattering oats used for human food was as early as 23 to 79 A.D. by the Roman writer Pliny (Coffman 1961).

**Figure 7.1** **(See color insert following page 114.)** Photographs of wild-weedy oat species. (Top) *A. murphyi* growing in southern Spain. (Bottom) *A. barbata* and *A. sterilis* growing with wild *Hordeum* in Izmir Province, Turkey.

## 7.2 SPECIATION IN *AVENA*

### 7.2.1 Introduction to the Genus

The genus *Avena* L. (Poaceae) belongs to the tribe Aveneae and is comprised of a polyploid species series with $x = 7$ and chromosome numbers of $2n = 2x = 14$, $2n = 4x = 28$, and $2n = 6x = 42$ (see Table 7.1 for a full list and designation of species). All of the representative species are

**Table 7.1  Taxonomic Classification in the Genus *Avena* L.**

| Section | Taxonomic Species | Chromosome No., Genomes |
|---|---|---|
| Avenotrichon (Holub) Baum | *A. macrostachya* Bal. ex Coss. et Dur. | 2n = 28, MMMM (?) |
| Ventricosa (Baum) | *A. clauda* Dur. | 2n = 14, CpCp |
| | *A. eriantha* Dur. | 2n = 14, CpCp |
| | *A. ventricosa* Bal. ex Coss. | 2n = 14, CvCv |
| Agraria (Baum) | *A. brevis* Roth. | 2n = 14, AsAs |
| | *A. hispanica* Ard. | 2n = 14, AsAs |
| | *A. nuda* L. | 2n = 14, AsAs |
| | *A. strigosa* Schreb. | 2n = 14, AsAs |
| Tenuicarpa (Baum) | *A. agadiriana* Baum et Fedak | 2n = 28 |
| | *A. atlantica* Baum et Fedak | 2n = 14, AsAs |
| | *A. barbata* Pott. ex Link. | 2n = 28, AABB |
| | *A. canariensis* Baum, Rajhathy et Sampson | 2n = 14, AcAc |
| | *A. damascena* Rajhathy et Baum | 2n = 14, AdAd |
| | *A. hirtula* Lag. | 2n = 14, AsAs |
| | *A. longiglumis* Dur. | 2n = 14, AlAl |
| | *A. lusitanica* (Tab. Mor.) Baum Comb. Nov. et Stat. Nov. | 2n = 14, AsAs |
| | *A. matritensis* Baum | 2n = 14, AsAs (?) |
| | *A. prostrata* Ladiz. | 2n = 14, ApAp |
| | *A. wiestii* Steudel. | 2n = 14, AsAs |
| Ethiopica (Baum) | *A. abyssinica* Hochst. | 2n = 28, AABB |
| | *A. vaviloviana* (Malz.) Mordv. | 2n = 28, AABB |
| Pachycarpa (Baum) | *A. insularis* Ladiz. | 2n = 28, AACC |
| | *A. maroccana* Gdgr. | 2n = 28, AACC |
| | (*A. magna* Murphy et Terrell | 2n = 28) |
| | *A. murphyi* Ladiz. | 2n = 28, AACC |
| Avena | *A. atherantha* Presl. | 2n = 42, AACCDD |
| | *A. byzantina* C. Koch | 2n = 42, AACCDD |
| | *A. fatua* L. | 2n = 42, AACCDD |
| | *A. hybrida* Peterm. | 2n = 42, AACCDD |
| | *A. occidentalis* Dur. | 2n = 42, AACCDD |
| | *A. sativa* L. | 2n = 42, AACCDD |
| | *A. sterilis* L. | 2n = 42, AACCDD |
| | *A. trichophylla* C. Koch | 2n = 42, AACCDD |

*Source*: Revised from Baum, B.R., *Oats: Wild and Cultivated, a Monograph of the Genus Avena L. (Poaceae)*, Supply & Services Canada, Ottawa, 1977, chap. 2; Leggett, J.M., in *Oat Science and Technology*, Agronomy Monograph 33, Marshall, H.G. and Sorrells, M.E., Eds., ASA and CSSA, Madison, WI, 1992, pp. 29–52.

autogamous annuals with the exception of *A. macrostachya*, which is a perennial, allogamous autotetraploid (Baum and Rajhathy 1976). Thirty-two taxonomic entities are generally recognized by oat workers, though there is some disagreement over classification of some species/taxa (Table 7.1). The generally accepted taxonomic names of the various morphological types of oat have been retained so that they can be readily identified. However, taxa that share common genomes and are interfertile (e.g., *A. sativa*, *A. fatua*, *A. sterilis*) are considered to constitute a single biological species (Ladizinsky 1988). One or more of the four basic genomes AA, BB, CC, and DD have been allocated to the various taxa based on cytological observations of meiotic chromosome pairing of inter- and intraspecific hybrids, karyotype analyses using C-banding, genomic *in situ* hybridization (GISH), and *in situ* hybridization with genome-specific DNA probes.

## 7.2.2  Diploids

There are 16 diploid *Avena* taxa with a chromosome number 2n = 2x = 14 (Table 7.1). These were divided into three sections by Baum (1977) and into two karyotypic groups, A and C. He

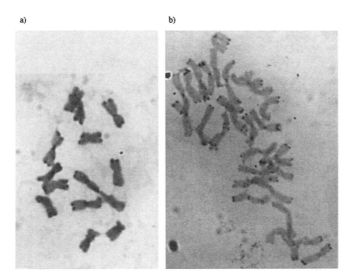

**Figure 7.2**  C-banded somatic metaphase chromosomes of *A. lusitanica* accessions: (a) CN 25777 (2n = 2x = 14) and (b) CN 25414 (2n = 4x = 28). (Accessions courtesy of Plant Gene Resources of Canada.)

classified the four cultivated diploid species as section Agraria (Baum), with the eight remaining A genome diploids being assigned to section Tenuicarpa (Baum), and the three C genome taxa assigned to section Ventricosa (Baum).

The A genome diploids are subdivided into two cytogenetic groups. The interfertile $A_sA_s$ genome (strigosa) group described by Rajhathy and Morrison (1959) and designated by Rajhathy (1961) contains the species *A. brevis, A. hirtula, A. hispanica, A. lusitanica, A. nuda, A. strigosa, A. wiestii*, and the more recently described *A. atlantica* (Baum and Fedak 1985a; Leggett 1987). The species *A. matritensis* currently has no karyotypic classification, though chromosome pairing of inter- and intraspecific hybrids (Leggett, unpublished data) indicates that it has the $A_s$ genome. All of these taxa have mostly metacentric to submetacentric chromosomes that are principally euchromatic. A number of accessions of these species, particularly *A. lusitanica*, are autotetraploid (Figure 7.2). Although hybrids within this group are fertile, disparities in marker distances between the *A. atlantica* × *A. hirtula* (614 cM, 192 cDNA markers; O'Donoughue et al. 1992) and *A. strigosa* × *A. wiestii* (880 cM, 181 mostly cDNA markers; Rayapati et al. 1994) maps indicate that minor chromosome divergence has occurred within this group.

The second set of A genome species has divergent karyotypes that deviate from one another and from the $A_sA_s$ genome. The species represented in this group are *A. canariensis* (Figure 7.3a; $A_cA_c$; Baum et al. 1973), *A. damascena* (Figure 7.3b; $A_dA_d$; Rajhathy and Baum 1972), *A. longiglumis* (Figure 7.3c; $A_lA_l$; Rajhathy and Morrison 1959), and *A. prostrata* ($A_pA_p$; Ladizinsky 1971b). The chromosomes of *A. canariensis* contain a relative abundance of pericentric heterochromatin (Figure 7.3; Fominaya et al. 1988a). The only fertile cross-combination among species of this group is *A. longiglumis* × *A. prostrata* (Ladizinsky 1973a).

The C genome diploid group contains three taxa and two genomic designations. The species *A. eriantha* (formerly known as *A. pilosa*) and *A. clauda* are interfertile, have the same karyotype, and share the genomic designation $C_pC_p$ (Rajhathy and Dyck 1963; Rajhathy 1966). The third species, *A. ventricosa*, forms sterile hybrids with *A. eriantha* and *A. clauda* and has the genome designation $C_vC_v$ (Figure 7.3d), though one accession of *A. ventricosa* with a single translocation has been observed (Rajhathy 1966, 1971; Rajhathy and Thomas 1967). In comparison with the A genome diploids, chromosomes of the C genome diploids are principally subtelocentric and contain abundant heterochromatin (Figure 7.3; Fominaya et al. 1988a).

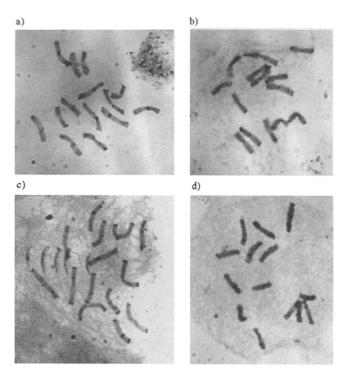

**Figure 7.3**    C-banded root-tip chromosomes of *Avena* diploids: (a) *A. canariensis* CN 23031, (b) *A. damascena* Cc7045, (c) *A. longiglumis* Clav 7087, and (d) *A. ventricosa* CN 21992. (Accessions courtesy of Plant Gene Resources of Canada, USDA–NPGS, and the BBSRC, U.K.)

### 7.2.3   Tetraploids

The tetraploid group of *Avena* (2n = 4x = 28) is represented by eight species: *A. abyssinica*, *A. agadiriana*, *A. barbata*, *A. insularis*, *A. macrostachya*, *A. maroccana*, *A. murphyi*, and *A. vaviloviana* (Table 7.1). The species *A. barbata*, *A. abyssinica*, and *A. vaviloviana* are traditionally classified in the same group due to their interfertility and share the genomic designation AABB (Rajhathy and Morrison 1959; Sadasivaiah and Rajhathy 1968). However, a recent study indicates that the Ethiopian endemic species *A. abyssinica* and *A. vaviloviana* (section Ethiopica Baum) have diverged from *A. barbata* for a class of structural repeats (Irigoyen et al. 2001).

Three tetraploid species comprise *Avena* section Pachycarpa Baum: *A. maroccana* (also known as *A. magna*), *A. murphyi*, and the recently discovered *A. insularis* (Baum 1977; Ladizinsky 1998). Although the three species have been given the designation AACC (Murphy et al. 1968), C-banding revealed significant morphological differences in their karyotypes, particularly among their euchromatic A genome chromosomes (Figure 7.4; Jellen and Ladizinsky 2000). C-banding and intraspecific hybrid pairing analyses revealed that minor chromosome rearrangements differentiate four populations of *A. insularis* collected in Tunisia and Sicily (Ladizinsky and Jellen 2003).

Section Tenuicarpa Baum contains *A. barbata* and *A. agadriana* (Baum and Fedak 1985b). The latter species has been shown to be closely related to the $A_s$ genome diploids at the gross DNA level from genomic *in situ* hybridization (GISH) studies, although chromosome pairing affinities indicate numerous chromosome rearrangements exist between the A genome diploid species and *A. agadiriana* (Leggett 1989a; Figure 7.4c).

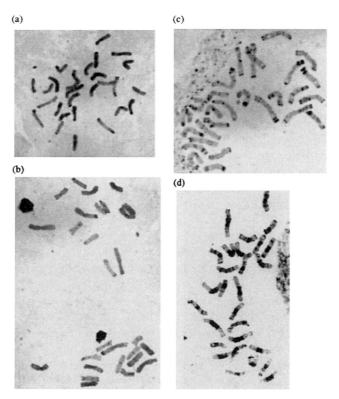

**Figure 7.4** C-banded root-tip chromosomes of *Avena* tetraploids: (a) *A. maroccana* CN 23060, (b) *A. insularis* from Bargou, Tunisia, (c) *A. agadiriana* CN 25837, and (d) *A. macrostachya* CN 24393. (Accessions courtesy of Plant Gene Resources of Canada and G. Ladizinsky, Hebrew University of Jerusalem.)

Chromosome pairing studies of hybrids between *A. macrostachya* and other species of oat have not provided a clear picture of genomic affinities, although the genomes of this autotetraploid are closer to the C genome diploids than to any of the A genome species (Leggett 1985; Pohler and Hoppe 1991). C-banding analysis revealed that most of the chromosomes have large blocks of pericentric heterochromatin, something not observed in any of the *Avena* diploids (Figure 7.4d). However, GISH studies indicated that the *A. macrostachya* genomes have a close affinity with the C genome diploid species at the gross DNA level, since probing *A. macrostachya* with total DNA from the C genome diploid *A. eriantha* resulted in a uniform label across all the *A. macrostachya* chromosomes, while hybridization with *A. strigosa* total DNA resulted in no detectable hybridization signal (Leggett and Markhand 1995b).

### 7.2.4 Hexaploids

Hexaploid *Avena* (2n = 6x = 42) is represented by eight taxa that share the genomic designation AACCDD (Table 7.1; Baum 1977; Leggett 1992). Intertaxa hybrids are fertile, although varying degrees of partial sterility are expected due to chromosome rearrangements (Ladizinsky 1970; McMullen et al. 1982; Singh and Kolb 1991). The taxonomic species include the following: *A. atherantha*, *A. byzantina*, *A. fatua*, *A. hybrida*, *A. occidentalis*, *A. sativa*, *A. sterilis*, and *A. trichophylla*. A major translocation involving chromosomes 7C and 17 differentiates *A. byzantina* and *A. sativa* (Jellen and Beard 2000).

## 7.3 INTERSPECIFIC HYBRIDIZATION

### 7.3.1  Introduction

An understanding of the evolutionary relationships in *Avena* is essential for oat geneticists, as these relationships may hinder the sexual transfer of genes into cultivated oat. Comparisons between the polyploid structure of oat and wheat must be treated with great caution. Whereas *Hordeum*, *Secale*, and *Triticum* are assigned taxonomically to the Triticeae tribe, *Avena* pertains to a different tribe, the Avenae (Clayton and Renvoize 1986).

The hexaploid cultivated oat is thought to have evolved via structural differentiation of chromosomes, hybridization, and chromosome doubling in a number of steps from diploids to tetraploids, and thence to hexaploids (see Rajhathy and Thomas 1974). Although this sequence is not clear, interspecies relationships are well documented, as outlined below.

### 7.3.2  Diploid × Diploid Hybrids

#### 7.3.2.1  Interspecific Hybridization in the A Genome Diploid Group

The $A_sA_s$ subgenome group (Table 7.1) taxa *A. strigosa*, *A. hirtula*, *A. wiestii*, and *A. atlantica* have similar karyotypes and are interfertile, their hybrids displaying mostly normal bivalent pairing (Holden 1966; Ladizinsky 1973b; Leggett 1987). Markhand and Leggett (1996) demonstrated that *A. lusitanica*, *A. hispanica*, and *A. matritensis* are homologous to *A. strigosa* at the gross DNA level using GISH. Ongoing chromosome pairing studies indicate that these species and *A. nuda* (not to be confused with hexaploid *A. sativa* ssp. *nuda* or naked oats) and *A. brevis* are closely related to each other and can be considered a single biological species complex (J.M. Leggett, personal observations).

The hybrid *A. strigosa* × *A. prostrata*, in contrast, exhibited a high frequency of multivalents and a low frequency of bivalents, indicating that there are large structural differences between the chromosomes of the two species (Ladizinsky 1973a). The formation of three trivalents and one pentavalent in a number of pollen mother cells (PMCs) is reminiscent of the chromosome pairing observed by Holden (1966) in the hybrid *A. strigosa* × *A. longiglumis*. Clearly, *A. strigosa* is separated from *A. prostrata* and *A. longiglumis* by at least five chromosome rearrangements. Ladizinsky (1973a) also reported the hybrid *A. longiglumis* × *A. prostrata* as forming up to four ring bivalents with a mean bivalent frequency of 5.74. This result, together with the partial self-fertility of this hybrid, indicates that there is a considerable degree of homology between their chromosomes.

As described by Leggett (1989b), the pattern of chromosome pairing in the hybrids *A. damascena* × *A. strigosa* and *A. strigosa* × *A. canariensis* was similar. The slightly higher frequency of univalents and the lower frequency of multivalents in the former hybrid probably result from a slightly larger translocated segment in the *A. canariensis* parent than in the *A. damascena* parent. The maximum chromosome pairing recorded in both hybrids indicates that *A. strigosa* differs from *A. canariensis* and *A. damascena* by at least three translocations.

Interestingly, hybrids between *A. canariensis* and *A. damascena*, which are morphologically diverse and geographically isolated, have regular bivalent pairing but are self-sterile (Leggett 1984a; Nishiyama et al. 1989). This is analogous to the situation in some hybrids between *Lolium* and *Festuca*, where there is complete and regular chromosome pairing but sterility of $F_1$ hybrids (Jauhar 1975a; for review, see Jauhar 1993).

The diploid hybrid *A. prostrata* × *A. canariensis* (Thomas and Leggett 1974) displayed similar meiotic chromosome pairing to the *A. prostrata* × *A. damascena* hybrid reported by Rajhathy and Baum (1972), and the apparent affinity of *A. prostrata* to *A. canariensis* and *A. damascena* indicates that these diploid species have likely arisen from a common ancestor.

Meiotic chromosome pairing in the hybrid *A. longiglumis* × *A. damascena* (Leggett 1984a) was similar to that described by Rajhathy and Baum (1972) for the hybrid *A. prostrata* × *A. damascena*, the largest configuration recorded being an octavalent, indicating three translocations involving four pairs of chromosomes. The number and frequency of multivalents in both hybrids effectively isolate the species, though the complete pairing is evidence of overall homology with rearrangement. Chromosome pairing in the hybrid *A. longiglumis* × *A. canariensis* was very similar to that of *A. longiglumis* × *A. damascena*, as might have been deduced from the affinity between *A. damascena* and *A. canariensis* (Leggett 1989b).

### 7.3.2.2  Interspecific Hybridization in the C Genome Diploid Group

We have a clearer understanding of relationships among the C genome taxa than among those of the A genome. The *A. eriantha* × *A. clauda* hybrid exhibited mostly normal disomic pairing, the only exception being that an occasional multivalent resulted from what was deemed to be a minor translocation (Rajhathy and Thomas 1967). These taxa are completely interfertile and, in this respect, are similar to the *A. strigosa* group of A genome diploids. Moreover, Rajhathy and Thomas (1967) and Ladizinsky (1968) showed that the main difference between these taxa — the unit of seed dispersal being the spikelet in *A. eriantha* and the individual floret in *A. clauda* — was under the control of one or two genes.

The third taxonomic entity belonging to the C genome diploids, *A. ventricosa*, has sufficiently different chromosome architecture from the *A. eriantha*–*A. clauda* complex to isolate these taxa from one another. The formation of multivalents and univalents (Rajhathy and Thomas 1967; Nishiyama and Yabuno 1975) indicates that a number of structural differences exist between these two taxonomic groups. It is therefore evident that *A. eriantha*–*A. clauda* and *A. ventricosa* represent distinct biological species.

### 7.3.3  Tetraploid × Tetraploid Hybrids

Species relationships in the tetraploid oats are somewhat more complicated due to considerable divergence. Traditionally they have been split into two main groups: the *barbata* group and the *maroccana*–*murphyi* complex, each with a separate phylogeny. *Avena agadiriana* and *A. macrostachya* are distinct from the other tetraploids. The recent discovery of *A. insularis* (Ladizinsky 1998) has shed further light on the relationships of the tetraploid and hexaploid oat.

### 7.3.3.1  Interspecific Hybridization in the Section Pachycarpa Tetraploid Group

*Avena maroccana*, *A. murphyi*, and *A. insularis* have the *sterilis*-type seed dispersal unit, and based on this alone, the former two species were thought to be closely related. However, these two species have unique karyotypes, are morphologically distinct, and are sexually isolated from one another because of habitat differences. Chromosome pairing in their hybrid (Ladizinsky 1971a) illustrates rearrangement, since multivalents and univalents were observed in addition to bivalents. According to Ladizinsky (1999), *A. maroccana*, *A. murphyi*, and *A. insularis* differ from each other by at least four chromosomal rearrangements.

### 7.3.3.2  Interspecific Hybridization in the A. barbata Tetraploid Group

The *A. barbata* group of tetraploids, which traditionally includes the section Tenuicarpa tetraploid as well as the section Ethiopica tetraploids *Avena vaviloviana* and *A. abyssinica*, are interfertile and have near-normal meiotic pairing. The occurrence of multivalents in these cross-combinations indicates that there is some structural differentiation between them, and occasional reduction in fertility has been observed (Holden 1966). In addition, Irigoyen et al. (2001) reported divergence for DNA repeats between *A. barbata* and the Ethiopian tetraploids.

### 7.3.3.3  *Interspecific Hybridization Involving Tetraploids in Different Groups*

Meiosis in the hybrids *A. maroccana* × *A. barbata* (Ladizinsky 1969) and *A. barbata* × *A. murphyi* was characterized by the formation of univalents and multivalents (J.M. Leggett, unpublished data). Together with the dissimilarities in karyotype and distinct morphologies, these data indicate that the AB and AC genome species evolved along different pathways.

Chromosome pairing in hybrids between *A. agadiriana* and *A. maroccana* indicates that there is only residual homology between the species, with 43 to 71% of chromosomes remaining unpaired in four separate hybrids. In contrast, the hybrid *A. agadiriana* × *A. barbata* had a mean of 6.96 chromosomes remaining unpaired, indicating greater homology (Leggett 1989a).

Analysis of pairing in the *A. macrostachya* × *A. murphyi* hybrid indicated only residual homology remains, since a trivalent was formed in 28% of pollen mother cells, whereas preferential bivalent pairing would be expected between the 14 homologous *A. macrostachya* chromosomes alone (Leggett 1985). Intriguingly, the *A. barbata* × *A. macrostachya* hybrid reported by Hoppe and Pohler (1989) had a mean of 12.2 bivalents, indicating that 5.2 bivalents must have resulted either from the pairing of *A. barbata* chromosomes or from mixed pairings of both species. This is especially noteworthy since the genomes of *A. barbata* are AABB and the genomes of *A. macrostachya* are apparently closer to the C genome (Leggett and Markhand 1995b), but this may be confounded by the operation of a pairing control gene in *A. macrostachya* (Leggett 1985).

### 7.3.4  Triploid Hybrids

Triploid hybrids between the diploid and tetraploid species are informative in that they can give some indication as to the possible role that the various diploid species may have played in the formation of the tetraploids, and subsequently the hexaploids.

### 7.3.4.1  *Hybrids Involving Diploids and the A. barbata Tetraploid Group*

It is clear from hybrid pairing analysis that the *A. strigosa* group of diploids, especially the morphologically similar *A. hirtula*, gave rise to these tetraploids (Ladizinsky 1973b). The chromosome pairing observed in *A. barbata* × *A. hirtula* hybrids (Holden 1966; Ladizinsky 1973b) consists of a mean of five bivalents. In contrast, the chromosome pairing recorded in the hybrids *A. barbata* × *A. longiglumis* (Holden 1966) and *A. barbata* × *A. prostrata* (Ladizinsky 1974) consisted of only 3.03 and 3.24 bivalents, respectively.

### 7.3.4.2  *Hybrids Involving Diploids and the Section Pachycarpa Tetraploid Group*

Chromosome pairing in the hybrids *A. strigosa* × *A. maroccana*, *A. maroccana* × *A. wiestii* (Sadanaga et al. 1968), and *A. hirtula* × *A. maroccana* (Rajhathy and Sadasivaiah 1969) is similar, which is expected since taxa of the same biological diploid species ($A_sA_s$) constitute one of the parents in each hybrid. The low association per chromosome in the *A. maroccana* × *A. strigosa* group hybrids led Ladizinsky and Zohary (1971) to conclude that there was substantial divergence between the A genome chromosomes of the two species, and that the proposal of Murphy et al. (1968) that *A. maroccana* contained the *A. strigosa* A genome should be disregarded. Ladizinsky (2000) exploited this heterology to create an amphiploid between *A. strigosa* and *A. maroccana*, from which he derived synthetic hexaploids that varied in their chromosome associations from 21 bivalents to 17 bivalents plus 2 quadrivalents. Ladizinsky also observed low levels of chromosome pairing in the triploid hybrids *A. strigosa* × *A. insularis* (Ladizinsky 1999) and *A. prostrata* × *A.*

*maroccana* (Ladizinsky 1974) and concluded that the residual homology between the chromosomes in both instances, combined with phenotypic dissimilarities between the species, ruled out *A. strigosa* and *A. prostrata* as the A genome sources of these two respective tetraploids.

Hybrids between *A. longiglumis* and *A. maroccana* or *A. murphyi* (Rajhathy 1971; Ladizinsky 1971a) showed an increase in chromosome pairing over the *A. maroccana* × *A. strigosa* hybrids. This led Rajhathy (1971) to conclude that *A. longiglumis* was the donor of the A genome of *A. maroccana*. Baum et al. (1973) suggested *A. canariensis* as the progenitor of *A. maroccana* based on geographic distribution and the bidentate lemma tip that the species share. Chromosome pairing of the hybrid, however, did not reflect a truly homologous relationship, with a mean of 12.3 chromosomes remaining unpaired (Leggett 1980).

It was hypothesized that one of the C genome diploids was the probable progenitor of the second genome of *A. maroccana*, but the hybrid *A. eriantha* × *A. maroccana* produced by Kummer and Miksch (1977) had such a low frequency of paired chromosomes that they proposed that the genomic designation of *A. maroccana* should revert to AAMM, as originally proposed by Sadanaga et al. (1968). Nevertheless, in spite of widespread structural rearrangement, *in situ* hybridization reveals that homology still exists at the DNA level between one set of chromosomes of the section Pachycarpa tetraploids and the C genome diploids (Fominaya et al. 1995; Leggett and Markhand 1995b).

### 7.3.5 Interspecies Hexaploid Hybrids

On first examination, it appears that genes may be easily transferred among the hexaploid oat species, since they are interfertile, and chromosome pairing is more or less normal. Where multivalents are formed, the effect of polyploid buffering frequently compensates for loss, and in subsequent generations, fertility is usually restored rapidly. An extreme example of this was reported by Leggett (1983) in the hybrid *A. hybrida* × *A. sativa*, in which six translocations involving 10 pairs of chromosomes were recorded, yet a fertility level of 71% was observed in the $F_1$ and plants from sterile to fully fertile were recovered in the $F_2$. Further consequences of translocations are considered in Section 7.5.

### 7.3.6 Diploid × Hexaploid Hybrids

Available data on hybrid pairing between diploids and hexaploids are inconclusive with respect to identification of the ancestors of the latter group. The *A. strigosa* × *A. sativa* hybrid observed by Marshall and Myers (1961) appears to have slightly higher chromosome pairing than the *A. longiglumis* × *A. sativa* hybrid reported by Thomas and Jones (1965), the *A. canariensis* × *A. sativa* and *A. prostrata* × *A. sativa* hybrids recorded by Thomas and Leggett (1974), or the *A. damascena* × *A. sativa* hybrid constructed by Leggett (1984a). However, when one considers that most of the bivalents were associated as rods, it is evident that these diploid species are equivalent in their relationship to the hexaploid *A. sativa*. The tetraploid hybrids involving C genome *A. eriantha* or *A. ventricosa* with *A. sativa* (Thomas and Rajhathy 1967; Thomas 1970) displayed minimal chromosomal homology. Clearly, the available cytological data with regard to hybrids between the diploids and the hexaploids do not point to any species as a direct ancestor, and they explain the difficulties encountered when utilizing the diploids as sources of genes for improving *A. sativa* cultivars.

### 7.3.7 Pentaploid Hybrids

#### 7.3.7.1 *Hybrids Involving Hexaploids and the A. barbata Tetraploid Group*

Gross differences in plant morphology, together with the cytological evidence obtained from a series of interspecific hybrids, indicate that the *A. barbata* group of tetraploids was not involved in the formation of the hexaploid species. The pentaploid hybrids *A. abyssinica* × *A. sativa*

(Rajhathy and Morrison 1960), *A. sterilis* × *A. barbata* (Ladizinsky 1969), and *A. barbata* × *A. sativa* exhibited chromosome pairing means of 16 to 21 univalents in meiotic metaphases (J.M. Leggett, unpublished data).

### 7.3.7.2   *Hybrids Involving Hexaploids and the Section Pachycarpa Tetraploids*

This group of tetraploids has been considered the most likely tetraploid ancestors of the hexaploids due to their morphological and karyotypic similarities to *A. sterilis*. Chromosome pairing at the meiotic metaphase in the hybrids *A. sterilis* × *A. maroccana*, *A. sativa* × *A. maroccana* (Sadanaga et al. 1968), and *A. murphyi* × *A. sterilis* (Ladizinsky 1971a) confirmed the affinity between the species and indicated that one of the genomes of the tetraploid was at least partially shared in common with one of the hexaploid genomes.

Since the discovery by Ladizinsky (1998) of *A. insularis* growing alone and in mixed stands with *A. sterilis* in Sicily in the mid-1990s, evidence is growing that this species is one of the actual progenitors of the hexaploids. In the same study, the partially fertile hybrid *A. sativa* × *A. insularis* demonstrated a 12% increase in chromosome pairing over any other such pentaploid combination. A comparison of C-banding patterns revealed striking similarities between *A. insularis* and *A. sterilis* for at least 10 pairs of chromosomes (Jellen and Ladizinsky 2000).

### 7.3.7.3   *Hybrids Involving Hexaploids and the Remaining Tetraploids*

The chromosome pairing recorded in the *A. macrostachya* × *A. sativa* hybrid is likely confounded by the interference of a pairing control gene (Leggett 1985). However, it appears that there is some residual homology between the species, since approximately 62% of the *A. sativa* chromosomes had paired, in addition to the expected seven bivalents from the *A. macrostachya* parent (due to autopolyploidy).

The Moroccan tetraploid *A. agadiriana* has morphological similarities to *A. canariensis* and *A. sterilis* and might therefore be considered a candidate progenitor of hexaploid oat. Chromosome pairing in the hybrid *A. sativa* × *A. agadiriana*, however, was the lowest recorded for any species hybrid in the genus and is similar to that found in polyhaploids of oat (Leggett 1989a). Such pairing failure could be due either to a pairing control mechanism that restricts the number of synaptic events or to heterology.

## 7.4 GENETIC CONTROL OF CHROMOSOME PAIRING

The problem underlying the transfer of alien variation to the cultivated oat species is invariably one of lack of homology between the chromosomes of the alien and *A. sativa* genomes. If the pairing control mechanism could be relaxed in some way and there were sufficient chromosomal synteny, then nonhomologous chromosomes might be induced to pair and recombine. Rajhathy and Thomas (1972) discovered that accession CW 57 of the diploid species *A. longiglumis* contains a gene that suppresses the homoeologous pairing restriction mechanism in *A. sativa*. These signals function similarly to the gene in *Triticum speltoides* that suppresses the effects of the *Ph1* locus of chromosome 5B in hexaploid wheat (Riley et al. 1961). However, in *A. sativa* no nullisome with an effect similar to chromosome 5B of wheat has been identified (Leggett et al. 1992).

Jauhar (1977) concluded in a review of chromosome pairing literature on *Avena* amphiploids and polyhaploids that a strong and complex genetic pairing regulation system was operative in the genus. For example, haploids derived from oat polyploids exhibit essentially no homoeologous pairing. This hypothesis was supported by electron micrographs of polyploid *Avena* synaptonemal complexes, which revealed that zygotene pairing was restricted exclusively to homologs (Jones et al. 1989). In addition, artificial decaploids derived from AACCDD × AABB species crosses do not contain multivalents (Thomas and Jones 1965). This observation was interpreted by Jauhar (1977) to mean that the increased copies of the putative pairing regulation genes present in these AAAABBCCDD decaploids heighten pairing regulation, to the extent that the tetrasomic A genome chromosome set always pair as bivalents. Although a preponderance of more recent evidence has supported the contention of Rajhathy and Thomas (1974) that structural rearrangement has been a primary force in *Avena* genome differentiation (see Section 7.5), genetically controlled pairing regulation remains an important consideration in alien introgression (see also Chapter 1 by Jauhar in this volume).

Utilization of the *A. longiglumis* CW 57 homoeologous pairing system is difficult. Because the suppressor gene is present in the diploid species *A. longiglumis*, and due to the sterility of *A. longiglumis* × *A. sativa* hybrids, production of the fertile amphiploid of this cross is a necessary early obstacle. The amphiploid can then be crossed either directly to the alien species, or preferably, to an addition or substitution line where the added or substituted chromosomes carry the gene that is the target for introgression. Selfing of the resultant hybrid and selecting for the desired character at each generation should eventually result in the production of a euploid plant carrying the requisite gene on what is essentially a chromosome from the cultivated species. The use of a disomic substitution line would be preferable, since the homoeolog of the alien chromosome would be present as a monosome in the $F_1$ hybrid, thus further encouraging the alien chromosome to pair with it in the absence of its true homolog.

Although this technique is complex and can incorporate a large portion of unwanted alien variation from the recombination of *A. longiglumis* chromosomes with those of *A. sativa*, it has been successfully used to incorporate genes conferring mildew resistance from the tetraploid species *A. barbata* (Thomas et al. 1980b) and from the diploid *A. prostrata* (Griffiths and Thomas 1982). In addition, an *A. maroccana* × *A. longiglumis* CW 57 cross was used to create a synthetic hexaploid germplasm, Amagalon, having a major crown rust resistance locus from the tetraploid parent (Rothman 1984).

## 7.5 CHROMOSOME REARRANGEMENTS

### 7.5.1 Background

Chromosome rearrangements might be expected between distantly related species where, through the course of evolution, the constituent genomes have continued to change. This is one of the main mechanisms for speciation. Lewis (1966) envisioned two paths that could lead to chromosome repatterning in speciation: (1) a gradual accumulation of chromosomal differences, or (2) saltational reorganization of chromosomes that would involve a rapid restructuring of the karyotype. The latter scenario might be induced by some catastrophic event, and its effect would be to isolate a very few individuals in an open and probably hostile environment, yet with no competition. Thus, the new species would not only survive but also begin a novel evolutionary pathway. It seems likely that a combination of these different mechanisms, coupled with timely hybridization and chromosome doubling, has given rise to the extant *Avena* species.

As mentioned repeatedly in the sections above, and as is demonstrated below, chromosome rearrangements such as translocations are common in the genus *Avena*. The weight of combined data from FISH, GISH, C-banding, and hybrid pairing experiments indicates that the A/B/D and C genomes experienced significant divergence, with minor divergence of the A/B/D and subgenomes in each group. In their monograph, Rajhathy and Thomas (1974) stated that "if gradual accumulation rather than saltational origin of the structural rearrangements is visualized, the major repatterning would have occurred earlier." This does not detract from the possible saltational origin of the structural chromosome rearrangements. It is possible that the A genome species developed initially via gradual accumulation of structural changes, after which a massive structural reorganization of the karyotype in one lineage occurred to produce the progenitor of the C genome diploids. This C genome progenitor would then have diverged to produce the *A. clauda–A. eriantha* complex and *A. ventricosa*.

A number of studies failed to distinguish the A, B, and D genome chromosome sets from one another in repeat sequence FISH (Aniniev et al. 2002; Katsiotis et al. 1997, 2000; Linares et al. 1999) or GISH (Chen and Armstrong 1994; Leggett and Markhand 1995a,b; Jellen et al. 1994; Yang et al. 1999) experiments (Figure 7.5). However, Linares et al. (1998) suggested that their discovery of a differential abundance of sequences homologous to a pAS120a satellite repeat sequence probe from *A. strigosa* was proof of minor divergence between the A and D genomes. Whereas this sequence hybridized well to all chromosomes of *A. strigosa*, *A. longiglumis*, and one genome of *A. sativa*, it did not hybridize to C genome chromosomes. In addition, hybridization signals were weak when pAS120a was probed *in situ* to several other A genome diploids or to *A. maroccana* and *A. murphyi*. Their results indicated that the *A. maroccana* and *A. murphyi* genome formulae should be CCDD rather than AACC, and that the diploids *A. canariensis* and *A. damascena* should be reconsidered D genome species. In another study using the same pAS120a satellite sequence as a probe, Irigoyen et

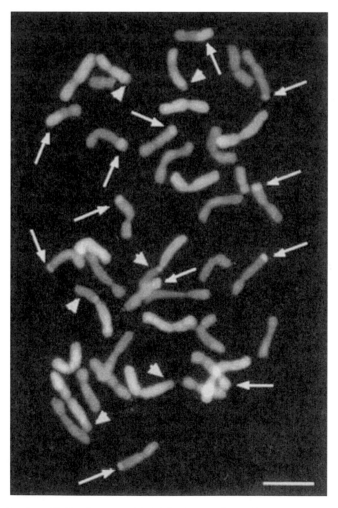

**Figure 7.5    (See color insert following page 114.)** Mitotic metaphase chromosome spread of *A. sativa* variety Sun II probed with rhodamine-labeled total genomic DNA derived from the C genome diploid *A. eriantha* and counterstained with DAPI. The pink fluorochrome indicates the C genome chromatin, while blue counterstain indicates chromatin from the A and D genomes. White arrows point to C genome translocations on A or D genome chromosomes, and yellow arrowheads point to A/D genome translocations on C genome chromosomes. Bar = 10 microns.

al. (2001) reported the ability to discriminate between A and B genome sets of chromosomes in *A. barbata* and *A. vaviloviana*. Although Linares et al. (1998) and Irigoyen et al. (2002) did not detect translocations between A and D genome chromosomes in *A. sativa* using the pAS120a probe *in situ*, Irigoyen et al. (2001) did detect A-B interchanges in *A. barbata* and *A. vaviloviana*. At the gross DNA level, it is evident that the A, B, and D genomes are very similar, since probing with total DNA extracted from any A genome diploid oat produces an even *in situ* hybridization signal across the chromosomes of any of these three genomes in the polyploids (Chen and Armstrong 1994; Jellen et al. 1994; Leggett and Markhand 1995a,b; Yang et al. 1999).

### 7.5.2 Detection of Translocations

A number of studies of meiotic chromosome pairing in hexaploid hybrids support the hypothesis that chromosomal rearrangement is a critical genome differentiation mechanism in *Avena*. One of the earliest comprehensive studies of meiotic irregularities in hexaploid oat was that of Joshi and Howard (1955), who observed varying degrees of multivalent and univalent formation at the first meiotic metaphase in winter × spring oat crosses. Because of these irregularities, the authors warned of possible problems from off-types, or duplicate-deficients (Dp-Df), being produced at high frequencies in breeding programs. Other reports of multivalent formation in intervarietal or intertaxa hexaploid oat crosses include those of Ladizinsky (1970, *A. byzantina*), McMullen et al. (1982, *A. sativa* × *A. sterilis*), and Singh and Kolb (1991, *A. sativa*). As a general rule, the more distantly related the two cultivars or taxa are, the higher the incidence of multivalents. However, the presence of multivalents in meiosis I may sometimes result from failure of the homoeologous pairing control mechanism rather than chromosome rearrangement, as was demonstrated in tall fescue interecotype hybrids by Jauhar (1974, 1975b). As another general rule, the smaller the rearrangement is, the less likely it is that it will be cytologically detectable using any of the techniques mentioned above.

When the C-banding technique began to be applied to oat chromosomes in the 1970s, researchers discovered that the C genome chromosomes were more heterochromatic than the A/B/D genomes (Yen and Filion 1977; Hutchinson and Postoyko 1986; Postoyko and Hutchinson 1986). However, C+ and C– banding regions at the termini of a number of chromosomes indicated the existence of large intergenomic A/D-C translocation segments (Fominaya et al. 1988b; Jellen et al. 1993a). These regions were confirmed to be intergenomic translocation segments using GISH in the hexaploids (Figure 7.5), *A. maroccana* (Leggett et al. 1994), and *A. murphyi* (Leggett and Markhand 1995b). Southern hybridization and comparative mapping can be used to detect smaller chromosome rearrangements not otherwise detectable by GISH (Rooney et al. 1994; Kianian et al. 1997; Fox et al. 2001; Portyanko et al. 2001; Wight et al. 2003).

### 7.5.3 Ancient Chromosome Structural Changes

In the hexaploid oat, it appears that there is a standard set of rearrangements that date back to before domestication. Most hexaploid oats carry the same translocation segments: two AA/DD-CC and three to five CC-AA/DD (Figure 7.5; Chen and Armstrong 1994; Jellen et al. 1994; Leggett and Markhand 1995a,b; Yang et al. 1999). When comparing C-banding patterns with GISH and FISH using genome-specific repetitive sequences, it is clear that these segments reside on chromosomes 1C, 7C, (8/9), 10, 12, (15), and 21 (Jellen et al. 1994; Yang et al. 1999; Irigoyen et al. 2002; Linares et al. 1992, 1996, 1999, 2000; Ananiev et al. 2002). These translocations are also present in accessions of the wild hexaploids *A. fatua* (Yang et al. 1999), *A. sterilis* (Zhou et al. 1999), and a number of varieties of *A. byzantina* and *A. sativa* (Markhand 1996; Jellen and Beard 2000).

An additional ancient reciprocal translocation involving chromosomes 7C and 17 differentiates landraces of Mediterranean *A. byzantina* from the *A. sativa* landraces of cooler climates (Jellen and Beard 2000). Zhou et al. (1999) reported the presence of this translocation in 80% of 96 *A.*

*sterilis* accessions. In addition, the translocation predominates in the floret-shattering hexaploids *A. fatua*, *A. hybrida*, and *A. occidentalis* (E. Jellen, personal observations).

### 7.5.4 Modern Chromosome Structural Changes

A number of studies have identified noninduced translocations that occurred in the recent past. In a study characterizing monosomics using the C-banding and RFLP techniques, Jellen et al. (1993b) identified five monosomics from the Kanota series having a very large reciprocal translocation between chromosomes 14 and 7C, but which was not present in the remaining lines. Although there is a small possibility that these differences are of ancient origin, it seems unlikely for two reasons. First, this translocation has not been observed in any of several hundred other hexaploid accessions examined using C-banding (Zhou et al. 1999; Jellen and Beard 2001). Second, the C-banding patterns of all the other chromosomes in these five lines closely resemble those of Kanota. Similarly, 15/18 Sun II-derived monosomic lines had a large translocation between chromosomes 14 and 3C that has not been observed in any other oat genotype, including disomic Sun II (Leggett and Markhand 1995a; Jellen et al. 1997).

Many other translocations have been identified in the oat hexaploids using GISH or C-banding. Markhand (1996) reported one accession of *A. sterilis* (Cc7079) having only four CC-AA/DD rearrangements. The old *A. sativa* variety Siberian had additional CC-AA/DD and AA/DD-CC rearrangements, both terminal in nature, while the variety Danish had six CC-AA/DD and one AA/DD-CC translocation, two of the CC-AA/DD translocations being present on the same chromosome, one being intercalary. C-banding analysis indicated that the AADD-CC translocation most likely is found on chromosome 2C (E.N. Jellen, personal observations). The AA/DD-CC translocation was different from the standard translocations and involved one of the small satellited (SAT) chromosomes that normally carry a CC genome terminal translocation on the long arm. Only the satellite and a tiny bit of the short arm of the AA/DD chromosome were translocated. It is likely that a reciprocal rearrangement including an inversion gave rise to these novel chromosome rearrangements. In another study, Jellen et al. (2003) found that *A. fatua* accession CN 25955 from Morocco had the ancestral 7C-17 translocation and an additional reciprocal translocation involving other portions of chromosomes 7C and 17 (Figure 7.6).

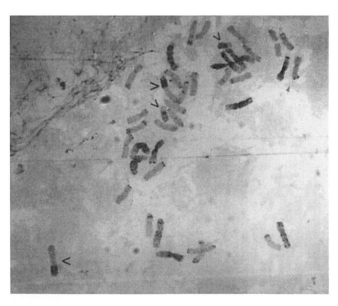

**Figure 7.6**   C-banded root-tip chromosomes of *A. fatua* accession CN 25955, having at least two intergenomic reciprocal translocations involving chromosomes 7C and 17 (arrowheads).

Additional evidence of ongoing genomic change in the genus *Avena* was observed in *A. agadiriana,* in which morphological and cytological differences led Leggett (1989a) to suggest that the different accessions were currently evolving from each other. These observations were confirmed by Jellen and Gill (1996), who described variation in C-banding karyotypes among five accessions of *A. agadiriana.*

A somewhat contradictory observation with regard to rapid genome change and the inherent implications for molecular mapping was made by Markhand (1996). He recorded an additional C genome translocation in an unidentified monosomic line (H-1/158). This translocation must have arisen during the previous sexual cycle, since no other seeds from the same parental plant contained it. While this could have been a unique event, it suggests that such intergenomic translocations occur at a fairly rapid rate.

From the evidence described above, it is clear that there are methods available to plant breeders to introgress desirable characters from all three *Avena* gene pools. However, they tend to be difficult and time-consuming, especially with regard to the secondary and tertiary gene pools. It is also apparent that while oat may be a more ancient allopolyploid than wheat, its genomes appear to be in a constant state of rearrangement. The isolation of a complete monosomic series, coupled with molecular genome analysis technologies, should help geneticists and breeders understand and manipulate the pairing control system of *Avena*, thus making gene transfer from the secondary and tertiary gene pools simpler.

## 7.6 GENE POOLS

### 7.6.1 Introgression

During introgression, elimination of alien chromosome segments due to poor recombination (linkage drag) is a major obstacle. However, it may also be difficult or impossible to eliminate entire alien chromosomes. Whether one or both of these obstacles is encountered is a function of the species involved and the gene pool to which it belongs. The wild species of *Avena* can be grouped into three gene pools, primary, secondary and tertiary, depending on the fertility of its interspecific hybrid with cultivated hexaploid oat *A. sativa*. Whereas the primary and tertiary gene pools for oat improvement are extensive and diverse, the secondary gene pool is relatively small and poorly represented in *ex situ* gene banks. In addition, as will be seen below, maize (*Zea mays* L.) should now be considered a member of the tertiary gene pool for oat improvement.

### 7.6.2 Introgressions Utilizing the Primary Gene Pool

The primary oat gene pool consists of all of the hexaploid *Avena* species. There are relatively minor restrictions on the flow of genes between the wild and cultivated species, and the desired traits can be recovered by a conventional backcrossing program. Thus, oat breeders have a truly remarkable array of landrace, weedy, and wild germplasm at their disposal, and there are many examples of introgressions from wild hexaploid species into the cultivated form leading to varieties. For example, Starter Spring oat, registered in 1990, has the following pedigree: Dal/3/Mn 67231/2/Diana/CI 8344/4/Noble (Stuthman et al. 1990). In this instance, the ancestor CI 8344 is a high-protein *A. sterilis* accession. Similarly, Ozark, with improved winter hardiness and released in 1989, was derived from a cross between the winter-susceptible Florida 501 × *A. sterilis* PI 29625 (Bacon 1991). The Spring oat Jay, developed by Purdue University Agricultural Research Programs and the USDA–ARS, has a complex pedigree that includes crown rust resistance from *A. sterilis* CIav 8079 through Iowa X434-II (Ohm et al. 2000). Likewise, a number of Brazilian varieties contain *A. sterilis* line ME 1563 or CRcpx in their lineage (L. Federizzi, personal communication).

Due to the relative ease of crossing and lack of barriers to recombination, chromosome engineering is not required for utilizing the hexaploid gene pool in oat breeding. However, linkage drag may be enhanced by the presence of translocations and other rearrangements, depending upon the diversity of the two parents involved in the cross. Linkage drag is also enhanced due to the requirement of eliminating undesirable alleles in three full genomes of recombining chromosomes.

### 7.6.3  Introgressions Utilizing the Secondary Gene Pool

Only three taxa, the tetraploids *A. murphyi, A. maroccana,* and *A. insularis,* constitute the secondary gene pool of cultivated hexaploid oat. Here, gene flow is partly restricted, in that $F_1$ hybrid seed can be fairly readily produced by mass pollination. Backcrossing these $F_1$ hybrids to the recurrent parent also requires repeat pollinations because of the inherent infertility; however, small numbers of seeds are produced since the hybrids are partially female fertile. Fertility levels improve with increased backcrossing.

Natural recombination can occur during the backcrossing procedure, so selection for the desired trait can be practiced directly. However, the chromosomes of the $F_1$ hybrid between the two species may not be sufficiently homologous, or even homoeologous, to pair at meiosis, thus preventing recombination. When such a situation arises, the normal procedure is to produce an addition line. As mentioned above, interspecies hybrids involving *A. insularis* have the greatest amount of chromosome pairing and are thus the most fertile.

There are fewer examples of the introgression of wild germplasm from the secondary gene pool, but recent examples are CDC Bell and a sister line, CDC Baler, forage oats developed at the Crop Development Centre, University of Saskatchewan (Rossnagel 1999). These lines were derived from the cross Av2401/2 × SO86044, where the Av2401/2 parent was an *A. sativa* breeding line derived from a cross between the tetraploid species *A. maroccana* and *A. sativa*, with the resulting pentaploid hybrid having been backcrossed to *A. sativa* (Thomas et al. 1980a).

In a novel series of experiments, Ladizinsky (1995) successfully introgressed genes for domestication traits from *A. sativa* into *A. maroccana* and *A. murphyi*. The purpose of this program was to derive a tetraploid oat cultigen having larger seed, higher seed protein, and higher yield in subtropical environments. Loci controlling the domestication traits were linked to each other and to a large heterochromatic knob marking the telomere of the wild version of chromosome 5C, long arm, when segregation was examined in a population derived from domesticated *A. maroccana* line Ba13-13 crossed with wild *A. maroccana* line 169 (Figure 7.7; E. Jellen, personal observations). Obstacles in the introgression process have included the replacement of undesirable genes in the *A. maroccana* parents for susceptibility to BYDV, day-length sensitivity, and tough rachilla, though considerable progress has been realized through the fourth backcross cycle as of the date of this printing (G. Ladizinsky, personal communication). Nevertheless, this *A. maroccana* improvement program demonstrates the relative ease with which genetic barriers between hexaploids and section Pachycarpa tetraploids can be overcome.

The secondary gene pool contains several desirable traits, although more extensive collections of *A. insularis* and *A. murphyi* will need to be made and evaluated. Valuable genes for pest protection include cereal cyst nematode resistance in *A. maroccana* and *A. murphyi* (P. Hagberg, personal communication). Known disease resistance genes include crown rust resistance in *A. maroccana,* including the major *Pc91* gene of the widely bred synthetic hexaploid Amagalon (Rothman 1984). In addition, *A. maroccana* possesses elevated seed protein content (Thomas et al. 1980a) and, along with the other section Pachycarpa tetraploids, increased seed size (Ladizinsky 1995).

### 7.6.4  Introgressions Utilizing the Tertiary Gene Pool

Gene flow in the tertiary group, which includes all the diploid species and the tetraploids other than section Pachycarpa, is very restricted, and thus is more difficult for the oat breeder to manage.

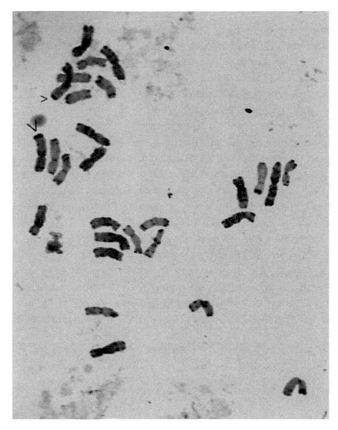

**Figure 7.7** C-banded root-tip chromosomes of $F_3$ *A. maroccana* plant BAM 25-1 (2n = 4x = 28), heterozygous for a 5C long-arm telomere knob. The plant was derived by a cross between domesticated *A. maroccana* line Ba13-13 (Ladizinsky 1995) and wild *A. maroccana* 169. Wild and domesticated portions of the homologous chromosomes are indicated by large and small arrowheads, respectively.

The initial problem is obtaining the interploidy hybrid, though this hurdle can usually be overcome through mass pollination and embryo rescue techniques. The tetraploid or pentaploid hybrids produced are invariably sterile, thus preventing backcrossing. The production of an amphidiploid through colchicine-mediated chromosome doubling usually restores sufficient fertility to facilitate backcrossing (using the hexaploid as the recurrent parent), resulting in the formation of alien chromosome addition lines.

Whole-chromosome addition lines can, however, prove to be problematic. For example, in some disomic addition lines the added chromosomes occasionally fail to pair in a small proportion of pollen mother cells (PMCs), which results in the production of balanced (3x) gametes. These balanced pollen grains are usually more competitive than the unbalanced gametes (3x + 1) due to certation, and thus participate in fertilization more frequently. Consequently, in each successive generation, increasingly higher proportions of euploid gametes are produced, eventually leading to the loss of the addition line. Such an addition decay was reported by Thomas et al. (1975) involving the addition of a pair of *A. barbata* chromosomes conferring mildew resistance to *A. sativa* cv. Manod. The increase in euploid gametes was so high in this instance that the addition line was not suitable for germplasm release.

The transmission of an alien chromosome in an addition line can be dependent upon which chromosome is involved, in a similar way to monosome transmission (Chang and Sadanaga 1964; Hacker 1965). Likewise, the fertility of chromosome addition lines can be dependent on which particular chromosome is involved in the addition. Thomas (1968) reported that two of the seven monosomic addition lines involving chromosomes Sat 2 and SM 5 added to the complement of *A. sativa* produced low fertility levels. Furthermore, the disomic addition of SM 5 had an even lower fertility, while the disomic addition of SM 6 had a fertility level some 68% lower than the monosomic addition line. In this instance, the sterility was apparently not related to meiotic disturbance and was assumed to have arisen as a consequence of genetic duplication or deleterious epistatic effects between *A. sativa* and *A. hirtula* genes. While the addition of certain chromosomes may result in morphological abnormalities, the addition of other chromosomes may have little or no phenotypic effect.

An alternative to addition lines is substitution lines, where the alien chromosome replaces a chromosome from the cultivated species (Figure 7.8). Such substitution lines can occur spontaneously during a backcrossing program designed to produce addition lines. This happens when there is meiotic instability leading to the loss of a cultivated chromosome. The problem with such random chromosome loss is that the substituted chromosome may not compensate for the loss of the *A. sativa* chromosome, which can result in instability, and hence loss of the stock.

If the *A. sativa* chromosome to be substituted is identified, then the appropriate monosome can be used as the female parent; but if it is not known, then the whole monosomic series must be crossed as female parents in order to derive the correct substitution. The greater the degree of homoeology between the substituted and replaced chromosomes, the better the chance will be for adequate compensation in replacing the lost chromosome. However, genic or structural differences that inevitably accumulate during the evolution of species, an abundance of which

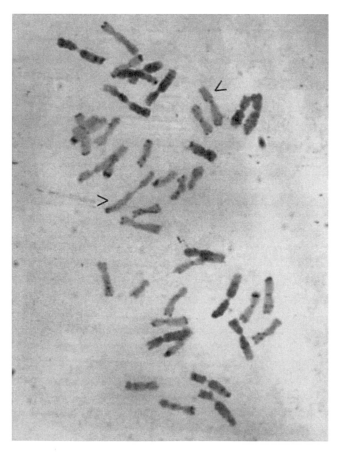

**Figure 7.8**   C-banded root-tip chromosomes of Wisconsin disomic substitution line DCS 1789 derived from crosses with a crown rust-resistant accession of *A. strigosa* (Dilkova et al. 2000). The pair of *A. strigosa* chromosomes substituted for chromosome 12 is indicated by arrowheads.

are found in polyploid *Avena* (see Section 7.5), can lead to a lack of compensation. Despite these obstacles, substitution lines in oats have been produced (Bhatti 1972; Sharma and Forsberg 1974; Powell 1980; Dilkova et al. 2000), which served not only to evaluate the ability of alien chromosomes to compensate for the loss of a crop chromosome but also as germplasms for further breeding efforts.

Some of the problems associated with whole-chromosome addition and substitution lines can be overcome by using the relevant telocentric chromosome line. In such cases, if the missing chromosome arm carries deleterious genes, the telocentric chromosome addition or substitution might have a considerable advantage over the whole-chromosome addition or substitution. Such a case was reported by Thomas et al. (1975), who observed that the pollen transmission of a ditelocentric addition line was greater than that of the disomic addition line.

The chances of natural recombination between added or substituted alien chromosomes and crop chromosomes in disomic addition or substitution lines depends on the degree of homology of the chromosomes involved. Since the *Avena* genomes have diverged considerably due to proliferation of genome-specific repeated DNA sequences (Ananiev et al. 2002; Katsiotis et al. 1997, 2000; Linares et al. 1998, 1999; Irigoyen et al. 2001) and chromosome rearrangement (see Section 7.5), the opportunity for recombination between native and alien chromosome segments is greatly reduced. Moreover, in these genetic stocks each chromosome has its true homolog with which to pair, and thus homoeologous and heterologous chromosome pairing is greatly reduced or precluded, especially in the presence of a homoeologous pairing suppression system. In such cases where natural recombination does not occur, there are other avenues available, such as the use of ionizing radiation or the exploitation of the homoeologous pairing induction system of *A. longiglumis* CW 57.

Three examples of introgression of characters from the diploids to the hexaploids are cited below. In the first two cases, it is not absolutely clear if the lines resulted from an addition, a substitution, or a natural recombination. The first example is the transfer of mildew resistance from the C genome species *A. eriantha* to the hexaploid reported by Hoppe and Kummer (1991). In the second example, the hybrids *A. strigosa* × *A. sativa* cv. Victory and *A. strigosa* × *A. sativa* cv. Abegweit were ancestors of the cultivars Hinoat, Gemini, and Foothill produced by Agriculture and Agri-Food Canada.

In the third example, *A. strigosa* × *A. sativa* crosses were used as sources of crown rust resistance in a complicated breeding program at the University of Wisconsin (Dilkova et al. 2000). One hexaploid line expressing resistance, N770-165-2-1, was derived from x-irradiation of one parent and carried a chromosome 6C-21 intergenomic reciprocal translocation (Figure 7.9). It is uncertain, however, where the resistance genes donated by *A. strigosa* reside in the genome of this line. A second line, DCS 1789, contained a pair of *A. strigosa* chromosomes instead of *A. sativa* chromosome 12 (Figure 7.8). In this case, since fertility and phenotype were normal, it was assumed that the *A. strigosa* chromatin is at least partly homoeologous, and therefore compensating, with respect to the missing genes from chromosome 12 (Dilkova et al. 2000).

The tertiary gene pool represents a rich reservoir of genetic diversity for improving cultivated oat. Mildew resistance genes have been discovered in *A. barbata*, *A. hirtula*, and *A. prostrata* (Thomas 1992). Crown rust resistance genes were identified by Marshall and Myers (1961) in *A. abyssinica* and *A. barbata*, and more recently by Aung et al. (1996) in *A. strigosa*. Stem rust resistance genes were found in accessions of *A. barbata* by Martens et al. (1980). Barley yellow dwarf virus (BYDV) resistance has been reported in *A. barbata*, *A. macrostachya*, and *A. strigosa* (Comeau 1984). Winter hardiness and perenniality genes are found in the Algerian montane species *A. macrostachya* (Baum and Rajhathy 1976). Oat quality may eventually be improved by the introgression of genes for high -glucan content from *A. damascena* or *A. atlantica* (Welch et al. 2000).

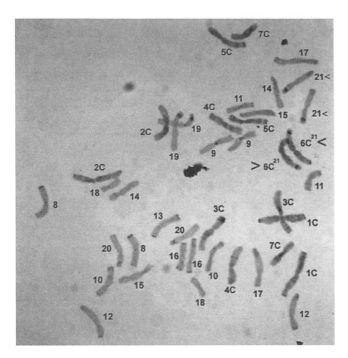

**Figure 7.9**    C-banded root-tip chromosomes of $F_4$ line SMIG-21 derived from the cross Sun II × N770-165-2-1. SMIG-21 has the genotype $6C^{21}/6C^{21}$(large arrowheads), 21/21 (small arrowheads), having four copies of most of the satellited short arm of 21. It is also deficient for the distal translocation segment from 6CS.

### 7.6.5    The Effect of Direction of a Cross and the Cytoplasm Donor

In producing hybrids between different *Avena* species, the direction of the cross is often critical. It is generally accepted that it is easier to produce hybrids between species of different ploidy levels by using the species of lower ploidy as the female parent, but this is not always true.

It has also been shown that the cytoplasm of the donor can be important. In alloplasmic lines derived from diploid × hexaploid crosses, Leggett (1984b) demonstrated that when the cytoplasm was derived from the C genome parent, subsequent progeny had much slower growth and a prolonged time to flowering relative to alloplasmic lines whose cytoplasm was derived from three different A genome parents. While these findings have little direct significance in terms of plant breeding, they demonstrate the differential effects of cytoplasms on quantitative trait expression in oat.

### 7.6.6    Oat–Maize Hybridization

An extreme example of chromosome addition lines in oat is the monosomic and disomic oat–maize addition lines developed at the University of Minnesota (Ananiev et al. 1997; Kynast et al. 2001; refer also to Chapter 5 in this volume). These materials were derived from an oat × maize wide crossing program initially conceived to produce haploid oat plants from which new oat monosomics could be isolated (Rines and Dahleen 1990; Jellen et al. 1997). Oat addition lines that are either disomic (for maize chromosomes 1 to 4, 6 to 7, and 9) or monosomic (for chromosomes 5, 8, and 10) have been isolated (Kynast et al. 2001). Moreover, all of the addition lines except 5 and 10 are sexually fertile and are capable of transmitting intact maize chromosomes (Kynast et al. 2001). Maize chromosome segments have been integrated into the oat genome using a radiation hybrid approach in these lines, as with maize chromosome 9 (Riera-Lizarazu et al. 2000). Although these stocks are of interest to maize molecular geneticists, they may also prove valuable in studying the effects of maize chromatin on oat agronomic performance.

## 7.7 DEVELOPMENT OF SUBARM ANEUPLOIDS

### 7.7.1 Introduction

Wheat geneticists have successfully exploited a natural chromosome breakage system by using a factor on chromosome 2C of *Aegilops cylindrica* to generate an extensive series of subarm deletion lines in wheat (Endo 1988; Endo and Gill 1996). These materials are being used to develop detailed, integrated cytogenetic maps of over 1000 DNA RFLP loci, as well as other DNA markers, genes, bacterial artificial chromosome (BAC) clones, and C-bands (Werner et al. 1992; Gill et al. 1993). The wheat cytogenetic maps are then used to formulate map-based cloning strategies and to study the distribution of recombination events along the physical wheat chromosomes. In addition, this deletion-generating system has been used to generate barley deletion and wheat–barley translocation stocks for genetic mapping (Serizawa et al. 2001).

Although such a system is not available in oat, it is possible to exploit the translocations present in the oat gene pool to generate subarm duplicate-deficient (Dp-Df) stocks. Although the unbalanced gametes produced from adjacent segregation of ring quadrivalent members in translocation het-erozygotes are not transmitted to progeny in most diploid species, Dp-Df progeny can be recovered in hexaploid oat due to genetic buffering (Phillips et al. 1971). As the examples below illustrate, these stocks have proven useful in one disease resistance study; they can be powerful tools for physical mapping of genetic markers; and, if the reciprocal translocation is large enough to identify using chromosome banding or it carries dominant phenotypic markers, the Dp-Df lines can be fairly easy to identify.

### 7.7.2 Duplicate-Deficient Segments and Crown Rust Resistance

Crown rust resistance genes *Pc-38*, *Pc-62*, and *Pc-63* are tightly linked, if not allelic (Harder et al. 1980). The oat cultivar Dumont and germplasm line Dif-63 carry *Pc-38* and *Pc-63*, respectively, although the segment carrying the crown rust resistance locus is involved in a reciprocal translocation in Dumont. Wilson and McMullen (1997a) manipulated this translocation difference to produce offspring with zero to four copies of the *Pc-63* gene. In a companion study, Wilson and McMullen (1997b) used hybrids derived from these lines to examine the effects of multiple copies of *Pc-62* and *Pc-63*, as well as a crown rust resistance suppressor gene closely linked to *Pc-38*, on resistance phenotypes. In the process, they determined that adjacent-1 (unbalanced) and alternate (balanced) gametes were transmitted at equal frequencies to selfed offspring.

### 7.7.3 Duplicate-Deficient Lines from Sun II × N770-165-2-1

Dilkova et al. (2000) reported an irradiation-induced translocation between chromosomes 6C and 21 in line N770-165-2-1. This reciprocal translocation is interesting because a large portion of the satellited arm of 21 is interchanged with a small part of the tip of the short arm of 6C. This line was subsequently crossed with the cultivar Sun II to produce a translocation heterozygote, and offspring from the new population of 82 individuals were allowed to self for several generations with single-seed descent (Jellen et al. 2003). Seedlings from at least three $F_4$ seed were then scored for chromosomes 6C and 21 using C-banding to approximately reconstruct the array of $F_3$ genotypes and to identify Dp-Df plants. Although this work is in progress, 44 lines have been scored, with 24 $F_3$s identified as heterozygous for the translocation, 9 homozygous normal (Sun II parental type), 10 homozygous translocation (N770-165-2-1 parent type), and 1 line fixed for the Dp-Df genotype

$6C^{21}/6C^{21}$, 21/21 (Figure 7.9). This line therefore contains a deficiency for the small terminal segment of 6CS and a large duplication for the short arm of 21, including the NOR/satellite region. Other aberrant $F_4$ plants included progeny of three different $F_3$s that produced $6C/6C^{21}$, $21^{6C}/21^{6C}$, a $6C^{21}$ trisomic, a $21^{6C}$ disomic, and a $6C/6C$, $21/21^{6C}$ (E. Jellen, unpublished results).

### 7.7.4    Segregation of 7C and 17 in Populations Derived from Translocation Heterozygotes

The initial oat mapping population derived from a Kanota × Ogle (KO) cross was segregating for the 7C-17 translocation contributed by Ogle. The main genetic manifestation of this segregation was a cluster of RFLP markers on KO linkage group 3, which corresponded to the translocation version of chromosome 7C due to inability of the Mapmaker linkage analysis program to accurately portray linkage relationships in the translocation quadrivalent (O'Donoughue et al. 1995; Fox et al. 2001). Among the recombinant inbred lines (RILs) derived from this cross, nine (KO 6, 7, 16, 17, 19, 21, 47, 71, and 82) were found to be Dp-Df for RFLP markers associated with chromosome 7C-17 segments (W. Wilson, E. Jellen, and M. Sorrells, unpublished data).

A second oat mapping population segregating for the 7C-17 translocation is derived from a Fulghum × Wintok (FW) cross. This population was initially produced by the oat breeding and genetics project at North Carolina State University to study inheritance of winter hardiness and crown freeze tolerance (P. Murphy and D. Livingston, personal communication). As data began to emerge associating presence of the 7C-17 translocation with oat genotypes from cooler climates (Jellen and Beard 2000; Zhou et al. 1999), however, the decision was made to screen a set of FW RILs for presence or absence of the translocation along with the cold stress tolerance traits. Regression analysis revealed a highly significant difference in cold tolerance between $F_{4:6}$ RILs containing the translocation and those having the normal forms of chromosomes 7C and 17 (Santos et al. 2005). Also of interest was the lack of Dp-Df lines, which suggests that Dp-Df stocks may be more difficult to produce in some genetic backgrounds (FW) than in others (KO).

## 7.8   INDUCED VARIATION

### 7.8.1   Ionizing Radiation

The use of ionizing radiation to cause chromosome breakage has been reported in many species, including oat (Aung et al. 1977; Dilkova et al. 2000; Riera-Lizarazu et al. 2000). The goal of such treatment is to produce translocations between nonhomologous chromosomes so that the new chromosome carries genes conditioning characters that are advantageous to the organism in question. As with the production of addition and substitution lines, there are problems associated with this technique.

Chromosome breakage from ionizing radiation occurs randomly. There is thus no guarantee that the chromosome segment having the desired genes will recombine, and even if it does, it is likely to harbor undesirable genes as well. In these lines, the added portion of alien chromosome must be able to compensate for that portion of the native chromosome that is lost. The target genes must also be fully expressed. The chance of obtaining the desired breakage and reunion of chromosomes is very small, and even if the desired genes are incorporated, other rearrangements may also have occurred, as was the case with the 6C-21 translocation induced by irradiation in Wisconsin line N770-165-2-1 (Figure 7.9; Dilkova et al. 2000).

### 7.8.2 Other Methods of Inducing Variation

Chemical induction may be the only practical method of generating a mutation of interest that is absent in the oat gene pool. Sodium azide and ethyl methanesulfonate (EMS) are the most commonly used (Rines 1985). Sodium azide was recently used to generate a short rachilla mutation in *A. sativa* (Burrows et al. 2002).

Transformation is another avenue that is available for improving *A. sativa*. Thus far, biolistic methodologies have proven effective in oat, and a number of researchers have transformed oat using a variety of genes (Somers et al. 1992; Zhang et al. 1999). One particularly interesting example was the introduction of coat protein genes of the P-PAV, MAV-PS1, or NY-RPV strains of BYDV to confer resistance to red leaf disease (McGrath et al. 1997). Another intriguing discovery was that oat transformation results in large-scale chromosome rearrangement (Svitashev et al. 2000).

## 7.9 FUTURE USES OF *AVENA* GENETIC RESOURCES

### 7.9.1 Adding Value to the Oat Crop

The value of the oat crop could be increased by exploiting the genetic variation in *Avena* relatives in selective breeding for a number of biochemical traits. Some of these traits, such as seed bran, oil, protein, antioxidant, and -glucan concentration, are potentially valuable in enhancing human nutrition. Elevating leaf and root saponin levels may enhance the plant's defensive capability against pathogens or pests while creating a market for the oat crop as a source of these compounds for commercial endeavors, including the biomedical industry. Quality and quantity of industrially important starches in oat grains may also be altered through selective breeding. Webster (1996) identified the development of two separate market classes of oat varieties as an important first step: one set for human food, perhaps with elevated -glucans and antioxidants, and the other category for animal feed, incorporating increased fat content.

A number of research groups are studying oat antioxidants for their potential in food stabilization and in enhancing human health. Several writers have recognized and discussed the importance of these compounds, as well as methodologies for measuring antioxidant levels in oat kernels and hulls (Lasztity et al. 1980; Collins 1986; Dimberg et al. 1993). *In vivo* studies of oat antioxidant effects by Rezar et al. (2003) showed that dietary supplementation with oats significantly decreased oxidative stress in swine. In addition, Liu et al. (2002) reported *in vitro* evidence for potential anti-inflammatory and antiatherogenic activities of oat phenolic avenanthramides. Lasztity et al. (1980) measured significant variation in tocopherol quantity and quality among oat cultivars, suggesting that variation should similarly be present in the *Avena* gene pool for other elements of the multi-faceted oat antioxidant complex (Webster 1996).

Another interesting area for potentially increasing the value of oat involves saponin production. Saponins are natural surfactants with potent antifungal properties and a variety of commercial uses (Osbourn 2003). The saponins include triterpenoidal and steroidal types, the former being common in dicots and the latter common in dicots as well as in some families of nongramineous monocots (Hostettman and Marston 1995). Unlike other cereal crops, however, oats synthesize several triterpenoid avenacins (Crombie and Crombie 1986) as well as steroidal avenacosides A and B (Tschesche et al. 1969; Tschesche and Lauven 1971). The role of root-produced avenacins in resisting attack by soil fungal pathogens was first described by Maizel et al. (1964) and is becoming increasingly clear, as at least seven nonallelic, saponin-deficient (*sad*) *A. strigosa* mutants susceptible to wheat take-all disease have been identified (Trojanowska et al. 2001), and natural variation for avenacin content among diploid *Avena* species has been documented (Osbourn et al. 1994). Genes for biosynthesis of avenacins could prove valuable in transferring resistance via transformation to cereals such as wheat that are susceptible to take-all disease (Osbourn 2003).

The soluble fraction of -glucan in oat kernels is another attractive trait for genetic manipulation, with higher levels being desirable for human-consumed oats due to its lowering effect on serum cholesterol (Braaten et al. 1994; Pomeroy et al. 2001), and reduced levels being desirable in feed oats (Peterson 2001). Peterson (1991) determined that oat -glucan concentration varies among cultivars and environments, but that heritability for this trait should be sufficiently high so that progress can be made in selection for high- or low--glucan cultivars. Within the wild *Avena* gene pool, Welch et al. (2000) reported groat -glucan levels ranging from 2.3 to 11.3% on a dry matter basis.

### 7.9.2   Expanding the Oat Crop into New Production Regions

There is considerable interest in extending the range of winter oats into colder climates through the incorporation of genes from the only perennial oat species, *A. macrostachya* (P. Murphy and D. Livingston, personal communication). Thus far, efforts have been unfruitful due to the apparent evolutionary distance between *A. sativa* and *A. macrostachya*, although further hybridization efforts using other *Avena* species in bridging crosses might prove fruitful.

A small number of studies suggest that the hardy perennial tall oatgrass (*Arrhenatherum elatius* L.) might be a close enough relative to warrant further investigation as a source of cold tolerance for improving the winter hardiness of oat. Gervais (1983) and Mitchell (2000) attempted intergeneric hybridization using oats, with the latter reporting a single putative Ogle × PI 253292 hybrid (0.02% success rate) that senesced on embryo rescue media before transplantation. Katsiotis et al. (1996, 2000) reported on repetitive DNA elements common to *Arrhenatherum* and the *Avena* A genome that were either absent, less abundant, or polymorphic in the C genome of *Avena*.

There has also been growing interest in expanding the range of oat into the tropics, for example, in Brazil (L. Federizzi, personal communication). Greater disease pressure, however, will require increased attention to incorporation of genes for horizontal rust resistance, slow-rusting phenotypes. Another intriguing possibility, with the advent of the oat–maize hybridization system and ongoing efforts to produce recombinant lines via irradiation (H. Rines, personal communication), is the incorporation of C4 pathway genes from maize.

## REFERENCES

Ananiev, E.V., Vales, M.I., Phillips, R.L., and Rines, H.W. 2002. Isolation of A/D and C genome specific dispersed and clustered repetitive DNA sequences from *Avena sativa*. *Genome* 45: 431–441.

Ananiev, E.V., Riera-Lizarazu, O., Rines, H.W., and Phillips, R.L. 1997. Oat-maize chromosome addition lines: a new system for mapping the maize genome. *Proc. Natl. Acad. Sci. U.S.A.* 94: 3524–3529.

Aung, T., Chong, J., and Leggett, J.M. 1996. The transfer of crown rust resistance gene Pc94 from a wild diploid to cultivated hexaploid oat. In *Proceedings of the 9th European and Mediterranean Cereal Rusts and Powdery Mildews Conference*, Lunteren, The Netherlands, p. 3.

Aung, T., Thomas, H., and Jones I.T. 1977. The transfer of the gene for mildew resistance from *Avena barbata* (4x) into the cultivated oat *A. sativa* by an induced translocation. *Euphytica* 26: 623–632.

Bacon, R.K. 1991. Registration of 'Ozark' oat. *Crop Sci.* 31: 1383.

Baum, B.R. 1977. *Oats: Wild and Cultivated, a Monograph of the Genus Avena L. (Poaceae)*. Supply & Services Canada, Ottawa, chap. 2.

Baum, B.R. and Fedak, G. 1985a. *Avena atlantica*, a new diploid species of the oat genus from Morocco. *Can. J. Bot.* 63: 1057–1060.

Baum, B.R. and Fedak, G. 1985b. A new tetraploid species of *Avena* discovered in Morocco. *Can. J. Bot.* 63: 1379–1385.

Baum, B.R. and Rajhathy, T. 1976. A study of *Avena macrostachya*. *Can. J. Bot.* 54: 2434–2439.

Baum, B.R., Rajhathy, T., and Sampson, D.R. 1973. An important new diploid *Avena* species discovered on the Canary Islands. *Can. J. Bot.* 51: 759–762.

Bhatti, I.M. 1972. Cytogenetic studies of the genomic structure in *Avena*. Ph.D thesis, University of Wales, Aberystwyth, U.K.

Braaten, J.T., Wood, P.J., Scott, F.W., Wolynetz, M.S., Lowe, M.K., Bradley-White, P., and Collins, M.W. 1994. Oat beta-glucan reduces blood cholesterol concentration in hypercholesterolemic subjects. *Eur. J. Clin. Nutr.* 48: 465–474.

Burrows, V.D., Konzak, C.F., McDiarmid, G., and Dey, J. 2002. A naked oat mutant with very short rachillas. *Can. J. Plant Sci.* 82: 83–84.

Chang, T.D. and Sadanaga, K. 1964. Breeding behavior, morphology, karyotype and interesting results of six monosomes in *Avena sativa* L. *Crop Sci.* 4: 609–613.

Chen, Q. and Armstrong, K. 1994. Genomic *in situ* hybridization in *Avena sativa*. *Genome* 37: 607–612.

Clayton, W.D. and Renvoize, S.A. 1986. *Genera Gramineum, Grasses of the World*, Kew Bulletin Additional Series XIII. Her Majesty's Stationary Office, London.

Coffman, F.A. 1961. *Oats and Oat Improvement*, F.A. Coffman, Ed. ASA, Madison, WI.

Collins, F.W. 1986. Oat phenolics: structure, occurrence and function. In *Oats: Chemistry and Technology*, F.H. Webster, Ed. AACC, St. Paul, MN, pp. 227–295.

Comeau, A. 1984. Barley yellow dwarf virus resistance in the genus *Avena*. *Euphytica* 33: 49–55.

Crombie, W.M.L. and Crombie, L. 1986. Distribution of avenacins A-1, A-2, B-1 and B-2 in oat roots: their fungicidal activity towards "take-all" fungus. *Phytochemistry* 25: 2069–2073.

Dilkova, M., Jellen, E.N., and Forsberg, R.A. 2000. C-banded karyotypes and meiotic abnormalities in germplasm derived from interploidy crosses in *Avena*. *Euphytica* 111: 175–184.

Dimberg, L.H., Theander, V., and Lingnert, H. 1993. Avenanthramides: a group of phenolic antioxidants in oats. *Cereal Chem.* 70: 637–641.

Endo, T.R. 1988. Induction of chromosomal structural changes by a chromosome of *Aegilops cylindrica* L. in common wheat. *J. Hered.* 79: 366–370.

Endo, T.R. and Gill, B.S. 1996. The deletion stocks of common wheat. *J. Hered.* 87: 295–307.

Fominaya, A., Vega, C., and Ferrer, E. 1988a. Giemsa C-banded karyotypes of *Avena* species. *Genome* 30: 627–632.

Fominaya, A., Vega, C., and Ferrer, E. 1988b. C-banding and nucleolar activity of tetraploid *Avena* species. *Genome* 30: 633–638.

Fominaya, A., Hueros, G., Loarce, Y., and Ferrer, E. 1995. Chromosomal distribution of a repeated DNA sequence from C-genome heterochromatin and the identification of a new ribosomal DNA locus in the *Avena* genus. *Genome* 38: 548–557.

Fox, S.L., Jellen, E.N., Kianian, S.F., Rines, H.W., and Phillips, R.L. 2001. Assignment of RFLP linkage groups to chromosomes using monosomic F1 analysis in hexaploid oat. *Theor. Appl. Genet.* 102: 320–326.

Gervais, C. 1983. Wide hybridization attempts in the tribe Avenae. *Nees. Ber. Schweiz. Bot. Ges. Bull. Soc. Bot. Suisse* 93: 195–212.

Gill, K.S., Gill, B.S., and Endo, T.R. 1993. A chromosome-region specific mapping strategy reveals gene-rich telomeric ends in wheat. *Chromosoma* 102: 374–381.

Griffiths, N.A.J. and Thomas, H. 1982. *Rpt. Welsh Plant Breed. Station*, Aberystwyth, Wales, pp. 145–146.

Hacker, J.B. 1965. The inheritance of chromosome deficiency in *Avena sativa* monosomics. *Can. J. Genet. Cytol.* 7: 316–327.

Hansen, J.R. and Renfrew, J.M. 1978. Palaeolithic-neolithic seed remains at Franchthi Cave, Greece. *Nature* 271: 349–352.

Harder, D.E., McKenzie, R.I.H., and Martens, J.W. 1980. Inheritance of crown rust resistance in three accessions of *Avena sterilis*. *Can. J. Genet. Cytol.* 22: 27–33.

Holden, J.W. 1966. Species relationships in the Avenae. *Chromosoma* 20: 75–124.

Hoppe, H.D. and Kummer, M. 1991. New productive hexaploid derivatives after introgression of *A. pilosa* features. *Vortr. Pflanzenzuchtg.* 20: 56–61.

Hoppe, H.D. and Pohler, W. 1989. Hybrids between *Avena barbata* and *A. macrostachya*. *Cereal Res. Commun.* 16: 231–235.

Hostettman, K.A. and Marston, A. 1995. *Saponins*. Cambridge University Press, Cambridge, U.K.

Hutchinson, J. and Postoyko, J. 1986. C-banding of *Avena* species. In *Genetic Manipulation in Plant Breeding, Proceedings of the International Symposium on. Eucarpia*, W. Horn, C.J. Jensen, W. Odenbach, and O. Schneider, Eds. Walter de Cruyter and Co., Berlin, pp. 157–160.

Irigoyen, M.L., Linares, C., Ferrer, E., and Fominaya, A. 2002. Fluorescence *in situ* hybridization mapping of *Avena sativa* L. cv. Sun II and its monosomic lines using cloned repetitive DNA sequences. *Genome* 45: 1230–1237.

Irigoyen, M.L., Loarce, Y., Linares, C., Ferrer, E., Leggett, M., and Fominaya, A. 2001. Discrimination of the closely related A and B genomes in AABB tetraploid species of *Avena*. *Theor. Appl. Genet.* 103: 1160–1166.

Jauhar, P.P. 1975a. Chromosome relationships between *Lolium* and *Festuca* (Gramineae). *Chromosoma* 52: 103–121.

Jauhar, P.P. 1975b. Genetic regulation of diploid-like chromosome pairing in the hexaploid species, *Festuca arundinacea* Schreb. and *F. rubra* L. (Gramineae). *Chromosoma* 52: 363–382.

Jauhar, P.P. 1977. Genetic regulation of diploid-like chromosome pairing in *Avena*. *Theor. Appl. Genet.* 49: 287–295.

Jauhar, P.P. 1993. *Cytogenetics of the Festuca-Lolium Complex*. Springer-Verlag, Berlin.

Jellen, E.N. and Beard, J. 2000. Geographical distribution of a chromosome 7C and 17 intergenomic translocation in cultivated oat. *Crop Sci.* 40: 256–263.

Jellen, E.N., Gardunia, B.W., Durrant, J.D., Jarvis, D., Ricks, S.J., Ames, D.C., Raymond, F.D., and Stevens, M.R. 2003. Genetic behavior of translocations, their geographic and taxonomic distribution, and facilitated translocation mapping using the QuadMap program in hexaploid oats (*Avena*). In *Plant and Animal Genome XI*, San Diego, January 11–14.

Jellen, E.N. and Gill, B.S. 1996. C-banding variation in the Moroccan oat species *Avena agadiriana* (2n = 4x = 28). *Theor. Appl. Genet.* 92: 726–732.

Jellen, E.N., Gill, B.S., and Cox, T.S. 1994. Genomic *in situ* hybridization differentiates between A/D- and C-genome chromatin and detects intergenomic translocations in polyploid oat species (genus *Avena*). *Genome* 37: 613–618.

Jellen, E.N. and Ladizinsky, G. 2000. Giemsa C-banding in *Avena insularis* Ladizinsky. *Genet. Res. Crop Evol.* 47: 227–230.

Jellen, E.N., Phillips, R.L., and Rines, H.W. 1993a. C-banded karyotypes and polymorphisms in hexaploid oat accessions (*Avena* spp.) using Wright's stain. *Genome* 36: 1129–1137.

Jellen, E.N., Rines, H.W., Fox, S.L., Davis, D.W., Phillips, R.L., and Gill, B.S. 1997. Characterization of 'Sun II' oat monosomics through C-banding and identification of eight new 'Sun II' monosomics. *Theor. Appl. Genet.* 95: 1190–1195.

Jellen, E.N., Rooney, W.L., Phillips, R.L., and Rines, H.W. 1993b. Characterization of the hexaploid oat *Avena byzantina* cv. Kanota monosomic series using C-banding and RFLP's. *Genome* 36: 962–970.

Jones, M., Rees, H., and Jenkins, G. 1989. Synaptonemal complex formation in *Avena* polyploids. *Heredity* 63: 209–219.

Joshi, A.B. and Howard, H.L. 1955. Meiotic irregularities in hexaploid oats. IV. Hybrids between *Avena sativa* (spring and winter varieties), *A. fatua, A. sterilis, A. byzantina* and *A. nuda*. *J. Agric. Sci.* 46:183–190.

Katsiotis, A., Hagidimitriou, M., and Heslop-Harrison, J.S. 1997. The close relationship between the A and B genomes in *Avena* L. (Poaceae) determined by molecular cytogenetic analysis of total genomic, tandemly and dispersed repetitive DNA sequences. *Ann. Bot.* 79: 103–109.

Katsiotis, A., Loukas, M., and Heslop-Harrison, J.S. 2000. Repetitive DNA, genome and species relationships in *Avena* and *Arrhenatherum* (Poaceae). *Ann. Bot.* 86: 1135–1142.

Katsiotis, A., Schmidt, T., and Heslop-Harrison, J.S. 1996. Chromosomal and genomic organization of *Ty1-copia*-like retrotransposon sequences in the genus *Avena*. *Genome* 39: 410–417.

Kianian, S.F., Wu, B.C., Fox, S.L., Rines, H.W., and Phillips, R.L. 1997. Aneuploid marker assignment in hexaploid oat with the C genome as a reference for determining remnant homoeology. *Genome* 40: 386–396.

Kummer, M. and Miksch, G. 1977. Successful hybridization between species *Avena pilosa* M. Bieb. and *Avena magna* Murphy et Terrel. *Cereal Res. Commun.* 5: 249–254.

Kynast, R.G., Riera-Lizarazu, O., Vales, M.I., Okagaki, R.J., Maquieira, S.B., Chen, G., Ananiev, E.V., Odland, W.E., Russell, C.D., Stec, A.O., Livingston, S.M., Zaia, H.A., Rines, H.W., and Phillips, R.L. 2001. A complete set of maize individual chromosome additions to the oat genome. *Plant Physiol.* 125: 1216–1227.

Ladizinsky, G. 1968. Cytogenetic and Evolutionary Study in the Genus *Avena*. Ph.D. thesis, The Hebrew University of Jerusalem, Rehovot, Israel.

Ladizinsky, G. 1969. New evidence on the origin of the hexaploid oats. *Evolution* 23: 676–684.

Ladizinsky, G. 1970. Chromosome rearrangements in the hexaploid oats. *Heredity* 25: 457–461.

Ladizinsky, G. 1971a. *Avena murphyi* a new tetraploid species from southern Spain. *Isr. J. Bot.* 20: 24–27.

Ladizinsky, G. 1971b. *Avena prostrata*: a new diploid species of oats. *Isr. J. Bot.* 20: 297–301.

Ladizinsky, G. 1973a. The cytogenetic position of *Avena prostrata* among the diploid oats. *Can. J. Genet. Cytol.* 15: 443–450.

Ladizinsky, G. 1973b. Genetic control of bivalent pairing in the *Avena strigosa* polyploid complex. *Chromosoma* 42: 105–110.

Ladizinsky, G. 1974. Cytogenetic relationships between the diploid oat *A. prostrata* and the tetraploids *A. barbata*, *A. magna* and *A. murphyi*. *Can. J. Genet. Cytol.* 16: 105–112.

Ladizinsky, G. 1988. Biological species and wild genetic resources in *Avena*. In *Proceedings of the 3rd International Oat Conference*, Lund, B. Mattsson, and R. Lyhagen, Eds., Svalof AB, Sweden, pp. 76–86.

Ladizinsky, G. 1995. Domestication via hybridization of the wild tetraploid oats *Avena magna* and *A. murphyi*. *Theor. Appl. Genet.* 91: 639–646.

Ladizinsky, G. 1998. A new species of *Avena* from Sicily, possibly the tetraploid progenitor of hexaploid oats. *Genet. Res. Crop Evol.* 45: 263–269.

Ladizinsky, G. 1999. Cytogenetic relationships between *Avena insularis* (2n = 28) and both *A. strigosa* (2n = 14) and *A. murphyi* (2n = 28). *Genet. Res. Crop. Evol.* 46: 501–504.

Ladizinsky, G. 2000. A synthetic hexaploid (2n = 42) oat from the cross of *Avena strigosa* (2n = 14) and domesticated *A. magna* (2n = 28). *Euphytica* 116: 231–235.

Ladizinsky, G. and Jellen, E.N. 2003. Cytogenetic affinities between populations of *Avena insularis* Ladizinsky from Sicily and Tunisia. *Genet. Res. Crop Evol.* 50: 11–15.

Ladizinsky, G. and Zohary. D. 1971. Notes on species delimitation, species relationships and polyploidy in *Avena* L. *Euphytica* 20: 380–395.

Lasztity, R., Berndorfer-Kraszner, E., and Huszar, M. 1980. On the presence and distribution of some bioreactive agents in oat varieties. In *Cereals for Food and Beverages. Recent Progress in Cereal Chemistry*, G. Inglett and L. Munck, Eds. Academic Press, New York, pp. 429–455.

Leggett, J.M. 1980. Chromosome relationships and morphological comparisons between the diploid oats *A. prostrata*, *A.canariensis* and the tetraploid *A. maroccana*. *Can. J. Genet. Cytol.* 22: 287–294.

Leggett, J.M. 1983. Chromosome relationships and morphological comparisons in the hybrids *Avena hybrida* × *A. sativa* and *A. hybrida* × *A. maroccana*. *Can. J. Genet. Cytol.* 25: 255–260.

Leggett, J.M. 1984a. Morphology and metaphase pairing in three *Avena* hybrids. *Can. J. Genet. Cytol.* 26: 641–645.

Leggett, J.M. 1984b. Cytoplasmic substitutions involving six *Avena* species. *Can. J. Genet. Cytol.* 26: 698–700.

Leggett, J.M. 1985. Interspecific hybrids involving the perennial oat species *Avena macrostachya*. *Can. J. Genet. Cytol.* 27: 29–32.

Leggett, J.M. 1987. Interspecific hybrids involving the recently described diploid taxon *Avena atlantica*. *Genome* 29: 361–364.

Leggett, J.M. 1989a. Inter- and intra-specific hybrids involving the tetraploid species *Avena agadiriana* Baum et Fedak sp. Nov. (2n = 4x = 28). In *Proceedings of the 3rd International Oat Conference*, Lund, B. Mattsson and R. Lyhagen, Eds., Svalof AB, Sweden, pp. 62–67.

Leggett, J.M. 1989b. Interspecific diploid hybrids in *Avena*. *Genome* 32: 346–348.

Leggett, J.M. 1992. Classification and speciation in *Avena*. In *Oat Science and Technology*, Agronomy Monograph 33, H.G. Marshall and M.E. Sorrells, Eds. ASA and CSSA, Madison, WI, pp. 29–52.

Leggett, J.M. and Markhand, G.S. 1995a. The genomic identification of some monosomics of *Avena sativa* L. cv. Sun II using genomic *in situ* hybridization. *Genome* 38: 747–751.

Leggett, J.M. and Markhand, G.S. 1995b. The genomic structure of *Avena* revealed by GISH. In *Kew Chromosome Conference IV*, P.E. Brandham and M.D. Bennett, Eds., London, pp. 133–139.

Leggett, J.M., Thomas, H.M., Meredith, M.R., Humphreys, M.W., Morgan, W.G., Thomas, H., and King, I.P. 1994. Intergenomic translocations and the genomic composition of *Avena maroccana* Gdgr. revealed by FISH. *Chromosome Res.* 2: 163–164.

Leggett, J.M., Thomas, H., and Naqvi, Z. 1992. Cytogenetic studies in *Avena*. In *Proceedings of the 4th International Oat Conference*, Vol. III, A.R. Barr, R.J. McLean, J.D. Oates, G. Roberts, J. Rose, K. Saint, and S. Tasker, Eds., Adelaide, Australia, pp. 123–127.

Lewis, H. 1966. Speciation in flowering plants. *Science* 152: 167–172.

Linares, C., Ferrer, E., and Fominaya, A. 1998. Discrimination of the closely related A and D genomes of the hexaploid oat *Avena sativa* L. *Proc. Natl. Acad. Sci. U.S.A.* 95: 12450–12455.

Linares, C., Irigoyen, M.L., and Fominaya, A. 2000. Identification of C-genome chromosomes involved in intergenomic translocations in *Avena sativa* L., using cloned repetitive DNA sequences. *Theor. Appl. Genet.* 100: 353–360.

Linares, C., Gonzalez, J., Ferrer, E., and Fominaya, A. 1996. The use of double fluorescence *in situ* hybridization to physically map the positions of 5S rDNA genes in relation to the chromosomal location of 18S-5.8S-26S rDNA and a C genome specific DNA sequence in the genus *Avena*. *Genome* 39: 535–542.

Linares, C., Loarce, Y., Serna, A., and Fominaya, A. 2001. Isolation and characterization of two novel retrotransposons of the Ty1-*copia* group in oat genomes. *Chromosoma* 110: 115–123.

Linares, C., Serna, A., and Fominaya, A. 1999. Chromosomal organization of a sequence related to LTR-like elements of Ty1-*copia* retrotransposons in *Avena* species. *Genome* 42: 706–713.

Linares, C., Vega, C., Ferrer, E., and Fominaya, A. 1992. Identification of C-banded chromosomes in meiosis and the analysis of nucleolar activity in *Avena byzantina* C. Koch cv. 'Kanota'. *Theor. Appl. Genet.* 83: 650–654.

Liu, L., Zubik, L., Marko, M.G., Collins, F.W., and Meydani, M. 2002. The antiatherogenic and anti-inflammatory potential of oat phenolics. *FASEB J.* 16: A223.

Maizel, J.W., Buckhardt, H.J., and Mitchell, H.K. 1964. Avenacin, an antimicrobial substance isolated from *Avena sativa*. Isolation and antimicrobial activity. *Biochemistry* 3: 424–426.

Markhand, G.S. 1996. Molecular Cytogenetic Studies of Genomic Structure in the Genus *Avena* L. Ph.D. thesis, University of Wales, Aberystwyth.

Markhand, G.S. and Leggett, J.M. 1996. The genomes of *A. lusitanica*, *A. hispanica* and *A. matritensis* confirmed using GISH. In *Proceedings of the 5th International Oat Conference and VIIth International Barley Genetics Symposium*, A. Slinkard, G. Scoles, and B. Rossnagel, Eds., Saskatoon, Saskatchewan, Canada, p. 347.

Marshall, H.G. and Myers, W.M. 1961. A cytogenetic study of certain interspecific *Avena* hybrids and the inheritance of resistance in diploid and tetraploid varieties to race of crown rust. *Crop Sci.* 1: 29–34.

Martens, J.W., McKenzie, R.I.H., and Harder, D.E. 1980. Resistance to *Puccinia graminis* Avenae and *P. coronata* Avenae in the wild and cultivated *Avena* populations of Iran, Iraq and Turkey. *Can. J. Genet. Cytol.* 22: 641–649.

McGrath, P.F., Vincent, J.R., Lei, C.H., Pawlowski, W.P., Torbert, K.A., Gu, W., Kaeppler, H.F., Wan, Y., Lemaux, P.G., Rines, H.R., Somers, D.A., Larkins, B.A., and Lister, R.M. 1997. Coat protein-mediated resistance to isolates of barley yellow dwarf in oats and barley. *Eur. J. Plant Pathol.* 103: 695–710.

McMullen, M.S., Phillips, R.L., and Stuthman, D.D. 1982. Meiotic irregularities in *Avena sativa* L./*A. sterilis* L. hybrids and breeding implications. *Crop Sci.* 22: 890–897.

Mitchell, C.C. 2000. Karyotypes of C-Banded *Arrhenatherum* Species and Intergeneric Hybridization Attempts Involving Oat. M.S. thesis, Brigham Young University, Provo, UT.

Murphy, H.C., Sadanaga, K., Zillinsky, F.J., Terrell, E.E., and Smith, R.T. 1968. *Avena magna*: an important new tetraploid species of oats. *Science* 159: 103–104.

Nishiyama, I. and Yabuno, T. 1975. Meiotic chromosome pairing in two interspecific hybrids and a criticism of the evolutionary relationship of diploid *Avena*. *Jap. J. Genet.* 50: 443–451.

Nishiyama, I., Yabuno, T., and Taira, T. 1989. Genomic affinity relationships in the genus *Avena*. *Plant Breed.* 102: 22–30.

O'Donoughue, L.S., Kianian, S.F., Rayapati, P.J., Penner, G.A., Sorrells, M.E., Tanksley, S.D., Phillips, R.L., Rines, H.W., Lee, M., Fedak, G., Molnar, S.J., Hoffman, D., Salas, C.A., Wu, B., Autrique, E., and Van Deynze, A. 1995. A molecular linkage map of cultivated oat. *Genome* 38: 368–380.

O'Donoughue, L.S., Wang, Z., Roder, M., Kneen, B., Leggett, M., Sorrells, M.E., and Tanksley, S.D. 1992. An RFLP-based linkage map of oats based on a cross between two diploid taxa (*Avena atlantica* × *A. hirtula*). *Genome* 35: 765–771.

Ohm, H.W., Cook, V.M., Shaner, G.E., Buechley, G.C., Sharma, H., Perry, K., Racliffe, R.H., and Cambron, S.E. 2000. Registration of 'Jay' spring oat. *Crop Sci.* 40: 569.

Osbourn, A.E. 2003. Saponins in cereals. *Phytochemistry* 62: 1–4.

Osbourn, A.E., Clarke, B.R., Lunness, P., Scott, P.R., and Daniels, M.J. 1994. An oat species lacking avenacins is susceptible to inflection by *Gaeumannomyces graminis* var. *tritici*. *Physiol. Mol. Plant Pathol.* 45: 457–467.

Peterson, D.M. 1991. Genotype and environment effects on oat beta-glucan concentration. *Crop Sci.* 31: 1517–1520.

Peterson, D.M. 2001. Oat antioxidants. *J. Cereal Sci.* 33: 115–129.

Phillips, R.L., Burnham, R.C., and Patterson, E.B. 1971. Advantages of chromosomal interchanges that generate haplo-viable deficiency-duplications. *Crop Sci.* 11: 525–528.

Pohler, W. and Hoppe, H.-D. 1991. Homeology between the chromosomes of *Avena macrostachya* and the *Avena* C genome. *Plant Breed.* 106: 250–253.

Pomeroy, S., Tupper, R., Cehun-Aders, M., and Nestel, P. 2001. Oat beta-glucan lowers total and LDL-cholesterol. *Aust. J. Nutr. Diet.* 58: 51–55.

Portyanko, V.A., Hoffman, D.L., Lee, M., and Holland, J.B. 2001. A linkage map of hexaploid oat based on grass anchor DNA clones and its relationship to other oat maps. *Genome* 44: 249–265.

Postoyko, J. and Hutchinson, J. 1986. The identification of *Avena* chromosomes by means of C-banding. In *Proceedings of the 2nd International Oats Conference*, D.A. Lawes and H. Thomas, Eds. Martinus Nijhoff Publ., Amsterdam, pp. 50–51.

Powell, W. 1980. Chromosome Manipulation in *Avena*. M.Sc. thesis, University of Wales, Aberystwyth.

Rajhathy, T. 1961. Chromosome differentiation and speciation in diploid *Avena*. *Can. J. Genet. Cytol.* 3: 372–377.

Rajhathy, T. 1966. Evidence and a hypothesis for the origin of the C genome of hexaploid *Avena*. *Can. J. Genet. Cytol.* 8: 774–779.

Rajhathy, T. 1971. The alloploid model in *Avena*. *Stadler Genet. Symp.* 3: 71–87.

Rajhathy, T. and Baum, B.R. 1972. *Avena damascena*: a new diploid oat species. *Can. J. Genet. Cytol.* 14: 645–654.

Rajhathy, T. and Dyck, P.L. 1963. Chromosomal differentiation and speciation in diploid *Avena*. II. Karyotype of *A. pilosa*. *Can. J. Genet. Cytol.* 5: 175–179.

Rajhathy, T. and Morrison, J.W. 1959. Chromosome morphology in the genus *Avena*. *Can. J. Bot.* 37: 331–337.

Rajhathy, T. and Morrison, J.W. 1960. Genome homology in the genus *Avena*. *Can. J. Genet. Cytol.* 2: 278–285.

Rajhathy, T. and Sadasivaiah. R.S. 1969. The cytogenetic status of *A. magna*. *Can. J. Genet. Cytol.* 11: 77–85.

Rajhathy, T. and Thomas, H. 1967. Chromosome differentiation and speciation in diploid *Avena*. III. Mediterranean wild populations. *Can. J. Genet. Cytol.* 9: 52–68.

Rajhathy, T. and Thomas, H. 1972. Genetic control of chromosome pairing in hexaploid oats. *Nat. New Biol.* 239: 217–219.

Rajhathy, T. and Thomas, H. 1974. *Cytogenetics of Oats*, Miscellaneous Publication of the Genetics Society of Canada 2. Genetics Society of Canada, Ottawa, Ontario.

Rayapati, P.J., Gregory, J.W., Lee, M., and Wise, R.P. 1994. A linkage map of diploid *Avena* based on RFLP loci and a locus conferring resistance to nine isolates of *Puccinia coronata* var. *avenae*. *Theor. Appl. Genet.* 89: 831–837.

Rezar, V., Pajk, T., Logar, R.M., Janezic, V.J., Salobir, K., Oresnik, A., and Salobir, J. 2003. Wheat bran and oat bran effectively reduce oxidative stress induced by high-fat diets in pigs. *Ann. Nutr. Metab.* 47: 78–84.

Riera-Lizarazu, O., Vales, M.I., Ananiev, E.V., Rines, H.W., and Phillips, R.L. 2000. Production and characterization of maize chromosome 9 radiation hybrids derived from an oat-maize addition line. *Genetics* 156: 327–333.

Riley, R., Kimber, G., and Chapman, V. 1961. Origin of genetic control of diploid like behaviour of polyploid wheat. *J. Hered.* 52: 22–25.

Rines, H.W. 1985. Sodium azide mutagenesis in diploid and hexaploid oats and comparison with ethyl methanesulphonate treatments. *Env. Exp. Bot.* 25: 7–16.

Rines, H.W. and Dahleen, L.S. 1990. Haploid oat plants produced by application of maize pollen to emasculated oat florets. *Crop Sci.* 30: 1073–1078.

Rooney, W.L., Jellen, E.N., Phillips, R.L., Rines, H.W., and Kianian, S.F. 1994. Identification of homoeologous chromosomes in hexaploid oat (*A. byzantina* cv. Kanota) using monosomics and RFLP analysis. *Theor. Appl. Genet.* 89: 329–335.

Rossnagel, B.G. 1999. CDC Baler. *Oat Newsletter* 45.

Rothman, P.G. 1984. Registration of four stem rust and crown rust resistant oat germplasm lines. *Crop Sci.* 24: 1217.

Sadanaga, K., Zillinsky, F.J., Murphy, H.C., and Smith, R.T. 1968. Chromosome associations in triploid, tetraploid and pentaploid hybrids of *Avena magna* (2n = 28). *Crop Sci.* 8: 594–597.

Sadasivaiah, R.S. and Rajhathy, T. 1968. Genome relationships in 4x *Avena*. *Can. J. Genet. Cytol.* 10: 655–659.

Santos, A.G., Livingston, D.P., Jellen, E.N., Wooten, D.R., and Murphy, J.P. 2005. A cytological marker associated with winterhardiness in oat. *Crop Sci.* 45 (in press).

Serizawa, N., Nasuda, S., Shi, F., Endo, T.R., Prodanovic, S., Schubert, I., and Kunzel, G. 2001. Deletion-based physical mapping of barley chromosome 7H. *Theor. Appl. Genet.* 103: 827–834.

Sharma, D.C. and Forsberg, R.A. 1974. Alien chromosome substitutions: a cause of instability for leaf rust resistance oats. *Crop Sci.* 14: 533–536.

Singh, R.J. and Kolb, F.L. 1991. Chromosomal interchanges in six hexaploid oat genotypes. *Crop Sci.* 31: 726–729.

Somers, D.A., Rines, H.W., Gu, W., Kaeppler, H.F., and Bushnell, W.R. 1992. Fertile, transgenic oat plants. *Bio/Technology* 10: 1589–1594.

Stuthman, D.D., Wilcoxson, R.D., and Rines, H.W. 1990. Registration of 'Starter' oat. *Crop Sci.* 30: 1365.

Svitashev, S., Ananiev, E., Pawlowski, W.P., and Somers, D.A. 2000. Association of transgene integration sites with chromosome rearrangements in hexaploid oat. *Theor. Appl. Genet.* 100: 872–880.

Thomas, H. 1968. The addition of single chromosomes of *A. hirtula* to the cultivated oat *A. sativa. Can. J. Genet. Cytol.* 10: 551–563.

Thomas, H. 1970. Chromosome relationships between the cultivated oat *Avena sativa* (6x) and *A. ventricosa* (2x). *Can. J. Genet. Cytol.* 12: 36–43.

Thomas, H. 1992. Cytogenetics of *Avena*. In *Oat Science and Technology*, Agronomy Monograph 33, H.G. Marshall and M.E. Sorrells, Eds. ASA and CSSA, Madison, WI, pp. 473–507.

Thomas, H., Haki, J.M., and Aurangzeb, S. 1980a. The introgression of characters of the wild oat *Avena magna* (2n = 6x = 42) into the cultivated oat *A. sativa* (2n = 6x = 42). *Euphytica* 29: 391–399.

Thomas, H. and Jones, M.L. 1965. Chromosomal differentiation in diploid species of *Avena*. *Can. J. Genet. Cytol.* 9: 154–162.

Thomas, H. and Leggett, J.M. 1974. Chromosome relationships between *Avena sativa* and the two diploid species *A. canariensis* and *A. prostrata. Can. J. Genet. Cytol.* 16: 889–894.

Thomas, H., Leggett, J.M., and Jones, I.T. 1975. The addition of a pair of chromosomes from the wild oat *Avena barbata* (2n = 28) into the cultivated oat *Avena sativa* L. (2n = 6x = 42). *Euphytica* 24: 717–724.

Thomas, H., Powell, W., and Aung, T. 1980b. Interfering with regular meiotic behaviour in *Avena sativa* as a method of incorporating the gene for mildew resistance from *A. barbata*. *Euphytica* 29: 635–640.

Thomas, H. and Rajhathy, T. 1967. Chromosome relationships between *Avena sativa* (6x) and *Avena pilosa* (2x). *Can. J. Genet. Cytol.* 9: 154–162.

Trojanowska, M.R., Osbourn, A.E., Daniels, M.J., and Threlfall, D.R. 2001. Investigation of avenacin-deficient mutants of *Avena strigosa*. *Phytochemistry* 56: 121–129.

Tschesche, R. and Lauven, P. 1971. Avenacosid B, ein sweites bisdesmosidisches Steroidsaponin aus *Avena sativa*. *Chem. Ber.* 104: 3549–3555.

Tschesche, R., Tauscher, M., Fehlhaber, H.W., and Wulff, G. 1969. Avenocasid A, ein bisdesmosidisches Steroidsaponin aus *Avena sativa*. *Chem. Ber.* 102: 2072–2082.

Webster, F.H. 1996. Oats. In *Cereal Grain Quality*, R.J. Henry and P.S. Kettlewell, Eds. Chapman & Hall, London, chap. 6.

Welch, R.W., Brown, J.C., and Leggett, J.M. 2000. Interspecific and intraspecific variation in grain and groat characteristics of wild oats (*Avena*) species: very high (1 3),(1 4)-b-D-glucan in an *A. atlantica* genotype. *J. Cereal Sci.* 31: 273–279.

Werner, J.E., Endo, T.R., and Gill, B.S. 1992. Toward a cytogenetically based physical map of the wheat genome. *Proc. Natl. Acad. Sci. U.S.A.* 89: 11307–11311.

Wight, C.P., Tinker, N.A., Kianian, S.F., Sorrells, M.E., O'Donoughue, L.S., Hoffman, D.L., Groh, S., Scoles, G.J., Li, C.D., Webster, F.H., Phillips, R.L., Rines, H.W., Livingston, S.M., Armstrong, K.C., Fedak, G., and Molnar, S.J. 2003. A molecular marker map in 'Kanota' × 'Ogle' hexaploid oat (*Avena* spp.) enhanced by additional markers and a robust framework. *Genome* 46: 28–47.

Wilson, W.F. and McMullen, M.S. 1997a. Recombination between a crown rust resistance locus and an interchange breakpoint in hexaploid oat. *Crop Sci.* 37: 1694–1698.

Wilson, W.F. and McMullen, M.S. 1997b. Dosage dependent genetic suppression of oat crown rust resistance gene *Pc-62*. *Crop Sci.* 37: 1699–1705.

Yang, Q., Hanson, L., Bennett, M.D., and Leitch, I.J. 1999. Genome structure and evolution in the allohexaploid weed *Avena fatua* L. *Genome* 42: 512–518.

Yen, S.-T. and Filion, W.G. 1977. Differential Giemsa staining in plants. V. Two types of constitutive heterochromatin in species of *Avena*. *Can. J. Genet. Cytol.* 19: 739–743.

Zhang, S., Cho, M.-J., Koprek, T., Yun, R., Bregitzer, P., and Lemaux, P.G. 1999. Genetic transformation of commercial cultivars of oat (*Avena sativa* L.) and barley (*Hordeum vulgare* L.) using *in vitro* shoot meristematic cultures derived from germinated seedlings. *Plant Cell Rep.* 18: 959–966.

Zhou, X., Jellen, E.N., and Murphy, J.P. 1999. Progenitor germplasm of domesticated hexaploid oat. *Crop Sci.* 39: 1208–1214.

Zohary, D. and Hopf, M. 1993. *Domestication of Plants in the Old World. The Origin and Spread of Cultivated Plants in West Asia, Europe and the Nile Valley*, 2nd Edition. Claredon/Oxford University Press, Oxford, UK.

# CHAPTER 8

# Utilization of Genetic Resources for Barley Improvement

**Ram J. Singh***

## CONTENTS

8.1 Introduction................................................................................................................234
8.2 Origin of Barley .......................................................................................................234
8.3 Taxonomy of Barley..................................................................................................235
8.4 Barley Germplasm Resources ...................................................................................235
8.5 Genomic Relationships in Barley .............................................................................241
    8.5.1 Diploid Species..............................................................................................241
    8.5.2 Tetraploid Species .........................................................................................241
    8.5.3 Hexaploid Species .........................................................................................242
8.6 Gene Pools of Barley ................................................................................................243
    8.6.1 Primary Gene Pool ........................................................................................243
    8.6.2 Secondary Gene Pool .....................................................................................243
    8.6.3 Tertiary Gene Pool .........................................................................................243
8.7 Germplasm Enhancement ..........................................................................................243
    8.7.1 Conventional Breeding ...................................................................................244
    8.7.2 Haploid Breeding...........................................................................................244
    8.7.3 Tetraploid Breeding .......................................................................................245
    8.7.4 Mutation Breeding..........................................................................................245
    8.7.5 Hybrid Barley Breeding .................................................................................245
    8.7.6 Wide Hybridization ........................................................................................246
        8.7.6.1 Production of Wheat–Barley Addition Lines ...........................246
        8.7.6.2 Production of Other Wheat-Barley Addition Lines...................249
    8.7.7 Somaclonal Variation.....................................................................................249
    8.7.8 Genetic Transformation .................................................................................249
    8.7.9 Exploitation of Apomixis ...............................................................................251
8.8 Conclusions................................................................................................................251
References ...........................................................................................................................251

---

*U.S. government employee whose work is in the public domain.

## 8.1 INTRODUCTION

Barley (*Hordeum vulgare* L., 2n = 14) is an important cereal used for food and feed. It is a rich source of fermentable sugars to produce malt for alcoholic beverages. Barley is one of the most ancient cereal crops, and archaeological study has documented that barley was cultivated as early as 7000 B.C. in Iran (Harlan 1979). In total production, barley is ranked fourth among world cereals after maize (*Zea mays* L.), rice (*Oryza sativa* L.), and wheat (*Triticum aestivum* L.). Barley is cultivated worldwide and is more tolerant to cold, drought, saline, and alkaline soils than other important cereals (Bothmer et al. 1995). It normally grows from 350 to 1500 m above sea level, although Tibetan barleys are cultivated from 2800 to 4050 m above sea level (Shao et al. 1982).

The objective of this chapter is to summarize information on origin, taxonomy, germplasm resources, genomic relationships, and germplasm enhancement of barley. Exploitation of apomixis and creation of genetic variation in barley through somaclonal and transformation methods will be discussed.

## 8.2 ORIGIN OF BARLEY

The center of origin of a crop can be defined as the region where its domestication first took place and where its wild ancestor and the derived cultivated species coexist (Molina-Cano et al. 1999). The origin of two-rowed and six-rowed barleys is controversial. Morphological, cytological, and genetic investigations have revealed that the cultivated two-rowed barley evolved from *Hordeum spontaneum* (C. Koch.) Thell. (Harlan 1979). The immediate ancestor of six-rowed cultivated barley is six-rowed wild-weedy *Hordeum agriocrithon* Aberg, and *H. agriocrithon* was derived from bottle-shaped wild barley (Shao et al. 1982). A recessive gene mutation in two-rowed barley resulted in the six-rowed form. Based on DNA markers closely linked to the *vrs1* locus (row-type gene), Tanno et al. (2002) concluded that six-rowed barleys were derived from two-rowed ones by a mutation at the *vrs1* locus. The dominant gene mutation from six-rowed to two-rowed is possible but is rare in nature. Production of tough-rachis cultivated barley from brittle-rachis wild form needs merely a mutation and fixation in the homozygous condition of one or two genes (Harlan 1979). Mutation and selection for the tough-rachis form by humans may have played a major role in the evolution of cultivated barley. Ladizinsky and Genizi (2001) simulated the evolution of the tough-rachis mutant from the brittle-rachis form. They proposed that the rate of inbreeding has the smallest effect; shifting cultivation has the minimal influence, but number of mutations and selection have played a major role in the establishment of domesticated barley. Based on allelic frequencies of esterase genes in wild and cultivated barley, multiple domestications for barley were suggested.

Harlan (1979) stated that the centers of origin of barley are diffuse in time and space and the question of the center of origin may never be solved. However, he concluded four centers of diversity:

1. The Ethiopian barleys are distinct, lack dormancy, isolated, and contain extreme diversity not recorded elsewhere. At one time Ethiopia was considered the primary center of origin.
2. The oriental barleys (Tibet, China, southern Japan, and Korea) are sharply distinct from other barleys.
3. Afghanistan and adjacent regions are not a center of diversity for barley.
4. Two-rowed barleys are grown in dry land farming areas of Iran, Iraq, and Turkey, while six-rowed barleys are grown under irrigation in the same regions.

Multiple centers of origin or diversity, a departure from traditional Vavilovian centers of diversity, for barley have been proposed from various studies (Peeters 1988; Cross 1992, 1994; Molina-Cano et al. 1999; Tanno et al. 2002). Multiple centers of diversity may arise through human migration dispersing the cultivated species away from its original center of origin. Examination of world germplasm collections and herbaria has revealed that not all centers of diversity represent center of origin (Cross 1992). A survey of a large worldwide barley germplasm collection has shown that material from the U.S. now contains the most variability, followed by that from Turkey, Japan, Russia, and China (Peeters 1988). Recently, Molina-Cano et al. (1999) proposed Morocco as a center of origin of barley and stated that the multicentric origin for the cultivated form started in Morocco and ended in Tibet.

The core collection (10% of the total) concept developed by Brown (1989) represents as much of the diversity (minimum repetitiveness) as possible in the crop. The core collection should be kept separately and not in the bulk population. It should be preserved as secondary sources. Liu et al. (1999) used six isozyme loci to detect genetic diversity in 350 East Asian accessions of the barley core collection and recorded genetic variation in both cultivars and landraces in different regions. Indian cultivars showed the highest diversity, followed by Korean and Chinese cultivars. Landraces from Bhutan and Nepal showed the least diversity. Cultivars carried genetic diversity higher than landraces within, as well as among, regions.

## 8.3 TAXONOMY OF BARLEY

The genus *Hordeum* L. belongs to the tribe Triticeae of the family Poaceae (Gramineae). Bothmer et al. (1995) divided species of *Hordeum* into four sections: (1) *Hordeum*, (2) *Anisolepis*, (3) *Critesion*, and (4) *Stenostachys*. The section *Hordeum* includes *H. vulgare* (cultivated barley), *H. bulbosum* L., and *H. murinum* L., and they are distributed in the Old World (northwest Europe, Greece, Turkey, Egypt, Iran, the Mediterranean region, Afghanistan to Kashmir) and consist of diploid (2n = 2x = 14), tetraploid (2n = 4x = 28), and hexaploid (2n = 6x = 42) cytotypes. Based on cytogenetic studies, Harlan (1965) proposed that *H. spontaneum* and *H. vulgare* could not be considered separate species because they cross readily, $F_1$ is fertile with regular pairing at meiosis, and the characters that separate them segregate in normal Mendelian fashion. The sections *Anisolepis*, *Critesion*, and *Stenostachys* contain 8, 7, and 14 species, respectively. Table 8.1 describes chromosome numbers, genomes, and geographical distributions of all species and subspecies. The distribution of *Hordeum* extends into arctic or subarctic regions in Siberia, Alaska, and southernmost Patagonia (Bothmer et al. 1995). The species of sections *Anisolepis* and *Critesion* belong to the New World, are wild, and carry diploid (2n = 14), tetraploid (2n = 28), and hexaploid (2n = 42) cytotypes. The species of section *Stenostachys* are distributed worldwide (Table 8.1). Based on geographical distribution of the genus *Hordeum*, most species inhabited in two major regions of the world: southwestern Asia and southern America (Bothmer and Jacobsen 1985; Table 8.1).

## 8.4 BARLEY GERMPLASM RESOURCES

The germplasm resources could be utilized to broaden the germplasm base of high-yielding improved varieties with resistance to abiotic and biotic stresses. Barley germplasm is enriched with improved cultivars, landraces, and wild species. Research institutes and international organizations maintain these materials (Table 8.2). Moseman and Smith (1985) classified barley germplasm collections as either base or working collections. Base collections are maintained in long-term storage for preservation. Working collections are divided into general and special collections and are available for distribution to scientists. Accessions in a general collection include cultivars, selections, breeding lines, and landraces, and they do not require special care for maintenance, and should be maintained in the gene banks, as this will result in a partial loss of the material (Zeven 1996). Farm maintenance would lead to a complete loss. *In situ* conservation may complement the *ex situ* method. This can preserve social and biological processes of evolution. The classical

**Table 8.1 *Hordeum* Species, 2n Chromosome Number, Genome, and Their Geographical Distribution**

| Species | Subspecies | 2n | Genome | Geographical Distribution |
|---|---|---|---|---|
| **Section *Hordeum*** | | | | |
| *H. vulgare* L. | *vulgare* | 14 | I | Temperate areas of the world |
| | *spontaneum* (C. Koch.) Thell. | 14 | I | Greece, Egypt, Iran, Afghanistan, western Pakistan, southern Tadzjikistan |
| | *agriochriton* (Åberg) Bowd. | 14 | I | Mediterranean, Middle East to 4500 m in Himalayas |
| *H. bulbosum* L. | | 14 | I | Greece, Egypt westwards |
| | | 28 | II | Greece and eastwards; locally southern Spain |
| *H. murinum* L. | *murinum* | 28 | YY | Northern and western Europe; from Atlantic to the Caucasus area |
| | *leporinum* (Link) Arcangeli | 28 | YY | Mediterranean region, eastward to Afghanistan |
| | | 42 | YYY | From Turkey eastward to Iran |
| | *glaucum* (Steudel) Tzvelev | 14 | Y | Southern Mediterranean region and eastward to Iran, Afghanistan, and Kashmir |
| **Section *Anisolepis*** | | | | |
| *H. pusillum* Nuttal | | 14 | H | Most of the U.S. except the westernmost part; few records from Canada and Mexico |
| *H. intercedens* Nevski | | 14 | H | Endemic to southwest California; adjacent Santa Barbara Islands; northwest Baja California of Mexico |
| *H. euclaston* Steudel | | 14 | H | Central Argentina, Uruguay, southern Brazil |
| *H. flexuosum* Steudel | | 14 | H | Buenos Aires, Argentina; scattered localities in adjacent areas, including Uruguay |
| *H. muticum* Presl | | 14 | H | Northwest Argentina, northeast Chile, Bolivia, Peru, Ecuador, Colombia |
| *H. chilense* Roemer & Schultes | | 14 | H | Central Chile; westernmost parts of the provinces of Neuquen and Rio Negro, Argentina |
| *H. cordobense* Bothmer, Jacobsen & Nicora | | 14 | H | Central and northern Argentina |
| *H. stenostachys* Godron | | 14 | H | Central and northern Argentina, Uruguay, southernmost Brazil |
| **Section *Critesion*** | | | | |
| *H. pubiflorum* Hooker f. | | 14 | H | Primarily in Tierra del Fuego and the Magellanes; province of Santa Cruz Argentina; Nuble in Chile |
| *H. halophilum* Grisebach | | 14 | H | From Tierra del Fuego northward on both sides of the Andes through Argentina and Chile; scattered in Bolivia and Peru |
| *H. comosum* Presl | | 14 | H | Andean areas in Chile and Argentina north to the province of Mendoza |
| *H. jubatum* L. | | 28 | HH | Mexico northward through U.S., Canada, and Alaska to eastern Siberia |

| Species | Subspecies | 2n | Genome | Distribution |
|---|---|---|---|---|
| H. arizonicum Covas | | 42 | HHH | Southern Arizona; few locations in the southeast part of California and northern Mexico |
| H. procerum Nevski | | 42 | HHH | Primarily in the provinces of Buenos Aires, Neuquen, Rio Negro, and La Pampa of central Argentina; rare in Chubut; single record from Uruguay |
| H. lechleri (Steudel) Schenck | | 42 | HHH | Chile and Argentina; Falkland Islands |

**Section Stenostachys**

| Species | Subspecies | 2n | Genome | Distribution |
|---|---|---|---|---|
| H. marinum Hudson | marinum | 14 | X | Mediterranean region, especially in the western part |
| | | 14 | X | Eastern Mediterranean to southwestern Asia |
| | gussoneanum (Parlatore) Thellung | 28 | XX | From Turkey to Afghanistan |
| H. secalinum Schreber | | 28 | HH | From southernmost Sweden and central Denmark along the Atlantic coast of Europe to Spain; scattered locations in the Mediterranean and inland Europe; a few locations in North Africa |
| H. capense Thunberg | | 28 | HH | Republic of South Africa and Lesotho |
| H. bogdanii Wilensky | | 14 | H | Central Asia, western Iran, Afghanistan, northwest Pakistan, northern India, southern Siberia, Mongolia, northern China |
| H. roshevitzii Bowden | | 14 | H | Southern Siberia, Mongolia, north-central China |
| H. brevisubulatum (Trinius) Link | brevisubulatum | 14, 28 | H,HH | Southeastern Siberia, Mongolia, northern China |
| | nevskianum (Bowden) Tzvelev | 14, 28 | H, HH | Nepal, northern Kashmir, Chitral in Pakistan, westernmost China, western Siberia, northeast Afghanistan |
| | turkestanicum (Nevski) Tzvelov | 28, 42 | HH, HHH | Central and northeast Afghanistan, Chitral in Pakistan, south Tadzjikistan, westernmost China |
| | violaceum (Boissier & Hohenacker) Tzvelev | 14, few 28 | H, HH | Central Turkey, Caucasus, the northern parts of western and eastern Azarbaijan in Iran, Alborz mountains |
| | iranicum Bothmer | Few 28, 42 | HH, HHH | Western and southern Iran in the Zagros, Alborz mountains |
| H. brachyantherum Nevski | brachyantherum | 28, 42 | HH, HHH | Western North America |
| | californicum (Covas & Stebbins) Bothmer, Jacobsen & Seberg | 14 | H | Southwestern California up to north of the Bay area; its distribution overlaps with that of subsp. brachyantherum, and in a few locations they grow together |
| H. depressum (Scribner & Smith) Rydberg | | 28 | HH | Western U.S. (central California, a few locations in the states of Oregon and Washington) |
| H. guatemalense Bothmer, Jacobsen & Jørgensen | | 28 | HH | The mountain region of Cuchumatanes in northern Guatemala (3000–3500 m) |
| H. erectifolium Bothmer, Jacobsen & Jørgensen | | 14 | H | A single locality in the western part of the province of Buenos Aires, Argentina |
| H. tetraploidum Covas | | 28 | HH | From the province of Mendoza to the province of Santa Cruz in Argentina |
| H. fuegianum Bothmer, Jacobsen & Jørgensen | | 28 | HH | Tierra del Fuego and a few locations in the region of Magellanes to southern Chile |

*(Continued)*

**Table 8.1** *Hordeum* **Species, 2n Chromosome Number, Genome, and Their Geographical Distribution** (continued)

| Species | Subspecies | 2n | Genome | Geographical Distribution |
|---|---|---|---|---|
| *H. parodii* Covas | | 42 | HHH | Northern Mendoza and southward of Santa Cruz in Argentina; a few scattered localities in Buenos Aires; in the region of Magellanes in south Chile and on the Chilean side of Tierra cel Fuego |
| *H. patagonicum* (Haumann) Covas | *patagonicum* | 14 | H | Southernmost part of the province of Chubut and along the coast of the province of Santa Cruz, Argentina |
| | *setifolium* (Parodi & Nicora) Bothmer, Giles & Jacobsen | 14 | H | Western part of the province of Chubut to northwest part of the province of Santa Cruz, Argentina |
| | *santacrucense* (Parodi & Nicora) Bothmer, Giles & Jacobsen | 14 | H | Southern part of the province of Santa Cruz, southward to the Strait of Magellan |
| | *mustersii* (Nicora) Bothmer, Giles & Jacobsen | 14 | H | |
| | *magellanicum* (Parodi & Nicora) Bothmer, Giles & Jacobsen | 14 | H | From southernmost Patagonia, with several inland localities to Tierra del Fuego, Argentina |

*Source:* Bothmer, R. von et al., in *Systematic and Ecogeographical Studies on Crop Genepools 7*, 2nd ed., International Plant Genetic Resources Institute, Rome, 1995.

Table 8.2  Barley Germplasm Collection at Six Major Centers of the World

| Species | PGRC | NSGC | NGB | ICARDA | VIR | NIAS |
|---|---|---|---|---|---|---|
| H. arizonicum | 7 | 1 | 6 | — | — | — |
| H. bogdanii | 18 | 16 | 35 | 1 | — | — |
| H. brachyantherum subsp. californicum | 23 | 2 | 11 | — | — | — |
| H. brachyantherum | 72 | 8 | 49 | 2 | — | — |
| H. brevisubulatum subsp. iranicum | — | 6 | — | — | — | — |
| H. brevisubulatum subsp. nevskianum | — | 1 | — | — | — | — |
| H. brevisubulatum subsp. turkestanicum | — | 1 | — | — | — | — |
| H. brevisubulatum subsp. violaceum | 17 | 21 | — | 4 | — | — |
| H. brevisubulatum | 11 | 9 | 7 | — | 10 | — |
| H. bulbosum | 527 | 167 | 15 | 19 | 38 | — |
| H. capense | 4 | 1 | 3 | — | — | — |
| H. fuegianum | — | — | 2 | — | — | — |
| H. chilense | 31 | 10 | 5 | 1 | — | — |
| H. comosum | 48 | 4 | 2 | 1 | — | — |
| H. cordobense | 1 | — | 5 | 2 | — | — |
| H. crachyantherum | — | — | 5 | — | — | — |
| H. depressum | 21 | — | 20 | 1 | — | — |
| H. enclaston | — | — | 2 | — | — | — |
| H. erectifolium | — | — | 1 | — | — | — |
| H. euclaston | 14 | — | 8 | 1 | — | — |
| H. flexuosum | 14 | — | 6 | 2 | — | — |
| H. fuegianum | — | — | 13 | — | — | — |
| H. geniculatum | — | — | — | 3 | 25 | — |
| H. guatemalense | — | — | 1 | — | — | — |
| H. halophilum | — | — | 4 | — | — | — |
| H. hoshevityvi | — | — | 2 | — | — | — |
| H. hybrid | 23 | 7 | — | — | — | — |
| H. intercedens | 17 | — | 3 | — | — | — |
| H. jubatum | 120 | 28 | 28 | 2 | 13 | — |
| H. lechleri | 59 | 2 | 34 | 2 | — | — |
| H. marinum | 19 | 10 | 99 | 32 | 5 | — |
| H. marinum subsp. gussoneanum | 69 | 13 | — | — | — | — |
| H. marinum subsp. marinum | 2 | — | — | 2 | — | — |
| H. murinum | 38 | 21 | 105 | 6 | 24 | 26 |
| H. murinum subsp. glaucum | 58 | 28 | — | 2 | — | — |
| H. murinum var. leporinum | 473 | 16 | — | 5 | — | — |
| H. mutabilis | — | — | 1 | — | — | — |
| H. muticum | 16 | 3 | 21 | 1 | — | — |
| H. nodosum | — | — | 3 | — | — | — |
| H. parodii | 70 | 2 | 33 | 1 | — | — |
| H. patagonicum magellanicum | 15 | — | — | — | — | — |
| H. patagonicum santacrucense | 1 | — | — | — | — | — |
| H. patagonicum setifolium | 7 | — | — | — | — | — |
| H. patagonicum | 49 | — | 62 | 1 | — | — |
| H. procerum | 9 | 3 | 7 | 2 | — | — |
| H. pubiflorum | 85 | — | 12 | 1 | — | — |
| H. pusillum | 25 | 10 | 13 | 2 | — | — |
| H. roshevitzii | 1 | 2 | 16 | 1 | — | — |
| H. secalinum | 1 | 4 | 4 | 3 | — | — |
| Hordeum spp. | 1277 | 39 | — | 6 | 48 | — |
| H. stenostachys | 53 | 12 | 23 | 4 | — | — |
| H. tetraploidum | — | — | 7 | — | — | — |

(Continued)

**Table 8.2   Barley Germplasm Collection at Six Major Centers of the World** (continued)

| Species | PGRC | NSGC | NGB | ICARDA | VIR | NIAS |
|---|---|---|---|---|---|---|
| *H. vulgare* subsp. *spontaneum* | 3875 | 1504 | 18 | 1696 | 150 | 16 |
| *H. vulgare* subsp. *agriochriton* | — | 1 | 45 | 9 | — | 2 |
| *H. vulgare* subsp. *vulgare* | 22,882 | 25,024 | 1804 | 23,639 | 17,197 | — |
| *H. vulgare* | 9684 | — | 141 | — | — | 6853 |
| *H. vulgare* var. *distichon* | — | — | — | — | — | 1996 |
| *Hordeum* mutant collection (Udda Lundquist) | — | — | 9647 | — | — | — |

*Source*: PGRC, www.agr.gc.ca/pgrc-rpc; NSGC, www.ars-grin.gov/npgs; NGB, www.ngb.se; ICARDA, www.icarda.cgiar.org; VIR, www.vir.nw.ru; NIAS, tnagamin@affrc.go.jp.

examples of *ex situ* preservation are potatoes (*Solanum* spp.) in the Andes of Peru, maize in southern Mexico, and wheat (*Triticum* spp.) in southern Turkey (Brush 1995). Maintenance of the accessions in special collections requires expertise in taxonomy, agronomy, physiology, genetics, and cytogenetics. Barley germplasm collections being maintained in many countries are available to scientists upon request. Major collections are located in the U.S., Canada, Sweden, Syria, Russia, and Germany. None of these major germplasm centers contains all of the wild *Hordeum* species (Table 8.2). Li et al. (1998) reported 18,838 accessions in the gene bank of China. Wild relatives constituted 2588 accessions.

A working barley collection is maintained by many barley-growing countries, particularly at the National Institute of Agrobiological Sciences (NIAS), Japan; Centro Internacional de Mejoramiento de Maiz Y Trigo (CIMMYT), Mexico; National Bureau of Plant Genetic Resources (NBPGR), New Delhi; Centre for Genetic Resources (CGN), the Netherlands; Research Institute of Crop Production (RICP), Prague, Czech Republic; Crop and Food Research (CGR), Christchurch, New Zealand; and Institut National de la Recherche Agronomique (INRA), Clermont-Ferrand, France (Table 8.3). Special working germplasm collections (genetic stocks; cytogenetic stocks — primary trisomics, tertiary trisomics, telotrisomics, acrotrisomics, duplications, and inversions; genetic male sterile lines; desynaptic mutants; and *Hordeum* species) are maintained at the National Small Grain Collection (NSGC) in Aberdeen, ID.

**Table 8.3   Working Barley Germplasm Depositaries in the World**

| Country | Number of Accessions |
|---|---|
| CIMMYT, Mexico (www.cimmyt.org) | |
|   *H. vulgare* | 14,986 |
| NBPGR, India (www.nbpgr.delhi.nic.in) | |
|   *H. vulgare* | 11,030 |
| CGN, The Netherlands (www.plant.wageningen-ur.nl/cgn) | |
|   *H. vulgare* | 3342 |
|   *H. spontaneum* | 15 |
|   *H. bulbosum* | 15 |
|   *H. murinum* | 27 |
|   *Hordeum* spp. | 37 |
| RICP, Prague, Czech Republic (www.vurv.cz) | |
|   *H. vulgare* (advanced cultivars) | 2939 |
|   *H. vulgare* (breeder's material) | 955 |
|   *H. vulgare* (local cultivars) | 99 |
|   *Hordeum* spp. | 81 |
|   Unknown | 411 |
| INRA-Amelioration & Sante des Plantes | |
|   *H.vulgare* | 6341 |
| C& FR, New Zealand (crossr@crop.cri.nz) | |
|   *H. vulgare* | 1436 |

## 8.5 GENOMIC RELATIONSHIPS IN BARLEY

Barley is one of the most extensively studied crop plants and is used in studies dealing with biosystematics, cytology, genetics, cytogenetics, mutation induction, population genetics, molecular cytogenetics, cell and tissue culture, genetic transformation, and plant breeding (Ramage 1985).

### 8.5.1 Diploid Species

Genomic relationships among barley species were established based on classical taxonomy, cytogenetics (karyotype analysis, crossability rate, meiotic chromosome pairing at metaphase I of meiosis in interspecific $F_1$ hybrids, $F_1$ hybrid viability, and fertility), and molecular methods (Bothmer et al. 1995; Singh 2003). Species carrying similar genomes hybridize readily and produce vigorous, fertile $F_1$ with regular chromosome pairing. Sometimes, parental species differ by chromosomal structural changes such as translocations or inversions. Unlike species are extremely difficult to hybridize, embryo abortion is common, and $F_1$ hybrids are weak, displaying either no or very little chromosome association at metaphase I of meiosis. This results in total seed sterility, although fertility is restored by chromosome doubling. Four basic genomes, I, H, X, and Y, have been identified in diploid barley (Table 8.1).

For the North and South American diploid and Asiatic diploid wild species, the genome symbol H has been assigned. North and South American diploid species exhibit high chromosome pairing (Table 8.4). Asiatic and North and South American diploids show intermediate chromosome pairing (Bothmer et al. 1986, 1995). This suggests some genomic differentiation between Old and New World diploid barley species. On the other hand, South and North American diploids possess marginal differentiation in their genomes, as bivalents ranged from 0 to 7, and a few hybrid combinations showed one quadrivalent configuration (Table 8.4). The South American taxa showed very high pairing in hybrids among themselves and almost normal pairing within any single combination between families (Bothmer et al. 1986). Mediterranean (*H. marinum*) and eastern Mediterranean to southwestern Asian (*H. marinum* subsp. *gussoneanum*) diploid barleys have very little genome homology with other diploid species (Bothmer et al. 1986). Based on this information, the genome symbol X has been assigned. *Hordeum murinum* subsp. *glaucum* has a geographical distribution from the southern Mediterranean region eastward to Iran, Afghanistan, to Kashmir. This is a unique species; its genomic relationship with other diploid species has not been established, and therefore it is assigned the genome symbol Y.

Molecular investigations support the classical taxonomy and cytogenetic information (Ørgaard and Heslop-Harrison 1994a,b; Svitashev et al. 1994; Ferrer et al. 1995; Taketa et al. 1999). The species that shared a basic genome expressed more similar hybridization patterns than species with different genomes (Ferrer et al. 1995). Taketa et al. (1999) studied comparative physical mapping of the rDNA (5S and 18S-25S) sites in nine wild *Hordeum* species. *H. vulgare* subsp. *spontaneum* carried similar composition of 5S and 18-S-25S rDNA sites. The closely related *H. bulbosum* L. (2n =14; I genome) showed one pair of 5S rDNA sites and one pair of 18S-25S rDNA sites on different chromosomes. Four diploid wild species, *H. marinum* Hudson (X genome), *H. glaucum* Steud. and *H. murinum* (Y genome), and *H. chilense* Brongn. (H genome), differed in the number (two or three pairs), location, and relative order of 5S. One or two major 18S-25S rDNA sites, but no minor 18S-25S rDNA sites, were recorded. They concluded that rDNA markers are an excellent marker to investigate chromosome evolution and phylogeny in the genus *Hordeum*.

### 8.5.2 Tetraploid Species

The majority of tetraploid species occur in the Old World (Table 8.1). Tetraploid cytotypes (2n = 28) barleys have been found for I, H, X, and Y genome species (Table 8.1). Tetraploid *H. bulbosum* (I genome) and *H. brevisubulatum* are true autotetraploids because a high frequency of quadrivalents

**Table 8.4  Interspecific *Hordeum* Hybrids with Chromosome Pairing at Diakinesis and Metaphase I**

| Hybrids (Genomes) | Average Chromosome Pairing (Range) | | | GAI | Authority |
|---|---|---|---|---|---|
| | I | II | III + IV | | |
| *Hordeum* (n = 7) | | | | | |
| *publiflorum* × *comosum* (H × H) | 0.24 (0–2) | 6.81 (5–7) | 0.05 (0–1)[a] | 0.98 | Bothmer et al. 1986 |
| *publiflorum* × *patagonicum* (H × H) | 0.19 (0–4) | 6.90 (5–7) | | 0.99 | Bothmer et al. 1986 |
| *comosum* × *euclaston* (H × H) | 0.40 (0–6) | 6.76 (4–7) | | 0.97 | Bothmer et al. 1986 |
| *patagonicum* × *flexuosum* (H × H) | 0.79 (0–4) | 6.60 (5–7) | | 0.94 | Bothmer et al. 1986 |
| *patagonicum* × *stenostachys* (H × H) | 0.43 (0–4) | 6.79 (5–7) | | 0.97 | Bothmer et al. 1986 |
| *bogdani* × *patagonicum* (I × H) | 7.38 (0–13) | 3.33 (1–7) | | 0.48 | Bothmer et al. 1986 |
| *brevisubulatum* × *brachyantherum* (I × H) | 10.24 (6–14) | 1.88 (0–4) | | 0.27 | Bothmer et al. 1986 |
| *brevisubulatum* × *brachyantherum* (I × H) | 3.20 (0–8) | 5.24 (3–7) | | 0.75 | Bothmer et al. 1986 |
| *marinum* × *bogdani* (X × H) | 13.71 (12–14) | 0.14 (0–1) | | 0.02 | Bothmer et al. 1986 |
| *marinum* × *brachyantherum* (X × H) | 13.10 (10–14) | 0.45 (0–2) | | 0.06 | Bothmer et al. 1986 |
| *bulbosum* × *patagonicum* (I × H) | 13.90 (12–14) | 0.06 (0–2) | | 0.01 | Bothmer et al. 1986 |
| *marinum* × *muticum* (X × H) | 13.00 (10–14) | 0.50 (0–2) | | 0.07 | Bothmer et al., 1986 |
| *roshevitzii* × *patagonicum* (I × H) | 13.42 (11–14) | 0.18 (0–1) | 0.03 (0–1) | 0.03 | Bothmer et al. 1986 |

[a]Includes V, VI (0.02).

and trivalents is observed at metaphase I (Bothmer et al. 1988b, 1995). Tetraploid species containing the H genome are autotetraploids, segmental allotetraploids, and genomic allotetraploids (Bothmer et al. 1995). The only tetraploid *H. marinum* subsp. *gussoneanum* with an X genome is allotetraploid because the chromosomes pair has 14 bivalents at metaphase I. However, observation of rDNA loci by fluorescent genomic *in situ* hybridization (fl-GISH) suggests an autoploid origin (Linde-Laursen et al. 1992). *H. murinum* subsp. *leporinum* (2n = 28; YY) is of autotetraploid origin, but may have a very strong diploidizing mechanism (Bothmer and Subrahmanyam 1988; Bothmer et al. 1995a). Analysis of rDNA sites in tetraploid *H. bachyantherum*, *H. chilense*, and *H. brevisubulatum* revealed their complex evolutionary history, probably involving the multiplication of minor rDNA sites.

## 8.5.3  Hexaploid Species

The majority of the hexaploid (2n = 42) barleys are wild and belong to the H genome species. They are allohexaploid (Rajhathy and Morrison 1961), and meiotic pairing is under genetic control (Gupta and Fedak 1985). These species are *H. arizonicum*, *H. procerum*, *H. lechleri*, *H. parodii*, and *H. brachyantherum* (distributed in the New World) and *H. turkestanicum* and *H. iranicum* (distributed in the Old World). Linde-Laursen et al. (1986) compared the Giemsa C-banding pattern of the North American *Hordeum* taxa — *H. pisillum*, *H. intercedens*, *H. brachyantherum* (2x, 4x, 6x), *H. jubatum*, *H. arizonicum*, *H. depressum* (2x, 4x) — and observed similar banding patterns.

As all taxa showed a low degree of banding pattern polymorphism, it was not sufficient to use C-banding patterns as a diagnostic marker.

## 8.6 GENE POOLS OF BARLEY

The wild relatives of a cultigen are rich reservoirs for traits of economic importance, and these traits are underexploited for developing improved varieties (Zamir 2001). Harlan and de Wet (1971) classified gene pools as primary, secondary, and tertiary based on crossability rate, hybrid viability, and meiotic chromosome pairing — the information that helps in their utilization for the improvement of crop plants.

### 8.6.1 Primary Gene Pool

The primary gene pool of cultivated barley includes landraces of various geographical regions and *H. vulgare* subsp. *spontaneum*. Both are cross-compatible and produce vigorous, viable, and fertile hybrids, facilitating transfer of desired traits from wild species to cultivated barley. Wild and cultivated barleys from several geographical areas constitute a highly diverse gene pool revealed by RFLP (Petersen et al. 1994). This study showed three distinct groups among barley cultivars (Nepal, Ethiopia, Europe) and two groups in wild barley (Iran, Turkey, Israel).

### 8.6.2 Secondary Gene Pool

The secondary gene pool of barley consists of all wild or weedy forms. Diploid and tetraploid *H. bulbosum* from a secondary gene pool for the cultivated barleys are used widely to produce haploid barley. Although *H. vulgare* and *H. bulbosum* contain the I genome, the $F_1$ hybrid is rescued by embryo culture because *H. bulbosum* chromosomes are eliminated during embryo development. The haploid embryos are doubled by colchicine treatment to generate homozygous barley lines for use directly in barley breeding programs. The gene exchange is feasible when the *H. bulbosum* genome is not eliminated during seed development. Chromosome elimination is dependent on the genotype and environment. Chromosome pairing in hybrids is normal, but seed fertility is low (Bothmer et al. 1995). Several abiotic and biotic traits of economic importance have been transferred to cultivated barleys from *H. bulbosum* (Pickering 2000).

### 8.6.3 Tertiary Gene Pool

The tertiary gene pool of barley is very large and includes all the species listed in Table 8.1 except *H. vulgare* subsp. *spontaneum* and *H. bulbosum*. The species in this gene pool are diverse genomically and grow in wide eco-geoclimatic conditions. Embryo rescue is needed to recover hybrids that are sterile. This restricts their rapid exploitation in barley improvement (Henry 2001).

Although there are numerous reports of interspecific and intergeneric hybrids produced involving *H. vulgare* and species of the tertiary gene pool (Fedak 1992), there has been no success in transferring genes from wild relatives of the tertiary gene pool into cultivated barley (Pickering 2000).

## 8.7 GERMPLASM ENHANCEMENT

Alien species have not been utilized for barley improvement because most of the wild *Hordeum* species are distantly related to the cultigen. Cultivated barley with 2n = 14 chromosomes is very sensitive to genetic imbalance and is naturally self-pollinated. Resistance to most pests and pathogens is under the control of single genes, and their action is different in new genetic backgrounds. Most of the identified resistance genes are effective for only a few years when used on a commercial basis. Thus, plant breeders are always looking for new sources of resistance in the primary, secondary, or tertiary gene pool. Several methods, described below, are used for varietal improvement of barley, particularly breeding for resistance to pests and pathogens, and improved brewing quality.

## 8.7.1  Conventional Breeding

Barley is an autogamous crop, and breeders have been confined to improve barleys for higher yields, agronomic traits such as kernel plumpness and straw strength, disease resistance, and malting quality by breeding techniques such as pedigree, backcrossing, bulk, male sterile-facilitated recurrent selection (MSFRS), and composite crosses (Anderson and Reinbergs 1985). Composite cross-breeding is the result of combining a number of single crosses into one large mixture or composite described in *One Man's Life with Barley* by Harlan (1957). Veteläinen (1994) suggested that a composite cross-population provides useful information on genetic conservation, natural selection, and dynamic source of genetic variability compared to traditional collection. Composite crosses in combination with MSFRS broadens the genetic base because of simultaneous input of parents and breakup of existing linkage blocks. This creates greater genetic variability than is available in other breeding methods (Anderson and Reinbergs 1985).

## 8.7.2  Haploid Breeding

Haploid ($2n = x = 7$) breeding in barley is ensued by producing double haploids through anther culture (Freidt and Foroughi-Wehr 1983), bulbosum system–chromosome elimination (Devaux et al. 1996), microspore culture (Kasha et al. 2001a,b), and haploid initiator gene (Hagberg and Hagberg 1980). Freidt and Foroughi-Wehr (1983) studied the field performance of androgenic doubled haploid spring barley from $F_1$ hybrids.

Haploids have been isolated in barley after interspecific crosses between *H. vulgare* and *H. bulbosum* (Choo et al. 1985). The genetics and mechanisms of barley haploid production have been extensively studied (Symko 1969; Bennett et al. 1976), including the genus *Hordeum* (Subrahmanyam 1977). The entire process of elimination of *bulbosum* chromosomes in interspecific hybrids of *H. vulgare* and *H. bulbosum* is under genetic and environmental control (Pickering 1984).

Hagberg and Hagberg (1987, 1991) identified a partially dominant haploid initiator gene (*hap*) in barley. The haploid initiator gene controls the abortion or survival of abnormal embryos and endosperms. Plants homozygous for the *hap* gene produce progeny that include 10 to 14% haploids. It is clearly shown in barley that male sperm nuclei reach the synergid cells about 1 h after pollination. One of the two male sperm nuclei reaches the two polar nuclei and forms a triploid endosperm. In plants of *hap/hap* and *hap/+* genotypes, the egg cell is not always reached by the other sperm nucleus. Thus, in a *hap/hap* plant, about half of the eggs stay unfertilized and some of these develop into haploid embryos. Using the *hap* system with marker genes, the breeders do not need to use embryo culture technique. However, they have to make a large number of crosses — a greater number than is needed using the bulbosum technique (Hagberg and Hagberg 1987).

The bulbosum technique for producing haploid barley is presently the most efficient and popular procedure (Pickering and Devaux 1992). A large number of haploids can be produced from any barley genotype, and frequency of albino plants is more rare than in the androgenesis system. The bulbosum system produces nearly all-haploid plants, while an average of 70% of the plants produced from androgenesis are spontaneously doubled haploids (Kasha et al. 1995). To produce doubled haploids, chromosomes are doubled by colchicine or nitrous oxide treatments (Jensen 1974; Subrahmanyam and Kasha 1975; Thiebaut and Kasha 1978; Thiebaut et al. 1979). Colchicine plus dimethyl sulfoxide (DMSO) was the most efficient treatment for doubling barley haploids during early stages of development. Haploid plants were treated at the two- to three-leaf stage for 5 h with a solution of 0.1% colchicine, and 2 to 4% DMSO resulted in 55.8% doubled sectors. Only 3% of the untreated plants set seeds, suggesting a low percentage of natural chromosomal doubling (Subrahmanyam and Kasha 1975). Pickering and Devaux (1992) compared several methods for producing haploids in barley and listed 12 barley cultivars produced and released from the bulbosum system. To date, 74 barley cultivars have been released by the bulbosum method (Devaux, personal communication, January 2003).

Utilization of haploidy breeding in barley has several potential advantages, such as it provides the quickest means of achieving homozygosity, produces uniform lines, provides a random sample of gametes, simplifies selection on the pure lines, fixes all genes for quantitative traits during breeding, and produces high-yielding and stable barley lines. The cost-effectiveness of haploidy breeding should be evaluated before initiating such a program (Kasha 1982).

### 8.7.3 Tetraploid Breeding

Autotetraploids (4x) occasionally appear spontaneously in natural diploid (2x) populations by nondisjunction either in somatic tissue (meristematic chromosome doubling) or in reproductive tissues by formation of unreduced gametes (DeWet, 1980). Autotetraploidy is successfully induced from temperature shocks, cell and tissue cultures, irradiations, and chemicals (colchicine, certain growth hormones, e.g., naphthalene–acetic acid) (Singh 2003).

Bender and Gaul (1966) suggested that diploidization of autotetraploids could be achieved artificially by an extensive genome reconstruction by means of induced chromosome mutations and gene mutations. Based on these assumptions, Freidt (1978) selected highly fertile lines derived from hybridization of five autotetraploid barley varieties previously irradiated up to seven times, but the results were negative. Seed fertility in tetraploid barley is related with the frequency of quadrivalent association (Fedak 1975). Autotetraploids do not breed true, but throw a low frequency of aneuploids in their progenies because of occasional 3:1 disjunction during anaphase I.

### 8.7.4 Mutation Breeding

Mutagenesis is a valuable tool to barley breeders in producing cultivars directly (instant cultivars) or indirectly via crosses with other improved genotypes (Anderson and Reinbergs 1985; Maluszynski et al. 1995). Induced mutants supplement genetic variability not available in the gene bank. Micke et al. (1985) listed 70 barley cultivars released by mutagenesis worldwide. These cultivars carried many economically valuable traits, such as erectoid growth habit, dwarf type, genetic male sterility, high yield, increased winter hardiness, high protein, earliness, short culm resulting in resistance to lodging, good grain and malting and brewing quality, and resistance to mildew, sprouting, stresses, covered and loose smut, and yellow rust.

Nawrot et al. (2001) selected 13 mutants with increased levels of tolerance to aluminum (Al) toxicity in $M_3$ generation after mutagenic treatment of four barley cultivars with N-methyl-N-nitroso urea (MNH) and sodium azide. Genetic studies demonstrated that a single recessive gene controlled Al tolerance in each mutant.

### 8.7.5 Hybrid Barley Breeding

A prerequisite for hybrid barley is to have genetic, cytoplasmic male sterility with restorer and maintainer lines. Barley being an autogamous crop requires floral structural modification for cross-pollination. Many sources of genetic (Hockett and Eslick 1968) and cytoplasmic (Ahokas 1983) male sterility have been identified, and their genetics have been established. Several sources of fertility restorer genes in barley have been found (Ahokas 1981), and these lines were stable at two different latitudes that helped hybrid barley production (Ahokas and Hockett 1981).

The *ms* genes cause complete breakdown of microsporogenesis, but megasporogenesis typically remains completely uninfluenced. By contrast, prophase I and prophase II meiotic mutants affect micro- and megasporogenesis, resulting in chromosomally imbalanced spores causing male and female sterility. On the other hand, both male and female spores are functionally normal in plants with self-incompatibility systems, but self-fertility is genotypically controlled. Compared to normal fertile plants, male sterile plants possess smaller, shrunken, and nondehiscent anthers.

Ramage (1965, 1991) proposed a scheme utilizing balanced tertiary trisomics (BTTs) for producing hybrid barley. Several modifications of the BTTs to produce hybrid barley were proposed, but commercialization of hybrid seed is still not economically feasible (Ramage 1991).

## 8.7.6 Wide Hybridization

Exotic germplasm has proven useful for increasing genetic variability in many crops, including barley, using conventional breeding methods or molecular techniques (Miflin 2000; Singh 2003). Exotic germplasm includes all germplasm that does not have immediate usefulness without hybridization and selection for adaptation for a given region (Hallauer and Miranda 1981; Singh 2003).

Utilization of exotic germplasm for broadening the germplasm base of Nordic barley is documented (Veteläinen et al. 1997). Studies on agronomic performance and adaptation revealed: (1) agronomically valuable genotypes can be produced through recombination using exotic germplasm, (2) incorporation of exotic material in a local genetic base is most successful, and (3) exotic germplasm has an effect on adaptation (Veteläinen 1996).

Alien genes are introgressed into cultigens by hybridization followed by backcrossing to the recurrent parent. *H. bulbosum* is a rich source of resistance to powdery mildew (Pickering et al. 1995) and helps in mapping genes conferring tolerance to boron toxicity (Jefferies et al. 1999), resistance to cereal cyst nematode (Kretschmer et al. 1997), and net blotch (Molnar et al. 2000). Some of the barriers, such as pollen tube inhibition, postfertilization endosperm breakdown, chromosome instability, chromosome pairing and crossing-over, hybrid infertility, and certation, prevent successful gene introgression between *H. vulgare* and *H. bulbosum* (Pickering 2000). Pickering et al. (1994) produced five of seven possible monosomic alien addition lines (MAALs) of *H. vulgare* × *H. bulbosum* from triploid hybrids between *H. vulgare* (2x) × *H. bulbosum* (4x).

Cultivated barley has been hybridized with wild species of the genus *Hordeum*, and other genera such as *Triticum*, *Secale*, *Aegilops*, and *Elymus*, in order to transfer traits of economic importance to barley and wheat (Fedak 1992; Bothmer et al. 1995). Intergeneric hybrids were totally sterile and did not progress beyond $BC_1$ and $BC_2$ generations (Fedak 1992). Islam and Shepherd (1990) reviewed published literature in wheat–barley hybridization.

### 8.7.6.1 Production of Wheat–Barley Addition Lines

Barley has been hybridized with diploid, tetraploid, and hexaploid wheat, and $F_1$ hybrids produced are euplasmic (wheat cytoplasm) and alloplasmic (barley cytoplasm). However, the only success story is from Islam and his colleagues (Islam et al. 1981; Shepherd and Islam 1988; Islam and Shepherd 1990). Six euplasmic disomic wheat–barley addition lines with 2n = 42 (wheat) + 1 (7H), 2 (2H), 3 (3H), 4 (4H), 6 (6H), or 7 (5H) were produced from 789 plants. Barley chromosomes were designated based on a homoeologous relationship with wheat chromosomes (in parenthesis) (Linde-Laursen 1997). These lines were generated through backcrossing to the recurrent (wheat) parent. Monosomic alien addition lines (MAALs) (2n = 42 wheat + 1 barley chromosome) were identified morphologically and cytologically. Disomic wheat–barley addition lines (DAALs) for chromosome 5 (1H) were male and female sterile (Islam et al. 1981). DAALs were isolated after selfing MAALs. A homoeologous relationship between barley and the wheat genome was established by isozyme banding pattern (Hart et al. 1980) and genetic studies (Shepherd and Islam 1988; Islam and Shepherd 1990). Koba et al. (1997) used the Japanese wheat cultivar

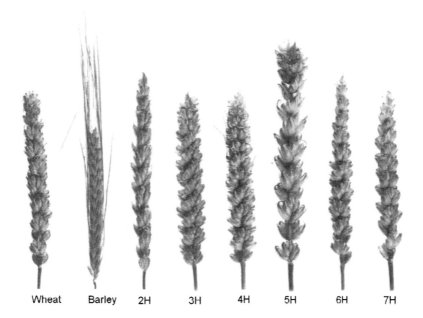

Wheat    Barley    2H    3H    4H    5H    6H    7H

**Figure 8.1**    **(See color insert following page 114.)** Spikes of wheat–barley disomics addition lines. From left to right: Chinese Spring wheat, Betzes barley, 2H (E), 3H (F), 4H (A), 5H (B), 6H (C), and 7H (D). (Homoeologous designation is from R. Islam, personal communication, 2003. Letters in parentheses are from Islam, A.K.M.R. et al., *Heredity*, 46, 161–174, 1981. With permission.)

Shinchunaga containing a crossability gene (*kr*) as female and obtained many hybrids. They identified only two wheat–barley DAALs and five translocated chromosome addition lines of common wheat. In a previous study, Koba et al. (1991) observed that barley chromosomes 1 (7H) and 5 (1H) were eliminated infrequently. This may be the reason for unsuccessful development of wheat–barley addition lines by other scientists.

### 8.7.6.1.1 Morphological Identification of Wheat–Barley Addition Lines

Islam et al. (1981) described morphological features of seven possible DAALs. The typical spike of wheat cv. Chinese Spring, barley cv. Betzes, and six DAALs is shown in Figure 8.1.

1. DAAL 1H (chromosome 5): Disomic alien addition line for 1H was not isolated because of sterility caused by extreme meiotic irregularities. Islam and Shepherd (2000) produced a fertile addition line involving barley chromosome 1H, which contains several genes for economic importance. The self-fertile stable disomic-monotelodisomic addition line carrying a pair of barley chromosomes 6H and a heteromorphic 1H/1HS pair was isolated (Figure 8.2; Islam and Shepherd 2000).
2. DAAL 2H (chromosome 2): Prostrate at juvenile stage, slender culms with narrow leaves and narrow, elongated florets with awnlets. Tsuchiya (1960) designated the plants carrying three dosages of chromosome 2 as slender.
3. DAAL 3H (chromosome 3): Similar to Chinese Spring, but the spikes are usually smaller and more dense. Triplo 3 (Pale) also carries dense spikes (Tsuchiya 1960).
4. DAAL 4H (chromosome 4): Thick culms and apically awnleted spikes often having supernumerary spikelets. Triplo 4, named Robust, has a thick stem and long, wide, thick, and dark-green leaves (Tsuchiya 1960).
5. DAAL 5H (chromosome 7): Thick culms and long lax spikes with larger florets and longer anthers. Triplo 7 has a short but lax spike, and large seeds (Tsuchiya 1960).

CS Wheat       1H/6H       1H/1HS

**Figure 8.2**    **(See color insert following page 114.)** Mature spike of Chinese Spring wheat (left), double monosomic wheat–barley addition line having barley 1H and 6H (center), and disomic-monotelo-disomic wheat–barley addition line having a pair of barley 6H and a heteromorphic 1H/1HS pair. (From Islam, A.K.M.R. and Shepherd, K.W., *Euphytica*, 111, 145–149, 2000. With permission.)

6. DAAL 6H (chromosome 6): Tapered spikes, with outwardly curved beaks on lemma. Tsuchiya (1960) named primary trisomic for Triplo 6 Purple. It is a major nucleolus organizer chromosome.
7. DAAL 7H (chromosome 1). Bushy at juvenile stage, many tillered with prolonged flowering habit and poor fertility. Tsuchiya (1960) designated the Triplo 1 primary trisomics as Bush. Triplo 1 has a bushy growth habit with many tillers and is partially fertile.

It can be concluded that an extra chromosome modifies plant morphology in wheat–barley disomics addition lines the same way it modifies in autotrisomics, with one exception. Morphologically, Triplo 5 closely resembles disomics. Tsuchiya (1960) designated this primary trisomics Pseudonormal. Chromosome 5 is the smallest chromosome in the barley complement. Taketa and Takeda (1997) examined the expression of seven dominant marker genes of barley in wheat–barley hybrids. They observed reduced expression or complete suppression of the dominant barley genes in the wheat–barley hybrids controlled by the wheat background.

### 8.7.6.1.2 Meiotic Behavior in Wheat–Barley Addition Lines

It is expected to form 22 bivalents at metaphase I of meiosis in wheat–barley disomic addition lines, but this association was not realized. The frequency of 22 bivalent formations varied from 76.9% (4H) to 91.3% (5H), but DAAL 1H was not examined (Islam et al. 1981). They also recorded no evidence of any meiotic pairing between wheat and barley chromosomes.

### 8.7.6.1.3 Seed Fertility and Stability in Wheat–Barley Addition Lines

Chinese Spring (CS) has 3.70 seeds per spikelet. Seed set per spikelet in six DAALs ranged from 1.68 (7H) to 3.62 (4H). The reduced seed fertility in DAAL 7H was attributed to nondehiscent anthers. Wheat–barley disomic addition lines were stable. The range of plants with 2 = 44 chromosomes was 73.3% (5H) to 97.5% (3H) in the selfed progeny of DAALs (Islam et al. 1981). However, the occurrence of monotelosomic and ditelosomic addition lines was frequently observed, and Islam and Shepherd (1990) isolated 12 of the 14 possible ditelosomic addition lines.

Induced meiotic pairing between wheat and barley chromosomes is feasible with the manipulation of the wheat gene *Ph1*, which restricts pairing to homologous chromosomes (Islam and Shepherd 1988, 1992). Murai et al. (2000) used allele-specific amplifications for identifying wheat–barley recombinant chromosomes.

### 8.7.6.1.4 Use of Wheat–Barley Addition Lines

Wheat–barley disomic addition lines were utilized to associate RFLP markers (Serizawa et al. 2001) and repetitive sequences (Tsujimoto et al. 1997) to barley chromosomes and to study the expression of dominant marker genes of barley (Taketa and Takeda 1997). A total of 23 cDNAs and two characterized genomic clones were mapped to chromosome arms by using wheat × barley telosomic addition lines (Cannell et al. 1992).

### 8.7.6.2 Production of Other Wheat–Barley Addition Lines

Six of the possible seven disomic addition lines of wheat (CS)–wild barley (*H. chilense* Brongn.) were produced (Miller et al. 1982). Cabrera et al. (1995) identified wheat–*H. chilense* addition lines by Giemsa C-banding and *in situ* hybridization. Hernández et al. (1999) used *H. chilense* chromosome-specific sequence-tagged site (STS) markers to identify and monitor introgression of *H. chilense* chromatin in bread and durum wheat.

Using wheat–barley and wheat–*H. chilense* disomic addition lines, Forster et al. (1990) located genes controlling tolerance to salt in barley and *H. chilense* chromosomes. Genes for positive effects for salt tolerance were located to chromosome 4H and 5H of barley and 1H[ch], 4H[ch], and 5H[ch] of *H. chilense*.

### 8.7.7 Somaclonal Variation

Somaclonal variation is a viable alternative to mutagenesis and a valuable tool for plant breeders to introduce variation (Skirvin et al. 2000). The genetic variation induced during culture is due to single nuclear gene mutations that exhibit Mendelian inheritance. Kole and Chawla (1993) isolated regenerated barley plants resistant to *Helminthosporium sativum* and high in yield. Bregitzer and Poulson (1995) reported negative somaclonal variation in regenerated barley.

Kihara et al. (1998) evaluated progeny of protoplast-derived barley for five agronomic traits (culm length, heading date, fertility, spike length, and spikelet density). They found that most of the progeny showed no abnormal traits and set seeds lower than control. Changes in some protoplast-derived plants may be culture induced and cultivar dependent. This suggests that somaclonal variation has failed to produce superior barley genotypes.

### 8.7.8 Genetic Transformation

The genetic transformation technology is a tool to integrate a trait of economic importance from distantly related species to a cultigen. The progress made in creating genetically modified organisms (GMOs) during the past 19 years (1986 to 2005) has been monumental (www.isb@vt.edu).

Fertile transgenic barley has been produced by *Agrobacterium*-mediated (Matthews et al. 2001), particle bombardment-mediated (Choi et al. 2002), electroporation-mediated (Ritala et al. 2002), and polyethyleneglycol (PEG)-mediated (Nobre et al. 2000) transformations. Several genes of economic importance have been inserted in the barley genome (Horvath et al. 2001; Manoharan and Dahleen 2002; Bregitzer and Tonks 2003). Manoharan and Dahleen (2002) used a commercial barley cv. Conlon to produce transgenic lines by particle bombardment of callus to isolate improved genotypes. They isolated 85 transgenic plants from 13 independent transformation events, and the progeny test revealed Mendelian inheritance for the transgenes. They are trying to incorporate disease and toxin resistance into improved cultivars (L.S. Dahleen, personal communication, 2003).

Choi et al. (2002) examined 19 transgenic lines produced in barley cv. Golden Promise; 8 lines were diploid (2n = 14) and 11 were tetraploid (2n = 28). Using FISH, they demonstrated integration of insert incorporated on barley chromosomes. Seven (37%) of 19 lines showed integration in the distal position of the chromosomes. Four lines (21%) had integration in telomeric regions, three (16%) in centromeric regions (Figure 8.3 top), three (16%) in satellite regions, and two (10%) in subtelomeric regions (Figure 8.3 bottom).

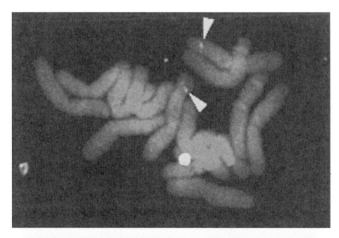

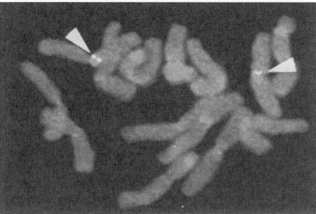

**Figure 8.3    (See color insert following page 114.)** FISH of inserted foreign genes (*sgfp* (S65T or *uidA*)) in metaphase chromosomes of transgenic barley plants. (Top) Two *sgfp* (S65T) signals in a homozygous T$_1$ diploid plant. (Bottom) Two *uidA* signals in a homozygous T$_1$ diploid plant. Arrowheads show signals. (Choi, H.W. et al., *Theor. Appl. Genet.*, 106, 92–100, 2002. With permission.)

Although stable transgenic barleys have been produced, none are yet grown commercially (P.H. Lemaux, L.S. Dahleen, and P. Bregitzer, personal communication).

### 8.7.9 Exploitation of Apomixis

Apomixis is a form of asexual reproduction in which seed is produced from the female gametophyte without fertilization. Methods for identification, characterization, and utilization of the apomictic trait in barley were developed (Kaji and Makino 1994; Jung et al. 1997). Possible occurrence of apomixis in complex interspecific hybrids of *Hordeum* was suspected (Bothmer et al. 1988a). An early embryogenesis gene, *EEA1*, has been cloned and characterized in a shrunken endosperm (*seg8*) barley mutant (Jung et al. 1997). Despite several reports on the occurrence of apomixis in barley, apomictic barley is not commercially feasible at this time.

## 8.8 CONCLUSIONS

Barley (*H. vulgare* L.) is an important cereal used for food, feed, and alcoholic beverages. In total world production, it is ranked fourth among cereals after maize, rice, and wheat. Barley has multiple centers of origin, grows worldwide from 350 to 4050 m above sea level, and evolved from *H. spontaneum*. Wild species of barley consist of diploid ($2n = 2x = 14$), tetraploid ($2n = 4x = 24$), and hexaploid ($2n = 6x = 42$) cytotypes. Major germplasm collections are maintained in the U.S., Canada, Sweden, Syria, Russia, and Germany. Barley is the most extensively studied crop and is used in studies dealing with biosystematics, cytology, genetics, cytogenetics, mutation, population genetics, molecular cytogenetics, cell and tissue culture, genetic transformation, and plant breeding.

## REFERENCES

Ahokas, H. 1981. Cytoplasmic male sterility in barley. X. Distribution of *msm1* fertility restoration ability in the wild progenitor of barley in Israel. *Ann. Bot. Fennici.* 18: 313–320.

Ahokas, H. 1983. Cytoplasmic male sterility in barley. 12. Associations between disease resistance and restoration of *msm1* fertility in the wild progenitor of barley. *Theor. Appl. Genet.* 65: 67–71.

Ahokas, H. and Hockett, E.A. 1981. Performance tests of cytoplasmic male-sterile barley at two different latitudes. *Crop Sci.* 21: 607–611.

Anderson, M.K. and Reinbergs, E. 1985. Barley breeding. In *Barley*, D.C. Rasmusson, Ed. American Society of Agronomy, Crop Science Society of America, Soil Science Society of America, Madison, WI, chap. 9, pp. 231–268.

Bender, K. and Gaul, H. 1966. Zur frage der diploidisierung autotetraploider gerste. *Z. Pflanzenzüchtg.* 56: 164–183 (in German with English summary.)

Bennett, M.D., Finch, R.A., and Barclay, I.R. 1976. The time rate and mechanism of chromosome elimination in *Hordeum* hybrids. *Chromosoma* 54: 175–200.

Bothmer, R. von et al. 1988a. Complex interspecific hybridization in barley (*Hordeum vulgare* L.) and the possible occurrence of apomixis. *Theor. Appl. Genet.* 76: 681–690.

Bothmer, R. von et al. 1995. An ecogeographical study of the genus *Hordeum*. In *Systematic and Ecogeographical Studies on Crop Genepools 7*, 2nd ed. International Plant Genetic Resources Institute, Rome.

Bothmer, R. von and Jacobsen, N. 1985. Origin, taxonomy, and related species. In *Barley*, D.C. Rasmusson, Ed. American Society of Agronomy/Crop Science Society of America/Soil Science of America, Madison, WI, pp. 19–56.

Bothmer, R. von, Flink, J., and, Landström T. 1986. Meiosis in interspecific *Hordeum* hybrids. I. Diploid combinations. *Can. J. Genet. Cytol* 28: 525–535.

Bothmer, R. von, Flink, J., and Landström, T. 1988b. Meiosis in interspecific *Hordeum* hybrids. IV. Tetraploid (4x × 4x) hybrids. *Genome* 30: 479–485.

Bothmer, R. von and Subrahmanyam, N.C. 1988. Assessment of chromosome associations in haploids and their parental accessions in *Hordeum*. *Genome* 30: 204–210.

Bregitzer, P. and Poulson. M. 1995. Agronomic performance of barley lines derived from tissue culture. *Crop Sci.* 35: 1144–1148.

Bregitzer, P. and Tonks, D. 2003. Inheritance and expression of transgenes in barley. *Crop Sci.* 43: 4–12.

Brown, A.H.D. 1989. Core collections: a practical approach to genetic resources management. *Genome* 31: 818–824.

Brush, S.B. 1995. *In situ* conservation of landraces in centers of crop diversity. *Crop Sci.* 35: 346–354.

Cabrera, A. et al. 1995. Characterization of *Hordeum chilense* chromosomes by C-banding and *in situ* hybridization using highly repeated DNA probes. *Genome* 38: 435–442.

Cannell, M. et al. 1992. Chromosomal assignment of genes in barley using telosomic wheat-barley addition lines. *Genome* 35: 17–23.

Choi, H.W., Lemaux, P.G., and Cho, M.-J. 2002. Use of fluorescence *in situ* hybridization for gross mapping of transgenes and screening for homozygous plants in transgenic barley (*Hordeum vulgare* L.). *Theor. Appl. Genet.* 106: 92–100.

Choo, T.M., Reinbergs, E., and Kasha, K.J. 1985. Use of haploids in breeding barley. *Plant Breed. Rev.* 3: 219–252.

Cross, R.J. 1992. A proposed revision of the IBPGR barley descriptor list. *Theor. Appl. Genet.* 84: 501–507.

Cross, R.J. 1994. Geographical trends within a diverse spring barley collection as identified by agro/morphological and electrophoretic data. *Theor. Appl. Genet.* 88: 597–603.

Devaux, P. et al. 1996. Doubled haploids in barley. In *V International Oat and VII International Barley Genetics Symposium*, G. Scoles and B. Rossnagel, Eds. University Extension Press, Saskatoon, Canada, pp. 213–222.

DeWet, J.M.J. 1980. Origins of polyploids. In *Polyploidy: Biological Relevance*, W.H. Lewis, Ed. Plenum Press, New York, pp. 3–15.

Fedak, G. 1975. Fertility and meiotic behaviour in tetraploid barley. *Can. J. Genet. Cytol.* 17: 121–123.

Fedak, G. 1992. Intergeneric hybrids with *Hordeum*. In *Biotechnology in Agriculture No. 5. Barley: Genetics, Biochemistry, Molecular Biology and Biotechnology*, P.R. Shewry, Ed. C.A.B. International, Wallingford, U.K., pp. 45–70.

Ferrer, E., Loarce, Y., and Hueros. G. 1995. Molecular characterization and chromosome location of repeated DNA sequences in *Hordeum* species and in the amphiploid tritordeum (× *tritordeum* Ascherson et Graebner). *Genome* 38: 850–857.

Forster, B.P. et al. 1990. Chromosome location of genes controlling tolerance to salt (NaCl) and vigour in *Hordeum vulgare* and *H. chilense*. *Heredity* 65: 99–107.

Freidt, W. 1978. Untersuchungen an autotraploiden gersten unter besonderer berücksichtigung der diploidisierung. I. Fertilität und vitalität und kornertrag. *Z. pflanzenzüchtg.* 81: 118–139 (in German with English summary.)

Freidt, W. and Foroughi-Wehr, B. 1983. Field performance of androgenetic doubled haploid spring barley from F$_1$ hybrids. *Z. Pflanzenzüchtg.* 90: 177–184.

Gupta, P.K. and Fedak, G. 1985. Genetic control of meiotic chromosome pairing in polyploids in the genus *Hordeum*. *Can. J. Genet. Cytol.* 27: 515–530.

Hagberg, A. and Hagberg, G. 1980. High frequency of spontaneous haploids in the progeny of an induced mutation in barley. *Hereditas* 93: 34–343.

Hagberg, A. and Hagberg, G. 1987. Production of spontaneously doubled haploids in barley using a breeding system with marker genes and the "hap"-gene. *Biol. Zent. Bl.* 106: 53–58.

Hagberg, A. and Hagberg, G. 1991. Production and analysis of chromosome duplications in barley. In *Chromosome Engineering in Plants: Genetics, Breeding, Evolution, Part A*, P.K. Gupta and T. Tsuchiya, Eds. Elsevier, Amsterdam, pp. 401–410.

Hallauer, A.R. and Miranda, Fo, J.B. 1981. *Quantitative Genetics in Maize Breeding*. Iowa State University Press, Ames.

Harlan, J.R. 1957. *One Man's Life with Barley*. Exposition Press, New York.

Harlan, J.R. 1965. The possible role of weed races in the evolution of cultivated plants. *Euphytica* 14: 173–176.

Harlan, J.R. 1979. On the origin of barley. In *Barley: Origin, Botany, Culture, Winter Hardiness, Genetics, Utilization, Pests*, USDA Agricultural Handbook 338. USDA, Washington, DC, pp. 10–36.

Harlan, J.R. and deWet, J.M.J. 1971. Toward a rational classification of cultivated plants. *Taxon* 20: 509–517.

Hart, G.E., Islam, A.K.M.R., and Shepherd, K.W. 1980. Use of isozymes as chromosome markers in the isolation and characterization of wheat-barley chromosome addition lines. *Genet. Res.* 36: 311–325.

Henry, R.T. 2001. Exploiting cereal genetic resources. *Adv. Bot. Res.* 34: 23–57.

Hernández, P. et al. 1999. Development and characterization of *Hordeum chilense* chromosome-specific STS markers suitable for wheat introgression and marker-assisted selection. *Theor. Appl. Genet.* 98: 721–727.

Hockett, E.A. and Eslick, R.F. 1968. Genetic male sterility in barley. I. Nonallelic genes. *Crop Sci.* 8: 218–220.

Horvath, H. et al. 2001. Stability of transgene expression, field performance and recombination breeding of transformed barley lines. *Theor. Appl. Genet.* 102: 1–11.

Islam, A.K.M.R. and Shepherd, K.W. 1988. Induced pairing between wheat and barley chromosomes. In *Proceedings of the 7th International Wheat Genetics Symposium*, T.E. Miller and R.M.D. Koebner, Eds. Bath Press, Avon, U.K., pp. 309–314.

Islam, A.K.M.R., Shepherd, K.W., and Sparrow, D.H.B. 1981. Isolation and characterization of euplasmic wheat-barley chromosome addition lines. *Heredity* 46: 161–174.

Islam, A.K.M.R. and Shepherd, K.W. 1990. Incorporation of barley chromosomes into wheat. In *Biotechnology in Agriculture and Forestry*, Y.P.S. Bajaj, Ed. Springer-Verlag, Heidelberg, Vol. 13, pp. 128–151.

Islam, A.K.M.R. and Shepherd, K.W. 1992. Production of wheat-barley recombinant chromosomes through induced homoeologous pairing. 1. Isolation of recombinants involving barley arms 3HL and 6HL. *Theor. Appl. Genet.* 83: 489–494.

Islam, A.K.M.R. and Shepherd, K.W. 2000. Isolation of a fertile wheat-barley addition line carrying the entire barley chromosome 1H. *Euphytica* 111: 145–149.

Jefferies, S.P. et al. 1999. Mapping of chromosome regions conferring boron toxicity tolerance in barley (*Hordeum vulgare* L.). *Theor. Appl. Genet.* 98: 1293–1303.

Jensen, C.J. 1974. Chromosome doubling techniques in haploids. In *Haploids in Higher Plant: Advances and Potential, Proceedings of the First International Symposium*, K.J. Kasha, Ed. Ainsworth Press, Guelph, Canada, pp. 153–190.

Jung, W., Skadsen, R.W., and Peterson, D.M. 1997. Cloning and characterization of an early embryogenesis gene, EEA1 (accession No. AF017430), from barley (PGR97-184). *Plant Physiol.* 115: 1730.

Kaji, R. and Makino, T. 1994. A proposed method for selecting obligate apomictic mutants. *Barley Genet. Newsl.* 23: 32.

Kasha, K.J. 1982. Production and applications of haploids in barley. *Actas V Congr. Latinoam. Genetica* 127–135.

Kasha, K.J. et al. 2001a. An improved *in vitro* technique for isolated microspore culture of barley. *Euphytica* 120: 379–385.

Kasha, K.J. et al. 2001b. Nuclear fusion leads to chromosome doubling during mannitol pretreatment of barley (*Hordeum vulgare* L.) microspores. *J. Exp. Bot.* 52: 1227–1238.

Kasha, K.J. et al. 1995. Production and application of doubled haploids in crops. In *Induced Mutations and Molecular Techniques for Crop Improvement*, IAEA-SM-340/9. International Atomic Energy Agency, Vienna, pp. 23–37.

Kihara, M. et al. 1998. Field performance of the progeny of protoplast-derived barley, *Hordeum vulgare* L. *Breed. Sci.* 48: 1–4.

Koba, T., Takumi, S., and Shimada, T. 1991. Efficient production of wheat-barley hybrids and preferential elimination of barley chromosomes. *Theor. Appl. Genet.* 81: 285–292.

Koba, T., Takumi, S., and Shimada, T. 1997. Isolation, identification and characterization of disomic and translocated barley chromosome addition lines of common wheat. *Euphytica* 96: 289–296.

Kole, P.C. and Chawla, H.S. 1993. Variation of *Helminthosporium* resistance and biochemical and cytological characteristics in somaclonal generations of barley. *Biol. Plant.* 35: 81–86.

Kretschmer, J.M. et al. 1997. RFLP mapping of the *Ha2* cereal cyst nematode resistance gene in barley. *Theor. Appl. Genet.* 94: 1060–1064.

Ladizinsky, G. and Genizi, A. 2001. Could early gene flow have created similar allozyme-gene frequencies in cultivated and wild barley? *Genet. Resourc. Crop Evol.* 48: 101–104.

Li, Y. et al. 1998. The use of genetic resources in crop improvement: lessons from China. *Genet. Resourc. Crop Evol.* 45: 181–186.

Linde-Laursen, I. 1997. Recommendations for the designation of the barley chromosomes and their arms. *Barley Genet. Newsl.* 26: 1–16.

Linde-Laursen, I. et al. 1992. Physical localization of active and inactive rRNA gene loci in *Hordeum marinum* spp. *gussoneanum* (4x) by *in situ* hybridization. *Genome* 35: 1032–1036.

Linde-Laursen, I., Bothmer, R. von, and Jacobsen, N. 1986. Giemsa C-banded karyotypes of *Hordeum* taxa from North America. *Can. J. Genet. Cytol.* 28: 42–62.

Liu, F., Bothmer, R. von, and Salomon, B. 1999. Genetic diversity among East Asian accessions of the barley core collection as revealed by six isozyme loci. *Theor. Appl. Genet.* 98: 1226–1233.

Maluszynski, M., Ahloowalia, B.S., and Sigurbjörnsson, B. 1995. Application of *in vivo* and *in vitro* mutation techniques for crop improvement. *Euphytica* 85: 303–315.

Manoharan, M. and Dahleen, L.S. 2002. Genetic transformation of the commercial barley (*Hordeum vulgare* L.) cultivar Conlon by particle bombardment of callus. *Plant Cell Rep.* 21: 76–80.

Matthews, P.R. et al. 2001. Marker gene elimination from transgenic barley, using co-transformation with adjacent 'twin T-DNAs' on a standard *Agrobacterium* transformation vector. *Mol. Breed.* 7: 195–202.

Micke, A., Maluszynski, M., and Donini, B. 1985. Plant cultivars derived from mutation induction or the use of induced mutants in cross breeding. *Mutat. Breed. Rev.* 3: 1–92.

Miflin, B. 2000. Crop improvement in the 21st century. *J. Exp. Bot.* 51: 1–8.

Miller, T.E., Reader, S.M., and Chapman V. 1982. The addition of *Hordeum chilense* chromosomes to wheat. In *Induced Variability in Plant Breeding.* EUCARPIA, Centre for Agricultural Publishing and Documentation, Wageningen, The Netherlands, pp. 79–81.

Molina-Cano, J.L. et al. 1999. Further evidence supporting Morocco as a center of origin of barley. *Theor. Appl. Genet.* 98: 913–918.

Molnar, S.J., James, L.E., and Kasha, K.J. 2000. Inheritance and RAPD tagging of multiple genes for resistance to net blotch in barley. *Genome* 43: 224–231.

Moseman, J.G. and Smith, D.H., Jr. 1985. Germplasm resources. In *Barley*, D.C. Rasmusson, Ed. American Society of Agronomy, Crop Science Society of America, Soil Science of America, Madison, WI, pp. 57–72.

Murai, K. et al. 2000. Barley allele-specific amplicons useful for identifying wheat-barley recombinant chromosomes. *Genes Genet. Syst.* 75: 131–139.

Nawrot, M., Szarejko, I., and Maluszynski M. 2001. Barley mutants with increased tolerance to aluminium toxicity. *Euphytica* 120: 345–356.

Nobre, J. et al. 2000. Transformation of barley scutellum protoplasts: regeneration of fertile transgenic plants. *Plant Cell Rep.*19: 1000–1005.

Ørgaard, M. and Heslop-Harrison, J.S. 1994a. Investigations of genome relationships between *Leymus, Psathyrostachys* and *Hordeum* inferred by genomic DNA: DNA *in situ* hybridization. *Ann. Bot.* 73: 195–203.

Ørgaard, M. and Heslop-Harrison, J.S. 1994b. Relationships between species of *Leymus, Psathyrostachys*, and *Hordeum* (*Poaceae, Triticeae*) inferred from Southern hybridization of genomic and cloned DNA probes. *Pl. Syst. Evol.* 189: 217–231.

Peeters, J.P. 1988. The emergence of new centers of diversity: evidence from barley. *Theor. Appl. Genet.* 76: 17–24.

Petersen, L., Østergård, H., and Giese, H. 1994. Genetic diversity among wild and cultivated barley as revealed by RFLP. *Theor. Appl. Genet.* 89: 676–681.

Pickering, R.A. 1984. The influence of genotype and environment on chromosome elimination in crosses between *Hordeum vulgare* L. × *Hordeum bulbosum* L. *Plant Sci. Lett.* 34: 153–164.

Pickering, R.A. 2000. Do the wild relatives of cultivated barley have a place in barley improvement? In *Barley Genetics VIII, Proceedings of the 8th International Barley Genetics Symposium*, Vol. I, S. Logue, Ed. Department of Plant Science, Waite Campus, Adelaide University, Australia, pp. 223–230.

Pickering, R.A. and Devaux, P. 1992. Haploid production: approaches and use in plant breeding. In *Biotechnology in Agriculture No. 5. Barley: Genetics, Biochemistry, Molecular Biology and Biotechnology*, P.R. Shewry, Ed. C.A.B. International, Wallingford, U.K., pp. 519–547.

Pickering, R.A. et al. 1995. The transfer of a powdery mildew resistance gene from *Hordeum bulbosum* L. to barley (*H. vulgare* L.) chromosome 2 (2I). *Theor. Appl. Genet.* 91: 1288–1292.

Pickering, R.A. et al. 1994. Characterization of progeny from backcrosses of triploid hybrids between *Hordeum vulgare* L. (2x) and *H. bulbosum* L. (4x) to *H. vulgare*. *Theor. Appl. Genet.* 88: 460–464.

Rajhathy, T. and Morrison, J.W. 1961. Cytogenetic studies in the genus *Hordeum*. V. *H. jubatum* and the new world species. *Can. J. Genet. Cytol.* 3: 378–390.

Ramage, R.T. 1965. Balanced tertiary trisomics for use in hybrid seed production. *Crop Sci.* 5: 177–178.

Ramage, R.T. 1985. Cytogenetics. In *Barley*, D.C. Rasmusson, Ed. American Society of Agronomy/Crop Science Society of America/Soil Science of America, Madison, WI, pp. 127–154.

Ramage, R.T. 1991. Chromosome manipulations in barley breeding. In *Chromosome Engineering in Plants: Genetics, Breeding, Evolution, Part A*, P.K. Gupta and T. Tsuchiya, Eds. Elsevier, Amsterdam, pp. 385–400.

Ritala, A. et al. 2002. Measuring gene flow in the cultivation of transgenic barley. *Crop Sci.* 42: 278–285.

Serizawa, N., Nasuda, S., and Endo, T.R. 2001. Barley chromosome addition lines of wheat for screening of AFLP markers on barley chromosomes. *Genes Genet. Syst.* 76: 107–110.

Shao, Q., Li, C.-S., and Baschan, C. 1982. Origin and evolution of cultivated barley: wild barley from western Szechuan and Tibet, China. *Barley Genet. Newsl.* 12: 37–42.

Shepherd, K.W. and Islam, A.K.M.R. 1988. Fourth compendium of wheat-alien chromosome lines. In *Proceedings of the 7th International Wheat Genetics Symposium*, T.E. Miller and R.M.D. Koebner, Eds. Bath Press, Avon, U.K., pp. 1373–1395.

Singh, R.J. 2003. *Plant Cytogenetics*, 2nd ed. CRC Press, Boca Raton, FL.

Skirvin, R.M. et al. 2000. Somaclonal variation: do we know what causes it? *AgBiotechNet* 2: 1–4.

Subrahmanyam, N.C. 1977. Haploidy from *Hordeum* interspecific crosses. I. Polyhaploids of *H. parodii* and *H. procerum. Theor. Appl. Genet.* 49: 209–217.

Subrahmanyam, N.C. and Kasha, K.J. 1975. Chromosome doubling of barley haploids by nitrous oxide and colchicine treatments. *Can. J. Genet. Cytol.* 17: 573–583.

Svitashev, S. et al. 1994. Phylogenetic analysis of the genus *Hordeum* using repetitive DNA sequences. *Theor. Appl. Genet.* 89: 801–810.

Symko, S. 1969. Haploid barley from crosses of *Hordeum bulbosum* (2x) × *Hordeum vulgare* (2x). *Can. J. Genet. Cytol.* 11: 602–608.

Taketa, S., Harrison, G.E., and Heslop-Harrison, J.S. 1999. Comparative physical mapping of the 5S and 18S-25S rDNA in nine wild *Hordeum* species and cytotypes. *Theor. Appl. Genet.* 98: 1–9.

Taketa, S. and Takeda, K. 1997. Expression of dominant marker genes of barley in wheat-barley hybrids. *Genes Genet. Syst.* 72: 101–106.

Tanno, K. et al. 2002. A DNA marker closely linked to the *vrs1* locus (row-type gene) indicates multiple origins of six-rowed cultivated barley (*Hordeum vulgare* L.). *Theor. Appl. Genet.* 104: 54–60.

Thiebaut, J. and Kasha, K.J. 1978. Modification of the colchicine technique for chromosome doubling of barley haploids. *Can. J. Genet. Cytol.* 20: 513–521.

Thiebaut, J., Kasha, K.J., and Tsai, A. 1979. Influence of plant development stage, temperature, and plant hormones on chromosome doubling of barley haploids using colchicines. *Can. J. Bot.* 57: 480–483.

Tsuchiya, T. 1960. Cytogenetic studies of trisomics in barley. *Jpn. J. Bot.* 17: 177–213.

Tsujimoto, H. et al. 1997. Identification of individual barley chromosomes based on repetitive sequences: conservative distribution of Afa-family repetitive sequences on the chromosomes of barley and wheat. *Genes Genet. Syst.* 72: 303–309.

Veteläinen, M. 1994. Exotic barley germplasm: variation and effects on agronomic traits in complex crosses. *Euphytica* 79: 127–136.

Veteläinen, M. 1996. Utilization of exotic germplasm in Nordic barley breeding and its consequences for adaptation. *Euphytica* 92: 267–273.

Veteläinen, M., Suominen, M., and Nissilä, E. 1997. Agronomic performance of crosses between Nordic and exotic barleys. *Euphytica* 93: 239–248.

Zamir, D. 2001. Improving plant breeding with exotic genetic libraries. *Nat. Rev.* 2: 983–989.

Zeven, A.C. 1996. Results of activities to maintain landraces and other material in some European countries *in situ* before 1945 and what we may learn from them. *Genet. Resour. Crop Evol.* 43: 337–341.

# CHAPTER 9

# Chromosome Mapping in Barley (*Hordeum vulgare* L.)

Jose M. Costa and Ram J. Singh

## CONTENTS

9.1   Introduction..................................................................................................258
9.2   Karyotype Analysis in Barley .....................................................................258
    9.2.1   Chromosome 1 (7H)....................................................................260
    9.2.2   Chromosome 2 (2H)....................................................................260
    9.2.3   Chromosome 3 (3H)....................................................................261
    9.2.4   Chromosome 4 (4H)....................................................................261
    9.2.5   Chromosome 5 (1H)....................................................................261
    9.2.6   Chromosome 6 (6H)....................................................................261
    9.2.7   Chromosome 7 (5H)....................................................................262
9.3   Development of a Cytogenetic Map of Barley by Chromosome Structural Changes........262
    9.3.1   Deficiencies................................................................................262
    9.3.2   Duplications................................................................................262
    9.3.3   Interchanges...............................................................................263
    9.3.4   Inversions...................................................................................263
9.4   Development of a Cytogenetic Map of Barley by Chromosome Numerical Changes..........264
    9.4.1   Primary Trisomics ......................................................................264
    9.4.2   Secondary Trisomics ..................................................................265
    9.4.3   Tertiary Trisomics ......................................................................265
    9.4.4   Telotrisomics..............................................................................265
    9.4.5   Acrotrisomics.............................................................................265
9.5   Molecular Maps of Barley ...........................................................................265
9.6   Utility of Genetic Maps of Barley ...............................................................266
9.7   Physical Mapping of Barley Chromosomes ................................................270
9.8   Comparative Mapping in the Grass Family..................................................273
9.9   Chromosome Evolution Mechanisms in Barley and Other Grasses ............275
9.10  Large-Insert Libraries of Barley Chromosomes .........................................275
9.11  Barley Expressed Sequence Tags.................................................................276
9.12  Conclusions..................................................................................................277
References ..............................................................................................................277

**Table 9.1   Comparison of the Genome Size of Important Crops**

| Plant Species | DNA Content[a] (Mb) | Number of n Chromosomes |
|---|---|---|
| Thale cress (*Arabidopsis thaliana*) | 120 | 5 |
| Rice (*Oryza sativa*) | 430–460 | 12 |
| Mungbean (*Vigna radiata*) | 500 | 11 |
| Sorghum (*Sorghum bicolor*) | 750–780 | 10 |
| Soybean (glycine max.) | 1100 | 20 |
| Tobacco (*Nicotiana tabacum*) | 1650 | 12 |
| Maize (*Zea mays*) | 2400–3200 | 10 |
| Barley (*Hordeum vulgare*) | 4900–5300 | 7 |
| Rye (*Secale cereale*) | 8000 | 7 |
| Oats (*Avena sativa*) | 11,300 | 21 |
| Hexaploid wheat (*Triticum aestivum*) | 16,000 | 21 |

[a]Arumuganathan and Earle (1991); Bennett and Smith (1976); Laurie and Bennett (1985).

## 9.1 INTRODUCTION

Barley is a widely studied model plant for chromosome mapping of the Triticeae. Some of the advantages of barley for genome mapping are that it is a diploid (2n = 14) and has seven large (6 to 8 μm) cytologically distinct chromosomes (Ramage 1985). Barley chromosomes contain approximately 5000 Mb DNA (Arumuganathan and Earle 1991; Bennett and Smith 1976). This makes barley the member of the widely cultivated Triticeae with the lowest amount of DNA, although it has 10 times more nuclear DNA than rice (*Oryza sativa* L.). Table 9.1 contains approximate nuclear DNA content and chromosome number of a selected group of plant species compared to barley.

The genomes of plant species are very different in terms of size, ploidy level, and chromosome number. Even plant species within the same family can have a large variation in DNA content. Rice and sorghum represent the low end of the spectrum for grasses (Poaceae), while oats and hexaploid wheat are at the high end.

Several recent reviews of mapping of the barley genome describe molecular markers and quantitative trait loci (QTL) assigned to barley chromosomes (Kleinhofs and Han 2002; Kleinhofs and Graner 2001; Hayes et al. 1996). In this review, we will focus on the history of gene mapping in barley and current comparative mapping with other grass species. The future trends in the genetic and physical mapping of barley will be discussed.

## 9.2 KARYOTYPE ANALYSIS IN BARLEY

The chromosome–linkage group relationship in barley has been established based on cytogenetics and molecular studies. Chromosome 5 is the smallest chromosome and chromosomes 6 and 7 are the nucleolus organizer chromosomes and are morphologically distinct. Based on conventional staining techniques, chromosomes 1 to 4 are difficult to distinguish (Figure 9.1). Barley cytogeneticists accepted the chromosome designation of Tjio and Hagberg (1951) until Tuleen (1973) and Künzel (1976) questioned the identity of chromosomes 1, 2, and 3. Both authors observed, based on the results of multiple translocation analysis, that chromosome 2 is the longest chromosome in the barley complement. The application of the Giemsa C-banding (Figure 9.2) and N-banding techniques helped to identify all the barley chromosomes not possible by conventional staining techniques (Singh 2003). Giemsa C-banding polymorphism among barley cultivars, however, may hinder precise chromosome identification. Furthermore, the combination of the aceto-carmine and the Giemsa staining techniques applied to the same cell has helped in developing karyotype analysis better than either technique alone.

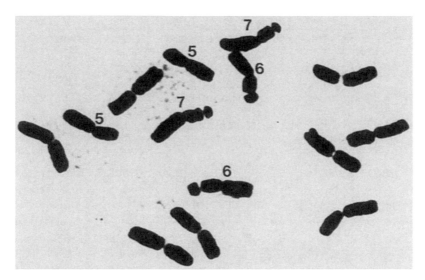

**Figure 9.1** Somatic mitotic metaphase chromosomes of barley after aceto-carmine staining showing 2n = 14 chromosomes. (From Singh, R.J., *Plant Cytogenetics*, 2nd ed., CRC Press, Boca Raton, FL, 2003. With permission.)

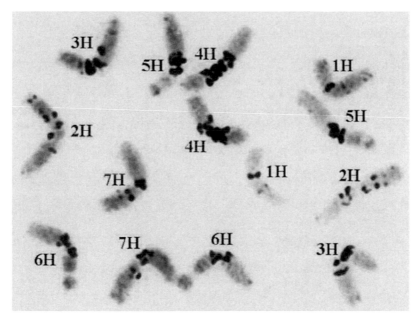

**Figure 9.2** Somatic mitotic metaphase chromosomes of barley after Giemsa C-banding. (From Singh, R.J. and Tsuchiya, T., *Z. Pflanzenzuchütg.*, 86, 336–340, 1981. With permission.)

Currently, molecular cytogenetic techniques are being used to construct extremely accurate karyotype and physical mapping of barley chromosomes. These are microdissection (Schondelmaier et al. 1993; Busch et al. 1995), polymerase chain reaction (PCR)-based methods (Sorokin 1994; Yoshino et al. 1998), sequential staining and imaging methods (Nakayama and Fukui 1995), fluorescence *in situ* hybridization (Fukui et al. 1994; Pedersen and Linde-Laursen 1995; Brown et al. 1999), and flow karyoptyping and sorting of mitotic chromosomes (Lysák et al. 1999).

Chromosome image analysis goes back to the early 1980s when computer systems were at the cradle stage to handle huge digital data of images, and only a few expensive image analyzing systems were available. In 1985, the first comprehensive chromosome image analyzing system

**Table 9.2   Nomenclature of Barley Aneuploids**

| Primary Trisomics | Telotrisomics | Acrotrisomics |
|---|---|---|
| Bush, 1 (7H) | lL; 1S (7HL; 7HS) | 1L$^{1S}$ (1HL$^{7HS}$) |
| Slender, 2 (2H) | 2L; 2S (2HL; 2HS) | − − |
| Pale, 3 (3H) | 3L; 3S (3HL; 3HS) | 3L$^{3S}$ (3HL$^{3HS}$) |
| Robust, 4 (4H) | 4L; 4S (4HL (deficient)) | 4L$^{4S}$ (4HL$^{4HS}$) |
| Pseudonormal, 5 (1H) | 5L; 5S (1HL; 1HS) | 5S$^{5L}$ (1HS$^{1HL}$) |
| Purple, 6 (6H) | 6S (6HS) | 6S$^{6L}$ (6HS$^{6HL}$) |
| Semierect, 7(5H) | 7S (5HS) | 7S$^{7L}$ (5HS$^{5HL}$) |

(CHIAS) with software fulfilling the basic requirements of cytologists and cytogeneticists was developed (Fukui 1986). The imaging method for barley chromosomes was improved by quantifying chromosome morphology, banding pattern (Fukui and Kakeda 1990), and differential condensation patterns of the metaphase chromosomes (Kakeda and Fukui 1994). Now personal computers with enough imaging capability allow image analysis in every cytology and cytogenetics laboratory. The basic points to keep in mind when using imaging methods are as follows:

1. Importance of the quality of the original chromosome images. No imaging method can create new information that is not originally included in the chromosome images.
2. The information of the images is reduced by each step of image manipulation. Imaging methods present the essence of the chromosome image of the original image as visible, and thus in a perceptible way.
3. Imaging methods can also display the image information as numerical data.

The standard and basic procedures for image analyses and the manuals can be obtained either by written form or via the Internet (http://mail.bio.eng.osaka-u.ac.jp/cell/).

The salient features of the seven barley chromosomes based on conventional and Giemsa C- and N-banding techniques (Singh and Tsuchiya 1982; Singh 2003) and homoeologous groups (in parentheses) (Linde-Laursen 1997; Costa et al. 2001) with wheat (Table 9.2) are described in the following.

## 9.2.1   Chromosome 1 (7H)

This is the third longest chromosome and is metacentric (Figure 9.3). Since both arms are almost equal in length, their designation in karyogram and idiogram is based on the morphological effects of telotrisomic plants, gene–chromosome arm relationships, and on the Giemsa N-banding pattern (Singh and Tsuchiya 1982; Singh 2003). The long arm contains only a centromeric heterochromatic band, and the short arm contains two intercalary bands in addition to a centromeric band. The band proximal to the centromere is darker than the one toward the telomere.

## 9.2.2   Chromosome 2 (2H)

This is the longest chromosome among the five nonsatellited chromosomes of barley complement. It carries its kinetochore at the median (l/s = 1.26) region (Figure 9.3). Tuleen (1973) and Künzel (1976) reported similar results by the reciprocal translocation method. The long arm is the third longest chromosome arm in the barley karyotype. The Giemsa N-banding pattern of 2HL indicates that this telocentric chromosome carries a centromeric and an intercalary band that is similar to the banding of the long arm of complete chromosome 2H. Two faint intercalary dots on each chromatid, however, are observed on the long arm at prometaphase.

Telo 2HS has a centromeric band, two intercalary bands, and a terminal band. The band proximal to the centromere is darker than the other two, and the terminal band is rather faint.

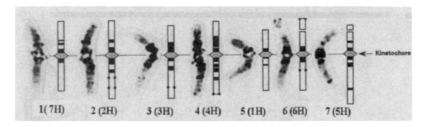

**Figure 9.3** Karyogram and idiogram of Giemsa N-banded chromosomes of barley. (From Singh, R.J. and Tsuchiya, T., *J. Hered.* 73, 227–229, 1982. With permission.)

### 9.2.3 Chromosome 3 (3H)

Tjio and Hagberg (1951) identified chromosome 3 as a median (arm ratio = 1.09) chromosome. If chromosomes 1 and 3 of Tjio and Hagberg (1951) are switched, their results will agree with those of Figure 9.3. Chromosome 3H shows a dark centromeric band. The band on the short arm appears as a large block at metaphase. The long arm has a dark interstitial band (close to the kinetochore) and a faint dot on each chromatid in the middle of the long arm (Figure 9.3). The morphological appearance of the telocentric chromosomes 3HL and 3HS indicate that they belong to the long and short arms of chromosome 3H, respectively.

### 9.2.4 Chromosome 4 (4H)

This chromosome contains kinetochore at the median region (l/s = 1.21). Conventional staining techniques do not distinguish chromosome 4H from chromosomes 1(7H), 2(2H), and 3 (3H). However, based on Giemsa C- and N-banding techniques, chromosome 4H is distinguished from the rest of the chromosomes because it is the most heavily banded in the barley complement; about 48% of the chromosome is heterochromatic. Sometimes it is difficult to locate the centromere position in condensed Giemsa banded metaphase chromosomes (Figure 9.2). However, the appearance of a diamond-shaped centromere position and the use of the aceto-carmine-stained Giemsa N-banding technique facilitates the precise localization of the kinetochore (Figure 9.3).

### 9.2.5 Chromosome 5 (1H)

This chromosome is the shortest among the five nonsatellited chromosomes of barley and has an arm ratio (1.42) similar to that of chromosome 3H. It has a centromeric band and an intercalary band on the long arm and a band on the short arm that is darker than those of the long arm (Figure 9.3).

### 9.2.6 Chromosome 6 (6H)

This chromosome has a larger satellite than chromosome 7 and has an arm ratio of 0.60 (without the satellite). Chromosome 6H shows a dark centromeric band in both arms and a faint dot on each chromatid on the telomere of the satellite at prometaphase. Telo 6HS is readily identified (Figure 9.3). Singh and Tsuchiya (1975) isolated, independently from other chromosomes, pachytene chromosome 6 associated with the nucleolus. Pachytene chromosomes of barley lack various distinctive landmarks, and they are about seven times longer than the somatic metaphase chromosomes.

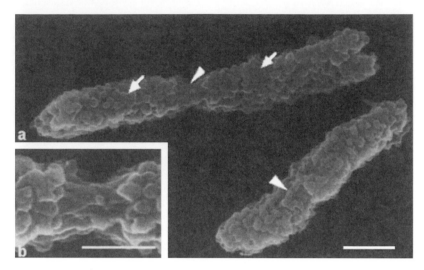

**Figure 9.4**   Scanning electron micrograph of barley chromosomes. (From Iwano, M. et al., *Chromo. Res.*, 5, 341–349, 1997. With permission.)

### 9.2.7   Chromosome 7 (5H)

This chromosome has the longest long arm in the barley karyotype and carries a submedian kinetochore. It shows an equally dense centromeric band at the distal portion of the long arm, and a faint intercalary band is observed in the short arm (Figure 9.3).

Iwano et al. (1997) studied the globular and fibrous structure in barley chromosomes by using high-resolution scanning electron microscopy. The surface of metaphase chromosomes was covered with globular protuberances. The centromeric region was smooth, but 20- to 40-nm fibers were folded compactly to form a higher-level organization surrounding the chromosomal axis (Figure 9.4).

## 9.3 DEVELOPMENT OF A CYTOGENETIC MAP OF BARLEY BY CHROMOSOME STRUCTURAL CHANGES

### 9.3.1   Deficiencies

Male and female spores of barley do not tolerate deficiency resulting in unviable plants. However, Finch (1983) identified a plant with 2n = 13 + 1 telo 6s), 6L missing, from the progeny of a plant 2n = 14 + 1 telo 4). Morphologically, the plant was dwarf, grew slowly, and had many tillers, and its leaves and spikes were deformed, highly sterile, and set two seeds. At meiotic metaphase I most of the sporocytes showed 6II + 1I. Deficiency was not transmitted as both seeds contained 2n = 14 chromosomes. Gecheff et al. (1994) identified a translocation line T-35 that contained homozygous deletion of rRNA genes residing in the nucleolus organizer region of 6H. Giemsa N-banding, *in situ* hybridization, and Southern blot analysis revealed an increased activity of the rRNA genes on 5H, the other nucleolus organizing region (NOR)-bearing chromosome. Deletion mapping in bread wheat can allow mapping markers with great precision (Endo and Gill 1996).

### 9.3.2   Duplications

An extra piece of chromosome segment either attached to the same homologous chromosome or transposed to one of the nonhomologous members of the genome is known as duplication

**Table 9.3  Chromosome Duplication Stocks**

| Chromosome | 7 (5H) | 6 (6H) | 5 (1H) | 4 (4H) | 3 (3H) | 2 (2H) | 1 (7H) | Sum of T-Breaks |
|---|---|---|---|---|---|---|---|---|
| 1 (7H) | 31 | 37 | 36 | 30 | 40 | 28 | — | Chromosome 1 (7H):202 |
| 2(2H) | 33 | 38 | 31 | 32 | 33 | — | | Chromosome 2 (2H):195 |
| 3(3H) | 33 | 27 | 28 | 35 | — | | | Chromosome 3 (3H):196 |
| 4(4H) | 29 | 28 | 35 | — | | | | Chromosome 4 (4H):185 |
| 5(1H) | 32 | 29 | — | | | | | Chromosome 5 (1H):187 |
| 6(6H) | 51 | — | | | | | | Chromosome 6 (6H):210 |
| 7(5H) | — | | | | | | | Chromosome 7 (5H): 209 |

Sum of T-breaks in collection: 1384

*Source*: Hagberg, A. and Hagberg, G., in *Chromosome Engineering in Plants: Genetics, Breeding, Evolution, Part A*, Gupta, P.K. and Tsuchiya, T., Eds., Elsevier, Amsterdam, 1991, pp. 401–410. With permission.

(Singh 2003). Hagberg and Hagberg (1991) reviewed the status of duplications in barley. Progeny of reciprocal translocations produces viable duplications in barley. The same pair of chromosomes must be involved in order to produce duplications, and most of the duplications have been produced in the barley cultivar Bonus. Table 9.3 lists the number of duplications available in barley. A total of 51 duplication lines produced in barley involved the short arms of chromosome 6 (6H) and 7 (5H). Both chromosomes have nucleolus organizers and can be easily identified. Furthermore, five duplications involving chromosome 5 (1H) and 6H and 5H were produced (Hagberg and Hagberg 1991).

Duplications in barley have a conspicuous morphological effect, and some, such as deficiency, are lethal. Duplications involving 1H are very weak or lost because of sterility problems. The majority of duplications for chromosomes 6H and 5H are also inferior morphologically to their parents. However, about 10% of the present duplication lines are equal or superior to their parents. Two duplication lines for proximal parts of chromosomes 6H and 5H showed increased grain yield over a 3-year period compared to their parents (Hagberg and Hagberg 1991).

Increased dosage of certain genes may enhance disease or insect resistance and enzyme activity in malting barley. Duplications would allow combinations of resistance genes that are now impossible to obtain because of tight linkage of gene loci (Ramage 1985).

### 9.3.3  Interchanges

Interchanges (translocations) are the result of the reciprocal exchange of terminal segments of nonhomologous chromosomes. Interchanges are induced spontaneously (Konishi and Lind-Laursen 1988) and by mutagens (Tjio and Hagberg 1951; Künzel et al. 1984). They were identified by Giemsa C-banding (Linde-Laursen 1988) and N-banding techniques (Xu and Kasha 1992), by partial pollen and seed sterility, and by genetic tests. Linde-Laursen (1988) localized breakpoints in 70 reciprocal translocations by Giemsa C-banding. All breakpoints were located in interband regions and were equally distributed among short and long arms. Interchanges in barley are used to isolate primary trisomics, identify and test the independence of linkage groups, associate new mutants to specific chromosomes, and conduct karyotype analysis (Singh 2003). This technique may be superior to the gene marker stock method because an interchange, in contrast to gene markers, usually does not affect the expression of other traits. Two hundred and forty translocation breakpoints of 120 T lines helped to construct physical restriction fragment length polymorphism (RFLP) maps of high resolution of seven barley chromosomes (Künzel et al. 2000).

### 9.3.4  Inversions

Paracentric inversions have been used to locate genes in specific segments of chromosome barley (Ekberg 1974).

## 9.4 DEVELOPMENT OF A CYTOGENETIC MAP OF BARLEY BY CHROMOSOME NUMERICAL CHANGES

### 9.4.1 Primary Trisomics

An individual with a normal chromosome complement plus an extra complete chromosome (2n = 2x + 1) is designated a primary trisomic, and the individual is called Triplo. Primary trisomics have been used extensively for the determination of gene–chromosome–linkage group relationships in several plant species (Singh 2003).

Generally, autotriploids, hypotriploids, and hypertriploids in barley are considered one of the best and most dependable sources for establishing primary trisomic series (see Singh 2003). A complete set of primary trisomics in barley has been isolated in a majority of diploid species from the progenies of autotriploids (3x) and autotriploid (3x) ∞ diploid (2x) crosses (Tsuchiya 1960, 1967). Another good source of primary trisomics is among the progenies of double trisomics or multiple trisomics. Tsuchiya (1960) obtained all seven primary trisomics from double and triple trisomics in barley. Occasionally, progenies of interchange heterozygotes produce primary trisomics due to a 3 and 1 chromosome disjunction at anaphase I from a non-co-oriented quadrivalent. The frequency of primary trisomics ranges from 0.94 to 5.65% (Ramage and Day 1960). The progenies of telotrisomics in barley produce primary trisomics with a low frequency, and the trisomic types are those to which telotrisomic belong (Singh and Tsuchiya 1975). The seven primary trisomics of barley have been classified into seven types based on distinct, easily noticeable morphological features (Chapter 7).

Theoretically, about 50% of primary trisomic plants are expected in the progenies of primary trisomics crossed as a female. However, this is rarely observed (Singh 2003). Trisomic seeds are thinner and later germinating, while larger seeds are generally disomic. Tsuchiya (1960) observed that small seeds showed poor germination but transmitted a considerably higher frequency of primary trisomic plants. The transmission of extra chromosomes through pollen in general is very low because in diploid species pollen with n + 1 chromosome constitution is unbalanced and generally unable to compete in fertilization, with pollen carrying the balanced, n, chromosome number.

In a critical combination, the genetic ratios are modified from 3:1 ($F_2$) and 1:1 ($BC_1$) to 17A:1a and 5A:1a, respectively (see Singh 2003). Primary trisomics are used to associate a marker gene with a particular chromosome, to determine chromosome–linkage group relationships, and to test the independence of linkage groups. Tsuchiya (1960) studied 10 marker genes belonging to four linkage testers with the seven primary trisomics of barley. The results indicated that chromosome 6 carried no genetic linkage group previously established, while two genetic linkage groups, III and VII, were located on chromosome 1. Thus, the linkage groups were reduced to six. However, several new genes, $e_c$ (early), $x_n$ (Xantha seedling), and $o$ (orange lemma), were associated with chromosome 6 by reciprocal translocations. Tsuchiya (1967) verified the association of gene $o$ with chromosome 6. Primary trisomics of barley have been used to locate qualitative traits to a particular chromosome. Shahla and Tsuchiya (1980) associated the gene $f3$ on chromosome 5 (1H) because $f3$ showed a trisomic ratio while it segregated in a disomic fashion with triplo 1 (7H), 2 (2H), 3 (3H), 4 (4H), 6 (6H), and 7 (5H).

McDaniel and Ramage (1970) distinguished primary trisomics of barley by disc electrophoresis of seed protein. A dosage effect of the extra chromosome differentiated alterations of protein bands. They suggested that the ultimate identification of a trisome is karyotype analysis, and protein and enzyme electrophoteric techniques should be used as an additional tool to identify aneuploidy.

### 9.4.2   Secondary Trisomics

A secondary trisomic contains an extra isochromosome. Tsuchiya (1960) identified only one secondary trisomic in the progeny of Triplo 2 (2H). Secondary trisomics resemble morphologically Triplo 2H.

### 9.4.3   Tertiary Trisomics

A tertiary trisomic contains an extra translocated chromosome consisting of parts of two nonhomologous chromosomes. Tertiary trisomics are isolated from interchanged heterozygotes due to a 3:1 segregation of a configuration of four chromosomes. A number of tertiary trisomics, primarily balanced tertiary trisomics, have been produced for use in commercial hybrid barley production (Ramage 1985). Tertiary trisomics are relatively vigorous, with good pollen and seed fertility. Some tertiary trisomics can be recognized morphologically, depending upon the composition of the tertiary chromosome.

### 9.4.4   Telotrisomics

A telocentric chromosome consists of a kinetochore (centromere) and one complete arm of a normal chromosome. Monotelotrisomics originate spontaneously, often in the progenies of barley primary, triploids, and novel compensating diploids (Tsuchiya 1991). The Giemsa N-banding pattern correctly identified and designated 11 telotrisomics of barley (Singh 2003).

Telotrisomics have been very useful in associating a gene with a particular arm of a chromosome in barley (Singh 2003). With the use of multiple marker stocks, centromere position and gene sequence on a linkage map can be determined. The principle of linkage studies with telotrisomics is different from that in simple primary trisomics. If a gene is not located on a particular arm of a chromosome, a disomic ratio is obtained for both disomic and trisomic portions (3:1::3:1). If a gene is on the telocentric chromosome, no recessive homozygote will be obtained in the telotrisomic portion, although the diploid portion will show a disomic ratio (3:1::4:0). A 7:1 ratio is expected in random chromosome segregation with a 50% female transmission rate of the telocentric chromosome. This ratio is narrowed to 5:1 when the female transmission rate approaches 33.3%.

### 9.4.5   Acrotrisomics

Trisomic plants carrying an extra acrocentric chromosome are designated acrotrisomics. In barley, six acrotrisomics (7HL[7HS], 3HL[3HS], 4HL[4HS], 1HS[1HL], 6HS[6HL], 5HS[5HL]) have been studied and identified morphologically and cytologically, and have been used in locating genes physically in chromosomes (Tsuchiya 1991). Acrotrisomics in barley have originated from the progeny of related primary trisomics. Precise identification of acrotrisomic plants in barley has been possible based on combining the aceto-carmine and Giemsa N-banding techniques (Singh 2003). All of the available four acrotrisomics of barley have been utilized in the physical localization of genes (Singh 2003).

## 9.5  MOLECULAR MAPS OF BARLEY

Barley genetics are unique because hundreds of phenotypically visible genetic mutations have been mapped (Franckowiak 1997). Characterization of the barley genome has evolved from initial linkage mapping to the current effort of physically mapping barley chromosomes. Genetic linkage mapping by recombination contributed to the development of the early chromosome maps (Robertson 1971). Cytogenetic linkage maps were developed when translocation and the primary trisomic methods were introduced (Tsuchiya 1991). Telotrisomic analysis further contributed in improving chromosomal maps by associating many genes with individual chromosome arms and locating centromere positions in the maps (Tsuchiya 1991).

More recently, a combination of cytogenetics with a molecular approach using polymerase chain reaction (PCR) has allowed the precise pinpointing of the location of translocations, leading to the creation of a physical map (Künzel et al. 2000). Additionally, sequencing of expressed sequence tags (ESTs) and large contiguous DNA fragments is contributing to characterize the barley genome (Smilde et al. 2001; Dubcovsky et al. 2001; Rostoks et al. 2002).

Since the early 1990s, a large number of molecular markers have been placed on the linkage maps of barley. Currently, over 1500 loci have been placed on these maps, providing a comprehensive catalog of markers and complete genome coverage. Several types of molecular markers, including restriction fragment length polymorphisms (RFLPs), random amplified polymorphic DNA (RAPD), sequence-tagged sites (STSs), simple sequence repeats (SSRs), and amplified fragment length polymorphisms (AFLPs), have been placed on these maps (Kleinhofs et al. 1993; Graner et al. 1991; Qi and Lindhout 1997; Costa et al. 2001). A database is maintained at http://barleygenomics.wsu.edu/, and mapped barley genes are periodically updated.

An aspect that may have lagged behind is the characterization of DNA-level variation and integration of molecular marker information with morphological marker maps of barley. Alleles for morphological trait loci may determine extreme phenotypes important for agricultural production (Robertson 1985) and may provide a starting point for comparative analysis of plant genomes. Several morphological traits of agronomic importance have been mapped using molecular markers such as the *denso* dwarfing gene (Laurie et al. 1993), the liguleless gene (Pratchett and Laurie 1994), and the *vrs1* locus (Komatsuda et al. 1997). The Oregon Wolfe barley (OWB) population (http://www.css.orst.edu/barley/WOLFEBAR/WOLFNEW.HTM) developed from the $F_1$ of a cross between dominant and recessive morphological marker stocks (Wolfe 1972), which showed an exceptional degree of phenotypic variation in a single reference population. A detailed molecular marker map of this population segregating for several morphological markers is available (Figure 9.5) and could serve as a framework for map integration, understanding genome organization, and map-based cloning of morphological trait loci (Costa et al. 2001). The OWB linkage map comprises a total linkage distance of 1387 cM. The average two-locus interval in the full map is 1.9 cM. The order of the markers and observed genetic distances are generally consistent with previously published molecular maps (Hayes et al. 1997; Kleinhofs et al. 1993; Qi et al. 1996).

## 9.6 UTILITY OF GENETIC MAPS OF BARLEY

Detailed genetic linkage maps may aid in map-based cloning of genes controlling morphological traits. For example, in the OWB population, seven markers (four AFLPs, one RFLP, and one SSR) cosegregated with the *rob* locus (Figure 9.5). These markers could be used as a starting point to identify genomic regions containing genes that determine morphological traits. Furthermore, Briggs (1991) indicated that a powerful method for the isolation of agronomically important genes from plants might be based on merging QTL maps. For example, the location of QTL for height in maize may coincide with that of dwarfing genes (Veldboom et al. 1994). A detailed genetic map of a population segregating for alleles at morphological trait loci could be used for testing a QTL in agronomically relevant populations (Robertson 1985). The Oregon Wolfe barley population is currently being used to test this hypothesis by mapping QTL for height, heading date, and other agronomic traits.

**Barley Chromosome 5  (1H)**

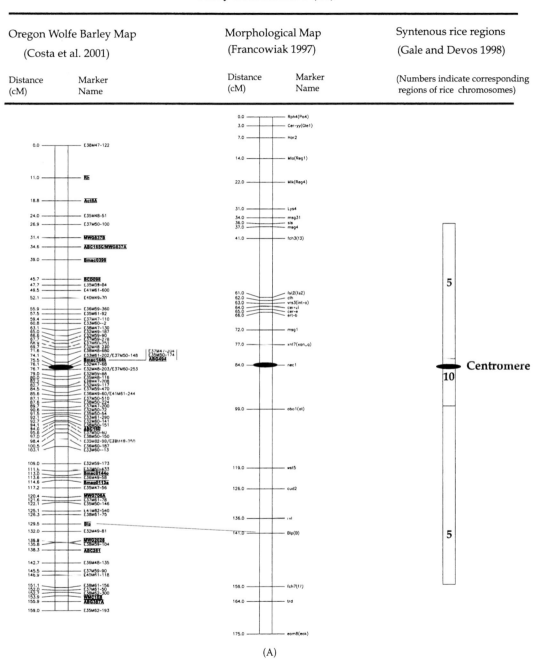

**Figure 9.5**  (A–G) Molecular and morphological marker maps of barley chromosomes, and syntenous rice chromosomal regions. (From Costa, J.M. et al., *Theor. Appl. Genet.*, 103, 415–424, 2001; Franck-owiak, J., *Barley Genet. Newsl.*, 26, 9–21, 1997; Gale, M. and Devos, K., *Science*, 282, 656–659, 1998. With permission.)

**Barley Chromosome 2  (2H)**

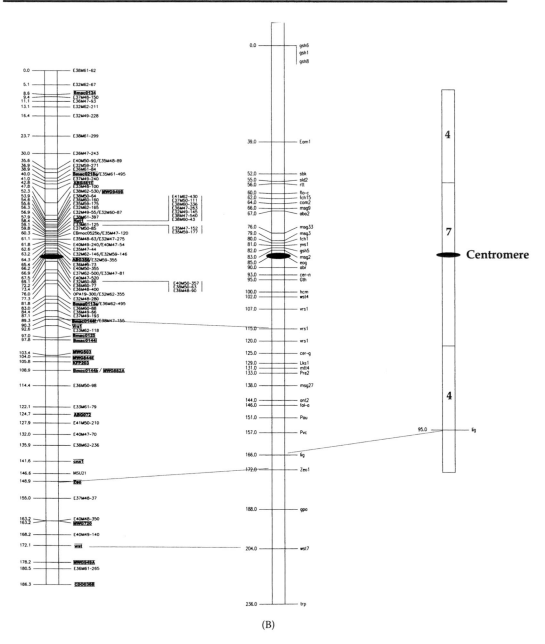

(B)

**Figure 9.5**     (*Continued*)

## Barley Chromosome 3 (3H)

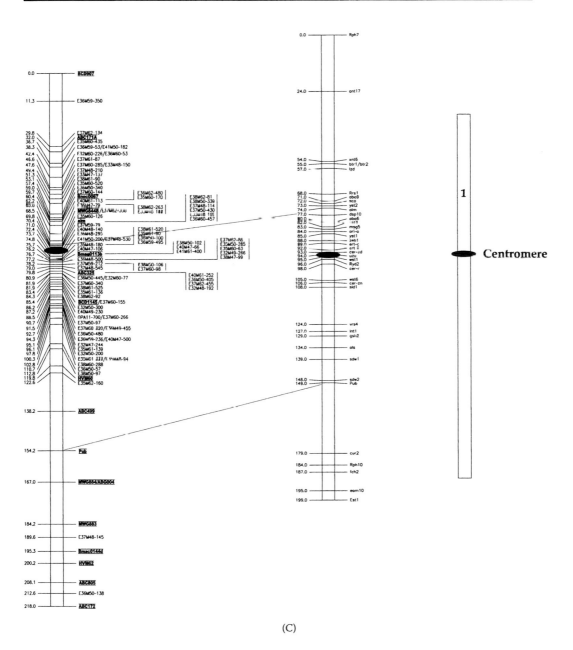

(C)

**Figure 9.5** (*Continued*)

**Barley Chromosome 4 (4H)**

| Oregon Wolfe Barley Map (Costa et al. 2001) | | Morphological Map (Francowiak 1997) | | Syntenous rice regions (Gale and Devos 1998) |
|---|---|---|---|---|
| Distance (cM) | Marker Name | Distance (cM) | Marker Name | (Numbers indicate corresponding regions of rice chromosomes) |

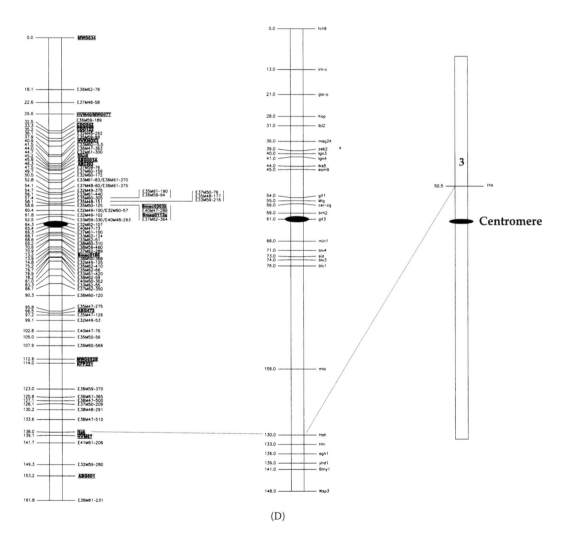

(D)

**Figure 9.5** (*Continued*)

## 9.7 PHYSICAL MAPPING OF BARLEY CHROMOSOMES

Translocations are useful to estimate the relationship between physical and genetic distances in barley. Künzel et al. (2000) used translocation stocks for precise location of the PCR marker with respect to the translocation. They observed high, medium, and low recombination regions in

**Barley Chromosome 7 (5H)**

| Oregon Wolfe Barley Map | Morphological Map | Syntenous rice regions |
|---|---|---|
| (Costa et al. 2001) | (Francowiak 1997) | (Gale and Devos 1998) |

| Distance (cM) | Marker Name | Distance (cM) | Marker Name | (Numbers indicate corresponding regions of rice chromosomes) |
|---|---|---|---|---|

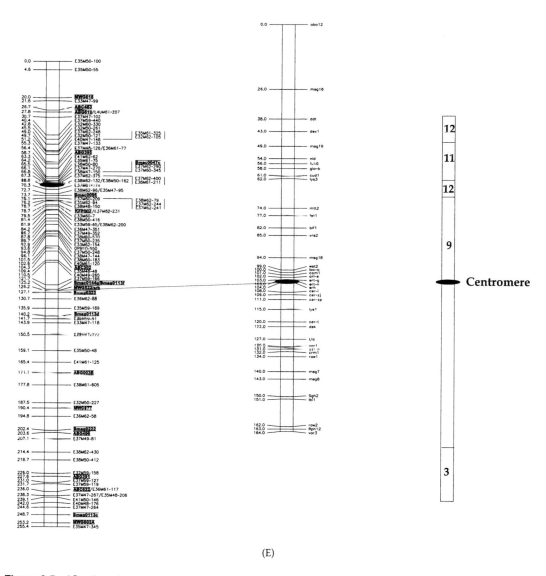

(E)

**Figure 9.5** *(Continued)*

barley chromosomes. The high and low recombination regions are presumed to be gene rich, while low recombination regions are considered to be gene poor (Barakat et al. 1997). Sequencing of a 60-kb region of the barley genome has revealed that there are indeed gene-rich islands (Panstruga et al., 1998). A density of approximately one gene per 20 kb was observed at the *mlo* and *Rar1*

**Barley Chromosome 6 (6H)**

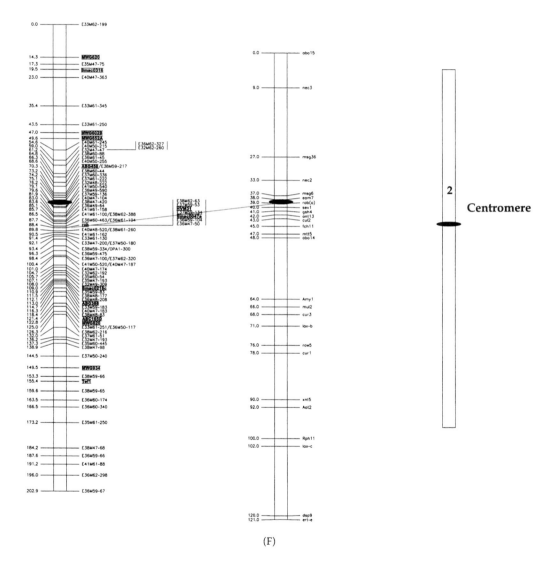

(F)

**Figure 9.5** (*Continued*)

loci (Panstruga et al. 1998; Shirasu et al. 2000), while a density of one gene per 4 to 5 kb was observed for the *Lrk* loci (Feuillet and Keller 1999). The gene density within gene-rich regions is similar to that observed in *Arabidopsis thaliana* (Bevan et al. 1998).

**Barley Chromosome 1 (7H)**

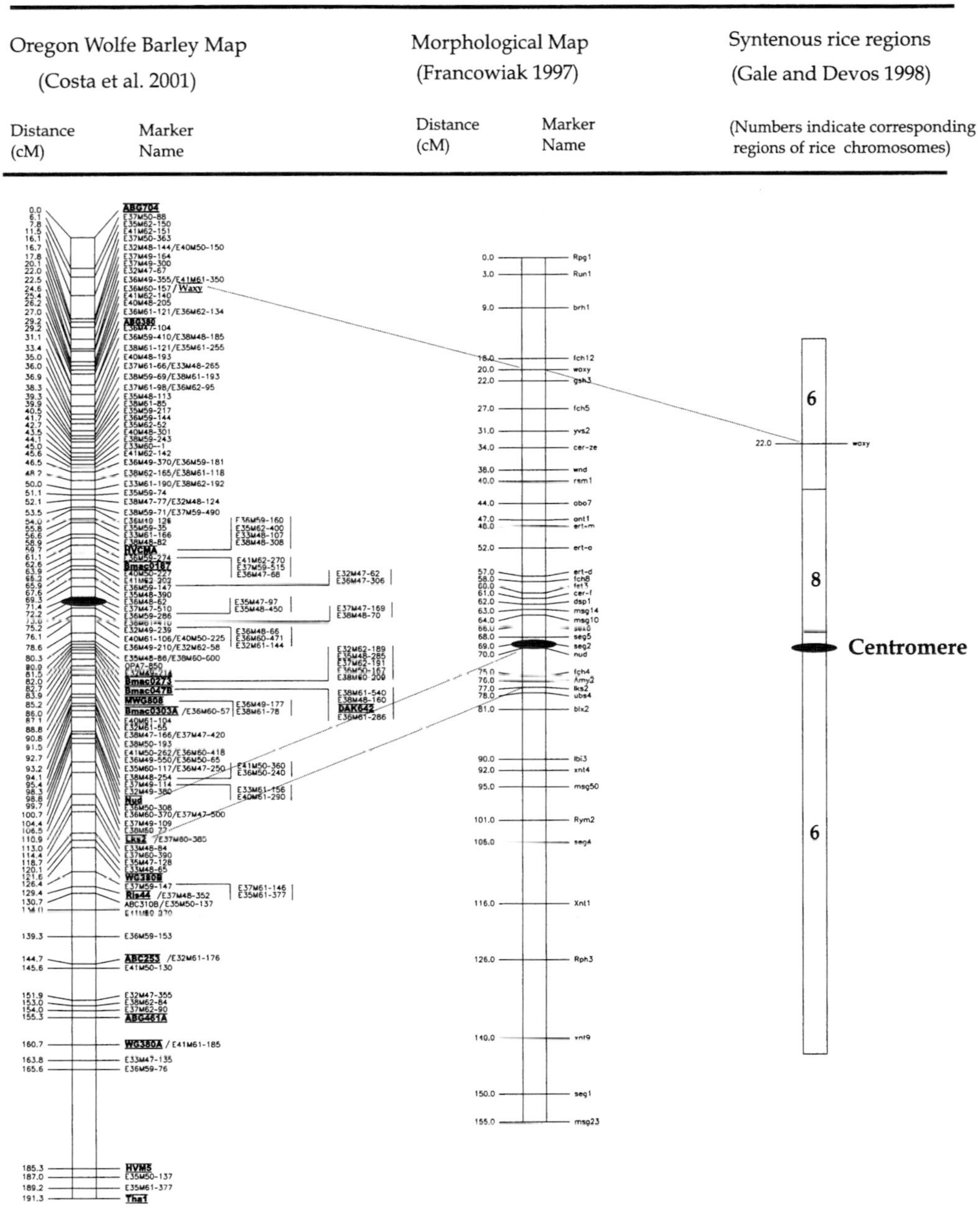

Oregon Wolfe Barley Map
(Costa et al. 2001)
Distance (cM)
Marker Name
Morphological Map
(Francowiak 1997)
Distance (cM)
Marker Name
Syntenous rice regions
(Gale and Devos 1998)
(Numbers indicate corresponding regions of rice chromosomes)
6
8
Centromere
6

(G)

**Figure 9.5** (*Continued*)

## 9.8 COMPARATIVE MAPPING IN THE GRASS FAMILY

The genomes of grasses and other important crop plants are different in size, ploidy level, and chromosome number (Table 9.1). Despite these significant differences, comparative mapping showed that the linear order (colinearity) of genetic markers and genes is relatively conserved among grass genomes, with very few chromosomal rearrangements (Van Deynze 1998). It is estimated that the seven chromosomes of barley may have similar gene content compared to other members of the Triticeae (rice, barley, wheat, maize, sorghum, and oats). This resource, called Gramene, focuses on rice but also provides information about maps, DNA sequences, genes, markers, QTL, and publications on other grasses, such as barley, and their relation with rice (Ware et al. 2002). These relationships among the genetic maps of grasses can be viewed at http://www.gramene.org. Currently, only a basic RFLP map derived from Steptoe by the Morex barley population (Kleinhofs et al. 1993) is represented in Gramene.

Rice has become the model system for the Poaceae because of its small genome size and commercial importance. Large efforts have been undertaken to obtain detailed genetic and physical maps of the rice genome, and a draft genomic sequence of rice is now available (Goff et al. 2002; Yu et al. 2002). Based on comparative mapping, the rice genome has extensive colinearity or synteny with other grasses, such as maize, sorghum, wheat, and barley (Gale and Devos 1998). The development of a large-insert library in barley (Yu et al. 2000) has allowed the study of large stretches of DNA sequences and has provided insight into the gene organization of barley. This allows direct comparison of corresponding regions in rice (Dubcovsky et al. 2001; Feuillet and Keller 1999, 2002). Data from comparative DNA sequencing among rice, barley, sorghum, and maize suggest that the difference in genome size between these organisms is primarily caused by multiple insertions of mobile genetic elements (retrotransposons) into noncoding regions, while the order of coding sequences is relatively well conserved (Dubcovsky et al. 2001; Sandhu and Gill 2002). Most of the detected rearrangements are small, involving inversion, duplication, translocation, or deletion of DNA segments that contain only one to three genes (Bennetzen and Ramkrishna 2002). These results indicate that thousands of small genetic rearrangements may have occurred in several grass lineages since their divergence from a common ancestor. These rearrangements have largely been missed by genetic mapping and may limit the use of the information collected from rice genome sequencing (Feuillet and Keller 2002).

Comparison of a large collinear region in barley (chromosome 5H) and rice (chromosome 3) by Dubcovsky et al. (2001) indicated that four predicted genes found in this region were conserved. These genes were present in the same orientation in rice; one was inverted and a second one was duplicated in tandem in barley but not in rice. Gene size and location and exon number were mostly conserved between barley and rice, but they also observed that these four genes were not colinear in Arabidopsis. In addition to the similarities within the exons of structural genes, a short stretch of homology was observed in the promoters and 3' untranslated regions. No extensive similarity was found in the intergenic regions. The larger distance between genes in barley was explained by the presence of transposable retroelements. Additionally, a set of rice expressed sequence tags (ESTs) previously mapped on rice chromosome 1 was used for comparative mapping in barley (Smilde et al. 2001). They observed that most clones showed distinct banding patterns in barley. Most polymorphic clones in a cross between Igri and Franka have been mapped, as predicted, to the corresponding barley chromosome 3H (Figure 9.5). Their comparative maps of barley chromosome 3H and rice chromosome 1 comprise 26 common markers covering more than 95% of the genetic length of both chromosomes. A 30-fold reduction of recombination was observed around the barley centromere, and synteny may be interrupted in this region.

Conservation of linkage is generally observed for many genes that have predominantly housekeeping functions (Peng et al. 1999). Some exceptions to the synteny or colinearity of genes of grasses, including barley, have been observed in genes for disease resistance (Leister et al. 1998). Disease resistance and their analog genes (RGAs) do not map in regions where they are expected based on syntenic relationships between barley and rice. It has been speculated that this may be due to the rapid evolution of disease resistance clusters, which is necessary to develop new specificities against the rapidly evolving pathogens (Hulbert et al. 2001). The degree of synteny of RGAs is clearly higher among more closely related species. RGA loci were identified in the vicinity of the barley leaf rust resistance loci *Rph4*, *Rph7*, and *Rph10*, whose orthologous RGAs were conserved between barley and wheat (Collins et al. 2001). Feuillet and Keller (2002) have extensively reviewed the progress and future prospects of comparative genomics in grasses.

To view the general relationships between barley and rice, the map of 713 molecular and 13 morphological markers from the OWB population, a map of barley morphological markers (Franckowiak 1997), and the approximate corresponding or syntenic chromosomal regions from rice (Gale and Devos 1998) are aligned in Figure 9.5. The chromosomal map of morphological markers is based on genetic linkage data, translocation, and telotrisomic analyses. The names of individual chromosomes and the order in which they are presented follow the Triticeae scheme for naming barley chromosomes (Linde-Laursen 1997). Revised locus symbols for morphological markers are based on a three-letter code (Franckowiak 1997).

## 9.9 CHROMOSOME EVOLUTION MECHANISMS IN BARLEY AND OTHER GRASSES

Sequence analyses in various species suggest uneven distribution of genes in the genomes of all grasses (Feuillet and Keller 1999; Shirasu et al. 2000). Comparison of the physical maps with genetic linkage maps showed that recombination in wheat and barley is confined to the gene-containing regions (Sandhu and Gill 2002). Gene-poor regions are composed of retrotransposon-like nontranscribing repeats and pseudogenes. In general, large-genome cereals such as barley and wheat have over 70% of the genome that is reiterated DNA (Feuillet and Keller 2002; Sandhu and Gill 2002). Simple repeats in the rye genome (found in tandem arrays) are in complex repeating units. This suggests that the unit amplified and then moved out into the rest of the genome. This basic unit is found in wheat and rye genomes, suggesting that it amplified before these two species diverged (Feuillet and Keller 2002; Sandhu and Gill 2002). In barley, repetitive sequences represent more than 70% of the genome (Vicient et al. 1999) but vary along the genome. For example, in a 60-kb region of the barley *mlo* locus that was sequenced, three genes were identified in addition to retrotransposons and complex repeat structures. The intergenic regions represented 25% of the DNA in this interval (Panstruga et al. 1998). In a similar study of the 66-kb *rar1* locus region in barley, a similar repeat structure accounted for 50% of the DNA sequences (Shirasu et al. 2000). The genome size difference between barley and rice can therefore be explained in part by the amplification of gene-poor regions. Preferential transposition to adjacent locations and the presence of vital genes flanking a gene-rich region may have restricted retrotransposon amplification to intergenic regions, resulting in tandem blocks of nontranscribing repeats. Insertional inactivation by adjoining retroelements and selection may have played a major role in stabilizing genomes (Sandhu and Gill 2002). The small nuclear genome of Arabidopsis, proposed as a model plant for genome studies, is drastically different from the genomes of grasses because it lacks the gene-empty regions separating gene clusters and does not have abundant transposon sequences in the intergenic regions (Bevan et al. 1998).

Initial sequencing information across grass species indicates that some gene families in the Triticeae have evolved by duplication and translocation events. For example, the *Lrk* gene family is duplicated in 1H and singly in 3H in wheat and barley (Feuillet and Keller 1999). It is hypothesized that the *Lrk* gene family evolved from a putative ancestor *Tak-Lrk* that translocated from chromosome 1H to 3H and duplicated. In rice and maize, it is duplicated only in 1H. Therefore, it is assumed that the *Lrk* gene family was originally on 1H and a copy moved to 3H (Feuillet and Keller 2002).

## 9.10 LARGE-INSERT LIBRARIES OF BARLEY CHROMOSOMES

Large DNA insert libraries, such as yeast artificial chromosome (YAC) or bacterial artificial chromosome (BAC), are essential tools for characterizing the genome of any organism and greatly facilitate map-based cloning of genes. In barley, initial efforts to develop these types of libraries involved using YAC clones (Kleine et al. 1993). Their use, however, was limited because of the presence of chimeric and unstable clones. BAC libraries are simpler to develop and tend to have a lower frequency of chimeric clones. Initially, a small BAC library that provided approximately one genome equivalent was developed by Lapitan et al. (1997). More recently, a large BAC library was constructed by Yu et al. (2000), which contains approximately six genome equivalents. This library provides a higher than 99% probability of recovering any specific sequence of interest. Initial efforts to screen this library indicated good genome coverage. Numerous BAC clones that represented more than 20 different genomic regions were detected using RGA probes (Yu et al. 2000). The availability of BAC libraries will be a valuable resource for future positional cloning of important genes in barley and for the study of gene distribution along the barley genome (Dubcovsky et al. 2001).

## 9.11 BARLEY EXPRESSED SEQUENCE TAGS

Expressed sequence tags (ESTs) are short (usually about 300 to 500 bp in length), single-pass cDNA sequence copies from mRNA. Typically, they are produced in large batches. They represent a snapshot of genes expressed in a given tissue or at a given developmental stage. They are tags (some coding, others not) of expression for a given cDNA library (Boguski et al. 1993). A large effort to identify and catalog ESTs is under way. By September 2003, there were 18,140,083 ESTs in the National Center for Biotechnology (NCBI) database, isolated mostly from humans and mouse. Barley researchers have generated numerous ESTs (over 300,000) from various tissues and developmental stages. A comparison of the progress of EST identification in barley and other plant species is presented in Table 9.4. To sort out the large amount of EST data into nonredundant sets of gene-oriented clusters, NCBI has devised an experimental system for automatically partitioning GenBank EST sequences into UniGenes. Each UniGene cluster contains sequences that represent a unique gene, as well as related information, such as the tissue types in which the gene was expressed.

Barley EST data can be utilized in a number of ways. This is generally termed data mining (Kantety et al. 2002). It consists of using the EST information databases to search for specific types of DNA sequences, such as simple repeats *in silico* (through computer searches). Kantety et al. (2002) searched the barley EST database for SSRs that can amplify across the grass genera for comparative mapping and genetics. They analyzed over 260,000 ESTs of barley, rice, maize, and wheat for their potential use in developing simple sequence repeat (SSR) markers. The frequency of SSR-containing ESTs (SSR-ESTs) of barley was 3.4%. Additionally, they also identified ESTs related to SSR-ESTs by BLAST analysis. The consensus and singleton sequences from each species were pooled and clustered to identify cross-species matches. Future functional analysis may reveal their role in plant metabolism and gene evolution (Kantety et al. 2002). Another use of the EST

**Table 9.4   Number of Expressed Sequence Tags (ESTs)
and Unigenes in the NCBI Database
(http://www.ncbi.nlm.nih.gov)**

| Plant Species | ESTs | Unigenes |
|---|---|---|
| Wheat (*Triticum aestivum*) | 449,979 | 22,841 |
| Barley (*Hordeum vulgare*) | 348,188 | 11,491 |
| Maize (*Zea mays*) | 362,510 | 13,137 |
| Thale cress (*Arabidopsis thaliana*) | 188,782 | 26,366 |
| Rice (*Oryza sativa*) | 260,783 | 31,080 |

*Note*: Information updated September 5, 2003.

data is to aid in identifying markers that map to a specific region of the barley genome by taking advantage of the knowledge of colinear regions between the rice and barley genomes (Figure 9.5). Once the region of the rice sequence is identified, anonymous EST clones from barley can be compared and matched with the homologous rice sequence (Goff et al. 2002; Yu et al. 2002). Previous attempts to use rice markers corresponding to a specific barley genome region exploiting synteny between barley and rice have only been partially successful (Kilian et al. 1995). One of the reasons for this partial success was the specific barley genomic region that was not sufficiently saturated with markers close enough to allow for chromosome landing within the average size of the BAC inserts available for barley (Tanksley et al. 1995; Yu et al. 2000). Instead, the potential of placing numerous barley ESTs based on the orthologous rice genomic region is very high because the number of deposited barley EST sequences in GenBank is high and constantly growing. Using this technique, Kudrna et al. (2002) mapped approximately 50% of anonymous barley ESTs to a specific target region in the barley genome.

A barley searchable genomics database has been established to manage the analysis of a large amount of genomic information that is currently being generated in barley through EST generation and sequencing, combined with the ever-increasing number of molecular markers (Druka et al. 2001). This database integrates data from the existing barley genetic and physical maps, mutant information, BAC library screening, EST sequencing, and mapping results (http://barleygenomics.wsu.edu/databases/databases.html).

## 9.12 CONCLUSIONS

Barley is a model plant species for genetic studies of cereals with a long and rich history of chromosome mapping. A large number of mutant genotypes have been characterized and mapped to barley chromosomes by aneuploidy. Numerous molecular markers are now available. Colinearity of genes has been observed along chromosomes of all the major cereals, with some exceptions. Although a small rice genome is being used as a source of probes and sequence information, there are similarities in transcribed regions, but not in untranscribed regions. Additionally, the use of rice as a model to support gene isolation from other grass genomes may be complicated by local rearrangements. Physical mapping of the barley genome is now under way by breakpoint translocations and direct sequencing of large fragments of genomic DNA. Several sequencing studies indicate that the barley genome has evolved by duplication, translocations, and insertion of retrotransposons that have generated a large proportion of repetitive DNA. Recent advances in genetic research of barley include the availability of large-insert DNA libraries, transposon tagging stocks, and sequencing of a large number of ESTs. These advances are generating an exciting pool of resources that will aid in fully characterizing barley chromosomes in the near future.

## REFERENCES

Arumuganathan, K. and Earle, E.D. 1991. Nuclear DNA content of some important plant species. *Plant Mol. Biol. Rep.* 9: 211–215.

Barakat, A., Carels, N., and Bernardi, G. 1997. The distribution of genes in the genomes of Gramineae. *Proc. Natl. Acad. Sci. U.S.A.* 94: 6857–6861.

Bennett, M.D. and Smith, J.B. 1976. Nuclear DNA amounts in angiosperms. *Phil. Trans. R. Soc. Lond. Biol.* 274: 227–274.

Bennetzen, J.L. and Ramakrishna, W. 2002. Numerous small rearrangements of gene content, order and orientation differentiate grass genomes. *Plant Mol. Biol.* 48: 821–827.

Bevan, M.I. et al. 1998. Analysis of 1.9 Mb of contiguous sequence from chromosome 4 of *Arabidopsis thaliana*. *Nature* 391: 485–488.

Boguski, M.S., Lowe, T.M., and Tolstoshev, C.M. 1993. dbEST: database for "expressed sequence tags." *Nat. Genet.* 4: 332–333.

Briggs, S.P. 1991. Identification and isolation of agronomically important genes from plants. In *Plant Breeding in the 1990s: Proceedings of the Symposium on Plant Breeding in the 1990s*, H.T. Stalker and J.P. Murphy, Eds. C.A.B. International, Wallingford, U.K.

Brown, S.E. et al. 1999. FISH landmarks for barley chromosomes (*Hordeum vulgare* L.). *Genome* 42: 274–281.

Busch, W. et al. 1995. Repeated DNA sequences isolated by microdisection. I. Karyotyping of barley (*Hordeum vulgare* L.). *Genome* 38: 1082–1090.

Collins, N. et al. 2001. Resistance genes analogs in barley and their relationship to rust resistance genes. *Genome* 44: 375–381.

Costa, J.M. et al. 2001. Molecular mapping of the Oregon Wolfe barleys: a phenotypically polymorphic doubled-haploid population. *Theor. Appl. Genet.* 103: 415–424.

Druka, A. et al. 2001. The Barley Genome Database at Washington State University: http://barleygenomics.wsu.edu/databases/databases.html. Verified October 5, 2004.

Dubcovsky, J. et al. 2001. Comparative sequence analysis of colinear barley and rice bacterial artificial chromosomes. *Plant Physiol.* 125: 1342–1353.

Ekberg, I. 1974. Cytogenetic studies of three paracentric inversions in barley. *Hereditas* 76: 1–30.

Endo, T.R. and Gill, B.S. 1996. The deletion stocks of common wheat. *J. Hered.*, 87: 295–307.

Feuillet, C. and Keller, B. 1999. High gene density is conserved at syntenic loci of small and large grass genomes. *Proc. Natl. Acad. Sci. U.S.A.* 96: 8265–8270.

Feuillet, C. and Keller, B. 2002. Comparative genomics in the grass family: molecular characterization of grass genome structure and evolution. *Ann. Bot.* 89: 3–9.

Finch, R.A. 1983. Deficiency of 6L in barley. *Barley Genet. Newsl.* 13: 2–3.

Franckowiak, J. 1997. Revised linkage maps for morphological markers in barley, *Hordeum vulgare*. *Barley Genet. Newsl.* 26: 9–21.

Fukui, K. 1986. Standardization of karyotyping plant chromosomes by a newly developed chromosome image analyzing system (CHIAS). *Theor. Appl. Genet.* 72: 27–32.

Fukui, K. and Kakeda, K. 1990. Quantitative karyotyping of barley chromosomes by image analysis methods. *Genome* 33: 450–458.

Fukui, K., Kamisugi, Y., and Sakai, F. 1994. Physical mapping of 5S rDNA loci by direct-cloned biotinylated probes in barley chromosomes. *Genome* 37: 105–111.

Gale, M.D. and Devos, K.M. 1998. Plant comparative genetics after 10 years. *Science* 282: 656–659.

Gecheff, K. et al. 1994. Cytological and molecular evidence of deletion of ribosomal RNA genes in chromosome 6 of barley (*Hordeum vulgare*). *Genome* 37: 419–425.

Goff, S.A. et al. 2002. A draft sequence of the rice genome (*Oryza sativa* L. ssp. *japonica*). *Science* 296: 92–100.

Graner, A. et al. 1991. Construction of an RFLP map of barley. *Theor. Appl. Genet.* 83: 250–256.

Hagberg, A. and Hagberg, G. 1991. Production and analysis of chromosome duplications in barley. In *Chromosome Engineering in Plants: Genetics, Breeding, Evolution, Part A*, P.K. Gupta and T. Tsuchiya, Eds. Elsevier, Amsterdam, pp. 401–410.

Hayes, P.M. et al. 1996. Barley genome mapping and its applications. In *Methods of Genome Analysis in Plants*, P.P. Jauhar, Ed., CRC Press, Boca Raton, FL. pp. 229–249.

Hayes, P.M. et al. 1997. Characterizing and exploiting genetic diversity and quantitative traits in barley (*Hordeum vulgare*) using AFLP markers. *J. Agr. Genomic* (http://www.cabi-publishing.org/gateways/jag/index.html, verified October 5, 2004).

Hulbert, S.H. et al. 2001. Resistance gene complexes: evolution and utilization. *Ann. Rev. Phytopathol.* 39: 285–312.

Iwano, M. et al. 1997. Globular and gibrous structure in barley chromosomes revealed by high-resolution scanning electron microscopy. *Chrom. Res.* 5: 341–349.

Kakeda, K. and Fukui, K. 1994. Dynamic changes in the morphology of barley chromosomes during the mitotic metaphase stage. *Jpn. J. Genet.* 69: 545–554.

Kantety, R.V. et al. 2002. Data mining for simple sequence repeats in expressed sequence tags from barley, maize, rice, and sorghum and wheat. *Plant Mol. Biol.* 48: 501–510.

Kilian, A. et al. 1995. Rice-barley synteny and its application to saturation mapping of the barley Rpg1 region. *Nucleic Acids Res.* 23: 2729–2733.

Kleine, M. et al. 1993. Construction of a barley (*Hordeum vulgare* L.) YAC library and isolation of a Hor1-specific clone. *Mol. Gen. Genet.* 240: 265–272.

Kleinhofs, A. and Han F. 2001. Molecular mapping of the barley genome. In *Barley Science: Recent Advances from Molecular Biology to Agronomy of Yeild and Quality.* Slafer, C.G., Araus, J.L., Savin, R., and Romagosa, I., Eds., Food Product Press, New York, pp. 31–45.

Kleinhofs, A. and Graner, A. 2001. An integrated map of the barley genome. In *DNA-Based Markets in Plants.* Phillips, R.L. and Vasil, I.K., Eds. Kluwer Academic Publishers, Dordrecht, The Netherlands, pp. 187–199.

Kleinhofs, A. et al. 1993. A molecular isozyme and morphological map of the barley (*Hordeum vulgare*) genome. *Theor. Appl. Genet.* 86: 705–712.

Komatsuda, T. et al. 1997. Identification of random amplified polymorphic DNA (RAPD) markers linked to the v locus of barley *Hordeum vulgare* L. *Theor. Appl. Genet.* 95: 637–642.

Konishi, T. and Linde-Laursen, I. 1988. Spontaneous chromosomal rearrangements in cultivated and wild barleys. *Theor. Appl. Genet.* 75: 237–243.

Kudrna, D.A. et al. 2002. Targeted saturation mapping of a high recombination region in barley using ESTs identified via synteny to rice. *Barley Genet. Newsl.* 32: 13–21.

Künzel, G. 1976. Indications for a necessary revision for the barley karyogramme by use of translocations. In *Barley Genetics III. Proceedings of the 3rd. International Barley Genetics Symposium,* H. Gaul, Ed., München, Thiemig, pp. 275–281.

Künzel, G., Gramatikova, M., and Hamann, S. 1984. Isolation of radiation-induced translocations in spring and winter barley. *Biol. Zbl.* 103: 649–653.

Künzel, G., Korzun, L., and Meister, A. 2000. Cytologically integrated physical restriction fragment length polymorphism maps of the barley genome based on translocation breakpoints. *Genetics* 154: 397–412.

Lapitan, N.L.V. et al. 1997. FISH physical mapping with barley BAC clones. *Plant J.* 11: 149–156.

Lauric, D.A. et al. 1993. Assignment of the *denso* dwarfing gene to the long arm of chromosome 3(3H) of barley by use of RFLP markers. *Plant Breed.* 111: 198–203.

Laurie, D.A. and Bennett, M.D. 1985. Nuclear DNA content in the genera *Zea* and *Sorghum*: intergeneric, interspecific and intraspecific variation. *Heredity* 55: 307–313.

Leister, D. 1998. Rapid reorganization of resistance gene homologues in cereal chromosomes. *Proc. Natl. Acad. Sci. U.S.A.* 95: 370–375.

Linde-Laursen, I. 1997. Recommendations for the designation of the barley chromosomes and their arms. *Barley Genet. Newsl.* 26: 1–3.

Lysák, M.A. et al. 1999. Flow karyotyping and sorting of mitotic chromosomes of barley (*Hordeum vulgare* L.). *Chromo. Res.* 7: 431–444.

McDaniel, R.G. and Ramage, R.T. 1970. Genetics of a primary trisomics series in barley: identification by protein electrophoresis. *Can. J. Genet. Cytol.* 12: 490–495.

Nakayama, S. and Fukui, K. 1995. Detection of sister chromatid exchange sites by sequential staining and imaging methods. *Jpn. J. Genet.* 70: 267–271.

Panstruga, R. et al. 1998. A continuous 60 kb genomic stretch from barley reveals molecular evidence for gene islands in a monocot genome. *Nucleic Acids Res.* 26: 1056–1062.

Pedersen, C. and Linde-Laursen, I. 1995. The relationship between physical and genetic distances at the *Hor1* and *Hor2* loci of barley estimated by two-color fluorescent *in situ* hybridization. *Theor. Appl. Genet.* 91: 941–946.

Peng, A.H. et al. 1999. 'Green revolution' genes encode mutant gibberellin response modulators. *Nature* 400: 256–261.

Pratchett, N. and Laurie, D.A. 1994. Genetic-map location of the barley developmental mutant liguleless in relation to RFLP markers. *Hereditas* 120: 35–39.

Qi, X. and Lindhout, P. 1997. Development of AFLP markers in barley. *Mol. Gen. Genet.* 254: 330–336.

Qi, X., Stam, P., and Lindhout, P. 1996. Comparison and integration of four barley genetic maps. *Genome* 39: 379–394.

Ramage, R.T. 1985. Cytogenetics. In *Barley,* D.C. Rasmusson, Ed. American Society of Agronomy/Crop Science Society of America/Soil Science of America, Madison, WI, pp. 127–154.

Ramage, R.T. and Day, A.D. 1960. Separation of trisomic and diploid barley seeds produced by interchange heterozygotes. *Agron. J.* 52: 590–591.

Robertson, D.W. 1971. Recent information of linkage and chromosome mapping. In *Barley Genetics. II. Proceedings of the 2nd International Barley Genetics Symposium*, R.A. Nilan, Ed. Washington State University Press, Pullman.

Robertson, D. 1985. A possible technique for isolating genic DNA for quantitative traits in plants. *J. Theor. Biol.* 117: 1–10.

Rostoks, N. et al. 2002. Genomic sequencing reveals gene content, genomic organization and recombination relationships in barley. *Funct. Integr. Genomics* 2: 51–59.

Sandhu, D. and Gill, K.S. 2002. Gene-containing regions of wheat and the other grass genomes. *Plant Physiol.* 128: 803–811.

Schondelmaier, J. et al. 1993. Microdissection and microcloning of the barley (*Hordeum vulgare* L.) chromosome 1HS. *Theor. Appl. Genet.* 86: 629–636.

Shahla, A. and Tsuchiya, T. 1980. Trisomic analysis of the gene (*f3*) for chlorina 3 in barley. *J. Hered.* 71: 359–361.

Shirasu, K. et al. 2000. A contiguous 66-kb barley DNA sequence provides evidence for reversible genome expansion. *Genome Res.* 10: 908–915.

Singh, R.J. 2003. *Plant Cytogenetics*, 2nd ed. CRC Press, Boca Raton, FL.

Singh, R.J. and Tsuchiya, T. 1975. Pachytene chromosomes of barley. *J. Hered.* 66: 165–167.

Singh, R.J. and Tsuchiya, T. 1981. Identification and designation of barley chromosomes by Giemsa banding technique: a reconsideration. *Z. Pflanzenzüchtg.* 86: 336–340.

Singh, R.J. and Tsuchiya, T. 1982. An improved Giemsa N-banding technique for the identification of barley chromosomes. *J. Hered.* 73: 227–229.

Smilde, W.D. et al. 2001. New evidence for the synteny of rice chromosome 1 and barley chromosome 3H from rice expressed sequence tags. *Genome* 44: 361–367.

Sorokin, A. et al. 1994. Polymerase chain reaction mediated localization of RFLP clones to microisolated translocation chromosomes of barley. *Genome* 37: 550–555.

Tanksley, S.D., Ganal, M.W., and Martin, G.B. 1995. Chromosome landing: a paradigm for map-based gene cloning in plants with large genomes. *Trends Genet.* 11: 63–68.

Tjio, J.H. and Hagberg, A. 1951. Cytological studies on some x-ray mutants of barley. *An. Estacion Exp. Aula Dei.* 2: 149–167.

Tsuchiya, T. 1960. Cytogenetic studies of trisomics in barley. *Jpn. J. Bot.* 17: 177–213.

Tsuchiya, T. 1967. The establishment of a trisomics series in a two-rowed cultivated variety of barley. *Can. J. Genet. Cytol.* 9: 667–682.

Tsuchiya, T. 1991. Chromosome mapping by means of aneuploids analysis in barley. In *Chromosome Engineering in Plants: Genetics, Breeding, Evolution, Part A.* P.K. Gupta and T. Tsuchiya, Eds. Elsevier, Amsterdam, pp. 361–384.

Tuleen, N.A. 1973. Karyotype analysis of multiple translocation stocks of barley. *Can. J. Genet. Cytol.* 15: 267–273.

Van Deynze, A.E. et al. 1995. Molecular-genetic maps for group 1 chromosomes of Triticeae species and their relation to chromosomes in rice and oat. *Genome* 38: 45–59.

Veldboom, L.R., Lee, M., and Woodman, W.L. 1994. Molecular marker-facilitated studies in an elite maize population. I. Linkage analysis and determination of QTL for morphological traits. *Theor. Appl. Genet.* 88: 7–16.

Vicient, C.M. et al. 1999. Structure, functionality, and evolution of the BARE-1 retrotransposon of barley. *Genetica* 107: 53–63.

Ware, D. et al. 2002. Gramene: a resource for comparative grass genomics. *Nucleic Acids Res.* 30: 103–105.

Wolfe, R.I. 1972. A multiple stock in Brandon, Canada. *Barley Genet. Newsl.* 2: 170.

Xu, J. and Kasha, K.J. 1992. Identification of barley chromosomal interchange using N-banding and *in situ* hybridization techniques. *Genome* 35: 392–397.

Yoshino, M., Nasuda, S., and Endo, T.R. 1998. Detection of terminal deletions in barley chromosomes by the PCR-based method. *Genes Genet. Syst.* 73: 163–166.

Yu, J. et al. 2002. A draft sequence of the rice genome (*Oryza sativa* L. ssp. *indica*). *Science* 296: 79–92.

Yu, Y. et al. 2000. A bacterial artificial chromosome library for barley (*Hordeum vulgare* L.) and the identification of clones containing putative resistance genes. *Theor. Appl. Genet.* 101: 1093–1099.

# CHAPTER 10

# Genetic Improvement of Pearl Millet for Grain and Forage Production: Cytogenetic Manipulation and Heterosis Breeding

Prem P. Jauhar,[*] Kedar N. Rai, Peggy Ozias-Akins, Zhenbang Chen, and Wayne W. Hanna

## CONTENTS

10.1  Introduction.................................................................................................282
10.2  Pearl Millet as a Poor Man's Crop ...........................................................282
10.3  Pearl Millet as a Research Organism.........................................................282
10.4  Origin and Taxonomy: Germplasm Resources ..........................................283
    10.4.1  Wild Relatives in the Primary Gene Pool...................................283
    10.4.2  Perennial Relatives in the Secondary Gene Pool ........................283
    10.4.3  Perennial Relatives in the Tertiary Gene Pool............................284
10.5  Chromosome Number and Genomic Evolution in the Genus *Pennisetum* .......................284
    10.5.1  Different Base Numbers: The Original Number .........................284
    10.5.2  Chromosome Pairing in Haploids: Implications on Genomic Evolution ..............285
10.6  Induced Polyploidy and Aneuploidy...........................................................286
10.7  Synthesis of Interspecific Hybrids: Genome Relationships .......................286
    10.7.1  Pearl Millet × Napier Grass Hybrids...........................................286
    10.7.2  Pearl Millet × Oriental Grass Hybrids.........................................286
    10.7.3  Pearl Millet × Fountain Grass Hybrids........................................288
    10.7.4  Pearl Millet × *P. schweinfurthii* Hybrids ...................................288
    10.7.5  Pearl Millet × *P. squamulatum* Hybrids.....................................288
10.8  Interspecific Hybridization and Breeding for Superior Fodder Traits .......289
10.9  Synthesis of Intraspecific Hybrids: Exploitation of Hybrid Vigor for Grain and Fodder Yield ...........................................................289
    10.9.1  Hybrid Options .............................................................................291
    10.9.2  Hybrid Parent Development ..........................................................292
        10.9.2.1  Cytoplasmic Male Sterility: Search and Utilization ...............292
        10.9.2.2  B-line Breeding .........................................................294
        10.9.2.3  Seed Parent Development ..........................................295

     10.9.2.4 Restorer Parent Development ................................................................295
    10.9.3 Hybrid Development and Testing ........................................................296
    10.9.4 Hybrids for Arid Conditions ..............................................................296
    10.9.5 Hybrid Parent Maintenance................................................................297
  10.10 Apomixis: Harnessing It for Heterosis Breeding...........................................297
    10.10.1 Incidence of Apomixis .....................................................................297
    10.10.2 Genetics of Apomixis.......................................................................298
    10.10.3 Transferring to Pearl Millet .............................................................298
    10.10.4 Possible Use of Apomixis to Develop Cultivars .............................299
  10.11 Direct Gene Transfer in Pearl Millet.............................................................300
  10.12 Conclusions and Perspectives ........................................................................301
  References ................................................................................................................302

## 10.1 INTRODUCTION

Pearl millet, *Pennisetum glaucum* (L.) R. Brown (= *Pennisetum typhoides* (Burm.) Stapf et Hubb.), is the most important member of the genus *Pennisetum* of the tribe Paniceae in the family Poaceae. The name *Pennisetum* was derived as a hybrid of two Latin words — *penna*, meaning feather, and *seta*, meaning bristle — and describes the typically feathery bristles of its species (Jauhar 1981a). Pearl millet is the sixth most important cereal crop in the world, ranking after wheat, rice, maize, barley, and sorghum. It is a valuable grain and fodder crop and is cultivated in many parts of the world, although in the U.S. it is grown primarily as a forage crop on less than 1 million ha. In tropical and warm-temperature regions of Australia and some other countries, it is also grown as a forage crop (Jauhar 1981a).

Pearl millet is an ideal organism for basic and applied research. In their extensive reviews, Jauhar (1981a) and Jauhar and Hanna (1998) compiled the available literature on cytogenetics and breeding of pearl millet and related species. This article covers some basic aspects of cytogenetics of pearl millet, its cytogenetic manipulation with a view to enrich it with alien genes, aspects of heterosis breeding facilitated by the cytoplasmic-nuclear male sterility (CMS) system and possibly by apomixis, and direct gene transfer into otherwise superior cultivars.

## 10.2 PEARL MILLET AS A POOR MAN'S CROP

Pearl millet is a dual-purpose crop used for grain and fodder and is grown primarily in Asia and Africa, where it occupies some 27 million ha (ICRISAT 1996). It is capable of growing on some of the poorest soils in dry, hot regions of Africa and Asia, where, as a poor man's source of dietary energy, it sustains a large proportion of the populace. It is also grown in other countries where, under relatively more favorable conditions, it provides grain for bullocks, dairy cows, and poultry. In Brazil, it occupies about 2 million ha and is mainly grown as a mulch crop in the soybean production system.

## 10.3 PEARL MILLET AS A RESEARCH ORGANISM

*Pennisetum* is a fascinating genus for conducting research on cytogenetic and evolutionary aspects. Pearl millet is the most important member of this genus. With $2n = 14$ large somatic chromosomes, it lends itself to investigation from the standpoints of classical and molecular cytogenetics, gene location by aneuploid analyses, and studies on haploidy and chromosome pairing. Its short life cycle, protogynous flowers, open-pollinated breeding system, and ability to set a large number of seeds per ear head make pearl millet highly suitable for intra- and interspecific hybridization. This breeding system facilitates the flow of genes between cultivated annual species and related wild species. Pearl millet's large chromosomes — larger than in most other species in the tribe Paniceae — and a distinctive pair of nucleolar organizers make it possible to study intergenomic and intragenomic chromosome pairing in interspecific hybrids (Jauhar 1968). Its outbreeding nature makes pearl millet an ideal crop for heterosis breeding (see Section 10.9).

## 10.4 ORIGIN AND TAXONOMY: GERMPLASM RESOURCES

It is generally agreed that pearl millet is of African origin, although the specific region where it originated is controversial. Harlan (1971) suggested the center of origin in a belt stretching from western Sudan to Senegal. However, based on the present-day distribution, Brunken et al. (1977) considered the Sahel zone of West Africa to be pearl millet's original home, the view favored by Clegg et al. (1984) based on chloroplast DNA studies. Based on the available evidence, Appa Rao and de Wet (1999) concluded that pearl millet originated in western Africa some 4000 years ago.

Over the decades, pearl millet has received several different taxonomic treatments, and hence different Latin names. Thus, it was treated as a constituent of at least six different genera, viz., *Panicum, Holcus, Alopecuros, Cenchrus, Penicillaria,* and *Pennisetum* (Jauhar 1981c). The name *Pennisetum typhoides* (Burm.) Stapf et Hubb., accepted by Bor (1960), was widely used by workers outside of the U.S. However, the name *Pennisetum glaucum* (L.) R. Brown, based on *Panicum glaucum* (L.) R. Brown, was adopted by Hitchcock and Chase (1951) in their *Manual of Grasses of the United States,* and hence accepted by American workers.

The need for collection and conservation of pearl millet germplasm for its improvement for present and future needs cannot be overemphasized. Appa Rao (1999) described the status of germplasm collections and genetic resources for pearl millet, particularly those at the International Crops Research Institute for the Semi-Arid Tropics (ICRISAT) in India. The ICRISAT collection includes the cultivated as well as weedy forms of pearl millet that belong to its primary gene pool.

### 10.4.1  Wild Relatives in the Primary Gene Pool

In the 32 wild annual relatives of pearl millet (Stapf and Hubbard 1934), there is a considerable variation in seed characters, as is generally observed between different cultivars. This variation must have been created by intercrossing between these diploid wild relatives and pearl millet, facilitated by the protogynous nature of the latter. Four taxa closely related to pearl millet, *Pennisetum americanum, Pennisetum nigritarum, Pennisetum echinurus,* and *Pennisetum albicauda,* were called allied species (Meredith 1955), and because they were interfertile with pearl millet, they were merged into a single species with pearl millet (Brunken et al. 1977).

These taxa form the primary gene pool of pearl millet, and may therefore be used as sources of desirable genes. Their easy crossability with pearl millet, coupled with regular chromosome pairing in the resulting hybrids, should facilitate gene introgressions. Because cultivated and wild pearl millet are generally sympatric, gene flow among them occurs, although it is very asymmetrical, with greater flow from wild pearl millet toward cultivated pearl millet (45%) than in the opposite direction (8%) (Couturon et al. 2003). Diploid species *Pennisetum violaceum* (Lam.) L. Rich and *Pennisetum mollissimum* Hochst. also fall in the primary gene pool of pearl millet (Jauhar and Hanna 1998) and are considered as subspecies of pearl millet (Martel et al. 1996).

### 10.4.2  Perennial Relatives in the Secondary Gene Pool

Elephant or Napier grass, *Pennisetum purpureum* Schum. (2n = 4x = 28; ÁÁBB), prized for its fodder for the wet tropical regions of the world, is a perennial relative of pearl millet. It is easily crossable with the latter, although the hybrids are predominantly sterile despite some degree of pairing between the parental chromosomes. Falling in the secondary gene pool of pearl millet,

Napier grass can be used for producing superior fodder hybrids with pearl millet (Gonzalez and Hanna 1984; Jauhar 1981a; Schank and Hanna 1995; Jauhar and Hanna 1998). Napier grass seems to be a valuable source of genetic variation for pearl millet (AA) because of the possibility of production in the former of monoploid gametes with either Á genome or B genome, which could be phenotypically expressed in pearl millet (Hanna 1990).

### 10.4.3  Perennial Relatives in the Tertiary Gene Pool

Other perennial species are crossable with pearl millet, although with some difficulty, resulting in highly sterile hybrids. These species include *Pennisetum squamulatum* Fresen. (2n = 6x = 54), a member of the tertiary gene pool and a source of genes for apomictic reproduction for pearl millet (see Section 10.10).

Although germplasm in the primary gene pool can be easily exploited for pearl millet improvement, species in the secondary and tertiary gene pools are also potential donors of desirable traits (Hanna 1986, 1990). Trispecific hybrids involving pearl millet, Napier grass, and *P. squamulatum* have also been produced for use as bridges to transfer germplasm across species (Dujardin and Hanna 1985).

## 10.5 CHROMOSOME NUMBER AND GENOMIC EVOLUTION IN THE GENUS *PENNISETUM*

### 10.5.1  Different Base Numbers: The Original Number

*Pennisetum* is a polybasic genus consisting of species with chromosome numbers as multiples of **5** (*P. ramosum* Hochst. (Schweinf); 2n = 10), **7** (*P. glaucum*; 2n = 14; *P. schweinfurthii* Pilger; 2n = 14; *P. purpureum*; 2n = 4x = 28), **8** (*P. massaicum* Stapf; 2n = 16, 32), and **9** (e.g., *P. orientale* L.C. Rich.; 2n = 18). The occurrence of cytotypes (intraspecific polyploidy) is a characteristic feature of perennial species (Jauhar 1981a). Thus, *P. orientale* has 2n = 18, 27, 36, 45, and 54, and *Pennisetum pedicellatum* Trin. has 2n = 36, 45, and 54. However, the annual species are devoid of such chromosomal races. For example, pearl millet and other annuals have only 2n = 14. Moreover, no chromosomal races have been reported in the annual or sometimes biennial species *P. ramosum* (2n = 10).

Another interesting feature is that the species with lower chromosome numbers have the larger chromosome sizes. Pearl millet, for example, has 2n = 14 (Rau 1929) but large chromosomes, considered larger than any other member of the tribe Paniceae (Avdulov 1931). *P. ramosum* probably has the largest chromosomes in the tribe, about 5% larger than those of pearl millet (Rangaswamy 1972; Jauhar 1981b). *P. ramosum* with x = 5 is the only species in the genus *Pennisetum* that possesses a lower haploid genome size (2.02 pg) than the x = 7 group. The mean DNA content per chromosome is almost the highest in this species (Martel et al. 1997). These authors found the genome size of *P. schweinfurthii* to be 2.49 pg, which is larger than other species of the genus. On the other hand, the species with higher chromosome numbers have strikingly smaller chromosomes. Thus, *P. orientale* (2n = 18) has considerably smaller chromosomes than pearl millet. Napier grass (2n = 4x = 28), an allotetraploid relative of pearl millet, also has smaller chromosomes than pearl millet. This size differential makes it possible to study intergenomic chromosome pairing in interspecific hybrids involving pearl millet (Jauhar 1968; Jauhar and Hanna 1998).

The occurrence of so many base numbers in *Pennisetum* is interesting. It is likely that the chromosome number of the cultivated species *P. glaucum* (2n = 2x = 14) was derived from a lower base number of x = 5. This is borne out by chromosome pairing in the haploid complement of pearl millet (Jauhar 1970a; Section 10.5.2) and by intergenomic and intragenomic chromosome pairing in interspecific hybrids with pearl millet (Jauhar 1968, 1981b; Section 10.7.1).

## 10.5.2   Chromosome Pairing in Haploids: Implications on Genomic Evolution

Pearl millet haploids ($2n = x = 7$; Figure 10.1a) generally form univalents during meiosis (Figure 10.1b) (Jauhar 1970a; Powell et al. 1975). An interesting feature was the formation of some rod and even ring bivalents, albeit with low frequency (Jauhar 1970a; Figure 10.1c and d). Chiasma terminalization in the ring bivalents was rapid, a characteristic feature of disomic pearl millet. The realization of a maximum of two bivalents per cell (Figure 10.1d) is attributable to homologies among four members of the complement that may have resulted from duplication during the course of evolution (Jauhar 1970a; Gill et al. 1973). It would appear, therefore, that the pearl millet complement has been derived from a base number of $x = 5$, making it a secondarily balanced species resulting from ancestral duplication of chromosomes. Corroborating evidence of the presence of duplicate loci came from the RFLP linkage maps of pearl millet (Liu et al. 1994). As outlined by Jauhar in Chapter 1 in this volume, several apparently diploid (or diploidized) cereal crops, like rice, maize, and sorghum, have in fact resulted from an ancestral round of polyploidy. Evolution in eukaryotes is known to be accompanied by gene duplication (Ohno 1970). It is believed that duplicated genetic material confers adaptive advantage, and according to Ohno's theory, having extra gene copies is essential for an organism to evolve. That the diversification of gene functions during the course of evolution requires prior gene duplication is further supported by recent work (Kellis et al. 2004).

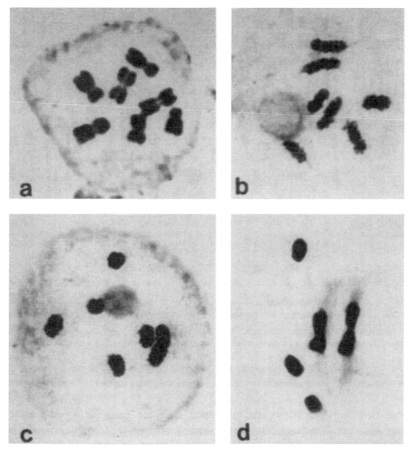

**Figure 10.1**   Chromosome pairing in pearl millet haploids ($2n = x = 7$). (a) Seven somatic chromosomes relatively similar in size. (b) Early meiotic prophase with seven univalents. (c) Diakinesis with 1 rod II + 5 I. (d) Meiotic metaphase I with 2 II + 3 I. (From Jauhar, P.P. and Joppa, L.R., in *Methods of Genome Analysis in Plants*, Jauhar, P.P., Ed., CRC Press, Boca Raton, FL, 1996, pp. 9–37.)

## 10.6 INDUCED POLYPLOIDY AND ANEUPLOIDY

Tetraploid (2n = 4x = 28) pearl millet has been produced by several workers (Krishnaswamy et al. 1950; see Jauhar 1981b for other references). Gill et al. (1969) and Jauhar (1970b) studied chromosome pairing in the raw ($C_O$) and advanced generation tetraploids. They noted a gradual shift from multivalents to bivalents — a sort of cytological diploidization in successive generations. This phenomenon was attributed to natural selection of genes that condition regular meiosis with predominance of bivalents (Jauhar 1970b).

Autotriploids (2n = 3x = 21) were produced by crossing synthetic autotetraploids with diploids (Gill et al. 1969; Jauhar 1970b). These triploids proved useful for producing a series of aneuploids. From the progeny of triploid × diploid crosses, several primary trisomics were isolated (Jauhar 1970b; Minocha et al. 1980) and used for assigning genes to various chromosomes (Minocha et al. 1980; see Jauhar and Hanna 1998).

## 10.7 SYNTHESIS OF INTERSPECIFIC HYBRIDS: GENOME RELATIONSHIPS

The degree of meiotic pairing will generally be in direct proportion to the degree of homology among parental chromosomes. Traditionally, therefore, the principal method of assessing genomic relationships among species has been the study of chromosome pairing in their hybrids (Jauhar and Joppa 1996). Several interspecific hybrids have been synthesized between pearl millet and other species in the genus *Pennisetum*. Because of marked size differences, chromosomes of pearl millet are easily distinguishable from those of other species, except *P. ramosum* and *P. schweinfurthii* (see Section 10.5.1), facilitating the study of pairing relationships. Knowledge of genomic relationships is useful in planning breeding strategies.

### 10.7.1  Pearl Millet × Napier Grass Hybrids

Interspecific triploid hybrids between diploid pearl millet (2n = 14; genome AA) and its allotetraploid relative Napier grass (2n = 4x = 28; genomes ÁÁBB) are among the most widely studied in the tribe Paniceae. Based on easily recognizable size differences among parental chromosomes, Jauhar (1968) was able to analyze both intergenomic and intragenomic chromosome pairing in the triploid hybrids (2n = 3x = 21; AÁB). He observed a range of zero to nine bivalents at meiotic metaphase I, with a mean of 5.3 II per cell. Most bivalents were formed between chromosomes of the A and Á genomes, and they were clearly heteromorphic because of the size differences of the parental chromosomes (Figure 10.2a and b). Intragenomic pairing (autosyndetic pairing) within the pearl millet complement (A genome) and within the Napier grass complement (ÁB genomes) also occurred, resulting in homomorphic bivalents because of the symmetrical karyotypes of the parental species. The formation of 7 I and 7 II in most cells showed that the genome in pearl millet was essentially homologous to one of the genomes, i.e., Á of Napier grass (Raman 1965). The formation of up to five heteromorphic bivalents in the AÁB hybrids indicated that the A and Á genomes are closely related, having probably arisen from a common progenitor with x = 5 chromosomes during the course of evolution (Jauhar 1968, 1981b; Jauhar and Hanna 1998). The occurrence of a species, *P. ramosum*, with 2n = 10 chromosomes supports this conclusion.

### 10.7.2  Pearl Millet × Oriental Grass Hybrids

Substantial size differences exist between chromosomes of pearl millet and those of diploid oriental grass (*P. orientale*; 2n = 18), the chromosomes of the latter being much smaller. Interspecific bivalents are therefore highly heteromorphic (Figure 10.2c and d). Patil and Singh (1964) and

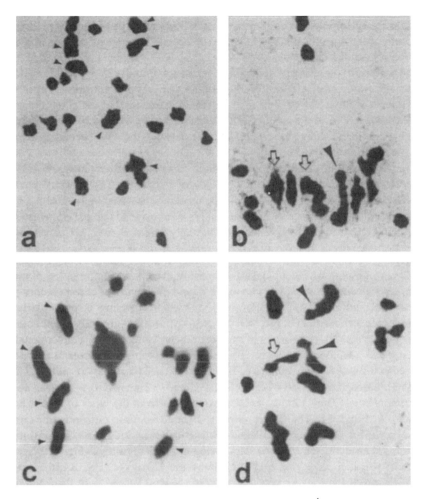

**Figure 10.2**  (a, b) Chromosome pairing in triploid hybrids (2n = 3x = 21; AÁB) between *P. glaucum* (2n = 14; AA) and *P. purpureum* (2n = 6x = 28; ÁÁBB). (a) Meiotic metaphase I showing 21 univalents, 17 large ones of *P. glaucum* (arrowheads) and 14 small from *P. purpureum*; note marked size differences among the parental chromosomes. (b) Metaphase I with 7 II + 7 I. The bivalents comprise two large symmetrical ones within the A genome (hollow arrows), one heteromorphic intergenomic bivalent between chromosomes of the A and Á genomes (solid arrow), and four intragenomic bivalents with the Á and B genomes. Note two large univalents of the A genome. (c, d) Chromosome pairing in interspecific hybrids (2n = 16) between *P. glaucum* (2n = 14) and *P. orientale* (2n = 18). (c) Diakinesis with 16 univalents, 7 large ones (arrowheads) of *P. glaucum* and 9 small from *P. orientale* (2n = 18). Note striking size differences among the parental chromosomes. (d) Meiotic metaphase I with two heteromorphic bivalents between *P. glaucum* and *P. orientale* chromosomes (solid arrows), and one autosyndetic bivalent within the *P. orientale* complement. (From Jauhar, P.P., *Adv. Agron.*, 34, 407–470, 1981b.)

Jauhar (1973, 1981b) studied chromosome pairing in the interspecific hybrids with 2n = 16. Based on both intergenomic and intragenomic chromosome pairing, Jauhar (1981b) inferred an ancestral relationship between the parental species. Hanna and Dujardin (1982) obtained hybrids (2n = 25) between diploid pearl millet and the tetraploid cytotype of *P. orientale* (2n = 36). The hybrids had 7 large pearl millet chromosomes and 18 small *P. orientale* chromosomes. Although the pearl millet chromosomes remained unpaired, the grass chromosomes paired mainly as bivalents (Dujardin and Hanna 1983), indicating the autotetraploid nature of the grass parent. The authors found that these hybrids or subsequent derivatives were either facultative or obligate apomicts. Dujardin and Hanna (1987) obtained partially fertile hybrid derivatives with 2n = 32 to use as possible bridges for germplasm transfer between pearl millet and *P. orientale*.

### 10.7.3 Pearl Millet × Fountain Grass Hybrids

Fountain grass, *Pennisetum setaceum* (Forsk.) Chiov., is a triploid (2n = 3x = 27), apomictic grass. Hanna (1979) synthesized interspecific hybrids using a male sterile line Tift 23 DA of pearl millet as a female parent and fountain grass as a male. Based on size differences, up to three chromosomes of pearl millet were found to associate with three chromosomes of fountain grass in the sterile hybrids.

### 10.7.4 Pearl Millet × *P. schweinfurthii* Hybrids

Hybrids between these two annual species with 2n = 14 chromosomes are vigorous, but mostly sterile. Low chromosome pairing (average of 0.48 to 1.97 bivalents in five hybrids), in conjunction with hybrid sterility, shows that the two species are not closely related (Hanna and Dujardin 1986).

### 10.7.5 Pearl Millet × *P. squamulatum* Hybrids

Hybrids between pearl millet and hexaploid *P. squamulatum* (2n = 6x = 54) (Figure 10.3) were produced by several workers (Patil et al. 1961; Jauhar 1981a,b; Dujardin and Hanna 1984). Dujardin and Hanna (1989) obtained partially fertile hybrids and suggested some homology among the parental chromosomes. Although these hybrids have some forage potential, they are not as high yielding as the pearl millet × Napier grass hybrids. The hexaploid *P. squamulatum* is apomictic and has been crossed to synthetic tetraploid pearl millet to transfer its apomixis to the latter (Hanna et al. 1989; see Section 10.8).

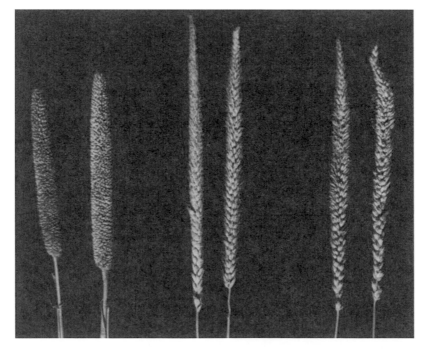

**Figure 10.3**    Heads of hybrids (2n = 41) between *P. glaucum* (2n = 14) and *P. squamulatum* (2n = 6x = 54). The latter has been used as a source of genes for apomixis for introduction into pearl millet (see Section 10.10).

## 10.8 INTERSPECIFIC HYBRIDIZATION AND BREEDING FOR SUPERIOR FODDER TRAITS

It is generally easier to produce interspecific hybrids using the protogynous pearl millet as the female parent. Pearl millet and Napier grass, both belonging to the section Penicillaria of the genus *Pennisetum*, are known to hybridize in nature to produce spontaneous hybrids (Stapf and Hubbard 1934). After Burton (1944) produced these interspecific hybrids in the U.S., they were produced in several other countries: India (Krishnaswamy and Raman 1949), South Africa (Gildenhuys 1950), Pakistan (Khan and Rahman 1963), Australia (Pritchard 1971), Sri Lanka (Dhanapala et al. 1972), and Nigeria (Aken'Ova and Chedda 1973). The main goal of crossing these species was to produce high-yielding, high-quality perennial fodder hybrids combining pearl millet's forage quality, non-shattering nature, and ability to establish readily, as well as the perennial, aggressive nature of Napier grass.

The hybrids generally show high heterosis for fodder yield and quality, and thus are high yielding and more acceptable than the Napier grass parent (Burton 1944; Krishnaswamy and Raman 1949; Patil 1963; Burton and Powell 1968; Hussain et al. 1968; Gupta 1974; Muldoon and Pearson 1977; Osgood et al. 1997). However, these hybrids are sterile and need to be propagated vegetatively, which puts a major limitation on their easy distribution to farmers. Seeds of the interspecific hybrid can be commercially produced in a frost-free environment by alternating rows of cytoplasmic-nuclear male sterile pearl millet with rows of Napier grass and by manipulating the planting dates of the parental line (Osgood et al. 1997).

Spontaneous chromosome doubling (amphidiploidy) has been observed in these hybrids (Jauhar and Singh 1969). Induced chromosome doubling in interspecific hybrids produced largely fertile amphidiploids ($2n = 6x = 42$; AAÁÁBB), which generally formed 21 II (Krishnaswamy and Raman 1954), resulting from preferential pairing between A-A, Á-Á, and B-B genome chromosomes. In view of the formation of 7 II + 7 I in the triploid hybrid (AÁB), some quadrivalents or trivalents would be expected in the derived amphidiploids, but they occur rarely, if at all. Jauhar (1981b) therefore postulated the possibility of some sort of genetic control on diploid-like pairing in the amphidiploids that makes them seed fertile. Hanna et al. (1984) observed that the triploid AÁB hybrids on average had higher dry matter yields and protein content than the hexaploids AAÁÁBB. However, the amphidiploids have excellent forage potential, and they can be propagated vegetatively as well as by seed (Schank and Hanna 1995).

## 10.9 SYNTHESIS OF INTRASPECIFIC HYBRIDS: EXPLOITATION OF HYBRID VIGOR FOR GRAIN AND FODDER YIELD

The utilization of heterosis or hybrid vigor is one of the most efficient means of crop yield improvement and has resulted in phenomenal increases in yields in open-pollinated crops like corn (see Chapter 5 in this volume). Pearl millet, a highly diverse crop with a predominantly cross-pollinated breeding system, displays high degrees of heterosis for grain (Figure 10.4) and fodder (Figure 10.5) yield and other agronomic traits. By crossing a dwarf cytoplasmic-nuclear male sterile single-cross $F_1$ hybrid Tift 8593 and dwarf pollinator Tift 383, Hanna et al. (1997) produced a dwarf leafy forage hybrid Tifleaf 3. This three-way hybrid is a very popular forage hybrid in the U.S.

An extensive survey of pearl millet literature showed single-cross hybrids having an average of 40% better-parent heterosis, and some of the crosses having more than 400% better-parent heterosis for grain yield (Virk 1988). This comparison, however, is more of academic interest. Because pearl millet is a highly cross-pollinated crop in which open-pollinated varieties (OPVs) represent its natural cultivar state, the assessment of yield advantage of hybrids over OPVs is of real practical significance. Ouendeba et al. (1993) studied heterosis and combining ability among five African landraces and observed that better-parent heterosis for grain yield ranged from 25 to

**Figure 10.4**  A field stand of a high-yielding single-cross dual-purpose pearl millet hybrid developed at ICRISAT, Hyderabad, India, and widely adopted by farmers in India. Note gigantic ear heads and excellent leafiness.

80%. Similarly, Ali et al. (2001) found significant heterosis for grain yield when they evaluated 11 medium- to late-maturity populations and their diallel crosses in five environments in India.

Availability of cytoplasmic-nuclear male sterility (CMS) made commercial production of hybrids possible. The development of CMS lines by Burton (1958, 1965a,b) in the U.S. and by Athwal (1965) in India greatly helped in the commercial exploitation of hybrid vigor. The remarkable speed with which Indian breeders developed high-yielding grain hybrids using CMS lines was described as "one of the most outstanding plant breeding success stories of all time" (Burton and Powell 1968). The first CMS-based single-cross hybrid HB 1 was shown to yield twice as much as the popular landrace-based OPVs (Athwal 1965). This finding played a catalytic role in the rapid development of the grain hybrid seed industry in India.

During the initial phase of pearl millet improvement programs at ICRISAT, the largest pearl millet program in the world, greater emphasis was placed on population improvement and OPV development. Results showed that some of the highest-yielding improved OPVs yielded 85% of the commercial single-cross hybrids. However, these comparisons were often made between the two types of cultivars that were of differing maturities. Over several years of yield trials in the All India Coordinated Pearl Millet Improvement Project (AICPMIP), it has been observed that single-cross hybrids often yield about 25 to 30% more than OPVs of comparable maturity. This order of grain yield advantage of hybrids has proved reasonably attractive for greater investment in hybrid research by both the public and private sector organizations in India. Consequently, there has been a dramatic rise in cultivar diversity and adoption as reflected in more than 70 hybrids under cultivation during the year 2003 on more than 4.5 million ha of the total of about 9.5 million ha

**Figure 10.5** A high-yielding experimental top-cross forage hybrid of pearl millet developed at ICRISAT, Hyderabad, India.

of pearl millet area in India. Figure 10.4, for example, shows a single-cross dual-purpose pearl millet hybrid developed at ICRISAT and widely adopted by farmers in India from 1988 to 2001. The only other country where pearl millet single-cross grain hybrids are grown for grain production on a limited scale is the U.S. While single-cross grain hybrids will become a reality in Africa in the long-term, the short- and medium-term prospects are more for other hybrid types, which are currently becoming feasible with the availability of suitable CMS sources.

## 10.9.1 Hybrid Options

The outbreeding system and availability of CMS makes it possible to develop various types of hybrids in pearl millet. These include single-cross hybrids, top-cross hybrids, three-way hybrids, three-way top-cross hybrids, and interpopulation hybrids (Talukdar et al. 1996, 1999). Single-cross hybrids are developed by crossing inbred seed parents (commercially known as male sterile, CMS, or A-lines) and inbred pollen parents (commercially known as restorer lines in the case of grain hybrids). For top-cross hybrids, the female parent is an A-line as in single-cross hybrids, but the male parent is an OPV, which is highly heterozygous as well as heterogeneous. Three-way hybrids are developed by crossing a male sterile single-cross $F_1$ hybrid with a restorer line (as in single-cross hybrids). In three-way top-cross hybrids, the female parent is a male sterile single-cross $F_1$ hybrid, while the male parent is an OPV (as in top-cross hybrids). In interpopulation hybrids, both parents are OPVs, with the female parent being a CMS-based male sterile population. Single-cross hybrids have the potential to achieve the highest levels of heterozygosity and genetic as well as phenotypic uniformity. At the other extreme, interpopulation hybrids are most heterogeneous genetically and vary most phenotypically. The other three hybrid forms will rank somewhere in between with respect to genetic heterogeneity and phenotypic variability.

The commercial viability of the hybrids depends on several factors. These include their relative grain yield advantage, acceptance by farmers and seed producers, feasibility of breeding seed parents and maintenance of hybrid parents, and research and development capacity of the National Agricultural Research Systems (NARS). The highest level of heterosis exploitation is possible in single-cross hybrids. It is for this reason, as well as for phenotypic uniformity, that only single-cross hybrids are the hybrid cultivar form currently under commercial grain production. Tifleaf 3 pearl millet, a three-way hybrid, is the leading forage cultivar in the U.S. (Hanna et al. 1997). Phenotypic uniformity of single-cross hybrids has its own special value for agencies involved in hybrid development and marketing, as it enables them to claim and protect their brand name and to maintain the purity of parental lines. Phenotypic uniformity is important for farmers since the seed and the grain crop conform to varietal description. However, the underlying genetic uniformity of single-cross hybrids has some associated disadvantages. The most significant of these is the enhanced vulnerability to certain diseases. For instance, single-cross hybrids of pearl millet have suffered from repeated downy mildew epidemics in India (Hash 1997). Single-cross hybrids also can be relatively more susceptible than OPVs to smut caused by *Moesziomyces penicillariae* (Bref) Vanky (Thakur 1989) and to ergot caused by *Claviceps fusiformis* Loveless (Thakur et al. 1983).

There is no theoretical basis for assuming that other hybrid types can equal the yield level of single-cross hybrids. Other criteria, however, would make them acceptable to seed industry and farmers under some agricultural situations. For instance, productivity at the seed production stage is important to the seed-producing farmers and agencies, and performance at the hybrid crop stage to the pearl millet farmers and consumers. Thus, resorting to hybrids based on male sterile $F_1$ seed parents (i.e., three-way hybrids) and male sterile population seed parents (i.e., interpopulation hybrids) can lead to a dramatic increase in productivity at the hybrid seed production stage. For top-cross hybrids also, higher productivity of the OPV (used as the male parent) will allow an increase in the proportion of land planted to the A-line and a decrease in that allocated for the pollen parent, thus increasing the hybrid seed yield per unit area. The use of male sterile $F_1$s and populations (as the female parent) in hybrid breeding has other advantages related to flowering manipulation and disease management (Talukdar et al. 1999; Rai et al. 2000a,b). Being genetically heterogeneous, three-way, top-cross, and interpopulation hybrids will be less vulnerable to diseases. It has been found that OPVs cultivated on large scales (i.e., 0.3 to 0.8 million ha) and for over 15 years in India did not register any decline in their downy mildew (DM) resistance levels, while most of the single-cross hybrids become susceptible after about 5 years of cultivation (Thakur, unpublished results).

Studies in western and central Africa show that top-cross and interpopulation hybrids can outyield the popular OPVs by 14 to 59% (Table 10.1). This represents a substantial advantage in favor of hybrid technology development. A top-cross hybrid (ICMH 312) developed by ICRISAT gave as much grain yield as a high-yielding and widely cultivated hybrid (ICMH 451) of comparable height and maturity in the All India Coordinated trials conducted over 3 years (Talukdar et al. 1999). It has been suggested that top-cross hybrids may provide a rapid means to combine the adaptation of local landraces or landrace-based improved populations (used as male parents) with higher seed yield potential of improved genotypes used as female parents (Mahalakshmi et al. 1992).

## 10.9.2  Hybrid Parent Development

### 10.9.2.1  *Cytoplasmic Male Sterility: Search and Utilization*

Cytoplasmic-nuclear male sterility (CMS) holds the key to seed parent development. Seed parent research starts with the search for commercially viable CMS sources. Since CMS results from a disharmonious interaction between the genetic factors in the nucleus and in the cytoplasm,

**Table 10.1 Grain Yield Advantage of Top-Cross and Interpopulation Hybrids over Open-Pollinated Varieties in Various Trials of Pearl Millet[a]**

| Hybrid/Location | No. of Hybrids in Trial | Best OPV | Yield Advantage over Best OPV (%)[a] |
|---|---|---|---|
| Top-cross hybrid Cinzana, Kolo, Sadore, Tara | 4 | CIVT | 14–38 |
| Lucydale, Makoholi | 100 | ICMVF 86415 | 38–52 |
| Interpopulation hybrid Bambey (2 years) | 35 | Souna II | 27–59 |
| Sadore, Bengou (2 years) | 10 | $P_3$ Kolo | 32–45 |

[a] Range for four top-ranking hybrids in the trial.

Source: Rai, K.N. et al., in *Proceedings of the International Conference on Genetic Improvement of Sorghum and Pearl Millet*, INTSORMIL Publication 97-5, 1997, pp. 71–83.

**Table 10.2 Pollen Shedders (PS) and Selfed Seed Set Distribution in Samples of Nonshedding (Non-PS) Plants of Three Isonuclear A-Lines of Pearl Millet: Range across Seven (Year × Season) Environments at Patancheru**

| A-Line | Pollen Shedders (%) | Percent Non-PS Plants in Selfed Seed Set Class | | | |
|---|---|---|---|---|---|
| | | 0 | 1–5 | 6–20 | 21–50 |
| $81A_1$ | 0.0–0.6 | 93.4–99.4 | 0.3–5.2 | 0.0–1.4 | Nil |
| $81A_4$ | Nil | 97.4–100.0 | 0.0–2.6 | Nil | Nil |
| $81A_5$ | Nil | 98.7–100.0 | 0.0–1.3 | Nil | Nil |

Source: Modified from Rai, K.N. et al., *Euphytica*, 121, 107–114, 2001a.

there is greater likelihood of finding it in segregating generations derived from crosses involving genetically diverse parents. Two CMS sources, designated $A_1$ (Burton 1958) and $A_3$ (Athwal 1966), were identified in populations of such crosses. Since pearl millet is a highly cross-pollinated crop allowing for extensive intercrossing between different genotypes under unprotected conditions, there are chances of finding CMS in landraces. The $A_2$ CMS source (Athwal 1961) and several unclassified CMS sources (Appa Rao et al. 1989) have been derived from diverse germplasm accessions.

Extending this logic further, one can expect broad-based gene pools to be the ideal germplasm to search for CMS, as exemplified by the $A_{egp}$ source identified in an early gene pool (Sujata et al. 1994). Several male sterile plants were identified in a large-seeded gene pool, of which one represented a CMS system distinctly different from all the reported sources and hence was designated as $A_5$ (Rai 1995). CMS sources have also been identified in populations derived from crosses between *P. glaucum* ssp. *violaceum* (= *monodii*), a wild relative of pearl millet, as female parents and cultivated pearl millet as male parents. Thus, a germplasm accession from Senegal was identified as a source of $A_v$ CMS (Marchais and Pernes 1985), and another accession from Senegal was a source of $A_4$ CMS (Hanna 1989).

Although several CMS sources have been identified in pearl millet, all are not commercially viable. The commercial viability of a CMS system is influenced by the stability of male sterility, frequency of its maintainers and restorers (in the case of grain hybrids) in germplasm and breeding materials, the extent of their male sterility expression in varying production environments, and the nature of genetic inheritance (including the effect of genetic background on the expression of male sterility) and character association. Research at ICRISAT shows that male sterility of A-lines with $A_4$ and $A_5$ cytoplasm is more stable than those with the current commercially used $A_1$ cytoplasm (Table 10.2). It has also been observed that A-lines with the $A_{egp}$ cytoplasm have more stable male

**Table 10.3 Effect of A$_1$ Cytoplasm on Grain Yield in Pearl Millet at Two Diverse Locations in India**

| Location | Year | Mean Grain Yield (t ha$^{-1}$) of Hybrids[a] on: | | | | |
|----------|------|------|------|------|------|------|
|          |      | 81A$_1$ | 81B | ICMA$_1$ 88004 | ICMB 88004 | SE± |
| Patancheru | 1990 | 3.06 | 2.63 | 3.15 | 2.69 | 0.166 |
|          | 1991 | 3.67 | 3.38 | 3.42 | 3.23 | 0.072 |
| Hisar    | 1990 | 2.26 | 2.20 | 2.41 | 2.28 | 0.059 |
|          | 1991 | 3.11 | 3.05 | 2.79 | 2.64 | 0.099 |

[a] Mean of four hybrids developed by crossing four common pollinators on each A-line.

sterility than those with the A$_1$ cytoplasm (K.N. Rai, unpublished results). Results from top-cross hybrids made with eight diverse populations show that the frequency of maintainers in germplasm and breeding materials is likely to be highest in A-lines with the A$_5$ cytoplasm (77 to 99%), followed by those with the A$_4$ (33 to 75%) and A$_1$ (22 to 49%) cytoplasm. Based on these two criteria, A-line breeding efficiency is likely to be highest with the A$_5$ cytoplasm, followed by the A$_4$ and then the A$_1$ cytoplasm. Since restorer frequency is the converse of maintainer frequency, it would mean that the frequency of restorers in germplasm and breeding materials would be lowest for the A$_5$ CMS, followed by that for the A$_4$ and A$_1$. Nevertheless, restorers of all three CMS sources have been found in a wide range of breeding populations.

Considering the association of Texas CMS with southern leaf blight epidemic on maize hybrids in the U.S. (Scheifele et al. 1970), and viewing the DM epidemic on Tift 23A$_1$-based pearl millet hybrids in India (Dave 1987), there has been concern over the possible association between this CMS source and DM susceptibility. However, it has been established that the A$_1$ CMS source is not associated with DM susceptibility (Anand Kumar et al. 1983; Yadav et al. 1993). The Tift 23 CMS-based hybrids are relatively more susceptible than the OPVs to ergot and smut. However, the greater susceptibility of hybrids is not due to the cytoplasm per se; rather, it is due to CMS-mediated male sterility (Rai and Thakur 1995, 1996). Preliminary research indicates that A$_1$ CMS-based hybrids may have some yield advantage over those based on the fertile cytoplasm of the counterpart B-lines or maintainer lines, and that this advantage may be environment dependent. For instance, evaluation of isonuclear hybrids showed that the mean grain yield of A$_1$ hybrids was 10 to 15% higher than the mean grain yield of hybrids based on their counterpart B-lines at Patancheru (southern India), while at Hisar (northern India), the A$_1$ hybrids had only 2 to 6% higher grain yield than the B-line hybrids (Table 10.3). Similar information with respect to character association of other CMS sources is not available.

### 10.9.2.2  B-line Breeding

B-line breeding constitutes the most critical part of seed parent development, as these lines must satisfy numerous requirements related to yield potential, adaptation and quality traits, and their sterility maintenance ability. Various breeding methods like population improvement, pedigree or pedigree bulk breeding, and backcross breeding have all been applied in B-line breeding of pearl millet, but pedigree breeding continues to be the most widely used breeding method. Maturity and several agronomic and quality traits such as plant height, growth habit (erect, semierect, and spreading type), lodging resistance, panicle size and compactness, exsertion, tip sterility, and seed size, shape, and color are highly heritable traits, amenable to effective visual selection, both at the plant and at the progeny (selfed as well as crossed) levels. Within-progeny selection for DM resistance has been shown to be much less effective than between-progeny selection; hence, very high emphasis is placed on between-progeny selection (Weltzien and King 1995), and the use of these approaches resulted in the development of a large number of B-lines with high levels of resistance to multiple pathotypes (Rai et al. 2001b).

Where more than one crop season in a year is possible, evaluation of sterility reaction of breeding lines in test crosses made on A-lines, for discarding purposes, can be done either in the rainy or dry summer season. However, since the sterility of A-lines and hybrids has been shown to be accentuated in the hot dry summer season (Rai and Hash 1990; Rai et al. 1996), greater reliance for screening for sterility can be placed when evaluated in the rainy season. Sterility evaluation can be done either on the basis of pollen shed or on the basis of seed set under selfing. Since the latter is more reliable, especially when done in the rainy season, it is advisable to use this to validate the results of evaluations based on pollen shed.

Evaluation of general combining ability (GCA) of potential B-lines to select those for conversion into A-lines is perhaps the least understood subject matter on which there is little unanimity. The problem arises from numerous issues, such as (1) inbreeding/selection stage appropriate for GCA evaluation, (2) the nature of the testers and its implications in determining the plot size of the crosses (test crosses or top crosses), (3) number of locations/environments in which the crosses need to be evaluated to derive valid estimates of GCA, and the access of breeding programs to these environments, (4) relative size of the GCA vs. specific combining ability (SCA) effects, and (5) the effectiveness of resource use in GCA evaluation of lines in relation to their numbers vs. use of this resource in evaluation of a larger number of progenies for performance per se. Khairwal and Singh (1999) summarized the results of a large number of studies and observed that both GCA and SCA play significant roles in determining grain yield in pearl millet, and the overriding importance of one over the other is highly influenced by the materials under study and perhaps also by the environments.

Results from several studies (Rai and Virk 1999) showed that performance per se of the lines is either positively correlated with GCA for grain yield or that there is no correlation between the two, implying that high-yielding lines can be as good general combiners as any other lines. These results have important implications in B-line breeding in that the promising seed parents will not be those that produce only high-yielding hybrids but those that additionally have high yield per se because the latter has a bearing on the seed production economy.

### 10.9.2.3  Seed Parent Development

The prospective B-lines for conversion into A-lines are usually identified at the $F_5/F_6$ stage of inbreeding. By this time, the lines will have been evaluated for performance per se and agronomic traits for 1 to 2 years, each during the rainy and summer seasons (where two evaluation seasons per year are possible — as in southern India), and for DM resistance at least once during the inbreeding/selection process. Since the lines become substantially uniform by this stage of inbreeding, bulk pollen from a B-line is used for crossing onto an A-line and for advancing backcrosses (BCs).

### 10.9.2.4  Restorer Parent Development

The general breeding procedure for restorer lines is similar to those for the B-lines except that the selection criteria (yield and agronomic traits) are relatively less stringent because the primary role of an R-line is to provide abundant pollen during hybrid seed production and produce high-yielding hybrids. This relaxation can, however, be traded off with selection for good general combining ability.

Besides the restorers themselves being good pollen producers, hybrids of good restorers should also produce profuse pollen. Whether the extent of pollen production has even a small degree of negative correlation with grain yield in environments free of the ergot and smut is still to be investigated. It seems likely that profuse pollen production in hybrids will confer some protection from ergot and smut infection (Rai and Thakur 1995, 1996). It is important to note that the pollen production ability and male fertility restoration of hybrids is determined in the target environments, using both pollen shedding and selfed seed set criteria. For instance, it has been observed that hybrids having excellent pollen production during the rainy season become sterile during the summer season under high temperatures that may prevail at flowering time. Thus, hybrids targeted for cultivation during the summer season should be evaluated for their fertility restoration during the summer season.

### 10.9.3  Hybrid Development and Testing

Hybrids are developed for high yield (grain and fodder as the case may be) with desirable adaptation and quality traits. Before making hybrids, it is assumed that both parental lines to be used in hybrids have met the basic agronomic, adaptation, and quality criteria, and that the pollinator has been tested *a priori* for its high level of fertility restoration. Stability of fertility restoration in pearl millet is influenced by genetic backgrounds of the hybrids, and more so by the environmental conditions, so it is important that the evaluation of this trait is done in the environment for which the hybrid is targeted.

Flowering synchrony between the seed and the pollen parents is an important factor determining the choice of the parental lines of hybrids. From the viewpoint of cost-effective seed production, it is desirable that both parents of hybrids have similar flowering time in the seed production environments. In practice, however, it is rarely achieved, in which case, staggered plantings of male parents (pollen parents) and female parents (seed parents) are required, adding to the cost of seed production. In India, up to 20 days of staggering of male and female parents of hybrids have been found to be manageable.

Plant height is another important factor determining the choice of hybrid parents. There are two aspects of it. The first issue deals with seed production, in which case the height of the male parent should be no less than that of the female parent. This allows for the free flow of pollen across the rows and thus cuts down the seed production cost by way of (1) better seed set, and consequently higher hybrid seed yield per unit area, and (2) greater female:male ratio, which also leads to higher hybrid seed yield. The second issue deals with the height of the hybrids, which becomes very important with regard to lodging, the relative value of grain vs. fodder, and the practical aspects of harvesting, where it is done manually.

Once the above criteria for selecting the potential parental lines of hybrids are satisfied, the next step is to produce hybrid seed and test hybrid performance. The numbers game is important at this stage: the larger the number of hybrids under evaluation, the higher the probability of identifying superior hybrids, when other aspects of parental diversity and combining ability are similar. In case the hybrids are in the hundreds, the first evaluation can be done in an unreplicated augmented design, with the principal criteria being yield potential (visual score or visual score combined with measured yield), confirmed fertility restoration, and agronomic traits relevant to farmers' acceptance. This is followed by replicated trials of selected hybrids at multiple locations — the number of replications and locations are primarily determined by the number of hybrids in the trial, which has an impact on the time frame for prerelease yield evaluations. In India, normally 1 to 2 years of yield evaluations of hybrids in trials managed by the originating centers, followed by 3 years of multilocational testing under the AICPMIP, is a standard protocol to identify hybrids for release. Following this protocol, more than 80 hybrids have been released from 1965 to 2003 in India.

### 10.9.4  Hybrids for Arid Conditions

In India, pearl millet hybrids have been widely adopted because of their high yields and resistance to diseases (Kelley et al. 1996). The hybrids show a high yield potential only under optimal conditions. Their adoption rates are relatively lower in the drier regions of western parts of India, especially in desert areas of western Rajasthan because of inadequate grain yields under drought conditions. In these dry areas, farmers improve their locally adapted landraces by crossing them with modern cultivars with high yield potential (Weltzien 2000; vom Brocke et al. 2002). Suitable hybrids for these harsh conditions may be developed by top crossing locally adapted landraces on suitable male sterile lines (Yadav et al. 2000), and work on development of suitable hybrids is in progress (Presterl and Weltzien 2003).

### 10.9.5  Hybrid Parent Maintenance

The highly cross-pollinated nature of pearl millet makes it prone to contamination from wind-borne pollen that can travel long distances and remain viable much longer than sorghum and maize pollen. Pollen shedders in A-lines that arise due to mutational changes (mostly in the cytoplasm) further complicate the maintenance of their genetic purity. Thus, maintenance of the purity of parental lines becomes an integral part of hybrid development.

The nucleus seed of parental lines of hybrids should be produced in the off-season (where such facilities exist) and at one third of the plant population recommended for planting of the commercial grain crop. This maximizes the individual plant expression and hence enables effective roguing of off-type plants and undesirable phenotypic deviants (Chopra et al. 1999). It is recommended that nucleus seed of the parental lines be produced in isolation plots that should be about 1500 m away from other genotypes of cultivated pearl millet and cross-compatible wild species. The seed should normally be produced once every 5 years with a target of about 30 kg of seed (same as recommended for OPVs), and this should be equally divided into six lots for future use in breeder seed production (lots 1 to 4), nucleus seed regeneration (lot 5), and a backup stock (lot 6) (Andrews and Harinarayana 1984).

### 10.10 APOMIXIS: HARNESSING IT FOR HETEROSIS BREEDING

Apomixis is a reproductive mechanism that allows a plant to produce carbon copies of itself through progenies derived from either selfed or open-pollinated seed. Therefore, if introduced into superior hybrids, apomixis would make them true breeding. The three main mechanisms of apomixis — apospory, diplospory, and adventitious embryony — have been comprehensively described in the literature (Nogler 1984; Asker and Jerling 1992; Koltunow 1993). Only apospory will be treated in this article, as it is relevant to *Pennisetum*.

### 10.10.1  Incidence of Apomixis

Apospory occurs in many *Pennisetum* species belonging to the tertiary gene pool of pearl millet, almost invariably in polyploid cytotypes (Jauhar 1981a). Diploid individuals from *Pennisetum* accessions have always been described as obligately sexual, as is generally the case in most plant groups (Jauhar 2003). The mode of reproduction of a plant can be determined by examination of cleared ovules or by progeny analysis. Sexually derived progeny are variable since *Pennisetum* species outcross and are heterozygous. Cleared ovules from an apomictic individual may contain only aposporous embryo sacs, or both aposporous and meiotically derived embryo sacs, in facultative apomicts. In near-obligate apomicts, however, meiosis typically aborts and the embryo sacs develop directly from a somatic cell of the nucellus. The aposporous embryo sacs can be easily distinguished from meiotically derived embryo sacs by their lack of antipodals. Seed development in the apomict proceeds only after pollination because fertilization of the central cell is required for endosperm formation even though the egg cell is parthenogenetic, and thus the embryo is solely of maternal origin.

Interest in apomixis has increased during the past 20 years because of (1) discoveries of facultative apomixis in sexual species, and of sexual plants in apomictic species, (2) the accumulation of new information on genetic control of apomixis, and (3) the availability of molecular tools for research on apomixis (Grimanelli et al. 2001; Grossniklaus et al. 2001; Roche et al. 2001; Spillane et al. 2001). The evolution of a broader understanding of apomixis could impact cultivar development in pearl millet through exploitation of heterosis (Hanna 1995).

### 10.10.2    Genetics of Apomixis

It has long been known that apomixis is under genetic control. However, the genetics of this important trait are difficult to study because sexual members and their apomictic counterparts are generally not available in the same species. Crosses will therefore need to be made between sexual and apomictic plants from different species. Studies on inheritance of apomixis are further complicated by the presence of facultative apomicts, the need to use an apomict as a male parent, and the lack of $F_2$ segregating populations.

In spite of these limitations, some understanding of the genetics of apomixis has been developed (Nogler 1984; Asker and Jerling 1992), and it is generally believed to be simple genetic control (Hanna 1995) that could improve its chances of being transferred to a crop species.

### 10.10.3    Transferring to Pearl Millet

The crossability of wild, apomictic *Pennisetum* species with sexual cultivated pearl millet has met with varied success ranging from no seed formation to recovery of infertile or fertile $F_1$ hybrids (Jauhar 1981a; Dujardin and Hanna 1989). Crosses between induced tetraploid sexual pearl millet and hexaploid apomictic *P. squamulatum* have been the most fertile and have shown the highest level of expression of apospory in crosses and backcrosses. Such crosses have been useful for studying the inheritance of apospory, and it has been shown that the trait is dominant and is transmitted as a single-dose allele (Ozias-Akins et al. 1998). However, backcrosses with the two-species hybrid rapidly declined in male fertility. It was therefore necessary to use another sexual hexaploid interspecific hybrid (pearl millet × *P. purpureum*) as a bridge for backcrossing (Dujardin and Hanna 1984; Hanna et al. 1992). DNA amplification-based molecular markers have been shown to cosegregate with apomixis in backcrosses (Ozias-Akins et al. 1993) and the $F_1$ hybrids (Ozias-Akins et al. 1998). These markers have proved useful for physical identification of the chromosome associated with apospory (Ozias-Akins et al. 2003; Akiyama et al. 2004).

The backcrossing program at Tifton, GA, where the aposporous mechanism is being transferred from *P. squamulatum* to cultivated pearl millet, has progressed to the $BC_8$ generation. A bottleneck was encountered at the $BC_3$ generation where only a single apomictic plant that shed pollen was found among the 1053 individuals screened (Dujardin and Hanna 1989). This plant contained 29 chromosomes, which is only one more than in the tetraploid pearl millet recurrent parent. However, molecular analysis showed that more than one linkage group from the apomictic parent remained in $BC_3$ (Ozias-Akins et al. 1993). These results were confirmed by direct visualization of alien chromosomes in the $BC_3$ line (Goel et al. 2003). This was accomplished by using fluorescent genomic *in situ* hybridization (fl-GISH), where total DNA from *P. squamulatum* was labeled and used as a probe onto chromosome spreads of $BC_3$. Unlabeled pearl millet DNA was used to block hybridization of any common sequences. In both mitotic and meiotic spreads, 3 chromosomes of 29 hybridized across their entire length with *P. squamulatum* DNA. These three chromosomes were observed to segregate randomly during meiosis, which could explain the low transmission rate of the trait (2 to 5%). The same technique was used to examine the chromosomal constitution of more advanced backcross generations, and the apomictic backcross plants were observed to have one to three chromosomes from *P. squamulatum* (Figure 10.6).

The single chromosome that was invariably associated with apomixis could be detected by two additional types of probes, both containing apomixis-linked molecular markers. The first probe consisted of pooled DNA amplification-based markers, some of which had been shown to contain

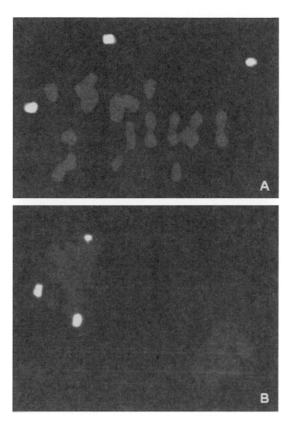

**Figure 10.6**  Fluorescent genomic *in situ* hybridization (FI-GISH) with labeled total genomic DNA from *P. squamulatum*. Chromosomal spreads from apomictic backcross lines at meiosis were hybridized with fluorescently labeled DNA from the apomictic parent and unlabeled DNA from the recurrent, sexual parent (*P. glaucum*). The three alien chromosomes are unpaired at meiosis (A) and can migrate either to the same pole (B) or opposite poles.

repetitive DNA (Ozias-Akins et al. 1998). This probe hybridized across a distinct region terminating the short arm of a single chromosome (Goel et al. 2003). The same region also was detected by probing with entire bacterial artificial chromosome clones that had been isolated from a library constructed from an $F_1$ polyhaploid genotype (Roche et al. 2002). The presence of this single chromosome was sufficient to confer on a host plant the ability to form aposporous embryo sacs and to produce asexual seeds. Plants from the most advanced backcross generation, $BC_8$, are morphologically similar to cultivated pearl millet, but seed set in obligate apomicts is usually less than 15%.

### 10.10.4   Possible Use of Apomixis to Develop Cultivars

The potential value of apomixis in producing new cultivars in cultivated pearl millet is high. Apomixis has many benefits, such as the ability to maintain heterozygosity through seed, thereby eliminating the need to have two parental lines and distance isolation to produce commercial hybrids. It also provides the opportunity to use individual apomictic plants (regardless of heterozygosity) from crosses of genetically diverse parents as cultivars, eliminating the need to maintain inbreds and cytoplasmic-nuclear male sterile lines, thereby simplifying the process of hybrid production. Apomixis would be especially beneficial to developing countries by making superior hybrid cultivars available at an economical price. Apomixis would also be beneficial in maintaining heterosis in forage hybrids through seed.

Regarding the prospect for release of a commercial pearl millet hybrid, several problems still need to be addressed. First, although cultivated pearl millet is diploid, present research indicates that apomixis may need to be used in tetraploid pearl millet because this reproductive mechanism is mainly expressed in polyploids. Excellent seed fertility is possible in hybrids of tetraploid pearl millet, so this would be a viable approach. Crossing schemes for using apomixis in plant breeding, which could be adapted to tetraploid pearl millet, have been discussed previously (Hanna 1995; Savidan 2000). Second, apomixis is currently incorporated into tetraploid pearl millet as alien chromosome addition/substitution lines; e.g., an entire alien chromosome still remains in even the most advanced backcrosses. Consequently, male fertility is reduced. Since this chromosome does not pair with any of the pearl millet chromosomes, it is unlikely that further integration can be achieved by recombination. It is possible that irradiation could be used to fragment the apomixis-associated chromosome, resulting in a translocation, and that a translocated chromosome could perhaps segregate more regularly during meiosis and result in a higher level of male fertility. Third, seed set in the apomictic backcrosses typically is less than 15%.

The best evidence for the cause of low seed set has come from the examination of endosperm development in $BC_3$ (Morgan et al. 1998). Aposporous embryo sacs of *Pennisetum* contain only four nuclei, one of which becomes the egg. The other three nuclei may be variably incorporated into two synergids and a uninucleate central cell or one synergid and a binucleate central cell. If the binucleate central cell is fertilized by a haploid sperm, the consequence would be an alteration in the maternal-to-paternal genome ratio in the endosperm from 2:1 to 4:1. The latter ratio results in endosperm degeneration in many grasses. In *P. squamulatum*, uninucleate central cells are predominant, whereas in the backcross lines, binucleate central cells are preponderant. However, flow cytometric analysis of the DNA content of developing endosperm showed that only triploid endosperm was present in viable seeds. Thus, if the binucleate central cells are fertilized, they probably do not support functional endosperm development.

Although complete introgression of apomixis into sexual pearl millet continues to be plagued with problems, the genetic materials produced have contributed enormously to our understanding of apomictic reproduction in this genus at the developmental, cellular, and molecular levels. Molecular cytogenetic characterization of these lines has allowed the identification of a segment of a single chromosome that is most likely sufficient for the transmission of this reproductive trait. Whether near-obligate apomixis can yet be achieved in an agronomically acceptable genetic background remains to be determined.

## 10.11 DIRECT GENE TRANSFER IN PEARL MILLET

Most genetic improvement of pearl millet has been accomplished by traditional breeding. Heterosis breeding has proven particularly fruitful in increasing the grain yield of pearl millet (Section 10.9). Interspecific hybridization has also been instrumental in breeding superior fodder and forage plants (Section 10.8). Genetic transformation could be used as a supplementary tool for asexually incorporating certain desirable traits into otherwise superior pearl millet cultivars, as has been done in other cereal crops (Jauhar and Chibbar 1999; Repellin et al. 2001; Jauhar and Khush 2002).

Although *in vitro* regeneration protocols, a prerequisite for genetic transformation, were established by several workers (Vasil and Vasil 1981; Morrish et al. 1990; Oldach et al. 2001), reports of genetic transformation of pearl millet are limited. Using microprojectile bombardment, Devi and Sticklen (2002) showed transient expression of the reporter gene, -glucuronidase (GUS), in shoot-tip clump cultures of pearl millet. Girgi et al. (2002) were able to produce herbicide-resistant pearl millet at a frequency of 0.18% using microprojectile bombardment of immature embryos. Using immature inflorescences, Goldman et al. (2003) were able to produce herbicide-resistant pearl millet plants at a frequency of 1 to 28 per successful bombardment, and achieved a frequency of cotransformation with green fluorescent protein gene ranging from 5 to 85%. Transgenic approaches to combat diseases like downy mildew are in progress in some laboratories.

## 10.12 CONCLUSIONS AND PERSPECTIVES

The world population is growing at an alarming rate of more than 2%, the growth being even more in the poor, malnourished parts of Africa and Asia. It may become very difficult to meet the ever-expanding demand for food, especially for this large segment of the underprivileged society. Approximately one sixth of the world population lives in the semiarid tropics comprising parts of Africa, Asia, and Latin America — the regions characterized by high temperatures, poor and depleted soils, and limited and erratic rainfall. Pearl millet is a dual-purpose crop, providing fodder for cattle and also serving as a poor man's bread. It is the staple food for and provides sustenance to a large segment of people in these impoverished regions. And this cereal has a remarkable capacity to grow in some of the poorest soils in the chronically drought-prone regions. The need for genetic improvement of pearl millet cannot be overemphasized, and it is a challenging task for breeders, cytogeneticists, biotechnologists, agronomists, and farmers.

Traditional breeding has played a major role in producing superior cultivars. Being allogamous, pearl millet is highly suited for and responds very well to heterosis breeding. Single-cross hybrids yield about 25 to 30% more than open-pollinated varieties of comparable maturity and are therefore widely adopted. Thus, in 2001 more than 70 hybrids were under cultivation on about 6 million ha of the total 10 million ha of pearl millet area in India. Exploitation of hybrid vigor will continue to be an important strategy for improving both grain and forage yields. Planting superior hybrids in the vast pearl millet growing areas of Africa, Asia, and South America would result in dramatic increases in grain yields. If apomixis were introduced in hybrids with desired heterozygosity and the right gene combinations, it would be possible to perpetuate hybrid vigor over extended periods, without having to produce hybrid seed and distribute to farmers year after year. Research in this area is in progress.

It is also very important to broaden the genetic base of hybrids to confer on them built-in insurance against future diseases. Hybrids with a narrow genetic base would be vulnerable to new pathotypes that may arise in the future. A potent strategy to avoid such an eventuality would be to produce genetically broad-based male sterile lines using germplasm resources with disease resistance. Such male sterile lines have indeed been developed at ICRISAT in India for use in synthesis of hybrids. These lines provide both high grain yields and resistance to downy mildew, a major disease caused by *Sclerospora graminicola* (Sacc.) Schroet. It has also been shown that top-cross and interpopulation hybrids may combine the adaptation of local landraces with higher seed yield potential of improved male sterile genotypes used as female parents, and such hybrids would be less vulnerable to downy mildew as well as the two floral diseases (ergot and smut).

Being the staple food of a large section of the population in Africa and Asia, pearl millet is the main source of dietary protein. Nutritional enhancement of its grain should therefore be an important goal of pearl millet breeding. With the genetic upgrading of its protein content and amino acid balance, pearl millet will be better able to feed the underprivileged sector of the society. Germplasm sources such as golden millet with β-carotene levels comparable to those in golden rice have been identified, thus paving the way for breeding hybrids with enhanced levels of this provitamin A.

Pearl millet is an ideal organism for both basic and applied research — the studies that have improved our knowledge of gene mapping and genome evolution. It is the only cereal that reliably produces both grain and fodder under some of the harshest environments. It is encouraging to see that genetic manipulation of the pearl millet genome has resulted in numerous superior hybrids. Modern tools of biotechnology could help incorporate value-added traits, including perhaps the apomictic mode of reproduction and disease resistance, into otherwise superior cultivars. A comprehensive effort should be made, using all available tools, to bring about genetic enrichment of pearl millet so that it may continue to play an important role in the welfare of the poor in the semiarid tropics of Africa and Asia.

## REFERENCES

Aken'Ova, M.E. and Chedda, H.R. 1973. Interspecific hybrids of *Pennisetum typoides* S. & H. × *P. purpureum* Schum. for forage in the hot humid lowland tropics of West Africa. *Nigerian Agric. J.* 10: 82–90.

Akiyama, Y., Conner, J.A., Goel, S., Morishige, D.T., Mullet, J.E., Hanna, W.W., and Ozias-Akins, P. 2004. High-resolution physical mapping in *Pennisetum squamulatum* reveals extensive chromosomal heteromorphism of the genomic region associated with apomixis. *Plant Physiol.* 134: 1733–1741.

Ali, A.M., Hash, C.T., Ibrahim, A.E.S., and Raj, A.G.B. 2001. Population diallel of elite medium- and long-duration pearl millet composites. I. Populations and their $F_1$ crosses. *Crop Sci.* 41: 705–711.

Anand Kumar, K., Jain, R.P., and Singh, S.D. 1983. Downy mildew reactions of pearl millet lines with and without cytoplasmic male sterility. *Plant Dis.* 67: 663–665.

Andrews, D.J. and Harinarayana, G. 1984. *Procedures for Seed Production of Pearl Millet Varieties*, Information Bulletin 16. ICRISAT, Patancheru, India.

Appa Rao, S. 1999. Genetic resources. In *Pearl Millet Breeding*, I.S. Khairwal, K.N. Rai, D.J. Andrews, and G. Harinarayana, Eds. Science Publishers, Enfield, NH, pp. 49–81.

Appa Rao, S. and de Wet, J.M.J. 1999. Taxonomy and evolution. In *Pearl Millet Breeding*, I.S. Khairwal, K.N. Rai, D.J. Andrews, and G. Harinarayana, Eds. Science Publishers, Enfield, NH, pp. 29–47.

Appa Rao, S., Mengesha, M.H., and Rajagopal Reddy, C. 1989. Development of cytoplasmic male-sterile lines of pearl millet from Ghana and Botswana germplasm. In *Perspectives in Cytology and Genetics*, G.K. Manna and U. Sinha, Eds. Kalyani, India, pp. 817–823.

Asker, S. and Jerling, L. 1992. *Apomixis in Plants*. CRC Press, Boca Raton, FL.

Athwal, D.S. 1961. Recent developments in the breeding and improvement of bajra (pearl millet) in the Punjab. *Madras Agric. J.* 48: 18–19 (abstract).

Athwal, D.S. 1965. Hybrid bajra-1 marks a new era. *Indian Farm.* 15: 6–7.

Athwal, D.S. 1966. Current plant breeding research with special reference to *Pennisetum*. *Indian J. Genet. Plant Breed.* 26A: 73–85.

Avdulov, N.P. 1931. Karyo-systematic investigations in the family Gramineae. *Bull. Appl. Bot. Genet. Plant Breed.*, Suppl. 43 (Russian).

Bor, N.L. 1960. *Grasses of Burma, Ceylon, India and Pakistan (excluding Bambuseae)*. Pergamon Press, London.

Brunken, J.N., de Wet, J.M.J., and Harlan, J.R. 1977. The morphology and domestication of pearl millet. *Econ. Bot.* 31: 163–174.

Burton, G.W. 1944. Hybrids between Napier grass and cattail millet. *J. Hered.* 35: 227–232.

Burton, G.W. 1958. Cytoplasmic male sterility in pearl millet (*Pennisetum glaucum*) (L.) R. Br. *Agron. J.* 50: 230–231.

Burton, G.W. 1965a. Pearl millet Tift 23A released. *Crops Soils* 17: 19.

Burton, G.W. 1965b. Male-sterile pearl millet Tift 18A released. *Crops Soils* 18: 19.

Burton, G.W. and Powell, J.B. 1968. Pearl millet breeding and cytogenetics. *Adv. Agron.* 20: 49–89.

Chopra, K.R., Chopra, R., Rabbani, G., and Thimaiah, K.K. 1999. Seed production. In *Pearl Millet Breeding*, I.S. Khairwal, K.N. Rai, D.J. Andrews, and G. Harinarayana, Eds. Science Publishers, Enfield, NH, pp. 445–478.

Clegg, M.T., Rawson, J.R.Y., and Thomas, K. 1984. Chloroplast DNA variation in pearl millet and related species. *Genetics* 106: 449–461.

Couturon, E., Mariac, C., Bezançon, G., Lauga, J., and Renno, J.-F. 2003. Impact of natural and human selection on the frequency of the F1 hybrid between cultivated and wild pearl millet (*Pennisetum glaucum* (L.) R. Br.). *Euphytica* 133: 329–337.

Dave, H.R. 1987. Pearl millet hybrids. In *Proceedings of the International Pearl Millet Workshop*. ICRISAT, Patancheru, Andhra Pradesh, India, pp. 121–126.

Devi, P. and Sticklen, M. 2002. Culturing shoot-tip clumps of pearl millet [*Pennisetum glaucum* (L.) R. Br.] and optimal microprojectile bombardment parameters for transient expression. *Euphytica* 125: 45–50.

Dhanapala, S.B., Siriwardene, J.A. de S., and Pathirana, K.K. 1972. NB 21 — a new hybrid Napier. *Ceylon Vet. J.* 20: 77.

Dujardin, M. and Hanna, W.W. 1983. Meiotic and reproductive behavior of facultative apomictic BC1 offspring derived from *Pennesitum americanum-P. orientale* interspecific hybrids. *Crop Sci.* 23: 156–160.

Dujardin, M. and Hanna. W. 1984. Cytogenetics of double cross hybrids between *Pennisetum americanum-P. purpureum* amphiploids and *P. americanum* × *Pennisetum squamulatum* interspecific hybrids. *Theor. Appl. Genet.* 69: 97–100.

Dujardin, M. and Hanna, W.W. 1985. Cytology and reproductive behavior of pearl millet-Napier grass hexaploids × *Pennisetum squamulatum* trispecific hybrids. *J. Hered.* 76: 382–384.

Dujardin, M. and Hanna, W.W. 1987. Inducing male fertility in crosses between pearl millet and *Pennisetum orientale* Rich. *Crop Sci.* 27: 65–68.

Dujardin, M. and Hanna, W. 1989. Crossability of pearl millet with wild *Pennisetum* species. *Crop Sci.* 29: 77–80.

Gildenhuys, P.J. 1950. A new fodder grass — A promising cross between babala and Napier fodder. *Farm. S. Afr.* 15: 189–191.

Gill, B.S., Minocha, J.L., Gupta, D., and Kumar, D. 1969. Chromosome behaviour and seed setting in autotetraploid pearl millet. *Indian J. Genet. Plant Breed.* 29: 462–467.

Gill, B.S., Sharma, H.L., and Dhesi, J.S. 1973. Cytomorphological studies of haploid pearl millet. *Cytologia* 38: 411–416.

Girgi, M., O'Kennedy, M.M., Morgenstern, A., Mayer, G., Lörz, H., and Oldach, K.H. 2002. Transgenic and herbicide resistant pearl millet (*Pennisetum glaucum* L.) R. Br. via microprojectile bombardment of scutellar tissue. *Mol. Breed.* 10: 243–252.

Goel, S., Chen, Z., Conner, J.A., Akiyama, Y., Hanna, W.W., and Ozias-Akins, P. 2003. Delineation by FISH of a single hemizygous chromosomal region associated with aposporous embryo sac formation in *Pennisetum squamulatum* and *Cenchrus ciliaris*. *Genetics* 163: 1069–1082.

Goldman, J.J., Hanna, W.W., Fleming, G., and Ozias-Akins, P. 2003. Fertile transgenic pearl millet [*Pennisetum glaucum* (L.) R. Br.] plants recovered through microprojectile bombardment and phosphinothricin selection of apical meristem-, inflorescence-, and immature embryo-derived embryogenic tissues. *Plant Cell Rep.* 21: 999–1009.

Gonzalez, B. and Hanna, W.W. 1984. Morphological and fertility responses in isogenic triploid and hexaploid pearl millet × Napier grass hybrids. *J. Hered.* 75: 317–318.

Grimanelli, D., Leblanc, O., Perotti, E., and Grossniklaus, U. 2001. Developmental genetics of gametophytic apomixis. *Trends Genet.* 17: 597–604.

Grossniklaus, U., Nogler, G., and van Dijk, P.J. 2001. How to avoid sex: the genetic control of gametophytic apomixis. *Plant Cell* 13: 1491–1497.

Gupta, V.P. 1974. Inter- and intra-specific hybridization in forage plants: genus *Pennisetum*. In *Breeding Researches in Asia and Oceania*, S. Ramanujam and R.D. Iyer, Eds. Indian Society of Genetics and Plant Breeding, New Delhi, pp. 162–172.

Hanna, W.W. 1979. Interspecific hybrids between pearl millet and fountain grass. *J. Hered.* 70: 425–427.

Hanna, W.W. 1986. Utilization of wild relatives of pearl millet. In *Proceedings of the International Pearl Millet Workshop*, ICRISAT, April 7–11.

Hanna, W.W. 1989. Characteristics and stability of a new cytoplasmic-nuclear male sterile source in pearl millet. *Crop Sci.* 29: 1457–1459.

Hanna, W.W. 1990. Transfer of germplasm from the secondary to the primary gene pool in *Pennisetum*. *Theor. Appl. Genet.* 80: 200–204.

Hanna, W. 1995. Use of apomixis in cultivar development. *Adv. Agron.* 54: 333–350.

Hanna, W.W. and Dujardin, M. 1982. Apomictic interspecific hybrids between pearl millet and *Pennisetum orientale* L.C. Rich. *Crop Sci.* 22: 857–859.

Hanna, W.W. and Dujardin, M. 1986. Cytogenetics of *Pennisetum schweinfurthii* Pilger and its hybrids with pearl millet. *Crop Sci.* 26: 449–453.

Hanna, W.W., Dujardin, M., and Monson, W.G. 1989. Using diverse species to improve quality and yield in the *Pennisetum* genus. In *Proceedings of the XVI International Grasslands Congress*, Nice, France, October 1–8, pp. 403–404.

Hanna, W., Dujardin, M., Ozias-Akins, P., and Arthur, L. 1992. Transfer of apomixis in *Pennisetum*. In *Proceedings of the Apomixis Workshop*, Atlanta, GA, pp. 30–33.

Hanna, W.W., Gaines, T.P., Gonzalez, B., and Monson, W.G. 1984. Effect of ploidy on yield and quality of pearl millet × Napier hybrids. *Agron. J.* 76: 969–971.

Hanna, W.W., Hill, G.M., Gates, R.N., Wilson, J.P., and Burton, G.W. 1997. Registration of 'Tifleaf 3' pearl millet. *Crop Sci.* 37: 1388.

Harlan, J.R. 1971. Agricultural origins: centers and noncenters. *Science* 174: 468–474.

Hash, C.T. 1997. Research on downy mildew of pearl millet. In *Proceedings of Integrating Research Evaluation Efforts*, M.C.S. Bantilan and P.K. Joshi, Eds. International Crops Research Institute for the Semi-Arid Tropics (ICRISAT), Patancheru, Andhra Pradesh, India, pp. 121–128.

Hitchcock, A.S. and Chase, A. 1951. *Manual of the Grasses of the United States*, 2nd ed., U.S. Department of Agriculture Miscellaneous Publication 200. USDA, Washington, DC.

Hussain, A., Ullah, M., and Ahmad, B. 1968. Napier bajra hybrid as compared with the parents-nutritive value based on chemical composition. *W. Pak. J. Agric. Res.* 6: 69–72.

International Crops Research Institute for the Semi-Arid Tropics (ICRISAT). 1996. Improving the Unimprovable: Succeeding with Pearl Millet. *ICRISAT Report*, May 1996.

Jauhar, P.P. 1968. Inter- and intra-genomal chromosome pairing in an interspecific hybrid and its bearing on basic chromosome number in *Pennisetum. Genetica* 39: 360–370.

Jauhar, P.P. 1970a. Haploid meiosis and its bearing on phylogeny of pearl millet, *Pennisetum typhoides* Stapf et Hubb. *Genetica* 41: 532–540.

Jauhar, P.P. 1970b. Chromosome behaviour and fertility of the raw and evolved synthetic tetraploids of pearl millet, *Pennisetum typhoides* Stapf et Hubb. *Genetica* 41: 407–424.

Jauhar, P.P. 1973. Inter- and intra-genomal chromosome relationships in a *Pennisetum* hybrid. *Genetics* 74: 126–127.

Jauhar, P.P. 1981a. *Cytogenetics and Breeding of Pearl Millet and Related Species*. Alan R. Liss., New York.

Jauhar, P.P. 1981b. Cytogenetics of pearl millet. *Adv. Agron.* 34: 407–470.

Jauhar, P.P. 1981c. The eternal controversy on the Latin name of pearl millet. *Indian J. Bot.* 4: 1–4.

Jauhar P.P. 2003. Formation of 2n gametes in durum wheat haploids: sexual polyploidization. *Euphytica* 133: 81–94.

Jauhar, P.P. and Chibbar, R.N. 1999. Chromosome-mediated and direct gene transfers in wheat. *Genome* 42: 570–583.

Jauhar, P.P and Hanna, W.W. 1998. Cytogenetics and genetics of pearl millet. *Adv. Agron.* 64: 1–26.

Jauhar, P.P. and Joppa, L.R., 1996. Chromosome pairing as a tool in genome analysis: merits and limitations. In *Methods of Genome Analysis in Plants*, P.P. Jauhar, Ed. CRC Press, Boca Raton, FL, pp. 9–37.

Jauhar, P.P. and Khush, G.S. 2002. Importance of biotechnology in global food security. In *Food Security and Environment Quality: A Global Perspective*, R. Lal, D.O. Hansen, N. Uphoff, and S. Slack, Eds. CRC Press, Boca Raton, FL, pp. 107–128.

Jauhar, P.P. and Singh, U. 1969. Amphidiploidization induced by decapitation in an interspecific hybrid of *Pennisetum. Curr. Sci.* 38: 420–421.

Kelley, T.G., Partasarathy Rao, E., Weltzien R., and Purohit, M.L. 1996. Adoption of improved cultivars of pearl millet in an arid environment: straw yield and quality considerations in western Rajasthan. *Expl. Agric.* 32: 161–171.

Kellis, M., Birren, B.W., and Lander, E.S. 2004. Proof and evolutionary analysis of ancient genome duplication in the yeast *Saccharomyces cerevisiae. Nature* 428: 617–624.

Khairwal, I.S. and Singh, S. 1999. Quantitative and qualitative traits. In *Pearl Millet Breeding*, I.S. Khairwal, K.N. Rai, D.J. Andrews, and G. Harinarayana, Eds. Science Publishers, Enfield, NH, pp. 119–157.

Khan, M.-D. and Rahman, H. 1963. Genome relationship and chromosome behaviour in the allotriploid hybrid of *Pennisetum typhoides* and *P. purpureum. W. Pak. J. Agric. Res.* 1: 61–65.

Koltunow, A.M. 1993. Apomixis: embryo sacs and embryos formed without meiosis or fertilization in ovules. *Plant Cell* 5: 1425–1437.

Krishnaswamy, N. and Raman, V.S. 1949. A note on the amphidiploid of the hybrid of *Pennisetum typhoides* Stapf and Hubbard × *P. purpureum* Schumach. *Curr. Sci.* 18: 15–16.

Krishnaswamy, N. and Raman, V.S. 1954. Studies on the interspecific hybrid of *Pennisetum typhoides* Stapf and Hubb. × *P. purpureum* Schumach. III. The cytogenetics of the colchicine-induced amphidiploid. *Genetica* 27: 253–272.

Krishnaswamy, N., Raman, V.S., and Nair, N.H. 1950. An autotetraploid in the pearl millet. *Curr. Sci.* 19: 252–253.

Liu, C.J., Witcombe, J.K., Pittaway, T.S., Nash, M., Hash, C.T., Busso, C.S., and Gale, M.D. 1994. An RFLP-based genetic map of pearl millet (*Pennisetum glaucum*). *Theor. Appl. Genet.* 89: 481–487.

Mahalakshmi, V., Bidinger, F.R., Rao, K.P., and Raju, D.S. 1992. Performance and stability of pearl millet topcross hybrids and their variety pollinators. *Crop Sci.* 32: 928–932.

Marchais, L. and Pernes, J. 1985. Genetic divergence between wild and cultivated pearl millets (*Pennisetum typhoides*). I. Male sterility. *Z. Pflanzenzüchtg.* 95: 103–112.

Martel, E., De Nay, D., Siljak-Yakovlev, S., Brown, S., and Sarr, A. 1997. Genome size variation and basic chromosome number in pearl millet and fourteen related *Pennisetum* species. *J. Hered.* 88: 139–143.

Martel, E., Ricroch, A., and Sarr, A. 1996. Assessment of genome organization among diploid species (2n = 2x = 14) belonging to primary and tertiary gene pools of pearl millet using fluorescent *in situ* hybridization with rDNA probes. *Genome* 39: 680–687.

Meredith, D., Ed. 1955. *The Grasses and Pastures of South Africa*. Cape Times Ltd., Parrow CP, South Africa.

Minocha, J.L., Gill, B.S., Sharma, H.L., and Sidhu, J.L. 1980. Cytogenetic studies on primary trisomics in pearl millet. In *Trends in Genetical Research in Pennisetums*, V.P. Gupta and J.L. Minocha, Eds. Punjab Agricultural University, Ludhiana, India, pp. 129–132.

Morgan, R.N., Ozias-Akins, P., and Hanna, W.W. 1998. Seed set in an apomictic BC$_3$ pearl millet. *Int. J. Plant Sci.* 159: 89–97.

Morrish, F.M., Hanna, W.W., and Vasil, I.K. 1990. The expression and perpetuation of inherent somatic variation in regenerants from embryogenic cultures of *Pennisetum glaucum* (L.) R. Br. (pearl millet). *Theor. Appl. Genet.* 80: 409–416.

Muldoon, D.K. and Pearson, C.J. 1977. Hybrid *Pennisetum* in a warm temperate climate: regrowth and standover forage production. *Aust. J. Exp. Agric. Anim. Husb.* 17: 277–283.

Nogler, G.A. 1984. Gametophytic apomixis. In *Embryology of Angiosperms*, B.M. Johri, Ed. Springer-Verlag, Berlin.

Ohno, S. 1970. *Evolution by Gene Duplication*. Springer-Verlag, Berlin.

Oldach, K.H., Morganstern, A., Rother, S., Girgi, M., O'Kennedy, M., and Lörz, H. 2001. Efficient *in vitro* plant regeneration from immature zygotic embryos of pearl millet (*Pennisetum glaucum* (L.) R. Br.) and *Sorghum bicolor* (L.) Moench. *Plant Cell Rep.* 20: 416–421.

Osgood, R.V., Hanna, W.W., and Tew, T.L. 1997. Hybrid seed production of pearl millet × Napier grass triploid hybrids. *Crop Sci.* 37: 998–999.

Ouendeba, B., Ejeta, G., Nyquist, W.E., Hanna, W.W., and Kumar, A. 1993. Heterosis and combining ability among African pearl millet landraces. *Crop Sci.* 33: 735–739.

Ozias-Akins, P., Akiyama, Y., and Hanna, W.W. 2003. Molecular characterization of the genomic region linked with apomixis in *Pennisetum/Cenchrus*. *Funct. Integr. Genomics* 3: 94–104.

Ozias-Akins, P., Lubbers, E.L., Hanna, W.W., and McNay, J.W. 1993. Transmission of the apomictic mode of reproduction in *Pennisetum*: co-inheritance of the trait and molecular markers. *Theor. Appl. Genet.* 85: 632–638.

Ozias-Akins, P., Roche, D., and Hanna, W.W. 1998. Tight clustering and hemizygosity of apomixis-linked molecular markers in *Pennisetum squamulatum* implies genetic control of apospory by a divergent locus which may have no allelic form in sexual genotypes. *Proc. Natl. Acad. Sci. U.S.A.* 95: 5127–5132.

Patil, B.D. 1963. Two new Pusa Napiers. *Indian Farm.* 12: 20–23.

Patil, B.D, Hardas, M.W., and Joshi, A.B. 1961. Auto-alloploid nature of *Pennisetum squamulatum*. *Fresen. Nat.* 189: 419–420

Patil, B.D. and Singh, A. 1964. An interspecific cross in the genus *Pennisetum* involving two basic numbers. *Curr. Sci.* 33: 161–162.

Powell, J.B., Hanna, W.W., and Burton, G.W. 1975. Origin, cytology, and reproductive characteristics of haploids in pearl millet. *Crop Sci.* 15: 389–392.

Presterl, T. and Weltzien, E. 2003. Exploiting heterosis in pearl millet for population breeding in arid environments. *Crop Sci.* 43: 767–776.

Pritchard, A.J. 1971. Hybrid between *Pennisetum typhoides* and *P. purpureum* as a potential forage crop in south-eastern Queensland. *Trop. Grassl.* 6: 35–39.

Rai, K.N. 1995. A new cytoplasmic-nuclear male sterility system in pearl millet. *Plant Breed.* 114: 445–447.

Rai, K.N., Anand Kumar, K., Andrews, D.J., Gupta, S.C., and Ouendeba, O. 1997. Breeding pearl millet for grain yield and stability. In *Proceedings of the International Conference on Genetic Improvement of Sorghum and Pearl Millet*, INTSORMIL Publication 97-5, pp. 71–83.

Rai, K.N., Anand Kumar, K., Andrews, D.J., and Rao, A.S. 2001a. Commercial viability of alternative cytoplasmic-nuclear male sterility system in pearl millet. *Euphytica* 121: 107–114.

Rai, K.N., Andrews, D.J., and Rao, A.S. 2000a. Feasibility of breeding male-sterile populations for use in developing inter-population hybrids of pearl millet. *Plant Breed.* 119: 335–339.

Rai, K.N., Chandra, S., and Rao, A.S. 2000b. Potential advantages of male-sterile $F_1$ hybrids for use as seed parents of three-way hybrids in pearl millet. *Field Crops Res.* 68: 173–181.

Rai, K.N. and Hash, C.T. 1990. Fertility restoration in male sterile × maintainer hybrids of pearl millet. *Crop Sci.* 30: 889–892.

Rai, K.N. and Thakur, R.P. 1995. Ergot reaction of pearl millet hybrids affected by fertility restoration and genetic resistance of parental lines. *Euphytica* 83: 225–231.

Rai, K.N. and Thakur, R.P. 1996. Smut reaction of pearl millet hybrids affected by fertility restoration and genetic resistance of parental lines. *Euphytica* 90: 31–37.

Rai, K.N., Thakur, R.P., Rao, V.P., and Rao, A.S. 2001b. Genetic resistance of pearl millet male-sterile lines to diverse Indian pathotypes of *Sclerospora graminicola*. *Plant Dis.* 85: 621–626.

Rai, K.N. and Virk, D.S. 1999. Breeding methods. In *Pearl Millet Breeding*, I.S. Khairwal, K.N. Rai, D.J. Andrews, and G. Harinarayana, Eds. Science Publishers, Enfield, NH, pp. 195–212.

Rai, K.N., Virk, D.S., Harinarayana, G., and Rao, A.S. 1996. Stability of male-sterile sources and fertility restoration of their hybrids in pearl millet. *Plant Breed.* 115: 494–500.

Raman, V.S. 1965. Progress of cytogenetic research in Madras state. In *Advances in Agricultural Sciences and Their Applications*, S. Krishnamurthi, Ed. Agricultural College Research Institute, Coimbatore, India, pp. 122–143.

Rangaswamy, S.R.S. 1972. Cytological studies on diploid and polyploid taxa of the genus *Pennisetum* Rich. *Genetica* 43: 257–273.

Rau, N.S. 1929. On the chromosome numbers of some cultivated plants of south India. *J. Indian Bot. Sci.* 8: 126–128.

Repellin, A., Båga, M., Jauhar, P.P., and Chibbar, R.N. 2001. Genetic enrichment of cereal crops via alien gene transfer: new challenges. *Plant Cell Tissue Organ Cult.* 64: 159–183.

Roche, D.R., Conner, J.A., Budiman, M.A., Frisch, D., Wing, R., Hanna, W.W., and Ozias-Akins, P. 2002. Construction of BAC libraries from two apomictic grasses to study the microcolinearity of their apospory-specific genomic regions. *Theor. Appl. Genet.* 104: 804–812.

Roche, D., Hanna, W., and Ozias-Akins, P. 2001. Is supernumerary chromatin involved in gametophytic apomixis of polyploid plants? *Sex. Plant Reprod.* 13: 343–349.

Savidan, Y. 2000. Apomixis: genetics and breeding. In *Plant Breeding Reviews*, J. Janick, Ed. John Wiley & Sons, New York, pp. 13–86.

Schank, S. and Hanna, W. 1995. Usage of *Pennisetum* in Florida and the tropics. In *Proceedings of the International Conference on Livestock in the Tropics*, Gainesville, FL, May 7–10, pp. 1–13.

Scheifele, G.L., Whitehead, W., and Rowe, C. 1970. Increased susceptibility of southern leaf spot (*Helminthosporium maydis*) in inbred lines and hybrids of maize with Texas male-sterile cytoplasm. *Plant Dis. Rep.* 54: 501–503.

Spillane, C., Steimer, A., and Grossniklaus, U. 2001. Apomixis in agriculture: the quest for clonal seeds. *Sex. Plant Reprod.* 14: 179–187.

Stapf, O. and Hubbard, C.E. 1934. *Pennisetum*. In *Flora of Tropical Africa*, Vol. 9, Part 6, D. Prain, Ed. Reeve & Co. Ltd., Ashford, Kent, England, pp. 954–1070.

Sujata, V., Sivaramakrishnan, S., Rai, K.N., and Seetha, K. 1994. A new source of cytoplasmic male sterility in pearl millet: RFLP analysis of mitochondrial DNA. *Genome* 37: 482–486.

Talukdar, B.S., Khairwal, I.S., and Singh, R. 1999. Hybrid breeding. In *Pearl Millet Breeding*, I.S. Khairwal, K.N. Rai, D.J. Andrews, and G. Harinarayana, Eds. Science Publishers, Enfield, NH, pp. 269–301.

Talukdar, B.S., Singh, S.D., and Prakash Babu, P.P. 1996. Prospects of topcross hybrids in increasing and stabilizing grain yield in pearl millet. *Crop Improv.* 23: 247–250.

Thakur, R.P. 1989. Flowering events in relation to smut susceptibility in pearl millet. *Plant Pathol.* 38: 557–563.

Thakur, R.P., Williams, R.J., and Rao, V.P. 1983. Control of ergot in pearl millet through pollen management. *Ann. Appl. Biol.* 103: 31–36.

Vasil, V. and Vasil, I.K. 1981. Somatic embryogenesis and plant regeneration from tissue cultures of *Pennisetum americanum* and the *Pennisetum americanum* × *Pennisetum purpureum* hybrid. *Am. J. Bot.* 68: 864–872.

Virk, D.S. 1988. Biometrical analysis in pearl millet: a review. *Crop Improv.* 15: 1–29.

vom Brocke, K., Presterl, T., Christinck, A., Weltzien R.E., and Geiger, H.H. 2002. Farmers' seed management practices open up new base populations for pearl millet breeding in a semi-arid zone of India. *Plant Breed.* 212: 36–42.

Weltzien, R.E. 2000. Supporting farmers' genetic resources management. Experiences with pearl millet in India. In *Encouraging Diversity. The Conservation and Development of Plant Genetic Resources*, C. Almekinders and E. de Boef, Eds. Intermediate Technology Publ., London, pp. 189–193.

Weltzien, R.E. and King, S.B. 1995. Recurrent selection for downy mildew resistance in pearl millet. *Plant Breed.* 114: 308–312.

Yadav, O.P., Manga, V.K., and Gupta, G.K. 1993. Influence of A1 cytoplasmic substitution on the downy mildew incidence of pearl millet. *Theor. Appl. Genet.* 87: 558–560.

Yadav, O.P., Weltzien-Rattunde, E., Bidinger, F.R., and Mahalakshmi, V. 2000. Heterosis in landrace-based topcross hybrids of pearl millet across arid environments. *Euphytica* 112: 285–295.

CHAPTER **11**

# Sorghum Genetic Resources, Cytogenetics, and Improvement

Belum V.S. Reddy, S. Ramesh, and P. Sanjana Reddy

## CONTENTS

11.1 Introduction ........................................................................................310
11.2 Taxonomy, Origin, and Domestication .........................................311
   11.2.1 Taxonomy .................................................................................311
   11.2.2 Origin ........................................................................................312
   11.2.3 Domestication .........................................................................312
11.3 Breeding Behavior and Pollination Control ................................313
11.4 Genetic Resources .............................................................................313
   11.4.1 Importance and Need for Conservation ............................313
   11.4.2 Status of Genetic Resources ................................................314
      11.4.2.1 ICRISAT, Patancheru, India ...............................314
      11.4.2.2 U.S. ..............................................................................315
      11.4.2.3 Africa ..........................................................................315
      11.4.2.4 China ...........................................................................315
   11.4.3 Maintenance of Genetic Resources .....................................315
   11.4.4 Core Collection .......................................................................316
   11.4.5 Evaluation, Characterization, and Documentation of Genetic Resources .............316
   11.4.6 Utilization of Genetic Resources ........................................317
11.5 Genetic Variability for Qualitative and Quantitative Traits ......318
   11.5.1 Morphological/Phenotypic Level ........................................318
   11.5.2 Biochemical Level ..................................................................319
   11.5.3 DNA Level ...............................................................................319
11.6 Genetics and Cytogenetics ..............................................................319
   11.6.1 Genetics ....................................................................................319
      11.6.1.1 Genetics of Morphological and Resistant Traits ......319
      11.6.1.2 Male Sterility .............................................................320
   11.6.2 Cytogenetics ............................................................................322
      11.6.2.1 Karyotype ...................................................................322
      11.6.2.2 Euploid Variation .....................................................322
      11.6.2.3 Aneuploid Variation .................................................322

          11.6.2.4 Apomixis ................................................................................322
11.7 Sorghum Improvement ..............................................................................323
      11.7.1 Environmental Response Characteristics ................................................323
      11.7.2 Conversion Programs........................................................................323
      11.7.3 Breeding Concepts and Breeding Material..........................................324
          11.7.3.1 Breeding Concepts ...........................................................324
          11.7.3.2 Breeding Materials ...........................................................325
      11.7.4 Adaptation and Productivity Enhancement..........................................325
      11.7.5 Trait-Based Breeding.........................................................................326
          11.7.5.1 Grain Yield and Adaptation ..............................................326
          11.7.5.2 Forage and Feed...............................................................326
          11.7.5.3 Resistance to Foliar, Stem, and Head Feeding Insects ..........328
          11.7.5.4 Disease Resistance ...........................................................329
          11.7.5.5 *Striga* Resistance.............................................................330
          11.7.5.6 Drought Tolerance.............................................................332
          11.7.5.7 Soil Chemical Toxicity Tolerance.......................................334
          11.7.5.8 Increased Nutritional Quality.............................................336
          11.7.5.9 Increased Micronutrient Density .......................................338
11.8 New Tools for Sorghum Improvement ......................................................339
      11.8.1 Farmers' Participatory Approach........................................................339
      11.8.2 Biotechnology...................................................................................340
          11.8.2.1 DNA Marker Technology ..................................................340
          11.8.2.2 Genetic Transformation Technology....................................344
          11.8.2.3 Transgenics and Conventional Breeding Integrated Technology............344
11.9 Summary....................................................................................................345
Acknowledgments ................................................................................................346
References .............................................................................................................346

## 11.1 INTRODUCTION

Sorghum (*Sorghum bicolor* (L.) Moench) is the world's fourth major cereal crop in terms of production, and fifth in acreage following wheat, rice, maize, and barley. Sorghum is mostly grown in the semiarid tropics (SAT) of the world as a subsistence dry-land crop by resource-limited farmers under traditional management conditions, thereby recording low productivity compared to the U.S. and Mexico. India grows the largest acreage of sorghum in the world, followed by Nigeria and Sudan, and produces the second largest tonnage after the U.S., with Nigeria being the third largest producer. In most of the regions of India, it is cultivated both as a rainy- and postrainy-season crop.

The yield and quality of sorghum produced worldwide is influenced by a wide array of biotic and abiotic constraints. Table 11.1 outlines the major biotic factors affecting sorghum production. The parasitic weed *Striga* (*S. asiatica*, *S. densiflora*, *S. hermonthica*) is prevalent in many regions of Africa and India and significantly impacts sorghum production. Abiotic constraints to sorghum production include soil properties (acidity and associated $Al^{3+}$ toxicity, salinity), extremes in temperature, and drought.

Grain sorghum quality is determined by several factors, such as visual quality, nutritional quality (including whole grain, protein, and starch digestibility; nutrient bioavailability), antinutritional factors such as tannins, processing characteristics, cooking quality, and consumer acceptability. Hulse et al. (1980) published an exhaustive review of chemical composition and nutritive value of sorghum grain. In India, grain produced in the postrainy season is highly valued for human food, while the grain produced in the rainy season has poor quality due to grain mold incidence.

**Table 11.1 Major Biotic Factors Affecting Global Sorghum Production**

| Insects | Diseases |
|---|---|
| Shoot fly (*Atherigona soccata* Rondani) | Grain mold (several genera) |
| Stem borer (*Chilo partellus*) | Charcoal rot (*Macrophomina phaseolina*) (Tassi) Goidanich |
| Midge (*Contarinia sorghicola* Coq.) | Downy mildew (*Peronosclerospora sorghi*) |
| Head bug (*Eurystylus oldi* Poppius) | Anthracnose (*Colletotrichum graminicola*) G.W. Wils. |
| Greenbug (*Schizaphis graminum*) (Rondani) | Rust (*Puccinia purpurea*) |
| Armyworms (*Spodoptera exempta*) | Leaf blight (several genera) |
| Locusts (several genera) | Several viral diseases |

Traditional foods made from sorghum include unfermented and fermented breads, porridges, couscous, boiled rice-resembling foods, and snacks, as well as alcoholic beverages. Sorghum blended with wheat flour is used to produce baked products, including yeast-leavened pan, hearth and flat breads, cakes, muffins, cookies, biscuits, and flour tortillas (Badi et al. 1990). Hard endosperm sorghum is used extensively in Southeast Asia for noodles and related products (Murty and Kumar 1995).

Grain sorghum is used to make products such as potable alcohol, malt, beer, liquids, gruels, starch, adhesives, core binders for metal casting, ore refining, and grits as packaging material. Grains are a rich and cheap source of starch and have applications in the food, pharmaceutical, textile, and paper industries. Malt drinks and malt cocoa-based weaning food and baby food industries are popular in Nigeria.

Sorghum grain is used for animal feed in the U.S., China, and South America, while in Africa and Asia the grain is used as human food (either directly or after brewing), with the dried stalks used for fodder, building materials, and fuel (Chandel and Paroda 2000). Sorghum is also grown for forage and is commonly grown in northern India and western Africa and fed to animals as a green chop, silage, or hay. Sweet sorghum is used to a limited extent in producing sorghum syrup and jaggery in India, and as a source of biofuel-alcohol in Brazil. Sweet sorghum stems are sold for chewing in many places of Africa and India.

Sorghum grain is one of the major ingredients in swine, poultry, and cattle feed in the western hemisphere (Bramel-Cox et al. 1995); however, demand for grain sorghum in poultry feed depends largely on the price of maize. More research is needed for its inclusion in poultry and dairy feed rations (Kleih et al. 2000).

Sorghum therefore assumes greater importance in the economics of several countries in Africa and Asia, largely inhabited by resource-limited farmers. Although wide ranges of traditional, novel, and industrially important products are produced from sorghum, the quality of products, such as ethanol, extrusion breakfast foods, etc., needs to be improved by utilizing the variability present in the crop.

## 11.2 TAXONOMY, ORIGIN, AND DOMESTICATION

### 11.2.1 Taxonomy

Linnaeus first described sorghum in 1753 under the genus *Holcus*. In 1794, Moench defined the genus *Sorghum* (from *Holcus*) to contain two species (Celarier 1959; Clayton 1961). Since then, *Sorghum* has undergone many taxonomic treatments to become the genus recognized today. Many authors have discussed the systematics, origin, and evolution of sorghum since Linnaeus (de Wet and Huckabay 1967; de Wet and Harlan 1971; Doggett 1988; Dahlberg 2000). Although sorghum is a diploid (or diploidized) cereal, it seems to have resulted from an ancestral round of polyploidy (see Chapter 1 by Jauhar in this volume).

**Table 11.2 The 15 Races of Cultivated *Sorghum bicolor* subsp. *bicolor***

| Basic Races | Intermediate/Hybrid Races |
|---|---|
| 1. Race *bicolor* (B) | 6. Race *guinea–bicolor* (GB) |
| 2. Race *guinea* (G) | 7. Race *caudatum–bicolor* (CB) |
| 3. Race *caudatum* (C) | 8. Race *kafir–bicolor* (KB) |
| 4. Race *kafir* (K) | 9. Race *durra–bicolor* (DB) |
| 5. Race *durra* (D) | 10. Race *guinea–caudatum* (GC) |
| | 11. Race *guinea–kafir* (GK) |
| | 12. Race *guinea–durra* (GD) |
| | 13. Race *kafir–caudatum* (KC) |
| | 14. Race *durra–caudatum* (DC) |
| | 15. Race *kafir–durra* (KD) |

Sorghum is classified under the family Poaceae, tribe Andropogoneae, subtribe Sorghinae, and genus *Sorghum* Moench (Clayton and Renvoize 1986). Garber (1950) and Celarier (1959) divided the genera into five subgenera: *Eu-sorghum*, *Chaetosorghum*, *Heterosorghum*, *Para-sorghum*, and *Stiposorghum*. *Chaetosorghum* and *Heterosorghum* are found in single species primarily in Australia and the South Pacific. *Para-sorghum* includes seven species found in the eastern hemisphere and Central America, while *Stiposorghum* contains 10 species endemic to Australia (Dahlberg 2000). Lazarides et al. (1991) present an excellent overview of the species in each of these four subgenera.

The basic chromosome number of sorghum is five, although striking differences in chromosome number, modes of origin, chromosome size, and geographic distribution of species are observed. Celarier (1959) hypothesized five as the lowest chromosome number in the *Para-sorghum* and *Stiposorghum* species, with polyploid proposed as autopolyploidy building by units of 10 (i.e., 2n = 10, 20, and 30).

*S. bicolor* spp. *bicolor* contains all the cultivated sorghums and are described as annual, with stout culms up to 5 m tall, often branched, and frequently tillering (Doggett 1988). Harlan and de Wet (1972) developed a simplified classification of cultivated sorghum into 5 basic and 10 hybrid races (Table 11.2) that proved to be of real practical utility for sorghum researchers. The 15 races of cultivated sorghum are identified by mature spikelets alone, although head type is sometimes helpful (Figure 11.1). The classification is based on five fundamental spikelet types — race *bicolor*, *guinea*, *caudatum*, *kafir*, and *durra* — and is detailed in Harlan and de Wet (1972).

The International Plant Genetic Resources Institute (IPGRI) Advisory Committee on Sorghum and Millets germplasm has accepted and recommended this classification to be used in describing sorghum germplasm (IBPGR/ICRISAT 1980).

## 11.2.2  Origin

The origin and early domestication of sorghum is hypothesized to have taken place around 5000 to 8000 years ago in northeastern Africa or at the Egyptian–Sudanese border (Mann et al. 1983; Wendorf et al. 1992). There is no argument against the African origin of sorghum, with the largest diversity of cultivated and wild sorghum also found there (de Wet 1977; Doggett 1988; Kimber 2000). The secondary center of origin of sorghum is the Indian subcontinent, with evidence for early cereal cultivation discovered at an archaeological site in western parts of Rojdi (Saurashtra) dating back to about 4500 before 1950 A.D. (Vavilov 1992; Damania 2002).

## 11.2.3  Domestication

Sorghum, a grass of the steppes and savannas of Africa, was probably domesticated in the northeast quadrant of Africa, an area that extends from the Ethiopia–Sudan border westward to Chad (Doggett 1970; de Wet et al. 1976). From there, it spread to India, China, the Middle East, and Europe soon after its domestication (Doggett 1965). The great diversity of *S. bicolor* has been created through disruptive selection (i.e., selection for extreme types) and by isolation and recombination in the extremely varied habitats of northeast Africa and the movement of people carrying the species throughout the continent (Doggett 1988). It has been diversified into a food source, sugar source, and construction material. Harvesting of entire panicles of sorghum by people altered the selection process (Kimber 2000). The basic morphological difference between a domesticated and a wild sorghum is the presence or absence of an abscission zone at the rachis, panicle, or spikelet nodes (Harlan et al. 1973). The process of domestication involved a change in several characteristics of the plant. A tough primary axis (rachis) and persistence of sessile spikelet were introduced early in the process. It is likely that the transformation of a loose and open inflorescence into a compact type involved several changes: (1) an increase in the number of branches per node, (2) an increase in the number of branches per primary inflorescence branch, and (3) a decrease in the internode length on the rachis. An increase in seed size was also probably a product of domestication, the seed becoming large enough to protrude from the glume (House 1985).

## 11.3 BREEDING BEHAVIOR AND POLLINATION CONTROL

Genetic variability is a prerequisite for any plant breeding program, with selfing and crossing the essential tools to direct and control this genetic variation. The knowledge of breeding behavior is therefore essential to decide the method used to breed cultivars, including pure lines, populations, and hybrids (House 1985).

Sorghum is primarily a self-pollinated crop; however, outcrossing does occur. In varieties with compact or semicompact panicles, selfing can be up to 90 to 95%, with 5 to 10% outcrossing (occurring more frequently at the tips of the panicles) (Doggett 1988). Varieties with loose or open panicles have higher rates of outcrossing: 30 to 60% (House 1985). In nature, the rate of outcrossing is affected by the wind as stigmas are most receptive during the first 3 to 5 days after their emergence, but can remain receptive up to a week or more after anthesis, depending upon temperature and humidity (Doggett 1988).

Hybridization, or crossing, of sorghum on a field scale is made feasible through genetic, cytoplasmic, and cytoplasmic-genetic male sterility systems (Stephens 1937; Ayyangar and Ponnaiya 1939; Stephens and Holland 1954). Limited scale crossing can be carried out through (1) emasculation with hot water and plastic bag technique or (2) hand emasculation and pollination techniques. Emasculation using hot water and plastic bags is cumbersome and requires a lot of preparation. It is safer to use hand emasculation, which can easily be done by unskilled staff with some training (House 1985).

## 11.4 GENETIC RESOURCES

### 11.4.1 Importance and Need for Conservation

Plant genetic resources are defined as the "genetic material of plants that is of value as a resource for the present and future generations of people" (IPGRI 1993). The importance of genetic resources was recognized at the intergovernmental platform under the umbrella of Food and Agricultural Organization (FAO) of the United Nations as the "common heritage of mankind," which should be made available without restriction (FAO 1983). Genetic resources have evolved as a product of domestication, intensification, diversification, and improvement through conscious and unconscious selection by countless generations of farmers. These landraces and improved cultivars provide the basic and strategic raw material for crop improvement the world over for present and future generations (Rai 2002).

Sorghum genetic resources

**Figure 11.1    (See color insert following page 114.)** Genetic variability in sorghum is shown by a wide range
of panicle colors.

Sorghum has an immense range of genetic resources (Figure 11.1), with much of the genetic
variability available in the African regions where domestication first occurred, and in the Asian
regions of early introduction. In Africa, the genetic variability occurs as cultivated species, wild
crop progenitors, and wild species (de Wet and Harlan 1971; Gebrekidan 1982). Landraces and
wild relatives of cultivated sorghum from the centers of diversity have been rich sources of resistance
to new pathogens, insect pests, and other stresses, such as high temperature and drought, as well
as sources of traits to improve food and fodder quality, animal feed, and industrial products.
However, this natural genetic diversity is by the adoption of improved varieties. To prevent the
extinction of landraces and wild relatives of cultivated sorghum, the collection and conservation
of sorghum germplasm was accelerated about four decades ago. Since then, germplasm collection
and conservation have become integral components of several crop improvement programs at both
national and international levels (Rosenow and Dalhberg 2000).

## 11.4.2  Status of Genetic Resources

At the global level, sorghum germplasm collections consist of approximately 168,500 accessions; the largest collection (21% of global total) is held at the International Crops Research Institute
for the Semi-Arid Tropics (ICRISAT), Patancheru, India. The total accessions consist of 18%
landraces/old cultivars, 21% advanced cultivars/breeding lines, 60% mixed categories of unknown
material, while very few are wild relatives (Chandel and Paroda 2000). In this section, we will
discuss the status of germplasm maintained at the ICRISAT, Patancheru, India, and at the National
Plant Germplasm System (NPGS) in the U.S., as well as Africa and China, as these countries have
large crop improvement programs (Rosenow and Dalhberg 2000).

### 11.4.2.1  ICRISAT, Patancheru, India

The Rockefeller Foundation in the Indian Agricultural Research Program initiated a sorghum
collection in 1960 (Murty et al. 1967; House 1985). A total of 16,138 accessions was assembled
from different countries, and international sorghum (IS) numbers were assigned to them. In 1976,
ICRISAT added its sorghum germplasm to the world collection in accordance with the recommen-

dations made by the Advisory Committee on Sorghum and Millet Germplasm sponsored by the International Board for Plant Genetic Resources (now the International Plant Genetic Resources Institute (IPGRI)) (IBPGR 1976; Prasada Rao et al. 1989).

At present, ICRISAT maintains 36,774 accessions from 90 countries, representing about 80% of the variability present in the crop (Eberhart et al. 1997). Most of the collections originated from developing countries in the semiarid tropics, with about 60% originating from India, Ethiopia, Sudan, Cameroon, Swaziland, and Yemen, the largest collection being from India. Landraces contribute 84% of the total collection compared with wild species, which contribute only 1%. The germplasm maintained at ICRISAT are classified into five races — *bicolor, guinea, caudatum, kafir,* and *durra* — and their derivatives, predominantly represented by the three basic races: *durra* (21.8%), *caudatum* (20.9%), and *guinea* (13.4%). The intermediate races *durra–caudatum* (12.1%), *guinea–caudatum* (9.5%), and *durra–bicolor* (6.6%) are common. India, Uganda, and Zimbabwe have all 5 basic and 10 intermediate races (Gopal Reddy et al. 2002).

### 11.4.2.2 U.S.

The U.S. Department of Agriculture (USDA) began the collection and distribution of sorghum around 1905. The Texas Agricultural Experimental Station was selected as the first station to work on sorghum in collaboration with the USDA (Quinby 1974). A total of 42,221 germplasm accessions are currently maintained at the National Plant Germplasm System (NPGS) (Dahlberg and Spinks 1995).

### 11.4.2.3 Africa

Ethiopia is a rich center of diversity of sorghum (Rosenow and Dahlberg 2000). The Ethiopian Sorghum Improvement Project (ESIP) began the collection, evaluation, documentation, and conservation of germplasm in the early 1970s, and through the early 1980s had amassed a collection of approximately 5500 accessions (Doggett 1988). It is estimated that through continued research efforts, the germplasm have grown to roughly 8000 collections (Rosenow and Dalhberg 2000). The distinct types of sorghum from Ethiopia are (1) *zera zeras,* (2) *durras,* and (3) *durra–bicolor* derivatives. The *zera zeras* have been extremely useful in providing germplasm for the improvement of food-type sorghums. Sudanese landrace collections were assembled at Tozi Research Station in Sudan in the 1950s, with the entire collection made available to the Rockefeller Foundation Project in India (Rosenow and Dahlberg 2000). The *caudatum* race dominates in Sudan, and Sudanese sorghums have been very useful as sources of drought resistance (Rosenow et al. 1999).

### 11.4.2.4 China

An extensive collection of sorghum landraces has been undertaken in China, with 12,836 germplasm accessions preserved in the National Genetic Germplasm Resources Bank, with 10,414 of these registered as genetic resources (Qingshan and Dahlberg 2001). Of these, 9652 accessions are local varieties, improved varieties, and strains that originated from 28 provinces and municipal and autonomous regions. These accessions are further classified according to use: 9859 sorghum accessions are food types, 394 varieties are for fodder or craft use, and 125 varieties are classified for use in sugar production. Only a limited number of local Chinese cultivars are conserved in sorghum collections at ICRISAT or the U.S. national collection, making these collections important plant genetic resources (Qingshan and Dahlberg 2001).

### 11.4.3 Maintenance of Genetic Resources

The largest sorghum collections are conserved and maintained at ICRISAT, the National Seed Storage Laboratory, Fort Collins, CO, and the USDA–ARS Plant Genetic Resources Conservation

Unit (PGRCU), Griffin, GA. Several countries maintain their own collections within their national plant genetic resource collections. Major field regeneration takes place at ICRISAT and at the USDA–ARS Tropical Agriculture Research Station, Mayagüez, Puerto Rico (Rosenow and Dahlberg 2000). Collections maintained at ICRISAT are regenerated in the postrainy season by selfing about 20 representative panicles from each line. Seed is harvested in equal quantities from these panicles and mixed to a bulk of about 500 g. The seed is preserved in aluminum cans in a medium-term storage facility (4°C and 20% relative humidity). Freshly rejuvenated accessions with 100% viability and 5 ± 1% seed moisture content are conserved in long-term storage (–20°C).

## 11.4.4  Core Collection

Sorghum scientists have developed subsets of the total base collection of sorghum in the past; however, these subsets were formed on a location-specific basis and did not represent the true genetic variation within the base collection (Prasada Rao et al. 1995). Therefore, the concept of core collection was developed (Brown 1989). A core collection consists of a limited set of accessions derived from an existing germplasm collection, chosen to represent the genetic and geographical spectrum of the whole collection. The rationale behind a core collection is the maintenance of as much representative genetic diversity as possible within a smaller, more manageable core collection (Dahlberg and Spinks 1995). A core collection was set up at ICRISAT by stratifying the total base collection geographically and taxonomically into subgroups within regions. Accessions in each subgroup were further clustered into closely related groups based on the principal component analysis (PCA) of the agronomic evaluation data. Representative accessions from each cluster were drawn in proportion to the total number of accessions present in that subgroup. Thus, the core collection of 3475 accessions set up at ICRISAT represents approximately 10% of the total world collection (Prasada Rao and Ramanatha Rao 1995). The core collection is an economical, practical, and effective method for conservation, maintenance, and utilization of the germplasm (Eberhart et al. 1997).

The U.S. set up a core collection of more than 200 accessions from the 42,221 accession base collection at the USDA in Mayagüez, Puerto Rico. The core collection represents genes for plant height, maturity, drought resistance, pericarp color, greenbug and aphid resistance, and downy mildew resistance (Dahlberg and Spinks 1995).

## 11.4.5  Evaluation, Characterization, and Documentation of Genetic Resources

A prerequisite for the efficient utilization of germplasm is that it must be properly evaluated, characterized, and documented onto a workable retrieval system so that any group of entries carrying any desired characteristics can be easily retrieved and used in breeding programs (Gebrekidan 1982). At ICRISAT, a total of 29,180 sorghum accessions have been characterized for 23 important morphoagronomic characters from the list of sorghum descriptors (IBPGR/ICRISAT 1980) during rainy and postrainy seasons. The range of variabilities available in cultivated races and their wild relatives is extensive, and the extreme types are so different as to appear to be separate species (Table 11.3) (Prasada Rao et al. 1995).

Characterization and passport data are documented using the ICRISAT Data Management Retrieval System (IDMRS) program and have been converted to the System 1032 (a Relational Database Management Software) for faster and more efficient management (Prasada Rao et al. 1995). ICRISAT maintains sorghum germplasm that has been evaluated and known to contain resistance to insect pests such as shoot fly, stem borer, midge, and head bug; resistance to diseases such as grain mold, anthracnose, rust, and downy mildew; resistance to the parasitic weed Striga; and important traits like glossiness, pop, sweet stalk, and scented types.

Approximately 50% of the total U.S. sorghum germplasm accessions have been characterized for 39 agronomic descriptors, with 21,661 accessions located at the USDA–ARS S-9 site in Griffin,

**Table 11.3 Range of Variation for Selected Traits of Sorghum Accessions Maintained at ICRISAT, Patancheru, India**

| Descriptor | Range of Variation | |
|---|---|---|
| Days to 50% flowering | 36 | 199 |
| Plant height (cm) | 55 | 655 |
| Pigmentation | Tan | Pigmented |
| Mid-rib color | White | Brown |
| Peduncle exertion (cm) | 0 | 55 |
| Head length (cm) | 2.5 | 71 |
| Head width (cm) | 1 | 29 |
| Head compactness and shape | Very loose stiff branches | Compact oval |
| Glume color | Straw | Black |
| Glume covering | Fully covered | Uncovered |
| Grain color | White | Dark brown |
| Glume size (mm) | 1 | 7.5 |
| 100-seed mass (g) | 0.58 | 8.56 |
| Endosperm texture | Completely starchy | Completely corneous |
| Threshability | Easy to thresh | Difficult to thresh |
| Luster | Lustrous | Nonlustrous |
| Subcoat | Present | Absent |

*Source*: Prasada Rao, K.E., Gopal Reddy, V., and Stenhouse, J.W., *Int. Sorghum Millets Newsl.*, 36, 15–19, 1995.

GA (Dahlberg and Spinks 1995). Passport and characterization data for U.S. sorghum germplasm accessions are documented on the Germplasm Resources Information Network (GRIN) and are also available through the sorghum curator located in Mayagüez, Puerto Rico.

### 11.4.6 Utilization of Genetic Resources

Only a small fraction of the total available sorghum collection could be fully utilized by breeders at any one time, as crop improvement programs are interested in germplasm that carry special desirable characters that are highly important at any particular point of time (Gebrekidan 1982). Early work on utilization of sorghum germplasm was confined to pure line selection within the cultivated landrace population in Africa and India that resulted in improved cultivars, some of which continue to be widely grown. Selection within dwarf populations was adopted, followed by exploitation of cytoplasmic male sterility that permitted the production of commercial hybrids (Dahlberg et al. 1997). Crossing or backcrossing between adapted introductions and local germplasm has been used to derive improved self-pollinated varieties and parental lines (Prasada Rao et al. 1989). Useful traits, such as increased seed number, larger panicles, greater total plant biomass, drought tolerance, disease resistance, greater plant height, longer maturity, greater leaf area indices, increased green leaf retention, and greater partitioning of dry matter have contributed to increased yield (Miller and Kebede 1984).

Utilization has been primarily limited to agronomically important and, in some cases, wild sources of germplasm. For example, use of *zera zera* sorghum has become widespread in the development of new, superior hybrids because of superior yield potential and grain quality (Duncan et al. 1991). The classic example of germplasm utilization in sorghum has been the Texas A&M–USDA Sorghum Conversion Program. To date, 673 converted lines have been released globally (listed in Duncan and Dahlberg 1993; Rosenow et al. 1995; TAES and USDA–ARS 1996).

The successful introgression of resistance to midge and downy mildew has greatly stabilized sorghum production in Australia and Argentina. Considerable opportunities remain for exploiting the collections to improve sorghum production globally. For example, over 340 accessions of sorghum belonging to subgeneric sections *Chaetosorghum*, *Heterosorghum*, *Stiposorghum*, *Parasorghum*, and *Eu-sorghum* have been evaluated for resistance to shoot fly at ICRISAT. Seven accessions with high levels of resistance, and some cases close to immunity, were found (Nwanze et al. 1995). Transfer of this high level of resistance to cultivated sorghum could greatly improve productivity of late sown crops in Africa and Asia, where shoot fly is a major production constraint (Dahlberg et al. 1997).

Since its establishment in 1972, ICRISAT has made efforts to enhance yield levels and identify resistance sources and use them to develop varieties and seed parents. The major germplasm sources utilized so far in varietal improvement include temperate lines from the U.S., *zera zera* lines from Ethiopia and Sudan, and some lines of Indian origin. Many stable resistant sources for shoot fly and stem borer have been identified in Africa, India, and the U.S. and have been used in both Indian and ICRISAT programs to confer resistance. ICRISAT developed an improved midge resistant line (ICSV 197 (SPV 694)), while germplasm lines with resistances to multiple diseases, listed in Table 11.1, are currently being used in breeding programs. ICRISAT used striga resistant germplasm to develop striga-resistant variety SAR 1. Nearly 1300 germplasm lines and 332 breeding lines were screened for early- and mid-season drought stresses by ICRISAT, identifying eight early-season and terminal drought-tolerant germplasm lines and nine mid-season drought-tolerant germplasm lines.

High-lysine sorghum lines have been developed from Ethiopian germplasm lines. Some promising high-lysine derivatives with shriveled and plump grain were obtained. Several sweet-stalked sorghum lines have been identified with the sweet-stalked trait currently being incorporated into an elite agronomic background. Many lines with desirable forage attributes have been identified, including germplasm with high-quality parameters and low hydrocyanic acid (HCN) and tannin content (Vidyabhushanam et al. 1989).

## 11.5 GENETIC VARIABILITY FOR QUALITATIVE AND QUANTITATIVE TRAITS

Genetic variability in *S. bicolor* has been assessed by researchers at three levels: (1) morphological/phenotyping, (2) biochemical (e.g., allozyme patterns), and, more recently, (3) DNA/nucleotide (variation in nucleotide sequence).

### 11.5.1 Morphological/Phenotypic Level

Genetic variability at the phenotypic level has been assessed using univariate and multivariate statistical tools. Univariate analysis evaluates factors such as range and variance and has been used extensively to evaluate sorghum. A vast reservoir of variability exists for several traits of interest in sorghum (Wenzel 1994; Geng 1994; Haussmann et al. 2001). This germplasm serves as a raw material for the genetic enhancement/prebreeding of varieties and hybrids the world over.

In sorghum, multivariate statistical tools such as Mahalanobi's $D^2$ statistic (1936), principal component analysis (Pearson 1901), and factor and canonical analysis (Spearman 1904) have been used to differentiate the genotypes/genetic resources. These methods were first used as a measure of genetic divergence to classify germplasm resources in crops in the Biometrical Genetics Unit of the Indian Agricultural Research Unit during the 1960s (Arunachalam et al. 1998). The potential of these methods was demonstrated in a reclassification of the species variability in the subgeneric section *Eu-sorghum* based on herbarium specimen measurements that replaced the classification made by Snowden (1936) and Chandrasekharaiah et al. (1969). By using Mahalanobi's $D^2$ statistic and canonical and factor analysis, the genetic diversity within sorghum was evaluated, and a world collection of genetic stocks of sorghum was cataloged and classified by grouping the genotypes into different clusters (Murty and Arunachalam 1967; Murty et al. 1967). In general, divergence analysis can be used to identify suitable parents for realizing heterotic $F_1$s, which in turn can be exploited commercially or used to derive superior recombinant inbred lines for further selection. This was demonstrated theoretically by Cress (1966): the higher the genetic divergence between the parents, the higher the heterosis of the $F_1$s would be. Vast literature is available to endorse this assumption, although a few contradictory results are also reported. Nevertheless, several researchers have carried out divergence and clustering analysis in sorghum. Ayana and Bekele (1999), Barthate et al. (2000), Kadam et al. (2001), Narkhede et al. (2001), Singh et al. (2001), and Umakanth et al. (2002) are some of the recent researchers who have attempted to assess diversity in sorghum germplasm using multivariate methods.

## 11.5.2 Biochemical Level

Genotypic differences in sorghum can also be determined using isozyme variation that evaluates enzymes with the same catalytic activity, but with different molecular weight and mobility in an electric field. The difference in the enzyme mobility is caused by point mutations resulting in amino acid substitution, so that isozymes reflect the products of different alleles rather than genes (Chahal and Gosal 2002). This principle has been exploited in plant breeding for quantifying genetic variability, characterizing germplasm, and identifying varieties/hybrids. While Arti (1993) and Reddy and Jacobs (2000) have used soluble proteins and isozyme markers, respectively, Schertz et al. (1990) and Aldrich et al. (1992) have used allozymes to assess genetic variability of sorghum germplasm. However, these biochemical markers are not widely used in quantifying the genetic divergence in sorghum, as the number of useful and easily assayable isozymes is a limiting factor.

## 11.5.3 DNA Level

Variation in nucleotide sequence has been exploited to assess the genetic diversity in sorghum germplasm. DNA markers such as restriction fragment length polymorphism (RFLP), random amplified polymorphic DNA (RAPD), sequence characterized amplified region markers (SCARs), simple sequence repeats (SSRs), and others have been used to assess and characterize genetic variability in sorghum genetic resources. Cui et al. (1995), Ahnert et al. (1996), Yang et al. (1996), Menkir et al. (1997), Dean et al. (1999), Ayana et al. (2000a,b), and Thimmaraju et al. (2000) are some of the latest studies where DNA markers have been used to quantify sorghum germplasm diversity. However, this is only illustrative and not meant to be an exhaustive list. RFLP and RAPD have been extensively used, with all studies indicating that DNA markers are effective in detecting the genetic variability in sorghum germplasm.

# 11.6 GENETICS AND CYTOGENETICS

## 11.6.1 Genetics

Knowledge of the nature of genetic control of both qualitative and quantitative characters of agronomic importance is fundamental in systematic and rapid improvement of these traits (House 1985). Excellent reviews on the genetics of various traits are found in Doggett (1988), Murty and Rao (1997), and Rooney (2000). The genetics of traits important to sorghum production are presented in this section.

### 11.6.1.1 Genetics of Morphological and Resistant Traits

Prominent among those who contributed to the genetics and breeding of sorghum in the early years were G.N. Rangaswamy Ayyangar and his colleagues, Rao (1972) in India, and J.R. Quinby and his colleagues in the U.S. (House 1985). They studied the inheritance of various morphological characters following Mendelian segregation. Quinby and Karper (1954) have shown that four recessive nonlinked brachytic dwarfing genes control height. Quinby et al. (1973) have shown that duration of growth and floral initiation are controlled by four loci, both dominant and recessive alleles at the

**Table 11.4 Genetics of Disease Resistance in Sorghum**

| Disease | Genetic Nature | Number of Alleles |
|---|---|---|
| Kernel smut | Incomplete dominant | 3 |
| Head smut | Dominant | — |
| Milo disease | Susceptibility is partially dominant | 1 |
| Anthracnose | Dominant, cytoplasm not significant | 1 |
| Rust | Dominant | 1 |
| Leaf blight | Susceptibility (in sudangrass) is dominant | 1 |
| Stay-green trait | Dominant at least with E 36-1 hybrids | — |
| Downy mildew | Dominant | >2 |

*Source*: House, L.R., *A Guide to Sorghum Breeding*, 2nd ed., International Crops Research Institute for the Semi-Arid Tropics, Patancheru, India, 1985.

maturity loci. Most tropical landraces/varieties are dominant at all four loci (four major genes *Ma1*, *Ma2*, *Ma3*, and *Ma4* with 13, 13, 16, and 12 different alleles, respectively, influencing days to maturity), but a recessive allele at the *Ma1* locus will cause them to be much less photoperiod sensitive and apparently less responsive to temperature variations, i.e., result in temperature zone adaptation. Other traits that received major attention were plant, glume, and grain color traits.

With increased interest in economic traits, including resistance to various yield-limiting factors, and with the development of novel procedures in quantitative genetics, many studies focused on the genetics of resistance to various traits and grain yield. The genetics of resistance to most diseases caused by fungi, bacteria, and viruses are in general simple inheritance of dominant alleles. Table 11.4 details genetics of the resistance to major diseases of sorghum. Grain mold resistance, on the other hand, is complex, with resistant (red-grained) hybrids produced by crossing susceptible red-grained female parents and white-grained restorer lines. It was established that flavan 4-ols at moderate levels in red-grained restorers were not sufficient enough to cause resistance in the parental lines, but were inherited in the $F_1$ hybrids and complemented to result in resistant hybrids (Reddy et al. 1992).

In contrast to diseases, the genetics of insect resistance is complex. Four insects are recognized as important pests throughout Asia, Africa, and India: shoot fly, stem borer (*Chilo* spp. and *Busseola* spp.), midge, and head bug (*Calocoris* spp.). Nonpreference mechanism is the predominant form of resistance and is quantitatively inherited mainly through additive gene action (Sharma et al. 1977).

Rana et al. (1980) reported that the $F_1$ is almost intermediate between the two parents for shoot fly resistance; however, resistance was found to be partially dominant under low to moderate shoot fly pressures. Resistance to stem borer is conferred by both tolerance and antibiosis, with primary damage explained by additive (A) and A × A interactions, and secondary damage controlled by A and nonadditive gene interactions (Rana and Murty 1971; Jotwani 1976). Resistance to both midge and head bug are predominantly under the control of additive gene action (Sharma et al. 1996; Ratnadass et al. 2002).

### 11.6.1.2 Male Sterility

Two types of male sterility are widely used in sorghum improvement programs: (1) genetic male sterility (GMS) and (2) cytoplasmic-nuclear male sterility (CMS).

#### 11.6.1.2.1 Genetic Male Sterility

Genetic male sterility is expressed in sorghum in many ways. In all cases, a recessive allele in homozygous condition (designated with alleles $ms_1$ to $ms_7$ and *al*) confers male sterility. Genetic male sterility is discussed in detail in Doggett (1988), Murty and Rao (1997), and Rooney (2000).

**Table 11.5 Male Sterility-Inducing Cytoplasms of Sorghum**

| Cytoplasm Fertility Group | Identity | Source Line | |
|---|---|---|---|
| | | Race | Origin |
| $A_1$ | Milo | D | — |
| | IS 6771C | G-C | India |
| | IS 2266C | D | Sudan |
| | IS 6705C | G | Burkina Faso |
| | IS 7502C | G | Nigeria |
| | IS 3579C | C | Sudan |
| | IS 8232C | (K-C)-C | India |
| | IS 1116C | G | India |
| | IS 7007C | G | Sudan |
| $A_2$ | IS 1262C | G | Nigeria |
| | IS 2573C | C | Sudan |
| | IS 2816C | C | Zimbabwe |
| $A_3$ | IS 1112C | D-(DB) | India |
| | IS 12565C | C | Sudan |
| | IS 6882C | K-C | U.S. |
| $A_4$ | IS 7920C | G | Nigeria |
| $9_E$ | IS 7218 | | Nigeria |
| | IS 112603C | G | Nigeria |
| $A_5$ | IS 7506C | B | Nigeria |
| $A_6$ | IS 1056C | D | India |
| | IS 2801C | D | Zimbabwe |
| | IS 3063C | D | Ethiopia |

*Source*: Schertz, K.F., in *Use of Molecular Markers in Sorghum and Pearl Millet Breeding for Developing Countries, Proceedings of an ODA Plant Sciences Research Conference*, J.R. Witcombe and R.R. Duncan, Eds., Norwich, U.K., March 29–April 1, 1993, pp. 35–37. With permission.

### 11.6.1.2.2   Cytoplasmic-Nuclear Male Sterility

The discovery of the male sterility resulting from the interaction of cytoplasmic and nuclear genes by Stephens and Holland (1954) laid the foundation and revolutionized the development of hybrid cultivar and hybrid seed production technology. The original source of the cytoplasm was the milo race, which induced male sterility in the nuclear background of the *kafir* race, and is designated as the $A_1$ cytoplasm. Since then, several sources and types of male sterile-inducing cytoplasms ($A_1$ to $A_6$) have been discovered and are reported in Table 11.5. In all these cytoplasms, a single/oligo recessive gene in the nucleus and sterile cytoplasm induces male sterility. These male sterile cytoplasms are differentiated based on the inheritance patterns of their fertility restoration, which is unclear, but dependent on the specific cytoplasm and nuclear combinations. Fertility restoration is controlled by a single gene in some combinations (e.g., $A_1$) but is controlled by two or three genes when the same nuclear genotype interacts with a different cytoplasm (Schertz 1994).

Although diverse male sterile cytoplasms have been identified, by far, only the milo cytoplasm ($A_1$) male sterility system is widely used because the hybrids based on this cytoplasm produce sufficient heterosis (20 to 30%) over the best available pure lines in sorghum. Although $A_2$ cytoplasm is as good as $A_1$ cytoplasm for mean performance as well as heterosis for economic traits such as grain yield, days to 50% flowering, and plant height, it is not popular, as the anthers in $A_2$ male steriles, unlike the $A_1$ male steriles, mimic the fertile or maintainer lines and lead to difficulties in monitoring the purity of hybrid seed production. Milo restorers need to be diversified in a *guinea* background to further enhance the yield advantage in hybrid development. So, there is a need to identify and breed for high-yielding nonmilo cytoplasm restorers. Based on $A_2$ CMS systems, only one hybrid, Zinza 2, has been developed and released in China for commercial cultivation (Liu Qing Shan et al. 2000).

## 11.6.2 Cytogenetics

*Sorghum bicolor* has a haploid chromosome number of 10, and it is classified as a diploid (2n = 2x = 20). Most species within *Sorghum* are diploid (2n = 20), but several species, most notably *Sorghum halapense,* are tetraploid (2n = 4x = 40). As the basic chromosome number in the Sorghastrae is five, it has been hypothesized that sorghum may be of tetraploid origin (Rooney 2000). Earlier studies on the meiotic chromosome pairing analysis did not provide evidence for the tetraploid origin of *S. bicolor* (Brown 1943; Endrizzi and Morgan 1955), and the information on the existence of homologous segments in the chromosomes of *S. bicolor* is poor; therefore, the chromosomes were regarded as distinct. Recent studies provide limited evidence about tetraploid origin of sorghum (Gomez et al. 1998). The molecular marker mapping studies of the genome by Chittenden et al. (1994), Pereira et al. (1994), and Dufour et al. (1996) demonstrated duplicated loci on the map, suggesting that sorghum has tetraploid origin. However, Subudhi and Nguyen (2000) contended that these evidences of tetraploidy are not satisfactory. They argued that in both analyses, the duplicated loci found on the mapped genome are only to an extent of 8 and 11%, respectively (Chittenden et al. 1994; Dufour et al. 1996). In a recent study, Peng et al. (1999) concluded that there is not enough evidence for tetraploidy origin of sorghum. Therefore, the cultivated sorghum could be considered a diploid from the perspective of genome organization (Subudhi and Nguyen 2000).

### 11.6.2.1 Karyotype

Analysis of sorghum chromosomes has been difficult, due to similarities in chromosome size and structure (Doggett 1988); nevertheless, several researchers have attempted to describe the karyotype of sorghum (Garber 1950; Celarier 1959; Gu et al. 1984; Mohanty et al. 1986; Yu et al. 1991).

### 11.6.2.2 Euploid Variation

Euploid variants such as haploids, triploids, and tetraploids occur naturally and can be induced. Autotetraploid sorghums have an increased grain size, which sparked interest in their development for grain production (Doggett 1962). The initial limitation in autotetraploid development was the high level of sterility; however, selection for improved fertility was successful, with resulting fertility levels near that of diploids (Doggett 1962; Luo et al. 1992). However, further research in this direction stalled, resulting in the nonrealization of tetraploid sorghum for grain production.

### 11.6.2.3 Aneuploid Variation

Aneuploids are organisms with more or less than an integral multiple of the haploid chromosome number. The most common form of aneuploidy observed in sorghum is trisomy, followed by translocation.

### 11.6.2.4 Apomixis

Apomixis in sorghum occurs when the embryo forms by apospory from a somatic cell in the nucellus. About 25% of the progeny developed apomictically can occur at a significantly higher percentage (Hanna and Schertz 1970; Rao and Murty 1972; Murty and Rao 1977). In most cases, apospory is the primary mechanism of apomixis in sorghum, although diplospory may occur (Murthy et al. 1979). Apomixis provides a mechanism to perpetuate a high-performing hybrid through self-pollination. Obligate apomixis is necessary to develop such a system, but all the reports of apomixis in sorghum involve only facultative apomixis, and efforts to increase the frequency of apomicts have not been successful (Reddy et al. 1980). To utilize facultative apomixis, the use of "vybrids" has been proposed (Murthy 1986).

## 11.7 SORGHUM IMPROVEMENT

### 11.7.1 Environmental Response Characteristics

Sorghum originated near the equator in northeastern Africa and is sensitive to day length and temperature (Miller 1982). It is classified as a short-day type with a critical photoperiod of 12 hours; however, some cultivars are classified into groups with higher critical photoperiods, including a class that is photoperiod insensitive (Miller et al. 1968). The sorghum plant differentiates from vegetative to reproductive growth when the day length becomes short (12 hours) and the rains diminish in tropical areas of Africa and India. At ICRISAT, it was shown that some high-yielding cultivars (CSH 1 and ICSV 112) with wide adaptability did not delay in flowering during rainy season, even when exposed to a day length of 17 hours (Alagarswamy, personal communication).

Information on the temperature effects on sorghum is scanty. Germination base temperature can differ within species and may vary from 4.6 to 16.5°C. Lines and hybrids designated as tropically adapted have lower germination base temperatures than the lines and hybrids designated as temperately adapted (Thomas 1980). High-yielding hybrids developed in temperate zones but grown in tropical environments have shown lower yields (Miller 1982).

### 11.7.2 Conversion Programs

Sorghum is a native of tropical Africa, with most of the species diversity found in the indigenous sorghums in tropical Africa and Asia. Most of these sorghums are landraces that do not flower in temperate zones in countries such as the U.S. Growing these photoperiod-sensitive lines as a summer crop in these temperate regions (when the day length is longer than 13 hours) makes it difficult to evaluate and use them for breeding. A dramatic decline in the expression of many growth-related characteristics is observed when photoperiod-sensitive germplasm is grown under continuous short days. Thus, photoperiod-insensitive germplasm has found much wider use in breeding programs (Rai et al. 1999). Many breeders recognized the positive correlation between height and yield in sorghum and developed lines as tall as possible to withstand the hazards of production (Miller 1982). However, maximum productivity is generally seen at about 1.75 to 1.80 m height and flowering at 68 to 70 days (Rao and Rana 1982). Height and maturity are controlled by four to eight genes, which when selected for can obtain lines with the desired height and maturity.

The major objective of the conversion program was to provide new sources of genetic diversity of exotic sorghums that have desirable plant height and maturity that are usable in sorghum improvement programs in Texas (Rosenow and Dahlberg 2000). The scheme essentially involved backcrossing early and dwarf $F_3$s at Mayaguez, Puerto Rico (selected from $F_2$s grown at Texas A&M), to the landrace. This was repeated four times before the final cross was made at Mayaguez using the landraces as the female to capture the landrace cytoplasm. This program had nearly 1279 lines that contributed to the breeding programs, not only in the U.S. but also in India and other places (Miller 1980). BTx 622 and BTx 623 are examples of lines that have made significant contributions to various seed parent development programs in several parts of the world (Reddy and Stenhouse 1994b; Rooney and Smith 2000).

In 1979, ICRISAT initiated a program to convert tall, late-flowering *zera zera* landraces (from the Ethiopia–Sudan border) and *kauras* and *guineenses* (from Nigeria). Several short and early-flowering lines were selected, with the converted lines of a height and maturity able to be cultivated under tropical, short-day conditions, and in temperate zones such as the U.S. The converted lines have been very useful in enhancing the use of exotic sorghum germplasm, broadening the genetic diversity, and providing new sources of desirable traits to sorghum researchers (Rosenow et al. 1997).

## 11.7.3 Breeding Concepts and Breeding Material

Breeding methods in sorghum improvement at various centers have taken into account the geographical mandate of individual research organizations, their materials and manpower resources, and short-term and long-term goals with respect to increasing productivity and genetic diversification (Rai et al. 1999). Breeding methods used in sorghum improvement are those developed for self-pollinated crops such as pure line selection, pedigree breeding, backcross breeding, population improvement, and hybrid breeding.

### 11.7.3.1 Breeding Concepts

Several breeding concepts are employed to utilize germplasm to develop high-yielding varieties and hybrid parents with resistance to biotic and abiotic stresses. Tropical germplasm are important as sources of dominant alleles for yield and plant height, while temperate germplasm are important for dominant alleles for earliness/maturity. The use of tropical × temperate crosses has produced several high-yielding varieties with desirable plant height (2.0 to 2.5 m) and maturity (100 to 110 days) (Rao and Rana 1982). Exotic germplasm are rich sources of genes for sorghum breeding, with crosses of exotic × exotic parents generally the most rewarding compared with local × local crosses (House et al. 1997). The most common breeding concept employed to introgress important traits is backcrossing, and it is most effective for simply inherited qualitative traits.

The type of restorer line used in breeding significantly impacts heterosis, or vigor, in the resulting hybrids. *Guinea* restorer lines contributed to the highest heterosis and grain yield per se in hybrids, followed by *caudatum* restorer lines. It was proposed that further gains can be made by making use of *guinea* sorghums after overcoming accompanying problems such as clasping of glumes to the grain in hybrids of *caudatum–kafir* male steriles and *guinea* restorers (Reddy and Prasada Rao 1993). Postrainy season adapted landrace germplasm possess excellent adaptive characteristics for the prevalent moisture-limiting conditions, with resulting landrace hybrids containing almost all the characteristics of the landraces preferred by farmers, as well as 15% superiority in grain yield over cultivated landraces (ICRISAT 1995).

Breeding methods for drought resistance and yield potential have been established by Reddy (1986) by selecting breeding materials for specific traits, such as emergence under crust, seedling drought recovery, and grain yield under drought-prone and yield potential areas for early-stage drought; drought recovery and grain yield under drought-prone and yield potential areas under mid-season drought; and stay-green, nonlodging, and grain yield under drought-prone and yield potential areas for terminal drought.

To date, most sorghum released globally were derived from the pedigree breeding program rather than from the population improvement programs at ICRISAT, indicating that for a short specific adaptation, pedigree selection appears to be more appropriate. It is therefore evident that the targeted gene pool approach is appropriate for a program that aims at a broad geographic mandate (Reddy et al. 2004a). Breeding schemes involving simultaneous selection for resistance and grain yield and converting the maintainer selections into male sterile lines have been used effectively to develop male sterile lines for resistance to pests and diseases in the shortest possible period of four years (Reddy et al. 2004a).

Breeding for insect resistance targets multiple traits. For stem borer resistance, the independence of antibiosis and the difference in patterns of inheritance of resistance to flower and peduncle damage and dead heart formation must be considered. The traits foliar and stem damage and the percentage of dead hearts were targeted with stem borer-resistant genotypes identified (Singh and Rana 1994). For shoot fly, the most important factor is to select for resistant germplasm in the season for which the material is intended (Jayanthi Kamala 1997). In selecting for resistance and increased yield, a multiple selection approach is employed with resistance selected on a family basis, and then selecting individuals within the resistant family for yield (ICRISAT 1995). In combining resistance characters that are simply inherited with grain yield, multiple crosses (three or four way) are as effective as single crosses. However, the selection for resistance of quantitatively inherited traits, such as resistance to stem borer or shoot fly, is not effective in four-way crosses (Reddy 1993).

### 11.7.3.2 Breeding Materials

The breeding materials so far exploited rather extensively in sorghum breeding programs in India, other tropical areas of the world, and the southern U.S. are derivatives of the *zera zera* group (Reddy and Stenhouse 1994a; Rosenow and Dalhberg 2000). Extensive use of the *zera zera* group of converted sorghums has contributed to produce disease resistance, yield potential, and grain quality of U.S. hybrid sorghum (Rosenow and Dalhberg 2000). Sorghum breeders in both public and private sectors throughout the world have been making extensive use of released, partially and completely converted, tall, late-maturing photoperiod-sensitive tropical exotic sorghum for conversion into short, photoperiod-insensitive, early types (Duncan et al. 1991). In the U.S., *kaura*, an African introduction, a source of yellow endosperm, was extensively used in sorghum improvement prior to the development of the conversion programs (Rosenow and Dalhberg 2000).

In India, *zera zera* lines have been extensively used to develop hybrids. This coupled with the selection for high grain yield, white grain color, medium grain mass, and optimum plant height and maturity resulted in materials looking alike, and this necessitated diversification (Reddy and Stenhouse 1994a).

### 11.7.4 Adaptation and Productivity Enhancement

Most of the varieties before 1960 in India were the result of pure line selection. The hybridization program was limited to improving grain yield and stalk juiciness. Breeders in the U.S. also aimed to improve lines for grain yield in the given location with short height suitable for combined harvest. By altering a maturity gene, grain yield could be enhanced substantially (Quinby and Karper 1946; Maunder 1972).

The discovery of genetic male sterility in sorghum opened up opportunities for recombination. Of the several genes reported to induce genetic male sterility, only two genes were widely used in population improvement, as they are stable over a range of environments (Reddy and Stenhouse 1994a; Murty and Rao 1997). This, in combination with various mating systems and reciprocal recurrent selection methods in exploiting additive (A), A × A and some epistatic genetic variations led many breeders to propose/take up the population improvement methods in the 1960s (Comstock and Robinson 1952; Doggett 1972; Eberhart 1972; Maunder 1972). These breeding enhancements led to the development of several populations in East and West Africa supported by funding from the International Development Research Council (IDRC) (Doggett 1972; Gardner 1972).

The sorghum improvement program at ICRISAT was initiated in 1972 with population improvement using the male sterility-inducing $ms_3$ gene following recurrent selection procedures to breed for wider adaptability. By 1980, the emphasis shifted to specific adaptation, and several specific disease- and pest-resistant gene pools with $ms_3/ms_7$ male sterile genes, following elaborate half-sib/$s_1$/$s_2$ testing procedures, were developed. Since the 1990s, trait-specific gene pool improvement using male sterile genes following simple mass selection alternating with recombination methods has become a corner stone in the development of diverse breeding materials (Reddy et al. 2004a). To date, 19 populations have been developed at ICRISAT using $v$ and $ms_7$ genes, into which 501 diverse germplasm accessions were introgressed.

Hybrids have distinct yield superiority (by at least 48%) and wide adaptability compared to the best available varieties in India and the U.S. (Maunder 1972; Rao 1972). Due to their wide adaptability, the All India Coordinated Sorghum Improvement Project (AICSIP) formulated the national policy of releasing hybrids for more than one state in India (Murty 1991, 1992; Murty and Rao 1997). Thus, the concept of wide adaptability of hybrids supported by the movement of the materials across continents was firmly established by the year 1985.

It was realized quite quickly that the ICRISAT-bred materials, although widely adaptable, did not adapt to drought environments in West Africa, or in high-altitude areas of Ethiopia, Kenya, etc. The ICRISAT drought resistance breeding program clearly established that drought is highly specific and that breeding for drought resistance therefore depends upon the stage of the crop at which the drought occurs, inferring that breeding for the traits that contribute to specific drought resistance should be carried out to realize further productivity gains in the target regions (Reddy 1986). This realization led to the establishment of different regional centers by ICRISAT in Africa, such as Niamey, Niger; SADC-ICRISAT, Bulawayo, Zimbabwe; and ICRISAT Sorghum and Pearl Millet East African Program, Nairobi, Kenya; as well as the Latin American Sorghum Improvement Program, El Baton, Mexico (Reddy and Stenhouse 1994a). These programs led to the release of head bug-resistant varieties across several countries.

The development and commercial cultivation of hybrid sorghums led to the improvement in grain yield by over 300% from 1950 to 1990 in the U.S. Later improvements were in terms of smaller yield increments, and enhanced disease and insect resistance and grain quality (Rooney and Smith 2000). Introduction of several seed parents of hybrid grain sorghum from the U.S. paved the way for hybrid development in Australia with several hybrids released during the 1960s. Maximum yield advance due to breeding during the 30 years since 1960 varied from 0.3 to 0.6% per year. However, the grain yield of subsequently released hybrids in 1986 was equal to the boot midge-susceptible hybrids in the absence of midge in Queensland (Henzell 1992). A further increase in yield was evident in nonsenescent hybrids AQL 40/RQL 36 and AQL 41/RQL 36, which averaged 111.4 and 113.1%, respectively, of the mean yield of three check hybrids (Texas 610SR, pride, and E57+). The increased yield potential of these hybrids was due to their late maturity and therefore greater biomass (Miller 1992).

Exploitation of heterosis was the main method of yield improvement in China. Through a program of selecting the best introduced seed parents, and then evaluating Chinese-bred pollinators, a series of hybrids were developed beginning in the 1970s. A progressive increase in yield (3.75 t ha$^{-1}$ in 1992–94 compared to 1.01 t ha$^{-1}$ in 1962–64) was accompanied by an improvement in quality traits. The increase in yield due to utilization of hybrids is estimated at 30 to 40%, with the remaining improvement being due to better cultivation conditions (Zhen Yang 1997).

### 11.7.5 Trait-Based Breeding

#### 11.7.5.1 Grain Yield and Adaptation

Breeding for high and stable grain yield with an improved harvest index continues to gain the top priority in sorghum (Rai et al. 1999). The use of adaptive trait breeding has been successful in achieving stable and productive cultivars. The identification and utilization of such traits as non-senescence, control of apical dominance, green bug and midge resistance, lodging resistance, tillering control, and resistance to downy mildew, head smut, foliar, and other diseases, have led to improvement in yield and its stability in many sorghum production areas (Miller et al. 1997).

#### 11.7.5.2 Forage and Feed

Forage is defined as "food for animals especially when taken by browsing or grazing" (*Webster's* 1986). Sorghum grown for forage in the tropics is often tall (2 to 3 m) but can be the very same dwarf hybrid grown for grain production and may or may not include grain fraction of the plant (Pedersen and Fritz 2000). However, from a definition point of view, the forage sorghum includes cultivars ranging from silage sorghum hybrids to varieties to *Sorghum sudanense* (2n = 10) and sorgo–sudangrass hybrids to sudangrass varieties (2n = 10) and hybrids (Kalton 1988). The uses of forage sorghum varies; in India it is commonly used as green fodder in northern states, and as stover after the grain harvest in southern states (House et al. 2000), while in the U.S., it is primarily used as silage, and to a limited extent, as pasture and hay for livestock (Kalton 1988).

Forage sorghums make efficient use of soil moisture by resorting to a semidormant state during stress, with a rapid response to moisture and a wide range of adaptability (Hanna and Cordona 2001). They also make efficient use of soil fertility by producing a higher number of tillers and biomass (Ahlrichs et al. 1991). Owen and Moline (1970) proposed a general classification of the forage sorghums based upon their use: (1) pasture using Columbus grass (*Sorghum almum*); (2) pasture and green chop or hay using Johnson grass (*Sorghum halepense*, 2n = 40), sudangrass (*S. sudanense*, 2n = 10), or sudangrass hybrids; (3) green chop or hay using sorghum–sudangrass hybrids; (4) silage using sorgos and canes; and (5) dual purposes (i.e., for production of grain and stover) using grain sorghums or single- or two-gene dwarf sorghums.

The primary objective of most forage breeding programs was forage yield (Kalton 1988). However, after 1988, efforts were directed toward improving quality traits. Although most forage quality traits like *in vitro* dry matter digestibility (IVDMD), crude protein (CP), neutral detergent fiber (NDF), and acid detergent fiber (ADF) appear to be quantitatively inherited, several simply inherited qualitative traits such as plant height, maturity, juiciness, sweetness, plant color, stay-green trait, etc., have a significant impact on forage quality (Bramel-Cox et al. 1995; Pederson 1997). Improvements in forage quality hold the greatest opportunities in improving forage sorghum (Hanna and Cordona 2001).

Owen and Moline (1970) concluded that the stage of maturity was the most important factor influencing the quantity and quality of forage produced. From heading to the ripe-seed stage, forage sorghums generally declined in protein content, crude fiber, and ash. An increase in lignin content followed by variable sugar content was observed with stage of maturity in grain sorghum. The sudangrass and sorghum–sudangrass hybrids were the most affected by maturity, with reduced protein content.

Four basic height genes are related to the dwarfing of the sorghum plant (Quinby and Karper 1954). It has been shown that dwarf sorghums containing one, two, or three of these dwarfing genes have no difference in leaf percentage; however, the percentage stalk vs. head was significantly affected. The one-dwarf type had 75% stalks and 7% heads, the two-dwarf type had 56% stalks and 23% heads, and three-dwarf type had 36% stalks and 43% heads (Gourley and Lusk 1978).

The usefulness of forage incorporated into diets of ruminants and forage used as biomass for biological conversion to liquid fuels is limited by the quantity of lignin present (Cherney et al. 1991). Hence, there is worldwide interest in improving the quality of forage through breeding to reduce or alter lignin content by incorporating brown mid-rib (*bmr*) trait in plants (Cherney 1990). *Bmr*, a single-gene trait, has the greatest forage quality, and Kalton (1988) recognized that great improvement was possible with increased incorporation of the *bmr* trait into all types of forage sorghum. Brown mid-rib mutant, discovered in maize in 1924 by Jorgenson (1931), was later induced in sorghum (Porter et al. 1978). Lignin concentrations in the *bmr* mutant are consistently lower than their normal counterparts in sorghum, while *in vitro* digestibility of *bmr* genotypes has been consistently higher than normal. Brown mid-rib mutant, differing in quantity and quality of lignin from normal genotypes, offers an opportunity to increase the overall digestion of plant fiber, which is a major constituent of forage crops, comprising 30 to 80% of their dry matter (Cherney et al. 1991). Several sorghum lines with high biomass were selected for the *bmr* trait at ICRISAT, with several white-grained B-lines, a red-grained B-line, and two red-grained varieties identified.

Stay-green is another simply inherited trait that is known to improve forage quality. Stay-green, governed by a recessive allele, not only slows down senescence but also arrests the decline in protein content of the aging leaves (Humphreys 1994). It also contributes to terminal drought and charcoal rot resistance (Rosenow and Clark 1995). Information on stay-green hybrid parental lines developed at ICRISAT is available at http:/www.icrisat.org/text/research/grep/homepage/sorghum/breeding/main.htm.

Kalton (1988) proposed that an ideal silage sorghum would include traits such as red seed, yellow endosperm, absence of testa layer, brown mid-rib, tan plant color, juicy stalk, moderate to low hydrocyanic acid (HCN), high IVDMD, good protein content, good leafiness, and green leaf retention.

ICRISAT has developed a strong forage sorghum improvement program that has developed a diversified set of hybrid parents, grain and dual-purpose varieties, and a population improvement program to improve sudangrass sorghums. Breeding has focused on high biomass, grain yield and stem sugar content, *bmr* lines, and grain types that tiller under stress conditions such as drought and stem borer resistance (Reddy et al. 1994). This program has resulted in the production of several lines with high tillers, and the identification of sweet-stalk lines, useful for developing ratoon and multicut sweet sorghum and dual-purpose and forage varieties and hybrids.

The objectives in the use of sorghum grain for livestock feed are different in ruminants (cattle, sheep, and goats) than in nonruminants (swine, poultry, and fish). In ruminants, the rumen microflora can upgrade poor-quality proteins and nonprotein nitrogen to the protein quality of the microflora itself, making protein levels less important, and as such, sorghum grain is primarily a source of starch. In nonruminants, the protein quality and quantity of sorghum grain are important because in sorghum-based diets it can contribute more than one third of the dietary crude protein for chicks and more than one half of the dietary crude protein for growing and finishing pigs (Bramel-Cox et al. 1995; Hancock 2000).

Of the several fairly simply inherited traits, such as pericarp color and endosperm color, type and texture differences in endosperm characteristics affect nutritional value in contrast to pericarp color (McCollough et al. 1972; Noland et al. 1977). Myer and Gorbet (1983) compared sorghum grain with waxy and normal starch types and low, medium, and high tannin content in nursery pigs. As tannin content increased, rate and efficiency of grain decreased, but sorghum with the waxy starch type was no better nutritionally than sorghum with normal starch type. There was a lack of improvement in energy digestibility and utilization in swine fed with floury and waxy endosperm sorghum, with the very soft or floury sorghums also having reduced yield and poor weathering ability.

Selection for improved *in vitro* protein (digestibility of pepsin) was found to be associated with lower yield and late maturity; however, the use of *in vitro* digestibility in conjunction with yield and maturity date has the potential to genetically improve grain sorghum as a feed grain for livestock, even though the environment will still have a major effect on nutritional quality. Unlike selection for floury and waxy endosperm texture, a selection index involving *in vitro* digestibility, grain yield, and maturity would result in genetic material to breed hybrids with acceptable yield, maturity, and improved nutritional value (Bramel-Cox et al. 1990, 1995).

### 11.7.5.3  *Resistance to Foliar, Stem, and Head Feeding Insects*

Foliage pests of grain sorghum in the U.S. are green bug (*Schizaphis graminum*), yellow sugarcane aphid (*Sipha flava*), chinch bug (*Blissus leucopterus*), fall armyworm (*Spodoptera frugiperda*), corn leaf aphid (*Rhopalosiphum maidis*), and Banks grass mite (*Oligonychus pratensis*). Biotype development has been a problem only with green bug, the key foliage pest of sorghum in the U.S. Foliage pests of grain sorghum outside the U.S. are shoot fly (*Atherigona soccata*), spotted stem borer (*Chilo partellus*), and sugarcane aphid (*Melanaphis sacchari*) (Peterson et al. 1997). Midge (*Contarinia sorghicola*) is a serious head pest common in India, eastern Africa, and Australia, with head bug (*Eurystylus oldi*) another important pest in India and western and central Africa.

**Figure 11.2** **(See color insert following page 114.)** (Left) Technique to screen sorghum lines resistant to midge. (Right) An improved sorghum cultivar resistant to midge produced by ICRISAT.

The lack of elite insect-resistant varieties and hybrids is due to low resistance levels when transferred into agronomically improved sorghum. When resistance is high, progenies are agronomically undesirable (Nwanze et al. 1995). In addition, resistance to some insects is quantitatively inherited and difficult to incorporate into elite, high-yielding varieties or hybrids (Peterson et al. 1997).

Development of improved germplasm or varieties resistant to shoot fly, stem borer, and midge is the example for the success of conventional breeding programs in Africa and India (Sharma 1993). Use of wild sorghums *Sorghum versicolor* and *Sorghum arundinacium* to obtain qualitatively different sources of resistance is the future objective. Absolute resistance to shoot fly was noticed in wild relatives *Sorghum dimidiatum* and *Sorghum australiense*, with efforts currently under way to exploit and introgress these using molecular markers.

Several midge-resistant varieties and hybrid parents have been developed at ICRISAT using pedigree selection and backcross techniques; however, these are mostly based on a single source of resistance that loses its resistance at the high altitudes and low temperatures of eastern Africa. In western and central Africa, advanced midge-resistant lines bred at Patancheru are crossed to locally adapted elite materials to obtain improved midge-resistant cultivars (Figure 11.2).

Breeding efforts for resistance to stem borer by ICRISAT and western and central Africa are focused on *C. partellus* and *B. fusca,* with several stem borer-resistant male sterile lines, pollinators, and varieties developed. In western and central Africa, a random mating population with sources of resistance to *B. fusca* and adapted high-yielding lines are being developed and pursued through recurrent selection procedures.

### 11.7.5.4 Disease Resistance

Table 11.2 lists major diseases. Disease management through genetic manipulation has been the most effective means of reducing losses in many crop species. However, managing all diseases by genetic means is neither feasible nor possible (Thakur et al. 1997). Breeding for host plant resistance is one of several methods of protecting the crop. Therefore, before a resistance breeding program begins, it must be determined that (1) the disease is of sufficient economic importance, (2) sufficient information is available on the nature of the host–pathogen system and on screening

**Figure 11.3   (See color insert following page 114.)** (Left) Screening of grain mold in sorghum by sprinkler irrigation method. (Right) An ICRISAT public sector-bred grain mold-resistant sorghum variety.

techniques, (3) well-defined sources of resistance are available, and (4) the expected economic output will occur within a given time.

Breeding for disease resistance in sorghum began about three decades ago (Rosenow 1992; Mukuru 1992). Usually an agronomically good, high-yielding cultivar is crossed to other parents with disease resistance, good grain quality, local adaptation, etc. Disease control through host genetic manipulation is difficult and has been slower for charcoal rot and grain mold, in which gene effects are small, compared to downy mildew, anthracnose, and leaf blight, in which gene effects are large (Thakur et al. 1997).

Current breeding methods at ICRISAT include both pedigree and population improvement for grain mold-resistance breeding (Figure 11.3). Pedigree selection using artificial screening for grain mold-resistance has resulted in high-yielding lines and hybrid parents with white grain types. A random mating population with white grains and *guinea*-type panicle and glume traits is being improved at ICRISAT. Male sterile lines with white, red, and brown grain colors and with resistance to grain mold have been developed. The possible roles of antifungal proteins that inhibit the growth of grain mold fungi are being investigated at ICRISAT and Texas A&M University. Several male sterile lines resistant to anthracnose (*Colletotrichum graminicola*) have been developed at ICRISAT.

### 11.7.5.5   Striga Resistance

Witchweed (*Striga* spp.) is endemic to subtropical regions of the world and comprises about 36 species, of which around 31 occur in Africa (Raynal-Roques 1987). Five species of *Striga* attack cultivated cereals: *Striga hermonthica* (Del.) Benth, *Striga aspera* (Willd) Benth, *Striga forbesii* Benth, *Striga asiatica* (L) Kuntze, and *Striga densiflora* Benth (Doggett 1984). Only three species are of widespread economic importance in Africa: *Striga asiatica* and *S. hermonthica*, which attack cereals, and *Striga gesnerioides*, which parasitize legumes, principally cowpea (*Vigna unguiculata*) (Doggett 1988).

*S. asiatica* occurs mainly in southern and central Africa, while *S. hermonthica* predominates in the semiarid zones of tropical eastern, central, and western Africa, where it attacks food crops such as sorghum, pearl millet (*Pennisetum glaucum*), maize (*Zea mays*), upland rice (*Oryza sativa*), sugarcane (*Saccharum officinarum*), and several wild grasses (Tarr 1962; Doggett 1988). *S. her-*

*monthica and S. asiatica* are obligate parasitic weeds of sorghum with significant economic importance. The yield losses from damage are often significant, with estimates varying from 10 to 70%, depending on crop cultivator and degree of infestation (Doggett 1988).

Characterization of germplasm, development of simple and efficient screening techniques, and a well-planned selection strategy for yield and other traits of importance in subsistence agriculture are essentials for breeding *Striga* resistance. Given the widespread *Striga* problem and the opportunities for natural and deliberate selection in environments where the host and parasite have coevolved, there has been suprisingly low genetic variability for *Striga* resistance in sorghum germplasm (Ejeta et al. 1997).

### 11.7.5.5.1 *Striga Screening Techniques*

Evaluation of host plant resistance to *Striga* in *Striga*-infested plots has been unsuccessful due to the inconsistent nature of *Striga* infestation both within the same field and among different fields across years, and due to complex interactions among the host, parasite, and environmental factors that affect the establishment of the parasite (Ejeta and Bulter 1993a; Volger et al. 1995).

Rapid laboratory screening procedures that predict field resistance to *Striga* on a per plant basis have been developed, with the double-plot technique and agar gel assay useful in screening large numbers of breeding progenies (Parker et al. 1977; Hess et al. 1992). The assay correlated well with the reported field resistance of the cultivars tested (Hess et al. 1992; Volger et al. 1995). This assay proved that *Striga* resistance in SRN 39, a superior source of field resistance against *Striga*, was primarily due to low production of germination stimulants and led to the release of this line for commercial cultivation in *Striga*-endemic areas of Sudan (Hess and Ejeta 1992; Volger 1992). This line is used extensively as a source of *Striga* resistance in a sorghum improvement program at Purdue University and at ICRISAT (ICRISAT 1982; Volger et al. 1995).

### 11.7.5.5.2 *Striga Breeding Strategies*

Both interspecific variability among *Striga* spp. and intraspecific variation for aggressiveness must be taken into account when breeding for *Striga* resistance (Ramaiah 1987; Ejeta et al. 1992). In order to obtain stable and polygenic resistance, breeding materials should be evaluated at multiple locations with different *Striga* populations or host-specific races (Ramaiah 1987). In addition to multilocational testing, breeding strategies should endeavor to:

1. Characterize crop germplasm, search for sources of resistance and tolerance in elite material, or improve currently available sources of resistance for agronomic performance
2. Include wild relatives with superior resistance in the breeding program
3. Pyramid resistance genes to obtain more durable and stable polygenic resistance
4. Transfer resistance genes into productive, well-adapted genotypes
5. Combine lines with different resistance mechanisms to form hybrids or synthetics to increase durablility of the resistance (Ramaiah 1987; Ejeta et al. 1992; Ejeta and Bulter 1993b; Burner et al. 1995; Haussmann et al. 2000a; Kling et al. 2000)

Other important considerations include careful definition of the target environment, farmer participation in identification of adapted parents for use in a backcross program, and determination of the most important region-specific selection traits, such as grain color and quality, plant height, maturity, photoperiod sensitivity, and disease resistance (Rattunde et al. 2000). The optimal genetic structure of the culitivar (i.e., degree of heterozygosity and heterogeneity) will also depend on the target environment, with the potential merit of heterozygous sorghum cultivars having significantly greater yield (18%) under *Striga* infestation than their parental lines (Haussmann et al. 2000a,b; Kling et al. 2000; Hess and Ejeta 1992).

Although several *Striga*-resistant lines were identified from the germplasm based on extensive laboratory and field screening at ICRISAT, many of them could not be used in the breeding programs due to their undesirable agronomic base. However, some germplasm lines were used in *Striga* resistance breeding, and several *Striga*-resistant male sterile lines have been developed. The *Striga*-resistant variety SAR 1 developed at ICRISAT was released for cultivation in *Striga*-endemic areas. Several *Striga*-resistant varieties have been identified in eastern, southern, and western Africa, Nigeria, and Sudan (Bebawi 1981; Obilana 1983; Ramaiah 1987; Riches et al. 1987; Doggett 1988; Obilana et al. 1991, Mbwaga and Obilana 1994; Hess et al. 1996). However, less than 50% of these varieties remain in use today, as some of them have been used as parents in developing the new and improved resistant lines.

In general, conventional approaches for breeding durable *Striga* resistance have not been successful due to (1) species-specific resistance against *Striga*; (2) intraspecific or physiological variants of *Striga*; (3) paucity of resistant genes in crop germplasm (Ejeta et al. 1991, 1997); (4) difficulty, expense, and occasional unreliability of testing for *Striga* resistance; (5) quarantine of the parasite; and (6) recessive character of some resistance genes (Haussmann et al. 2000a). A mix of conventional and marker technology approaches may have to be employed in the future in breaking down *Striga* resistance into simpler components that can be exploited for developing crop genotypes with durable resistance (Ejeta et al. 1997; Haussmann et al. 2000a).

### 11.7.5.6  *Drought Tolerance*

Drought stress is a major constraint to sorghum production worldwide. Although sorghum possesses excellent drought resistance compared to most other crops, improving its drought resistance would increase and stabilize grain and food production in low-rainfall, harsh environmental regions of the world (Rosenow et al. 1997). Drought resistance is the phenotypic expression of a number of morphological characteristics and physiological mechanisms, including drought escape, dehydration avoidance, and dehydration tolerance (Ludlow 1993). Therefore, drought resistance in sorghum is a complex trait affected by several interacting plant and environmental factors.

The response of sorghum to soil moisture stress is largely determined by the stage of growth at which moisture stress occurs, with two distinct responses to drought stress identified. The preflowering (mid-season) drought response is expressed when plants are stressed during panicle differentiation prior to flowering, while the postflowering drought response is expressed when moisture stress occurs during the grain filling stage (Rosenow and Clark 1995). Two further responses were described by ICRISAT: where emergence under crust and high temperature prevailed in drought-prone environments and wherein the lines with long mesocotyl emerge earlier than others; and early-stage (120-day-old seedlings) drought recovery where distinct differences among the genotypes are observed (Reddy 1985).

Research at ICRISAT screened and identified progenies with high levels of preflowering stress recovery, which were further selected for grain yield alternatively under mid-season drought and yield potential environments in India. When tested in drought-prone environments in Africa, these lines had greater stability and grain and biomass yield than other varieties. Similarly, promising lines with resistance to seedling emergence under crust and high temperature were also identified (ICRISAT 1982, 1986, 1987).

Symptoms of postflowering drought stress susceptibility include premature plant (leaf and stem) death or plant senescence, stalk collapse and lodging, charcoal rot (*Macrophomina phaseolina*), and a significant reduction in seed size, particularly at the base of the panicle. Tolerance is indicated when plants remain green and fill grain normally. Such green stalks also have good resistance to stalk lodging and charcoal rot. Such cultivars are referred to as having the stay-green trait.

Stay-green is as an important postflowering drought resistance trait. Genotypes possessing stay-green are able to maintain a greater green leaf area under postflowering drought than their senescent counterparts (Rosenow et al. 1997). Recent studies have shown that leaves stay green not only because of small sink demand but also due to higher leaf nitrogen status and transpiration efficiency, resulting in maintenance of photosynthetic capacity and, ultimately, higher grain yield and lodging resistance (Borrell and Douglas 1997; Borrell et al. 1999, 2000b; Borrell and Hammer 2000). Greater green leaf area duration during grain fill appears to be a product of different combinations of three distinct factors: green leaf area at flowering, time of onset of senescence, and subsequent rate of senescence. All are independently inherited and easily combined in breeding programs (Van Oosterrom et al. 1996; Borrell et al. 2000a). Stay-green hybrids produce significantly greater total biomass after anthesis, retain greater stem carbohydrate reserves, maintain greater grain growth rates, and have significantly greater yields under terminal drought stress than related but senescent hybrids (Henzell et al. 1997; Borrell et al. 1999, 2000b). Several physiological traits, such as heat tolerance, desiccation tolerance, osmotic adjustments, rooting depth, and epicular wax, are known to improve drought resistance (Downes 1972; Sullivan 1972; Sullivan and Ross 1979; Turner 1979; Jordan and Monk 1980; Kramer 1980; Jordan and Sullivan 1982; Peacock and Sivakumar 1987; Levitt 1992; Krieg 1993; Ludlow 1993; Van Oosterrom et al. 1996; Henzell et al. 1997). Although screening techniques based on these traits for drought resistance have been reported, little if any progress using specific physiological traits has been documented, partly because interaction of various physiological mechanisms involved in drought tolerance is still poorly understood (Christiansen and Lewis 1982; Garrity et al. 1982; Seetharama et al. 1982; Blum 1983, 1987; Jordan et al. 1983; Ejeta 1987; Bonhert et al. 1995).

The stay-green trait has been successfully used in Australia to develop postflowering drought stress resistance and lodging resistance in parental lines and commercial hybrids. Conventional breeding for stay-green is primarily based on two sources, B 35 and KS 19, of Ethiopian and Nigerian origin, respectively (Henzell et al. 1984, 1992a,b; Henzell and Hare 1996; Rosenow et al. 1997; Mahalakshmi and Bidinger 2002). KS 19 has been commercially used primarily in the breeding program of the Queensland Department of Primary Industries and Fisheries (QDPIF), while B 35 is widely used in both public and private sector breeding programs in the U.S. (Henzell and Hare 1996). B 35 has provided the major and best sources of stay-green used in the QDPIF program, and produced the key line QL 41, with high levels of stay-green expression. Crosses between QL 41 and sorghum midge-resistant lines formed the basis of the female stay-green and midge-resistant gene pool in the QDPIF programs, although less progress has been made in developing such germplasm in the male population (Rosenow et al. 1983; Henzell et al. 1997).

Germplasm and breeding lines tolerant to specific drought environments have been identified at ICRISAT; 36 lines with a stay-green score ranging from 1 to 2 on a scale of 1 to 5 (where 1 = more green and 5 = least green) and a plant agronomic score ranging from 1 to 3 on a scale of 1 to 5 (where 1 = very good and 5 = poor) were selected. For details see ICRISAT's webpage: http:/www.ICRISAT.ORG/Text/research/grep/homepage/sorghum/breeding/main.htm.

Most commercial sorghum hybrids in the U.S. possess good tolerance to preflowering drought stress; however, only a few have good postflowering resistance (Nguyen et al. 1997). In spite of the availability of simply inherited stay-green trait associated with terminal drought tolerance, progress in enhancing postflowering drought resistance is slow because the expression of this trait is strongly influenced by environmental factors and because of the limited number of stay-green sources currently used in sorghum breeding programs (Henzell and Hare 1996; Henzell et al. 1997).

Stay-green is expressed best in environments in which the crop is dependent on stored soil moisture but where this is sufficient to meet only a part of the transpiration demand. Sufficient expression of the trait for selection is thus dependent upon the occurrence of a prolonged period of drought stress of sufficient severity during the grain filling period to accelerate normal leaf senescence but not of sufficient magnitude to cause premature death of the plants (Henzell and Hare 1996). Because of this precise requirement for the trait expression, field environments do not offer ideal conditions for selection, and therefore, identification of quantitative trait loci (QTL) conferring stay-green trait and the molecular markers tightly linked to these QTL will provide powerful tools to enhance drought resistance (Henzell and Hare 1996; Crasta et al. 1999; Xu et al. 2000).

### 11.7.5.7  Soil Chemical Toxicity Tolerance

The demand for cereal grains in tropical environments characterized by soils that impose mineral stresses has mandated additional breeding research to adapt sorghum to these environments (Gourley et al. 1997a). Soil acidity (and associated $Al^{3+}$ toxicity) and salinity are probably the most important constraints to sorghum productivity in tropical environments, with an estimated 72.4 billion ha (18%) of the world's soils classed as acidic and 0.9 billion ha classed as saline and sodic (Vose 1987; Gourley et al. 1997a). These problematic soils cause more acute crop production constraints for resource-poor tropical farmers in developing countries than for temperate zone farmers in developed countries. However, improvement in nutrient use efficiency and tolerance to toxicities would benefit all farmers.

#### 11.7.5.7.1  Inheritance and Breeding for Soil Acidity Tolerance

Tolerance to soil acidity and $Al^{3+}$ toxicity in sorghum is controlled by a few dominant genes under additive and nonadditive action (Flores et al. 1991; Adamou et al. 1992; Zake et al. 1992; Maciel et al. 1994a; Gourley et al. 1997a). Soil acidity stress factors vary with location, soil depth, rainfall, temperature, effective cation exchange capacity (ECEC), natural content of essential elements, level of toxic ions, p-fixation capacity, and amount and quality of organic matter (OM) (Gourley et al. 1997a). These factors combined with a poor correlation of results obtained in nutrient culture for acidity tolerance in field or greenhouse studies are the causes of complexity of breeding for tolerance to soil acidity (Horst 1985; Marschner 1991). Nevertheless, much progress has been made since the EMBRAPA sorghum for tolerance to acid soils and International Sorghum and Millets (INTSORMIL) sorghum acid soil breeding project were initiated in Columbia in 1981 (Schaffert et al. 1975).

Many sorghum lines have been identified with good levels and substantial genetic variability in $Al^{3+}$ toxicity tolerance (de Andrade Lima et al. 1992; Maciel et al. 1992, 1994b; Gourley et al. 1997a). More than 6000 sorghum genotypes from the world collection were screened at Quilichao, Colombia, with around 8% found to tolerate 65% $Al^{3+}$ saturation, with a few of these genotypes able to produce greater than 2 t ha$^{-1}$ of grain (Gourley 1988). Many of these highly tolerant genotypes from the world collection originated in acid soil areas in Nigeria, Uganda, or Kenya and were classified as *caudatum* or *caudatum* hybrid races. The open-panicled *guinea* race and the hybrid *guinea–bicolor* lines had a higher overall percentage of acid-tolerant sorghum entries than those of other races and hybrids evaluated (Gourley 1988).

The INTSORMIL and EMBRAPA projects used a pedigree breeding method to identify Al-tolerant plants and screened the $F_2$ plants at 65 and 45% Al saturation, respectively. The resulting tolerant lines had yields between 2.6 and 4.6 t ha$^{-1}$, with the INTSORMIL lines shorter and earlier than the EMBRAPA lines (Gourley et al. 1997a). About 170 grain sorghum inbreds with tolerance to Al saturation levels of tropical acid soils developed by pedigree breeding have been released as germplasm by the Mississippi Agricultural and Forestry Experiment Station (MAFES) (Gourley et al. 1997b).

ICRISAT and Centro International de Agriculture Tropical (CIAT) screened large numbers of grain sorghum lines (male sterile, restorer, and forage lines) over four consecutive seasons under varied $Al^{3+}$ concentrations and identified high-yielding male sterile lines (MS), restorer lines, and forage sorghums tolerant to $Al^{3+}$ that have been distributed to various agencies in the region (ICRISAT, NARS, CIAT 1997). Evaluation of these high-yielding breeding materials in multiple locations resulted in the identification of many MS, R, B, and forage lines with wide adaptability (regression coefficient between 0 and 1) (Reddy et al. 1998; Reddy and Rangel 2000).

Four of the nearly 200 sorghum hybrids evaluated at Matazul (60% $Al^{3+}$ and 4.6% organic matter) were found to have outstanding tolerance to soil acidity, and they were also less susceptible to leaf diseases, were greener at the time of maturity, and were also taller than the control Real 60 (ICRISAT 2000). Hybrids therefore hold promise for improving the sustainability of acid savannas (Reddy et al. 2004a,b).

### 11.7.5.7.2 Inheritance and Breeding for Soil Salinity Tolerance

Studies on genetic basis of variability for salinity tolerance in sorghum are rather limited (Azhar and McNeilly 1989; Haggag et al. 1993; de la Rosa Ibarra and Maiti 1994; Fernandes et al. 1994; Igartua et al. 1994; Jiqing Peng et al. 1994; Maiti et al. 1994; Richter et al. 1995). The genetic control of salinity tolerance appears to be complex, with both additive and dominant gene effects important in controlling the expression of salt tolerance; however, the effect of dominant genes appears to be most important (Igartua et al. 1994).

Early screening for salt stress-tolerant sweet sorghum genotypes can be done *in vivo* by (1) growing the seedlings on sand in polystyrene containers and (2) growing the seedlings in spectrophotometer cuvettes (Montemurro et al. 1994). Genotypic response to salt stress is similar in both methods, and classification of sweet sorghum genotypes for salt tolerance was effectively demonstrated. A third method was suggested by Hassanein and Azab (1990) where seeds of sorghum genotypes are grown in water (control) and NaCl solution (ECE = 12, 16, or 20 mhos/cm).

### 11.7.5.7.3 Biochemical Basis of Resistance to Salinity

The concentration of biochemical components such as proline and hydrocyanic acid (HCN) increases with an increase in salinity levels in resistant genotypes but not in the susceptible genotypes of sorghum at the seedling stage (Richards and Dennett 1980; de al Rosa Ibarra and Maiti 1994). The detailed study of resistance mechanism of these lines would aid in incorporation of resistance in elite agronomic background.

Attempts to breed sorghum for salinity tolerance is rather limited. The major problem breeding sorghum for abiotic stresses such as salinity is the choice of optimal selection environments due to the high levels of both spatial and temporal variation in salinity-affected fields (Richards and Dennett 1980; Richards et al. 1987; Igartua 1995). Spatial variation occurs horizontally and vertically on very small scales, and therefore, most salinity-affected lands are actually comprised of many microenvironments, ranging from low to high salinity in the same field (Igartua 1995).

Testing genotypes across a broad range of salinity levels shows that genotype × salinity level interactions are commonly large and significant in sorghum, forcing the plant breeder to decide whether to work over the whole target environment (breeding for wide adaptation) or subdivide it into more homogeneous subenvironments (breeding for specific adaptation) (Azhar and McNeilly 1987; Igartua 1995). This decision depends mainly on the relative sizes of the genotype × year and genotype × location interaction (Austin 1993).

The target environments, both for breeding for saline soils and for wide adaptation, are actually a population of many possible environments, for which there exists a significant component of genotype × environement interaction (Igartua 1995). Three environment selection strategies have been proposed: (1) make selection in a stressful environment, (2) select under optimum growing conditions, and (3) use a combination of both the approaches; i.e., select materials that perform well under both stress and nonstress conditions (Calhoun et al. 1994). Option 3 appears to be the best, as separate selection for distinct environments is not possible, and greater accuracy of selections is achieved over more environments used (Igartua 1995). Selection of breeding populations for salinity tolerance over three salinity levels — low, high, and average of the two extreme levels — instead of the whole range of salinity levels, has demonstrated the usefulness of option 3 to select for improved yield in grain sorghum under saline soil conditions (Maas and Hoffman 1977; Igartua 1995).

Salinity causes more serious damage to sorghum at the seedling emergence stage than at any other stage (Jiqing Peng et al. 1994). Therefore, enhancing the salinity tolerance of grain sorghum at the germination emergence stage should be one of the breeding objectives sought for areas affected by soil salinity (Hassanein and Azab 1990). The effectiveness of divergent selection for tolerance to salinity at the germination emergence stage in grain sorghum has been demonstrated, where subsets of the progenies showing best and least tolerance to salinity were selected and recombined to constitute the next cycle. Evaluation of the original and four subpopulations under saline conditions evaluated in the laboratory and the field for germination emergence ability revealed the effectiveness of selection in separating the subpopulations (Igartua and Gracia 1998).

### 11.7.5.8  Increased Nutritional Quality

While mostly rural poor in semiarid developing countries consume sorghum as human food, it is used as animal feed in developed countries and may in the future be utilized more for feed in developing countries (Hamaker and Axtell 1997). The starch content of whole sorghum is about 70%, and its protein content is about 11% (flour weight basis), which is higher than in maize (Klopfenstein and Hoseney 1995).

The nutritional quality of sorghum is poor compared to other cereals, mainly due to the predominance of storage proteins, i.e., prolamins (*Kafirins*), which are known to be extremely low in the essential amino acid lysine, rich in leucine, and have low protein digestibility (which is lowered when the grain is cooked) (Deosthale et al. 1972; Maclean et al. 1983; Rao et al. 1984a; Magnavaca et al. 1993). People who depend on sorghum in their diet often develop pellagra mainly on account of a high leucine and isoleucine ratio (Gopalan and Srikantia 1960).

Sorghum is usually rich in glutamic acid, leucine, alanine, proline, and aspartic acid, with the level of amino acid highly positively correlated with protein content (Waggle and Deyoe 1966). Similar protein content of immature and mature grain, coupled with higher content of lysine, aspartic acids, and glycine and much lower glutamic acid, proline, and leucine, makes immature grain more nutritionally balanced than mature grain in terms of amino acid composition (Deyoe et al. 1970). Albumin and globulin, glutein, and prolamins are the best, intermediate, and poorest proteins, respectively, in terms of nutrition (Virupaksha and Sastry 1969; Wall and Blessin 1970).

Sorghum proteins have reduced digestibility due to (1) protein cross-linking, which lowers protein solubility, (2) a stronger association of proteins with undigested fiber components, and (3) the presence of a high proportion of peripheral endosperm with high levels of protein (Rooney and Plugfelder 1986). Although nitrogen fertilization can increase protein, lysine, and tryptophan yields per unit area, such an increase is associated with decreased concentrations of these amino acids in the grain (Deosthale et al. 1972).

Total lysine content and total biological value of sorghum protein can be compensated by the addition of synthetic amino acid, legumes, leafy vegetables and other cereal grains, and fermentation of sorghum grains (Pushpamma et al. 1970; Au 1979). Similarly, protein digestibility can be improved by decortication and extrusion of grain and by cooking in the presence of a reducing agent (Maclean et al. 1983; Hamker et al. 1987). However, genetic enhancement is the most sustainable and economical option to increase lysine content and improve protein digestibility.

### 11.7.5.8.1  Genetic Variability for Protein Content and Amino Acid Composition

Considerable variability has been reported for protein content as well as essential amino acid levels such as lysine, methionine, tryptophan, histidine, arginine, aspartic acid, threonine, serine, glutamic acid, proline, glycine, alanine, cystine, valine, methionine, isoleucine, tyrosine, and phenylalanine (Virupaksha and Sastry 1968, 1969; Deosthale et al. 1970; Reich and Atkins 1971; Mali and Gupta 1974; Nanda and Rao 1975a, 1975b).

### 11.7.5.8.2    Genetics of Protein Content and Its Digestibility

While protein content of sorghum grain is governed by additive gene action, essential amino acids such as lysine and methonine are controlled by both additive and nonadditive gene action. Another essential amino acid, tryptophan, appears to be governed by nonadditive gene action (Nanda and Rao 1975a, 1975b; Rana and Murty 1975; Singhania et al. 1979; Nayeem and Bapat 1984; Chinna and Phul 1986; Mallick et al. 1988). While the high lysine natural mutant gene (*hl*) is monogenic recessive (Singh and Axtell 1973), lysine content is inherited as a single gene with partial dominance in p-721 Q, a chemically induced high lysine mutant of sorghum (Mohan 1975; Axtell et al. 1979). Similarly, protein digestibility in a sorghum line, p-851171 derived from p-721 Q, is inherited as a simple Mendelian trait (Oria et al. 2000; Axtell 2001).

### 11.7.5.8.3    High Protein Digestibility and High Lysine Content

The identification of naturally high lysine Ethiopian sorghum mutants from the world sorghum collection, a chemically induced high-lysine mutant, and the recent identification of a sorghum line with high protein digestibility levels surpassing that of maize facilitated the prospects for combining high nutritional quality and grain yield (Singh and Axtell 1973; Mohan 1975; Oria et al. 2000).

The two high-lysine Ethiopian lines, IS 11758 and IS 11167, have exceptionally high lysine contents, are photoperiod sensitive, tall, and late, and have shrivelled or dented seeds. Their acceptance is limited due to many problems associated with their opaque kernel, reduced grain yield, slow drying in the field, increased susceptibility to molds and insects, and the tendency of the seed to crack when mechanically harvested (Rao et al. 1984a; Ejeta and Axtell 1990). Similarly, the chemically induced high-lysine strain p-721 Q has soft kernel and floury endosperm with reduced yielding ability (Axtell 2001; Rao et al. 1984b; Asante 1995).

Earlier attempts to select agronomically desirable recombinants that are dwarf to medium plant height, are early maturing, and are relatively photoperiod insensitive with superior protein quality from various cross-combinations of the very tall, late, photoperiod-sensitive Ethiopian high-lysine parents were not fruitful (Rao et al. 1983b). Although it was possible to select plump, corneous grain types with the high-lysine trait, as breeding continued and generations advanced, there was a reduction in the frequency of plants with high levels of lysine. Selection during 1994 from a high-lysine population developed at Purdue University in the U.S. had 0.30 to 0.49% lysine (flour basis), compared to normal sorghum cultivars that contain about 0.24% lysine, with grain yield in this group ranging from 3 to 4 t ha$^{-1}$ (Hamaker and Axtell 1997). Establishment of a negative correlation between protein and lysine content in several segregating generations of Ethiopian high-lysine × agronomically superior varieties forced the breeders to improve lysine at moderate protein levels (Rao et al. 1983a, 1984b).

Ejeta and Axtell (1990) were able to select modified endosperm of p-721 opaque (high-lysine mutant) with vitreous kernels similar to normal types. But vitreous phenotypes have been detected in an advanced generation ($F_6$) of breeding. However, crosses between p-721 Q (high protein digestibility) and other elite lines resulted in improved yields (Axtell et al. 1979). The recent identification of a sorghum line, p-851171 (a derivative of p-721 Q), with protein digestibility levels surpassing that of maize raises the hopes of improving protein digestibility in sorghum (Oria et al. 2000). However, to date, no studies have examined the association between protein digestibility and lysine concentration in crosses involving p-851171 (Axtell 2001).

Recent development of a rapid protein digestibility assay for identifying highly digestible sorghum lines will help in screening a large number of breeding lines for protein digestibility (Aboubacar et al. 2002). Sorghum germplasm with high protein digestibility as high or higher than maize or other staple cereals have been identified through Purdue/INTSORMIL research. In addition, lines having vitreous kernels with good food grain and processing properties have been identified and are available for breeding programs at Purdue University (Axtell 2001).

### 11.7.5.8.4   Sorghum Whole-Grain Digestibility and Its Improvement

Whole-grain digestibility of sorghum is sometimes a problem; however, local processing techniques convert sorghums into digestible foods. Development of highly digestible sorghum is desirable, but in many environments, the highly digestible sorghums are predigested by molds and insects in the field and have quite low yields. Therefore, highly digestible types may be limited to production in extremely dry environments, where the grain is not subjected to humidity after maturity. It is difficult to improve digestibility without enhancing the susceptibility of the grain to deterioration; thus, efforts to enhance digestibility of sorghum must be done with care (Rooney et al. 2003).

### 11.7.5.8.5   Tannins and Other Polyphenols and Their Genetic Manipulation

Contrary to the belief of many scientists and nutritionists that all sorghums contain tannins, sorghums without a pigmented testa do not contain tannins (Bulter 1990a). The tannin (brown) sorghums have a very definite pigmented testa (caused by combination of dominant $B_1$-$B_2$-$S$ genes) with levels of condensed tannins and may offer resistance to birds and grain mold (Tipton et al. 1970; McMillian et al. 1972; Bulter 1990b; Rooney et al. 2003). The rate of preharvest germination is significantly lower for most high-tannin sorghums (Asante 1995; Harris and Burns 1970; Chavan et al. 1980).

Huge losses in yield due to bird damage in nonresistant sorghum hybrids have been reported in the U.S. (Anonymous 1967; Harris 1969; Tipton et al. 1970). Hence, the beneficial effects of tannins in conferring bird resistance continue to be important in bird-affected areas of the world, particularly Africa (Bulter 1990b). However, experience in Africa has shown that birds will eat high-tannin sorghums when alternative food sources (such as white-grained sorghum) are not available. Nutritionally, tannins interact with and precipitate protein during processing. The level of tannins in high-tannin sorghums is enough to cause significant antinutritional effects, especially if the diet is inadequate in protein (Bulter 1990b). Adding extra protein to the ration overcomes the effects of the tannins. The tannin sorghums decrease feed efficiency by about 10% when fed to livestock. The tannin sorghums have high antioxidant activities and may be a very important source of nutraceuticles. Thus, we might someday use the sorghums with a pigmented testa and dominant spreader gene as potent, more efficient sources of antioxidants than fruits or berries (Rooney et al. 2003).

### 11.7.5.9   Increased Micronutrient Density

Micronutrient malnutrition, primarily the result of diets poor in bioavailable vitamins and minerals, causes blindness and anemia (even death) in more than half of the world's population, especially women and preschool children. Two micronutrients, iron (Fe) and zinc (Zn) and provitamin A (-carotene) are recognized by the World Health Organization (WHO) of the United Nations as limiting. Deficiency for Fe, Zn, and -carotene is highest in South and Southeast Asia and sub-Saharan Africa. These are also the regions where sorghum is cultivated and consumed as food by a large number of people. Therefore, biofortification of sorghum provides the needed micronutrients and -carotene by taking advantage of the consistent daily consumption of large amounts of sorghum food by people.

By breeding sorghum for high levels of micronutrients and -carotene in grain, the plant breeding strategy seeks to develop sorghum with inherent fortification. Plant breeding holds great promise for making a significant, low-cost, and sustainable contribution to reducing micronutrient and -carotene deficiencies without resorting to programs that depend on behavioral change in people who consume sorghum as a staple food. However, this approach will be successful only if farmers are willing to adopt such varieties, if the edible parts of these varieties are palatable and acceptable to consumers, and if the incorporated micronutrients can be absorbed by the human body (Bouis et al. 2000). The fundamental assumption in proposing a plant breeding option to overcome micronutrient malnutrition is that nutrient density traits must be delivered in cultivars of the highest yield. In order to have maximum impact, top-yielding lines are needed to convince local farmers to grow them when the target consumer is in no position to pay a higher price for quality. To do this, a major increase in breeding costs will be necessary in order to maintain progress in yield concurrently (Graham et al. 1999).

Studies on Fe and Zn bioavailability in rats fed with genetically enriched beans and rice have indicated that enriching these staples with Fe and Zn provides significantly more bioavaiable Fe and Zn (Platt 1962). If this is true (optimistically presumed true) with genetically enriched sorghum with Fe and Zn, then consumption of such enriched sorghum would amount to a 50% increase in the dietary intake of Fe and Zn among several million families consuming sorghum as a staple food, and hence should support breeding sorghum with improved Fe and Zn micronutrient contents.

Sorghum is richer in Zn and Fe than most other cereals, such as pearl millet, maize, rice, and wheat (Platt 1962). Considering the prospects of large genetic variability and presumably simpler genetic inheritance for Zn and Fe, as evidenced from other cereals like rice, wheat, and maize, these mineral nutrients levels can be further improved through concerted breeding efforts (Banziger and Long 2000; Gregorio et al. 2000; Monasterio and Graham 2000). A fairly good variability coupled with additive genetic inheritance and moderate heritability of -carotene levels brightens the prospects of improving this trait in sorghum (Nanda and Rao 1974, 1975a,b).

## 11.8 NEW TOOLS FOR SORGHUM IMPROVEMENT

### 11.8.1 Farmers' Participatory Approach

The development of cultivars with farmers' preferred traits through farmers' participation in crop breeding provides an option in the development of cultivars adapted to a wide range of environments (Gowda et al. 2000). Farmer participatory approaches to breeding of improved crop cultivars are categorized into participatory varietal selection (PVS) and participatory plant breeding (PPB). Both PVS and PPB are relatively new approaches for crop improvement (Maurya et al. 1988; Witcombe et al. 1996). PVS involves farmers selecting suitable varieties from finished or near-finished breeding material/lines arising from on-station crop improvement programs through a process of evaluation in their fields and under their own management. PPB involves farmers selecting from the breeder-developed segregating material and is a logical extension of PVS. A detailed methodology of these two approaches and their comparison to conventional breeding methods can be found in Witcombe et al. (1996).

A higher adoption rate of varieties developed using farmer participation is expected as the varieties are better suited for resource-poor farmers, and they facilitate a liberal release system through horizontal spread from farmer to farmer (Rana et al. 1998). While the long-term aim of PVS is to increase biodiversity, where indigenous variability is high, it can also reduce biodiversity (Witcombe et al. 1996). PPB has a greater effect in increasing biodiversity, although its impact may be limited to smaller areas, with PPB a dynamic form of *in situ* genetic conservation (Witcombe et al. 1996). PPB has been used to a rather limited extent compared to PVS. In Ethiopia, yield potential of landraces of *Sorghum* has been enhanced by mass selection by farmers in collaboration with the scientists from the Plant Genetic Resources Center (Worede and Mekbib 1993).

The farmer participatory approach was used to make crosses between *rabi* landraces and improved varieties for the postrainy season in India. These crosses were made to compare the efficiency of conventional breeding and PPB (Reddy et al. 2004a), and participatory varietal selection facilitated the release of the variety SPV 1359 for postrainy season cultivation in Maharashtra and Karnataka States in India during the year 1999–2000.

Participatory sorghum improvement has been used in Africa focusing on two key objectives: (1) modification of the priorities and objectives of sorghum improvement research for Mali to better meet farmers' needs and preferences and (2) farmer assessment of specific new varieties of sorghum in a wide range of production zones (Rattunde 2002). With an objective of developing diversified sorghum populations and lines that incorporate farmer-preferred plant and grain traits, the program succeeded in identifying several genotypes that possessed suitable traits for plant height, maturity duration, resistance to drought, grain (size and color), yield ability, and actual grain yield, with four varieties enjoying wide farmer preference (Chintu 2000).

Improved sorghum cultivars developed by ICRISAT in collaboration with the Institute for Agricultural Research (IAR) were evaluated using farmer participation, with the cultivars outyielding local varieties. The farmers accepted these improved cultivars because of their terminal drought tolerance, easy threshability, bold grain, medium height, and succulent and palatable fodder (Tabo et al. 1999).

## 11.8.2  Biotechnology

Traditional plant breeding has made significant contributions to sorghum improvement; however, they have been slow in targeting complex traits like grain yield, grain quality, drought tolerance, cold tolerance, and resistances to many pests and diseases. For efficient genetic management of such traits, biotechnology offers new and potentially powerful tools to plant breeders. Of the wide range of biotechnological tools, DNA marker technology and genetic transformation have wide application in crop improvement programs across the globe and are discussed in this section.

### 11.8.2.1  DNA Marker Technology

DNA markers have the potential to enhance the operation of plant breeding programs through a number of ways, including fingerprinting of elite genetic stocks, assessment of genetic diversity, addressing genome evolution, phylogeny relevant to germplasm management, increasing the efficiency of selection for difficult traits through their tight linkages with DNA markers, and making environment-neutral selection to map-based cloning (Ejeta et al. 2000; Subudhi and Nguyen 2000). The long-term utility of marker-assisted selection in sorghum improvement is likely to be jointly determined by the identification and mapping of phenotypes with a direct impact on productivity and quality, but which are difficult to study and manipulate by classical means (Paterson 1994). The integration of marker-assisted selection (MAS) into conventional plant breeding promises a more rapid pyramiding of desirable genes into improved cultivars to provide durable resistance (Haussmann et al. 2000b).

Traits that are suitable for molecular breeding include tolerance to drought and heat and resistances to *Striga*, grain mold, downy mildew, stem borer, green bug/aphid, shoot fly, midge, head bug, and chinch bug, as well as grain hardiness and stover quality. The priority traits have been identified as drought, *Striga*, shoot fly, and downy mildew resistance (Witcombe and Duncan 1994).

The most fundamental step required for a detailed genetic study and marker-assisted breeding approach in any crop is the construction of linkage maps (Tanksley et al. 1989). With DNA marker technology, screening the plants using several markers for different pathogens simultaneously is possible without the need to inoculate the pathogens (Lu 1994). However, expression of such resistance genes under variable field environments needs to be tested. Sorghum genome mapping based on DNA markers began in the early 1990s, and since then, many sorghum linkage maps have been constructed (Subudhi and Nguyen 2000, for review). Mapping of genes/QTL of agronomic importance, such as grain yield components, drought and cold tolerance, photoperiod response, resistance to shoot fly, stem borer and head bug, downy mildew, grain mold, and *Striga*, is described below.

### 11.8.2.1.1 Grain Yield Component Traits

Breeding for grain yield per se is difficult as it has low heritability and is governed by polygenes. Little is known about the number and location of genes and their interactive effects on the expression of the component traits. High-density DNA marker maps provide this opportunity by making it possible to identify, map, and measure their effects by detecting QTL. Considerable progress has been made in this direction, and several yield components, such as kernel weight panicle$^{-1}$, threshing (%), dehulling yield (%), panicle length, tiller number, flowering or maturity, number of seed branches panicle$^{-1}$, 100/1000 kernel weight, number of kernel panicle$^{-1}$, and seed size, have been mapped (Lin et al. 1995; Pereira et al. 1995; Paterson et al. 1998; Rami et al. 1998; Crasta et al. 1999; Deu et al. 2000; Hart et al. 2002). Grain quality attributes like kernel flouriness, kernel friability, kernel hardiness, amylose content (%), protein content, and lipid content (%); fodder quality traits like stay-green; and juicy mid-rib have also been mapped (Tuinsta et al. 1996, 1997; Rami et al. 1998; Crasta et al. 1999; Deu et al. 2000; Subudhi et al. 2000; Tao et al. 2000; Xu et al. 2000; Haussmann et al. 2002). Depending on their relative effects and position, many QTL could be used as targets for marker-assisted selection and provide opportunity for accelerating breeding programs (Subudhi and Nguyen 2000).

### 11.8.2.1.2 Drought Tolerance

QTL studies have identified several genomic regions of sorghum associated with pre- and postflowering drought tolerance (Tuinsta et al. 1996, 1997, 1998; Crasta et al. 1999; Ejeta et al. 2000; Xu et al. 2000; Kebede et al. 2001). The molecular genetic analysis of QTL influencing stay-green trait, an important postflowering drought resistance, resulted in the identification of up to four QTL (Tao et al. 2000; Xu et al. 2000; Haussmann et al. 2002). Three of the four stay-green QTL showed consistency across different genetic backgrounds and environments, with the QTL Stg-2 expected to increase our understanding of stay-green trait, leading to either marker-assisted introgression of this QTL into an elite agronomic background or map-based cloning to genetically engineer this locus into improved cultivars (Subudhi et al. 2000).

ICRISAT recently mapped several new QTL for the drought tolerance/stay-green trait from two *Striga* resistance mapping populations derived from an agronomically elite, *Striga*-susceptible, stay-green parent E 36-1 that was not detected in previous studies based on sources B 35 and SC 5. To date, there are now three available stay-green sources (B 35, SC 56, and E 36-1) for which QTL have been mapped and identified on all 10 sorghum linkage groups. ICRISAT has initiated a marker-assisted backcross program to introgress QTL for the stay-green trait from sources B 35 and E 36-1 into a diverse range of farmer-accepted sorghum cultivars adapted to diverse agroecologies in tropical Asia, Africa, and Latin America, with lines developed from this marker-assisted backcross program available shortly (ICRISAT 2002).

### 11.8.2.1.3 Cold Tolerance

Tolerance to early-season cold temperature is needed in much of the sorghum production areas of the U.S. Important advantages of this trait are attributed to seedling vigor, resulting in greater biomass and grain yield in cold and dry environments. Studies of Ejeta et al. (2000) suggest the presence of undetected QTL for early-season cold tolerance, highlighting the need for exhaustive detection of all relevant QTL by generating a series of well-controlled and -characterized phenotypic data.

### 11.8.2.1.4 Striga Resistance

Efforts are currently under way to identify and map genes for qualitative and quantitative resistance to *Striga* in three sorghum mapping populations derived from three crosses (Bennetzen et al. 2000; Ejeta et al. 2000; Haussmann et al. 2000c). By generating a dense linkage map using RFLP markers, Ejeta et al. (2000) were able to map the locus for one of the better-characterized mechanisms of resistance to *Striga*, viz., the low germination stimulant (*lgs*). They also placed a putative QTL for *Striga* resistance using phenotypic data from field evaluation of mapping population against *Striga hermonthica* and *Striga asiatica*. Single-marker analysis detected six QTL for resistance to *S. hermonthica* (accounted for 37% of the variation in resistance) and five QTL for resistance to *S. asiatica* (accounted for 49% of the variation in resistance).

Three QTL for *Striga* resistance from source N 13 have been identified that offer some degree of stability across subsets of the mapping populations and *S. hermonthica* strains of the parasitic weed from Eastern and Western Africa in a collaborative study between ICRISAT and the University of Hohenheim, Germany. However, map coverage is incomplete, and the single recessive gene for low levels of *Striga* seed germination stimulant does not yet cosegregate with any of the 10-molecular marker-based linkage groups. Attempts are under way to extend the linkage maps of these two populations before completing QTL analysis while also initiating marker-assisted back-cross transfer of the three resistance QTL from the source to a number of farmer-accepted sorghum cultivars of African origin (ICRISAT 2002).

The identification of individual genes or QTL for *Striga* resistance and their transfer into adapted cultivars will also allow the evaluation of any costs of *Striga* resistance, i.e., whether resistance is associated with any yield drag. Such costs of resistance might have been one of the reasons for slow breeding process in the past (Haussmann et al. 2000b).

### 11.8.2.1.5 Photoperiod Response

Photoperiod sensitivity appears to be a key feature matching flowering time to the length of the rainy season, and it plays an important role in securing the level and quality of harvest (Deu et al. 2000). Photoperiod response in cereals is determined by three main components (Major 1980): (1) basic vegetative phase (BVP) — defined as the shortest possible time for floral initiation when the plants are not responsive to changes in photoperiod; (2) photoperiod sensitivity (PS) — which expresses the varietal linear response to flowering time as plants respond to day-length changes; and (3) minimum optimal photoperiod (MOP) — defined as the photoperiod threshold beyond which the vegetative period is influenced by changes in day length.

Four major gene loci controlling flowering time have been identified in sorghum ($Ma_1$ to $Ma_4$) (Quinby 1973). Recently, Aydin et al. (1997) showed that other maturity loci ($MA_5$ and $MA_6$) might be involved in floral initiation of ultralate sorghum genotypes in the U.S. Lin et al. (1995) also identified another QTL for flowering time that is located on linkage group H of the linkage map described by Pereira et al. (1994). The genotypes involved in these studies were day-neutral and quantitatively short-day sorghums able to flower in the U.S. In more photoperiod-sensitive sorghum, one QTL has been detected on linkage group H for PS that explained 18% of total phenotypic variation. A second QTL was detected on the same linkage group for the direct measurement of photoperiod response (Trouche et al. 1998). QTL have also been detected and mapped for BVP and PS components of photoperiod response (Deu et al. 2000).

### 11.8.2.1.6 Disease Resistance Traits

*11.8.2.1.6.1 Grain Mold* — It is unlikely that traditional breeding methods will achieve high-yielding, bold, white-grained sorghum with complete protection against grain mold (Reddy et al. 2000). New biotechnology techniques such as QTL analyses and marker-assisted selection provide new opportunities to enhance grain mold resistance. To date, 12 QTL have been detected that explain 10 to 33.8% of phenotypic variation in grain mold incidence; however, these QTL are unstable over different environmental conditions, indicating that genotype by environment interaction is critical in grain mold resistance (Rami et al. 1998; Rooney and Klein 2000; Klein et al.

2001). While two grain mold QTL are related to phenotypic or kernal traits, the remaining three QTL were not associated; however, several possible traits that influence these QTL were predicted, including higher levels of antifungal proteins in resistant compared to susceptible lines (Rodriguez-Herrera et al. 1999; Rooney and Klein 2000).

*11.8.2.1.6.2 Downy Mildew* — Existing techniques for breeding varieties with resistance to the three pathotypes of downy mildew (*Peronosclerospora sorghi* (Weston and Uppal) C.G. Shaw) involve screening varieties after inoculation with the pathogen (Craig 1987; Sifuentes and Frederiksen 1988). Although this technique is excellent, it is costly and time-consuming and depends upon suitable environmental conditions for accuracy (Gowda et al. 1994).

The availability of tightly linked genetic markers for resistance genes will permit multigenic resistance to downy mildew. RFLP markers linked to different resistance genes were identified in the mid-1990s, and using marker-assisted selection (MAS), it should be possible to pyramid these genes to confer resistance to the three downy mildew pathotypes into agronomically acceptable cultivars (Gowda et al. 1994, 1995; Magill et al. 1997; Thakur et al. 1997).

Anthracnose (*Colletotrichum graminicola*) and head smut (*Sporosorium reilianum*) are globally important diseases for which RFLP markers closely linked to resistance genes have been identified. MAS will make possible the introgression of these resistance genes into agronomically elite cultivars (Oh et al. 1993; Rosenow 1994). Charcoal stalk rot is important in the postrainy season sorghum in India and other countries and is closely related to soil moisture deficit only during the grain fill stage. Considerable progress has been made toward charcoal stalk rot resistance (Rosenow 1994).

### 11.8.2.1.7 Insect Resistance Traits

*11.8.2.1.7.1 Shoot Fly Resistance* — Nonavailability of sources with absolute resistance to shoot fly in cultivated sorghum, coupled with the complex nature of the inheritance, has resulted in slow and inefficient progress in shoot fly resistance breeding using conventional approaches. The identification of genomic regions for each of the known components of shoot fly resistance, as well as markers tightly linked to these regions, will allow MAS breeding or map-based cloning to introgress or incorporate disease resistance genes into elite cultivars. Sajjanar (2002) identified eight QTL for shoot fly resistance components, with one major QTL for glossiness (phenotypic expression of 34.3 to 46.5% in the three screening environments, with highest expression in postrainy season) identified that was linked with dead hearts (%) under high shoot fly pressure. This QTL may be a useful target for MAS for shoot fly resistance in sorghum.

At ICRISAT, QTL mapping is under way for shoot fly resistance using recombinant inbred line (RIL) populations, with a linkage map with reasonable genome coverage constructed identifying six QTL in at least two screening environments. The phenotypic variance explained by each of these QTL ranged from 62.9% for glossiness to 4.5% for seedling vigor, offering potential in MAS breeding for resistance (ICRISAT 2002).

*11.8.2.1.7.2 Sorghum Midge Resistance* — A poor understanding of resistance to sorghum midge has slowed progress in host plant resistance to sorghum midge, as there are three different mechanisms of midge resistance involved in insect–host plant interactions: antixenosis, antibiosis, and tolerance (Kogan and Ortman 1978). Very recently, QTL associated with two of the mechanisms of midge resistance, antixenosis and antibiosis, were identified from a relatively large recombinant inbred (RI) population through accurate glass house screening (Tao et al. 2003). Two regions on separate linkage groups (A&G) were associated with antixenosis and explained 34.5% of the variation in the difference of egg and pupal counts in the RI population. The identification of genes for different mechanisms of midge resistance will be particularly useful for exploring new sources of midge resistance and for gene pyramiding of different mechanisms in sorghum breeding through marker-assisted selection.

The only program to find molecular markers putatively linked with midge resistance is in Australia, with two QTL explaining 27% of the variation in ovipositional antixenosis (nonpreference), while one QTL explained 34.5% of variation in antibiosis from a study of a RIL population (Tao et al. 1996, Henzell et al. 2002).

*11.8.2.1.7.3 Sorghum-Spotted Stem Borer Resistance* — QTL mapping is currently under way for resistance to several species of sorghum stem borers using RILs derived from three populations, with mapping of SSRs achieved with a reasonably good coverage of the sorghum genome (ICRISAT 2002).

### 11.8.2.2 Genetic Transformation Technology

The process of introduction, integration, and expression of foreign genes in the host is called genetic transformation. Combined use of recombinant DNA technology, gene transfer methods, and tissue culture techniques has led to the efficient production of transgenics in a wide array of crop plants. Unlike conventional plant breeding, in transgenesis, only the cloned genes of agronomic importance are introduced into the plants without linkage drag from the donor. This approach has the potential to serve as an effective means of removing certain specific defects of an otherwise well-adapted cultivar, which is difficult using conventional breeding approaches (Chahal and Gosal 2002).

Compared to all other major food crops, sorghum transformation is in its infancy, with much technical progress yet to be achieved (Seetharama et al. 2003). Successful transformation of sorghum was first achieved in the early 1990s, with highly advanced transformation technology now available that is capable of transforming at least 150 kbp of foreign DNA into the sorghum genome (Hamilton et al. 1996; Seetharama et al. 2003). At ICRISAT, sorghum transformation has been achieved and efforts are continuing to develop procedures for *Agrobacterium*-mediated transformation. Transgenic sorghums have been developed with resistance to stem borer that have been molecularly characterized and are undergoing insect bioassays. Preliminary results have indicated that 9 of 11 transgenic plants expressed a good level of protection against stem borer damage (ICRISAT 2002). Although transgenic sorghum has been developed, the absence of pleiotropic effects must be assured, and consideration of biosafety issues is important before useful transgenic plants can be commercialized.

### 11.8.2.3 Transgenics and Conventional Breeding Integrated Technology

Hybrid seed will be the final delivery vehicle of transgenic technology to the farmers, and as such, integration of conventional breeding with transgenic research will be required (Seetharama et al. 2003). A case study with special reference to sorghum is detailed in Reddy and Seetharama (2001). Hybrid sorghum production offers an opportunity to separately pyramid two sets of traits in different parents: the efficiency of this approach is not yet adequately tested in any crop; however, it is likely that most of the transgenes will be first introduced into the proper female parent (A-line), and most of the natural resistance may be bred into the male set (restorer lines).

The questions related to and public acceptance of transgenic crops are not unique to sorghum. Considering the fact that sorghum is a poor man's food crop and that only a small proportion of production enters the global market, public institutions will have a greater say and responsibility in this respect (Reddy et al. 2001). Conversely, sorghum farmers in developing countries may not be able to use transgenic sorghum, as the focus of transgenic research may be toward industrial and export purposes. All these issues have to be considered on a case-by-case basis, and once the extent of economic and social benefits is weighed against the acceptable magnitude of risk, if any, decisions will be made.

## 11.9 SUMMARY

Sorghum, the fourth major cereal crop of the world in production and fifth in acreage after wheat, rice, maize, and barley, is mostly grown by resource-limited farmers under minimal/traditional management conditions in the semiarid tropics of the world as a subsistence crop, thereby recording low productivity compared to other developed regions of the world. The yield and quality of sorghum produce are influenced by a wide array of biotic (shoot fly, stem borer, head bug, aphid, armyworms, and locusts among insects; grain mold, charcoal rot, downy mildew, anthracnose, rust, and leaf blight among diseases; and *S. asiatica*, *S. densiflora*, and *S. hermonthica*, a parasitic weed) and abiotic (problematic soils like saline and acidic, temperature extremities and drought) constraints.

Sorghum, with its primary center of origin in Africa and secondary center of origin in the Indian subcontinent, probably domesticated in the northeast quadrant of Africa, from where it spread to India, China, the Middle East, and Europe soon after its domestication. ICRISAT, which is a major repository of world sorghum germplasm, has the responsibility of preservation, maintenance, characterization, and distribution.

The genetic variability available in cultivated races and their wild relatives is extensive, and the extreme types are so different as to appear as separate species. The major germplasm sources utilized so far in varietal improvement include temperate lines from the U.S., *zera zera* lines from Ethiopia and Sudan, and some lines of Indian origin.

High-lysine sorghum lines from Ethiopia were used in the breeding program for transferring this trait to a desirable agronomic background. Conversion of tall, late, or nonflowering (in the U.S.) sorghum germplasm from tropics into short, early-photoperiod-insensitive forms jointly by the Texas Agricultural Experiment Station and USDA, and conversion of tall, late-flowering *zera zera* (from the Ethiopia–Sudan border) landraces and *kauras* and *guineenses* (from Nigeria) by ICRISAT resulted in enhanced use of exotic sorghum germplasm, broadening of genetic diversity, and provision of new sources of desirable traits to sorghum researchers. The breeding materials so far exploited rather extensively in sorghum improvement programs in India, other tropical areas of the world, and the southern U.S. belonged to derivatives of the *zera zera* group. Extensive use of the *zera zera* group of converted sorghums has made a major contribution to disease resistance, yield potential, and grain quality of U.S. hybrid sorghums. In the U.S., *Kaura* (origin: Nigeria), African introductions, and the source of yellow endosperm were extensively used in sorghum improvement prior to the development of the conversion programs.

Two types of male sterility, GMS and CMS, are widely used in sorghum improvement programs. Of the several GMS-inducing alleles (such as $ms_1$, $ms_2$, $ms_3$, $ms_4$, $ms_5$, $ms_6$, $ms_7$, and $al$), $ms_3$ followed by $ms_7$ have been widely used in population improvement, as they are stable over a range of environments. Of the several different sources of CMS, only $A_1$ cytoplasm has been extensively used in the development of commercial sorghum hybrids worldwide. However, recent evidences point out the potential of $A_2$ cytoplasm, especially in the development of hybrids. Extensive research is under way for the development of $A_2$ cytoplasm-based hybrids at ICRISAT and through Indian programs. In China, $A_2$ cytoplasm-based hybrids are already in commercial cultivation. Breeding methods used in sorghum improvement world over are those developed for self-pollinated crops, such as pure line selection, pedigree breeding, backcross breeding, population improvement (using GMS), and hybrid breeding.

Conventional approaches to enhance resistance to diseases such as grain mold, downy mildew, anthracnose, and head smut; insect pests like shoot fly and stem borer; endemic weed, *Striga*; and terminal drought have met with partial success. The identification of genomic regions affecting each of these resistance traits as well as markers tightly linked to these regions would greatly facilitate the exploitation of MAS breeding or map-based cloning to introgress or incorporate resistance genes into elite cultivars. While considerable progress has been made in combining enhanced levels of host plant resistance (HPR) to sorghum midge with local adaptation in the breeding programs at ICRISAT, Texas A&M University, and in Australia through conventional breeding approaches, application of MAS is further expected to hasten this progress.

While attempts to enhance salinity tolerance of sorghum in saline areas are limited, much progress has been achieved in breeding sorghum for soil acidity tolerance with and without $Al^{3+}$ toxicity. Future breeding programs should aim at specific adaptations in view of significant genotype $\times Al^{3+}$ toxicity interaction variation and variation in $Al^{3+}$ saturation from location to location. Further, the possibility of different resistance mechanisms due to varying levels of $Al^{3+}$ saturation necessitates the study of resistance mechanisms and their genetics in the selected sorghum lines. The considerable variability in soil salinity levels both spatially and temporally, and with much variation within the same plot, also warrants breeding for wide adaptation for salinity tolerance in sorghum.

The identification of natural, exceptionally high lysine Ethiopian mutants from the world sorghum collection, a chemically induced high-lysine mutant, and, recently, a sorghum line with high protein digestibility levels in plump grain background surpassing that of maize, as well as identification of molecular markers linked to high protein digestibility or high lysine in high-lysine sorghum lines, facilitated the prospects for combining high protein quality and protein digestibility with grain yield.

It is expected that there will be greater use of genomic tools for germplasm management and gene manipulation through marker-assisted selection, as more and more useful traits are tagged in the near future. Sorghum, as a drought-tolerant crop species with a small genome size, is an excellent model for the investigation of genes involved in drought tolerance and plant adaptation to harsh climate conditions. It can be transformed with a number of candidate genes to enhance resistance to biotic and abiotic stress factors. With the advent of sophisticated transformation technology, it should be possible to transfer resistance genes from the secondary gene pool into cultivated sorghums, which was not possible hitherto due to crossing barriers. Enormous potential exists for further improvement of grain yield and boldness with high levels of Fe, Zn, and vitamin A, resistance to grain mold and shoot fly, and tolerance to abiotic stresses, including problematic soils, considering the vast variability left untapped in this crop following specific adaptation approaches.

## ACKNOWLEDGMENTS

A financial grant from Suri Sehgal Foundation, India, in partial support of this work and encouragement received from Dr. C.L.L. Gowda, Global Theme Leader, Crop Improvement, ICRISAT, to carry out this work are gratefully acknowledged. We are indebted to Dr. Sally Dillon, Australian Tropical Crops and Forages Collection, Queensland Department of Primary Industries and Fisheries, Australia, for his critical review and editing of the manuscript and providing invaluable suggestions and comments to improve the quality of the manusript.

## REFERENCES

Aboubacar, A., Axtell, J.D., Huang, C.P., and Hamaker, B.R. 2002. A rapid protein digestibility assay for identifying highly digestible sorghum lines. *Cereal Chem.* 78: 160–165.

Adamou, M., Gourley, L.M., Watson, C.E., Mclean, S.D., and Goggo, A.S. 1992. Evaluation of combining ability of acid soil tolerant sorghum germplasm in Niger. In *Agronomy Abstracts*, November 1–6, Minneapolis, p. 88.

Ahlrichs, S.L, Duncan, R.R., Ejeta, G.E., Hill, P.P., Baligar, V.C., Wright, R.J., and Hanna, W.W. 1991. Pearl millet and sorghum tolerance to aluminum in acid soil. In *Plant Soil Interactions at Low pH*, R.J. Wright, Ed. Kluwer Academic Publishers, Dordrecht, The Netherlands, p. 197.

Ahnert, D., Lee, M., Austin, D.F., Woodman, W.L., Openshaw, S.J., Smith, J.S.C., Porter, K., and Dalton. 1996. Genetic diversity among the elite *Sorghum* inbred lines assessed with DNA markers and pedigree information. *Crop Sci.* 36: 1385–1392.

Aldrich, P.R., Doebley, J., Schertz, K.F., and Stec, A. 1992. Patterns of allozyme variation in cultivated and wild *Sorghum bicolor. Theor. Appl. Genet.* 85: 451–460.

Anonymous. 1967. Bird resistance sorghum where damage to corn is severe. *Crops Soils* 19: 21–22.

Arti, M. 1993. Water-soluble proteins of Indian sorghum cultivars. *Sorghum Newsl.* 34: 53.

Arunachalam, V., Prabhu, K.V., and Sujatha, V. 1998. Multivariate methods of quantitative evaluation of crop improvement: conventional and molecular approaches revisited. In *Crop Productivity and Sustainability: Shaping the Future, Proceedings of 2nd International Crop Science Congress*, New Delhi, India, V.L. Chopra, R.B. Singh, and A. Varma, Eds., pp. 793–807.

Asante, S.A. 1995. Sorghum utility and utilization. *Afr. Crop Sci. J.* 3: 231–240.

Au, P. 1979. Study of Fermentation of Grain Sorghum to Improve Its Amino Acid and Vitamin Content. M.S thesis, University of Missouri–Columbia, Columbia.

Austin, R.B. 1993. Augmenting yield-based selection. In *Plant Breeding Principles and Prospects*, M.D. Hayward, N.O. Bosemark, and I. Romagosa, Eds. Chapman & Hall, London, pp. 391–405.

Axtell, J.D. 2001. *Breeding Sorghum for Increased Nutritional Value*, 2001 Annual Report. INTSORMIL Sorghum/Millet Collaborative Support Program (CRSP), pp. 67–73.

Axtell, J.D., Van Scoyoc, S.W., Christensen, P.J., and Ejeta, C. 1979. Current status of protein quality improvement in grain sorghum. In *Seed Protein Improvement in Cereals and Grain Legumes*. IAEA, Vienna, pp. 357–366.

Ayana, A. and Bekele, E. 1999. Multivariate analysis of morphological variation in sorghum [*Sorghum bicolor* (L.) Moench] germplasm from Ethiopia and Eritrea. *Genet. Resour. Crop Evol.* 46: 273–284.

Ayana, A., Bekele, E., and Bryngelsson, T. 2000a. Genetic variation in wild sorghum [*Sorghum bicolor* spp. *verticilliflorum* (L.) Moench] germplasm from Ethiopia assessed by random amplified polymorphic DNA (RAPD). *Hereditas* 3: 249–254.

Ayana, A., Bekele, E., and Bryngelsson, T. 2000b. Genetic variation of Ethiopian and Eritrean sorghum [*Sorghum bicolor* (L.) Moench] germplasm assessed by random amplified polymorphism DNA (RAPD). *Genet. Resour. Crop Evol.* 47: 471–482.

Aydin, S., Rooncy, W.L., and Miller, F.R. 1997. Identification and characterization of the Ma5 and Ma6 maturity loci in sorghum. In *Proceedings of the International Conference on Genetic Improvement of Sorghum and Pearl Millet*, Lubbock, TX, September 22–27, 1996, pp. 641–642.

Ayyangar, G.N.R. and Ponnaiya, B.W.X. 1939. The occurrence and inheritance of earheads with empty anther sacs in sorghum. *Curr. Sci.* 8: 116.

Azhar, F.M. and McNeilly, T. 1987. Variability for salt tolerance in *Sorghum bicolor* (L.) Moench under hydroponic condition. *J. Agron. Crop Sci.* 159: 269–277.

Azhar, F.M. and McNeilly, T. 1989. Heritability estimates of variation for NaCl tolerance in *Sorghum bicolor* (L.) Moench seedlings. *Euphytica* 43: 69–72.

Badi, S., Pedersen, B., Monowar, L., and Eggum, B.O. 1990. The nutritive value of new and traditional sorghum and millet foods for Sudan. *Plant Foods Hum. Nutr.* 40: 5–19.

Banziger, M. and Long, J. 2000. The potential for increasing the iron and zinc density of maize through plant-breeding. *Food Nutr. Bull.* 21: 397–400.

Barthate, K.K., Patil, J.V., and Thete, R.V. 2000. Genetic divergence in sorghum under different environments. *Indian J. Agric. Res.* 34: 85–90.

Bebawi, F.F. 1981. Response of sorghum cultivars and *Striga* population to nitrogen fertilization. *Plant Soil* 59: 261–267.

Bennetzen, J.L., Gong, F., Xu, J., Newton, C., and de Oliveira, A.C. 2000. The study and engineering of resistance to parasitic weed *Striga* in rice, sorghum and maize. In *Breeding for Striga Resistance in Cereals, Proceedings of a Workshop*, B.I.G. Haussmann, D.E. Hess, M.L. Koyama, L. Grivet, H.F.W. Rattunde, and H.H. Geiger, Eds., Ibadan, Nigeria, August 18–20, 1999, pp. 197–205.

Blum, A. 1983. Genetic and physiological relationships in plant breeding for drought resistance. *Agric. Water Manage.* 7: 195–202.

Blum, A. 1987. Genetic and environmental considerations in the improvement of drought stress avoidance in sorghum. In *Food Grain Production in Semi-Arid Africa, Proceedings of the International Drought Symposium*, Nairobi, Kenya, May 19–23, 1983, pp. 91–99.

Bonhert, H.J., Nelson, D.E., and Jenson, R.G. 1995. Adaptations to environment stresses. *Plant Cell* 7: 1099–1111.

Borrell, A.K., Bidinger, F.R., and Sunitha, K. 1999. Stay-green associated with yield in recombinant inbred sorghum lines varying in rate of leaf senescence. *Int. Sorghum Millets Newsl.* 40: 31–33.

Borrell, A.K. and Douglas, A.C.L. 1997. Maintaining green leaf area in grain sorghum increased nitrogen uptake under post-anthesis drought. *Int. Sorghum Millets Newsl.* 38: 89–91.

Borrell, A.K. and Hammar, G.L. 2000. Nitrogen dynamics and the physiological basis of stay-green in sorghum. *Crop Sci.* 40: 1295–1307.

Borrell, A.K., Hammar, G.L., and Douglas, A.C.L. 2000a. Does maintaining green leaf area in sorghum improve yield under drought? I. Leaf growth and senescence. *Crop Sci.* 40: 1026–1037.

Borrell, A.K., Hammar, G.L., and Henzell, R.O. 2000b. Does maintaining green leaf area in sorghum improve yield under drought? II. Dry matter production and yield. *Crop Sci.* 40: 1037–1048.

Bouis, H.E., Graham, R.D., and Welch, R.M. 2000. The Consultative Group on International Agricultural Research (CGIAR) micro nutrient project: justification and objectives. *Food Nutr. Bull.* 21: 374–381.

Bramel-Cox, P.J., Lauver, M.A., and Witt, M.E. 1990. Potential gain from selection in grain sorghum for higher protein digestibility. *Crop Sci.* 30: 521–524.

Bramel-Cox, P.J., Kumar, K.A., Hancock, J.D., and Andrews, D.J. 1995. Sorghum and millets for forage and feed. In *Sorghum and Millets: Chemistry and Technology*, D.A.V. Dendy, Ed. American Association of Cereal Chemists, St. Paul, MN, pp. 325–364.

Brown, A.H.D. 1989. The case for core collection. In *The Use of Plant Genetic Resources*, A.H.D. Brown, O.H. Frankel, D.R. Marshall, and J.T. Williams, Eds. Cambridge University Press, Cambridge, U.K., pp. 136–156.

Brown, M.S. 1943. Haploid plants in sorghum. *J. Hered.* 34: 163–166.

Bulter, L.G. 1990a. *Tannins and Other Phenols: Effects on Sorghum Production and Utilization*, INTSORMIL Annual Report (a technical research report of the grain sorghum/pearl millet collaborative research support program (CRSP)). CRSP, Lincoln, NE, pp. 140–144.

Bulter, L.G. 1990b. The nature and amelioration of the antinutritional effects of tannins in sorghum grain. In *Sorghum Nutritional Quality: Proceedings of an International Conference*, G. Ejeta, E.T. Mertz, L.W. Rooney, R. L'Schaffert, and J. Yohe, Eds., Purdue University, West Lafayette, IN, February 26–March 1, pp. 191–205.

Burner, D.K., Kling, J.G., and Singh, B.B. 1995. *Striga* research and control, a perspective from Africa. *Plant Dis.* 79: 652–660.

Calhoun, D.S., Gebeyehu, C., Miranda, A., Rajaram, S., and Van Ginkel, M. 1994. Choosing evaluation environments to increase grain yield under drought conditions. *Crop Sci.* 34: 673–678.

Celarier, R.P. 1959. Cytotaxonomy of the Andropogonea. III. Sub-tribe sorgheae, genus sorghum. *Cytologia* 23: 395–418.

Chahal, G.S. and Gosal, S.S. 2002. *Principles and Procedures of Plant Breeding: Biotechnological and Conventional Approaches*. Narosa Publishing House, New Delhi, p. 486.

Chandel, K.P.S. and Paroda, R.S. 2000. *Status of Plant Genetic Resources Conservation and Utilization in Asia-Pacific Region*, Regional Synthesis Report 32, Asia-Pacific Association of Agricultural Institutions, FAO Regional Office for Asia and the Pacific, Bangkok.

Chandrasekharaiah, S.R., Murty, B.R., and Arunachalam, V. 1969. Multivariate analysis of divergence in the genus *Eu-sorghum*. *Proc. Natl. Inst. Sci. India B* 35: 172–195.

Chavan, J.K., Kadam, S.S., and Salunkhe, D.K. 1980. Changes in tannin free amino acids, reducing sugars, and starch during seed germination of low and high tannin cultivars of sorghum. *J. Food Sci.* 46: 638–639.

Cherney, J.H. 1990. Normal and brown-midrib mutations in relation to improved ligno cellulose utilization. In *Microbial and Plant Opportunities to Improve Lignocellulose Utilization by Ruminants*, D.E. Akin and L.G. Lungdahl, Eds. Elsevier, Amsterdam, pp. 205–214.

Cherney, J.H., Cherney, D.J.R., Akin, D.E., and Axtell, J.D. 1991. Potential of brown-midrib, low lignin mutants for improving forage quality. *Adv. Agron.* 46: 157–198.

Chinna, B.S. and Phul, P.S. 1986. Heterosis and combining ability studies for protein, lysine and tryptophan in sorghum [*Sorghum bicolor* (L) Moench]. *Genet. Agrar.* 40: 405–414.

Chintu, E.M. 2000. Development of sorghum varieties through participatory plant breeding in Malawi. *Int. Sorghum Newsl.* 41: 23–24.

Chittenden, L.M., Schertz, K.F., Lin, V.R., Wing, R.A., and Paterson, A.H. 1994. A detailed RFLP map of *Sorghum bicolor* × *S. propinquum*, suitable for high density mapping, suggests ancestral duplication of sorghum chromosomes or chromosomal segments. *Theor. Appl. Genet.* 87: 925–933.

Christiansen, M.N. and Lewis, C.F. 1982. *Breeding Plants for Less Favourable Environments*. John Wiley & Sons, New York.

Clayton, W.D. 1961. Proposal to conserve the generic name *Sorghum* Moench (Gramineae) versus *Sorghum* adans (Gramineae). *Taxonomy* 10: 242.

Clayton, W.D. and Renvoize, S.A. 1986. *Genera Graminum Grasses of the World*, Kew Bulletin Addition Series XIII. Royal Botanic Gardens, Kew, London, pp. 338–345.

Comstock, R.E. and Robinson, H.F. 1952. Genetic parameters, their estimation and significance. In *Proceedings of the Sixth International Grass Lands Congress*, pp. 284–291.

Craig, J. 1987. Tiered temperature system for producing and storing conidia of *Peronosclerospra sorghi*. *Plant Dis.* 71: 365–368.

Crasta, O.R., Xu, W., Rosenow, D.T., Mullet, J.E., and Nguyen, H.T. 1999. Mapping of post-flowering drought resistance traits in grain sorghum: association of QTLs influencing premature senescence and maturing. *Mol. Gen. Genet.* 262: 579–588.

Cress, C.E. 1966. Heterosis of the hybrid related to gene frequency differences between two populations. *Genetics* 53: 269–274.

Cui, Y.X., Xu, G.W., Magill, C.W., Schertz, K.F., and Hart, G.E. 1995. RFLP based assay of *Sorghum bicolor* (L.) Moench genetic diversity. *Theor. Appl. Genet.* 90: 787–796.

Dahlberg, J.A. 2000. Classification and characterization of sorghum. In *Sorghum, Origin, History, Technology and Production*, Wiley Series in Crop Science, C.W. Smith and R.A. Frederiksen, Eds. John Wiley & Sons, New York, pp. 99–130.

Dahlberg, J.A., Hash, C.T., Kresovich, S., Maunder, B., and Gilbert, M. 1997. Sorghum and pearl millet genetic resources utilization. In *Proceedings of the International Conference on Genetic Improvement of Sorghum and Pearl Millet*, September 22–27, 1996, Lubbock, TX, pp. 42–54.

Dahlberg, J.A. and Spinks, M.S. 1995. Current status of the U.S. sorghum germplasm collection. *Int. Sorghum Millets Newsl.* 36: 4–12.

Damania, A.B. 2002. The Hindustan centre of origin of important plants. *Asian Agri-Hist.* 6: 333–341.

de Andrade Lima, M.M., Maciel, G.A., Tabosa, J.N., Tavares, J.A., and Neto, C. de A. 1992. Advanced grain sorghum trial for aluminium tolerance. *Sorghum Newsl.* 33: 58.

de la Rosa Ibarra, M. and Maiti, R.K. 1994. Morphological and biochemical basis of resistance of glossy sorghum to salinity at seedling stage. *Int. Sorghum Millets Newsl.* 35: 118–119.

de Wet, J.M.J. 1977. Domestication of African cereals. *Afr. Econ. Hist.* 3: 15.

de Wet, J.M.J. and Harlan, J.R. 1971. The origin and domestication of *Sorghum bicolor*. *Econ. Bot.* 25: 128–135.

de Wet, J.M.J., Harlan, J.R., and Price, E.G. 1976. Variability in *Sorghum bicolor*. In *Origins of African Plant Domestication*, J.R. Harlan, J.M.J. de Wet, and A.B.C. Stemler, Eds. The Mountain Press, The Hague, pp. 453–463.

de Wet, J.M.J. and Huckabay, J.P. 1967. The origin of *Sorghum bicolor*. II. Distribution and domestication. *Evolution* 211: 787–802.

Dean, R.E., Dahlberg, J.A., Hopkins, M.S., Mitchell, S.E., and Kresovich, S. 1999. Genetic redundancy and diversity among 'orange' accessions in the U.S. National *Sorghum* collection as assessed with simple sequence repeat (SSR) markers. *Crop Sci.* 39: 1215–1221.

Deosthale, Y.G., Mohan, V.S., and Rao, V. 1970. Varietal differences in protein, lysine and leusine content of grain sorghum. *J. Agric. Food Chem.* 18: 644–646.

Deosthale, Y.G., Nagarajan, V., and Vesweswar Rao, K. 1972. Some factors influencing the nutrient composition of sorghum grain. *Indian J. Agric. Sci.* 42: 100–108.

Deu, M., Grivet, L'Trouche, G., Barro, C., Ratnadass A., Diabate, M., Hamada, A., Fliedel, G., Rami, J.F., Grenier, C., Hamon, P., Glaszmann, J.C., and Chantereau, J. 2000. Use of molecular markers in the sorghum breeding program at CIRAD. In *Application of Molecular Markers in Plant Breeding*, training manual on seminar held at IITA, B.I.G. Haussmann, H.H. Greiger, D.E. Hess, C.T. Hash, and P. Bramel-Cox, Eds., Ibadan, Nigeria, August 16–17, 1999.

Deyoe, C.W., Shoup, F.K., Miller, G.D., Bathurst, J., Laing, D., Standford, P.E., and Murphy, L.S. 1970. Amino acid composition and energy value of immature sorghum grain. *Cereal Chem.* 47: 363–368.

Doggett, H. 1988. *Sorghum*, 2nd ed. Longmans Scientific and Technical Publishers, London, U.K.

Doggett, H. 1962. Tetraploid hybrid sorghum. *Nature* 196: 755–756.

Doggett, H. 1965. The development of cultivated sorghum. In *Crop Plant Evolution*, I. Hutchinson, Ed. Cambridge University Press, Cambridge, U.K., pp. 50–69.

Doggett, H. 1970. *Sorghum*. Longmans Green, London.

Doggett, H. 1972. The important sorghum in East Africa. In *Sorghum in the Seventies*, N.G.P Rao and L.R. House, Eds. Oxford and IBH Publishing Co., New Delhi, pp. 47–59.

Doggett, H. 1984. *Striga*: its biology and control — an overview. In *Striga: Biology and Control*, E.S. Ayensu, H. Doggett, R.D. Keynes, J. Marton-Lefevre, L.J. Musselman, C. Parker, and A. Pickering, Eds. ICSU Press, Paris, p. 27.

Downes, R.W. 1972. Discussion: physiological aspects of sorghum adaptation In *Sorghum in the Seventies*, N.G.P Rao and L.R. House, Eds. Oxford and IBH Publishing Co., New Delhi, pp. 256–274.

Dufour, P., Grivet, L., D'Hont, A., Deu, M., Trouche, G., Glaszmann, J.C., and Haman, P. 1996. Comparative genetic mapping between duplicated segments on maize chromosomes 3 and 8 and homeologous regions in sorghum and sugarcane. *Theor. Appl. Genet.* 92: 1024–1030.

Duncan, R.R., Bramel-Cox, P.J., and Miller, F.R. 1991. Contributions of introduced sorghum germplasm to hybrid development in the U.S. In *Use of Plant Introductions in Cultivar Development*, Part 1, CSSA Special Publication 117, H.L. Shands and L.E. Wiesner, Eds. Crop Science Society of America, Madison, WI, pp. 69–102.

Duncan, R.R. and Dahlberg, J.A. 1993. Cross-reference of PI/IS/SC numbers from U.S. conversion program. *Sorghum Newsl.* 34: 72–80.

Eberhart, S.A. 1972. Techniques and methods for more efficient population improvement in sorghum. In *Sorghum in the Seventies*, N.G.P. Rao and L.R. House, Eds. Oxford and IBH Publishing Co., New Delhi, pp. 195–213.

Eberhart, S.A., Bramel-Cox, P.J., and Prasada Rao, K.E. 1997. Preserving genetic resources. In *The Proceedings of the International Conference on Genetic Improvement of Sorghum and Pearl Millet*, Lubbock, TX, September 22–27, 1996.

Ejeta, G. 1987. Breeding sorghum hybrids for irrigated and rainfed conditions in Sudan. In *Food Grain Production in Semi-Arid Africa, Proceedings of the International Drought Symposium*, Nairobi, Kenya, May 19–23, 1986, pp. 121–130.

Ejeta, G. and Axtell, J.D. 1990. Development of hard endosperm high lysine sorghum lines. In Sorghum Nutritional Quality, Proceedings of an International Conference, G. Ejeta et al., Eds., Purdue University, West Lafayette, IN, February 26–March 1, pp. 126–141.

Ejeta, G. and Bulter, L.G. 1993a. Host-parasite interactions throughout the *Striga* life cycle and their contributions to *Striga* resistance. *Afr. Crops Sci. J.* 1: 75–80.

Ejeta, G. and Bulter, L.G. 1993b. Host-plant resistant to *Striga*. In *International Crop Science I*, D.R. Buxton et al., Eds. International Crop Science Congress, Crop Science Society of America, Madison, WI, pp. 561–569.

Ejeta, G., Bulter, L.G., and Babiker, A.G. 1992. *New Approaches to the Control of Striga, Striga Research at Purdue University*, Research Bulletin 991. Agricultural Experiment Station, Purdue University, West Lafayette, IN, 1992.

Ejeta, G., Bulter, L.G., Hess, D.E., Obilana, T., and Reddy, B.V.S. 1997. Breeding for *Striga* resistance in sorghum. In *Proceedings of the International Conference on Genetic Improvement of Sorghum and Pearl Millet*, Lubbock, TX, September 22–27, 1996, pp. 504–516.

Ejeta, G., Bulter, L.G., Hess, D.E., and Vogler, R.K. 1991. Genetic and breeding strategies for *Striga* resistance in sorghum. In *Proceedings of the Fifth International Symposium on Parasitic Weeds*, Ransom J.K. et al., Eds., Nairobi, Kenya, pp. 539–544.

Ejeta, G., Goldsbrough, P.B., Tunistra, M.R., Grote, E.M., Menkir, A., Ibrahim, Y., Cisse, N., Weerasuriya, Y., Melake-Berhan, A., and Shaner, C.A. 2000. Molecular marker applications in sorghum. In *Application of Molecular Markers in Plant Breeding*, training manual on seminar held at IITA, B.I.G. Haussmann, H.H. Geiger, D.E. Hess, C.T. Hash, and P. Bramel-Cox, Eds., Ibadan, Nigeria, August 16–17, 1999.

Endrizzi, J.E. and Morgan, D.T., Jr. 1955. Chromosomal interchanges and evidence for duplication in haploid *Sorghum vulgare*. *J. Hered.* 46: 201–208.

FAO. 1983. *International Undertaking on Plant Genetic Resources*. FAO of the United Nations, Rome.

Fernandes, M.B., Castro, J.R., de Fernandes, V.L.B., Aquino, B.F., de Alves, J.S., and Gois, F.C. de. 1994. Evaluation of forage sorghum cultivars in salt affected soils of Rio Grande do Norte, Brazil. *Pesquisa Agropecuaria Brasileira* 29: 25–261.

Flores, C.I., Clark, R.B., Pedersen, J.F., and Gourley, L.M. 1991. Leaf mineral element concentrations in sorghum (*Sorghum bicolor*) hybrids and their parents grown at varied aluminium saturations on an ultisol. In *Plant Soil Interactions at Low pH*, R.J. Wright, V.C. Baligar, and R.P. Murrmann, Eds. Kluwer Academic Publishers, Dordrecht, The Netherlands, pp. 1095–1104.

Garber, E.D. 1950. Cytotaxonomy studies in the genus sorghum. *Univ. Calif. Publ. Bot.* 23: 283–362.

Gardner, C.O. 1972. Development of superior populations of sorghum and their role in breeding program. In *Sorghum in the Seventies*, N.G.P. Rao and L.R. House, Eds. Oxford and IBH Publishing Co., New Delhi, pp. 180–183.

Garrity, D.P., Sullivan, C.Y., and Ross, W.M. 1982. Alternative approaches to improving grain sorghum productivity under drought stress. In *Drought Resistance in Crops with Emphasis in Rice*. International Rice Research Institute, Manila, Philippines, pp. 339–356.

Gebrekidan, B. 1982. Utilization of germplasm in sorghum improvement. In *Proceedings of the International Symposium on Sorghum*, L.R. House, L.K. Mughogho, and J.M. Peacock, Eds., ICRISAT, Patancheru, India, November 2–7, 1981, 1: 335–345.

Geng, B.R. 1994. Study on some hereditary properties of sudangrass (*Sorghum sudanense*). *Grass. China* 3: 58–61.

Gomez, M.I., Islam-Faridi, M.N., Zwick, M.S., Czeschin, D.G., Jr., Wing, R.A., Stelly, D.M., and Price, J.H. 1998. Tetraploid nature of *Sorghum bicolor* (L.) Moench. *J. Hered.* 89: 188–190.

Gopal Reddy, G.N., Kameshwara Rao, N., Reddy, B.V.S., and Prasada Rao, K.E. 2002. Geographic distribution of basic and intermediate races in the world collection of sorghum germplasm. *Int. Sorghum Millet Newsl.* 43: 15–17.

Gopalan, C. and Srikantia, S.G. 1960. Leucine and pellagra. *Lancet* 1: 954–957.

Gourley, L.M. 1988. Breeding sorghum for acid soils of the humid tropics. In *Africaland: Land Development and Management of Acid Soils in Africa II*, M. Lathan, Ed. Inter Board for Soils Research and Management (IBSRAM), Bangkok, Thailand, pp. 261–273.

Gourley, L.M. and Lusk, J.W. 1978. Genetic parameters related to sorghum silage quality. *J. Dairy Sci.* 61: 1821–1827.

Gourley, L.M., Watson, C.E., Goggi, A.S., and Axtell, J.D. 1997b. Grain sorghum inbreds tolerant to tropical acid soils released as germplasm. *Int. Sorghum Millets Newsl.* 38: 93–94.

Gourley, L.M., Watson, C.E., Schaffert, R.E., and Payne, W.A. 1997a. Genetic resistance to soil chemical toxicities and deficiencies. In *International Conference on Genetic Improvement of Sorghum and Pearl Millet*, Lubbock, TX, September 22–27, 1996, pp. 461–480.

Gowda, B.T.S., Halaswamy, B.H., Seetharam, A., Virk, D.S., and Witcombe, J.R. 2000. Participatory approach in varietal improvement: a case study in finger millet in India. *Curr. Sci.* 79: 366–368.

Gowda, P.S.B., Magill, C.W., and Frederiksen, R.A. 1994. Tagging downy mildew resistance genes in sorghum. In *Use of Molecular Markers in Sorghum and Pearl Millet Breeding for Developing Countries, Proceedings of an ODA Plant Sciences Research Program Conference*, J.R. Witcombe and R.R. Duncan, Eds., Norwich, U.K., March 29–April 1, 1993.

Gowda, P.S.B., Xu, G.W., Frederiksen, R.A., and Magill, C.W. 1995. DNA markers for downy mildew resistance genes in sorghum. *Genome* 38: 823–826.

Graham, R.D., Senadhira, D., Beebe, S.E., Iglesias, C., and Ortiz-Monasterio, I. 1999. Breeding for micronutrient density in edible portions of staple food crops: conventional approaches. *Field Crops Res.* 60: 57–80.

Gregorio, G.B., Senadhira, D., Htut, H., and Graham, R.D. 2000. Breeding for trace mineral density in rice. *Food Nutr. Bull.* 21: 382–386.

Gu, M.H., Ma, H.T., and Liang, G.H. 1984. Karyotype analysis of seven species in the genus sorghum. *J. Hered.* 75: 196–202.

Haggag, M.E., Shafey, S.A., and Mousa, M.E. 1993. Variation in salinity tolerance among forage sorghum: sudan hybrids. *J. Agric. Sci.* 18: 1597–1608.

Hamaker, B.R. and Axtell, D. 1997. Nutritional quality of sorghum. In *Proceedings of the International Conference on Genetic Enhancement of Sorghum and Pearl Millet*, Lubbock, TX, September 22–27, 1996, pp. 531–538.

Hamilton, C.M., Frary, A., Lewis, C., and Tanksley, S.D. 1996. Stable transfer of intact molecular weight DNA into plant chromosomes. *Proc. Natl. Acad. Sci. U.S.A.* 93: 9975–9979.

Hamker, B.R., Kirleis, A.W., Bulter, L.G., Axtell, J.D., and Mertz, E.T. 1987. Improving the *in vitro* protein digestibility of sorghum with reducing agents. *Proc. Natl. Acad. Sci. U.S.A.* 84: 626–628.

Hancock, J.D. 2000. Value of sorghum and sorghum Co products in diets for livestock. In *Sorghum: Origin, History, Technology and Production*, C.W. Smith and R.A. Frederiksen, Eds. John Wiley & Sons, New York, p. 731.

Hanna, W.W. and Cordona, S.T. 2001. Pennisetums and sorghum in integrated feeding systems in the tropics. In *Tropical Forage Plants: Development and Use*, A.S. Rios and W.D. Pitman, Eds. CRC Press, Boca Raton, FL, pp. 193–200.

Hanna, W.W. and Schertz, K.F. 1970. Inheritance and trisome linkage of seedling characters in *Sorghum bicolor* (L.) Moench. *Crop Sci.* 10: 441–443.

Harlan, J.R. and deWet, J.M.J. 1972. A simplified classification of cultivated sorghum. *Crop Sci.* 12: 172–176.

Harlan, J.R., de Wet, J.M.J., and Price, E.G. 1973. Comparative evolution of cereals. *Evolution* 27: 311–351.

Harris, H.B. 1969. Bird resistance in grain sorghum. *Proc. 24th Ann. Corn Sorghum Res. Conf.* 24: 113–122.

Harris, H.B. and Burns, R.E. 1970. Influence of tannin content on pre-harvest and seed germination in sorghum. *Agron J.* 62: 835–836.

Hart, G.E., Schertz, K.F., Peng, Y., and Syed, N.Y. 2002. Genetic mapping of *Sorghum bicolor* (L.) Moench: QTLs that control variation in tillering and other morphological characters. *Theor. Appl. Genet.* 3: 1232–1242.

Hassanein, A.M. and Azab, A.M. 1990. Test for salt tolerance in grain sorghum. *Bull. Fac. Agric.* 41: 265–276.

Haussmann, B.I.G., Hess, D.E., Reddy, B.V.S., Welz, H.G., and Geiger, H.H. 2000b. Analysis of resistance to *Striga hermonthica* in diallel crosses of sorghum. *Euphytica* 116: 33–40.

Haussmann, B.I.G., Hess, D.E., Reddy, B.V.S., Mukuru, N., Seetharama, N., Kayentao, S.Z., Welz, H.G., and Geiger, H.H. 2000c. QTL for *Striga* resistance in sorghum populations derived from IS 9830 and N 13. In *Breeding for Striga Resistance in Cereals, Proceedings of a Workshop*, B.I.G. Haussmannn, M.L. Koyama, L. Grivet, H.F. Rattunde, D.E. Hess, Eds., Ibadan, Nigeria, August 18–20, 1999.

Haussmann, B.I.G., Hess, D.E., Sissoko, L., Kayentao, M., Reddy, B.V.S., Welz, H.G., and Geiger, H.H. 2001. Diallel analysis of sooty stripe resistance in sorghum. *Euphytica* 122: 99–104.

Haussmann, B.I.G., Hess, D.E., Welz, H.G., and Geiger, H.H. 2000a. Improved methodologies for breeding *Striga* resistant sorghums. *Field Crops Res.* 66: 195–211.

Haussmann, B.I.G., Mahalakshmi, V., Reddy, B.V.S., Seetharama, N., Hash, C.T., and Geiger, H.H. 2002. QTL mapping of stay-green in two sorghum recombinant inbred populations. *Theor. Appl. Genet.* 106: 143–148.

Henzell, B., Jordan, D., Tao, Y., Hardy, A., Franzmann, B., Fletcher, D., MacCosker, T., and Bunker, G. 2002. Grain sorghum breeding for resistance to sorghum midge and drought. In *Plant Breeding for the 11th Millennium, Proceedings on the 12th Australian Plant Breeding Conference*, McComb, J.A., Ed., Perth, Western Australia, September 15–20, pp. 81–86.

Henzell, R.G. 1992. Grain sorghum breeding in Australia: current status and future prospects. In *Proceedings of the Second Australian Sorghum Conference*, M.A. Foale, R.G. Henzell, and P.N. Vance, Eds., Gatton, Australia, February 4–6, pp. 70–82.

Henzell, R.G., Brengman, R.L., Fletcher, D.S., and McCosker, A.N. 1992a. The release of Q140 and Q141, two-grain sorghum B-lines with a high level of nonsenescence and low levels of sorghum midge resistance. In *Proceedings of the Second Australian Sorghum Conference*, M.A. Foale, R.G. Henzell, and P.N. Vance, Eds., Gatton, Australia, February 4–6, pp. 360–366.

Henzell, R.G., Brengman, R.L., Fletcher, D.S., and McCosker, A.N. 1992b. Relationship between yield and nonsenescence (stay-green) in some grain sorghum hybrids grown under terminal drought stress. In *Proceedings of the Second Australian Sorghum Conference*, M.A. Foale, R.G. Henzell, and P.N. Vance, Eds., Gatton, Australia, February 4–6, pp. 355–358.

Henzell, R.G., Dodman, R.L., Done, A.A., Brengman, R.L., and Meyers, R.E. 1984. Lodging, stalk rot and root rot in sorghum in Australia. In *Sorghum Root and Stalk Rots: A Critical Review, Proceedings of the Consultative Group Discussion of Research Needs and Strategies for Control of Sorghum Root and Stalk Rot Diseases*, L.K. Mughogho, Ed., Bellagio, Italy, November 27–December 2, 1983, pp. 225–236.

Henzell, R.G., Hammar, G.I., Borrell, A.K., McIntyre, C.L., and Chapman, S.C. 1997. Research on drought resistance in grain sorghum in Australia. *Int. Sorghum Millets Newsl.* 38: 1–8.

Henzell, R.G. and Hare, B.W. 1996. Sorghum breeding in Australia: public and private endeavors. In *Proceedings of the Third Australian Conference*, Tamworth, NSW Australia, February 20–22, pp. 159–171.

Hess, D.E. and Ejeta, G. 1992. Inheritance of resistance to *Striga* in sorghum genotype SRN 39. *Plant Breed.* 109: 233–241.

Hess, D.E., Ejeta, G., and Butler, L.G. 1992. Selecting sorghum genotypes expressing a quantitative biosynthetic trait that confers resistance to *Striga*. *Photochemistry* 31: 493.

Hess, D.E., Obilana, A.B., and Grard, P. 1996. *Striga* research at ICRISAT. In *Advances in Parasitic Plant Research: Proceedings of the Sixth International Parasitic Weed Symposium*, Cordoba, Spain, M.T. Moreno, J.I. Cnbero, D. Berner, D. Joel, L.J. Musselman, and C. Parker, Eds., pp. 827–834.

Horst, W.J. 1985. Quick screening of cowpea (*Vigna unguiculata*) genotypes for aluminium tolerance in an aluminium: treated acid soil. *Z. Pflanzenernaehr. Bodenkd.* 148: 335–348.

House, L.R. 1985. *A Guide to Sorghum Breeding*, 2nd ed. International Crops Research Institute for the Semi-Arid Tropics, Patancheru, India.

House, L.R., Gomez, M., Murty, D.S., Sun, Y., and Verma, B.N. 2000. Development of some agricultural industries in several African and Asian countries. In *Sorghum: Origin, History, Technology and Production*, C.W. Smith and R.A. Frederiksen, Eds. John Wiley & Sons, New York, pp. 131–190.

House, L.R., Verma, B.N., Ejeta, G., Rana, B.S., Kapran, I., Obilana, A.B., and Reddy, B.V.S. 1996. Developing countries breeding and potential of hybrid sorghum. In *The Proceedings of the International Conference on Genetic Improvement of Sorghum and Pearl Millet*, Lubbock, TX, September 22–27, pp. 84–96.

Hulse, J.H., Laing, E.M., and Pearson, O.E. 1980. *Sorghum and the Millets: Their Composition and Nutritive Value*. Academic Press, New York, pp. 530–592.

Humphreys, M.O. 1994. Variation in carbohydrate and protein content of rye grasses: potential for genetic manipulations, 5–Oct. 1994. In *Breeding for Quality: Proceedings of the 19th Eucarpia Fodder Crops Section Meeting*, D. Reheul and A. Ghesquiere, Eds., Merelbeke, Belgium, pp. 165–172.

IBPGR. 1976. *Proceedings of the Meeting of the Advisory Committee on Sorghum and Millet Germplasm*, October 3–7. ICRISAT, Hyderabad, India.

IBPGR/ICRISAT. 1980. *Sorghum Descriptors*. IBPGR, Rome.

ICRISAT. 1982. *Annual Report, 1981*. ICRISAT, Patancheru, India.

ICRISAT. 1986. *Annual Report, 1985*. ICRISAT, Patancheru, India.

ICRISAT. 1987. *Annual Report, 1986*. ICRISAT, Patancheru, India.

ICRISAT. 1995. Restorers. In *ICRISAT Asia Region Annual Report 1992*. ICRISAT, Patancheru, India.

ICRISAT. 2000. *A Research and Network Strategy for Sustainable Sorghum Production Systems for Latin America*, 9th season report, January–June 2000. ICRISAT, Patancheru, India (limited distribution).

ICRISAT. 2002. *Archival Report*. ICRISAT, Patancheru, India.

ICRISAT, NARS, CIAT. 1997. *A Research and Network Strategy for Sustainable Sorghum Production Systems for Latin America*, Third Report. ICRISAT, Patancheru, India.

Igartua, E. 1995. Choice of selection environment for improving crop yields in saline areas. *TAG* 91: 1016.

Igartua, E. and Gracia, M.P. 1998. Divergent selection for salinity tolerance at the germination-emergence stage in grain sorghum. *Maydica* 43: 161.

Igartua, E., Gracia, M.P., and Lasa, J.M. 1994. Characterization and genetic control of germination emergence responses of grain sorghum to salinity. *Euphytica* 76: 185.

IPGRI. 1993. *Diversity for Development*. IPGRI, Rome.

Jayanthi, P.D.K. 1997. Genetics of Shoot Fly Resistance in Sorghum Hybrids of Cytoplasmic Male Sterile Lines. Ph.D. thesis, Acharya N.G. Ranga Agricultural University, Rajendranagar, Hyderabad, India.

Jordan, W.R., Dugas, W.A., and Stenhouse, P.J. 1983. Strategies for crop improvement for drought prone regions. *Agric. Water Manage.* 7: 281–299.

Jordan, W.R. and Monk, R.L. 1980. Enhancement of drought resistance of sorghum: progress and limitations. In *Proceedings of the 35th Annual Corn and Sorghum Research Conference*. Annual Seed Trade Association, Chicago, pp. 185–204.

Jordan, W.R. and Sullivan, C.Y. 1982. Reaction and resistance of grain sorghum to heat and drought. In *Sorghum in the Eighties*. ICRISAT, Hyderabad, India, pp. 131–142.

Jorgenson, L.R. 1931. Brown-midrib in maize and its linkage relations. *J. Am. Soc. Agron.* 23: 549–557.

Jotwani, M.G. 1976. Host plant resistance with special reference to sorghum. *Proc. Natl. Acad. Sci. U.S.A.* 46: 42.

Kadam, D.E., Patil, F.B., Bhor, T.J., and Harer, P.N. 2001. Genetic diversity studies in sweet sorghum. *J. Maharshtra Agric. Univ.* 26: 140–143.

Kalton, R.R. 1988. Overview of forage sorghums. *Proc. Ann. Corn Sorghum Res. Conf.* 43: 1–12.

Kebede, H., Subudhi, P.K., Rosenow, D.T., and Nguyen, H.T. 2001. Quantitative trait influencing drought tolerance in grain sorghum (*Sorghum bicolor* L. Moench). *Theor. Appl. Genet.* 103: 266–276.

Kimber, C.T. 2000. Origins of domesticated sorghum and its early diffusion to India and China. In *Sorghum: Origin, History, Technology and Production*, C.W. Smith and R.A. Frederiksen, Eds. John Wiley & Sons, New York, 2000, pp. 3–98.

Kleih, U., Bala Ravi, S., and Rao, B.D., 2000. Industrial utilization of sorghum in India. In *Sorghum Utilization and the Livelihoods of the Poor in India*, A.J. Hall and B. Yoganand, Eds. ICRISAT, Patancheru, India, pp. 73–78.

Klein, R.R., Rodriguez-Herrera, R., Schlueter, J.A., Klein, P.E., Yu, Z.H., and Rooney, W.L. 2001. Identification of genomic regions that affect grain mold incidence and other traits of agronomic importance in sorghum. *Theor. Appl Genet.* 102: 307–319.

Kling, J., Fajemisin, J.M., Badu-Apraku, B., Diallo, A., Menkir, A., and Melake-Berhan, A. 2000. *Striga* resistance breeding in maize. In *Breeding for Striga Resistance in Cereals, Proceedings of a Workshop*, B.I.G. Haussmann, D.E. Hess, M.L. Koyama, L. Grivet, H.F.M. Rattunde, and H.H. Geiger, Eds., Ibadan, Nigeria, August 18–20, 1999, pp. 103–118.

Klopfenstein, C.F. and Hoseney, R.C. 1995. Nutritional Properties of sorghum and the millets. In *Sorghum and Millets: Chemistry and Technology*, D.A.V. Dendy, Ed. American Association of Cereal Chemists, St. Paul, MN, pp. 125–168.

Kogan, M. and Ortman, E.E. 1978. Antixenosis: a new term proposed to replace painters 'non-preference' modality of resistance. *Bull. Entomol. Soc. Am.* 24: 175–176.

Kramer, P.J. 1980. Drought stress and the origin of adaptations. In *Adaptation of Plants to Water and High Temperature Stress*, N.C. Turner and P.J. Karmer, Eds. Wiley Interscience, New York, pp. 7–20.

Krieg, D.R. 1993. Stress tolerance mechanisms in above ground organs. In *Proceedings of a Workshop on Adaptation of Plants to Soil Stresses*, Lincoln, NE, August 1–4, 1993, pp. 65–79.

Lazarides, M., Hacker, J.B., and Andrew, M.H. 1991. Taxonomy, cytology and ecology of indigenous Australian sorghums (*Sorghum* Moench: Andropogoneae: poaceae). *Aust. Syst. Bot.* 4: 591–635.

Levitt, J. 1992. *Responses of Plants to Environmental Stresses*. Academic Press, New York.

Lin, Y.R., Schertz, K.F., and Paterson, A.H. 1995. Comparative analysis of QTLs affecting plant height and maturity across the Poaceae, in reference to an interspecific sorghum population. *Genetics* 140: 391–411.

Lu, Q. 1994. RFLP techniques and sorghum breeding. In *Use of Molecular Markers in Sorghum and Pearl Millet Breeding for Developing Countries, Proceedings of an ODA Plant Sciences Research Programme Conference*, J.R. Witcombe and R.R. Duncan, Eds., Norwich, U.K., March 29–April 1, 1993, pp. 15–16.

Ludlow, M.M. 1993. Physiological mechanisms of drought resistance. In *Proceedings of Symposium on Application and Prospects of Biotechnology*, Mabry, T.J., Nguyen, H.T., and Dixon, R.A., Eds., Lubbock, TX, November 5–7, 1992, pp. 11–34.

Luo, Y.W., Yen, X.C., Zhang, G.Y., and Liang, G.H. 1992. Agronomic traits and chromosomal behavior of autotetraploid sorghums. *Plant Breed.* 109: 46–53.

Maas, E.V. and Hoffman, G.S. 1977. Crop salt tolerance: current assessment. *J. Irrig. Drainage Div. ASCE* 103: 115–134.

Maciel, G.A., de Andrade Lima, M.M., Duncan, R.R., de Franca, J.G.E., and Tabosa, J.N. 1994a. Combining ability of aluminium tolerance in grain sorghum lines. *Int. Sorghum Millets Newsl.* 35: 73–74.

Maciel, G.A., de Andrade Lima, M.M., Santos, J.P.O., Tabosa, J.N., and de Franca, J.G.E. 1994b. Screening new grain sorghum introductions for aluminium tolerance. *Int. Sorghum Millets Newsl.* 35: 117–118.

Maciel, G.A., Tavares, J.A., de Andrade Lima, M.M., de Tabosa, J.N., and Neto, C.A. 1992. Preliminary evaluation of grain sorghum for aluminium tolerance in northeastern Brazil. *Sorghum Newsl.* 33: 57.

Maclean, W.C., Lopez de Romana, G., and Graham, G.G. 1983. The effects of decortication and extrusion on the digestibility of sorghum by preschool children. *J. Nutr.* 113: 2171.

Magill, C.W., Boora, K., Sunitha Kumari, R., Osorio, J., Oh, B.J., Gowdsa, B., Cui, Y., and Frederiksen, R. 1997. Tagging sorghum genes for disease resistance: expectations and reality. In *International Conference on Genetic Improvement of Sorghum and Pearl Millet*, Rosenow et al., Eds. Lubbock, TX, September 22–27, 1996, pp. 316–325.

Magnavaca, R., Lakins, B.A., Schaffert, R.C., and Lopes, M.A. 1993. Improving protein quality of maize and sorghum. In *International Crop Science I*, D.R. Buxton, R.A. Shibles, R.A. Forsberg, B.L. Blad, K.H. Asay, G.H. Paulsen, and R.F. Wilson, Eds. Crop Science Society of America, Madison, WI, pp. 649–653.

Mahalakshmi, V. and Bidinger, F.R. 2002. Evaluation of stay-green sorghum germplasm lines at ICRISAT. *Crop Sci.* 42: 965–974.

Mahalanobi, P.C. 1936. On the generalised distance in statistics. *Proc. Natl. Inst. Sci. India* B2: 49–55.

Maiti, R.K., de la Rosa, M. and Alicia, L. 1994. Evaluation of several sorghum genotypes for salinity tolerance. *Int. Sorghum Millets Newsl.* 35: 121.

Major, D.J. 1980. Photoperiod response characteristics controlling flowering of nine crop species. *Can. J. Plant Sci.* 60: 777–784.

Mali, P.C. and Gupta, Y.P. 1974. Chemical composition and protein quality of improved Indian varieties of *Sorghum Vulgare* pers. *Indian J. Nutr. Diet.* 11: 289–295.

Mallick, A.S., Gupta, M.P., and Pandey, A.K. 1988. Combining ability of some quality traits in grain sorghum [*Sorghum bicolor* (L) Moench]. *Indian J. Genet.* 48: 63–68.

Mann, J.A., Kimber, C.T., and Miller, F.R. 1983. The origin and early cultivation of sorghums in Africa. *Tex. Agric. Exp. Stn. Bull.* 1454.

Marschner, H. 1991. Mechanisms of adaptation of plants to acid soils. *Plant Soil* 134: 1–20.

Maunder, A.B. 1972. Objectives and approaches to grain and forage sorghum improvement in the Americas. In *Sorghum in the Seventies*, N.G.P. Rao and L.R. House, Eds. Oxford and IBH Publishing Co., New Delhi, pp. 60–100.

Maurya, D.M., Bottrall, D.M., and Farrington, J. 1988. Improved livelihood, genetic diversity and farmers' participation: strategy for rice breeding in rain fed areas of India. *Exp. Agric.* 24: 311–320.

Mbwaga, A.M. and Obilana, A.J. 1994. Distribution and host specificity of *Striga asiatica* and *Striga hermonthica* in cereals in Tanzania. *Int. J. Pest Manage.* 39: 449–451.

McCollough, R.L., Drake, C.L., Roth, G.M., Brent, B.E., Riley, R.G., and Schalles, R.R. 1972. Several reports on the nutritive value of hybrid sorghum grains. *Kan. Agric. Exp. Stn. Bull.* 577: 15–17.

McMillian, W.W., Wiseman, B.R., Burns, R.E., Harris, H.B., and Greene, G.L. 1972. Bird resistance in diverse germplasm of sorghum. *Agron J.* 64: 821–822.

Menkir, A., Goldsbrough, P., and Ejeta, G. 1997. RAPD based assessment of genetic diversity in cultivated races of sorghum. *Crop Sci.* 37: 564–569.

Miller, F., Muller, N., Monk, R., Murty, D.S., and Obilana, A.B. 1997. Breeding photoperiod insensitive sorghums for adaptation and yield. In *International Conference on Genetic Improvement of Sorghum and Pearl Millet*, Lubbock, TX, September 22–27, 1996, pp. 59–65.

Miller, F.R. 1980. The breeding of sorghum. *Tex. Agric. Exp. Stn.* 1451: 128–136.

Miller, F.R. 1982. Genetic and environmental response characteristics of sorghum. In *Sorghum in the Eighties*. ICRISAT, Patancheru, India, pp. 393–402.

Miller, F.R. 1992. Improvements in sorghum during the last 25 years. In *Proceedings of the Second Australian Sorghum Conference*, M.A. Foale, R.G. Henzell, and P.N. Vance, Eds., Gatton, Australia, February 4–6, pp.

Miller, F.R., Barnes, D.K., and Cruzado, H.J. 1968. Effect of tropical photoperiods on the growth of *Sorghum bicolor* (L.) Moench, when grown in 12 monthly plantings. *Crop Sci.* 8: 499–502.

Miller, F.R. and Kebede, Y. 1984. Genetic contributions to yield gains in sorghum, 1950 to 1980. In *Genetic Contributions to Yield Gains of Five Major Crop Plants*, CSSA Special Publication 7, W.R. Fehr, Ed. CSSA, ASA, Madison, WI, pp. 1–14.

Mohan, D.P. 1975. Chemically Induced High Lysine Mutant in *Sorghum bicolor* (L.) Moench. Ph.D. thesis, Purdue University, Lafayette, IN.

Mohanty, B.D., Maiti, S., and Ghosh, P.D. 1986. Establishment of karyotype in *Sorghum bicolor* through somatic metaphase and pachytene analysis. In *Perspectives in Cytology and Genetics*, Proceedings of the Fifth All India Congress of Cytology and Genetics, Vol. 5, G.K. Manna and U. Sinha, Eds., Bhubaneshwar, India, October 7–10, 1984, pp. 559–563.

Monasterio, I. and Graham, R.D. 2000. Breeding for trace minerals in wheat. *Food Nutr. Bull.* 21: 392.

Montemurro, F., Rigoldi, M.P., Sunseri, F., and Vanadia, S. 1994. Early screening methodologies for selecting salt stress tolerant sweet sorghum [*Sorghum bicolor* (L.) Moench]. *Rivista Agron.* 28: 179–183.

Mukuru, S.Z. 1992. Breeding for grain mold resistance. In *Sorghum and Millet Diseases, A Second World Review*, W.A.J. de Milliano, R.A. Frederiksen, and G.D. Bengston, Eds. ICRISAT, Patancheru, India, pp. 273–285.

Murthy, U.R. 1986. Apomixis: achievements, problems and future prospects. In *Advanced Methods in Plant Breeding*. Oxford and IBH, New Delhi.

Murthy, U.R., Schertz, K.F., and Bashaw, E.C. 1979. Apomictic and sexual reproduction in sorghum. *Indian J. Genet. Plant Breed.* 39: 271–278.

Murty, B.R. and Arunachalam, V. 1967. Factor analysis of genetic diversity in the genus sorghum. *Indian J. Genet.* 27: 123–135.

Murty, B.R., Arunachalam, V., and Saxena, M.B.L. 1967. Cataloguing and classifying a world collection of genetic stocks of sorghum. *Indian J. Genet.* 27A: 1–312.

Murty, D.S. and Kumar, K.A. 1995. Traditional uses of sorghum and millets. In *Sorghum and Millets: Chemistry and Technology*, D.A.V. Dendy, Ed. American Association of Cereal Chemists, St. Paul, MN, pp. 185–221.

Murty, U.R. and Rao, N.G.P. 1997. Sorghum. In *Genetics, Cytogenetics and Breeding of Crop Plants*, Vol. 2, *Cereal and Commercial Crops*, P.N. Bahl, P.M. Salimath, and A.K. Mandal, Eds. Oxford & IBH Publishing Co. Pvt. Ltd., New Delhi, pp. 197–239.

Murty, U.R. 1991. National Programme on Sorghum Research in India. Paper presented at the consultative meeting to consider establishment of Regional Sorghum Research Network for Asia, Patancheru, India, September 16–19.

Murty, U.R. 1992. *ICAR_ICRISAT Collaborative Research Projects: Sorghum Progress Report*. National Research Center on Sorghum 27.

Murty, V.R. and Rao, N.G.P. 1977. Sorghum. In *Genetics, Cytogenetic and Breeding of Crop Plants*, Vol. 2, P.N. Bahl, P.M. Salimath, and A.K. Mandal, Eds. Oxford and IBH Publishing Co. Pvt. Ltd, New Delhi, pp. 197–239.

Myer, R.O. and Gorbet, D.W. 1985. Waxy vs. normal grain sorghums with varying tannin contents in diets for young pigs. *Anim. Feed Sci. Technol. Amsterdam* 12(3): 179–186.

Nanda, G.S. and Rao, N.G.P. 1974. Gene action for vitamins in grain sorghum. *Crop Improve.* 1: 53–60.

Nanda, G.S. and Rao, N.G.P. 1975a. Gene action for content of amino acid in grain sorghum. *Indian J. Genet.* 35: 395–398.

Nanda, G.S. and Rao, N.G.P. 1975b. Genetic analysis of some exotic ∞ Indian crosses in sorghum. IX. Nutritional quality and its association with grain yield. *Indian J. Genet.* 35: 131–135.

Narkhede, B.N., Akade, J.H., and Awari, V.R. 2001. Genetic diversity in *rabi* sorghum local types [*Sorghum bicolor* (L.) Monech]. *J. Maharashtra Agric. Univ.* 25: 245–248.

Nayeem, K.A. and Bapat, D.R. 1984. Combining ability in grain sorghum. *Indian J. Genet.* 44: 353–357.

Nguyen, H.T., Xu, W., Rosenow, D.T., Mullet, J.F., and McIntyre, L. 1997. Use of biotechnology in sorghum drought resistance breeding conference on genetic improvement of sorghum and pearl millet. In *International Conference on Genetic Improvement of Sorghum and Pearl Millet*, Lubbock, TX, September 22–27, 1996, pp. 412–424.

Noland, P.R., Campbell, D.R., Sharp, R.N., and Johnson, Z.B. 1977. Influence of pericarp and endosperm color and type on digestibility of grain sorghum by pigs. *Anim. Feed Sci. Technol.* 2: 219–224.

Nwanze, K.F., Seetharama, N., Sharma, H.C., and Stenhouse, J.W. 1995. Biotechnology in pest management: improving resistance to sorghum to insect pests. *Afr. Crop Sci. J.* 3: 209–215.

Obilana, A.B. 1983. *Striga* studies and control in Nigeria. In *Proceedings of the Second International Workshop on Striga*, K.V. Ramaiah and M.J. Vasudeva Rao, Eds., Quagadougou, Upper Volta, Burkina Faso, October 5–8, 1981, pp. 87–98.

Obilana, A.B., de Milliano, W.A.J., and Mbwaga, A.M. 1991. *Striga* research in sorghum and millets in southern Africa: status and host plant resistance. In *Proceedings of the Fifth International Symposium of Parasitic Weeds*, J.K. Ransom, L.J. Musselman, A.D. Worsham, and C. Parker, Eds., Nairobi, Kenya, June 24–30, pp. 435–441.

Oh, B.J., Gowda, P.S.B., Xu, G.W., Frederiksen, R.A., and Magill, C.W. 1993. Tagging *Acremonium* wilt, downy mildew and head smut resistance genes in sorghum using RFLP and RAPD markers. *Sorghum Newsl.* 34: 34.

Oria, M.P., Hamaker, B.R., and Axtell, J.D. 2000. A highly digestible sorghum cultivar exhibits a unique folded structure of endosperm protein bodies. *Proc. Natl. Acad. Sci. U.S.A.* 97: 5065–5070.

Owen, F.G. and Moline, W.J. 1970. Sorghum for forage. In *Sorghum Production and Utilization*, J.S. Wall and W.M. Ross, Eds. AVI Publication Co., Westport, CT, pp. 382–415.

Parker, C., Hitchock, A.M., and Ramaiah, K.V. 1977. The germination of *Striga* species by crop root exudates, techniques for selecting resistant crop cultivars. In *Proceedings of the 6th Asian-Pacific Weed Science Society Conference*, Indiana, Indonesia, 1: 67.

Paterson, A.H. 1994. Status of genome mapping in sorghum and prospects for marker-associated selection in sorghum improvement. *Int. Sorghum Millet Newsl.* 35: 89–91.

Paterson, A.H., Schertz, K.F., Lin, Y., and Li, Z. 1998. Case history in plant domestication: sorghum an example of cereal evolution. In *Molecular Dissection of Complex Traits*, A.H. Paterson, Ed. CRC Press, Boca Raton, FL.

Peacock, J.M. and Sivakumar, M.V.K. 1987. An environmental physiologist's approach to screening for drought resistance to sub-Saharan Africa. In *Food Grain Production in Semi-Arid Africa, Proceedings of International Drought Symposium*, Nairobi, Kenya, May 19–23, 1986, pp. 101–120.

Pearson, K. 1901. On the lines and planes of closest fit to a system of points in a space. *Philos.Mag.* 2: 557–572.

Pederson, J.F. 1997. Breeding sorghum and pearl millet for forage and fuel. In *International Conference on Genetic Improvement of Sorghum and Pearl Millet*, Lubbock, TX, September 22–27, 1996, pp. 539–548.

Pedersen, J.F. and Fritz, J.O. 2000. Forages and fodder. In *Sorghum: Origin, History, Technology and Production*, W.C. Smith and R.A. Frederiksen, Eds. John Wiley & Sons, New York, pp. 797–810.

Peng, J., Lill, H., Li, J., and Tan, Z. 1994. Screening Chinese sorghum cultivars for tolerance to salinity. *Sorghum Millets Newsl.* 35: 124.

Peng, Y., Schertz, K.F., Cartinhour, S., and Hart, G.E. 1999. Comparative genome mapping of *Sorghum bicolor* (L.) Moench using an RFLP map constructed in a population of recombinant inbred lines. *Plant Breed.* 118: 225–235.

Pereira, M.G., Ahnert, D., Lee, M., and Klier, K. 1995. Genetic mapping of quantitative trait loci for panicle characteristics and seed weight in sorghum. *Rev. Bras. Genet.* 18: 249–257.

Pereira, M.G., Lee, M., Bramel-Cox, P., Wordman, W., Docbley, J., and Whitkus, R. 1994. Construction of an RFLP map in sorghum and comparative mapping in maize. *Genome* 37: 236–243.

Peterson, G.C., Reddy, B.V.S., Youm, O., Teetes, G.I., and Lambright, L. 1997. Breeding for resistance for foliar- and stem-feeding insects of sorghum and pearl millet. In *International Conference on Genetic Improvement of Sorghum and Pearl Millet*, Lubbock, TX, September 22–27, 1996, pp. 281–302.

Platt, B.S. 1962. *Tables of Representative Values of Food Commonly Used in Tropical Countries*, Medical Research Council Special Report Series 302 (revised edition of SRS 253). H.M. Stationery Office, London.

Porter, K.S., Axtell, J.D., Lechtenberg, V.L., and Colenbrander, V.F. 1978. Phenotype, fiber composition and *in vitro* dry matter disappearance of chemically induced brown-midrib (bmr) mutants of sorghum. *Crop Sci.* 18: 205–208.

Prasada Rao, K.E., Gopal Reddy, V., and Stenhouse, J.W. 1995. Sorghum genetic resources at ICRISAT Asia Center. *Int. Sorghum Millets Newsl.* 36: 15–19.

Prasada Rao, K.E., Mengesha, M.H., and Reddy, V.G. 1989. International use of sorghum germplasm collection. In *The Use of Plant Genetic Resources*, A.H.D. Brown, O.H. Frankel, D.R. Marshall, and J.T. Williams, Eds. Cambridge University Press, Cambridge, U.K., pp. 150–167.

Prasada Rao, K.E. and Ramanatha Rao, V. 1995. Use of characterization data in developing a core collection of sorghum. In *Core Collection of Plant Genetic Resources*, T. Hodgkin, H.D. Brown, J.L. Hinthum, and E.A.V Morales, Eds. John Wiley & Sons, Chichester, U.K., pp. 109–111.

Pushpamma, P., Ratnakumari, H., and Geervani, P. 1970. Nutritional quality of sorghum and legume based food mixture for infants and preschool children II. *Nutr. Rep. Int.* 19: 643–649.

Qingshan, L. and Dahlberg, J.A. 2001. Chinese sorghum genetic resources. *Econ. Bot.* 55: 401–425.

Quinby, J.R. 1973. The genetic control of flowering and growth in sorghum. In *Advances in Agronomy*, N. Brady, Ed. Academic Press, New York, pp. 125–162.

Quinby, J.R. 1974. *Sorghum Improvement and the Genetics of Growth*. Texas A&M University Press College Station, Lubbock, Texas.

Quinby, J.R., Hesketh, J.D., and Voigt, R.L. 1973. Influence of temperature and photoperiod on floral initiation and leaf number in sorghum. *Crop Sci.* 13: 243–246.

Quinby, J.R. and Karper, R.E. 1946. Heterosis in sorghum resulting from the heterozygous condition of a single gene that affects duration of growth. *Am. J. Bot.* 33: 716–721.

Quinby, J.R. and Karper, R.E. 1954. Inheritance of height in sorghum. *Agron J.* 46: 212–216.

Rai, M. 2002. Genetic resources and intellectual property rights in agriculture perspective. *Indian J. Pulses Res.* 15: 1–18.

Rai, K.N., Murty, D.S., Andrews, D.J., and Bramel-Cox, P.J. 1999. Genetic enhancement of pearl millet and sorghum for the semi-arid tropics of Asia and Africa. *Genome* 42: 617–628.

Ramaiah, K.V. 1987. Breeding cereal grains for resistance to witch weed. In *Parasitic Weeds in Agriculture*, Vol. 1, L.J. Musselman, Ed. CRC Press, Boca Raton, FL, pp. 227–242.

Rami, J.F., Dufour, P., Trouche, G., Fliedel, G., Mestress, C., Davrieux, F., Blanchard, P., and Hamon, P. 1998. Quantitative trait loci for grain quality, productivity, morphological and agronomical traits in sorghum (*Sorghum bicolor* L. Moench). *Theor. Appl. Genet.* 97: 605–616.

Rana, B.S., Jotwani, M.G., and Rao, N.G.P. 1980. Inheritance of host plant resistance to sorghum shoot fly. In *Insect Science and Its Application*, IC1PE, Nairobi, Kenya, May 4–8.

Rana, B.S., Kaul, S.L., Chari, A., Reddy, B.V.S., WitCombe, J.R., and Virk, D.S. 1998. Farmer's participatory varietal selection for improving *rabi* sorghum productivity in India. In *Participatory Plant Improvement: Proceedings of the MSSRF-ICRISAT Workshop 1998*, Chennai, India, pp. 31–37.

Rana, B.S. and Murty, B. R. 1971. Genetic analysis of resistance to stem borer in sorghum. *Indian J. Genet.* 31: 521–529.

Rana, B.S. and Murty, B.R. 1975. Heterosis and components of genetic variation for protein and lysine content in some grain sorghums. *Theor. Appl. Genet.* 45: 225–230.

Rao, N.G.P. 1972. Sorghum breeding in India: recent developments. In *Sorghum in the Seventies*, N.G.P. Rao and L.R. House, Eds. Oxford and IBH Publishing Co., New Delhi, pp. 101–142.

Rao, V.J.M., Deosthale, Y.G., Rana, B.S., Vidyasagar Rao, K., and Rao, N.G.P. 1983a. Genetic analysis of some exotic × Indian crosses in sorghum. XXXI. Nutritional quality in grain sorghum: variability for protein, lysine and leucine. *Indian J. Genet.* 43: 380–382.

Rao, V.J.M., Deosthale, Y.G., Rana, B.S., Vidyasagar Rao, K., and Rao, N.G.P. 1983b. Genetic analysis of some exotic × Indian crosses in grain sorghum. XXXII. Nutritional quality in grain sorghum: behaviour of crosses involving between high lysine sorghum and agronomically superior types. *Indian J. Genet.* 43: 395–403.

Rao, V.J.M., Mehta, S.L., and Rao, N.G.P. 1984a. Breeding sorghums for high protein content and quality. *Indian J. Genet.* 44: 305–313.

Rao, N.G.P. and Murthy, U.R. 1972. Further studies on obligate apomixis in grain sorghum, *Sorghum bicolor* (L.) Moench. *Indian J. Genet.* 32: 379–383.

Rao, N.G.P. and Rana, B.S. 1982. Selection in temperate and tropical crosses of sorghum. In *Sorghum in the Eighties, Proceedings of International Symposium on Sorghum*, 2–7 Nov. 1981, ICRISAT, Patancheru, India, pp. 403–420.

Rao, V.J.M., Rana, B.S., Reddy, B.B., and Rao, N.G.P. 1984b. Genetic upgrading of protein quality in sorghum. In *Nutritional and Processing Quality of Sorghum*, D.K. Salunkhe, J.K. Chavan, and S.J. Jadhav, Eds. Oxford and IBH Publishing Co., New Delhi, pp. 67–90.

Ratnadass, A., Chantereau, J., Coulibaly, M.F., and Cilas, C. 2002. Inheritance of resistance to the panicle-feeding bug (*Eurystylus oldi*) and the sorghum midge (*Stenodiplosis sorghicola*) in sorghum. *Euphytica* 123: 131–138.

Rattunde, E.W. 2002. Participatory breeding with sorghum in Mali: statistical and analytical aspects. In *Quantitative Analysis of Data from Participatory Methods in Plant Breeding*, M.R. Bellon and J. Reeves, Eds. CIMMYT, Mexico, D.F., pp. 140–141.

Rattunde, H.F.W., Obilana, A.B., Haussmann, B.I.G., Reddy, B.V.S., and Hess, D.E. 2000. In *Breeding for Striga Resistance in Cereals, Proceedings of a Workshop*, B.I.G. Haussmann, D.E. Hess, M.L. Koyama, L. Grivet, H.F.W. Rattunde, and H.H. Geiger, Eds., Ibadan, Nigeria, August 18–20, p. 85.

Raynal-Roques, A. 1987. The genus *Striga* (Scrophulariaceae) in western and central Africa: a survey. In *Parasitic Flowering Plants*, H. Weber and W. Forstreuter, Eds. Marburg, Germany, p. 675.

Reddy, B.V.S. 1985. Relatorio final de consultoria. IICA/EMBRAPA/IPA, p. 83.

Reddy, B.V.S. 1986. Genetic improvement for drought resistance in sorghum: a plant breeder's viewpoint. In *Genetic Improvement of Drought Resistance, Proceedings of a Discussion Series of the Drought Research Seminar Forums*, ICRISAT, Patancheru, India, pp. 28–32.

Reddy, B.V.S. 1993. Varietal improvement: breeding *Striga* resistant seed parents. In *Cereal Program, ICRISAT, Annual Report 1992*. ICRISAT, Patancheru, India.

Reddy, B.V.S., Bandyopadhyay, R., Ramaiah, B., and Ortiz, R. 2000. Breeding grain mold resistant sorghum cultivars. In *Technical and Institutional Options for Sorghum Grain Mold Management: Proceedings of an International Consultation*, A. Chandrashekar, R. Bandyopadhyay, and A.J. Hall, Eds., May 18–19. ICRISAT, Patancheru, India, pp. 195–224.

Reddy, B.V.S., Hall, A.J., and Rai, K.N. 2001. The long road to partnership: private support of public research on sorghum and pearl millet. In *Sharing Perspectives on Public-Private Sector Interaction: Proceedings of a Workshop*, A.J. Hall, B. Yoganand, V. Rasheed Sulaiman, and N.G. Clark, Eds., April 10. ICRISAT, Patancheru, India, pp. 27–34.

Reddy, B.V.S., Hash, C.T., Stenhouse, J.W., Nigam, S.N., Singh, L., and Van Rheenen, H.A. 1994. *Crop Improvement for Livestock Crop Residue Feed at ICRISAT Asia Center*. pp. 85–92.

Reddy, B.V.S., Mughogho, L.K., and Jambunathan, R. 1992. Breeding grain mold resistance seed parents and hybrids. In *Cereals Program, Annual Report, 1991*. ICRISAT, Patancheru, pp. 28–29.

Reddy, B.V.S. and Prasada Rao, K.E. 1993. *Varietal Improvement: Genetic Diversification in Cereals Program*, ICRISAT Annual Report 1992. ICRISAT, Patancheru, pp. 48–51.

Reddy, B.V.S., Prakasha Rao, P., Deb, U.K., Stenhouse, J.W., Ramaiah, B., and Ortiz., R. 2004a. Global sorghum genetic enhancement processes at ICRISAT. In Sorghum Genetic Enhancement: Research Process, Dissemination and Impacts, M.C.S. Bantilan, U.K. Deb, C.L.L. Gowda, B.V.S. Reddy, A.B. Obilana, and R.E. Evenson, Eds. ICRISAT, Patancheru, pp. 65–102.

Reddy, B.V.S. and Rangel, A.F. 2000. Genotype (G) × environment (E) interactions in sorghum in acid-soils of the oriental Llanos of Colombia. In *A Research and Network Strategy for Sustainable Sorghum and Pearl Millet Production Systems for Latin America: Proceedings of the Workshop*, Villavicencio, Meta, Colombia, November 24–26, 1998. ICRISAT, Patancheru, pp. 46–51.

Reddy, B.V.S., Rangel, A.F., and Iglesaias, C. 1998. Latin American Sorghum Project: some results. In *Proceedings of the First International Symposium on Sorghum*, Rio Bravo, Tamaulipas, Mexico, May 27–30, pp. 70–79.

Reddy, B.V.S., Rangel, A.F., Ramaiah, B., and Ortiz, R. 2004b. A research and network strategy for sustainable sorghum production systems for Latin America. In *Sorghum Genetic Enhancement: Research Process, Dissemination and Impacts*, M.C.S. Bantilan, U.K. Deb, C.L.L. Gowda, B.V.S. Reddy, A.B. Obilana, and R.E. Evenson, Eds. ICRISAT, Patancheru, pp. 139–148.

Reddy, B.V.S. and Seetharama, N. 2001. Sorghum improvement: a case study for integrating traditional breeding and transgenic research methods. In *Sorghum: Tissue Culture, Transformation and Genetic Engineering*, N. Seetharama and I.D. Godwin, Eds. ICRISAT and Oxford Publishers, India.

Reddy, B.V.S. and Stenhouse, J.W. 1994a. Sorghum improvement for semi-arid tropics region: past current and future research thrusts in Asia. *PKV Res. J.* 18: 155–169.

Reddy, B.V.S. and Stenhouse, J.W. 1994b. Improving Postrainy Season Sorghum, a Case for Landrace Hybrids Approach. Invited paper presented at All India Coordinated Sorghum Improvement Project (AICSIP), Pantnagar, India, April 18–20.

Reddy, C.S., Schertz, K.F., and Bashaw, E.C. 1980. Apomictic frequency in sorghum R 473. *Euphytica* 29: 223–226.

Reddy, N.P.E. and Jacobs, M. 2000. Polymorphism among kakirins and esterases in normal and lysine-rich cultivars of *Sorghum. Indian J. Genet.* 60: 159–170.

Reich, V.H. and Atkins, R.E. 1971. Variation and interrelationship of protein, oil content and seed weight in grain sorghum. *Iowa State J. Sci.* 46: 13–22.

Richards, R.A. and Dennett, C.W. 1980. Variation in salt concentration in a wheat field. *Soil Water* 44: 8–9.

Richards, R.A., Dennett, C.W., Qualset, C.O., Epstein, E., Norlyn, J.D., and Winslow, M.D. 1987. Variation in yield of grain and biomass in wheat, barley and triticale in a salt-affected field. *Field Crops Res.* 15: 277–287.

Riches, C.R., De Milliano, W.A.J., Obilana, A.T., and House, L.R. 1987. Witch weeds (*Striga* spp.) of sorghum and pearl millet. In *Third Regional SADCC/ICRISAT Sorghum and Millet Improvement Project Workshop*, Lusaka, Zambia, September 20–25, 1986.

Richter, C., Heiligtag, B., Gertling, M., and Abdullahzadeh, A. 1995. Salt tolerance of different varieties of *Sorghum bicolor* and *Vicia faba*. *Tropenlandwirt* 96: 141–152.

Rodriguez-Herrera, R., Waniska, R.D., and Rooney, W.L. 1999. Antifungal proteins and grain mold resistance in sorghum with a non-pigmented testa. *J. Agric. Food Chem.* 47: 4802–4806.

Rooney, W.L. 2000. Genetics and cytogenetics. In *Sorghum, Origin, History, Technology and Production*, Wiley Series in Crop Science, C.W. Smith and R.A. Frederiksen, Eds. John Wiley & Sons, New York, pp. 261–307.

Rooney, L.W., Hamker, B.R., and Botorou, Q. 2003. Grains in West Africa: Processing and Product Marketing through Value Added Supply Chain Management. Concept note paper presented in the expert meeting on alternative uses of sorghum and pearl millet in Asia, Patancheru, India, July 1–4.

Rooney, W.L. and Klein, R.R. 2000. Potential of marker-assisted selection for improving grain mold resistance in sorghum. In *Technical and Institutional Options for Sorghum Grain Mold Management: Proceedings of an International Consultation*, A. Chandrashekar, R. Bandyopadhyay, and A.J. Hall, Eds., Patancheru, India, May 18–19, pp. 183–194.

Rooney, L.W. and Plugfelder, R.L. 1986. Factors affecting starch digestibility with special emphasis on sorghum and corn. *J. Anim. Sci.* 63: 1607.

Rooney, W.L. and Smith, C.W. 2000. Techniques for developing new cultivars. In *Sorghum, Origin, History, Technology and Production*, Wiley Series in Crop Science, C.W. Smith and R.A. Frederiksen, Eds. John Wiley & Sons, New York, pp. 329–347.

Rosenow, D.T. 1992. Using germplasm from the world collection in breeding for disease resistance. In *Sorghum and Millet Diseases: A Second World Review*, W.A.J. de Milliano, R.A. Frederiksen, and G.D. Bengston, Eds. ICRISAT, Patancheru, India, pp. 319–324.

Rosenow, D.T. 1994. Evaluation of drought and disease resistance in sorghum for use in molecular marker-assisted selection. In *Use of Molecular Markers in Sorghum and Pearl Millet Breeding for Developing Countries, Proceedings of an ODA Plant Sciences Research Programs Conference*, J.R. Witcombe and R.R. Duncan, Eds., Norwich, U.K., March 29–April 1, 1993, pp. 27–31.

Rosenow, D.T. and Clark, L.E. 1995. Drought and lodging resistance for a quality sorghum crop. In *Proceedings of the Fiftieth Annual Corn and Sorghum Industry Research Conference*, Chicago, December 6–7, pp. 82–97.

Rosenow, D.T. and Dalhberg, J.A. 2000. Collection, conversion and utilization of sorghum. In *Sorghum, Origin, History, Technology and Production*, Wiley Series in Crop Science, C.W. Smith and R.A. Frederiksen, Eds. John Wiley & Sons, New York, pp. 309–328.

Rosenow D.T., Dahlberg, J.A., Paterson, G.C., Clark, L.E., Sotomayor-R'os, A., Miller, F.R., Hamburger, A.J., Madera-Torres, P., Quiles-Belen, A., and Woodfin, C.A. 1995. Release of 50 converted sorghum lines and 253 partially converted sorghum bulks. *Int. Sorghum Millets Newsl.* 36: 19–31.

Rosenow, D.T., Ejeta, G., Clark, L.E., Grilbert, M.L., Henzell, R.G., Borrell, A.K., and Muchow, R.C. 1997. Breeding for pre- and post-flowering drought stress resistance in sorghum. In *International Conference on Genetic Improvement of Sorghum and Pearl Millet*, Lubbock, TX, September 22–27, 1996, pp. 400–411.

Rosenow, D.T., Quisenberry, J.E., Wendt, C.W., and Clark, L.E. 1983. Drought tolerant sorghum and cotton germplasm. In *Plant Production and Management under Drought Conditions*, J.F. Stone and W.O. Willis, Eds. The Netherlands Elsevier Science Publishers B.V., Amsterdam, pp. 207–222.

Rosenow, D.T., Woodfin, C.A., Clark, L.E., and Sij, J.W. 1999. Drought resistance in exotic sorghum. In *Agronomy Abstracts*. Agronomy Society of America, Madison, WI, p. 166.

Sajjanar, G.M. 2002. Genetic Analysis and Molecular Mapping of Components of Resistance to Shoot Fly [*Atherigona soccata* Rond.] in Sorghum [*Sorghum bicolor* (L.) Moench]. Ph.D. thesis, University of Agricultural Sciences, Dharwad, India, 2002.

Schaffert, R.E., McCrate, A.J., Trevisan, W.L., Bueno, A., Meira, J.L., and Rhykerd, C.L. 1975. Genetic variation in *Sorghum bicolor* (L.) Moench for tolerance to high levels of exchangeable aluminium in acid soils of Brazil. In *Proceedings of Sorghum Workshop*, University of Puerto Rico, Mayaguez, Puerto Rico, pp. 151–160.

Schertz, K.F. 1994. Male sterility in sorghum: its characteristics and importance. In *Use of Molecular Markers in Sorghum and Pearl Millet Breeding for Developing Countries, Proceedings of an ODA Plant Sciences Research Conference*, J.R. Witcombe and R.R. Duncan, Eds., Norwich, U.K., March 29–April 1, 1993, pp. 35–37.

Schertz, K.F., Stec, A., and Deobley, J.F. 1990. *Isozyme Genotypes of Sorghum Lines and Hybrids in the United States*, Texas Agricultural Experimental Station 1719, Lubbock, TX, p. 15.

Seetharama, N., Mythili, P.K., Rani, T.S., Harshvardhan, D., Ranjani, A., and Sharma, H.C. 2003. Tissue culture and alien gene transfer in sorghum. In *Plant Genetic Engineering*, Vol. 2, *Improvement of Food Crops*, P.K. Jaiwal and R.P. Sing, Eds. Sci Tech Publishing LLC, pp. 235–265.

Seetharama, N., Reddy, B.V.S., Peacock, J.K., and Bidinger, F.R. 1982. Sorghum improvement for drought resistance. In *Drought Resistance in Crop Plants with Emphasis on Rice*. IRRI, Philippines, pp. 317–356.

Shan, L.Q., Ai, P.J., Yin, L.T., and Yao, Z.F. 2000. New grain sorghum cytoplasmic male-sterile A2V4A and F1 hybrid Jinza no.12 for northwest China. *Int. Sorghum Millets Newsl.* 41: 31–32.

Sharma, H.C. 1993. Host plant resistance to insects in sorghum and its role in integrated pest management. *Crop Prot.* 12: 11–34.

Sharma, H.C., Abraham, C.V., Vidyasagar, P., and Stenhouse, J.W. 1996. Gene action for resistance in sorghum to midge, *Contarinia Sorghicola*. *Crop Sci.* 36: 259–265.

Sharma, G.C., Jotwani, M.G., Rana, B.S., and Rao, N.G.P. 1977. Resistance to sorghum shoot fly [*Atherigona Soccata* Rondani] and its genetic analysis. *J. Ent. Res.* 1: 1–12.

Sifuentes, J. and Frederiksen, R.A. 1988. Inheritance of resistance to pathotypes 1, 2 and 3 of *Peronosclerospora sorghi* in sorghum. *Plant Dis.* 72: 332–333.

Singh, B.V. and Rana, B.S. 1994. Influence of varietal resistance on disposition and larval development of stalk borer, *Chilo Partellus* Swinhoe and Its relationship to field tolerance in sorghum. *Insect Sci. Application* 5: 287–296.

Singh, G., Singh, H.C., Ramakrishna, and Singh, S. 2001. Genetic divergence in *Sorghum bicolor* (L.) Moench. *Ann. Agric. Res.* 22: 229–231.

Singh, R. and Axtell, J.D. 1973. High lysine mutant gene (*hl*) that improves protein quality and biological value of grain sorghum. *Crop Sci.* 13: 535.

Singhania, D.L., Deosthale, Y.G., and Rao, N.G.P. 1979. A study of gene action for protein and lysine content in sorghum (*Sorghum bicolor* (L) Moench). *Indian J. Hered.* 11: 25–34.

Snowden, J.D. 1936. *The Cultivated Races of Sorghum*. Allard and Co., London.

Spearman, C. 1904. General intelligence objectively determined and measured. *Am. J. Psychol.* 25: 201–293.

Stephens, J.C. 1937. Male sterility in sorghum: its possible utilization in production of hybrid seed. *J. Am. Soc. Agron.* 29: 690–696.

Stephens, J.C. and Holland, R.F. 1954. Cytoplasmic male-sterility for hybrid sorghum seed production. *Agron J.* 46: 20–23.

Subudhi, P.K. and Nguyen, H.T. 2000. New horizons in biotechnology. In *Sorghum, Origin, History, Technology and Production*, Wiley Series in Crop Science, C.W. Smith and R.A. Frederiksen, Eds. John Wiley & Sons, New York, pp. 349–397.

Subudhi, P.K., Rosenow, D.T., and Nguyon, H.T. 2000. Quantitative trait loci for the stay-green trait in sorghum (*Sorghum bicolor* (L.) Moench): consistency across genetic backgrounds and environments. *Theor. Appl. Genet.* 101: 733–741.

Sullivan, C.Y. 1972. Mechanisms of heat and drought resistance in grain sorghum and methods of measurements. In *Sorghum in the Eighties*, N.G.P. Rao and L.R. House, Eds. Oxford and IBH Publishing Co., New Delhi, pp. 247–263.

Sullivan, C.Y. and Ross, W.M. 1979. Selecting for drought and heat resistance in grain sorghum. In *Stress Physiology in Crop Plants*, H. Mussell and R.C. Staples, Eds. Wiley Interscience, New York, pp. 263–281.

Tabo, R., Ogungible, A.O., Gupta, S.E., and Ajayi, O. 1999. Participatory evaluation of sorghum cultivars in northern Nigeria. *Int. Sorghum Millet Newsl.* 40: 36–38.

TAES and USDA–ARS. 1996. Release of converted exotic lines from the world sorghum collection.

Tanksley, S.D., Young, N.D., Paterson, A.H., and Bonierbale, M.W. 1989. RFLP mapping in plant breeding: new tool for an old science. *Biotechnology* 7: 257–264.

Tao, Y.Z., Hardy, A., Drenth, J., Henzell, R.G., Franzmann, B.A., Jordan, D.R., Bulter, D.G., and McIntyre, C.L. 2003. Identification of two different mechanisms for sorghum midge resistance through QTL mapping. *Theor. Appl. Genet.* 107: 116–122.

Tao, Y.Z., Henzell, R.G., Jordan, D.R., Butler, D.G., Kellu, A.M., and McIntyre, C.L. 2000. Identification of genomic regions associated with stay-green in sorghum by testing RILs in multiple environments. *Theor. Appl. Genet.* 100: 1125–1232.

Tao, Y.Z., McIntyre, C.L., and Henzell, R.G. 1996. Applications of molecular markers to Australian sorghum breeding programs. I. Construction of a RFLP map using sorghum recombinant inbred lines. In *Proceedings of the Third Australian Sorghum Conference*, M.A. Foale, R.G. Henzell, and J. Kneipp, Eds., Tamworth, Australia, February 20–22, pp. 443–450.

Tarr, S.A.J. 1962. *Diseases of Sorghum, Sudangrass and Brown Corn.* The Commonwealth Mycological Institute, Kew, Surrey, U.K., p. 380.

Thakur, R.P., Frederiksen, R.A., Murty, D.S., Reddy, B.V.S., Bandyaophayay, R., Giorda, L.M., Odvody, G.N., and Claflin, L.E. 1997. Breeding for disease resistance in sorghum. In *International Conference on Genetic Improvement of Sorghum and Pearl Millet*, Lubbock, TX, September 22–27, 1996, pp. 303–315.

Thimmaraju, R., Krishna, T.G., Kuruvinashetti, M.S., Ravikumar, R.L., and Shenoy, V.V. 2000. Genetic diversity among sorghum genotypes assessed with RAPD markers. *Karnataka J. Agric. Sci.* 13: 564–569.

Thomas, G.L. 1980. Thermal and Photo Thermal Effects on the Growth and Development of Diverse Grains Sorghum Genotypes. Ph.D. dissertation, Texas A&M University, College Station, 1980.

Tipton, K.W., Floyd, E.H., Marshall, J.G., and McDevitt, J.B. 1970. Resistance of certain grain sorghum hybrids to bird damage in Louisiana. *Agron. J.* 62: 211–213.

Trouche, G., Rami, J.E., and Chantereau, J. 1998. QTLs for photoperiod response in sorghum. *Int. Sorghum Millets Newsl.* 39: 94–96.

Tuinsta, M.R., Grote, E.M., Goldsbrough, P.B., and Ejeta, G. 1996. Identification of quantitative trait loci associated with pre-flowering drought tolerance in sorghum. *Crop Sci.* 36: 1337–1344.

Tuinsta, M.R., Grote, E.M., Goldsbrough, P.B., and Ejeta, G. 1997. Genetic analysis of post-flowering drought tolerance and components of grain development in *Sorghum bicolor* (L.) Moench. *Mol. Breed.* 3: 439–448.

Tuinsta, M.R., Grote, E.M., Goldsbrough, P.B., and Ejeta, G. 1998. Evaluation of near-isogenic sorghum lines contrasting for QTL markers associated with drought tolerance. *Crop Sci.* 38: 825–842.

Turner, N.C. 1979. Drought resistance and adaptation to water deficits in crop plants. In *Stress Physiology in Crop Plants*, H. Mussel and R.C. Staples, Eds. Wiley Interscience New York, pp. 343–372.

Umakanth, A.V., Madhusudhana, R., Madhavilatha, K., Hema Kumar, P., and Kaul, S. 2002. Genetic architecture of yield and its contributing characters in postrainy-season sorghum. *Intl Sorghum Millet Newsl.* 43: 37–39.

Van Oosterrom, E.J., Jayachandran, R., and Bidinger, F.R. 1996. Diallel analysis of the stay-green trait and its components in sorghum. *Crop Sci.* 36: 549–555.

Vavilov, N.I. 1992. *Origin and Geography of Cultivated Plants*, V.F. Dorofeev, Ed. Cambridge University Press, Cambridge, U.K., p. 332.

Vidyabhushanam, R.V., Rana, B.S., and Reddy, Belum V.S. 1989. Use of Sorghum Germplasm and Its Impact on Crop Improvement in India. Paper presented at Collaboration on Genetic Resources: Summary Proceedings of a Joint ICRISAT/NBPGR (ICAR) Workshop on Germplasm Exploration and Evaluation in India, Patancheru, India, November 14–15, 1988.

Virupaksha, T.K. and Sastry, L.V.S. 1968. Studies on the protein content and amino acid composition of some varieties of grain sorghum. *J. Agric. Food Chem.* 16: 199–203.

Virupaksha, T.K. and Sastry, L.V.S. 1969. Alcohol soluble proteins of grain sorghum. *Cereal Chem.* 46: 284–293.

Volger, R.K. 1992. Genetic Control of Low Stimulant Production and Its Potential as a Predictor of Field Resistance against *Striga asiatica* in Sorghum. M.Sc. thesis, Purdue University, West Lafayette, IN, 1992.

Volger, R.K., Ejeta, G., and Bulter, L.G. 1995. Integrating biotechnological approaches for the control of *Striga. Afr. Crop Sci. J.* 3: 217–222.

Vose, P.B. 1987. Genetic aspects of mineral nutrition: progress to date. In *Genetic Aspects of Plant Mineral Nutrition*, W.H. Gabelman and B.C. Loughman, Eds. Martinus Nijhoff Publishers, Dordrecht, The Netherlands, pp. 3–13.

Waggle, D.H. and Deyoe, C.W. 1966. Relationship between protein level and amino acid composition of sorghum grain. *Food Stuffs* 38: 18–19.

Wall, J.S. and Blessin, C.W. 1970. Composition of sorghum plant and grain. *Sorghum Production and Utilization*, J.S. Wall and W.M. Ross, Eds., AVI Pub., Westport, CT, pp. 118–166.

*Webster's Ninth New Collegiate Dictionary.* 1986. Merriam-Webster, Springfield, MA.

Wendorf, F., Close, A.E., Schild, R., Wasylikowa, K., Housley, R.A., Harlan, R.A., and Krolik, H. 1992. Saharan exploitation of plants 8000 years B.P. *Nature* 359: 721–724.

Wenzel, W.G. 1994. Tillering, panicle mass and threshability as components of yield in grain sorghum. *Int. Sorghum Millets Newsl.* 35: 72–73.

Witcombe, J.R. and Duncan, R.R. 1994. Use of molecular markers in sorghum and pearl millet breeding for developing countries. In *Proceedings of an ODA Plant Sciences Conference*, Norwich, U.K. March 29–April 1, 1993.

Witcombe, J.R., Joshi, J.R., Joshi, K.D., and Sthapit, B.R. 1996. Farmer participatory crop improvement. *Exp. Agric.* 32: 453–460.

Worede, M. and Mekbib, H. 1993. Linking genetic resource conservation to farmers in Ethiopia. In *Cultivating Knowledge, Genetic Diversity, Farmer Experimentation and Crop Research*, W. de Boef, K. Amanor, and K. Wellard, Eds. International Technology Publications, London, pp. 78–84.

Xu, W.W., Subudhi, P.K., Crasta, O.R., Rosenow, D.T., Mullet, J.E., and Nguyen, H.T. 2000. Molecular mapping of QTLs conferring stay-green in grain sorghum (*Sorghum bicolor* (L.) Moench). *Genome* 43: 461–469.

Yang, W., de Oliveira, A.C., Godwin, L., Schertz, K.C., and Bennetgen, J.L. 1996. Comparison of DNA marker technologies in characterizing plant genome diversity: variability in Chinese *Sorghums*. *Crop Sci.* 36: 1669–1676.

Yang, Z. 1997. Sorghum breeding research in China. *Int. Sorghum Millets Newsl.* 38: 15–18.

Yu, H., Liang, G.H., and Kofoid, K.D. 1991. Analysis of C-banding chromosome patterns of sorghum. *Crop Sci* 31: 1524–1527.

Zake, V.M., Watson, C.E., Jr., and Gourley, L.M. 1992. In *Proceedings of Workshop on Adaptation of Plants to Soil Stresses*, Lincoln, NE, August 1–4, 1993, pp. 80–99.

# CHAPTER 12

# Rye (*Secale cereale* L.): A Younger Crop Plant with a Bright Future

**Rolf Schlegel**

## CONTENTS

12.1  Introduction..........................................................................................................366
12.2  Botany..................................................................................................................367
   12.2.1  Gross Morphology.....................................................................................367
   12.2.2  Root System ..............................................................................................367
   12.2.3  Seeds ..........................................................................................................368
12.3  Physiology ...........................................................................................................368
12.4  Cytology ..............................................................................................................368
   12.4.1  Chromosome Number ...............................................................................368
   12.4.2  Molecular Structure of Genome................................................................368
   12.4.3  Primary Trisomics and Telotrisomics ......................................................369
   12.4.4  B Chromosomes ........................................................................................370
   12.4.5  Karyotype ..................................................................................................370
   12.4.6  Reciprocal Translocations ........................................................................372
   12.4.7  Neocentric Activity...................................................................................373
   12.4.8  Haploid Rye...............................................................................................373
   12.4.9  Homoeology ..............................................................................................374
   12.4.10 Taxonomy, Cytotaxonomy, and Origin....................................................375
   12.4.11 Alien Introgression ...................................................................................376
   12.4.12 DNA Transfer ...........................................................................................377
12.5  Genetics ...............................................................................................................377
   12.5.1  General.......................................................................................................377
   12.5.2  Chromosomal and Regional Localization of Genes and Markers .......................378
12.6  Cytoplasmic Male Sterility and Restorer............................................................378
12.7  Breeding...............................................................................................................378
   12.7.1  Diploid Rye ...............................................................................................379
       12.7.1.1  Population Breeding...................................................................379
       12.7.1.2  Synthetics ..................................................................................380
       12.7.1.3  Hybrid Breeding........................................................................381

12.7.2  Tetraploid Rye ...................................................................................................382
12.8  Rye Cropping ...........................................................................................................384
    12.8.1  Seeding .........................................................................................................385
    12.8.2  Soil Type.......................................................................................................385
    12.8.3  Water.............................................................................................................385
    12.8.4  Micronutrients...............................................................................................385
    12.8.5  Fertilization...................................................................................................386
    12.8.6  Temperature ..................................................................................................386
    12.8.7  Susceptibility and Resistance.......................................................................386
    12.8.8  Growth Regulators.........................................................................................387
    12.8.9  Incorporation in Crop Rotation....................................................................387
    12.8.10 Allelopathic Effects ......................................................................................387
    12.8.11 Volunteering..................................................................................................387
12.9  Rye as a Donor for Genetic Improvement of Wheat .............................................387
    12.9.1  Genome Additions.........................................................................................387
    12.9.2  Chromosome Additions.................................................................................388
    12.9.3  Chromosome Substitutions............................................................................388
    12.9.4  Chromosome Translocations .........................................................................388
12.10  Conclusion...............................................................................................................388
References .............................................................................................................................389

## 12.1 INTRODUCTION

Rye (*Secale cereale* L.) is a cereal that played a major role in feeding European populations throughout the Middle Ages, owing to its considerable winter hardiness. Recently the world production amounts to about 30 million tons (MT). The cultivated rye resulted from crossbreeding between *Secale vavilovii* Grossh. and the perennial species, *Secale anatolicum* Boiss. and *Secale montanum* Guss. It is part of the quite young cultivated plants and called a secondary crop, which originated as a weed in emmer and barley fields of the Near East. First cultivation began in Persia, Central Anatolia, and north of the Black Sea region about 3000 years ago. The domestication probably happened at several locations but, presumably, within the general area defined below. Rye grains found in Neolithic sites in Austria and Poland are considered to be of wild origin. The earliest seed of cultivated rye in central Europe came from the Hallstatt period, 1000 to 500 B.C. From there, the cropping of rye moved northwest toward Sweden from 2500 to 2000 B.C. During the 16th century rye cultivation subsequently increased, and at the beginning of the 20th century it succeeded even wheat in acreage.

Original rye was a meter-high grass. The long straw was used in particular for roofing of buildings. Meanwhile, rye became a modern crop plant with all the technological and agronomic advantages. There is a subsequent increase in rye grain yield caused by improvement of agronomy and new (hybrid) varieties.

Rye has a low acreage and low production on a worldwide basis (see Figure 1.1 and Figure 1.2 in Chapter 1 in this volume). Although the rye acreage decreased by more than one half during the last four decades, the cool temperate zones of Europe remained the major growing areas. About 94% of the world production is harvested in Russia (37% of the total acreage), Belarus (9% of the total acreage), Poland (22% of the total acreage), and Germany (9% of the total acreage) (Table 12.1). Acreage in the U.S. has been decreasing as well. All rye of the U.S. is winter rye, mostly used for grain production. The leading states are South Dakota, Georgia, Minnesota, and North Dakota.

About 50 to 75% of the yearly harvest is used for bread making, resulting in rich, dark bread that holds its freshness for about a week. The rest is used for feeding and for the production of alcohol.

**Table 12.1 Largest Rye-Producing Countries in the World (1996 to 2001)**

| Country | 1996 | 1997 | 1998 | 1999 | 2000 | 2001 |
|---|---|---|---|---|---|---|
| Russia | 5,934 | 7,478 | 3,269 | 4,782 | 5,400 | 6,000 |
| Poland | 5,652 | 5,299 | 5,664 | 5,181 | 4,003 | 4,921 |
| Germany | 4,214 | 4,580 | 4,775 | 4,329 | 4,208 | 5,151 |
| Belarus | 1,794 | 1,788 | 1,384 | 929 | 1,239 | 1,500 |
| Ukraine | 1,092 | 1,348 | 1,140 | 919 | 968 | 1,500 |
| Total | 22,930 | 24,987 | 20,823 | 19,964 | 19,694 | 22,718 |

*Note*: Data are in thousands of tonnes.

*Source*: FAO (http://faostat.fao.org/faostat), 2002.

## 12.2 BOTANY

Rye is, as wheat, barley, and oats, a member of the grass family, the Poaceae (= Gramineae), subfamily Pooideae, tribe Triticeae. The divergence of the wheat and rye lineages from the Pooideae happened about 7 million years ago. The common name is rye, feral rye, or cereal rye — in the following named as rye. The use of the name *cereal rye* reduces confusion with ryegrasses (*Lolium* spp.) in the English eloquent countries. The scientific name of cultivated rye is *Secale cereale* L. Internationally, rye is called Roggen (German), Centeno (Spanish), rosh (Russian), Segale comune (Italian), zyto (Polish), Seigle (French), råg (Swedish), Rug (Danish), Ruis (Finish), Rogge (Dutch), and Rúgur (Icelandic). It is a typical allogamous plant species showing a high degree of self-incompatibility.

### 12.2.1 Gross Morphology

Rye is an erect annual grass with flat blades and dense spikes. The habit resembles that of wheat, but usually taller. It has the longest stems of all cultivated small grains (culms of 50 (minimum 35) to 120 (maximum 300) cm), and these provide most of the photosynthetic area. The average height of older ryes is 150 cm; e.g., the variety Viatka is 190 cm; Wojcieszyckie, 160 cm; Dankowske Zlote, 145 cm; and Kustro, 135 cm. Modern, high-yielding varieties are much shorter, in particular hybrid varieties. Their long compact ears, possessing seeds of equal size, distinguish the latter.

During grain formation, stems with sheaths account for 60 to 80% of the total plant area. As compared to wheat, the flag leaf is smaller and less important in photosynthesis (blades of 4 to 12 mm wide). Leaves (stems and spikes) have a bluish color by their waxy surface. The spike is longer and more slender (spike of 4.5 (minimum 2) to 12 (maximum 19) cm) somewhat nodding when mature.

The inflorescence and seeds are as follows: spikelets usually are two flowered, solitary, placed flatwise against the rachis, the rachilla disarticulating above the glumes and produced beyond the upper floret as minute stipes. The glumes are narrow, rigid, acuminate, or subulate pointed. Lemmas are broader, sharply keeled, five nerved, and ciliate on the keel and exposed margins, tapering into a long awn. The lemma and palea that enclose the floret are free threshing, and the lemma often bears barbs on the keel and has an awn of intermediate length. Anthers are about 7 mm long, with light yellow to purple color. The rye grain is longer and more slender than that of wheat, and it sometimes shows a pale but often greenish seed coat. The pale grain is genetically coded by recessive gene loci. Purple seeds were found in mutant plants (Ruebenbauer et al. 1983).

### 12.2.2 Root System

Rye has the best-developed root system among annual cereal crops; as with other grasses, the system is fibrous, with no defined taproot. The extensive root system, including its specific rhizosphere, enables it to be the most drought-tolerant cereal crop and makes it among the best green manures for improving soil structure. Field studies revealed that decomposition of incorporated [14]C-labeled rye residue was accelerated through having rye plants growing in the soil. This was believed to be due to the rhizosphere microbial complex. Comparison of shoot and root dry weight and soil moisture during progressive growth stages for wheat, triticale, and rye in both greenhouse pot and field studies demonstrated significantly higher root dry weight in rye when grown in pots, but wheat showed greater root mass, from depths of 20 to 50 cm, in the field studies. Root growth is greatest from the seedling to the flag leaf for all three cereals. Rye generally roots to a depth of 90 to 230 cm; thus, it is deeper rooting than other small grains.

### 12.2.3  Seeds

Rye seeds may germinate in storage or even while still in the spike (preharvest sprouting). Seeds will germinate at temperatures as low as 3 to 5°C, but the optimal range is 25 to 31°C. In most northern countries, rye can be established when seeded as late as October 1. The minimal temperature for germination ranges from 1 to 2°C. In the autumn, rye grows more rapidly than wheat, oat, or various other annual grasses. Although rye is usually regarded as a winter crop, several spring-sown varieties are available. As a long-day plant, flowering is induced by 14 hours of daylight at a temperature of 5 to 10°C. Short-day length can cause rye plants to remain vegetative for up to 7 and more years. This behavior is used for maintenance or multiplication of specific genotypes by cloning under short-day conditions.

The crude protein content is the lowest among the cereals (8 to 10% of the dry matter), while the starch content ranges with 50 to 70% of dry matter, again between barley and wheat. The fat content is about 2% in seed. Pentosans play a particular role in rye utilization. Their content is between 9 to 10% of the dry matter, depending on the variety. About 25% of the pentosans are water soluble. Pentosans decrease digestion when fed to animals. The lysine content is about 3.7%, the methionine content is 3.6%, and the threonine content is about 3.0% of the mean crude protein content. Thus, rye shows the highest lysine content among the cereals (wheat 2.5%, barley 3.5%, triticale 3.0%).

## 12.3 PHYSIOLOGY

The maturation date of rye varies according to soil moisture, but vegetative growth stops once reproduction begins. In general, rye matures earlier than oat. During seed set, leaf blades provide 15 to 20% of the photosynthetic area, which is much lower than for maize, wheat, and oat. Stems and sheaths have lower rates of photosynthesis. For winter rye, photosynthetic area decreases rapidly after seed setting and does not achieve a plateau near the maximum, as seen with other grains.

## 12.4 CYTOLOGY

### 12.4.1  Chromosome Number

The chromosome number of rye is $2n = 2x = 14$. The genome formula of diploid rye is given as $R^{cer}R^{cer}$.

### 12.4.2  Molecular Structure of Genome

The 1C DNA content of the rye genome is 9.5 pg (Bennett and Smith 1976). The 1C DNA content of an individual rye chromosome ranges from about 1.2 to 1.4 pg. Chromosome length and DNA content are directly proportional. Only 10 to 20% of the genome can be assigned biochemically to the major part of the genome, which belongs to the repeated sequence category. The kinetic analysis of genome organization has revealed that repeated sequences are interspersed among unrepeated sequences. The discovery of a very rapidly reannealing class of DNA constitutes 4 to 10% of the genome, although it is believed to be composed largely of sequences capable of renaturation. This class contains long tandem arrays of simple, repeated sequences (Appels 1982). Ranjekar et al. (1974) were the first to demonstrate several buoyant density components in a fraction of DNA renaturing with a density of 0 to 0.1 (10 to 12% of the genome). However, the predominant component is a well-defined species at 1.702 g/cc in a CsCl gradient. Smith and Flavell (1977) considered this class of DNA to consist of mainly palindromic sequences, which are distributed in clusters throughout at least 30% of the genome. DNA with a mean fragment length of 500 bp was fractionated to allow recovery of very rapidly renaturing fractions ($C_ot$ = 0 to 0.2). This DNA contained several families of highly repeated sequences. Two of them were purified, which resulted in a fraction renaturing to a density of 1.701 g/cc and comprising 0.1% of the total genome, and the other polypyrimidin tract DNA, which comprised 0.1% of the genome. Other hybridization studies among cereals have shown that 22% of the DNA is species-specific repeated sequences (Rimpau et al. 1978).

### 12.4.3 Primary Trisomics and Telotrisomics

Primary trisomics of spontaneous origin in rye have been described as early as 1935 (Takagi 1935). A complete series of primary trisomics was established in 1962 (Kamanoi and Jenkins 1962); however, this set was lost later on. Two new sets of primary trisomics were produced from the winter rye varieties Danae by Mettin et al. (1972) and Heines Hellkorn by Zeller et al. (1977). The first set was derived from triploid × diploid crosses, and it was identified based on cytological (Figure 12.1) and morphological observations. The extra chromosome in primary trisomics transmitted about 15% through the egg cell and almost 0% through the pollen. Consequently, backcrosses to the female plants are needed in order to maintain primary trisomics.

The progeny of primary trisomics produced monotelocentric and other chromosome aberrations for almost all of the 14 chromosome arms (Schlegel and Sturm 1982; Sturm and Melz 1982; Melz and Schlegel 1985; Melz and Sybenga 1994; Benito et al. 1996).

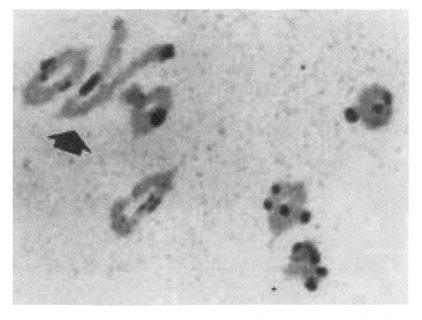

**Figure 12.1** Metaphase I configuration of primary rye trisomic 5R (2n = 2x = 15) with 6II + 1III after C-banding.

Until molecular studies became standard in gene mapping during 1990, primary trisomics and telotrisomics have been the most powerful tool for locating genes in rye (Sybenga 1965; Sturm 1978; Melz et al., 1984). Morphological or physiological characters were located to a particular chromosome, by using primary trisomics, based on modified ratios in $F_2$ and $BC_1$ populations (see Chapter 8).

### 12.4.4  B Chromosomes

B chromosomes of rye are very invasive. It is the only plant species where the B chromosomes undergo nondisjunction at the postmeiotic mitosis of both female and male gametophytes. It is controlled by a gene or gene complex on the B chromosome itself. By reduced fertility and persistent univalent formation, a high degree of chromosomal instability is found, which results in a highly polymorphic population.

The B chromosome of rye is relatively large. It amounts to about 10% of the DNA of the haploid A genome. It has a widespread global distribution and exhibits little structural variation. It is presumed to be derived largely or entirely from the A chromosome set based on similar DNA composition shown by *in situ* hybridization with genomic DNA or the rye-specific R173 repeat.

Despite general similarity, the B chromosome does not pair with any A chromosomes. However, translocations between B and A chromosomes were observed (Pohler and Schlegel 1990). Chromosome 3R was involved in the rearrangement. Schlegel and Pohler (1994) demonstrated that a terminal segment, including a large terminal heterochromatic block, was transferred to the terminal region of the long arm of the B chromosome. Although several authors assumed a monophyletic origin of rye B chromosomes in the distant past and a perpetuation in various populations, the results imply similar events also nowadays, which is partially supported by stable fragments occasionally revealed, as well as the presence of B chromosomes in numerous plant species.

Defined A-B translocations were thought to be a useful tool for gene amplification, chromosome identification, genetic investigation, and gene transfer experiment.

B chromosomes appear to be transcriptionally inactive, implying either loss of coding sequences or selective amplification of repetitive DNA in its evolution. By AFLP studies, it was demonstrated that DNA products specific to B-containing plants were seen at a frequency of about 1%, significantly less than that predicted on a mass basis. Most of the fragments were found to be highly repetitive and were shown by fluorescent *in situ* hybridization (FISH) to be distributed over both A and B genomes. Dispersed fragments fell into two groups: the first showed an equal density of distribution across both A and B genomes, consistent with earlier observations, while the second showed an increased concentration at a pericentric site on the long arm of the B. The redundancy of members of the second group suggests that this region is less complex than the average for the B and may represent a hot spot of sequence amplification.

The terminal heterochromatin of the B chromosome contains both A- and B-specific sequences (Houben et al. 1996). The B-specific D1100 family is the major repeat species located in the terminal heterochromatin. The most distinctive region of the B chromosome is a subtelomeric domain that contains an exceptional concentration of B-specific DNA sequences. At metaphase, this domain appears to be the physical counterpart of the subtelomeric heterochromatic regions of the standard B chromosome (Langdon et al. 2000).

### 12.4.5  Karyotype

Several attempts have been made to establish a common chromosome designation (Table 12.2), a general karyological characterization, and a standard karyotype in diploid rye. All efforts, however, did not completely account for the natural variation of chromosome morphology and structural heterozygosity in allogamous rye populations (Heneen 1962). To overcome the difficulties at least partially, the participants of the first and second Workshops on Rye Chromosome Nomenclature and Homoeology Relationships decided to consider the Imperial rye additions to the hexaploid

**Table 12.2 Comparison of Different Proposals for Designation of Rye Chromosomes**

| Author | Chromosome | | | | | | |
|---|---|---|---|---|---|---|---|
| Lewitzky 1931 | IV | V | I | III | VI | VII | II |
| Pathak 1940 | A | B | F | G | D | E | C |
| Tijo and Levan 1950 | I | II | V | III | IV | VI | VII |
| Oinuma 1952 | b | g | c | a | e | d | f |
| Lima de Faria 1952 | II | IV | III | I | VI | VII | V |
| Bhattacharyya and Jenkins 1960 | I | III | II | V | IV | VI | VII |
| Heneen 1962 | 1 | 2 | 3 | 4 | 5 | 6 | 7 |
| Vosa 1974 | I | IV | II | V | VI | III | VII |
| Verma and Rees 1974 | 2 | 3 | 5 | 1 | 4 | 6 | 7 |
| Vries and Sybenga 1976 | II | III | I | IV | VI | V | VII |
| Schlegel and Mettin 1982 | 2R | 7R | 3R | 4R | 5R | 6R | 1R |

**Table 12.3 Variation of Chromosome Length in Haploid Rye Rosta-7**

| Characteristics | Chromosome | | | | | | |
|---|---|---|---|---|---|---|---|
| | 1R | 2R | 3R | 4R | 5R | 6R | 7R |
| Satellite | 15.24 | | | | | | |
| SD | 1.40 | | | | | | |
| Short arm | 39.32 | 75.44 | 63.73 | 65.25 | 45.03 | 50.51 | 58.78 |
| SD | 0.40 | 0.63 | 0.65 | 1.26 | 1.30 | 0.99 | 0.76 |
| Long arm | 69.68 | 87.02 | 65.63 | 88.21 | 93.25 | 89.67 | 71.54 |
| SD | 1.26 | 1.15 | 1.26 | 4.70 | 1.18 | 0.82 | 4.10 |
| Total length | 124.24 | 162.46 | 129.36 | 153.47 | 138.29 | 140.19 | 130.33 |
| SD | 1.35 | 1.60 | 1.82 | 5.99 | 1.47 | 1.48 | 5.51 |
| Arm ratio | 1.27[a] | 1.15 | 1.03 | 1.35 | 2.07 | 1.78 | 1.22 |
| SD | 0.06 | 0.02 | 0.02 | 0.03 | 0.08 | 0.04 | 0.01 |

*Note*: Confidence limits at 1% of probability for total chromosome length = 6.30, for arm length = 3.64, and for the arm ratio = 0.06.

[a]Satellite included in the short arm.

wheat variety Chinese Spring as the standard rye chromosome set, although these chromosomes are not completely identical with those of the population of Imperial variety (Sybenga 1983).

Based on the agreements, a preliminary karyotype from C-banding patterns and a homoeologous designation of added rye chromosomes to wheat were proposed. The participants anticipated producing a standard diploid rye homozygous for genetic, biochemical, or molecular studies, which should be, in addition, available for cytogenetic test crosses. Therefore, Schlegel et al. (1987) established from the rye variety Petka a haploid, a dihaploid, and, finally, a tetrahaploid plant from the original haploid genome. This variety was chosen because of its spring growth habit, dominantly greenish grains. The variety Petka was bred in Petkus (Germany) and released in 1961. Its genetic background is related to the gene pool of Petkus rye, which has been used worldwide in breeding and research.

The data demonstrate remarkable differences in mean total length as well as relative arm length, which range from 124.24 units (chromosome 1R) to 162.46 units (2R) and from 45.03 units (chromosome arm 5RS) to 93.25 units (chromosome 5RL), respectively. The arm ratios vary from 1.02 (chromosome 3R) to 2.07 (chromosome 5R). The small sample standard deviations mean that most of the length differences or arm index variations are statistically significant (Table 12.3).

The karyogram drawn from the data has been used for detailed description using the C-banding pattern. Prominent blocks of telomeric heterochromatin are stained, which is a common pattern of diploid rye (Figure 12.2 and Figure 12.3). Rye chromosomes 2R, 3R, and 6R also show prominent N-bands near the centromeres (Schlegel and Gill 1984). Moreover, there is quite a good correlation between the distribution of heavy knobs of chromomeres described by Lima de Faria (1952) and C- or N-bands. Recently, tetrad-FISH analysis and linkage maps based on restriction fragment

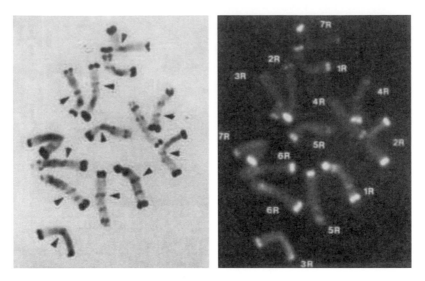

**Figure 12.2**   Somatic spread of diploid rye after sequential C-banding (left) and fluorescence DAPI staining (right).

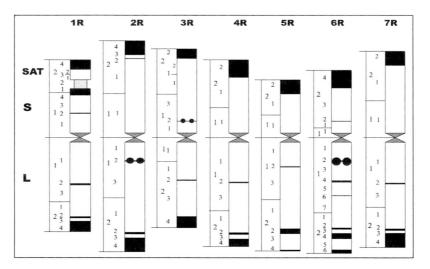

**Figure 12.3**   Standard karyogram established according to the karyological measurements as well as C- and N-band distribution.

length polymorphism (RFLP) markers clearly indicated that heterochromatin strongly suppresses recombination of entire chromosomal regions (Kagawa et al. 2002).

The band positions and band sizes are related to the relative arm length, so there are sufficient references for identifying each of the seven chromosomes individually. Applying the standard chromosome band nomenclature taken from Schlegel et al. (1986b), a specific reference karyogram was established excluding structural heterozygosity of the genome. This standard karyotype from a dihaploid and tetrahaploid progeny was proposed as general reference material in genetic and cytogenetic studies.

## 12.4.6   Reciprocal Translocations

Rye maintains chromosomal interchanges within populations in a more or less high frequency and complexity (Candela et al. 1979). Chromosomal interchanges are of particular importance both for evolutionary studies of rye and for several genetic and breeding applications. Until 1980, the nomenclature of rye chromosomes was confusing (Schlegel and Mettin 1982). Therefore, several efforts were made to designate rye chromosomes according to the homoeologous relationships within the Triticeae (Sybenga 1983). A translocation tester set of rye was established and was used for crossing with wheat–rye chromosome addition lines. In this way, for the first time, a complete series of reciprocal translocations was produced (Sybenga et al. 1985). Although several other interchanges have been established before and after those experiments (Sybenga and Wolters 1972; Vries and Sybenga 1976; Augustin and Schlegel 1983), this series became the key tester set for rye involving all seven chromosomes by interchanged chromosomes as follows: 1RS-4RL, 1RS-5RL, 1RS-6RS, 2RS-5RS, 2RL-6RL, 3RS-5RL, 4RL-5RL, and 5RL-7RS. They originated from irradiated pollen grains of Petkus rye crossed with several inbred rye lines.

Despite intensive utilization for gene mapping in rye (Vries and Sybenga 1976), reciprocal translocations were also used for diploidization experiments in tetraploid rye. However, the complex structural heterozygosity of reconstructed tetraploid karyotypes not only increased the preferential bivalent pairing but also decreased fertility.

### 12.4.7 Neocentric Activity

Neocentric activity is a newly derived kinetic activity outside the proper centromere. Kattermann (1939) first described this phenomenon in rye. Neocentromeres are rare in plants but less infrequent in meiosis. Three types of neocentromeres have been described in rye:

1. The neocentromere shows a stable structural differentiation in one end of a given chromosome. It is inherited as a Mendelian gene.
2. The neocentromeres are located at terminal regions of some chromosomes. They are mostly associated with distal heterochromatin, variable in activity and number among individuals and cells within an individual and under polygenic control (Viinikka 1985). They occur in both inbred lines and allogamous populations.
3. Schlegel (1987) described an additional type in haploid rye. A proximal constriction present on the long arm of chromosome 5R is co-oriented with the ordinary centromere. It behaves like a dicentric chromosome. However, molecular studies showed that the 5RL constriction lacks detectable quantities of two repetitive DNA sequences, CCS1 and the 180-bp knob repeat, present at cereal centromeres and neocentromeres (Manzanero et al. 2000).

### 12.4.8 Haploid Rye

Several workers have reported haploids and meiotic studies on haploids in rye. Associations of two or more nonhomologous chromosomes are often observed at metaphase I. Levan (1942) was the first to demonstrate statistically that chiasma formation between the seven chromosomes was not random. He suggested that one particular chiasma is formed at a quite high frequency, while the remaining arises at random. The mean chiasma frequency per pollen mother cell, as given by different authors, ranged from 0.03 to 0.44. Differentiation between true chiasma formation and secondary end-to-end attachments may be either by chromosome co-orientation or by chromatid bridges and acentric fragments during anaphase I. By testing random or non-random association of nonhomologous chromosomes in haploid, a statistically significant increase of chromosomes with heavy telomeres on both ends was found; i.e., chromosomes 1R, 2R, 3R, and 7R may show duplicated segments contributing to chromosome pairing in a haploid rye. No chiasma formation was observed in the heterochromatic telomeres. This indicates that this kind of repetitive DNA does not function as a homologous region contributing to crossing-over (Schlegel et al. 1987).

## 12.4.9  Homoeology

It is supposed that cereals within the subtribe Triticinae have a common origin and, likewise, a partial structural homology. It was first concluded by more or less good ability of chromosomes to substitute each other in interspecific hybrids (Gupta 1971; Koller and Zeller 1976; Schlegel 1990). Later, it was supported by intergeneric hybridization, particularly with wheat, that rye chromosomes can even show chiasmatic pairing with wheat chromosomes (Schlegel and Weryszko 1979). The comparatively high degree of wheat–rye chromosome pairing, and thus recombination, became of interest for plant breeders utilizing the incorporation of useful characteristics of rye species for wheat and triticale improvement. The frequency of wheat–rye pairing ranged from zero in crosses with *Secale silvestre* Host to 2.4% per pollen mother cell (PMC) in *S. montanum*.

Recent molecular maps show remarkable conservation of gene order among wheat, rye, barley, millet, rice, etc., disrupted only by a few gross interchromosomal rearrangements. Rye has diverged from wheat by at least 7 to 13 translocation events (Figure 12.4) after only 6 million years of divergence, while barley appears to reflect precise synteny with the basic wheat genome (Devos et al. 1993). Other authors provide evidence that some genomes fix rearrangements more readily than others. These different rates of species divergence through chromosomal rearrangement do not appear to correlate to the breeding system, because high levels of evolutionary translocations are found in both rye, an outbreeder, and, e.g., *Aegilops umbellulata* Zhuk., a predominantly self-pollinated species.

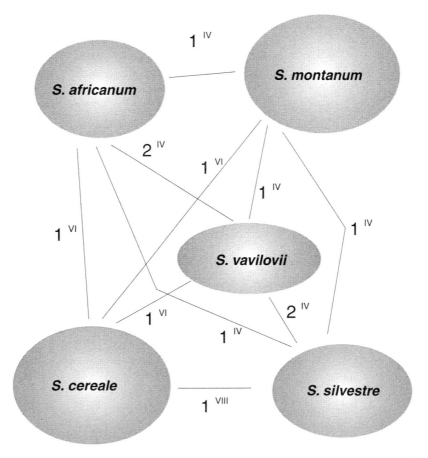

**Figure 12.4**  Schematic drawing of cytological differences between rye species by presence of interchanges (IV = quadrivalent, VI = hexavalent, VIII = octovalent).

### 12.4.10  Taxonomy, Cytotaxonomy, and Origin

Apart from artificial polyploids, all rye species are diploid (2n = 2x = 14). However, numerous reports about accessory or B chromosomes are available in both wild and cultivated *S. montanum* and *S. cereale*. Roshevitz (1947) recognized 14 species, but it is questionable whether all these have to be given specific rank. Two groups of species can be readily separated as being important in the evolution of the cultivated rye. First, there is a group of annual weeds, such as *Secale ancestrale* Zhuk., *Secale dighoricum* (Vavilov) Roshev., *Secale segetale* (Zhuk.) Roshev., and *Secale afghanicum* (Vavilov) Roshev., which cytologically resemble each other and cultivated rye (Schlegel and Weryszko 1979). They could be included as subspecies of *S. cereale*. This group is confined to agricultural areas, the weedy types being widespread in cereal crops in Iran, Afghanistan, and Transcaspia (Zohary 1971).

Second, there is an aggregate of wild perennial forms widely distributed from Morrocco through the Mediterranean area, Anatolia to Iraq and Iran. These have been separated into distinct species but are most probably described as variants of a single species, *S. montanum*. Members of this group are cytologically similar and interfertile. However, they differ from the *S. cereale* complex by two major reciprocal translocations involving three pairs of chromosomes. Stutz (1972) proposed a stepwise evolution of *S. cereale* from *S. montanum*. The annual species are supposed to be derived from the introgression of *S. montanum* into *S. vavilovii*. This annual self-fertile species was in turn derived from the annual *S. silvestre* as a consequence of chromosome translocation (Figure 12.5). Frederiksen and Petersen (1998) revised rye taxonomy. They recognized only three species of *Secale*. *S. silvestre*, *S. strictum*, and *S. cereale*. *S. strictum* has priority over *S. montanum*. It includes two subspecies: *strictum* and *africanum*. *S. cereale* also includes two subspecies, i.e., the cultivated taxa, marked by their tough rachises of ssp. *cereale*, and the wild or weedy taxa, showing fragile rachis of ssp. *ancestrale*.

Comparison of chloroplast DNA variation confirmed the distinctness of *S. silvestre* from the remaining taxa. This basic differentiation between *S. silvestre* and *S. montanum* or *S. strictum* took place during the Pliocene or later. In this context, it became clear that rye shows a purely maternal

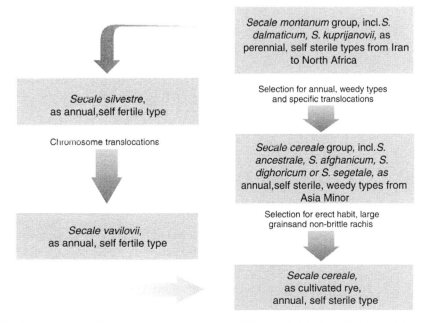

**Figure 12.5**  Proposed evolutionary origin of cultivated rye (*S. cereale* L.). [Modified from G.M. Evans 1976, in: N.W. Simmonds (ed.): *Evolution of Crop Plants*, Longman, London & New York, 1976.]

inheritance of chloroplasts (Corriveau and Coleman 1988), although Fröst et al. (1970) postulated a biparental mode of inheritance.

The most likely place of origin of the weedy rye is in central and eastern Turkey, northwest Iran, and Armenia. It is the area of maximum genetic diversity and coincides with the high degree of variability of the perennial *S. montanum*. The cold and harsh climate of this area possibly favored rye instead of the two main cereals, barley and wheat, which spread into this region. There is no doubt that under such conditions rye could have become established as a weed. Weedy annual races with brittle and semibrittle rachises were and still are colonizers.

## 12.4.11    Alien Introgression

In many taxa, introgression is a more or less frequent phenomenon, depending on the spontaneous tendency to interspecific hybridization. As hybridization played an important role in the evolution of plant speciation, it has been suggested that interspecific crosses will offer better results in taxa in which such evolutionary mechanisms have been common. Only a few experiments have been successful with diploid crops, although the improvement of their performance is as important or even more valuable than in polyploids.

By crossing of hexaploid wheat and diploid rye and subsequent backcrossing, several monosomic and monotelosomic rye–wheat chromosome additions were produced (Schlegel 1982; Schlegel et al. 1986a) using rye cytoplasm as background. However, during the first experiment alloplasmic rye–wheat additions were available. The hexaploid wheat Chinese Spring was used as a female.

Monosomic additions were maintained by backcrossing to the euploid rye parent and by subsequent microscopic screening. Because of missing male transmission of the extra chromosome, self-pollination did not result in disomic addition lines. Characteristic morphological and physiological distinctness could be detected depending on the alien chromosome added to rye. Since the population size was too small, no agronomic performance testing was possible.

In order to transfer wheat chromosomal segments or single genes into the rye complement, experiments were designed to take advantage of gametic selection after spike irradiation, which would facilitate recovery of desirable translocations. Thus, premeiotic spikes of several addition lines were exposed to x-rays of 1000 r dosage.

The resulting pollen grains were used to fertilize emasculated euplasmic spring rye plants. The resulting $F_1$ seedlings were screened by chromosomal banding techniques for the presence of alien chromatin and further characterized by meiotic analysis. A successful example of gene transfer could be demonstrated involving chromosome 6B of hexaploid wheat, var. Chinese Spring. Thirty-six treated spikes used for pollination produced 361 seeds for cytological screening. Since male transmission of the extra chromosome could not be fully excluded, all of the progenies were checked for somatic chromosome number. Among 188 plants analyzed, 169 (90%) individuals had 14 chromosomes, 1 (0.5%) 14+t[rye], 1 (0.5%) 13, 1 (0.5%) 28, and 16 (8.5%) 14+1[wheat]. Thus, the added wheat chromosome can be transmitted, at least to some extent, through the pollen cells, although previous studies of selfed alloplasmic 6B additions never yielded disomic progeny. By application of differential chromosome staining, all 15 chromosome plants were classified as rye plus 1 wheat chromosome 6B. However, among the 14 chromosome progeny, two types were found to have novel sites of N-bands, as compared to rye. In addition to the 13 normal rye chromosomes, one of the two plantlets showed in all somatic cells one prominent chromosome with a heavy N-band situated near the centromere. This band did not match any of the rye-specific N-bands. Its location and size resemble the heterochromatic block of chromosome 6B. More detailed studies revealed a dicentric structure of the rye–wheat translocation. This result provides clear evidence of transfer of alien genetic material to diploid rye. It is tolerated not only in the embryo but also by the adult plants. It demonstrates that wheat genes can compensate for the loss of rye genetic information (Schlegel and Kynast 1988).

## 12.4.12    DNA Transfer

Compared to other cereal and crop plants, rye was less accessible for DNA and gene transfer experiments. In addition, rye is known as one of the most recalcitrant species in tissue culture. Neither the tissue culture ability nor the regeneration ability is as good as in wheat, barley, rice, or other grasses. Merely a genotype of *S. vavilovii* (Grosse et al. 1996) showed a higher *in vitro* utilization than the average *S. cereale* genotypes tested (Flehinghaus et al. 1991). This strongly hindered the application the co-cultivation approach using *Agrobacterium tumefaciens*. The first breakthrough was obtained by Peña et al. (1987). Work on the development of the male germ line of rye showed that 14 days before first meiotic metaphase the archesporial cells are highly sensitive to caffeine and colchicine injected into the developing floral tillers. The authors considered that at this stage the archesporial cells might also be permeable to other molecules, such as DNA. Therefore, they injected DNA carrying a dominant selectable marker gene into rye plants. It was the plasmid pLGVneo1103, including the aminoglycoside phosphotransferase II gene (APH 3') under the control of the nopaline synthetase promoter. The plants were injected 2 weeks before meiosis. Because of the high degree of self-incompatibility of the JNK rye, seeds were produced by pairwise crossing of infected tillers.

From 3023 seeds screened for kanamycin resistance and derived from 98 plants, 7 seedlings remained green after 10 days' growth on kanamycin-containing medium. These apparently kanamycin-resistant plantlets were assayed for the presence of APH (3')II enzymatic activity. Two of the seven resistant seedlings, each resulting from an independent injection experiment, showed APH (3')II activity. This was the first unequivocal report on a gene transfer experiment in rye.

Recently, a reproducible transformation system for rye was established by using inbred lines with superior regeneration potential from tissue cultures. Biolistic parameters, such as tissue age during bombardment and microprojectile density, were compared in a multifactorial experiment. By using the selectable marker genes *bar* or *nptII*, transformation efficiency between 2 and 4% of the bombarded explants was found. A total of 37 independent transgenic rye plants were produced by biolistic gene delivery. For *Agrobacterium tumefaciens*-mediated gene delivery, different influencing factors were combined and led to morphologically normal and fertile transgenic plants at a frequency of 3.9% of the inoculated explants. Moreover, a selection system was also developed for direct production of transgenic plants. The high-molecular-weight glutenin subunit genes *Ax1*, *Dx5*, and *Dy10* from wheat were introduced into rye and their stable expression in endosperm of primary transformants of rye and in their segregating progeny demonstrated (Herzfeld 2002).

## 12.5 GENETICS

### 12.5.1   General

The development and utilization of new techniques in genetic analysis, karyotype identification, and chromosome manipulation, and the elucidation of homoeologous relationships in the Triticinae have greatly contributed to the present understanding of genetics and cytogenetics of rye. Knowledge of rye genetics, for a long period less developed as compared to other diploid crops (maize, barley, tomato, or pea), is progressing and has become useful in marker-aided selection. Although Jain (1960) in his review article stated that the genetics of rye had not received so much attention and that simple markers were hardly available, genetic and cytological markers for all rye chromosomes are now available in reasonable amounts.

Based on compilation of available data of genetics in rye, wheat, and triticale, as well as wheat–rye addition and substitution lines, a comprehensive presentation of gene designation, localization, and linkage relationships has been established. Since 1960 (Jain 1960), with updates in 1982, 1986, 1996, and 1997 (Schlegel and Mettin 1982; Schlegel et al. 1986b; Schlegel and

Melz 1996; Schlegel and Korzun 1997; Schlegal et al. 1997b), 1700 biochemical, molecular, and morphological markers have become available in rye. The number of markers mapped to rye chromosomes is about 80% molecular markers (Rogowsky et al. 1992; Quarrie et al. 1994), 12% biochemical markers (Fra-Mon et al. 1984; Hart and Gale 1988; McIntosh 1988; Hull et al. 1991, 1992), and 8% morphological markers. The best investigated rye chromosome is 1R, with about 350 mapped loci, followed by chromosomes 6R (280 loci), 5R (260 loci), 2R (215 loci), 4R (205 loci), 7R (190 loci), and 3R (160 loci), respectively. Meanwhile, markers for economically important traits, such as resistance and fertility traits, were applied in hybrid breeding of rye (Dreyer et al. 1996). Even first quantitative trait loci (QTL) were mapped on several rye chromosomes (Schlegel and Meinel 1994; Börner et al. 2000).

### 12.5.2  Chromosomal and Regional Localization of Genes and Markers

On the following pages, the localization of markers along the chromosomes and chromosomal regions are given. Since different authors published different results, often for the same marker, all available data have to be considered. Because of limited space in this paper, most of the mapping and linkage data from numerous authors are not included. Readers are encouraged to visit the author's website (www.desicca.de/plant_breeding), which is updated annually.

## 12.6 CYTOPLASMIC MALE STERILITY AND RESTORER

Schnell (1959) provided a first report on rye with cytoplasmic-genic pollen sterility. He found numerous offsprings showing pollen sterility in selfed lines of European rye varieties. However, inheritance of the character was not clear (Kobyljanskij 1969). More intensive crossing experiments demonstrated the presence of cytoplasmic-genic inheritance of male sterile plants in rye (Geiger and Schnell 1970). Systematic screening and genetic studies carried out at the University of Hohenheim, Germany, since 1966 among 46 European, 1 Argentinean, and 96 accessions from Iran yielded a cytoplasmic male sterile sample from Argentina that was designated Pampa cytoplasm. It became the main constituent for the German hybrid breeding program of rye, and it behaves analogous to the Texas cytoplasm in corn.

It is easy to find nonrestorer plants for reliable sterility maintenance, whereas fully effective fertility restorers occur rather seldomly. On the other hand, there are several other cytoplasms reacting in a similar but not in an identical manner as the Pampa cytoplasm, when tested for male sterility in crosses with specific pollinators. Adolf and Neumann (1981) and Steinborn et al. (1993) detected a second type of CMS cytoplasm. Compared to the Pampa type, it is classified as Vavilov type. The latter is difficult to maintain, while it was easy to find restorers for it, and vice versa for the Pampa cytoplasm.

Restorer genes were detected not only in exotic accessions but also in several European varieties. The frequency of gametes with restorer genes ranged between 10 and 20% across all accessions. During the last decade genes were mapped and described in detail (Glass et al. 1995; Dreyer et al. 1996; Greiger and Miedaner 1996; Miedaner et al. 2000).

Systems of chromosomal male sterility have not attained any practical meaning. Although there are *ms* genes, suitable markers, and balanced tertiary trisomics with very low male transmission (as in the balanced tertiary trisomic (BTT) system of barley), those prerequisites could not be combined in an efficient breeding program (Sybenga 1985).

## 12.7 BREEDING

Breeding methods have naturally been much influenced by the outbreeding nature of the crop. Rye has a gametophytic two-locus incompatibility system (Lundquist 1956). Early breeding approaches could best be described as forms of simple recurrent selection. Rye shows inbreeding depression, but inbred lines of acceptable vigor can be isolated and used in the construction of synthetic (and even hybrid) varieties, following suitable progeny tests for combining ability. The aims of recent breeding have been improvement of grain yield, improved stability, fast growth, fine stems, resistance to powdery mildew and brown rust, protein content, and quality, together with cold tolerance and shorter straws. Less problematic food and baking qualities and resistance to deformity largely determine the value of bread rye varieties. Varieties with plenty of leaf mass are particularly suited for use as green fodder.

## 12.7.1 Diploid Rye

Population breeding comprises the development of open-pollinated and synthetic varieties. In both cases the variety constitutes a panmictic population. The population is produced by random fertilization, at least in the final generation of seed production. The gametophytic self-incompatibility prevents self-fertilization under open pollination. It helps to avoid inbreeding depression.

### 12.7.1.1 Population Breeding

In population breeding, only self-incompatible varieties have usually been produced. An open-pollinated variety is the direct outcome of population improvement. When a breeding population has reached a per se performance level comparable to that of existing cultivars, it may be considered a new variety. Depending on the local statutory regulations, it can be released and registered as a variety.

Various selection procedures are being described for population improvement in rye (Ferwerda 1956; Laube and Quadt 1959; Sengbusch 1940; Vettel and Plarre 1955; Wolski 1975; Geiger 1982). Improvement of self-compatible populations aims at improving either the per se performance, the potential of the population for synthetic variety production, or both. In self-fertile material, selection exclusively aims at improving the potential of the population for hybrid variety production.

For all approaches the intra- and interpopulations' general combining ability of the parental units, i.e., plants, clones, pairs of plants, or pairs of clones, has to be improved. In addition, for hybrid breeding the mutational load of the populations has to be diminished in order to reduce inbreeding depression in line establishment. The different selection procedures can be operated in various ways according to the experimental facilities. Moreover, the procedures can be used simultaneously or successively in a given selection scheme. It follows the generalized scheme of population improvement of Halauer and Miranda (1981). The breeding procedure is divided into different selection cycles. Each cycle includes a parental unit (plants, clones, pairs of plants, or pairs of clones to be evaluated), a selection unit (plants, clones, or progeny that provide data used as the basis for selection), and a recombination unit (plants, clones, or progeny that finally recombine to form the improved population).

Cloning of parental material is necessary to achieve sufficient seeds for progeny tests in multilocation yield trials or even for microplots. The number of clones per genotype can be limited or even few when CMS testers are involved.

Selfing of heat-treated self-incompatible $S_0$ plants generally yields a few seeds only. In order to increase the number of seeds for yield trials, the $S_1$ lines have to be multiplied by free pollination under isolation cages or in isolation.

Considering a wide range of experiments, it is difficult to predict an optimum number of parental, selection, and recombination units and an optimum allocation of testing facilities. Even if just one selection procedure is taken into consideration, the optima may vary considerably depending on the underlying genetic and environmental factors.

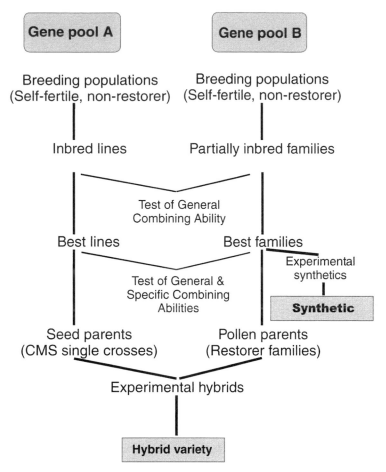

**Figure 12.6** Schematic drawing of an integrated breeding program for simultaneous development of hybrid and synthetic varieties in rye.

### 12.7.1.2 Synthetics

The term *synthetics* in rye is used to designate varieties that are produced by crossing *inter se* a number of selected parents with subsequent multiplication by open pollination under isolation. The parents can be clones, inbred families, or other genotypes. Long-term maintenance of clones is not feasible under practical conditions. However, their gametic arrays can be reproduced from $S_1$ families, which can be established by selfing under heat treatment (Wricke 1978).

The potential of a genotype as constituent of a synthetic variety is indicated by its general combining ability. Population improvement and the formation of synthetics ought to be organized as an integrated program. When a high testing accuracy is practiced in a population improvement, a selected fraction of the recombined material may serve as varietal parent without additional testing. The optimal number of parents is determined by almost the same parameters as the optimum population size in recurrent selection. Since synthetic breeding is not directed to selection with varieties but to creating better parents, intravarietal genetic variance is of minor concern, and the number of varietal parents can be chosen smaller than in long-term population improvement. Studies on rye revealed that the genetic variance among synthetics sharply decreases as the number of parents increases (Geiger et al. 1981). If several unrelated, well-combining and high-performing plant populations are at the breeder's disposal, the question arises whether a synthetic variety should be composed of parents from one single or from a certain number of such populations. Only in the case where a population is outstanding in both per se performance and variance of general combining ability is it likely to be better suited as a source for synthetics than any population set. In all other situations, the optimum has to be determined by predicting the expected performance of the best synthetic of each set of populations. Frequently, two or more populations turn out to be more promising than just one.

By the complex theoretical and practical situation, synthetic rye varieties have not gained acceptance at the seed market. If improvement of both hybrid and population varieties is conducted at the same breeding company, and integrated approach would be desirable to make more use of genetic resources, labor supply, and technical equipment, a general scheme for simultaneous development of hybrid and synthetic varieties in rye is proposed (Geiger 1982; also see Figure 12.6).

### 12.7.1.3  Hybrid Breeding

The main goal of hybrid breeding is a stable grain yield. This includes tolerance to drought and nutrient stress because rye is widely grown on poor, sandy soils where it has a higher relative performance than wheat and triticale. Rye can now compete with these cereals even on more fertile, productive soils. Since hybrids are genetically more uniform than population varieties, breeding for disease resistance is urgently needed, especially for those diseases that cannot be prevented by chemical means (Miedaner and Geiger 1999; Miedaner et al. 1995). Concerning quality, a large grain weight and resistance to preharvest sprouting are the most important features.

Hybrid breeding started in 1970 in Germany. The first three hybrid varieties, Forte, Aktion, and Akkord, were released in Germany in 1984. In multilocal testing trials they surpassed standard population varieties with approximately 10% in yield. Moreover, they showed shorter straw and better resistance to preharvest sprouting, but lower thousand-kernel weight and unchanged lodging resistance.

Today, more than 20 hybrid varieties are on the official list, occupying more than 60% of the total rye acreage. Some of these hybrids also are registered and distributed in Austria, Denmark, France, the U.K., Scandinavian countries, and the Netherlands. Independent hybrid rye breeding programs are running at present in Sweden and in Poland, and the first Polish hybrids were released in 1999. In addition, several programs are conducted in different areas of Russia. Outside Europe, the only hybrid rye breeding program is situated at the University of Sydney, Australia.

Rye is the only cross-pollinated species among the small grain cereals. Selfing is naturally prevented by an effective gametophytic self-incompatibility mechanism. Self-fertile forms have been found in several populations and are routinely used for developing inbred lines. Selfing results in strong inbreeding depression, and hybrids display a high heterosis. For grain yield, new data show a relative midparent heterosis of about 100%. Heterosis can be exploited only when preselected inbred lines of different gene pools are crossed to an $F_1$ hybrid. The use of cytoplasmic-genic male sterility is necessary. Male sterility is mainly caused by introgression in the Pampa cytoplasm of Argentinean rye that is environmentally highly stable. Pollen fertility in the hybrids is restored by the use of dominant nuclear-coded restorer genes from European or exotic populations.

Systematic search for gene pools with maximal heterosis revealed that two German populations, Petkus and Carsten, were particularly well matching. Inbred lines from the two pools are used successfully for the development of hybrids. A simplified scheme of hybrid rye breeding is shown in Figure 12.7. Seed parent lines are developed from Petkus and pollinator lines from the Carsten gene pool. Intensive selection for line performance is practiced in selfing generations $S_1$ and $S_2$.

Selfing is done by hand under isolation bags. After one or two stages of selection, the seed parent lines are transferred into the CMS-inducing Pampa cytoplasm by repeated backcrossing yielding in backcross progenies. $BC_1$ and $BC_2$ are subsequently crossed to parents from the opposite pool to select for test-cross performance, i.e., combining ability for grain yield. The reverse procedure is used for the pollinator lines. They are grown between isolation walls in adequate plots

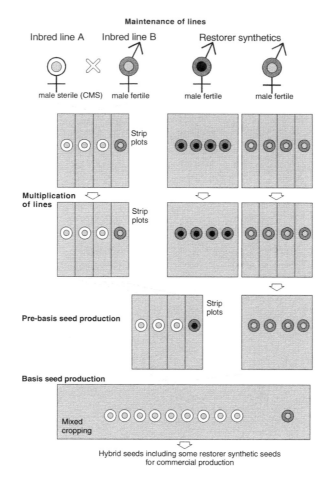

**Figure 12.7**  Schematic drawing of hybrid seed production in rye.

and crossed to CMS single crosses as testers. The test crosses are evaluated in multienvironmental trials with two replications.

Commercial hybrids are produced between CMS single crosses as seed parents and restorer synthetics as pollinator parents. The latter are mostly composed of two inbred lines crossed by hand and further multiplied by random open pollination. This complex type of hybrid (top-cross hybrid) needs several stages/years for production (Figure 12.7). The main advantages are a cheaper and more stable seed production with less risk for the breeder, a higher vigor of the hybrid seed, and an extended period of pollen shedding by the pollinator synthetic.

Substantial progress in grain yield has been achieved in hybrid rye breeding. In 1999, the best hybrid surpassed the best open-pollinated check by 20% in the official German trials. Over the years, progress was significantly higher for hybrid than for the population breeding. This clearly reflects that hybrid breeding is genetically the most efficient method available. Progress was also made for breeding rye with lodging resistance, leaf rust resistance, and bread-making quality. Hybrid rye has become more attractive to farmers than the traditional populations. Progress in hybrid breeding steadily increased this superiority since 1990 (Geiger 1990).

## 12.7.2  Tetraploid Rye

Most cultivated rye contains seven pairs of somatic chromosomes. Artificially produced tetraploid rye with 14 pairs of chromosomes is grown for seed production in limited amounts in Europe.

As a forage crop, it is grown on a large scale either individually or in mixtures with other forage crops in the U.S. and other countries. Tetraploid ryes often show perennial growth habit by introgression of *S. montanum*.

After discovery of the colchicine method for inducing polyploidy in 1937, development of polyploid crops was regarded as an unconventional technique to penetrate yield and other barriers in plant breeding. The two more or less universal effects of chromosomal doubling are increased cell size and decreased fertility. Consequently, crops that benefit most from increased cell size and suffer least from reduced fertility are inherently predisposed to benefit from polyploid breeding. Crops most amenable to improvement through chromosome doubling should have a low chromosome number, be harvested primarily for their vegetative parts, and be cross-pollinating. Two other conditions, the perennial habit and vegetative reproduction, have a bearing on the success of polyploid breeding by reducing a crop's dependence on seed production. From the 1950s to the 1970s, induced autopolyploidy in rye was considered to be an important breeding method. Rye showed good prerequisites for an autopolyploid crop. Russia, Poland, Germany, and Sweden spent considerable attention to the production of tetraploid rye.

Low fertility, seed shriveling, and aneuploid offsprings are believed to be influenced by irregular meiotic chromosome pairing. Two major approaches of chromosomal pairing regulation were investigated:

1. The research group of H. Rees at Aberystwyth (U.K.) (Hazarika and Rees 1967) favored an increased quadrivalent formation with convergent or parallel centromere co-orientation as a means of reduced aneuploidy and, consequently, improved fertility. A disjunction index (number of pollen mother cells without univalents and trivalents divided by the total number of pollen mother cells) was used as a measure for meiotic pairing regularity. They even demonstrated a positive correlation between the disjunction index and fertility.
2. Alternatively, Sybenga (1964) proposed a preferential bivalent pairing in order to reduce meiotic irregularities by induced allopolyploidization of an autotetraploid. Besides a genetic control of diploid-like chromosome pairing, a complex system of experimentally induced reciprocal translocations was intensively discussed. Despite strong efforts and tests over decades, no practical benefit could be achieved.

Schlegel (1976) introduced a method for mass production of autotetraploid rye by so-called valence crosses. Under microplot isolators, nonemasculated tetraploid genotypes (clonal plants) were crossed by spontaneous pollination with diploid genotypes (clonal plants). The tetraploid used as mother plants showed a recessive pale grain character, while the diploid used as male plants showed dominantly green seeds. In this way, green xenia could be selected among the pale grains after harvest of the mother plants. All the green xenia must be hybrids, either triploid or tetraploid. Microscopic chromosome counting can differentiate them. The triploids are produced from fusion of a diploid female gamete and a haploid male gamete, while the tetraploids are derived from a diploid female and an unreduced male gamete (Schlegel and Mettin 1975).

In this way, diploid genotypes showing high chiasma frequencies per PMC (>17 Xta/PMC) and other useful agronomic characters could be transferred into the so-called half-meiotic tetraploids. The method proved to be efficient for broadening genetic variability of tetraploid breeding material without deleterious effects of colchicine treatment. Until this time, all tetraploids available worldwide based on colchicine-induced genotypes were in a very limited number.

On the other hand, the higher chiasma frequency of the diploid parent did not significantly influence the number of multivalents of the tetraploids, despite the missing correlation between the quadrivalent frequency, aneuploid frequency, and fertility, at least in advanced breeding strains. However, depending on the karyological structure of the diploid parent, a more or less strong modification of the bivalent frequencies could be observed; i.e., the bigger the differentiation of the chromosome structure between the parental genotypes, the higher was the number of bivalents per PMC. Although just minor changes of the karyotype have been considered, they contributed

**Table 12.4 Utilization of Breeding Methods Considering the Reproduction System and Origin of Population**

| Reproduction | | Basic Population Derived from | | Breeding Method | |
|---|---|---|---|---|---|
| Autogamous | − | Crossing | + | Mass selection | − |
| Allogamous | + | Wild origin | − | Pedigree method | − |
| Anemophilous | + | Polyploidization | + | Combination breeding | + |
| Self-fertile | − | Induced mutation | + | Residue method | + |
| Self-sterile | + | | | Hybrid breeding | + |

**Table 12.5 Different Selection Procedures for Interpopulation Improvement in Self-Fertile Rye**

| Method | | Parental Unit | Selection Unit | Recombination Unit | Evaluation |
|---|---|---|---|---|---|
| Test crosses (CMS tester) | | Individual plants | Test crossing | $S_1$ line from self-fertile $S_0$ plant | Visual screening in one- or two-rowed observation plots |
| | | Individual clones | Test crossing | $S_1$ line from self-fertile $S_0$ plant | Regular yield testing |
| Inbred family selection | $S_1$ line from self-fertile $S_0$ plant | Individual plants | $S_1$ line from self-fertile $S_0$ plant | $S_1$ line from self-fertile $S_0$ plant | Visual screening in one- or two-rowed observation plots |
| | $S_1$ line from self-fertile $S_0$ plant | Individual clones | $S_1$ line from self-fertile $S_0$ plant | $S_1$ line from self-fertile $S_0$ plant | Microplot yield testing (1–2 m$^2$) |
| | $S_2$ line from self-fertile $S_0$ plant | Individual $S_1$ plants | $S_2$ line from self-fertile $S_0$ plant | $S_2$ line from self-fertile $S_0$ plant | Visual screening in one- or two-rowed observation plots |
| Combined test crosses of $S_1$ families | | Individual clones | Test cross of $S_1$ line from self-fertile $S_0$ plant | $S_1$ line from self-fertile $S_0$ plant | Regular yield testing or microplot yield testing (1–2 m$^2$) |

*Source*: Modified after Geiger, H.H., *Tag. Ber. Adl. Berlin*, 198, 305–332, 1982.

to a preferential bivalent pairing, obviously within the parental diploid genomes of both female and male donors.

Summarizing the data of over 50 years' research, neither an increased chiasma frequency or a higher frequency of quadrivalents per PMC nor a diploidization mechanism substantially contributed to breeding progress. Interchromosomal substitutions as a source of irregular chromosome pairing and aneuploidy, i.e., aneusomy, could be excluded as well (Schlegel et al. 1985). Gradual selection for yield, fertility, and low degree of shriveling was more successful than experimentally induced genetic or cytological changes.

## 12.8 RYE CROPPING

Rye may serve as grain, hay, pasture, cover crop, green fodder, and green manure. Several traits have to be considered not only in breeding but also for the technology depending on the utilization (Table 12.4 to Table 12.6).

Rye is a good pioneer crop for sterile soils. When used as a cover crop, it is grown for erosion control, to add organic matter, to enhance soil life, and for weed suppression. It may also stabilize and prevent leaching of excess soil or manure nitrogen. It has been used to protect soil from wind erosion in exposed areas and, with its tall stature, may be of some value in providing windbreaks. It is a good green manure because it produces large quantities of organic matter, but should be

**Table 12.6 Quality Characteristics of Rye Depending on the Type of Utilization**

| Traits | Type of Utilization | | | |
|--------|---------|----------------|--------------------|------------------|
| | Backing | Animal Feeding | Alcohol Production | Industrial Usage |
| Starch content | Neutral | Neutral | Important | Important |
| Protein content | Neutral | Important | Neutral | Neutral |
| α-Amylase | Negative | Neutral | Important | Important |
| Pentosan content | Important | Negative | Negative | Important |

used only in rotation with row crops because other grain crops are graded down in the market if they contain rye seed.

## 12.8.1 Seeding

Recommended seeding depths are 2.5 to 5.0 cm. The suggested seeding rates are 90 to 112 kg/ha. The optimal seed rate for hybrid material is about 150 seeds/m². When rye is seeded late, the rate should be increased up to 336 kg/ha to achieve rapid and complete vegetational cover and reduce erosion. To minimize erosion, a leaf area index of 1.0, i.e., complete cover, may be necessary.

## 12.8.2 Soil Type

Rye grows best on well-drained loam or clay loam soils, but even heavy clays, light sands, and infertile or poorly drained soils are feasible. It will grow on soils too poor to produce other grains or clover. On light sandy soils rye can produce more than 7 t/ha for several years when hybrid varieties are grown. In general, it is tolerant of different soil types. Rye is well known to tolerate acid soil. The range of best suitability is pH 5.0 to 7.0, but tolerance is between 4.5 and 8.0. Rye also shows high aluminum tolerance (Gallego et al. 1998). It even gives good yields on poor, sandy soils and does better than oat on sandy soil.

## 12.8.3 Water

Rye grows best with ample moisture, but in general it does better in low-rainfall regions than do legumes, and it can outyield other cereals on droughty, sandy, infertile soils. Its extensive root system enables it to be the most drought tolerant cereal crop, and its maturation date can change based on moisture availability. The structure of the rye plant enables it to capture and hold protective snow cover, which enhances winter hardiness. This snow retention might also be expected to enhance water availability. However, under intensive condition of rye production, a dense stand of plants in autumn can promote the growth of snow mold. Diploid varieties are more drought tolerant than tetraploid. Rye requires about 20 to 30% less water than wheat per unit of dry matter formation.

## 12.8.4 Micronutrients

In plant physiology, rye is recognized among cereals as a genotype showing high ability for uptake of micronutrients, e.g., under iron, copper, manganese, and zinc deficiency (Erenoglu et al. 1999). When grown in zinc-deficient calcareous soil in the field, the rye cultivars had the highest and the durum wheat the lowest zinc efficiency. Under zinc deficiency, rye had the highest rate of root-to-shoot translocation of zinc. The results indicated that high zinc efficiency of rye could be attributed to its greater zinc uptake capacity from soils. By utilization of wheat–rye addition lines it was demonstrated that genes on chromosome arms 1RS and 7RS are associated with high zinc efficiency (Schlegel et al. 1999).

Rye shows the ability to take up iron and copper under deficient conditions (Podlesak et al. 1990). As shown by Marschner et al. (1989) and Treeby et al. (1989), the release of phytosiderophores is the main mechanism of grasses to acquire iron and copper in the rhizosphere. An initial study demonstrated that rye secretes mainly 3-hydroxymugineic acid under iron deficiency conditions, but also mugineic acid and 2'-deoxymugineic acid (Mori et al. 1990). The latter authors found a clear correlation between high amount of those chelators and the presence of chromosome 5R in wheat–rye addition lines. The genes for high copper and iron efficiency were physically mapped on the distal region of chromosome 5RL by using a specific wheat–rye translocation line (Schlegel et al. 1993).

The different behavior of root uptake is not necessarily correlated with the concentration of micronutrients in the shoot (Schlegel et al. 1997a). Nevertheless, chromosomes 2R and 7R were associated with improved manganese and iron concentrations in the shoots, chromosome 1R with zinc, and chromosome 5R with copper concentration.

### 12.8.5 Fertilization

Rye even produces high yield under poor soil conditions and without or limited extra dressing. Intensive systems of rye production require additional nitrogen during the starting phase of vegetation. It promotes tillering and spikelet fertility. Depending on soil and variety, up to 30 to 50 kg/ha N can be recommended. Application of nitrogen during the period of shoot emergence and two-node stage influences the stand density. In order to utilize the yield potential, a third portion of nitrogen can be applied during the stage of spike emergence when sufficient water resources are available. The application of sludge is possible, however, in limited amounts. Precise calculations of mineral fertilizers have to be considered.

### 12.8.6 Temperature

Rye is grown in the cool temperate zones or at high altitudes. It is the most winter hardy of all small cereal grains. Its cold tolerance exceeds that of wheat, including the most hardy winter wheat varieties, and it is seldom injured by cold weather. Alien cytoplasm may even increase cold tolerance (Limin and Fowler 1984). Minimal temperatures for germinating rye seed have been variously given as 1 to 5°C. It grows better in cooler weather. Vegetative growth for rye requires a temperature of at least 4°C. It also can be incorporated earlier in the spring than other cereals. It is one of the best crops where fertility is low and winter temperatures are extreme.

### 12.8.7 Susceptibility and Resistance

Rye is afflicted by the following fungi: powdery mildew (*Erysiphe graminis*), ergot of rye (*Claviceps purpurea*), take-all of wheat (*Gaeumannomyces graminis*, *Ophiobolus graminis*), stalk smut (*Urocystis occulta*), stem rust (*Puccinia graminis* f. *secalis*), brown leaf rust (*Puccinia dispersa*), yellow rust (*Puccinia striiformis*, *Puccinia glumarum*), eyespot (*Cercosporella herpotrichoides*), snow mold (*Fusarium nivale*), fusariosis of rye spikes (*Fusarium* spp.), spot blotch (*Helminthosporium sativum*, *Bipolaris sorokiniana*), black mold (*Cladosporium herbarium*), anthracnose (*Colletotrichum graminicola*), septoria leaf blotch (*Septoria secalis*), and leaf blotch of rye (*Rhynchosporium secalis*). Viral diseases include barley yellow dwarf, wheat dwarf, soil-borne mosaic, and oat blue dwarf. Rye is susceptible to glyphosate and paraquat.

Rye may be attacked by nematodes, such as *Ditylenchus dipsaci*, *Anguina tritici*, and *Heterodera avenae*. Wheat, potato, turnip, lupin, alfalfa, and white mustard can be grown preceding rye to reduce *D. dipsaci*. Clean seed and crop rotation reduces *A. tritici*. The use of leguminous and root crops in rotation reduces *H. avenae*. Rye harbors particularly low densities of root lesion nematode (*Pratylynchus penetrans*).

### 12.8.8  Growth Regulators

The lodging resistance is reduced because of plant height of common rye varieties. Lodging significantly decreases grain yield and quality, despite more costly harvest. Even when seeding rate and nitrogen dressing are optimal, growth regulators can be applied twice for stem reduction, considering the specific characteristics of the variety and local and climatic conditions. Before and after application of growth regulators there should be sufficient soil moisture.

### 12.8.9  Incorporation in Crop Rotation

Rye is very suitable for several crop rotations showing high adaptability. The new hybrid varieties show high yield potential and, if an optimal cropping technique is used, outyield other cereals under comparable agronomic conditions. However, wheat is usually grown in the better soils, and it receives more attention within the crop rotation. Trials where rye is grown in monoculture or in rotation with winter wheat and winter barley showed the better yield performance. In some countries, it is sown with legumes or other grasses.

### 12.8.10  Allelopathic Effects

Rye produces several compounds that inhibit crops and weeds. Rye provides excellent weed suppression through both allelopathic and competitive mechanisms. Rye residues maintained on the soil surface release 2,4-dihydroxy-1,4(2H)-benzoxazin-3-one (DIBOA) and a breakdown product 2(3H)-benzoxazalinone (BOA), both of which are strongly inhibitory to germination and seedling growth of several dicot- and monocotyledonous plant species (Chase et al. 1991a,b). Further, microbially produced transformation products of BOA demonstrate several-fold increases in phytotoxin levels. Hence, a variety of natural products contribute to the herbicidal activity of rye residues. Several studies have demonstrated the allelopathic characteristics of rye residues and root exudates containing DIBOA and BOA. Experiments have shown marked reductions in germination and growth of several problem agronomic weeds, including barnyard grass (*Echinochloa crusgalli*), common lambs' quarters (*Chenopodium album*), common ragweed (*Ambrosia artemisiifolia*), green foxtail (*Setaria viridis*), and redroot pigweed (*Amaranthus retroflexus*).

### 12.8.11  Volunteering

Rye can become a weed through volunteering and sometimes escapes in waste places and fields; e.g., it is a common weed. In some areas, it grows even with only 15 to 20 l/m$^2$ of precipitation.

## 12.9 RYE AS A DONOR FOR GENETIC IMPROVEMENT OF WHEAT

The numerous crossing experiments and chromosomal manipulations between rye and wheat are exceptional within the plant kingdom and among the crop plants. Already in 1888, Rimpau (1891) produced the first intergeneric and partially fertile hybrid between hexaploid wheat and diploid rye. Since the first success, a wide range of various combinations of genomes, individual chromosomes, and even chromosome segments were established (Schlegel 1990).

### 12.9.1  Genome Additions

The basic aim of rye genome transfer into wheat was to combine the quality characteristics of wheat with the agronomic unpretentiousness of rye. From the many intergeneric hybridizations made between wheat and *Aegilops*, *Triticum*, *Haynaldia*, *Hordeum*, and *Agropyron* species, only the wheat–rye combination became of breeding and agronomic value. Hexaploid triticale (AAB-BRR) became a new, worldwide-grown, and man-made crop plant (Schlegel 1996). After a long period of cytogenetic research and breeding efforts, hexaploid triticale now enlarges the spectrum of cereal crops by high yield, good resistance against diseases, and improved protein content.

## 12.9.2 Chromosome Additions

Among more than 200 different wheat–alien chromosome addition lines, there are complete series of disomic wheat–rye additions, including the adequate disomic telocentric addition lines (Shepherd and Islam 1988). Individually added chromosomes are available from the rye varieties Imperial and King II. Incomplete sets exist from the varieties Dakold and Petkus Rye (Riley and Chapman 1958). Those addition lines were not only used for first genetic mapping studies in rye, but also as source for targeted gene transfer experiments from rye into wheat.

## 12.9.3 Chromosome Substitutions

In the past, wheat–rye or wheat–*Aegilops* chromosome substitutions were used for studies of homoeologous relationships between cereal genomes. First reports on spontaneous wheat–rye chromosome substitutions 5R (5A) were given in 1937 (Kattermann 1937) and 1947 (O'Mara 1947). Driscoll and Anderson (1967) reported the substitution of the wheat chromosomes 3A, 3B, 3D, and 1D by rye chromosome 3R. About 20 years later, 1R (1D) (Müller et al. 1989) and 1R (1A) substitutions (Koebner and Singh 1984) were produced in order to improve wheat for baking quality and resistance against diseases. Schlegel (1997) described other wheat–rye chromosome substitutions with breeding relevance.

## 12.9.4 Chromosome Translocations

Although a first experimental wheat–rye translocation (4B-2R) was produced in 1967 (Driscoll and Anderson 1967), introgression of rye genetic information into wheat became most famous by the spontaneous 1RS.1BL wheat–rye translocation (Mettin et al. 1973; Zeller 1973). Now more than 250 cultivars of wheat from all over the world contain this particular type of translocation (Schlegel 1997). The most important phenotypic deviation from common wheat cultivars is the so-called wheat–rye resistance, i.e., the presence of wide-range resistance to races of powdery mildew and rusts, which is linked with decreased bread-making quality, good ecological adaptability and yield performance (Schlegel and Meinel 1994), and better nutritional efficiency. The origin of alien chromosome was intensively studied for genetic and historical reasons. It turned out that four sources exist: two in Germany (most likely one source; Schlegel and Korzun 1997), one in the U.S., and one in Japan. The variety Salmon (1RS.1BL) is a representative of the latter, and the variety Amigo (1RS.1AL) is a representative of the penultimate group, while almost all remaining cultivars can be traced back to one or to the other German origin. Another wheat–rye translocation with breeding importance was found in the Danish variety Viking. It carries a 4B-5R interchange (Schlegel et al. 1993), causing high iron, copper, and zinc efficiency, compared to common wheat.

## 12.10 CONCLUSION

Rye (*S. cereale* L.), a secondary crop, has been cultivated for more than 5000 years. The domestication probably happened at several locations of Persia and Central Anatolia. Because of its winter hardiness and unpretentiousness, it became an important cereal crop in Europe, particularly during the cold periods of the Middle Ages. It is still grown on 9 million ha worldwide, mainly in Russia, Germany, Poland, Belarus, and Ukraine. Original rye was a meter-high grass. The long straw of landraces was used even for roofing of buildings. Meanwhile, rye became a modern crop with all the technological and agronomic advantages. There is a subsequent increase in rye grain yield caused by improvement of agronomy and new (hybrid) varieties.

Quite early rye became a subject of cytological research. The large chromosomes and the diploid genome made rye suitable for several microscopic studies. However, the allogamous flowering and genetic heterozygosity aggravated genetic analysis. Despite first approaches utilizing primary trisomics and reciprocal translocations, the breakthrough of rye genetics was associated with the latest molecular techniques. Currently there are more than 2000 biochemical, molecular, and morphological markers available. About 80% molecular markers, 12% biochemical markers, and 8% morphological traits have been located to the chromosomes. The most investigated chromosome of rye is chromosome 1R, followed by chromosomes 6R, 5R, 2R, 4R, 7R, and 3R. Cytological and genetic studies were always promoted by the dual role of rye: as an independent crop plant and as an important donor species for wheat improvement. Triticale, the first man-made crop, carries the diploid genome of rye. Some of the most successful wheat varieties of the world possess genes from chromosome 1R of rye.

## REFERENCES

Adolf, K. and Neumann, H. 1981. Probleme bei der Entwicklung und Nutzung eines Funktionssystems für die Hybridzüchtung beim Roggen. *Tag. Ber. Adl. Berlin* 191: 61–72.

Appels, R. 1982. The molecular cytology of wheat-rye hybrids. *Rev. Cytol.* 80: 93–132.

Augustin, C. and Schlegel, R. 1983. Chromosome manipulations. II. Production and characterization of interchromosomal translocations for genetic mapping in rye. *Arch Züchtungsforsch.* 12: 180–184.

Benito, C., Romero, M.P., Henriques-Gil, N., Llorente, F., and Figueiras, A.M. 1996. Sex influence on recombination frequency in *Secale cereale* L. *Theor. Appl. Genet.* 93: 926–931.

Bennett, M.D. and Smith, J.B. 1976. Nuclear DNA amounts on angiosperms. *Philos. Trans. R. Soc. Lond.* 274: 227–274.

Bhattacharyya, N.K. and Jenkins, B.C. 1960. Karyotype analysis and chromosome designation for *Secale cereale* L. "Dakold." *Can. J. Genet. Cytol.* 2: 168–277.

Börner, A., Korzun, V., Voylokov, A.V., Worland, A.J., and Weber, W.E. 2000. Genetic mapping of quantitative trait loci in rye (*Secale cereale* L.). *Euphytica* 116: 203–209.

Candela, M., Figueiras, A.M., and Lacadena, J.R. 1979. Maintenance of interchange heterozygosity in cultivated rye, *Secale cereale* L. *Heredity* 42: 283–289.

Chase, W.R., Nair, M.G., and Putnam, A.R. 1991a. 2,2'-oxo-1,1'-azobenzene: selective toxicity of rye (*Secale cereale* L.) allelochemicals to weed and crop species. II. *J. Chem. Ecol.* 17: 9–19.

Chase, W.R., Nair, M.G., Putnam, A.R., and Mishra, S.K. 1991b. 2,2'-oxo-1,1'-azobenzene: microbial transformation of rye (*Secale cereale* L.) allelochemical in field soils by *Acinetobacter calcoaceticus*. III. *J. Chem. Ecol.* 17: 1575–1584.

Corriveau, J.L. and Coleman, A.W. 1988. Rapid screening method to detect potential biparental inheritance of plastid DNA and results for over 200 angiosperm species. *Am. J. Bot.* 75: 1443–1458.

Devos, K.M., Atkinson, M.D., Chinoy, C.N., Francis, H.A., Harcourt, R.L., Koebner, R.M.D., Liu, C.J., Masojc, P., Xie, D.X., and Gale, M.D. 1993. Chromosomal rearrangements in the rye genome relative to that of wheat. *Theor. Appl. Genet.* 85: 673–680.

Dreyer, F., Miedaner, T., and Geiger, H.H. 1996. Chromosomale Lokalisation von Restorergenen aus argentinischen und iranischem Roggen (*Secale cereale* L.). *Vortr. Pflanzenzücht.* 32: 280–281.

Driscoll, C.J. and Anderson, L.M. 1967. Cytogenetic studies in Transec: a wheat-rye translocation line. *Can. J. Genet. Cytol.* 9: 375–380.

Erenoglu, B., Cakmak, I., Römheld, V., Derici, R., and Rengel, Z. 1999. Uptake of zinc by rye, bread wheat and durum wheat cultivars differing in zinc efficiency. *Plant Soil* 209: 245–252.

Ferwerda, F.P. 1956. Recurrent selection as a breeding procedure for rye and other cross-pollinated plants. *Euphytica* 5: 175–184.

Flehinghaus, T., Deimling, S., and Geiger, H.H. 1991. Methodical improvement in rye anther culture. *Plant Cell Rep.* 10: 397–400.

Fra-Mon, P., Salcedo, G., Aragoncillo, C., and Garcia-Olmedo, G. 1984. Chromosomal assignment of genes controlling salt-soluble proteins (albumins and globulins) in wheat and related species. *Theor. Appl. Genet.* 69: 167–173.

Frederiksen, S. and Peterson, G. 1998. A taxonomic revision of *Secale* (Triticeae, Poaceae). *Nordic J. Bot.* 18: 399–420.

Fröst, S., Vaivars, L., and Carlbom, C. 1970. Reciprocal extrachromosomal inheritance in rye (*Secale cereale* L.). *Hereditas* 65: 251–260

Gallego, F.J., Calles, B., and Benito, C. 1998. Molecular markers linked to the aluminium tolerance gene *Alt1* in rye. (*Secale creale* L.) *Theor. Appl. Genet.* 97: 1104–1109.

Geiger, H.H. 1982. Breeding methods in diploid rye (*Secale cereale* L.). *Tag. Ber. Adl. Berlin* 198: 305–332.

Geiger, H.H. 1990. Wege, Fortschritte und Aussichten der Hybridzüchtung. In *Pflanzenproduktion im Wandel*, G. Haug et al., Eds. VCH Verlag, Weinheim, Germany, pp. 41–47.

Geiger, H.H., Diener, C., and Singh, R.K. 1981. Influence of self-fertility on the performance of synthetic populations in rye (*Secale cereale* L.). In *Quantitative Genetics and Plant Breeding*, A. Gallais, Ed. pp. INRA Service des Publications, Versailles, France, 169–177.

Geiger, H.H. and Miedaner, T. 1996. Genetic basis and phenotypic stability of male fertility restoration in rye. *Vortr. Pflanzenzücht.* 35: 27–38.

Geiger, H.H. and Schnell, F.W. 1970. Cytoplasmic male sterility in rye (*Secale cereale* L.). *Crop Sci.* 10: 590–593.

Glass, C., Miedaner, T., and Geiger, H.H. 1995. Lokalisation von Restorergenen der Winterroggenlinie L18 mit RFLP-Markern. *Vortr. Pflanzenzücht.* 31: 52–55.

Grosse, B.A., Deimling, S., and Geiger, H.H. 1996. Kartierung von Genloci für Antherenkultureignung bei Roggen mittels molekularer Marker. *Vortr. Pflanzenzücht.* 32: 46–48.

Gupta, P.K. 1971. Homoeologous relationship between wheat and rye chromosomes. *Genetica* 42: 199–213.

Hart, G. and Gale, M.D. 1988. Guidelines for nomenclature biochemical/molecular markers. In *Proceedings of the 7th International Wheat Genetics Symposium*, Cambridge, England, pp. 1215–218.

Hazarika, M.H. and Rees, H. 1967. Genotypical control of chromosome behaviour in rye. X. Chromosome pairing and fertility in autotetraploids. *Heredity* 22: 317–332.

Heenen, W.K. 1962. Chromosome morphology in inbred rye. *Hereditas* 48: 182–200.

Herzfeld, J.C.P. 2002. Development of a Genetic Transformation Protocol for Rye (*Secale cereale* L.) and Characterisation of Transgene Expression after Biolistic or *Agrobacterium*-Mediated Gene Transfer. Ph.D. thesis, Martin Luther University, Halle-Wittenberg, Germany, pp. 1–74.

Houben, A., Kynast, R.G., Heim, U., Hermann, H., Jones, R.N., and Forster, J.W. 1996. Molecular cytogenetic characterisation of the terminal heterochromatic segment of the B-chromosome of rye (*Secale cereale*). *Chromosoma* 105: 97–103.

Hsam, S.L.K. and Zeller, F.J. 1996. Allelism of mildew resistance genes *Pm8* and *Pm17* in wheat-rye translocated chromosome T1BL.1RS. *Vortr. Pflanzenzücht.* 35: 186–187.

Hull, G.A., Halford, N.G., Kreis, M., and Shewry, P.R. 1991. Isolation and characterisation of genes encoding rye prolamins containing a highly repetitive sequence motif. *Plant Mol. Biol.* 17: 1111–1115.

Hull, G.A., Sabelli, P.A., and Shewry, P.R. 1992. Restriction fragment analysis of the secalin loci of rye. *Biochem. Genet.* 30: 85–97.

Jain, S.K. 1960. Cytogenetics of rye (*Secale* ssp.). *Bibliogr. Genet.* 19: 1–86.

Kagawa, N., Nagaki, K., and Tsujimoto, H. 2002. Tetrad-FISH analysis reveals recombination suppression by interstitial heterochromatin sequences in rye (*Secale cereale*). *Mol. Genet. Genomics* 267: 5–10.

Kamanoi, M. and Jenkins, B.C. 1962. Trisomics in common rye, *Secale cereale* L. *Seiken. Ziho.* 13: 118–123.

Kattermann, G. 1937. Zur Cytologie halmbehaarter Stämme aus Weizen-Roggen-Bastardierungen. *Züchter* 9: 196–199.

Kattermann, G. 1939. Ein neuer Karyotyp beim Roggen. *Chromosoma* 1: 284–299.

Kobyljanskij, V.D. 1969. Cytoplasmic male sterility in diploid rye. *Vestn. Selskokhooz. Nauk.* 6: 18–22 (in Russian).

Koebner, R.M.D. and Singn, N.K. 1984. Amelioration of the quality of a wheat-rye translocation line. In *Proceedings of the 4th Wheat Breeding Society of Australia*, Toowoomba, pp. 86–90.

Koller, D.L. and Zeller, F.J. 1976. The homoeologous relationships of rye chromosome 4R and 7R with wheat chromosomes. *Genet. Res.* 28: 177–188.

Laube, W. and Quadt, F. 1959. Roggen (*Secale cereale* L.). In *Züchtung der Getreidearten. Handbuch der Pflanzenzüchtung*, H. Kappert et al., Eds. Verl. P. Parey, Berlin, pp. 35–102.

Langdon, T., Seago, C., Jones, R.N., Ougham, H., Thomas, H., Forster J.W., and Jenkins, G. 2000. *De novo* evolution of satellite DNA on the rye B chromosome. *Genetics* 154: 869–884.

Levan, A. 1942. Studies on the meiotic mechanism of haploid rye. *Hereditas* 28: 6–11.

Lewitzky, G.A. 1931. The morphology of chromosomes. *Bull. Appl. Bot. Genet. Plant Breed.* 27: 19–74.

Lima de Faria, A. 1952. Chromosome analysis of the chromosome complement of rye. *Chromosoma* 5: 1–68.

Limin, A.E. and Fowler, D.B. 1984. The effect of cytoplasm on cold hardiness in alloplasmic rye (*Secale cereale* L.) and triticale. *Can. J. Genet. Cytol.* 26: 405–408.

Lundquist, A. 1956. Self-incompatibility in rye. I. Genetic control in the diploid. *Hereditas* 42: 293–348.

Manzanero, S., Puertas, M., Jimenez, G., and Vega, J.M. 2000. Neocentric activity of rye 5RL chromosome in wheat. *Chromosome Res.* 8: 543–554.

Marschner, H., Treeby, M., and Römheld, V. 1989. Role of root-induced changes in the rhizosphere for iron acquisition in higher plants. *Z. Pflanzenernährung. Bodenkunde.* 152: 197–204.

McIntosh, R.A. 1988. Catalogue of gene symbols for wheat. In *Proceedings of the 7th International Wheat Genetics Symposium*, Cambridge, England, pp. 1225–1323.

Melz, G., Neumann, H., Müller, H.W., and Sturm, W. 1984. Genetic analysis of rye (*Secale cereale* L.). I. Results of gene location on rye chromosomes using primary trisomics. *Genet. Polon.* 25: 112–115.

Melz, G. and Schlegel, R. 1985. Identification of seven telotrisomics of rye (*Secale cereale* L.). *Euphytica* 34: 361–366.

Melz, G. and Sybenga, J. 1994. The 3rd Workshop of Rye Genetics and Cytogenetics: revision and completion of the genetic map of rye (Gross Lüsewitz, Germany, October 25–28, 1993). *Genet. Polon.* 35: 131–132.

Mettin, D., Balkandschiewa, J., and Müller, F. 1972. Studies on trisomics of rye, *Secale cereale* L. *Tag. Ber. Akad. Landwirtsch. Wiss.* 119: 153–161 (in German).

Mettin, D., Blüthner, W.-D., and Schlegel, R. 1973. Additional evidence on spontaneous 1B/1R wheat-rye substitutions and translocations. In *Proceedings of the 4th International Wheat Genetics Symposium*, Columbia, MO, pp. 179–184.

Miedaner, T., Fromme, F.J., and Geiger, H.H. 1995. Genetic variation for foot-rot and *Fusarium* head-blight resistance among full-sib families of a self-incompatible winter rye (*Secale cereale* L.) population. *Theor. Appl. Genet.* 91: 862–868.

Miedaner, T., Glass, C., Dreyer, F., Wilde, P, Wortmann, H., and Geiger, H.H. 2000. Mapping of genes for male-fertility restoration in Pampa A CMS winter rye (*Secale cereale* L.). *Theor. Appl. Genet.* 101: 1226–1233.

Miedaner, T. and Geiger, H.H. 1999. Vererbung quantitativer Resistenzen gegen Pilzkrankheiten bei Roggen. *Vortr. Pflanzenzücht.* 46: 157–168.

Mori, S., Kishi-Nishizawa, N., and Fujigaki, J. 1990. Identification of rye chromosome 5R as carrier of the genes for mugineic acid synthetase and 3-hydroxymugineic acid synthetase using wheat-rye addition lines. *Jap. J. Genet.* 65: 343–353.

Müller, G., Vahl, U., and Wiberg, A. 1989. Die Nutzung der Antherenkulturmethode im Zuchtprozess von Winterweizen. II. Die Bereitstellung neuer Winterweizen-dh-Linien mit 1AL-1RS-Translokationen.*Plant Breed.* 103: 81–85.

Oinuma, T. 1952. Karyomorphology of cereals. Karyotype alteration of rye, *Secale cereale* L. *Biol. J. Okayama Univ.* 1: 12–71.

O'Mara, J.G. 1947. The substitution of a specific *Secale cereale* chromosome for a specific *Triticum aestivum* chromosome. *Genetics* 32: 99–100.

Pathak, G.N. 1940. Studies on the cytology of cereals. *J. Genet.* 39: 437–467.

Peña, A., de la, Lörz, H., and Schell, J. 1987. Transgenic rye plants obtained by injecting DNA into young floral tillers.*Nature* 325: 274–276.

Podlesak, W., Werner, T., Grün, M., Schlegel, R., and Hülgenhof, E. 1990. Genetic differences in the copper efficiency of cereals. In *Plant Nutrition: Physiology and Applications*, M.L. van Beusichem, Ed. Kluwer Academic Publishers, Amsterdam, pp. 297–301.

Pohler, W. and Schlegel, R. 1990. A rye plant with frequent A-B chromosome pairing. *Hereditas* 112: 217–220.

Quarrie, S.A., Steed, A., Semikhodsky, A., Calestani, C., Fish, L., Galiba, G., Sutka, J., and Snape, J.W. 1994. Identifying quantitative trait loci (QTL) for stress response in cereals. In *Ann. Rep. J. Innes Centre & Sainsbury Lab.* (Norwich, U.K.), pp. 25–26.

Ranjekar, P.K., Lafontaine, J.G., and Palotta, D. 1974. Characterization of repetitive DNA in rye (*Secale cereale*). *Chromosoma* 48: 427–440.

Riley, R. and Chapman, V. 1958. The production and phenotypes of wheat-rye chromosome addition lines. *Heredity* 12: 301–315.

Rimpau, W. 1891. *Kreuzungsprodukte landwirtschaftlicher Kulturpflanzen.* P. Parey Verlag, (Berlin), pp. 1–56.

Rimpau, J., Smith, D.B., and Flavell, R.B. 1978. Sequence organization analysis of the wheat and rye genomes by interspecies DNA/DNA hybridization. *J. Mol. Biol.* 123: 327–359.

Rogowsky, P.M., Shepherd, K.W., and Langridge, P. 1992. Polymerase chain reaction based mapping of rye involving repeated DNA sequences. *Genome* 35: 621–626.

Roshevitz, R.J. 1947. A monograph of wild, weedy and cultivated species of rye. *Act. Inst. Bot. Acad. Sci.* 6: 49–176.

Ruebenbauer, T., Kubara-Szpunar, L., and Pajak, K. 1983. Inheritance of violet coloration in rye kernels (*Secale cereale* L.). *Genet. Polon.* 24: 313–317.

Schlegel, R. 1976. The relationship between meiosis and fertility in autotetraploid rye, *Secale cereale* L. *Tag. Ber. Adl. Berlin* 143: 31–37.

Schlegel, R. 1982. First evidence for rye-wheat additions. *Biol. Zbl.* 101: 641–646.

Schlegel, R. 1987. Neocentric activity in chromosome 5R of rye revealed by haploidy. *Hereditas* 107: 1–6.

Schlegel, R. 1990. Efficiency and stability of interspecific chromosome and gene transfer in hexaploid wheat, *Triticum aestivum* L. *Kulturpflanze* 38: 67–78.

Schlegel, R. 1996. Triticale: today and tomorrow. In *Triticale: Today and Tomorrow, Developments in Plant Breeding 5,* H. Guedes-Pinto et al., Eds. Kluwer Academic Publ., Amsterdam, pp. 21–32.

Schlegel, R. 1997. Current list of wheats with rye introgressions of homoeologous group 1. *Wheat Inf. Serv.* 84: 64–69.

Schlegel, R., Cakmak, I., Torun, B., Eker, S., and Köleli, N. 1997a. The effect of rye genetic information on zinc, copper, manganese and iron concentration of wheat shoots in zinc deficient soil. *Cereal Res. Commun.* 25: 177–184.

Schlegel, R. and Gill, B.S. 1984. N-banding analysis of rye chromosomes and the relationship between N-banded and C-banded heterochromatin. *Can. J. Genet. Cytol.* 26: 765–769.

Schlegel, R. and Korzun, V. 1997. About the origin of 1RS.1BL wheat-rye chromosome translocations from Germany. *Plant Breed.* 116: 537–540.

Schlegel, R. and Kynast, R. 1988. Wheat chromosome 6B compensates genetic information of diploid rye, *Secale cereale* L. In *Proceedings of the 7th International Wheat Genetics Symposium,* Cambridge, U.K., pp. 421–426.

Schlegel, R., Kynast, R., and Schmidt, J.C. 1986a. Alien chromosome transfer from wheat into rye. In *Genetic Manipulation in Plant Breeding,* W. Horn et al., Eds. Gruyter Publ., Berlin, pp. 129–136.

Schlegel, R., Kynast, R., Schwarzacher, T., Römheld, V., and Walter, A. 1993. Mapping of genes for copper efficiency in rye and the relationship between copper and iron efficiency. *Plant Soil* 154: 61–65.

Schlegel, R. and Meinel, A. 1994. A quantitative trait locus (QTL) on chromosome arm 1RS of rye and its effect on yield performance of hexaploid wheat. *Cereal Res. Commun.* 22: 7–13.

Schlegel, R. and Melz, G. 1996. A comprehensive gene map of rye, *Secale cereale* L. *Vortr. Pflanzenzüchtg.* 35: 311–321.

Schlegel, R., Melz, G., and Korzun, V. 1997. Genes, marker and linkage data of rye (*Secale cereale* L.). 5th updated inventory. *Euphytica* 101: 23–67.

Schlegel, R., Melz, G., and Mettin, D. 1986b. Rye cytology and cytogenetics: current status. *Theor. Appl. Genet.* 72: 721–734.

Schlegel, R., Melz, G., and Nestrowicz, R. 1987. A universal reference karyotype in rye, *Secale cereale* L. *Theor. Appl. Genet.* 74: 820–826.

Schlegel, R. and Mettin, D. 1975. Studies of valence crosses in rye (*Secale cereale* L.). IV. The relationship between meiosis and fertility in tetraploid hybrids. *Biol. Zbl.* 94: 295–315.

Schlegel, R. and Mettin, D. 1982. The present status of chromosome recognition and gene localization in rye, *Secale cereale* L. *Tag. Ber. Akad. Landwirtsch. Wiss.* 198: 131–152.

Schlegel, R., Özdemir, A., Tolay, I., Cakmak, I., Saberi, H., and Atanasova, M. 1999. Localisation of genes for zinc and manganese efficiency in wheat and rye. In *Plant Nutrition: Molecular Biology and Genetics*, G. Gissel-Nielsen and A. Jensen, Eds. Kluwer Academic Publ., Amsterdam, pp. 417–424.

Schlegel, R. and Pohler, W. 1994. Identification of an A-B translocation in diploid rye. *Breed. Sci.* 44: 279–283.

Schlegel, R., Scholz, I., and Fischer, K. 1985. Aneusomy as source of meiotic disturbances in tetraploid rye? *Biol. Zbl.* 104: 375–384.

Schlegel, R. and Sturm, W. 1982. Meiotic chromosome pairing of primary trisomics of rye, *Secale cereale* L. *Tag. Ber. Akad. Landwirtsch. Wiss.* 198: 225–243.

Schlegel, R. and Weryszko, E. 1979. Intergeneric chromosome pairing in different wheat-rye hybrids revealed by the Giemsa banding technique and some implications on karyotype evolution in the genus *Secale*. *Biol. Zbl.* 98: 399–407.

Schnell, F.W. 1959. Roggen. In *Dreißig Jahre Züchtungsforschung*, W. Rudorf, Ed. G. Fischer Verl., Stuttgart, pp. 135–139.

Sengbusch, R. von. 1940. Pärchenzüchtung unter Ausschaltung von Inzuchtschäden. *Forschungsdienst* 10: 545–549.

Shepherd, K.W. and Islam, A.K.M.R. 1988. Fourth compendium of wheat-alien chromosome additions. In *Proceedings of the 7th International Wheat Genetics Symposium*, Cambridge, U.K., pp. 1373–1387.

Smith, D. B. and Flavell, R. B. 1977. Nucleotide sequence organization in rye genome. *Biochem. Biophys. Acta.* 474: 82–97.

Steinborn, R., Schwabe, W., Weihe, A., Adolf, K., Melz, G., and Börner, T. 1993. A new type of cytoplasmic male sterility in rye (*Secale cereale* L.): analysis of mitochondrial DNA. *Theor. Appl. Genet.* 85: 822–824.

Sturm, W. 1978. Identification of Trisomics in the Rye Variety "Esto" and Trisomic Analysis of Short Straw Gene *Hl*. Ph.D. thesis, Akad. Landwirtschaftswissenschaften. pp. 1–180 (in German).

Sturm, W. and Melz, G. 1982. Telotrisomics. *Tag. Ber. Akad. Landwirtsch. Wiss.* 198: 217–225.

Stutz, H.C. 1972. The origin of cultivated rye. *Am. J. Bot.* 59: 59–70.

Sybenga, J. 1964. The use of chromosomal aberrations in the autopolyploidization of autopolyploids. In *Proceedings of the Symposium IAEA/FAO*, Rome, pp. 741–768.

Sybenga, J. 1965. The quantitative analysis of chromosome pairing and chiasma formation based on the relative frequencies of MI configurations. II. Primary trisomics. *Genetica* 36: 339–350.

Sybenga, J. 1983. Rye chromosome nomenclature and homoeology relationships. 1st Workshop on Rye. *Z. Pflanzenzücht.* 90: 297–304.

Sybenga, J. 1985. Critical appraisal of balanced chromosomal ms systems in hybrid rye breeding. In *EUCARPIA Meeting of Cereal Section Rye*, Svalöv, Sweden, pp. 307–311.

Sybenga, J., Eden, J. van, Meijs, Q.G. van der, and Roeterding, B.W. 1985. Identification of the chromosomes of the rye translocation tester set. *Theor. Appl. Genet.* 69: 313–316.

Sybenga, J. and Wolters, Z.H.G. 1972. The classification of the chromosomes of rye (*Secale cereale* L.): a translocation tester set. *Genetica* 43: 453–464.

Takagi, F. 1935. Karyogenetical studies on rye. I. A trisomic plant. *Cytologia* 6: 496–501.

Tijo, J.N. and Levan, A. 1950. The use of oxyquinoline in chromosome analysis. *Anal. Est. Exp. Aula Dei.* 2: 21–64.

Treeby, M., Marschner, H., and Römheld, V. 1989. Mobilization of iron and other micronutrient cations from a calcareous soil by soil-borne, microbial and synthetic metal chelators. *Plant Soil* 114: 217–226.

Verma, S.G. and Rees, H. 1974. Giemsa staining and the distribution of heterochromatin in rye chromosomes. *Heredity* 32: 118–122.

Vettel, F. and Plarre, W. 1955. Mehrjährige Heterosisversuche mit Winterroggen. *Z. Pflanzenzücht.* 42: 233–247.

Viinikka, Y. 1985. Identification of the chromosomes showing neocentric activity in rye. *Theor. Appl. Genet.* 70: 66–71.

Vosa, C.G. 1974. The basic karyotype of rye (*Secale cereale*) analyzed with Giemsa and fluorescence method. *Heredity* 33: 403–408.

Vries, J.M. de and Sybenga, J. 1976. Identification of rye chromosomes: the Giemsa banding pattern and the translocation tester set. *Theor. Appl. Genet.* 48: 35–43.

Wolski, T. 1975. The use of pair crosses in rye breeding. *Hod. Rosl. Aklim. Nasienn.* 19: 509–520.

Wricke, G. 1978. Degree of self-fertilization under free pollination in rye populations containing a self-fertility gene. *Z. Pflanzenzücht.* 82: 281–285.

Zeller, F.J. 1973. 1B/1R wheat-rye chromosome substitutions and translocations. In *Proceedings of the 4th International Wheat Genetics Symposium*, Columbia, MO, pp. 209–221.

Zeller, F.J., Kimber, G., Gill, B.S. 1977. The identification of rye trisomics by translocations and Giemsa staining. *Chromosoma* 62: 279–289.

Zohary, D. 1971. Origin of southwest Asiatic cereals: wheats, barley, oats and rye. In *Plant Life of Southwest Asia*, P.H. Davis et al., Eds. Proceedings Royal Botanical Society of Edinburgh (U.K.), pp. 235–263.

# CHAPTER 13

# Triticale: A Low-Input Cereal with Untapped Potential

Tamás Lelley

## CONTENTS

13.1  Introduction..................................................................................................396
13.2  The Wheat–Rye Galaxy ...............................................................................397
13.3  History .........................................................................................................398
13.4  Cytogenetics ................................................................................................400
       13.4.1  Tetraploid Triticale .........................................................................402
13.5  Genetics .......................................................................................................404
13.6  Breeding.......................................................................................................406
       13.6.1  Yield.................................................................................................407
       13.6.2  Winter Hardiness .............................................................................407
       13.6.3  Lodging............................................................................................408
       13.6.4  Preharvest Sprouting .......................................................................408
       13.6.5  Abiotic Stress...................................................................................409
       13.6.6  Biotic Stress.....................................................................................410
       13.6.7  Hybrid Breeding ..............................................................................412
       13.6.8  Biotechnology..................................................................................413
               13.6.8.1  Molecular Markers and Mapping ......................................413
               13.6.8.2  Haploidy ............................................................................414
       13.6.9  Quality .............................................................................................415
               13.6.9.1  Use for Food ......................................................................415
               13.6.9.2  Use for Feed ......................................................................416
               13.6.9.3  Forage and Dual Use .........................................................416
13.7  Sources of Genetic Variation.......................................................................417
13.8  Final Conclusion..........................................................................................419
References ...............................................................................................................419

## 13.1 INTRODUCTION

Triticale is a truly man-made crop. It does not exist in nature, and it would not survive in the wild without human care. Triticale is a typically human inspiration. Knowing the undemanding, robust nature of rye being better than any other cereal in poorer environments, and the high yield potential and good quality of wheat, man wanted to create a less demanding, high-yielding, and good-quality plant, and he succeeded. Almost.

The more than 130 years of history of triticale is a didactic play of human determination. Again and again, when enthusiasm started to fade, innovative new ideas led to the next step of improvement in reducing the initial weaknesses of triticale, such as lateness, high sterility, too much plant height, and grain shriveling. Today's triticale varieties, under appropriate environmental conditions, successfully compete or even outcompete their parents, wheat and rye. According to the FAO database (FAOSTAT 2004), on 3,356,778 ha worldwide, some 13,784,447 t of triticale were harvested, with an average yield of 41,065 hg/ha. Data can be compared from countries with developed agriculture and where all three cereals — wheat, rye, and triticale — are grown on a large scale. Complementing these data with information about the prevailing environmental conditions where triticale is being grown, it appears that triticale found its place in growing areas that were less favorable for wheat and also not typical for growing rye (Figure 13.1).

But triticale is not only a successful new cereal. It is a wonderful object for studying the evolution of a polyploid species. Polyploidy — especially allopolyploidy, the merging of two different species into one genetic unit with new characters and extended adaptation — is one of nature's most elegant solutions to increase variability and fix heterosis. The genetic control of chromosome pairing in the wheat parent has had a stabilizing effect in combining the benefits of both polyploidy and hybridity, a combination that promotes the founding of successful species (see Chapter 1 by Jauhar in this volume). In nature, such a new creation has to be successful right at the beginning of its existence, not suppressed by the parents with which the new plant has to compete. Two complete sets of genes that are doing more or less the same in the progenitors have to find a way to cooperate within a short period. The study of polyploid evolution received a major boost from recent developments in molecular biology, which enables us to study the above-

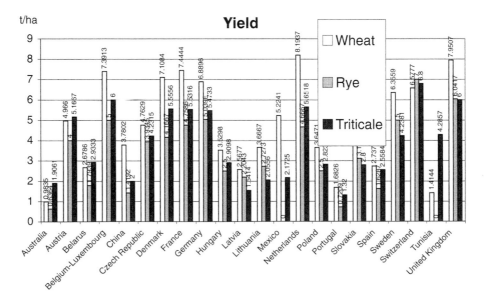

**Figure 13.1** Average yield per hectare of wheat rye and triticale in countries where all three crops are grown. (FAOSTAT, 2003.)

mentioned process of genome adjustment and fine-tuning by molecular means. It is not surprising that the whole complexity of the process of adjustment of gene and genome duplications is realized as myriad interactions, and the understanding of their mechanism, function, and significance is still in its infancy (Wendel 2000). Triticale offers a unique opportunity to watch closely what is taking place right at the beginning, and what might go wrong. It appears that the potential triticale offers for this field of research is not yet fully recognized.

Finally, the author is aware that only a small part of the available literature on triticale could be considered when composing this essay, and selection was clearly influenced by his own almost 30 years of involvement in triticale research.

## 13.2 THE WHEAT–RYE GALAXY

Rye (*Seacel cereale* L.) is wheat's (*Triticum aestivum* L.) closest relative among the major cereal genera (Flavell et al. 1977) and is used most intensively to extend the genetic variability of wheat via intergeneric recombination (Friebe et al. 1998). Attempts were even made to introduce wheat chromatin into diploid rye (Schlegel and Kynast 1988). Figure 13.2 shows the combinations in which rye chromatin was added to wheat and vice versa. While wheat–rye addition and substitution lines played an important role in determining homoeologous relationships between the two genera and were extensively used to search for useful genes in rye for wheat breeding, rye–wheat addition lines remain so far only an excellent demonstration of the power of cytogenetic techniques (Schlegel et al. 1986). From the many wheat–rye translocations produced, the 1BL.1RS transloca-

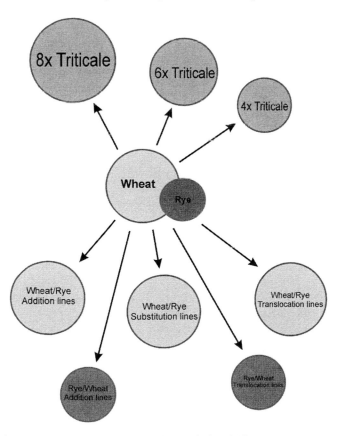

**Figure 13.2**  The wheat–rye galaxy demonstrating the mode in which wheat and rye chromatin have been combined into a functional genotype.

tion especially deserves mention, since more than 300 wheat cultivars around the world are known to possess this translocation (Rabinovitch 1998), and a large number of publications have been devoted to the role of this translocation in wheat breeding (Graybosch 2001; Lelley et al. 2004). All this work to combine wheat and rye chromatin started with the first creation of octoploid triticale in the 19th century. While this particular wheat–rye combination was dealt with in the previous chapter, essentially this whole chapter is devoted to hexaploid (6x) triticale, with a special section on the 4x triticale.

## 13.3 HISTORY

The Scottish amateur botanist A.S. Wilson is cited to have reported for the first time, in 1876, on successful fertilization of wheat with rye (Wilson 1876), followed by an American plant breeder, E.S. Carman, in 1884 (Carman 1884 from Leighty 1916). Although these plants were vigorous in growth, they remained completely sterile because of the lack of regular pairing of chromosomes during gamete formation of the essentially haploid $F_1$ hybrid, certainly not recognized in those days. But another peculiarity was also not known then, i.e., the presence in wheat of the dominant crossability genes *Kr1* and *Kr2*, which strongly reduce successful fertilization between wheat and rye, and which were not described until 1943 (Lein 1943). It has to be considered a lucky coincidence that in 1888, Rimpau, a German plant breeder, not only succeeded in producing a single $F_1$ wheat–rye hybrid, but this hybrid produced 12 seeds that gave fertile plants of uniform appearance (Rimpau 1891). This true breeding strain attracted great attention and is still maintained in the gene bank in Gatersleben, Germany, as Triticale Rimpau (accession number TCA 26). It is interesting to mention that the octoploid nature of this plant was cytologically demonstrated only in 1935 (Lindschau and Oehler 1935).

Intensive work with triticale in the 20th century started in Russia. In Saratov, a large-scale natural intergeneric hybridization was observed between wheat and rye (Meister 1921), from which later true breeding and fertile intermediate types were derived (Tiumiakov 1930). In 1930, the first botanical description of *Triticum secalotricum saratoviense* MEISTER was presented (Meister 1930). Lewitzky and Benetzkaya (1930) published the first drawings of the 56 somatic chromosomes of an octoploid triticale. They were the first to postulate an apogamous origin of the amphidiploid. Lebedeff, who produced amphidiploid wheat–rye hybrids in the Ukraine, reported the first aneuploidy in triticale. He was the first to articulate the special situation of the open-pollinating rye in a more or less inbreeding triticale by saying: "With respect to its hereditary complexity, which arose for rye in the amphidiploid, as a consequence homozygosity is to a degree increased, which corresponds to that which has been achieved by repeated inbreeding" (Lebedeff 1934, translated to English).

Arne Müntzing in Lund, Sweden, was one of the first to start intensive work on triticale in Europe (Müntzing 1939). Müntzing devoted much of his scientific career to the octoploid triticale. His review "Triticale Results and Problems" from 1979 is a most valuable source of information on the first 50 years of the history of triticale. It is worth recalling his statement made 24 years ago: "It is true that a large amount of work will be needed for the production of improved strains with higher value and usability than the material of triticale now available. Anyhow, it is justified to state that the threshold now has been passed and triticale is here to stay" (Müntzing 1979).

In the 1930s, work with octoploid triticale also started in Germany. New octoploid strains were produced and studied cytogenetically. They were intercrossed or crossed on primary wheat–rye hybrids to create genetic variability for selection. Triticale Rimpau was also used as a crossing partner (Oehler 1936; v. Berg and Oehler 1938; Katterman 1939).

The name *triticale* was first mentioned in the publication of Lindschau and Oehler (1935) and was credited to be coined by Tschermak, one of the three rediscoverers of Mendel's laws. The Latin generic name *Triticosecale* Wittmack was suggested by Baum (1971).

A major event in the history of triticale was the invention of the chromosome-doubling technique by using colchicine (Blakeslee and Avery 1937). Colchicine, an alkaloid of the autumn crocus *Colchicum autumnale* L., stops spindle formation and prevents the separation of sister chromatides and their migration to opposite poles. The resulting nucleus of a hitherto haploid (amphihaploid) cell becomes amphidiploid, giving rise to an amphidiploid tissue. Using this technique, new strains of triticale could be produced at will.

"It is possible that the interest in triticale as a potential new crop would have tapered off entirely if the efforts had been limited to octoploid material" (Müntzing 1979). But instead, at the end of the 1940s and into the 1950s a new era began, when hexaploid triticale entered the scene. Works by O'Mara (1948), Nakajima (1950), Kiss and Rédei (1953), and Sanches-Monge (1956) indicated that 6x triticale might be better in meiotic stability and fertility than octoploids. From then on, hexaploid triticale took over the lead, leaving China the only country where research and breeding efforts on octoploid triticale persisted until the 1990s. The reason why Chinese researchers took this position was twofold. Up to the 1960s, Chinese scientists produced 4700 primary strains of octoploid triticale, representing such a rich man-made resource for conventional crossbreeding that it was impossible to change the research priority. A second reason was the satisfactory (even better than wheat) bread-making quality of octoploid triticale, which made it well suited for home-made steamed bread consumed by farmers. The quality of hexaploid triticale was unsuitable for bread making (Bao and Yan 1993).

Meanwhile, on the hexaploid level a further milestone was the beginning of crosses between 6x and 8x triticale. In this way, a combination of the A and B genomes of tetraploid and hexaploid wheat has been achieved, leading to several superior properties of the new crop (Kiss 1966; Pissarev 1966). In 1968 two hexaploid varieties — numbers 57 and 64 — were released in Hungary; in 1969 these two varieties were grown on about 40,000 ha in Hungary (Kiss 1971). Crossing between hexaploid and octoploid triticale to produce the so-called secondary 6x triticale became one of the standard methods in triticale breeding.

In 1954 at the University of Manitoba, Winnipeg, the Rosner Research Chair was established with a generous private donation. The supported research led in 1969 to the first licensed triticale variety Rosner in Canada (Larter et al. 1970). In 1964 Borlaug initiated the triticale research program at International Maize and Wheat Improvement Center (CIMMYT) in Mexico, financially supported by the Rockefeller Foundation. In 1968 this program was taken over by F.J. Zillinsky. In 1971 the most comprehensive cooperative research program between CIMMYT and the University of Manitoba started, providing three entirely different climatic zones and two growing seasons per year for selection. In 1969 CIMMYT initiated an international triticale testing program, in which many countries around the world cooperate.

Another push in the history of triticale breeding was the detection of the first D genome chromosome substitutions in the R genome of hexaploid triticale. Such substitutions may occur following 6x triticale × 6x wheat crosses. The most prominent outcome of such a cross was the CIMMYT variety Armadillo, which was celebrated as a breakthrough in triticale breeding (Zillinsky and Borlaug 1971). Armadillo had a shorter straw, better fertility, higher yield, improved test weight, day-length insensitivity, early maturity, and good nutritional quality. By 1970 practically all the material in the CIMMYT triticale breeding program originated from crosses with Armadillo as one of the crossing parents (Zillinsky 1974). As was shown later in Armadillo, chromosome 2R was replaced by 2D (Gustafson and Zillinsky 1973). Meanwhile, many other positive results were reported of 6x triticale × 6x wheat crosses (e.g., Larter et al. 1968; Jenkins 1969; Popov and Tsvetkov 1975). The development of the Giemsa technique for differential chromosome staining opened a new horizon for triticale cytogenetics. Using this technique, Merker (1975) identified a number of D genome chromosome substitutions in the CIMMYT breeding material; a better understanding of the consequences of specific R/D substitutions could be expected. Time has shown, however, that to use the full potential of triticale, a complete rye genome is preferable. Today's triticale varieties are essentially free of any R/D substitutions.

The history of triticale is indeed a changing one. The idea to combine the favorable traits of the two species wheat and rye to produce a high-yielding, low-input nutritious cereal as animal feed or for human consumption still stimulates breeding efforts. New vistas, e.g., double haploid (DH) technology, molecular markers, and hybrid triticale, are being tested for use in practical breeding. The first international symposium devoted entirely to triticale was held in El Batan, Mexico, in 1973. The cereal section of EUCARPIA organized its first triticale meeting in 1973 in Leningrad, followed by a meeting in Radzików, Poland, in 1979, in Clermond-Ferrand, France, in 1984, and in Schwerin, then the German Democratic Republic, in 1988. The first International Triticale Symposium was held in Sydney, Australia, in 1986. Since then, triticale researchers and breeders have met regularly at 4-year intervals: in Passo-Fundo, Brazil (1990), in Lisbon, Portugal (1994), in Red Deer, Alberta, Canada (1998), and lastly in perhaps the most successful country for triticale breeding in the world, in Radzikow, Poland (2002).

Objectively seen, triticale has not yet fulfilled the probably overstated expectations. CIMMYT's Annual Report 2001–2002 states that "Triticale … is not widely sown, or even known, though it has been around since the 19th century, and CIMMYT has worked on it for more than 30 years. Originally promoted as a new grain for human consumption, triticale has made little headway against more established crops." Whether this is due to the still too-short evolutionary history of the plant, the larger-than-expected incompatibility of the R genome with the two wheat genomes A and B, the fact that nature has never created a competitive triticale, or the lack of a powerful lobby for triticale, these and other related questions will be discussed in the following pages of this chapter.

## 13.4 CYTOGENETICS

The role of chromosomes in heredity was first described in the milestone publication of Sutton in 1903. The assumption that true breeding stable wheat–rye hybrids represented an addition of the complete chromosome complement of the six wheat (AABBDD) and the two rye (RR) genomes, with the genome formula of AABBDDRR, became a reality by counting the number of chromosomes in root tips of the plants (Lewitzky and Benetzkaya 1930; Lindschau and Oehler 1935). It soon became clear that the chromosome number of octoploid triticale was not stable. A great part of the observed genetic variation was caused by aneuploidy, mainly by loss of chromosomes. Much of the early cytogenetic work on triticale was devoted to determining the aneuploid frequency of different strains under different conditions. Krolow (1962) reported, for example, that lines with the highest seed fertility had an average euploid frequency of 59.7%. Ten years later, Weimarck (1973) found in selected strains of hybrid derivatives an average euploid frequency of 69.3%, ranging from 40.0 to 84.2%.

A marked improvement in cytological stability of triticale was observed at the hexaploid level, AABBRR. Tsuchiya, studying advanced strains, found in 21 strains an average euploid frequency of 85% and in another six strains an average of 70%, the lowest being 57.2% (Tsuchiya and Larter 1969, 1971). The early work on the cytogenetics of triticale was excellently reviewed by Tsuchiya (1974).

It was known in the 1930s that in the first meiotic metaphase of octoploid triticale a number of unpaired chromosomes occur (Müntzing 1939). The average number of unpaired chromosomes, univalents, was found to be different between lines, and Bjurman (1958) described the phenomenon that the meiosis of the $F_1$ plants was much more irregular than that of the two parents.

The meiotic abnormalities found in 6x triticale were essentially the same as those found in octoploids. Univalents, rarely multivalents, laggards, and micronuclei varied in number in different strains, even between different plants of the same strain (Nakajima 1952; Merker 1971).

The occurrence of univalents in meiosis of triticale had been considered to be the result of asynchrony of the meiotic rhythm of wheat and rye in triticale, and consequently a failure of meiotic pairing of homologous chromosomes in zygotene and pachytene, i.e., asynapsis (Stutz 1962; Kempanna and Seetharam 1972). Differences observed in the duration of meiosis, particularly in zygotene and pachytene, between wheat, rye, and triticale were found by Bennett et al. (1971), who suggested that this asynchrony was likely a cause of failure of regular pairing and chiasma formation, especially between rye chromosomes in triticale. Tsuchiya (1970) observed, however, that in most cases chromosome pairing was complete at diakinesis with 21 bivalents, while in metaphase only half of the pollen mother cells (PMCs) showed complete pairing. Tsuchiya's observation was confirmed by the author of the present chapter. While in metaphase I of different 6x triticale genotypes the average univalent frequencies varied between 0.1 and 9.3, the salient feature of the meiosis of these different plants was the lack of univalents in diakinesis (Lelley 1974). This finding strongly suggested that meiotic disturbances in triticale, expressed in the formation of univalents with irregular distribution at anaphase, were caused mainly by lack of a sufficient number of crossing-over. Crossing-over is primarily necessary to maintain physical linkage between homologous chromosomes until the first meiotic metaphase. Further studies of the meiosis of $F_1$ hybrids between 6x and 8x triticale and population rye, with the genomic formulas of A-B-RR and A-B-D-RR, respectively, in which homologous pairing could occur only between rye chromosomes, revealed highly significant plant-to-plant differences in chiasma frequency. This observation is best explained by a polygenic control of meiosis in rye and a heterogeneity of the pollinating rye parents (Lelley 1975a). A polygenic control of chromosome pairing in rye was suggested by Rees and Thomson (1956). The introduction of the differential staining technique of chromosomes with Giemsa allowed us to show definitely that univalents and rod bivalents in triticale mainly occurred in the rye genome (Lelley 1975b). This phenomenon was first recognized for octoploid triticale by Pieritz (1970). Nevertheless, a number of reports established that cytological disturbances and seed fertility are not necessarily correlating phenomena (e.g., Merker 1973; Weimarck 1973), leading to the belief that they are the different expressions of a common underlying physiological disturbance of the wheat–rye amphidiploid.

The polygenic control of chromosome pairing in rye (Rees and Thompson 1956), however, offered an excellent opportunity to study the genetic interaction between wheat and rye in triticale on a quantitative character — the phenomenon of chromosome pairing — with high accuracy.

Very early in the history of triticale the assumption was articulated that one of the reasons for incongruity between wheat and rye as an amphidiploid could be the fundamentally different reproductive biology, and consequently different genetic structure of the two species. In triticale the autogamous character of wheat prevails, rendering the rye homozygous (Lebedeff 1934). Wheat, a highly self-pollinating species with its homozygous gene loci, relies mostly on additive gene action, whereas rye as an obligate cross-pollinator with efficient self-incompatibility mechanisms operates with dominance as the major source of genetic variance. In rye, the genetic variance with respect to meiotic behavior becomes visible if self-incompatibility can be overcome and inbred lines of rye are developed. Cytogenetic studies on inbred lines of rye bring to light a wealth of variation in cytological characters, e.g., chiasma frequency and distribution, sister chromatid exchange with u-type exchange, and formation of unreduced gametes (Lelley 1978a; Lein and Lelley 1986; Lelley et al. 1986).

The production of primary triticale using such inbred lines offers a unique opportunity to study the influence of a specific wheat background on the described cytological phenomena characterizing the inbred lines. Crossing of such highly homozygous inbred lines with pure lines of wheat, followed by chromosome doubling of the $F_1$ by colchicine, leads to an exact addition of the two genotypes. The expression of characters, especially meiotic characters, can be directly compared in the parents and in their amphidiploid. The use of a selected inbred line to produce triticale was first suggested by Lebedeff (1934) and Müntzing (1939), with the idea of avoiding the transfer of deleterious genes commonly present in rye populations. Accumulated experience, meanwhile, did not prove the benefit of inbred lines for the production of triticale (Kaltsikes 1973). Their use in studying the contribution of rye to the phenotypic expression of triticale, however, is of major importance

(Lelley 1978b). Inbred lines with well-defined chiasma frequencies were used by the author for the production of primary hexaploid triticale. Studying meiosis with the differential (Giemsa) chromosome staining technique clearly showed that two rye inbred lines exhibiting a high heterosis for chiasma frequency in their $F_1$ hybrid, as was indeed expected following Rees (1955), as the rye component of two hexaploid triticale with the same wheat component showed in their $F_1$ a significant reduction of chiasma frequency. The number of unpaired rye chromosomes increased dramatically in the hybrid. Since the wheat parents of the two triticales were identical, the wheat genome of the $F_1$ remained homozygous without change in its chiasma frequency (Lelley 1981). Obviously, the genetic potential of the two inbred lines to have a significantly increased chiasma frequency in their $F_1$ hybrid could not be realized with the wheat genomes in the background. In fact, a change for the worse has happened. In view of the earlier literature on the meiotic behavior of triticale hybrids (e.g., Bjurman 1958; Merker 1973; Pohler et al. 1978), this result is not surprising. Nonetheless, it clearly demonstrates the negative interactions between wheat and rye in triticale on a quantitative trait. In a follow-up study, six primary triticales were produced using two advanced durum genotypes and three inbred lines selected for hybrid rye production, i.e., with a good general combining ability for hybrid rye breeding. Comparison of the observed and expected chiasma frequency of parental and amphidiploid genotypes revealed a general tendency to a reduction of chiasma frequency in the amphiploid — up to 12.9%. Crosses were made between these six primary triticale genotypes, the three parental rye inbreds, and two parental wheat lines in all possible intraspecific combinations. The resulting 15 triticale hybrids belonged in three groups: (1) heterozygous only for the rye, (2) heterozygous only for the wheat, and (3) heterozygous for both wheat and rye genomes. A thorough cytological analysis of this material has proven the highly disturbed meiotic behavior of rye chromosomes in triticale $F_1$ hybrids, especially if both wheat and rye genomes were in the heterozygous state. This occurred in spite of the fact that the diploid rye $F_1$ hybrids showed a 15 to 22% heterosis for chiasma frequency. The least reduction in chiasma frequency was observed in hybrids heterozygous only for the wheat genome. Obviously, interactions between the wheat and rye genomes have negatively affected the meiotic behavior of the rye chromosomes, especially if they were in the heterozygous state (Jung et al. 1985).

## 13.4.1 Tetraploid Triticale

If hexaploid triticale is considered an excellent object of study of polyploid evolution with either classical or molecular tools, the more so should tetraploid triticale inspire the fantasy of research workers on this subject.

The story of tetraploid triticale is almost as old as that of hexaploid triticale. Early attempts to produce tetraploid triticale are reviewed by Scoles and Kaltsikes (1974). Krolow (1973) succeeded in synthesizing a tetraploid triticale using *Triticum tauschii* (Cross.) Schmal. (synonymous with *Aegilops squarrosa* L.) and *Secale sylvestre*; the amphidiploid, however, remained highly sterile. Sodkiewicz (1984) received a tetraploid from the cross of *Triticum monococcum* L. and *S. cereale*, but the plant could be maintained only vegetatively. Bernard and Bernard (1987) produced a few fertile tetraploids using *T. tauschii* and *S. cereale*. Finally, the most successful tetraploids turned out to be those with a diploid wheat genome representing a mixture of chromosomes of the A and B genomes and a complete diploid rye genome. This type of triticale was obtained first by Krolow (1973), through crossing hexaploid triticale with diploid rye, and repeated selfing of the A-B-RR hybrid. In such a hybrid the homologous pairing of the rye chromosomes leads to their regular distribution, whereas the wheat chromosomes will remain unpaired and their distribution will be random. Due to the random distribution of wheat chromosomes, a great number of aneuploid gametes, although with a complete rye genome, will occur. But there is a statistical chance for the occurrence of gametes having along with the haploid rye genome a complete A or B genome, or any of the 126 combinations of a seven-chromosome mixed genome, in which each homoeologous group is represented. Such a mixed wheat genome, for example, 1A, 2B, 3B, 4A, 5B, 6A, 7A, can be considered to be genetically balanced owing to the compensating ability of the homoeologous chromosomes (Sears 1966). Therefore, such a 14-chromosome gamete might survive and take part in fertilization. The frequency of such a functional gamete is certainly very low, and the number of tetraploid triticales, which require the union of two gametes with the same seven wheat chromosomes in their mixed genomes, remained limited. The progress in producing new tetraploids remained slow (Lukaszewski et al. 1984).

Compared with hexaploid or octoploid triticale, tetraploid triticale with a mixed wheat genome shows a remarkable meiotic stability and low aneuploid frequency (Krolow 1974; Hohmann 1984; Lukaszewski et al. 1987a). Moreover, tetraploid triticale crossed with octoploid enabled the incorporation of D genome chromosomes into hexaploid triticale (Krolow 1973). Finally, the production of tetraploid triticale with a mixed A and B genome vividly demonstrated a possible way of evolutionary emergence of new polyploid species, as has been suggested by Zohary and Feldman (1962).

Baum and Lelley (1988) found crossing hexaploid triticale with rye and backcrossing the $F_1$ as female with the hexaploid parent to be an efficient way to produce new tetraploids with a mixed diploid wheat genome. The rationale in this scheme is that, in theory, if an egg cell with a complete rye and a complete balanced wheat genome with 14 chromosomes in total is formed, and this egg cell meets a 21-chromosome gamete of the hexaploid parent, a plant with 35 chromosomes will result. When this plant is selfed, in meiosis all homologous chromosomes will pair, leaving seven chromosomes as univalents. The paired chromosomes will be distributed regularly, while unpaired chromosomes will be eliminated, leaving behind a viable 14-chromosome gamete. This is supposed to be true for micro- and macrosporogenesis. Upon selfing of such plants, 28-chromosome balanced tetraploids will emerge. Study of the chromosome number of the backcross progeny of such cross combinations revealed that 57% of the backcross seeds had 35 chromosomes, which must have arisen from the fertilization of 14-chromosome egg cells with 21-chromosome pollen. The progeny of selfing such 35-chromosome plants, which we later called founder plants, consisted of more than 25% tetraploid triticale. Thus, the backcrossing of an A-B-RR hybrid with its triticale parent as male turned out to be an efficient and rapid method of producing new 4x triticale genotypes (Baum and Lelley 1988). The chromosomal constitution of the mixed wheat genome could be determined either by C-banding (Lukaszewski et al. 1984) or by using chromosome arm specific restriction fragment length polymorphism (RFLP) probes (Lelley et al. 1995).

A specific use of tetraploid triticale was suggested by Krolow (1973). Through crossing octoploid with tetraploid triticale, chromosomes of the D-genome can be substituted for A-, B- and/or R-genome chromosomes. A first report on the effect on yield and quality parameters of 84 such D-chromosome substitution lines was presented by Kazman et al. (1993). The most interesting results concerned the substitution of 1D for either 1A, 1B or 1R, showing for the 1D/1A substitution a significant increase of sedimentation volume. Several of the lines exhibited a comparable or even higher sedimentation volume and protein content than the wheat varieties included in the experiment. Some of them surpassed in yield the triticale variety "Clercal."

The results of a thorough field experiment investigating the effect of 18 D-genome chromosome substitution lines were recently presented by Budak et al. (2004). The substitution lines were derived from the winter triticale variety "Presto". They found no grain yield increase associated with any of the tested D-genome chromosomes, but three substitutions decreased grain yield. Chromosome 5D increased or decreased kernel hardiness when substituted for 5R or 5A, respectively. Introgression of 1D and 6D unproved end-use quality. The authors conclude that, apart from beneficial effects of individual loci located on some D-genome chromosomes, no major benefit can be expected from D-genome chromosome substitutions.

The agronomic value of tetraploid triticale will probably never reach the level of making it attractive for agricultural practice. Nevertheless, its mere existence and successful application in extending the genetic variation of hexaploid triticale (Kazman and Lelley 1994), tetraploid wheat, and rye (Lapiski 2002) demonstrate evolutionary processes: how the genetic material can be shifted around between related species to create new variations for selection in a changing environment.

## 13.5 GENETICS

Polyploidy is recognized as one of the most important evolutionary events leading to an enlargement of genetic variation and increased potential of adaptation. It is estimated that over 70% of flowering plants had at least one event of polyploidization in their evolution (Wendel 2000). Allopolyploidy is the union of complete genomes of two (or more) related species in a common cytoplasm. The phenotypic expression of this plant is from then on determined by the joint action of homoeologous or orthologous genes of both genomes. Mutual disturbances up to complete incompatibilities can be the outcome. The fewer incompatibilities, the more harmonious the cooperation that occurs between those genes and the higher the chance of the new amphiploid being a successful competitor. Natural selection determines the value of a new amphiploid; the creation itself is a random process.

Our understanding of the organization and structure of eukaryotic genomes experienced a dramatic growth in recent years, giving new impetus to study the events following polyploidization. A number of mechanisms have been described, which can act after polyploidization to foster stabilization in the short term, e.g., harmonization of gene expression by gene silencing or gene activation, dosage compensation, transcriptional activation of retrotransposons, rapid sequence elimination, genomic rearrangements, and in evolutionary perspectives, e.g., elimination of genes through mutations, functional diversification (Matzke et al. 1999; Comai 2000; Wendel 2000; Pikaard 2001; Ozkan et al. 2001; Levy and Feldman 2002). In view of this new surge of interest in studying polyploid evolution, it is worth recalling results of earlier studies made on triticale mainly by means of classical genetics or cytogenetics.

When talking of triticale it has to be borne in mind that triticale as an amphidiploid was never found in nature, even though natural hybridization between wheat and rye is possible. Many wheat genotypes carry the recessive interspecific crossability alleles, easily producing $F_1$ hybrids with rye. Therefore, it has to be accepted that although primary triticale, similar to those found by Rimpau (1891), must have occurred repeatedly in nature, these plants did not survive. The more intriguing is the fact that today, under human care, triticale is grown worldwide on more than 3 million ha, and in some areas of the world it has become a regular constituent of the agroecosystem.

Gustafson and Flavell (1996) stated that rye gene expression in wheat background is totally unpredictable. McIntosh and Singh (1987) described a number of stem rust resistance genes of rye origin, well expressed in triticale. Other genes have no detectable expression in triticale. The nucleolus-organizing region (NOR) on the short arm of chromosome 1R is suppressed in triticale by either of the two wheat NOR loci (Gustafson and Flavell 1996). The lack of expression of a heat-shock protein in triticale was documented by Somers et al. (1992). Variation in expressing gliadin genes in primary triticale was demonstrated by Günther et al. (1996). The nonexpression of the enzyme 6-phosphogluconat-dehydrogenase in a specific primary triticale was reported by Lelley (1985). In another primary triticale Lelley (1996) detected a single gene with no visible effect on the carrier inbred rye, but with a detrimental effect on a primary triticale synthesized with this particular inbred line, causing delayed growth and strongly reduced fertility. Two genes causing hybrid necrosis in rye were first detected in triticale, where through interaction with the known hybrid necrosis genes in wheat they also produced hybrid necrosis (Jung and Lelley 1985a; Scoles 1985; Ren and Lelley 1988, 1989).

Producing primary triticale lines with defined (homozygous) wheat and rye parents allows the study of more complex agronomic traits by comparing the performance of the amphiploid with that of both parental genotypes (Lelley 1978b). In a diallel cross of six newly synthesized hexaploid triticale and through intraspecific hybridization of their two parental wheat and three rye genotypes, comparisons were made for morphological and yield characters among parents, $F_1$s and $F_2$s. The planted material was cytologically controlled, allowing data evaluation with and without aneuploids. Owing to genomic constitution of the triticale lines, hybrids could be subdivided according to heterozygosity only in the wheat, only in the rye, or in both genomes. Aneuploid frequency in $F_2$ generation was highest when both wheat and rye genomes were heterozygous and lowest when only the wheat genome was heterozygous. Data analysis revealed that allelic interactions within the genomes leading to heterosis in the hybrid of parental diploid rye were generally suppressed in the amphidiploid triticale by interactions between wheat and rye. These interactions had a clear negative effect on cytological phenomena, i.e., chiasma frequency leading to an increased number of univalents, but could be positive as well, e.g., for kernels per spike. This study showed a different contribution of the two parents to these interactions.

Heterozygosity in the rye genome appeared to be detrimental for triticale, whereas heterozygosity in the wheat genome only resulted in some positive interactions (Jung et al. 1985; Jung and Lelley 1985b). In a further study, six inbred lines of rye, selected for good general combining ability in a hybrid rye breeding program of the University of Hohenheim, Stuttgard, Germany, were used to produce primary hexaploid triticale in an orthogonal manner with three advanced *Triticum durum* Desf. and one *Triticum turgidum* L. breeding line. From the possible 24 combinations, 20 could be realized. The primary triticale lines with *T. turgidum* were made available to us by Dr. G. Oettler, University of Hohenheim. The unsuccessful attempt to synthesize the missing four genotypes indicated the total incompatibility of certain wheat–rye combinations. Performance trials were conducted under field conditions in 2 years and two locations. From a total of 7200 triticale plants planted in the field, the chromosome number could be established in 4113. General and specific combining abilities of the parental genomes of wheat and rye were calculated; the terms *general genome combining ability* (g.g.c.a.) and *specific genome combining ability* (s.g.c.a.) were introduced. While the average fertility of euploids over all genotypes was higher than that of all plants, no correlation between aneuploid frequency and fertility was found comparing each genotype separately, as has been found in our above-described study, as well as in several previous experiments (see Müntzing 1979). The data showed that in this set of primary triticale, the rye genome exerted very little effect on quantitative characters, except for height and spikelet number. The wheat component had a major effect on heading date, plant height, spikelet length, and their numbers. It also influenced floret fertility. Thousand-kernel weight (TKW) was consistently higher in triticale than in the parents because of nonadditive gene actions. Fertility of the triticale lines was well below the mid-parent value; the g.g.c.a. and s.g.c.a. ratios for the two parents indicated a significant role of nonadditive gene actions for this character contributed especially by the rye genome (Lelley and Gimbel 1989). The data supported our earlier findings concerning the predominant role of intergenomic interaction for kernel per spike and TKW (Jung and Lelley 1985b).

In a later study, however, Oettler et al. (1991) found in a similar experiment only 20% of the genetic variance to be caused by interaction effects. Experiments with primary triticale are certainly not representative of the behavior of advanced breeding material. As summarized by Gustafson (1983), meiotic instability can affect raw amphiploids, but not secondary triticale. Similarly, for other characters, crossing and selection will soon eliminate many of those alleles in wheat or rye responsible for negative interactions. Unfortunately, studies as described above are not possible with secondary material. On the other hand, as stated by Oettler (1998), even after decades of intensive breeding work triticale is still characterized by cytological instability. As pointed out earlier, amphiploids have to be competitive right at the beginning to avoid being eliminated by natural selection. The twice-as-high costs required for maintenance breeding of triticale compared with other cereals and the doubled margin of tolerance for triticale compared with other cereals in the process of varietal registration in Germany (Dr. B. Schinkel, Lochow-Petkus GmbH, personal communication, March 2003) indicate that those intergenomic interactions, made visible in the above-described studies, are still interfering with the performance of triticale. It remains to be seen whether further breeding work will be able to make up for an evolutionary coadaptation, which presumably played an essential role in the establishment of many agronomically successful natural allopolyploids, e.g., wheat, oat, rapeseed, cotton.

Nevertheless, primary triticale is still an excellent object to study those complex genetic and epigenetic interactions mentioned earlier, which may decide in the short term the fate of a newly produced amphidiploid. Unfortunately, despite the large amount of recent literature demonstrating a revived interest in the evolution of polyploids using molecular approaches, triticale is not on the list of model plants. The first promising experiments using primary triticale for studying genome interactions and genome evolution in allopolyploids were reported by Voylokov and Tikhenko (1998, 2002) and Ma et al. (2002).

## 13.6 BREEDING

Triticale, the synthetic intergeneric amphidiploid, entered agricultural practice on the hexaploid level, with four wheat and two rye genomes (AABBRR). Octoploids and tetraploids continue to be used only to introduce new genetic variation.

Wheat, including tetraploid wheat, is a self-pollinating crop, and this character prevails in triticale, although the rye component is a typical allogamous species. For breeding triticale, the methods of self-pollinated crops are applied, aiming at selection of positively transgressive pure lines from crosses of more or less pure lines. Two peculiarities, however, distinguish triticale breeding from breeding conventional self-pollinating crops, e.g., barley or wheat, which have to be considered before starting a breeding program.

1.  Every cross between two triticale lines will more or less affect the delicate genetic balance of the wheat and rye components, established through selection in the pure lines. A similar interference is not to be expected between the DD and AABB genomes in a wheat cross. In triticale, the created imbalance will, through segregation, result in phenotypic variation, which cannot be fixed, however attractive in rare cases it may be. Therefore, in early generations a breeder may prefer simple laborsaving mass selection instead of a pedigree treatment until a certain balance between wheat and rye is restored.
2.  Although considered to be a self-pollinating crop, triticale may outcross and interfere with development of pure lines. The outcrossing rate was measured by several researchers with different results. They all agree, however, that it varies depending on the genotype and environmental conditions and ranges between close to zero and 60% (Kiss 1970; Singh 1979; Sowa and Krysiak 1996; Hermann 2002).

Ingenuity is a basic requirement in the plant breeding profession. In the case of triticale it may be slightly more so. It is worth describing briefly the breeding method of one of the world's most successful triticale breeding programs — DANKO, breeder of the variety Lasko, which was registered in 1982. Although not recommended for Poland because of its insufficient winter hardiness, it was registered and grown in eight other European countries and in New Zealand. The variety Lasko and its principal breeder, Dr. T. Wolski, played an eminent role in the development of triticale breeding in the last 20 years. At present, triticale is grown in Poland on 1 million ha (FAOSTAT 2004). In Germany, where triticale was grown in 2004 on 507,000 ha, 6 out of 29 registered varieties were of Polish origin.

DANKO breeders use the pedigree method with single plant or ear selection, currently from the $F_3$, previously from the $F_4$ generation onward. Progenies are tested in unreplicated microtrials and are multiplied separately. Progenies are selected for yield and other characteristics. When sufficiently uniform, strains are bulked and tested in trials at three locations in three or four replications, in 5 m$^2$ plots, later in 10 m$^2$ plots at more locations (Banaszak and Marciniak 2002).

It is also worth considering the experiences of a German triticale breeder, because of the long tradition and high standard of plant breeding in that country. Owing to the relatively high outcrossing rate of triticale, compared with wheat, three or four more generations are needed to reach the level of homogeneity necessary for variety registration. Even though the German Federal Office of Varietal Registration applies a double margin in triticale registration trials for deviation of register characters — e.g., in drill plots, 10 deviating plants of 2000 will be tolerated instead of 5 for wheat or barley — breeders believe that progress is hampered by the "still too strict homogeneity requirements." Valuable material will be lost because it will never reach the required homogeneity. Therefore, accurate estimate of homogeneity and frequent reselection of lines are two major components of a successful breeding program. As a further consequence of instability, maintenance breeding of triticale lasts longer and is more expensive than that of wheat (Schachschneider 1996).

## 13.6.1 Yield

One of the major attractions of triticale is its high yield potential. It is the number one target of most breeding programs, and as Figure 13.1 shows, triticale has the potential to compete successfully with its two parents — wheat and rye.

According to Banaszak and Marciniak (2002), the number of ears per square meter is the most important yield component, followed by number of grains per ear and a medium TKW. Earlier, triticale varieties tended to produce shriveled grains with reduced hectoliter weight. Indeed, shriveled grain was a salient character of triticale for many years and much research was invested to clear up the causes. For a long while, the most appealing hypothesis seemed to be the frequent occurrence of anaphase bridges and aberrant nuclei in the coenocytic stage of endosperm development, caused by the failure of rye chromosomes to separate at their telomeres, where in rye a large amount of late-replicating DNA is located (Bennett 1977). Terminal heterochromatin of the rye chromosomes was first made responsible for cytological disturbances of triticale by Merker (1976) and Thomas and Kaltsikes (1976). Bennett (1977) proposed a causal link between grain shriveling and late-replicating segments of heterochromatin in rye telomeres, suggesting the reduction or elimination of these segments to improve grain quality of triticale. This causal link was questioned by Varghese and Lelley (1983), who found in triticale genotypes subjected to a disruptive selection for grain type no differences in C-banding pattern between sublines with plump and shriveled seeds. Seal (1986), making a similar observation, concluded that aberrant coenocytic nuclei may be one of many causes for grain shriveling in triticale, but cytological and physiological disturbances may also be a parallel consequence of earlier events. Meanwhile, intensive breeding efforts largely eliminated grain shriveling as a typical feature of triticale, but under certain circumstances shriveling still may occur, indicating once more the not yet complete harmony between the two parental genomes in the amphidiploid.

## 13.6.2 Winter Hardiness

While many octoploid triticale genotypes had *a priori* excellent winter hardiness, due to the winter hardiness of both their parental species, hexaploid triticale required much breeding effort to achieve an adequate winter hardiness, and most winter-hardy triticales do not reach the winter hardiness of bread wheat. Winter hardiness was first transferred into hexaploid triticale by crossing octoploid with hexaploid triticale. Using this method, Schulindin (1975) demonstrated that winter hardiness could be introduced in hexaploid triticale. More details of this work written in Russian are given in Müntzing (1979). As a report by Salo (1998) proves, winter hardiness of present-day triticale varieties can be sufficient to compete successfully with winter wheat in southern Finland. Winter hardiness is one of the most important breeding objectives of the DANKO breeding program. For selection, cool chambers are used as described by Banaszak et al. (1998).

### 13.6.3  Lodging

Although the introduction of hexaploid triticale led to a substantial reduction of plant height, lodging resistance has remained a major target of breeding efforts. Several attempts have been made to introduce dwarfing genes from wheat and rye into triticale, e.g., a recessive dwarfing gene from the rye variety Snoopy by Zillinsky and Borloug (1971), or another recessive dwarfing gene of a mutant rye line (Nalepa et al. 1980), with rather disappointing results, proving the general observation that recessive genes of rye are poorly expressed in triticale. The *H1* dominant dwarfing rye gene of Kobylyansky (1975) was introduced into several European triticale breeding programs, including the DANKO program (Wolski 1990a; Wolski and Gryka 1998). Gregory (1980, 1984) showed that the gibberellin-insensitive dwarfing genes of wheat, *Rht1* and *Rht2*, are expressed in triticale and therefore can be used to improve lodging resistance. The introduction of the *Rh3* dwarfing gene was disappointing because of pleiotropic effects causing excessive tillering and everlasting segregation often linked with reduced fertility. Börner and Melz (1988) described the two dominant semidwarf genes from rye (*Ddw1* and *Ddw2*) as gibberellic acid-sensitive, i.e., not detectable by testing seedlings with $GA_3$. Nevertheless, especially *Ddw1* localized on chromosome 5R by Sturm and Engel (1980), and most probably identical to *H1*, has been used successfully to improve lodging resistance of triticale (Wolski and Gryka 1996). Use of a $GA_3$-insensitive dominant dwarfing gene in rye, described by Jlibene and Gustafson (1992), has not yet been reported.

### 13.6.4  Preharvest Sprouting

Germination and growth of grains still in the ear before harvest may sharply decrease grain yield, vigor, milling, and baking quality. Under specific environmental circumstances it will become a serious problem of triticale. Moreover, preharvest sprouting appears to be the most recalcitrant negative characteristic of triticale. Very little breeding progress has been made so far. Sprouting occurs when dormancy of the grain is overcome or absent. It is stimulated by wet conditions combined with relatively high temperatures. Breeding against sprouting is difficult, because of the complexity of the trait, large genotype × environment interactions, and inefficiency of selection under natural conditions. Even by utilizing the excellent sprouting resistance of the rye variety Otello or of other resistant rye strains, no improvement could be achieved (Huskowska et al. 1985). The sprouting tolerance of the available triticale material originates from transgressions obtained in the progeny of triticale crosses.

Current protocols of identifying genotypes with reduced tendency for sprouting are based on two approaches: (1) wetting the spike under field or laboratory conditions, closely reflecting the field situation, or (2) germination test of threshed seeds (Wu and Carver 1999).

A widely used parameter for assessing the degree of sprouting is measuring the falling number, which is related to amylase activity. Falling number indicates starch degradation before the appearance of visible symptoms (Zanetti et al. 2000). Falling number is a simple and inexpensive selection method, but it does not always correlate sufficiently with sprouting resistance (Wgrzyn et al. 1998). The variety Pronto, for example, with the best sprouting resistance from the DANKO program, has a lower falling number than the variety Moreno, with the highest falling number in the program (Banaszak and Marciniak 2002).

While provoking sprouting is an active selection method allowing screening of segregating populations, measuring falling number alone is only suitable for differentiating genotypes when field conditions favor preharvest sprouting. In a study by Haesaert and De Baets (1996), 71 triticale genotypes were tested for level of preharvest sprouting by germinating hand-threshed seeds along with seeds in spike, by measuring falling number, and by field observations. Less than 10% of the genotypes could be classified as more or less resistant and were identified as potential genetic sources for breeding. They found good correlation between germination of threshed seeds and seeds in spike, and these data showed good agreement with field observations. They found, however, no association between germination data and falling number. Derycke et al. (2002) recommend testing the falling number after three different storage conditions of the seeds: (1) dry conditions, (2) moistened once, followed by dry conditions, and (3) constantly humid conditions. Then they suggest calculating a rank correlation for the three methods. Such experiments resulted in the highest correlation between field observations and the laboratory test.

Oettler (2002) found a short-cycle recurrent selection for falling number highly effective by increasing cycle means with minimal loss in genetic variation. A high response to selection and a wide range in cycle 4 indicated potential for further advancement.

Sodkiewicz (2003) reported the transfer and full expression of sprouting resistance genes of *T. monococcum* determined in hexaploid triticale mature spike sprouting.

## 13.6.5  Abiotic Stress

Physical and chemical factors of the environment that adversely affect yield are regarded in agriculture as abiotic stress. Conversely, resistance to environmental stress is manifested as the stability of yield in a changing environment, and the disparity between potential and actual yield is mainly explained by environmental stress. It has been estimated that generally 71% of yield reduction is caused by abiotic stress (Acevedo and Fereras 1993). Triticale is considered to be the most tolerant to abiotic stresses among cereals. It can be used directly in sustainable cropping systems as a low-input crop or indirectly as a donor of genes controlling important tolerance mechanisms (Anio 2002). The current tendency to a more sustainable agriculture means a move away from manipulating the environment to fit the plant's needs toward a genetic manipulation to tolerate the environment with reduced input.

At present, the most widespread strategy in breeding for abiotic stress tolerance is conducting multiyear and multilocation field trials, followed by statistical analysis, even though this empirical approach, based on yield in different stress environments, is relatively inefficient mainly due to large genotype × environment interactions. The use of molecular markers may bring a substantial improvement in efficiency. A summary of identified major genes and quantitative trait loci (QTL) related to resistance to abiotic stresses is given in Anio (2002). None of the 54 citations refer to triticale as the source of the character, and so far no triticale has been transformed with an abiotic stress resistance gene.

The synthetic amphidiploid triticale has the potential of both parents, and as reviewed by Jessop (1996), tolerance was found to low nutrient availability, drought, frost, soil acidity, aluminum and other element toxicities, and salinity. An illuminating example of how interactions of the wheat and rye genomes determine stress tolerance of triticale is its tolerance to aluminum toxicity. Soil acidity limits crop production in many parts of the world; e.g., in Poland over 60% of arable soils are acidic or very acidic, and such soils predominate also in the small grain-producing regions of southern Brazil. In such soils aluminum is regarded as the most important yield-limiting factor. Baier et al. (1998) described the visual evaluation of acidic soil tolerance in comparison with yield data through 3 years of 100 triticale cultivars, together with 4 wheat and 1 rye genotype in Passo Fundo, Brazil. They found good correlation between yield and visual scores of tolerance, indicating the possibility to select for aluminum tolerance by visual observation.

Rye is considered to be the most aluminum-tolerant cereal, followed by wheat. In triticale, however, the expression of rye genes is restricted by the wheat background: the lower the tolerance of the wheat, the lower the expression of the rye tolerance genes (Anio and Gustafson 1984). In wheat the tolerance genes reside on several chromosomes of the A and D genomes (Anio and Gustafson 1984; Anio 1990). In rye Anio and Gustafson (1984) found chromosomes 3R, 4R, and 6R, Gallego and Benito (1997) only 3R and 4R, to carry major resistance genes against aluminum toxicity. Lukaszewski (1990) produced two sets of chromosome substitutions in which, with very few exceptions, the homoeologous A, B, and R chromosomes of two triticale cultivars, Presto and Rhino, were replaced by the respective D genome chromosomes. This substitution series was used by Budzianowski and Wos (2002) to study the genetics of aluminum tolerance in triticale. In laboratory tests, 4-day-old seedlings were transferred for 48 hours to a nutrient solution with 10 or 20 mg Al/l. The study showed that the absence of any rye chromosome reduced the Al tolerance of triticale. Replacement of 3R by 3D caused a dramatic reduction of tolerance. In the cultivar Presto the replacement of 4R showed a similar effect; unfortunately, in the cultivar Rhino the 4D/4R substitution was missing. Thus, except for the tolerance coded by 6R, the results of Anio and Gustafson (1984) and Gallego and Benito (1997) have been proven. Gustafson and Ross (1990), studying the effect of all 42 chromosome arms of wheat on the expression of rye Al tolerance, found 18 wheat chromosome arms affecting the expression of rye Al tolerance, removal of 2DL, 6DL, and 5DS completely suppressing Al tolerance of rye. More data and an extensive discussion on aluminum tolerance of triticale were published by Budzianowski and Wos in 2004. The findings demonstrate the complexity of the genetics of aluminum tolerance of triticale, but also show how knowledge may be accumulated on the genetic control of resistance against an abiotic stress factor.

### 13.6.6  Biotic Stress

Similar to its high tolerance to abiotic stress, triticale also shows good tolerance to biotic stress. In fact, triticale is the healthiest among all cereals, and although biodiversity from naturally evolving populations does not exist for this young man-made plant, the whole gene pool of wheat and rye is available for successful breeding work as long as intergenomic interaction does not thwart expectations. So far, for example, it has not been possible to transfer the good scab resistance of some rye genotypes into triticale. *Fusarium* is one of the most serious pathogens of triticale, causing root and foot rots, leaf and head blight; the latter, Fusarium head blight (FHB), is the most significant because of severe quantitative losses and quality defects. It reduces yield, test weight, and germination capacity, and contaminates seeds with mycotoxins, most importantly desoxinivalenol (DON). But a large amount of genetic variation with respect to tolerance against the disease seems to exist in triticale. In Polish material a wide variation of tolerance in varieties and breeding lines was reported by Goral et al. (2002). A large-scale FHB resistance test, including 100 varieties, advanced breeding lines and genotypes collected from all across Europe, detected a variation between 0 and 60% yield reduction, and a large potential for resistance breeding against *Fusarium culmorum* (Vrolijk and Suijs 2002). This experiment, with small changes made in the collection, will be continued over the years to accumulate long-term data on resistance development. Comparing 10 wheat and 10 triticale varieties in an infection experiment with *Fusarium graminearum* through 3 years, Kleier and Fossati (1998) found significantly lower disease rating and TKW loss for triticale than for wheat. They found no variety resistant to snow mold (*Fusarium nivale, Typhula ishikariensis*, and *Typhula incarnata*) by naturally occurring infection. Diallel analysis of 10 winter triticale covering a wide range of resistance levels revealed the major importance of general combining ability (GCA) variance, suggesting the predominance of additive gene effects for resistance (Heinrich et al. 2002). However, a significant GCA × environment interaction was also detected, making multilocation testing of the breeding material necessary. Recently, Miedaner et al. (2004) studied the phenotypic and genotypic correlations among symptom ratings, content of Fusarium exoantigens (ExAg), which is a mixture of water soluble extracellular fungal products, and DON, in an experiment of six location-by-year combinations. A highly aggressive single-spore isolate of *Fusarium culmorum* was used to inoculate 55 winter triticale genotypes. A tight correlation, established between DON and ExAg content, is of special interest, because the costs for analyzing the latter are much less than for determining DON content. The high correlation (0.71) between rating Fusarium-damaged kernels and DON content in triticale led to the conclusion that reduced FHB symptoms should lead to a correlated selection response for lower DON content. Selection, at least in early generations, can be carried out by determining ExAg content.

One of the major reasons for the success of triticales in Germany (507,000 ha growing area) is a low-cost production technology based on its good resistance against fungal diseases and low fungicide requirement (Schinkel 2002). Disease score data of national yield trials between 1988 and 2001 of wheat, rye, and triticale were used to follow the development of disease susceptibility of triticale in comparison to its two parents. Triticale was found to be less susceptible to diseases than wheat and rye, the main disease at present in Germany being leaf blotch (*Stagonospora* = *Septoria nodorum*), where little improvement in resistance level was seen in recent years. While genetic variation for leaf blotch seems to be low, for brown rust a large genetic coefficient of variation promises good breeding success (Schinkel 2002).

There are few studies on the inheritance of resistance to *S. nodorum* in triticale. Head resistance of the variety Alamo was found to be determined by a single recessive gene (Wo 1996). In an incomplete diallel analysis of six winter triticale genotypes representing significantly different resistance levels to leaf blotch, Sowa et al. (1998a) tried to find out whether the parents could be characterized by their GCA and specific combining ability (SCA) effects using artificial inoculation. They found significant GCA and SCA effects, GCA effects having been larger in both $F_1$ and $F_2$. The reaction of $F_1$s from crosses of resistant and moderately resistant parents indicated that dominance was a major type of gene action conditioning resistance to *S. nodorum*. However, significant effects of reciprocal crosses were also found (Sowa et al. 1998a). In a further field study, Sowa et al. (1998b) used parents, $F_1$, $F_2$, and backcross populations of three medium resistant, three medium susceptible, and a highly susceptible genotype. After artificial inoculation with a mixture of 15 isolates of *S. nodorum*, they found a continuous variation of disease expression without any sign of individually acting major genes. The data suggested that the high level of resistance of some of the genotypes is probably controlled by several minor genes, implying that breeders could elevate the resistance level by using appropriate sources and selecting in later generations (Sowa et al. 1998b). An attempt to introduce resistance to *S. nodorum* from *T. monococcum* via a *T. monococcum–S. cereale* amphiploid into triticale is described by Arseniuk and Sodkiewicz (2002).

The superior rust resistance of triticale is partly due to resistance genes available in rye. Their functioning in wheat background is well documented. The variety Transec derived its leaf rust resistance from chromosome 2R of rye (Driscoll and Anderson 1967). Stem rust resistance was derived from chromosome 3R of Imperial rye through translocation (Steward et al. 1968). Several European wheat varieties derived their leaf and stem rust resistance from Petkus rye through the original 1B/1R substitution (Zeller 1973). Morrison et al. (1977) demonstrated that five triticale lines carried at least five genes for resistance to two prevalent races of wheat stem rust, indicating that a range of resistances against this pathogen is available to the triticale breeder. On the other hand, McIntosh et al. (1983) described in Australia a mutant stem rust pathotype, which was virulent on triticale — attacking ca. 70% of triticale lines emerging from the CIMMYT program — showing that the fate of resistance genes in triticale is not different from that of resistance genes in other cereals. The option to transfer alien resistance genes into triticale was recently demonstrated by Sodkiewicz and Strzembicka (2004), who reported on the introgression into triticale of complete and partial resistance against *Puccinia triticina* from *T. monococcum*.

Triticale is highly tolerant to powdery mildew (*Erysiphe graminis*) owing to its rye component. At least three major resistance genes have been identified and localized in rye on chromosomes 1R, 2R, and 6R (Heun and Friebe 1990). *Pm7* has been located on the long arm of chromosome 5R (Heun and Friebe 1990). *Pm8* and *Pm17* were found on the short arm of chromosome 1R and appear to be allelic (Friebe et al. 1989; Heun et al. 1990; Hsam and Zeller 1997), and *Pm20* was found on the long arm of chromosome 6R (Friebe et al. 1994). Although many rye-derived resistance genes have been identified, few have made a significant contribution to wheat breeding because they were mostly introduced through noncompensating wheat rye translocations (Friebe et al. 1996). This restraint certainly does not apply to triticale.

Finally, an intriguing case of reaction to a disease of the two parental species in triticale is that to barley yellow dwarf virus (BYDV), rye — especially winter rye — being very tolerant and durum wheat very sensitive. Moreover, it appears that resistance to BYDV confers on triticale not only tolerance against the virus, but BYDV tolerance seems to be related to tolerance to many biotic and abiotic stresses (St.-Pierre et al. 1998), probably due to genes that have a pleiotropic effect controlling reaction to other stress. However, BYDV tolerance genes of rye are not expressed in every wheat genotype (Bizimungu et al. 1996; Comeau and Arseniuk 1998). In fact, it is very likely that a whole range of useful genes is present in triticale, which remain inactive due to negative interactions with genes in the wheat genome. Improving triticale may thus come from optimizing intergenomic interactions (Comeau and Arseniuk 1998).

### 13.6.7  Hybrid Breeding

The commercial success of various hybrid crops inspires the ongoing discussion on the use of heterosis in triticale breeding as well. The accomplishments of rye hybrid varieties clearly underline the high hybrid potential of the rye component of triticale. The partially open-pollinating character of triticale has also for a long time stimulated thoughts about hybrid breeding. Contraindications of hybrid production are cytological disturbances known to occur in $F_1$ hybrids, but it was observed that heterosis can have a masking effect on cytological disturbances; i.e., plump seeds are developed although the embryo is aneuploid (Struss and Röbbelen 1989).

One of the key issues for commercial production of hybrids is the expected amount of heterosis. Little information is available on this topic in the literature. The first trial with advanced breeding material in two environments was reported by Barker and Varughese (1992). Eight complete triticale lines and 28 crosses were tested at two sites with full irrigation and under drought stress. Yield under irrigation showed a mean percentage heterosis over the high parent of 7.2% with a maximum of 19.7%, while under drought the average heterosis was 25.2% with a maximum of 68.2%. The GCA effects were highly significant for all traits studied, i.e, yield, test weight, and height, under both moisture regimes; highly significant SCA was found only for yield. The GCA/SCA sum of squares was lowest for yield, indicating more nonadditive effects for yield in this material. Further microplot studies (e.g., Mangat and Dhindsa 1995; Grzesik and Wgrzyn 1998) also reported considerable heterosis on yield components. In drilled plots, the first results on $F_1$ heterosis for grain yield were presented by Pfeiffer et al. (1998) in spring triticale and by Oettler et al. (1998) in winter triticale. In both studies heterosis for grain yield of up to 20% above the high parent was found. Pfeiffer at al. (1998) found higher average mid-parent heterosis for substituted (2D/2R) × complete R genome hybrids (10.7%) when compared with hybrids having only complete R genomes (8.6%).

These data strongly suggested a substantial gain in genetic yield potential based on existing genetic variation using triticale hybrids in commercial production. In a further study Oettler et al. (2001) investigated heterosis for eight agronomic traits in six $F_1$ and $F_2$ hybrids in drilled plots of 2.5 or 3 $m^2$ in three environments. Average heterosis compared with the mid-parent value was found to be 10.5% for the $F_1$ and 5.0% for the $F_2$ generation. Kernels/spike and 1000-kernel weight contributed positively to heterosis, whereas spikes per square meter showed negative heterosis. Negative heterosis was found in both $F_1$ and $F_2$ generations for falling number. This study also attested to a likelihood of up to 20% heterosis for grain yield. Oettler et al. (2003) presented data on 24 $F_1$ hybrids produced by a chemical hybridizing agent and grown in 5 $m^2$ plots in 1 year at two locations and at two levels of nitrogen application. The regression of hybrid performance on mid-parent values for grain yield was similar for the two N levels, i.e., 0 and 90 kg/ha; mid-parent heterosis for grain yield (q/ha) was found to be higher without (10.1%) than with (9.5%) N fertilizer, supporting the often articulated assumption that hybrid production of triticale is especially suitable for less favorable growing conditions (see also Barker and Varughese 1992, mentioned above.) Maximum relative grain yield heterosis of 20% at N0 and 25% at N90 kg/ha indicates a considerable potential for increasing yield by systematic search for heterotic groups and testing parents for combining ability. The study revealed GCA as the main variance component, underlining the importance of additive effects as was found earlier by Barker and Varughese (1992) and Dhindsa et al. (1998). However, in all experiments where combining ability was studied in several environments, the importance of GCA × environment interaction and the necessity for testing hybrids under several environmental conditions was stressed (e.g., Mangat and Dhindsa 1995; Dhindsa et al. 1998; Oettler et al. 2003).

A prerequisite for economic production of hybrid triticale is a stable system to handle male sterility (MS) and fertility restoration. There are two options for male sterility induction: use of gametocide or maternally inherited male sterility (CMS). The application of presently available chemical agents to induce MS is restricted owing to potential health hazards and environmental effects. On the other hand, several authors reported that replacement of cultivated triticale cytoplasm by the cytoplasm of related species may result in male sterility (Tsunewaki et al. 1984; Cauderon et al. 1985; Nalepa 1990; Warzecha et al. 1996). Sanches-Monge (1975) reported for the first time that hexaploid triticale with *Triticum timopheevi* (Zhuk.) cytoplasm was completely male sterile. No negative effects associated with *timopheevi* cytoplasm in triticale have been reported so far. It is widely known that *T. timopheevi* cytoplasm was used very intensively for inducing male sterility in wheat. Hybrid wheat production failed because of insufficient male fertility restoration. In contrast, in triticale the majority of tester pollen parents restore male fertility. In several hundred tested genotypes only 3 to 5% did not restore male sterility (Warzecha and Salak-Warzecha 2002). The two genes *Rfc3* and *Rfc4* restoring pollen fertility in *T. timopheevi* cytoplasm reside in chromosomes 6R and 4R, respectively (Curtis and Lukaszewski 1993).

## 13.6.8  Biotechnology

### 13.6.8.1  *Molecular Markers and Mapping*

Molecular markers are revolutionary new tools for plant breeding. Many different applications are possible, including studying genetic relationships, tracing alien chromosome substitutions, and tagging economically important genes and QTL for marker-aided selection. Genetic diversity in 127 European winter triticale genotypes was estimated using simple sequence repeat (SSR) and amplified fragment length polymorphism (AFLP), as well as pedigree information (coancestry coefficient) by Tams et al. (2004 and 2005). For both, line and hybrid breeding, information about genetic diversity within a germplasm pool is essential for section of crossing partners and to establish heterotic groups for hybrid triticale. The investigation has clearly shown the superiority of marker data compared to pedigrees. Also, the comparison of the data obtained by the two different marker types are enlightening. No clear grouping of triticale germplasm was detected, indicating the lack of germplasm management, which would, however, be desirable for hybrid breeding.

Probably the first molecular map of triticale was presented by González et al. (2002a). They used randomly amplified polymorphic DNA (RAPD), random amplified microsatellite polmorphism (RAMP), amplified fragment length polymorphism (AFLP), and SSR markers on 73 doubled haploid (DH) lines obtained by anther culture from the cross Torote × Presto. The map consists of 385 markers, 29% of them showing a distorted segregation in the DH population, 67% of these belonging to Presto. The complete map length was calculated to be 2647.3 cM, with an average marker distance of 6.9 cM. Most markers (47.1%) were placed on the R genome, the fewest (15.3%) on the A genome. The first use of this map was the identification and location of several QTL affecting the successive phases of androgenesis in triticale, i.e., production of embryos, regeneration capacity, and proportion of green plants (González et al. 2002b).

An AFLP- and RFLP-analysis of primary triticales and complete sets of wheat-rye addition lines, together with the parental genotypes (Ma et al. 2002 and 2004), provide an exciting insight into the evolutionary processes in triticale, following polyploidization. It is demonstrated that genomic sequence changes, primarily sequence losses, are very common in triticale, and these changes occur predominantly in the rye genome. The degree of variation was much higher in triticale than it was found earlier in wheat or in *Brassica*. This is attributed to the difference in the parental relationship of the species involved, triticale being no an interspecific but an intergeneric amphiploid. The study also showed that wheat genome conservation was much higher, in octoploid than in hexaploid triticale. AFLP band conservation was much higher, if the band was present in both progenitors, indicating that alteration occurred more frequently in regions with less similarity of the parents. Variation was found to be much higher in regions containing repetitive sequences compared to low-copy sequences. RFLP data indicated that changes in expressed sequences were also dramatically higher in rye than wheat, 61.6 versus 2.7% respectively. Both marker types produced new bands not present in either of the parents. The possibility of a concerted evolution is suggested.

### 13.6.8.2 Haploidy

The development of near-homozygous lines after crossing requires up to seven selfing generations. Production of haploids followed by chromosome doubling by colchicine leads to complete homozygosity in one generation after crossing. In plant breeding, haploid technology is a great help in many situations, e.g., developing inbred lines, producing a DH population for QTL analysis, achieving homogeneity for varietal registration, or simply for varietal development.

Ya-Ying et al. (1973) reported the first haploid triticale plants produced by anther culture. Later, several studies showed the influence of factors affecting the development of microspores into fertile plants, i.e., plant genotypes (Sosinova et al. 1981; Charmet and Bernard 1984), developmental stage of the microspore (Bernard 1980; Kozdój and Zimny 1993), effect of cold treatment of anthers (Sosinova et al. 1981; Charmet and Bernard 1984), growth or stress conditions of the donor plants (Bernard 1977, 1980; Immonen et al. 1998), culture medium and medium supplements (Bernard 1977; Sosinova et al. 1981; Davies et al. 1998; Verzea and Ittu 1998; Immonen and Robinson 2000), and seasonal effects (Kozdój and Zimny 1993). Microspores in the uninuclear stage from cold-pretreated anthers were taken. After isolating microspores, Sayed-Tabatabaei (1998) reported on the possibility of obtaining 200 green plants per spike.

Another avenue to produce doubled haploid triticale is crossing it with maize. There is much information in the literature on the success of wheat × maize, but very little on triticale × maize crosses. Rogalska and Mikulski (1996) reported on 3484 flowers of eight different triticale genotypes pollinated with maize, producing 108 grains. Dissecting 35 embryos yielded altogether 23 DH plants, all originating from a single triticale genotype. Iganaki et al. (1998) made maize crosses with 20 triticale genotypes. Pollinating a total of 200 spikes, they obtained 62 plants with mainly 21 and a few with 22 chromosomes. A higher frequency of embryos was formed with 2D and 4D chromosome substitution lines. They showed that finding the proper plant growth regulators and pollen source might be crucial for haploid production using triticale × maize crosses. Alternative pollen sources, i.e., *Tripsacum dactyloides* L. or teosinte (*Zea maize* L. ssp. *mexicana*) may increase embryo formation. Wedzony et al. (1998) crossed five triticale varieties and five $F_1$ hybrids with a pollen mixture of inbred and open-pollinating maize genotypes, and applied several auxin analogs, including dicamba and picloram, after pollination. Dicamba stimulated growth of the ovaries significantly more than picloram, and both were more efficient in growth stimulation than the other tested growth regulators. The most efficient concentration appeared to be 75 or 100 mg/l solution of picloram or dicamba producing on average between 17.1 and 25.1 embryos per 100 pollinated florets. The frequency of excised embryos did not differ between genotypes, making trical × maize crossings an efficient alternative to another culture to produce doubled haploids in triticale. In a very recent publication Pratap et al. (2005) describe the effect of nine different Gramineae genera (*Z. mays, Sorghum bicolor, Pennisetum americanum, Setaria italica, Festuca arundinacea, Imperata cylindrica, Cynodon dactylon, Lolium temulentum* and *Phalaris minor*) on haploid induction in eight triticale × triticale and 15 triticale × wheat crosses. In each case, except for one triticale × wheat cross, embryo formation and green haploid plantlet regeneration was obtained, when pollination was done with maize and *I. cylindrica*. But the latter outperformed significantly the former in embryo formation and regeneration. In case of triticale hybrids no significant difference was observed between maize and *I. cylindrica*, indicating a surprisingly high potential of the latter to produce haploids in triticale and triticale ×wheat hybrids presenting itself as an alternative to maize.

## 13.6.9 Quality

The biological value of triticale makes it suitable for use as animal feed and for human consumption. The first comprehensive review on this issue was published by Hulse and Laing (1974), concerning mainly its proteins compared with both parents. They concluded that compared to wheat, advanced triticale lines are at least equal in protein content but superior in biological quality. Triticale has a higher content of the essential amino acid lysine than wheat. The considerable literature devoted to the processing and supplementation of wheat and wheat flour is largely relevant to triticale. The evidence accumulated up to that point suggested that healthy triticale is equal, and possibly superior to, wheat in producing most farm animal diets. Since triticale tolerates environments unfavorable to wheat, it was considered a promising addition to the world's cereal crops for both human food and animal feed. While in recent years a significant increase in the growing area of triticale has been observed, especially in Europe, it is mainly used as animal feed. Its utilization as a food grain remains rather limited. The reasons are associated with grain composition factors, breeding priorities, region-specific grain preferences of consumers, competitiveness with other grains, marketing, and processing aspects (Peña 1996).

### 13.6.9.1 Use for Food

One major limiting factor in the utilization of triticale grain as food is its high alpha amylase activity (AAA) prior to harvest, which is caused by wet conditions. It significantly alters the functional properties of starch, reducing its water-binding capacity and eventually the quality of bread. On the other hand, high AAA can be advantageous in the production of triticale malt, which can be used as an additive in the food industry or in brewing. However, a slightly higher proteolytic activity, resulting in high levels of solubilized protein, causes problems of fermentation, storage, and color of the beer (Gupta et al. 1985). For malting triticale, the process of germination should be shorter, and at lower temperatures and relative humidity than generally applied for malting barley (Beirão da Costa and Cabo Verde 1996).

Pentosans (arabinos + xylose) are cell wall constituents and play a major role in determining the viscous properties of rye dough. They are important for baking quality — they determine dough yield, stability, and volume, and partially influence bread loaf volume and crumb texture. Rye dough viscosity is influenced largely by its water-soluble pentosan content. Proteins are less important in rye dough than they are in wheat dough (Drews and Seibel 1976). The soluble pentosan content of triticale grain and flour is similar to that of wheat and much lower than that of rye (Saini and Henry 1989). Therefore, triticale doughs in rye bread production will have inferior dough viscosity properties and baking quality, compared to those made with 100% rye flour. Triticale flour may be used as a substitute for wheat or rye flour in mixed wheat–rye or rye–wheat breads, respectively (Peña 1996).

Storage proteins in wheat interact to form gluten. Gluten quality and quantity are responsible for the viscoelastic properties of dough that enable the production of leavened and unleavened breads. The storage protein content of triticale is considerably lower than that of wheat, and only a part of it is capable of forming gluten. Another part, secalins, inherited from rye, does not form gluten. These differences in amount and composition of storage proteins result in inferior bread-making quality of triticale compared to wheat. Although genetic variability for gluten content exists, the highest gluten content in triticale is still 10 to 15% below that of wheat (Peña 1996). In bread wheat, the *Glu-D1* locus on chromosome 1D is considered to be the most influential for bread-making quality. Hexaploid triticale does not contain the D genome, but attempts have been made to transfer, either by translocation or by whole-chromosome substitution the *Glu-D1* locus into triticale (Lukaszewski et al. 1987b; Lukaszewski and Curtis 1992; Kazman and Lelley 1994; Lukaszewski 1998; Lafferty and Lelley 2001). While final conclusions from these studies were not yet derived, the generally observed increase of Zeleny sedimentation value following incorporation of the *Glu-D1* locus in hexaploid triticale indicates that this approach may create new genetic variation for selecting triticale lines with substantially improved bread-making quality. Recently, three germplasms of hexaploid triticale with introgressed wheat storage protein loci, *Glu-D1*, *Gli-D1*, and *Glu-D3*, were registered in the U.S. Preliminary tests indicate significant improvement in all bread-making quality parameters (Lukaszewski 2003).

### 13.6.9.2  Use for Feed

The major use of triticale at present is for livestock feed, primarily for pigs and poultry. It serves as a main energy source in their diet. When studying the relevant literature on the nutritive value of triticale as animal feed, it is necessary to keep in mind that since the first triticale varieties were released more than 30 years ago, large progress in kernel characteristics has been achieved. Work carried out until the early 1980s necessarily used seeds of varieties with often shriveled grains. The degree of shriveling became worse under unfavorable conditions during grain-filling time. Shriveling resulted in lowered volume weight and starch content but led to comparatively high protein content. Therefore, large variation in degree of shriveling could be the cause for contradictory results of earlier investigations on the nutritive value of triticale.

According to the review by Hughes and Choct (1999), the apparent metabolizable energy (AME) content of triticale, which is the energy content of grains for use in broiler chicken and laying hen diets, is comparable with that of wheat and barley but is lower than that of sorghum. Hughes and Cooper (2002) showed that the AME of seven triticale varieties grown under different environmental conditions, including drought stress, never fell below 13 MJ/kg dry matter, as has been observed frequently with wheat in Australia over the past 20 years.

Boros (2002) studied the physicochemical characteristics of 16 triticale strains and varieties of Polish origin. His data indicate that recent triticales are very similar to wheat in viscosity (V) of the grain water extract (WEV) and the content of soluble dietary fiber (SDF) and soluble arabinoxilans (SAX). Broiler chickens are especially sensitive to soluble and viscous arabinoxilans, which constitute the major fraction of dietary fiber in rye and are responsible for its low nutritive value but are not expressed to the same extent in triticale. The TKW and test weight of triticale were found, however, to be lower than those of the check wheat variety Almari. The study concludes that without contamination with rye, present-day triticale can be used as the only cereal component in the diet of chickens.

As stated by Van Barneveld and Cooper (2002), when pig diets are formulated to supply equal levels of digestible amino acids and digestible energy, the performance of growing pigs fed triticale as the sole cereal base is equal or superior to the performance observed when wheat-based diets are fed. When compared with wheat, barley, and sorghum, the triticale amino acids digestible in the small intestine are similar to those measured in wheat and sorghum and superior in most cases to that measured in barley. Furthermore, as was shown by Myer (2002), triticale is an effective replacement for maize in diets not only for growing pigs but also for early weaned pigs (3 to 8 weeks old).

### 13.6.9.3  Forage and Dual Use

Many investigations were made into the potential of triticale as a forage crop. Bishnoi et al. (1978) found that triticale has a higher forage yield and protein content than wheat, rye, and barley. The use of triticale as a whole-crop silage has been reported to provide a promising alternative to other cereals (Gregory 1987; Bocchi et al. 1996; Sun et al. 1996; Stallknecht and Wichman 1998).

Its superior biomass production in comparison to other cereals has been proven, for example, for the northern region of Mexico (Lozano et al. 1998) or for eastern Algeria (Benbelkacem 2002). A recent experiment in Belgium carried out through four growing seasons, however, demonstrated a strong influence of year, location, and genotype on yield and quality parameters. It was found, in general, that values for digestibility and net energy for lactation of whole plant silage were lower than values common for forage maize. But a higher proportion of spikes in the total dry matter increased digestibility and net energy for lactation. Moreover, the more environmentally friendly production system, compared with the intensive cultivation of maize in monoculture in Belgium, might outweigh the shortcomings of triticale (Haesaert et al. 2002).

The concept of using triticale as a dual-purpose crop, first for grazing or harvesting as green fodder and then allowing regeneration for grain production, has been highlighted in several studies (Andrews et al. 1991; Scott and Hines 1991; Garcia del Moral 1991; Royo et al. 1993). In an experiment studying eight triticale genotypes with a fodder oat variety as a control, Singh et al. (1996) found distinct genetic differences for most of the investigated characters, particularly number of tillers per meter after fodder harvest (70 days after sowing). As found by other workers also, the postcut crop showed a clear height reduction with a corresponding increase in lodging resistance. One genotype was found as a "near ideal dual type" with adequate fodder yield (72% of control) without significant reduction in grain yield. It appears that for optimum dry matter and grain yields in dual-purpose use, knowledge of the apical development patterns of the plants is essential. In a study of Boujenna et al. (1996), two cutting treatments were compared with an uncut control: one cutting made at Zadok's stage 30 (C30, pseudostem erection) and one at C31 (first node detectable). The study concluded that forage removal lengthens the spikelet initiation period but reduces the duration from the terminal spikelet stage to anthesis. Cutting at stage C31 reduced significantly both the primordial initiation rate and the number of spikelets per ear, which did not occur when cutting was made at stage C30.

## 13.7 SOURCES OF GENETIC VARIATION

Triticale, the synthetic wheat–rye amphidiploid, does not have wild relatives or landraces as sources of genetic variation for use in its improvement. Despite the lack of diversity created by evolution, there is a broad spectrum of possibilities to introduce genetic variation into triticale from its two parents and their relatives. However, in applied breeding, where short-term aims prevail, the crossing of established varieties is customary to avoid larger interference with the delicate balance between the wheat and rye genomes as established during the course of selection. It has to be kept in mind that even after decades of intensive breeding, triticale is still characterized by cytological instabilities, "which continue to be a problem" (Oettler 1998). Whether the persisting cytological instability is a symptom of lack of evolutionary coadaptation or a sign of intrinsic differences between the two parental species, which prevented the natural emergence of triticale as a competitive new allopolyploid, is not known at present. It might be revealed by future studies at the molecular level.

Synthesizing primary hexaploid and octoploid triticale from selected parents appears to be a straightforward approach in creating new genetic variation. General genome combining ability for both tetraploid wheat and rye was detected by Lelley and Gimbel (1989) using 20 primary hexaploid triticales in comparison with their parents, but data have indicated that while for certain characters the wheat genome could have a definite effect, e.g., on plant height, the statistical additive effect of the rye genome was low, and therefore, preselection of a rye genome cannot be recommended. Performance of the primary triticale could not be predicted from the performance of either parents (Geiger et al. 1993). Moreover, primary triticales usually exhibit considerable cytological instability and are more or less sterile. Their agronomic performance is generally low. Crossing them with each other or with advanced triticale genotypes will lead to completely new interactions between the parental genomes and dissolve any quantitative influence of the original rye genotype.

Present-day triticale varieties, having the highest economic impact, are the product of numerous cycles of hybridization, mostly between secondary hexaploid genotypes, followed by a rigorous and long-lasting selection. Originally, the term *secondary* triticale was coined for progenies of crosses between hexaploid × octoploid triticale. Later, however, it became customary to call all triticale genotypes secondary that were not the immediate descendants of a wheat × rye cross followed by chromosome doubling. Of 19 released cultivars, Wolski (1990b) in his winter triticale breeding program developed seven from two- and five from three-way crosses between hexaploid triticale genotypes. In spring triticale of CIMMYT's breeding program, Varughese et al. (1996) also found the highest benefit of such crosses, as measured by numbers of lines that reached an advanced stage. Measured by cluster analysis, greater useful variation in morphological and agronomic traits was found among winter than spring triticale (Royo et al. 1995). For resistance breeding sufficient genetic variability was found for selection against *Septoria nodorum*, Fusarium head blight, and barley yellow dwarf virus (Collin et al. 1990; Arseniuk et al. 1994; Ittu et al. 1996; Maier and Oettler 1996). On the other hand, crosses between hexaploid triticale contributed little to improvement of preharvest sprouting, still one of the most recalcitrant shortcomings of present-day triticale varieties (Pfeiffer 1994; Haesaert and De Baets 1996).

For long-term progress in triticale breeding, in order to expand the genetic variability without immediate practical implications, a vast amount of genetic variation in wheat and rye and in their relatives is available to breeders. Hybridizing hexaploid triticale with hexaploid wheat followed by backcrossing with triticale leads to the introduction of allelic variation of the A and B genomes of hexaploid wheat into the A and B genomes of triticale, which originate from tetraploid wheat. The crossing scheme was recommended by Barker et al. (1989). Experimental data supporting the usefulness of triticale × wheat crosses were presented by Lelley (1992). From the 19 varieties, Wolski (1990b) produced four originated from (triticale × wheat) × triticale crosses.

Triticale × wheat crosses may lead to D/R substitutions and to so-called segmental triticale. The most famous of such substitutions was the CIMMYT variety Armadillo, in which chromosome 2R was replaced by 2D (Gustafson and Zillinsky 1973). The high-yield capacity of this genotype was attributed to the presence of 2D. Armadillo was a highly preferred crossing partner of CIM-MYT's triticale breeding program. In the 1977–78 international triticale screening nursery ca. 40 lines having the complete rye complement and ca. 300 having one or two D/R substitutions were included (Zillinsky 1985). However, time has shown that the benefit of D/R substitution has been overestimated, and present-day varieties regularly possess the complete rye genome. Through the development of tetraploid triticale with a mixed diploid A/B genome, a new approach was opened up to substitute D genome chromosomes for their homoeologs in the A and B genomes of hexaploid triticale (Kazman and Lelley 1994). The substitution of chromosome 1A by 1D leads to the introduction of the storage protein locus *Glu-D1* with its alleles, responsible for high bread-making quality of hexaploid wheat, into hexaploid triticale with the prospect of improving its bread-making quality (Lafferty and Lelley 2001).

The value of (triticale × rye) × triticale crosses is more questionable. Only one of Wolski's 19 triticale varieties had rye in its pedigree (Wolski 1990b). The experiment reported by Lelley (1992) showed that the intergenomic interaction was more affected by variation in the rye genome of (triticale × rye) × triticale crosses than by variation in the wheat genome of (triticale × wheat) × triticale crosses. It appears that genes in the rye genome are more often suppressed in their expression in triticale than those of wheat; for example, the excellent preharvest sprouting resistance of the rye variety Otello was not expressed in triticale (Huskowska et al. 1985).

Crossing triticale with relatives of wheat and rye also has been attempted. Bernard and Gay (1985) crossed triticale with *Aegilops ventricosa* Tausch to introduce eyespot resistance (*Cercosporella herpatrichodes*) into triticale, producing only a few plants with poor agronomic performance. Gruszecka et al. (1996) crossed *Aegilops crassa* Boiss., *Aegilops juvenalis* (Thell.) Eig.,

*Aegilops triaristata* Willd., and *Aegilops squarrosa* with hexaploid triticale to introduce new resistance genes against pests and diseases, but the $F_1$ plants showed severe meiotic disturbances and low pollen viability. Similarly, cytological disturbances prevented the utilization of traits in triticale from *T. monococcum* (Sodkiewicz and Tomczak 1996). Induced mutagenesis was also attempted in triticale, using chemical agents, irradiation, or somaclonal variation. However, the products of these experiments rarely reached the stage of field-grown trials.

For short-term benefits in practical breeding programs Oettler (1998) suggests crossing advanced triticale genotypes, being the most promising strategy, followed by crosses between primary and secondary genotypes, and finally (triticale × wheat) × triticale crosses. For middle- or long-term benefits, new primary hexaploid and octoploid triticales have to be synthesized, which may include wild rye species (Attia and Lelley 1986).

## 13.8 FINAL CONCLUSION

In the introduction to this chapter I stated that by producing triticale, "man wanted to create a less demanding, high-yielding, and good-quality plant, and he succeeded. Almost." Why only almost? Are not 3 million ha of triticale grown worldwide sufficient proof of the success and vitality of this plant? Does triticale still show any marked differences in its adaptation, compared with other, natural polyploid crops? If so, what are those differences? Triticale is the only synthetic polyploid produced commercially, mainly for seed yield. Seeds are the immediate results of the production of competitive fertile gametes and a successful fertilization. A well-developed seed is a composite of a functional embryo with a balanced chromosome number and an adequate-size endosperm, forming the sink for deposition of starch and proteins to nurture the embryo until photosynthesis of the first leaves takes over. All these developments are based on a delicately balanced genetic program. In triticale, meiosis leading to euploid gametes and mitosis producing an adequate number of endosperm cells are today still prone to disturbances, owing to the genotype or environmental conditions. The reasons for these phenomena are not yet fully understood.

The success of triticale largely depends on steadfast breeding efforts. Varieties successfully competing in the agricultural arena with other cereals are coming from breeders and only very rarely from scientific laboratories. Decades of crossing and selection accumulated the experience for choosing the right crossing partners and determining selection intensities. Breeders learned quickly to recover those rare combinations of wheat and rye genotypes with the least disturbed intergenomic interactions but ever-improving agronomic characteristics. Man has succeeded once more to prevail against nature, at least temporarily.

We learned in recent years that polyploidy requires the fast adjustment of expression of orthologous genes, previously responsible for the phenotypic expression of the plant at the diploid level. A number of mechanisms may act and interact during this process. A quasi-natural selection, as breeders perform it, is at present probably the most efficient way to push ahead this process of genome adjustment for an improved genotypic expression.

Economic circumstances may promote triticale production. However, the recent success of triticale, especially in Europe, is based to a great extent on the agronomic value of the plant. We hope that its low production costs, robustness, high yield potential, and high nutritive value will eventually open the way to less developed production systems, for which it is predestined, as envisaged by so many triticale workers.

## REFERENCES

Acevedo, E. and Fereras, E. 1993. Resistance to abiotic stresses. In *Plant Breeding: Principles and Prospects*, M.D. Hayward, N.O. Bosemark, and I. Romagosa, Eds. Chapman & Hall, New York, pp. 406–421.

Andrews, A.C., Wright, R., Simpson, P.G., Jessop, R., Reeves, S., and Wheeler, J. 1991. Evaluation of new cultivars of triticale as dual-purpose forage and grain crops. *Aust. J. Exp. Agric.* 31: 769–775.

Aniol, A. 1990. Genetics of aluminum tolerance in wheat (*Triticum aestivum* L. Thell). *Plant Soil* 123: 223–227.

Aniol, A. 2002. Environmental stress in cereals: an overview. In *Proceedings of the 5th International Triticale Symposium*, E. Arseniuk and R. Osiski, Eds., Radzikow, Poland, 1: 112–121.

Aniol, A. and Gustafson, J.P. 1984. Chromosomal location of genes controlling aluminum tolerance in wheat, rye and triticale. *Can. J. Genet. Cytol.* 26: 701–705.

Arseniuk, E., Czembor, H.J., Scharen, A.L., Sowa, W., Krysiak, H., and Zimny, J. 1994. Response of triticale, wheat, and rye germplasm lines to artificial inoculation with *Stagonospora* (=*Septoria*) *nodorum* under controlled environmental and field conditions. *Triticale Topics* 12: 13–23.

Arseniuk, E. and Sodkiewicz, W. 2002. Study of phenotypic traits of partial resistance to *Stagonospora nodorum* in winter triticale introgressive lines, commercial cultivars and dihaploid lines. In *Proceedings of the 5th International Triticale Symposium*, E. Arseniuk and R. Osiski, Eds., Radzikow, Poland, 1: 163–177.

Attia, T. and Lelley, T. 1986. Production and cytology of 6x triticale containing wild rye species. *Egypt. J. Genet. Cytol.* 15: 285–296.

Baier, A.C., de Sousa, C.N.A., and Wiethölter, S. 1998. Tolerance of triticale to acid soil. In *Proceedings of the 4th International Triticale Symposium*, P. Juskiw and C. Dyson, Eds., Red Deer, Alberta, Canada, 2: 285–288.

Banaszak, Z. and Marciniak, K. 2002. Wide adaptation of DANKO triticale varieties. In *Proceedings of the 5th International Triticale Symposium*, E. Arseniuk and R. Osinski, Eds., Radzikow, Poland, 1: 217–222.

Banaszak, Z., Marciniak, K., Wolski, T., and Szolkowski, A. 1998. Breeding for winterhardiness in the DANKO triticale program. In *Proceedings of the 4th International Triticale Symposium*, P. Juskiw and C. Dyson, Eds., Red Deer, Alberta, Canada, 2: 94–96.

Bao, W. and Yan, Y. 1993. Octoploid triticale in China. *Adv. Sci. China Biol.* 3: 55–76.

Barker, T.C. and Varughese, G. 1992. Combining ability and heterosis among eight complete spring hexaploid triticale lines. *Crop Sci.* 32: 340–344.

Barker, T.C., Varughese, G., and Metzger, M. 1989. Alternative backcross methods for introgression of variability into triticale via interspecific hybrids. *Crop Sci.* 29: 963–965.

Baum, B.R. 1971. The taxonomic and cytogenetic implications of the problem of naming amphidiploids of *Triticum and Secale*. *Euphytica* 20: 302–306.

Baum, M. and Lelley, T. 1988. A new method to produce 4x triticales and their application in studying the development of a new polyploid plant. *Plant Breed.* 100: 260–267.

Beirão da Costa, M.L. and Cabo Verde, M.J. 1996. Triticale malting: an evaluation of characteristics and production. In *Triticale: Today and Tomorrow*, H. Guedes-Pinto, N. Darvey, and V.P. Carnide, Eds. Kluwer Academic Publ., Dordrecht, The Netherlands, pp. 763–769.

Benbelkacem, A. 2002. Development and use of triticale (× *Triticosecale* Wittmack) in eastern Algeria. In *Proceedings of the 5th International Triticale Symposium*, E. Arseniuk and R. Osiski, Eds., Radzikow, Poland, 1: 283–286.

Bennett, M.D. 1977. Heterochromatin, aberrant endosperm nuclei and grain shrivelling in wheat-rye genotypes. *Heredity* 39: 411–419.

Bennett, M.D., Champan, V., and Riley, R. 1971. The duration of meiosis in pollen mother cells of wheat, rye and triticale. *Proc. R. Soc. Lond. B* 178: 259–275.

v. Berg, K.H. and Oehler, E. 1938. Untersuchungen über die Cytogenetik amphidiploider Weizen-Roggen-Bastarde. *Züchter* 10: 226–238.

Bernard, S. 1977. Etude de quelques facteurs contribuant a la réussite de l'andronese par culture d'antheres *in vitro* chez le Triticale hexaploide. *Ann. Plantes* 27: 629–635.

Bernard, S. 1980. *In vitro* androgenesis in hexaploid triticale: determination of physical conditions increasing embryoid formation and green plant production. *Z. Pflanzenzüchtg.* 85: 308–321.

Bernard, S. and Bernard M. 1987. Creating new forms of 4x, 6x and 8x primary triticale associating both complete R and D genomes. *Theor. Appl. Genet.* 74: 55–59.

Bernard, S.M. and Gay, G. 1985. Introduction of *Aegilops ventricosa* germplasm into hexaploid triticale. In *Genetics and Breeding of Triticale*, EUCARPIA meeting. INRA, Clermont-Ferrand, France, pp. 221–225.

Bishnoi, U.R., Chitapong, I., Hughes, I., and Nishimura, J. 1978. Quantity and quality of triticale and other small grain silage. *Agron. J.* 70: 439–441.

Bizimungu, B., Collin, J., St. Pierre, C.-A., and Comeau, A. 1996. Evaluation of new winter octoploid triticales for their reaction to single and combined infections with BYDV and *Typhula ishikariensis*. In *Triticale: Today and Tomorrow*, H. Guedes-Pinto, N. Darvey, and V.P. Carnide, Eds. Kluwer Academic Publ., Dordrecht, The Netherlands, pp. 541–547.

Bjurman, B. 1958. Note on the frequency of univalents in some strains of triticale and their hybrids. *Hereditas* 44: 189–192.

Blakeslee, A.F. and Avery, A.G. 1937. Methods of inducing doubling of chromosomes in plant. *J. Hered.* 28: 392–411.

Bocchi, S., Lazzaroni, G., and Maggiore, T. 1996. Evaluation of triticale as a forage plant through the analysis of cinetic of some qualitative parameters from stem elongation to maturity. In *Triticale: Today and Tomorrow*, H. Guedes-Pinto, N. Darvey, and V.P. Carnide, Eds. Kluwer Academic Publ., Dordrecht, The Netherlands, pp. 827–834.

Börner, A. and Melz, G. 1988. Response of rye genotypes differing in plant height to exogenous gibberellic acid application. *Arch. Züchtungsforsch.* 18: 79–82.

Boros, D. 2002. Physico-chemical quality indicators suitable in selection of triticale for high nutritive value. In *Proceedings of the 5th International Triticale Symposium*, E. Arseniuk and R. Osiski, Eds., Radzikow, Poland, 1: 239–244.

Boujenna, A., Ramos, J.M., Yanez, J.A., and Garzía del Moral, L.F. 1996. Apical development in triticale grown for dual purpose in a Mediterranean environment. In *Triticale: Today and Tomorrow*, H. Guedes-Pinto, N. Darvey, and V.P. Carnide, Eds. Kluwer Academic Publ., Dordrecht, The Netherlands, pp. 873–877.

Budak, H., Baenzinger, P.S., Beecher, B.S., Graybosch, R.A., Campbell, B.T., Shipman, M.J., Erayman M., and Eskridge, K.M. 2004. The effect of introgression of wheat D-genome chromosomes into "Presto" triticale. Euphytica 137: 261–270.

Budzianowski, G. and Wos, H. 2002. The effect of single "D"-genome substitutions on aluminum tolerance of winter and spring triticale. In *Proceedings of the 5th International Triticale Symposium*, E. Arseniuk and R. Osiski, Eds., Radzikow, Poland, 2: 189–196.

Budzianowski, G. and Wos, H. 2004. The effect of single D-genome chromosomes on aluminum tolerance of triticale. *Euphytica* 137: 165–172.

Carman, E.S. 1884. *Rural New Yorker*, August 30.

Cauderon, Y., Cauderon, A., Gay, G., and Roussel, J. 1985. Alloplasmic lines and nucleocytoplasmic interactions in triticale. In *Genetics and Breeding of Triticale*, EUCARPIA meeting. INRA, Clermont-Ferrand, France, pp. 177–192.

Charmet, G. and Bernard, S. 1984. Diallel analysis of androgenic plant production in hexaploid triticale (× *Triticosecale*, Wittmack). *Theor. Appl. Genet.* 69: 55–61.

Collin, J., Comeau, A., and St.-Pierre, C.A. 1990. Tolerance to barley dwarf virus in triticale. *Crop. Sci.* 30: 1008–1014.

Comai, L. 2000. Genetic and epigenetic interactions in allopolyploid plants. *Plant Mol. Biol.* 43: 387–399.

Comeau, A. and Arseniuk, E. 1998. Disease resistance and tolerance in triticale: an evolutionary viewpoint. In *Proceedings of the 4th International Triticale Symposium*, P. Juskiw and C. Dyson, Eds., Red Deer, Alberta, Canada, 1: 124–147.

Curtis, C.A. and Lukaszewski, A.J. 1993. Localization of genes in rye that restore male fertility to hexaploid wheat with *timopheevi* cytoplasm. *Plant Breed.* 11: 106–114.

Davies, P.A., Cooper, K.V., Morton, S., and Zanker, T.P. 1998. Triticale doubled haploids in South Australia. In *Proceedings of the 4th International Triticale Symposium*, P. Juskiw and C. Dyson, Eds., Red Deer, Alberta, Canada, 2: 61–65.

Derycke, V., Haesaert, G., Latré, J., and Struik, P.C. 2002. Relation between laboratory sprouting resistance test and field observations in triticale (× *Triticosecale* Wittmack) genotype. In *Proceedings of the 5th International Triticale Symposium*, E. Arseniuk and R. Osiski, Eds., Radzikow, Poland, 1: 123–133.

Dhindsa, G.S., Maini, G., Nanda, G.S., and Singh, G. 1998. Combining ability and heterosis for yield and its components in triticale. In *Proceedings of the 4th International Triticale Symposium*, P. Juskiw and C. Dyson, Eds., Red Deer, Alberta, Canada, 2: 116–118.

Drews, E. and Seibel, W. 1976. Bread-baking and other uses around the world. In *Rye: Production, Chemistry, and Technology*, W. Buschuk, Ed. American Association of Cereal Chemists, St. Paul, MN, pp. 127–178.

Driscoll, C.J. and Anderson, L.M. 1967. Cytogenetic studies of Transec wheat-rye translocation line. *Can. J. Genet. Cytol.* 9: 375–380.

FAOSTAT. 2004. Agricultural data. http://faostat.fao. org.

Flavell, R.B., Rimpau, J., and Smith, D.B. 1977. Repeated sequence DNA relationships in four cereal genomes. *Chromosoma* 63: 205–222.

Friebe, B., Heun, M., and Bushuk, W. 1989. Cytological characterization, powdery mildew resistance and storage protein composition of tetraploid and hexaploid 1BL/1RS wheat-rye translocation lines. *Theor. Appl. Genet.* 78: 425–432.

Friebe, B., Heun, M., Tuleen, N., Zeller, F.J., and Gill, B.S. 1994. Cytogenetically monitored transfer of powdery mildew resistance from rye into wheat. *Crop Sci.* 34: 621–625.

Friebe, B., Jiang, J., Raupp, W.J., McIntosh, R.A., and Gill, B.S. 1996. Characterization of wheat-alien translocations conferring resistance to diseases and pests: current status. *Euphytica* 91: 59–87.

Friebe, B., Raupp, W.J., and Gill, B.S. 1998. Alien sources for disease and pest resistance in wheat improvement. In *Current Topics in Plant Cytogenetics Related to Plant Improvement*, T. Lelley, Ed. Universitätsverlag, Vienna, pp. 63–71.

Gallego, F.J. and Benito, C. 1997. Genetic control of aluminum tolerance in rye (*Secale cereale* L.). *Theor. Appl. Genet.* 95: 393–399.

Garcia del Moral, L.F. 1991. Leaf area, grain yield and yield components following forage removal in triticale. *J. Agron. Crop Sci.* 168: 100–102.

Geiger, H.H., Oettler, G., Marker, R., Utz, H.F., and Wehmann, F. 1993. Evaluating rye inbred lines per se and in euplasmic and alloplasmic crosses for triticale suitability. *Crop Sci.* 33: 725–729.

González, J.M., Jouve, N., Gustafson, J.P., and Muniz, L.M. 2002a. A genetic map of molecular markers in × *Triticosecale* Wittmack. In *Proceedings of the 5th International Triticale Symposium*, E. Arseniuk and R. Osiski, Eds., Radzikow, Poland, 2: 85–93.

González, J.M., Jouve, N., and Muiz, L.M. 2002b. Practical application of a molecular genetic map in triticale: locating QTL for androgenesis. In *Proceedings of the 5th International Triticale Symposium*, E. Arseniuk and R. Osiski, Eds., Radzikow, Poland, 2: 95–101.

Goral, T., Cichy, H., Busko, M., and Perkowski, J. 2002. Resistance to head blight and mycotoxin concentrations in kernels of Polish winter triticale lines and cultivars inoculated with *Fusarium culmorum*. In *Proceedings of the 5th International Triticale Symposium*, E. Arseniuk and R. Osiski, Eds., Radzikow, Poland, 2: 85–93.

Graybosch, R.A. 2001. Uneasy unions: quality effects of rye chromatin transfers to wheat. *J. Cereal Sci.* 33: 3–16.

Gregory, R.S. 1980. A technique for identifying major dwarfing genes and its application in a triticale breeding programme. *Hodowla Roslin* 24: 407–418.

Gregory, R.S. 1984. A technique for identifying major dwarfing genes in triticale. *Z. Pflanzenzüchtg.* 92: 177–184.

Gregory, R.S. 1987. Triticale breeding. In *Wheat breeding*, F.G.H. Lupton, Ed. Chapman & Hall, London, pp. 269–286.

Gruszecka, D., Tarkowski, C., Stefanowska, G., and Marciniak, K. 1996. Cytological analysis of F1 *Aegilops* hybrids with *Triticosecale*. In *Triticale: Today and Tomorrow*, H. Guedes-Pinto, N. Darvey, and V.P. Carnide, Eds. Kluwer Academic Publ., Dordrecht, The Netherlands, pp. 195–201.

Grzesik, H. and Wegrzyn, S. 1998. Heterosis and combining ability in some varieties of triticale. In *Proceedings of the 4th International Triticale Symposium*, P. Juskiw and C. Dyson, Eds., Red Deer, Alberta, Canada, 2: 129–133.

Günther, T., Hesemann, C.U., and Oettler, G. 1996. Gel electrophoretic gliadin patterns of euplasmic and alloplasmic primary triticale and the corresponding wheat parents. In *Triticale: Today and Tomorrow*, H. Guedes-Pinto, N. Darvey, and V.P. Carnide, Eds. Kluwer Academic Publ., Dordrecht, The Netherlands, pp. 211–216.

Gupta, N.K., Singh, T., and Bains, G.S. 1985. Malting of Triticale. Effect of Variety, Steeping Moisture, Germination and Giberellik Acid. *Brewers Digest*, March 24–27.

Gustafson, J.P. 1983. Cytogenetics of triticale. In *Cytogenetics of Crop Plants*, M.S. Swaminathan, P.K. Gupta, and U. Sinha, Eds. Macmillan India Ltd., Bombay pp. 225–250.

Gustafson, J.P. and Flavell, R.R. 1996. Control of nucleolar expression in triticale. In *Triticale: Today and Tomorrow*, H. Guedes-Pinto, N. Darvey, and V.P. Carnide, Eds. Kluwer Academic Publ., Dordrecht, The Netherlands, pp. 119–125.

Gustafson, J.P. and Ross, K. 1990. Control of alien gene expression for aluminum tolerance in wheat. *Genome* 33: 9–12.

Gustafson, J.P. and Zillinsky, F.J. 1973. Identification of D-genome chromosomes from hexaploid wheat in a 42-chromosome triticale. In *Proceedings of the 4th International Wheat Genetics Symposium*, Columbia, MO, pp. 225–231.

Haesaert, G. and De Baets, A.E.G. 1996. Preharvest sprouting resistance in triticale: preliminary results. In *Triticale: Today and Tomorrow*, H. Guedes-Pinto, N. Darvey, and V.P. Carnide, Eds. Kluwer Academic Publ., Dordrecht, The Netherlands, pp. 615–622.

Haesaert, G., Derycke, V., Latré, J., Debersaque, F., D'hooghe, K., Coomans, D., and Rombouts, G. 2002. A study on triticale (× *Triticosecale* Wittmack) for whole plant silage in Belgium. In *Proceedings of the 5th International Triticale Symposium*, E. Arseniuk and R. Osiski, Eds., Radzikow, Poland, 1: 261–269.

Heinrich, N., Miedaner, T., and Oettler, G. 2002. Diallel analysis of Fusarium head blight resistance in winter triticale. In *Proceedings of the 5th International Triticale Symposium*, E. Arseniuk and R. Osiski, Eds., Radzikow, Poland, 2: 251–253.

Hermann, M. 2002. Close range outcrossing in triticale. In *Proceedings of the 5th International Triticale Symposium*, E. Arseniuk and R. Osiski, Eds., Radzikow, Poland, 2: 351–355.

Heun, M. and Friebe, B. 1990. Introgression of powdery mildew resistance from rye into wheat. *Phytopathology* 80: 242–245.

Heun, M., Friebe, B., and Bushuk, W. 1990. Chromosomal location of the powdery mildew resistance gene of Amigo wheat. *Phytopathology* 80: 1129–1133.

Hohmann, U. 1984. Cytology and fertility of primary and secondary tetraploid triticale of advanced generations. In *Proceedings of the EUCARPIA Meeting on Triticale*, Clermont-Ferrand, France, pp. 267–275.

Hsam, S.L.K. and Zeller, F.J. 1997. Evidence of allelism between genes *Pm8* and *Pm17* and chromosomal location of powdery mildew and leaf rust resistance genes in the common wheat cultivar 'Amigo'. *Plant Breed.* 116: 119–122.

Hughes, R.J. and Choct, M. 1999. Chemical and physical characteristics of grains related to variability in energy and amino acid availability in poultry. *Aust. J. Agric. Res.* 50: 689–703.

Hughes, R.J. and Cooper, K.V. 2002. Nutritive value of triticale for broiler chickens is affected by variety, weather condition and growth site. *Proc. Aust. Poultry Sci. Symp.* 14: 131–134.

Hulse, J.H. and Laing, E.M. 1974. Nutritive Value of Triticale Protein. International Development Research Centre (IDRC), Ottawa, Canada.

Huskowska, T., Wolski, T., and Ceglinska, A. 1985. Breeding of winter triticale for improvement of grain characters. In *Proceedings of the EUCARPIA Meeting in Genetics and Breeding of Triticale*, M. Bernard and S. Bernard, Eds., Clermont-Ferrand, France, pp. 641–649.

Iganaki, M.N., Mergoum, M., Pfeiffer, W.H., Mujeeb-Kazi, A., and Lukaszewski, A.J. 1998. Crossability of triticale (× *Triticosecale* Wittmack) with maize and the effect of D-genome chromosomes. In *Proceedings of the 4th International Triticale Symposium*, P. Juskiw and C. Dyson, Eds., Red Deer, Alberta, Canada, 2: 71–75.

Immonen, S., Merjankari, I., Tauriainen, A., and Hovinen, S. 1998. Androgenesis in triticales of diverse origin, impact of stress. In *Proceedings of the 4th International Triticale Symposium*, P. Juskiw and C. Dyson, Eds., Red Deer, Alberta, Canada, 1: 38–40.

Immonen, S. and Robinson, J. 2000. Stress treatments and ficoll for improving green plant regeneration in triticale anther culture. *Plant Sci.* 150: 77–84.

Ittu, M., Ittu, G., and Saulescu, N.N. 1996. Screening for *Fusarium* scab resistance in triticale. In *Triticale: Today and Tomorrow*, H. Guedes-Pinto, N. Darvey, and V.P. Carnide, Eds. Kluwer Academic Publ., Dordrecht, The Netherlands, pp. 527–533.

Jenkins, B.C. 1969. History of the development of some promising hexaploid triticales. *Wheat Inf. Serv.* 28: 18–20.

Jessop, R.S. 1996. Stress tolerance in newer triticales compared to other cereals. In *Triticale: Today and Tomorrow*, H. Guedes-Pinto, N. Darvey, and V.P. Carnide, Eds. Kluwer Academic Publ., Dordrecht, The Netherlands, pp. 419–427.

Jlibene, M. and Gustafson, J.P. 1992. The identification of a gibberellic acid insensitive gene in *Secale cereale*. *Plant Breed.* 108: 229–233.

Jung, C. and Lelley, T. 1985a. Hybrid necrosis in triticale caused by gene-interaction between its wheat and rye genomes. *Z. Pflanzenzüchtg.* 94: 344–347.

Jung, C. and Lelley, T. 1985b. Genetic interactions in wheat and rye genomes in triticale. 2. Morphological and yield characters. *Theor. Appl. Genet.* 70: 427–432.

Jung, C., Lelley, T., and Röbbelen, G. 1985. Genetic interactions in wheat and rye genomes in triticale. 1. Cytological results. *Theor. Appl. Genet.* 70: 422–426.

Kaltsikes, P.J. 1973. Univalency in triticale. In *Triticale — Proc. Int. Symp.*, IDRC-024e, R. MacIntyre and M. Campbell, Eds. El Batan, Mexico, pp. 159–167.

Katterman, G. 1939. Über heterogenomische amphidiploide Weizen-Roggen-Bastarde. *Z. Pflanzenzüchtg.* 23: 179–209.

Kazman, E. and Lelley, T. 1994. Rapid incorporation of D-genome chromosomes into A and/or B genomes of hexaploid triticale. *Plant Breed.* 113: 89–98.

Kazman, E., Lelley, T. and Röbbelen, G., 1993. Die Backqualität von 6x-triticale: Möglichkeiten chromosomaler Manipulation. Bericht 44. Arbeitstagung der AG Saatzuchtleiter, Gumpenstein, Austria, 37–45.

Kempanna, C. and Seetharam, A. 1972. Studies into meiotic stability, pollen and seed fertility in triticales. *Cytologia* 37: 327–333.

Kiss, Á. 1966. Neue Richtung in der Triticale-Züchtung. *Z. Pflanzenzüchtg.* 55: 309–329.

Kiss, Á. 1970. Spontaneous crossing between hexaploid triticale Rosner and triticale no. 64. *Wheat Inf. Serv.* 31: 24–25.

Kiss, Á. 1971. Origin of the preliminary released Hungarian hexaploid varieties, nos. 57 and 64. *Wheat Inf. Serv.* 32: 20–22.

Kiss, Á. and Rédei, G. 1953. Experiments to produce rye-wheat (triticale). *Acta Agron. Acad. Scientiarum Hung.* 3: 257–276.

Kleier, G. and Fossati, D. 1998. Disease reaction of triticale and wheat varieties. In *Proceedings of the 4th International Triticale Symposium*, P. Juskiw and C. Dyson, Eds., Red Deer, Alberta, Canada, 2: 223–226.

Kobylyansky, V.D. 1975. Effect of the dominant character of short straw on some quantitative characters of winter rye. *Hod. Rosl. Aklim. Nas.* 19: 495–501.

Kozdój, J. and Zimny, J. 1993. Microspore development stages in chilled and unchilled anthers of Triticale (× *Triticosecale* Wittmack). *Bull. Pol. Acad. Sci.* 93: 108–116.

Krolow, K.-D. 1962. Aneuploidie und Fertilität bei amphidiploiden Weizen-Roggen-Bastarden (Triticale). I. Aneuploidie und Selektion auf Fertilität bei oktoploiden Triticale-Formen. *Z. Pflanzenzüchtg.* 48: 177–197.

Krolow, K.-D. 1973. 4x triticale production and use in triticale breeding. In *Proceedings of the 4th International Wheat Genetics Symposium*, Missouri Agricultural Experiment Station, Columbia, pp. 691–696.

Krolow, K.-D. 1974. Research work with 4x triticale in Germany (Berlin). In *Proce. Inter. Symp. IDRC-024*, R. MacIntyre and M. Campbell, Eds., El Batan, Mexico, pp. 51–60.

Lafferty, J. and Lelley, T. 2001. Introduction of high molecular weight glutenin subunits 5 + 10 for the improvement of bread-making quality of hexaploid triticale. *Plant Breed.* 120: 33–37.

Lapiski, B. 2002. Application of tertaploid triticale in improvement of related crop species. In *Proceedings of the 5th International Triticale Symposium*, E. Arseniuk and R. Osiski, Eds., Radzikow, Poland, 1: 71–77.

Larter, E.N., Shebeski, L.H., McGinnis, R.C., Evans, L.E., and Kaltsikes, P.J. 1970. Rosner, a hexaploid triticale cultivar. *Can. J. Plant Sci.* 50: 122–124.

Larter, E.N., Tsuchiya, T., and Evans, L. 1968. Breeding and cytology of triticale. In *Proceedings of the 3rd International Wheat Genetics Symposium*, Canberra, pp. 213–221.

Lebedeff, V.N. 1934. Neue Fälle der Formierung von Amphidiploidie in Weizen-Roggen-Bastarden. *Z. Pflanzenzüchtg.* 19: 509–525.

Leighty, C.E. 1916. Carman's wheat-rye hybrids. *J. Hered.* 7: 420–427.

Lein, A. 1943. Die genetische Grundlage der Kreuzbarkeit zwischen Weizen und Roggen. *Z. indukt. Abstamm. Vererbungslehre* 81: 28–61.

Lein, V. and Lelley, T. 1986. A separate control for frequency and within bivalent distribution of chiasmata in rye (*Secale cereale* L.). *Genome* 29: 419–424.

Lelley, T. 1974. Desynapsis as a possible source of univalents in metaphase I of triticale. *Z. Pflanzenzüchtg.* 73: 249–258.

Lelley, T. 1975a. Genetic control of pairing of rye chromosomes in triticale. *Z. Pflanzenzüchtg.* 75: 24–29.

Lelley, T. 1975b. Identification of univalents and rod bivalents in tritical with Giemsa. *Z. Pffanzensüchtg.* 75: 252–256.

Lelley, T. 1978a. Genetic control of chiasma frequency and distribution in rye *Secale cereale. Can. J. Genet. Cytol.* 20: 471–474.

Lelley, T. 1978b. Specific interaction for meiotic regulation of wheat and rye in triticale, consequences for triticale breeding. In *Proceedings of the 5th International Wheat Genetics Symposium*, S. Ramanujam, Ed., New Delhi, pp. 1213–1217.

Lelley, T. 1982. Meiotic behaviour of the rye genome in triticale. In *Proceedings of the International Symposium Section on Mutation and Polyploidy of EUCARPIA in Induced Variability in Plant Breeding*, C. Broertjes, Ed., Wageningen, The Netherlands, pp. 101–105.

Lelley, T. 1985. Triticale breeding a new approach. In *Proceedings of the EUCARPIA Meeting in Genetics and Breeding of Triticale*, M. Bernard and S. Bernard, Eds., Clermont-Ferrand, France, pp. 135–143.

Lelley, T. 1992. Triticale, still a promise? *Plant Breed.* 109: 1–17.

Lelley, T. 1996. The verdict of triticale: a critical view. In *Triticale: Today and Tomorrow*, H. Guedes-Pinto, N. Darvey, and V.P. Carnide, Eds. Kluwer Academic Publ., Dordrecht, The Netherlands, pp. 49–55.

Lelley, T. and Gimbel, E. M. 1989. "Genome combining ability" of wheat and rye in triticale. *Plant Breed.* 102: 273–280.

Lelley, T., Machmoud, A.S., and Lein, V. 1986. Genetics and cytology of unreduced gametes in rye (*Secale cereale* L.). *Genome* 29: 635–638.

Lelley, T., Kazman, E., Devos, K.M., and Gale, M.D. 1995. Use of RFLPs to determine the chromosome composition of tetraploid triticale (A/B)(A/B)RR. *Genome* 38: 250–254.

Lelley, T., Eder, C., and Grausgruber, H. 2004. Influence of 1BL.1RS wheat-rye chromosome translocation on genotype by environment interaction. *J. Cereal Sci.* 39: 313–320.

Levy, A.A. and Feldman, M. 2002. The impact of polyploidy on grass genome evolution. *Plant Physiol.* 130: 1587–1593.

Lewitzky, G.A. and Benetzkaya, G.K. 1930. Cytological investigation of constant intermediate rye-wheat hybrids. In *Proceedings of the All-Union Congress of Genetics and Breeding,* Leningrad, pp. 345–352 (in Russian with English summary).

Lindschau, M. and Ochler, E. 1935. Untersuchungen am konstant intermediären additiven Rimpau'schen Weizen-Roggen-Bastard. *Züchter* 7: 228–233.

Lozano, A.J., Zamora, V.M., Solís, H.D., Mergoum, M., and Pfeiffer, W.H. 1998. Triticale forage production and nutritional value in the northern region of Mexico. In *Proceedings of the 4th International Triticale Symposium*, P. Juskiw and C. Dyson, Eds., Red Deer, Alberta, Canada, 2: 259–267.

Lukaszewski, A.J. 1990. Development of aneuploid series in hexaploid triticale. In *Proceedings of the 2nd International Triticale Symposium*, Passo Fundo, Brazil, pp. 397–400.

Lukaszewski, A.J. 1998. Improvement of breadmaking quality of triticale through chromosome translocations. In *Proceedings of the 4th International Triticale Symposium*, P. Juskiw and C. Dyson, Eds., Red Deer, Alberta, Canada, 1: 102–110.

Lukaszewski, A.J. 2003. Registration of three germplasms of hexaploid triticale with introgressions of wheat storage protein loci from chromosome 1D of bread wheat. *Crop Sci.* 43: 2316.

Lukaszewski, A.J., Apolinarska, B., and Gustafson, J.P. 1987b. Introduction of the D-genome chromosomes of bread wheat into hexaploid triticale with a complete rye genome. *Genome* 29: 425–430.

Lukaszewski, A.J., Apolinarska, B., Gustafson, J.P., and Krolow, K.-D. 1984. Chromosome constitution of tetraploid triticale. *Z. Pflanzenzüchtg.* 93: 222–236.

Lukaszewski, A.J., Apolinarska, B., Gustafson, J.P., and Krolow, K.-D. 1987a. Chromosome pairing and aneuploidy in tetraploid triticale. I. Stabilized karyotypes. *Genome* 29: 554–561.

Lukaszewski, A.J. and Curtis, C.A. 1992. Transfer of the *Glu-D1* gene from chromosome 1D of bread wheat to chromosome 1R in hexaploid triticale. *Plant Breed.* 109: 203–210.

Ma, X.-F., Rodriguez Milla, M.A., and Gustafson, J.P. 2002. AFLP-based genome studies of triticale following polyploidization. In *Proceedings of the 5th International Triticale Symposium*, E. Arseniuk and R. Osiski, Eds., Radzikow, Poland, 1: 87–94.

Ma, X.-F., Fang, P. and Gustafson J.P. 2004. Polploidization-induced genome variation in tritical. *Genome* 47: 839–848.

Maier, F.J. and Oettler, G. 1996. Genetic variation for head blight resistance in triticale caused by *Fusarium graminearum* isolates of different deoxynivalenol production. *Euphytica* 89: 387–394.

Mangat, G.S. and Dhindsa, G.S. 1995. Combining ability studies in spring × winter triticale crosses for yield and its components over environments. *Cereal Res. Commun.* 23: 73–78.

Matzke, M.A., Mittelsten Scheid, O., and Matzke, A.J. 1999. Rapid structural and epigenetic changes in polyploid and aneuploid genomes. *Bioessays* 21: 761–767.

McIntosh, R.A. and Singh, S.J. 1987. Rusts: real and potential problems for triticale. In *Proceedings of the 1st International Triticale Symposium*, Sydney, N.L. Darvey, Ed., Australian Institute of Agricultural Science, pp. 199–207.

McIntosh, R.A., Luig, N.H., Milne, D.L., and Cusick, J. 1983. Vulnerability of triticale to wheat stem rust. *Can. J. Plant Pathol.* 5: 62–69.

Meister, G.K. 1921. Natural hybridization of wheat and rye in Russia. *J. Hered.* 12: 467–470.

Meister, G.K. 1930. The present purposes of the study of interspecific hybrids. In *Proceedings of the USSR Congress of Genetic Plant and Animal Breeding*, 2: 27–43 (in Russian with English summary).

Merker, A. 1971. Cytogenetic investigations in hexaploid triticale. I. Meiosis, aneuploidy and fertility. *Hereditas* 68: 281–290.

Merker, A. 1973. Cytogenetic investigations in hexaploid triticale. II. Meiosis, aneuploidy and fertility in F1 and F2. *Hereditas* 73: 285–290.

Merker, A. 1975. Chromosome composition of hexaploid triticales. *Hereditas* 80: 41–52.

Merker, A. 1976. The cytogenetic effect of heterochromatin in hexaploid triticale. *Hereditas* 83: 215–222.

Miedaner, T., Heinrich, N., Schneider, B., Oettler, G., Rohde, S. and Rabenstein, F. 2004. Estimation of deoxynivalenol (DON) content by symptom rating and exoantigen content for resistance selection in wheat and triticale. Euphytica 139: 123–132.

Morrison, R.J., Larter, E.N., and Green, G.J. 1977. The genetics of resistance to *Puccinia graminis tritici* in hexaploid triticale. *Can J. Genet. Cytol.* 19: 683–693.

Müntzing, A. 1939. Studies on the properties and the ways of production of rye-wheat amphidiploids. *Hereditas* 25: 387–430.

Müntzing, A. 1979. Triticale results and problems. *Z. Pflanzenzüchtg.* 10: 1–130.

Myer, R.O. 2002. Triticale grain in young pig diets. In *Proceedings of the 5th International Triticale Symposium*, E. Arseniuk and R. Osiski, Eds., Radzikow, Poland, 1: 271–276.

Nakajima, G. 1950. Genetical and cytological studies in breeding of amphidiploid types between *Triticum* and *Secale*. I. The external characters and chromosomes of fertile F1 *T. turgidum* (n = 14) ∞ *S. cereale* (= 7) and its F2 progenies. *Jap. J. Genet.* 25: 139–148.

Nakajima, G. 1952. Genetical and cytological studies in the breeding of amphidiploid types between *Triticum* and *Secale*. III. A. The maturation division in PMC's of F2 plants having 2n = 42 chromosomes raised from *T. turgidum* (n = 14) and *S. cereale* (n = 7). *Bot. Mag.* 66: 288–294.

Nalepa, S. 1990. Hybrid triticale: present and future, In *Proceedings of the 2nd International Triticale Symposium*, Passo Fundo, Rio Grande do Sul, Brasil, pp. 402–407.

Nalepa, S., Grzesik, H., and Pilch, J. 1980. A study of hybrids between triticale, hexa- and dwarf rye. *Hod. Rosl. Aklim. Nas.* 24: 531–541.

Oehler, E. 1936. Heutiger Stand der Weizen-Roggenkreuzungen in Deutschland. *Umschau* 51: 1-4.

Oettler, G. 1998. Creating genetic variability in triticale and its potential for breeding. 1. Agronomic traits. In *Proceedings of the 4th International Triticale Symposium*, P. Juskiw and C. Dyson, Eds., Red Deer, Alberta, Canada, 1: 1–12.

Oettler, G. 2002. Improving falling-number by recurrent selection in winter triticale. In *Proceedings of the 5th International Triticale Symposium*, E. Arseniuk and R. Osiski, Eds., Radzikow, Poland, 1: 253–259.

Oettler, G., Wehmann, F., and Utz, H.F. 1991. Influence of wheat and rye parents on agronomic characters in primary hexaploid and octoploid triticale. *Theor. Appl. Genet.* 81: 401–405.

Oettler, G., Becker, H.C., Hoppe, G., and Wahle, G. 1998. Heterosis for yield and yield components in multi-location trials of winter triticale. In *Proceedings of the 4th International Triticale Symposium*, P. Juskiw and C. Dyson, Eds., Red Deer, Alberta, Canada, 2: 151–155.

Oettler, G., Becker, H.C., and Hoppe, G. 2001. Heterosis for yield and other agronomic traits of winter triticale F1 and F2 hybrids. *Plant Breed.* 120: 351–353.

Oettler, G., Burger, H., and Melchinger, A.E. 2003. Heterosis and combining ability for grain yield and other agronomic traits in winter triticale. *Plant Breed.* 122: 318–321.

O'Mara, J.G. 1948. Fertility in allopolyploids. *Rec. Genet. Soc. Am.* 15: 52 (abstract).

Ozkan, H., Levy, A.A., and Feldman, M. 2001. Allopolyploidy-induced rapid genome evolution in the wheat (*Aegilops-Triticum*) group. *Plant Cell* 13: 1735–1747.

Peña, R.J. 1996. Factors affecting triticale as a food crop. In *Triticale: Today and Tomorrow*, H. Guedes-Pinto, N. Darvey, and V.P. Carnide, Eds. Kluwer Academic Publ., Dordrecht, The Netherlands, pp. 753–761.

Pfeiffer, W.H. 1994. Triticale improvement strategies at CIMMYT: existing genetic variability, and its implication to projected genetic advance. In *Proceedings of the 5th Portuguese Triticale Conference, Elvas*, Portugal, p. 9.

Pfeiffer, W.H., Sayre, K.D., and Mergoum, M. 1998. Heterosis in spring triticale hybrids. In *Proceedings of the 4th International Triticale Symposium*, P. Juskiw and C. Dyson, Eds., Red Deer, Alberta, Canada, 1: 86–91.

Pieritz, W.J. 1970. Elimination von Chromosomen in amphidiploiden Weizen-Roggen-Bastarden (Triticale). *Z. Pflanzenzüchtg.* 56: 90–109.

Pikaard, C.S. 2001. Genomic change and gene silencing in polyploids. *Trends Genet.* 17: 675–677.

Pissarev, V. 1966. Different approaches in triticale breeding. *Hereditas*, Suppl. 2: 279–290.

Pohler, W., Kistner, G., Kison, H.-U., and Szigat, G. 1978. Meioseuntersuchungen an Triticale V. Meioseverhalten, Pollenvitalität und Fertilität von Triticale F1 Bastarden und deren Eltern. *Biol. Zbl.* 97: 453–470.

Popov, P. and Tsvetkov, S. 1975. Development of winter primary and secondary triticale (2n = 42). In *Triticale: Studies and Breeding, Proceedings of the International Symposium 1973*, Leningrad, pp. 151–157.

Rabinovitch, S.V. 1998. Importance of wheat-rye translocations for breeding modern cultivars of *Triticum aestivum* L. *Euphytica* 100: 323–340.

Rees, H. 1955. Heterosis in chromosome behaviour. *Proc. R. Soc. Lond. B* 144: 150–159.

Rees, H. and Thomson, J.B. 1956. Genotypic control of chromosome behaviour in rye. III. Chiasma frequency in heterozygoutes and homozygoutes. *Heredity* 10: 409–424.

Ren, Z.L. and Lelley, T. 1988. Genetics of hybrid necrosis in rye. *Plant Breed.* 100: 173–180.

Ren, Z.L. and Lelley, T. 1989. Hybrid necrosis in triticale and the expression of necrosis genes in allopolyploids. *Theor. Appl. Genet.* 76: 742–748.

Rimpau, W. 1891. Kreuzungsprodukte landwirtschaftlicher Kulturpflanzen. *Landwirtschaftl. Jahrb.* 20: 335–371.

Rogalska, S.M. and Mikulski, W. 1996. Induction of haploids in triticale (× *Triticosecale* Witt.) by crossing it with maize (*Zea mays*). In *Triticale: Today and Tomorrow*, H. Guedes-Pinto, N. Darvey, and V.P. Carnide, Eds. Kluwer Academic Publ., Dordrecht, The Netherlands, pp. 379–382.

Royo, C., Montesinos, E., Molina-Cano, J.L., and Serra, J. 1993. Triticale and other small grain cereals for forage and grain in Mediterranean condition. *Grass For. Sci.* 48: 11–17.

Royo, C., Soler, C., and Romagosa, I. 1995. Agronomical and morphological differentiation among winter and spring triticales. *Plant Breed.* 114: 413–416.

Saini, H.S. and Henry, R.J. 1989. Fractionation and evaluation of triticale pentosans: comparison with wheat and rye. *Cereal Chem.* 66: 11–14.

Salo, Y. 1998. Triticale cultivation in the Nordic marginal area. In *Proceedings of the 4th International Triticale Symposium*, P. Juskiw and C. Dyson, Eds., Red Deer, Alberta, Canada, 2: 340–343.

Sanches-Monge, E. 1956. Studies on 42-chromosome triticale. I. The production of the amphidiploids. *An. Aula Dei* 4: 191–207.

Sanches-Monge, E. 1975. Hexaploid triticale with different cytoplasm. In *Proceedings of the International Symposium 1973*, Leningrad, pp. 175–180.

Sayed-Tabatabaei, B.E., Patel, M., and Darvey, N.L. 1998. Isolated microspore culture in triticale. In *Proceedings of the 4th International Triticale Symposium*, P. Juskiw and C. Dyson, Eds., Red Deer, Alberta, Canada, 2: 82–84.

Schachschneider, R. 1996. Einige Erfahrungen aus 12 Jahren Triticale-Züchtung. In *Triticale, Züchtung Erzeugung und Verwertung*, Vorträge für Pflanzenzüchtung, 34: 278–291.

Schinkel, B. 2002. Triticale still a healthy crop? In *Proceedings of the 5th International Triticale Symposium*, E. Arseniuk and R. Osiski, Eds., Radzikow, Poland, 1: 157–162.

Schlegel, R., Kynast, R., and Schmidt, J.-C. 1986. Alien chromosome transfer from wheat into rye. In *Genetic Manipulation in Plant Breeding*. Walter de Gruyter, Berlin, pp. 129–136.

Schlegel, R. and Kynast, R. 1988. Wheat chromosome 6B compensates genetic information of diploid rye, *Secale cereale* L. In *Proceedings of the 7th International Wheat Genetics Symposium*, T.E. Miller and R.M.D. Koebner, Eds., Cambridge, U.K., pp. 421–426.

Schulindin, A.F. 1975. Genetical grounds of synthesis in various triticales and their improvement by breeding. In *Triticale: Studies and Breeding, Proceedings of the International Symposium 1973*, Leningrad, pp. 53–69.

Scoles, G.J. 1985. A gene for hybrid necrosis in an inbred line of rye (*Secale cereale* L.). *Euphytica* 34: 207–212.

Scoles, G.J. and Kaltsikes, P.J. 1974. The cytology and cytogenetics of triticale. *Z. Pflanzenzüchtg.* 73: 13–43.

Scott, W.R. and Hines, S.E. 1991. Effect of grazing on grain yield of winter barley and triticale: the position of the apical dome relative to the soil surface. *New Zealand J. Agric. Res.* 34: 177–184.

Seal, A.G. 1986. Causes of grain shriveling in triticale. In *Proceedings of the 1st International Triticale Symposium*, Sydney, N.L. Darvey, Ed., Australian Institute of Agricultural Science, pp. 31–34.

Sears, E.R. 1966. Nullisomic-tetrasomic combinations in hexaploid wheat. In *Chromosome Manipulations and Plant Genetics*, R. Riley and K.R. Lewis, Eds., Oliver and Boyd, Edinburgh, pp. 29–45.

Singh, J., Dhindsa, G.S., Nanda, G.S., and Batta, R.K. 1996. Prospects of triticale as a dual purpose crop. In *Triticale: Today and Tomorrow*, H. Guedes-Pinto, N. Darvey, and V.P. Carnide, Eds. Kluwer Academic Publ., Dordrecht, The Netherlands, pp. 867–871.

Singh, V.P. 1979. Extent of natural crossing in triticale. *Crop Improv.* 6: 63–65.

Sodkiewicz, W. 1984. Amphiploid *Triticum monococum* L. × *Secale cereale* L. (AARR): a new form of tetraploid triticale. *Cereal Res. Commun.*, 12: 1–2.

Sodkiewicz, W. 2003. Diploid wheat: *Triticum monococcum* as a source of resistance genes to preharvest sprouting of triticale. *Cereal Res. Commun.* 30: 323–328.

Sodkiewicz, W. and Strzembicka, A. 2004. Application of *Triticum monococcum* for the improvement of triticale resistance to leaf rust (*Puccinia triticina*). *Plant Breed.* 123: 39–42.

Sodkiewicz, W. and Tomczak, M. 1996. Variation of some physiological and spike characters affecting the reproductive behaviour of introgressive triticale lines with *T. monococcum* genetic information. In *Triticale: Today and Tomorrow*, H. Guedes-Pinto, N. Darvey, and V.P. Carnide, Eds. Kluwer Academic Publ., Dordrecht, The Netherlands, pp. 867–871.

Somers, D.J., Gustafson, J.P., and Filion, W.G. 1992. The influence of the rye genome on expression of heat-shock proteins in triticale. *Theor. Appl. Genet.* 34: 845–848.

Sosinova, A., Lukjanjuk, S., and Ignatova, S. 1981. Anther cultivation and induction of haploid plants in triticale. *Z. Pflanzenzüchtg.* 86: 272–285.

Sowa, W. and Krysiak, H. 1996. Outcrossing in winter triticale, measured by occurrence of tall plants. In *Triticale: Today and Tomorrow*, H. Guedes-Pinto, N. Darvey, and V.P. Carnide, Eds. Kluwer Academic Publ., Dordrecht, The Netherlands, pp. 593–596.

Sowa, W., Wgrzyn, S., Arseniuk, E., and Krysiak, H. 1998a. Diallel analysis of leaf resistance to *Septoria nodorum* Berk. in crosses of winter triticale (× *Triticosecale* Wittmack). In *Proceedings of the 4th International Triticale Symposium*, P. Juskiw and C. Dyson, Eds., Red Deer, Alberta, Canada, 2: 231–234.

Sowa, W., Wgrzyn, S., Arseniuk, E., and Krysiak, H. 1998b. Heritability and quantitative determination of the gene action of *Septoria nodorum* resistance in seven winter triticales (× *Triticosecale* Wittmack). In *Proceedings of the 4th International Triticale Symposium*, P. Juskiw and C. Dyson, Eds., Red Deer, Alberta, Canada, 2: 235–238.

Stallknecht, G.F. and Wichman, D.M. 1998. The evaluation of winter and spring triticale (× *Triticosecale* Wittmack) for grain and forage production under dryland cropping in Montana, USA. In *Proceedings of the 4th International Triticale Symposium*, P. Juskiw and C. Dyson, Eds., Red Deer, Alberta, Canada, 2: 272–274.

Steward, D.M., Gilmore, D.M., and Ausemus, E.T. 1968. Resistance to *Puccinia graminis* derived from *Secale cereale* incorporated to *Triticum aestivum. Phytopathology* 58: 508–511.

St.-Pierre, C.A., Rioux, S., and Comeau, A. 1998. Yield stability of wheat through genetic resistance to biotic and abiotic stresses. In *Proceedings of the 9th International Wheat Genetics Symposium*, Vol. 1, A.E. Slinkard, Ed., Saskatoon, Saskatchewan, pp. 49–52.

Struss, D. and Röbbelen, G. 1989. Cytological irregularities masked by heterozygosity in triticale. *Plant Breed.* 102: 265–272.

Sturm, W. and Engel, K.H. 1980. Trisomanalyse des Alleles HI für Kurzstrohigkeit bei *Secale cereale* L. *Arch. Züchtungsforsch.* 10: 31–35.

Stutz, H.C. 1962. Asynchronous meiotic chromosome rhythm as a cause of sterility in triticale. *Genetics* 47: 988.

Sun, Y.S., Xie, Y., Wang, Z.Y., Hai, L., and Chen, X.Z. 1996. Triticale as forage in China. In *Triticale: Today and Tomorrow*, H. Guedes-Pinto, N. Darvey, and V.P. Carnide, Eds. Kluwer Academic Publ., Dordrecht, The Netherlands, pp. 879–886.

Sutton, W.S. 1903. The chromosomes in heredity. *Biol. Bull. Mar. Boil. Lab. Woods Hole* 4: 231–248.

Tams, S.H., Bauer, E., Oettler, G., and Melchinger, A.E. 2004. Genetic diversity in European winter triticale determined with SSR markers and coancenstry coefficient. *Theor. Appl. Genet.* 108: 1385–1391.

Tams, S.H., Melchinger, A.M. and Bauer, E, 2005. Genetic similarity among European winter triticale elite germplasms assessed with AFLP and comparisons with SSR and pedigree data. *Plant Breed.* 124: 154–160.

Thomas, J.B. and Kaltsikes, P.J. 1976. The genomic origin of unpaired chromosomes in triticale. *Can. J. Genet. Cytol.* 18: 687–700.

Tiumiakov, N. 1930. Fertility and comparative morphology of rye-wheat hybrid of balanced type. In *Proceedings of the USSR Congress of Genetic Plant and Animal Breeding*, 2: 497 508.

Tsuchiya, T. 1970. Chromosome pairing at diakinesis in hexaploid triticale. *Wheat Inf. Serv.* 31: 22–23.

Tsuchiya, T. 1974. Cytological stability of triticale. In *Triticale: First Man-Made Cereal*, C.C. Tsen, Ed. American Association of Cereal Chemistry, St. Oaul, MN, pp. 62–89.

Tsuchiya, T. and Larter, E.N. 1969. Chromosomal stability in some hexaploid strains of triticale. *Crop Sci.* 9: 235–236.

Tsuchiya, T. and Larter, E.N. 1971. Further results on chromosomal stability of hexaploid triticale. *Euphytica* 20: 591–596.

Tsunewaki, K., Iwanaga, M., Maekawa, M., and Tsuji, S. 1984. Production and characterization of alloplasmic lines of a triticale 'Rosner'. *Theor. Appl. Genet.* 68: 169–177.

Van Barneveld, R.J. and Cooper, K.V. 2002. Nutritional quality of triticale for pigs and poultry. In *Proceedings of the 5th International Triticale Symposium*, E. Arseniuk and R. Osiski, Eds., Radzikow, Poland, 1: 277–282.

Varghese, J.P. and Lelley, T. 1983. Origin of nuclear aberrations and seed shriveling in triticale: a re-evaluation of the role of C-heterochromatin. *Theor. Appl. Genet.* 66: 159–167.

Varughese, G., Pfeiffer, W.H., and Pena, R.J. 1996. Triticale: a successful alternative crop (part 2). *Cereal Foods World* 41: 635–645.

Verzea, M. and Ittu, Gh. 1998. Anther culture response of some Romanian winter triticale genotypes. In *Proceedings of the 4th International Triticale Symposium*, P. Juskiw and C. Dyson, Eds., Red Deer, Alberta, Canada, 2: 85–87.

Voylokov, A.V. and Tikhenko, N.D. 1998. Identification and localization of rye polymorphic genes specifically expressed in triticale. In *Proceedings of the 4th International Triticale Symposium*, P. Juskiw and C. Dyson, Eds., Red Deer, Alberta, Canada, 1: 290–296.

Voylokov, A.V. and Tikhenko, N.D. 2002. Triticale as a model for study genome interaction and genome evolution in allopolyploid plants. In *Proceedings of the 5th International Triticale Symposium*, E. Arseniuk and R. Osiski, Eds., Radzikow, Poland, 1: 63–70.

Vrolijk, A.J.J. and Suijs, L.W.M. 2002. Fusarium in triticale. In *Proceedings of the 5th International Triticale Symposium*, E. Arseniuk and R. Osiski, Eds., Radzikow, Poland, 2: 263–265.

Warzecha, R. and Salak-Warzecha, K. 2002. Hybrid triticale: prospects for research and breeding. Part II. Development of male sterilirty. In *Proceedings of the 5th International Triticale Symposium*, E. Arseniuk and R. Osiski, Eds., Radzikow, Poland, 1: 193–198.

Warzecha, R., Salak-Warzecha, K., and Staszewski, Z. 1996. CMS system in hexaploid triticale. In *Triticale: Today and Tomorrow*, H. Guedes-Pinto, N. Darvey, and V.P. Carnide, Eds. Kluwer Academic Publ., Dordrecht, The Netherlands, pp. 225–232.

Wedzony, M., Marcinska, I., Ponitka, A. Slusarkiewicz-Jarzina, A. and Wozna, J. 1998. Production of doubled haploids in triticale (× *Triticosecale* Wittm.) by means of crosses with maize (*Zea mays* L.) using picloram and dicamba. *Plant Breed.* 117: 211–215.

Wgrzyn, S., Gut, M., Grzesik, H., Bichonski, A., and Strus, M. 1998. Resistance to pre-harvest sprouting and other traits of winter triticale. In *Proceedings of the 4th International Triticale Symposium*, P. Juskiw and C. Dyson, Eds., Red Deer, Alberta, Canada, 2: 354–356.

Weimarck, A. 1973. Cytogenetic behaviour in octoploid triticale. I. Meiosis, aneuploidy and fertility. *Hereditas* 74: 103–118.

Wendel, J.F. 2000. Genome evolution in polyploids. *Plant Mol. Biol.* 42: 225–249.

Wilson, A.S. 1876. On wheat and rye hybrids. *Trans. Proc. Bot. Soc. Edinburgh* 12: 286–288.

Wolski, T. 1990a. Impact of semidwarf rye germplasm on triticale improvement. In *Proceedings of the 2nd International Triticale Symposium*, Passo Fundo, Brazil, pp. 144–149.

Wolski, T. 1990b. Winter triticale breeding. In *Proceedings of the 2nd International Triticale Symposium*, Passo Fundo, Brazil, pp. 41–48.

Wolski, T. and Gryka, J. 1996. Semidwarf winter triticale. In *Triticale: Today and Tomorrow*, H. Guedes-Pinto, N. Darvey, and V.P. Carnide, Eds. Kluwer Academic Publ., Dordrecht, The Netherlands, pp. 581–587.

Wolski, T. and Gryka, J. 1998. Further progress in semidwarf winter triticale program. In *Proceedings of the 4th International Triticale Symposium*, P. Juskiw and C. Dyson, Eds., Red Deer, Alberta, Canada, 2: 163–166.

Wos, H. 1996. Inheritance of head resistance of winter triticale to *Phaeospharia nodorum*. In *Triticale: Today and Tomorrow*, H. Guedes-Pinto, N. Darvey, and V.P. Carnide, Eds. Kluwer Academic Publ., Dordrecht, The Netherlands, pp. 535–539.

Wu, J. and Carver, B.F. 1999. Sprout damage and preharvest sprout resistance in hard white winter wheat. *Crop Sci.* 39: 441–447.

Ya-Ying, W., Ching-San, S., Ching-Chu, W., and Nan-Fen, C. 1973. The induction of pollen plantlets of Triticale and *Capsicum annum* from anther culture. *Science Sinica* 16: 147–151.

Zanetti, S., Winzeller, M., Keller, M., Keller, B., and Messmer, M. 2000. Genetic analysis of pre-harvest sprouting resistance in a wheat × spelt cross. *Crop Sci.* 40: 1406–1417.

Zeller, F.J. 1973. 1B/1R wheat-rye substitutions and translocations. In *Proceedings of the 4th International Wheat Genetics Symposium*, E.R. Sears and L.M.S. Sears, Eds., Columbia, MO, pp. 209–221.

Zillinsky, F.J. 1974. The triticale improvement program at CIMMYT. In *Triticale IDRC-024e, Proceedings of the International Symposium*, El Batan, Mexico, pp. 155–157.

Zillinsky, F.J. 1985. Triticale: an update on yield, adaptation and world production. In *Triticale*, CSSA Special Publication 9. CSSA, Madison, WI, pp. 1–7.

Zillinsky, F.J. and Borloug, N.E. 1971. Progress in developing triticale as an economic crop. *CIMMYT Res. Bull.* 17: 1–27.

Zohary, D. and Feldman, M. 1962. Hybridization between amphidiploids and the evolution of polyploids in the wheat (*Aegilops-Triticum*) group. *Evolution* 16: 44–61.

# Index

## A

Abiotic stresses, 80
    engineering tolerance in wheat breeding, 108–109
    germplasm development in maize, 180–182
    introgression for rice tolerance to, 136–137
    in triticale, 409–410
Acid soil tolerance, in maize, 182
Acrotrisomics, in barley, 265
Adultery, plant improvement facilitated by, 11
Advanced-backcross progenies, in rice, 133–134
African rice, introgression into Asian rice, 137
Alien addition lines
    monosomic in rice, 131
    oat-maize, 167–169
    rice monosomic, 130–131
    wheat-rye, 397
Alien gene transfer
    in bread wheat, 74–79
    cereal crop enrichment through, 18
    hybrid synthesis step in, 34–35
    hybridization with homologous genomes, 35–36
    impact from documented, 80–81
    impact from undocumented, 81–82
    mapping in durum wheat, 42–43
    molecular mapping in rice, 133, 140–142
    multiple combinations in durum wheat, 47–48
    strategies in rice, 132–133
    transfer into cultivated wheats, 33–34
Alien introgression, molecular characterization of, 142
Allopolyploid wheats
    conservation of intergenomic relatedness, 30–31
    evolutionary pathways, 29
    induced haploidy, 32–33
    mechanisms of diploidization, 31–32
    triticale, 396, 404
Allopolyploidy, 5, 7
    in polyploid wheats, 5–6
Alpha amylase activity (AAA), in triticale, 415
Aluminum toxicity tolerance
    molecular mapping for rice, 141–142
    in rye, 409
    in triticale, 409–410
Amino acid composition, sorghum variability for, 336–337
Amphiploidy, 5
Amplified fragment length polymorphisms (AFLPs), 101
    analysis in rice, 122

    in triticale, 413
Aneuploidy
    in bread wheat, 66
    and maize chromosome number manipulations, 163
    nomenclature in barley, 260
    in oats, 221–222
    in pearl millet, 286
    in rice, 130
Anther culture, in rice, 144–145
Apomixis
    in barley, 251
    cultivar development via, 299–300
    genetics of, 298
    incidence in pearl millet, 297–298
    induction via genetic engineering, 150
    in maize, 189
    mutagenesis-induced in rice, 149–150
    in pearl millet, 297–300
    for rice germplasm enhancement, 148–150
    screening wild rice germplasm for, 148–149
    in sorghum, 322–323
    transferring to pearl millet, 298–299
Apparent metabolizable energy (AME), triticale content, 416
Arid conditions, pearl millet hybrids for, 296–297
Asian rice, introgression from African rice into, 137
Asymmetric synthetic genomes, bread wheat, 67–68
Australian National Wheat Molecular Marker Program (NWMMP), 102

## B

B-line breeding, in pearl millet, 294–295
Backcrossing
    in bread wheat hybridization, 65
    meiotic analyses of, 78
Bacterial artificial chromosomes (BACs), 106
Bacterial blight
    gene introgression for rice resistance to, 136
    gene mapping for resistance in rice, 141
Banks grass mite resistance, in sorghum, 328
Barley, vii
    acrotrisomics in, 265
    apomixis in, 251
    chromosome 1, 260, 273
    chromosome 2, 260, 268
    chromosome 3, 261, 269
    chromosome 4, 261, 270

chromosome 5, 261, 267
chromosome 6, 261, 272
chromosome 7, 262, 271
chromosome deficiencies, 262
chromosome duplications, 262–263
chromosome evolution mechanisms, 275
chromosome image analysis in, 259–260
chromosome interchanges, 263
chromosome inversions, 263
chromosome mapping in, 257–258
conventional breeding, 244
cytogenetic manipulation, 16–17
cytogenetic mapping of, 262–263, 264–265
diploid species, 241
expressed sequence tags in, 276–277
gene pools, 243
genetic resources for improvement, 233–234
genetic transformation in, 249–251
genomic relationships, 241–243
geographical distribution, 236–238
germplasm collection centers, 239–240
germplasm enhancement, 243–251
germplasm resources, 235–240
haploid breeding, 244–245
hexaploid species, 242–243
historical origins, 234–235
hybrid breeding, 245–246
interspecific hybrids, 242
karyotype analysis in, 258–260
large-insert chromosome libraries, 275–276
molecular maps, 265–266
mutation breeding, 245
physical chromosome mapping, 270–273
primary gene pool, 243
primary trisomics, 264
relationships to rice, 275
secondary gene pool, 243
secondary trisomics, 265
self-pollination in, 17
somaclonal variation in, 249
somatic mitotic metaphase chromosomes, 259
taxonomy, 235
tertiary gene pool, 243
tertiary trisomics, 265
tetraploid breeding, 245
tetraploid species, 241–242
utility of genetic maps, 266–270
wide hybridization, 246–249
working germplasm depositary, 240
Barley yellow dwarf virus resistance
   in bread wheat, 80
   in oats, 219
   in triticale, 411–412
Base chromosome numbers
   changing through gene introgressions, 10
   pearl millet genome, 284
Beta-carotene, WHO guidelines, 338
Biotechnology, vii
   applications to sorghum improvement, 340–344
   triticale applications, 413–415
Biotic stresses, 80

in triticale, 410–412
Black mold resistance, in rye, 386
Blast resistance, gene mapping in rice, 141
BPH resistance, gene mapping in rice, 141
Bread-making quality, 83, 108
   rye, 366
   triticale, 399, 415
Bread wheat, vii
   1B1R translocation, 80
   cytogenetic manipulation and alien gene transfer, 74–79
   durum wheat as forerunner of, 6
   evolutionary origin of, 4, 7
   futuristic anticipation, 82–86
   gene pyramiding in, 84–85
   genetic diversity and distribution, 62–63
   genetic resources for improvement of, 61–62
   high pairing, 75
   impact from documented transfer, 80–81
   impact from undocumented transfer, 81–82
   intergeneric hybridization, 74
   intergenetic hybrids production, 63–64
   interspecific hybridization, 68–74
   molecular markers and genomics, 99
   origin and evolution scheme, 69
   primary gene pool, 63
   secondary gene pool, 63
   synthetic hexaploid, 70
   T1AL1RS translocation, 80
   tertiary gene pool, 63
   wide cross germplasm in, 79–80
Bridge crosses, in bread wheat, 69
Brown rust resistance
   in rye, 379, 386
   in triticale, 411
Bulbosum method, 16

### C

Carbohydrates, in cereal crops, 4
Cardiovascular disease, whole grains and prevention of, vii, 4
Cereal crops, vii
   climate conditions and, 5
   cytogenetic architecture, 1–4
   nutritional value, 4–5
   perspectives and challenges, 17–19
   as primary food source, 2
   total world production of, 2, 3
Cereal genomes, diversity of origin, 14–15
Cholesterol lowering, via high-glucan oats, 224
Chromosome breakage
   in bread wheat, 75
   in oats, 222–223
Chromosome doubling technique, 399
Chromosome engineering, 3
   durum wheat genome, 27–29
   in rice, vii
Chromosome evolution mechanisms, 275
Chromosome image analysis, in barley, 259–260
Chromosome libraries, barley, 275–276
Chromosome mapping, in barley, 257–258, 270–273

Chromosome number
  barley, 236–238
  manipulations in maize, 163
  pearl millet genome, 284–285
  rice, 119
  rye, 368
Chromosome pairing
  breeding implications, 10
  control mechanisms in polyploid grasses, 8
  gene introgressions and changes in base chromosome
        numbers, 10
  genetic control in oats, 210–211
  in pearl millet haploids, 285
  use in genome analysis, 9–10
    cytogenetic and evolutionary implications, 9–10
    genetic control of, 6–7, 7–8
Chromosome rearrangements
  ancient chromosome structural changes, 213–214
  modern chromosome structural changes, 214–215
  in oats, 211–215
  translocation detection, 213
Chromosome segment substitution lines, in rice, 137–138
Chromosome structural changes, in barley, 262–263
Chromosome translocations, 18
  from intergeneric bread wheat stocks, 77
  rice, 129
  whole-arm translocations in durum wheat, 39–41
CIMMYT
  See also International Maize and Wheat Improvement
        Center, vii
  maize gene bank, 171
  maize inbreds released by, 176
  synthetic hexaploid wheat potential at, 82
  triticale research program at, 399
CMS sources, introgression in rice, 136
Cold tolerance, 83
  in sorghum, 341
Comparative mapping
  in grass family, 273–275
  in wheat, 102–105
Conventional breeding, 3
  barley, 244
  integration with transgenics, 344
  in pearl millet, 301
  slowness of improvements through, 109
Copper, rye takeup of, 385
Corn leaf aphid resistance, in sorghum, 328
Crop pest resistance, in maize, 179
Crown rust resistance, in oats, 218, 219, 221
Cultivated rice. See Rice
Cytogenetic architecture
  maize, 161–162
  triticale, 400–404
Cytogenetic manipulation
  barley, 16–17
  in bread what, 85
  bread wheat, 74–79
  in diploid cereals, 15
  maize, 15, 159–161, 162–170
  in oat improvement, 199–200
  oats, 12

pearl millet, 16
  in pearl millet, 281–282
  of polyploid cereal crops, 11
  rice, 15, 128–132
  and rice germplasm enhancement, 115–118
  rye, 16–17
  sorghum, 16–17
  specificity of wheat chromosome pairing, 12
  triticale, 16–17
  wheat, 11–12
Cytogenetic mapping
  in barley, 262–263
  via chromosome numerical changes, 264–265
Cytogenetics, vii
  of diploid cereals, 12–14
Cytoplasmic male sterility
  in pearl millet hybridization, 290, 292–294
  in rye, 378
  and rye hybridization, 17
  in sorghum, 320–321, 345
  transfer in rice, 147
  in triticale, 413

D

D-genome chromosomal segments, 45–47
DANKO triticale breeding program, 406
Dietary protein
  in cereal crops, 4
  cereal crops as third world source of, vii,2
  maize breeding for, 15, 159–161
  in rye, 379
  sorghum trait-based breeding for, 328
Diploid cereals
  barley, 241
  cereal genome diversity and synteny, 14–15
  cytogenetic makeup and ancient polyploid origin,
        12–14
  cytogenetic manipulation and breeding in, 15–17
  genomic evolution, 12
  oats, 202–203
  rice, 149
  rye, 379
Direct gene transfer
  in bread wheat, 69, 71
  in durum wheat, 48
  in pearl millet, 300–301
Disease resistance
  bacterial blight in rice, 136
  engineering in wheat breeding, 108
  enhancing in maize, 177–178
  grassy stunt virus in rice, 134–135
  sorghum, 319–320
  in sorghum, 342–343, 345
  tungro disease in rice, 135–136
DNA marker technology, and sorghum improvement,
        340–344
DNA transfer, in rye, 377
Doubled haploidy (DH)
  in bread wheat breeding, 85
  triticale applications, 400

Downy mildew resistance, in sorghum, 343
Drought tolerance
　bread wheat, 82
　in maize, 180, 181
　in sorghum, 332–334, 341, 345
　sorghum breeding methods for, 324
　sorghum molecular breeding for, 340
　in triticale, 409
　via conventional wheat breeding, 108
Durum wheat, vii
　chromosome engineering in, 27–29
　direct gene transfer in, 48
　engineering with targeted introgressions, 39–48
　evolution of, 29–33, 30
　as forerunner of bread wheat, 6
　hybrid synthesis in, 34–35
　molecular markers and genomics, 99
　multiple combinations of alien segments, 47–48
　origin and evolution of, 69
　reducing targeted chromosomal region size, 43–44
　transfer of alien genes into, 33–48
　transfer of chromosomal segments, 41–47
　transfers containing D-genome chromosomal segments, 45–47
　value of mapping alien recombinant chromosomes, 43–43
　whole-arm translocations in, 39–41
　wild relatives as sources of desirable genes, 33

**E**

Embryo rescue, in rice, 134
Emmer, 6
Ergot of rye, 386
Erosion control, use of rye for, 384
ESTs
　in barley, 276–277
　mapping on wheat chromosome bins, 104
European corn borer (ECB) resistance, in maize, 171
Evolutionary pathways, allopolyploid wheats, 29–33
Expressed sequence tags (ESTs). *See* ESTs

**F**

Fall armyworm resistance, in sorghum, 328
Fluorescence in situ hybridization (FISH)
　analysis in rice, 130
　in bread wheat hybridization, 77, 79
　durum wheat, 31
　maize chromosome hybrid lines, 170
　pearl millet apomictic backcross lines, 299
　in transgenic barley plants, 250
Fodder yield
　pearl millet intraspecific hybridization for, 289–291
　sorghum trait-based breeding for, 326–328
　triticale, 416
Fountain grass, hybridization with pearl millet, 288
Frost tolerance, in triticale, 409
Functional genomics, 19
　in rice, 152
Fungal disease resistance, in triticale, 411

Fusarium head blight (FHB), 108
　in rye, 386
　triticale resistance to, 410

**G**

Gene introgressions, and changes in base chromosome numbers, 10
Gene pools
　barley, 243
　oats, 215
Gene pyramiding, 101
　in bread wheat, 84–85
Gene recombination, in bread wheat hybridization, 81
Genetic diversity
　bread wheat, 62–63
　as prerequisite for durability of resistance, 62
　research in maize, 181–182
　and rice genetic erosion, 123–124
　sorghum, 323–324
　in sorghum, 318–319
　in triticale, 417–419
Genetic engineering, 18
　induction of apomixis via, 150
　in maize, 171
　in wheat, 99–100
Genetic enhancement, maize, 159–161
Genetic erosion, rice, 123–124
Genetic male sterility
　sorghum, 320
　in sorghum, 345
Genetic maps. *See also* Molecular mapping
　utility in barley, 266–267
Genetic resources, vii
　for barley improvement, 233–234
　for bread wheat improvement, 61–62
　conserving, viii
　maintenance of sorghum, 315–316
　rice, 123–125
　sorghum, 313–318
Genetic transformation
　in barley, 249–251
　in sorghum, 344
Genome analysis
　usefulness of chromosome pairing in, 9–10
　wheat, 100
Genome size, rice, 15
Genome splitting, 67
Genomic evolution, diploid/diploidized cereals, 12–17
Genomic in situ hybridization (GISH), 44, 144
　characterization of rice parental genomes, MAALs, and homeologous pairing, 142–144
　oat-maize hybrids, 168
　wide cross progeny characterization in rice, 133
Genomic reconstruction, induced haploidy in, 32–33
Germplasm
　barley, 235–240
　characterizing in maize, 173
　collection and conservation, vii
　conservation in rice, 124
　exploration and collection of rice, 123

need for free technology sharing, 88
  pearl millet, 283–284
  wide cross in bread wheat, 79–80
Germplasm conservation, in maize, 171–173
Germplasm enhancement
  apomixis for, 148–150
  in barley, 243–251
  in rice via cytogenetic manipulation, 115–118, 132–144
  through rice cell and tissue culture, 144–147
  via induced mutations in rice, 147–148
Golden rice
  comparison with golden millet, 301
  vitamin A and iron content of, 18
Grain mold resistance, in sorghum, 330, 342–343
Grain quality improvement, in wheat breeding, 108
Grain shriveling, in triticale, 407
Grains, as world food source, vii
Grass family, comparative mapping in, 273–275
Grassy stunt virus, gene introgression in rice, 134–135
Green bug resistance, in sorghum, 328
Green Revolution, vii,3, 17

H

Haploid breeding
  barley, 244–245
  rice, 130
  rye, 373
  triticale, 414–415
Head scab resistance
  in bread wheat hybrids, 76, 80, 82
  sorghum breeding for, 328–329
Heat tolerance, sorghum molecular breeding for, 340
Heterosis breeding
  in maize, 15, 161
  in pearl millet, 15, 281–282, 289–291, 297–300
  in triticale, 411–412
  triticale as elegant solution in, 396
Heterotic patterns, in maize breeding, 173
Hexaploidy
  in barley, 242–243
  in oats, 205
High-glucan content
  in oats, 219
  strategies in oat breeding, 224
Historical origins
  barley, 234–235
  bread wheat, 4, 7
  cultivated rice, 119
  durum wheat, 69
  oats, 201–202
  pearl millet, 283–284
  rye, 366, 375–376
  sorghum, 312
  triticale, 398–400
Homeologous pairing
  characterization through GISH, 142–144
  as key to gene transfer, 10and durum wheat
    hybridization efforts, 36–37
  suppression in wheat, 31, 100

Homologous pairing, and hybridization in durum wheat, 35–36
Host plant resistance (HPR), in sorghum, 345
Hybrid breeding
  in barley, 245–246
  in rye, 381–382
  in triticale, 412–413
Hybrid parent development, in pearl millet, 292–296
Hybrid parent maintenance, in pearl millet, 297
Hybrid production, in rice, 147
Hybrid synthesis
  in alien gene transfer into wheat, 34–35
  with durum wheat mutants lacking Ph1, 37–39
  and Ph1 regulation in homeologous pairing, 36–37
  in rice, 133
Hybrid vigor, in pearl millet, 289–291

I

ICARDA. See International Center for Agricultural
    Research in the Dry Areas (ICARDA)
ICRISAT
  midge resistance program for sorghum, 329
  pearl millet hybridization program at, 290, 291
  research in sorghum drought tolerance, 332
  sorghum breeding for soil acidity tolerance, 334
  sorghum projects, 313–314, 340, 346
Inbred progenitors, maize, 174
Induced haploidy, in wheat, 32–33
Induced mutations, and germplasm enhancement in rice, 147–148
Induced polyploidy, in pearl millet, 286
Induced variation, in oats, 222–223
Insect resistance
  in maize development, 179–180
  in sorghum, 343–344
  sorghum breeding methods for, 324–325
Institute for Agricultural Research (IAR), sorghum
    improvement projects, 340
Intergeneric hybridization, 100
  bread wheat, 74
  in rice, 146
  as source for what-alien chromosome translocations, 77
Intergenetic bread wheat hybrids
  asymmetric synthetic genomes, 67–68
  cytology, 65–66
  maintenance of, 66–67
  phenology of, 65
  production in bread wheat, 63–64
International bread wheat screening nursery (IBWSN), 83
International Center for Agricultural Research in the Dry
    Areas (ICARDA), vii
International Maize and Wheat Improvement Center
    (CIMMYT), vii
International Rice Gene Bank (IRG), 124
International Triticale EST Cooperative, 102–103
International Triticale Mapping Initiative (ITMI), 72
International What Genetic Symposia, 76
Interspecies hexaploid hybrid, in oats, 209
Interspecific hybridization, 100
  barley, 242

bread wheat, 68–74
in oats, 201–210
in pearl millet, 286–288
in rice, 133–134
for superior fodder traits in pearl millet, 2898
Intraspecific hybrids, in pearl millet, 289–297
Introgressions, 10. *See also* Gene introgressions
from AA genome wild rice species, 134
for abiotic stress tolerance in rice, 136–137
from African rice to Asian rice, 137
from distant rice genomes, 142
durum wheat genome engineering with, 39–48
evaluation for transfer of target traits, 133
genes from distantly related rice genomes, 138–140
molecular characterization of, 133
in oats, 215
in rice, 133
in rye, 376
for tungro disease resistance in rice, 135–136
yield-enhancing loci in rice, 138
Ionizing radiation, and induced variation in oats, 222–223
Iron
rye takeup of, 385
World Health Organization guidelines, 338

**K**

Karyotype analysis
in barley, 258–262
rye, 370–372

**L**

Leaf blotch resistance, in rye, 386
Low nutrient tolerance, in triticale, 409
Lysine content, in sorghum, 337–338, 345

**M**

MAALs, 144
GISH characterization of, 142–144
production in rice, 133–134
Maize, vii,13
acid soil tolerance in, 182
apomixis in, 189
B-A translocations, 165–166
breeding for morphological traits, 172
breeding for nutritional quality, 182–189
breeding for stress tolerance, 177–182
characterizing and using inbred progenitors, 175
chromosomal manipulations in oat background,
167–170
chromosomal rearrangement manipulations, 163
crop pest resistance in, 179
crossbred performance, 174
crossing triticale with, 414
cytogenetic architecture, 161–162
cytogenetic manipulation and breeding, 15, 162–170
developing potentially useful breeding/germplasm
methodologies, 173
development of inbred progenitors, 174

development of QPM donor stocks, 184–185
development of soft opaques, 183
drought-tolerant germplasm, 181
early breeding work, 171–173
early-maturity germplasm in, 172
enhancing disease resistance in, 177–178
expanded QPM development efforts, 185–186
First-generation mutant problems, 183–184
FISH analysis of hybrid lines, 170
genetic engineering in, 171
genetic enhancement of, 159–161, 170–171
genetic modifiers, 184
germplasm conservation in, 171–173
germplasm development for abiotic stresses, 180–182
hybrid development ant testing, 175
hybrid options, 174–175
hybrid-oriented source germplasm and hybrid
development, 173–177
hybridization with oats, 220–221
improving nutritional quality, 173
inbred line evaluation nurseries, 175
inbreeding tolerance, 174
insect resistant germplasm in, 179–180
inversions, 166–167
mitotic metaphase chromosome spread of, 161
ploidy level manipulations of, 162
reciprocal translocations in, 164–165
reducing plant height, 172
shifts in recurrent selection procedures, 173–174
subchromosome fragment stocks, 169–170
value-added traits in, 170–171
waterlogging-tolerant germplasm, 181–182
Maize inbreds, release of, 176–177
Maize testers research, 176
Map-based cloning, in wheat, 106–107
Marker-assisted selection (MAS), 101
Meiotic behavior, in wheat-barley addition lines, 248
Micronutrient uptake, in rye, 385
Micronutrients
in rye, 385–386
in SH bread wheat hybrids, 83
WHO malnutrition guidelines, 338
Midge resistance, in sorghum, 329, 343–344, 345
Mildew resistance, in wheat, 108
Millet. *See* Pearl millet
Molecular breeding, in sorghum, 340
Molecular mapping
in barley, 265–266
barley chromosomes, 267, 269, 270
of introgressed alien genes in rice, 140–142
in triticale, 413–414
Molecular markers
and pest/disease resistance in wheat, 18
in wheat, 99–100, 100–102
Monosomic alien addition lines, in rice, 130–131
Mutagenesis-induced apomixis, in rice, 149–150
Mutation breeding, barley, 245

**N**

Napier grass, hybridization with pearl millet, 286

National Seed Storage Laboratory (NSSL), 125
Nematode resistance, in rye, 386
Nutritional quality
    maize breeding for, 173, 182–189
    sorghum breeding for increased, 336

## O

Oat-maize addition lines, 167–169, 220–221, 224
    subarm aneuploid development, 221–222
Oats, taxonomy, vii, 202
    ancient chromosome structural changes, 213–214
    chromosome rearrangements in, 211–215
    crosses with maize, 167
    crown rust resistance in, 218, 219
    cytogenetic makeup of hexaploid, 7
    cytogenetic manipulation of, 12, 199–200
    detection of translocations, 213
    diploid diploid hybrids, 206–207
    diploid-hexaploid hybrids, 209
    diploids, 202–203
    duplicate deficient segments and crown rust resistance, 221
    effect of cross direction and cytoplasm donor, 220
    expanding crop into new regions, 224
    future value-added traits, 223–224
    gene pools, 215–221
    genetic control of chromosome pairing, 210–211
    genus introduction, 201–202
    hexaploids, 205
    high-glucan content in, 219
    history of, 200–201
    induced variation, 222–223
    interspecies hexaploid hybrids, 209
    interspecific hybridization, 206–210
    ionizing radiation for chromosome breakage, 222–223
    modern chromosome structural changes, 214–215
    pentaploid hybrids, 209–210
    primary gene pool introgressions, 215–216
    secondary gene pool introgressions, 216
    speciation, 201–205
    stem rust resistance in, 219
    tertiary gene pool introgressions, 216–219, 219
    tetraploid-tetraploid hybrids, 207–208
    tetraploids, 264–265
    translocation heterozygotes, 222
    triploid hybrids, 208–209
Organelle recombinants, in rice, 147
Oriental grass hybrids, with pearl millet, 286–287

## P

*P. schweinfurthil,* hybridization with pearl millet, 288
*P. squamulatum,* hybridization with pearl millet, 288
Paracentric inversions, maize, 166–167
Parental genome characterization, in wide hybrids of rice, 143
PCR-based markers, 101
Pearl millet, vii
    apomixis in, 297
    B-line breeding in, 294–295

base chromosome number, 284
chromosome number, 284–285
chromosome pairing in haploids, 285
cytoplasmic male sterility in, 292–294
direct gene transfer in, 300–301
fodder trait breeding of, 289
fountain grass hybridization with, 288
gene transference to, 298–299
genetic improvement for grain and forage production, 281–282
hybrid development and testing, 296
hybrid options, 291–292
hybrid parent development, 292–296
hybrid parent maintenance, 297
hybrids for arid conditions, 296–297
incidence of apomixis in, 297–298
interpopulation hybrids, 291
interspecific hybrids, 286–288, 289
as major source of dietary protein in Africa/Asia, 301
napier grass hybrids with, 286
oriental grass hybrids, 286–287
*p. schweinfurthil* hybridization with, 288
*p. squamulatum* hybridization with, 288
perennial relatives in secondary gene pool, 283–284
as poor man's crop, 282
primary gene pool, 283
ranking in world grain production, 282
as research organism, 282–283
restorer parent development in, 295–296
as secondarily balanced species, 13
seed parent development in, 295
single-cross hybrids, 291
tertiary gene pool, 284
three-way hybrids, 291
top-cross hybrids, 291
wild relatives in primary gene pool, 283
Pentaploid hybrids, in oats, 209–210
Pericentric inversions, maize, 166–167
Pest resistance
    engineering in wheat breeding, 108
    in maize, 171
    in rye, 386
*Ph1* regulation, 6
    durum wheat mutations lacking, 37–39
    suppression of homeologous chromosome pairing by, 9–10
    and suppression of homeologous pairing in wheat, 31
    in wheat, 100
    and wheat hybridization with homeologous genomes, 36–37
Photoperiod response, in sorghum breeding, 342
Plant height reductions, maize, 172
Ploidy levels
    maize manipulations of, 162
    in wheat, 31–32
Pollen germination assay, maize, 165
Polyphenols, genetic manipulation in sorghum, 338
Polyploid cereals
    cytogenetic architecture, 5
    cytogenetic makeup of hexaploid oat, 7
    cytogenetic manipulation, 11–12

durum wheat, 6
evolutionary origin of bread wheat, 7
genetic control of chromosome pairing, 6–7
pearl millet, 286
polyploid wheats, 5–6
triticale as model of evolutionary study, 396
Polyploid wheat, and evolution by allopolyploidy, 5–6
Polyploidy, as dominant factor in plant speciation, 5
Powdery mildew resistance
in rye, 379, 386
in SH wheat hybrids, 83
in triticale, 411
Preharvest sprouting, in triticale, 408–409
Primary gene pool
in barley, 243
bread wheat, 63
in oats, 215–216
pearl millet, 283
Primary trisomics
in barley, 264
rice, 130
Productivity enhancement
sorghum breeding for, 325–326
*vs.* single trait improvement, 80
Protein content
genetics of, 337
sorghum variability for, 336–337
Protoplast fusion, intergeneric somatic rice hybrids from, 146

**Q**

Quality protein maize (QPM), 183
donor stock development, 184–185
expanded development efforts, 185–186
gene pools and populations, 187
germplasm performance, 186
hybrid development and testing, 186–189
prominent varieties, 188
recent releases, 188
superior tropical hybrids, 187

**R**

Radical wheat breeding, 49
Randomly amplified polymorphic DNA (RAPD), 413
Reciprocal crossing, in bread wheat, 74
Reciprocal translocations, in maize, 164–165
Red oat, 7
Restorer parent development
in pearl millet, 295–296
in rye, 378
in triticale, 413
Restriction fragment length polymorphisms (RFLPs), 101
rice analysis via, 122
Rice, vii
anther culture in, 144–145
chromosome number, 119
chromosome segment substitution lines in, 137–138
comparative genetic maps, 131–132
cytogenetic manipulation, 15

cytogenetics, 128–129
development of doubled haploids, 137
diploid species, 149
ex situ germplasm conservation, 124–125
exploration/conservation of genetic resources, 123–125
functional genomics, 152
genetic engineering to induce apomixis, 150
genetic enhancement via transformation, 150–151
genetic erosion, 123–124
genomic composition, 119
genomic relationships, 127–128
germplasm conservation, 124
germplasm enhancement via cell and tissue culture, 144–147
germplasm enhancement via cytogenetic manipulation, 115–118
germplasm enhancement via induced mutations, 147–148
haploids, Tripoli's, aneuploids, 130
homology with wheat, 105
hybrid production in, 147
identification and introgression of yield-enhancing species alleles, 138
introgression from distantly related genomes, 138–140
as model cereal crop, 15, 118, 274
molecular approaches to genomics, 128
molecular characterization of alien introgression, 142
molecular mapping of introgressed alien genes, 140–142
*O. bracyantha,* 127
*O. meyeriana* complex, 126–127
*O. officinalis* complex, 126
*O. ridleyi* complex, 126
*O. sativa* complex, 125–126
*O. schlechteri,* 127
organelle recombinants in, 147
origin of cultivated, 119
pachytene karyotype, 129
polyphyletic origin of, 121–123
production, area, and productivity, 117
related genera, 127
relationships to barley, 275
as secondarily balanced species, 14
seed protein analysis, 128
in situ germplasm conservation, 125
small genome size, 274
somaclonal variation in, 145
somatic cell hybridization in, 146–147
somatic karyotype, 129
species distribution, 119
strategies for alien gene transfer, 132–133
taxonomy, 125–128
tetraploid species, 149
transfer of cytoplasmic male sterility in, 147
translocations, 129
trisomics, 130–132
unknown complex, 127
wild progenitors of cultivated, 118–121
Rockefeller Foundation, support for triticale breeding, 399
Rye, vii, 365–366
alien introgression, 376

allelopathic effects, 387
aluminum toxicity tolerance, 409
B chromosomes in, 370
botany of, 367–368
breeding, 378–384
chromosomal and regional localization of genes and markers, 378
chromosome additions, 388
chromosome number, 368
chromosome substitutions, 388
chromosome translocations, 388
as closest relative to wheat, 397–398
cropping, 384–387
cross-pollination in, 381
crossing of hexaploid wheat with, 376
cytogenetic manipulation, 16–17
cytological differences between species, 374
cytology, 368–377
cytoplasmic male sterility in, 378
cytotaxonomy, 375–376
decreasing acreage of, 366
diploid breeding, 379
DNA transfer, 377
as donor for genetic wheat improvement, 387–388
erosion control with, 384
fertilization, 386
gene designation, 377
genetics, 377–378
genome additions, 387–388
gross morphology of, 367
growth regulators, 387
haploid, 373
homeology, 374
hybrid breeding, 381–382
inbreeding depression in, 379
incorporation in crop rotation, 387
integrated breeding program schematic, 380
karyotype, 370–372
micronutrient content, 385–386
molecular structure of genome, 368–369
neocentric activity in, 373
origins, 375–376
physiology of, 368
polygenic control of chromosome pairing in, 401
population breeding, 379
primary trisomics and telotrisomics, 369–370
proposals for chromosome designations, 371
reciprocal translations, 372–373
restorer lines, 378
root system, 367–368
seed production, 382
seeding, 385
seeds, 368
selection procedures in self-fertile, 384
self-incompatibility mechanism, 381
soil type, 385
somatic spread of diploid, 372
susceptibility and resistance in, 386
synthetics, 380–381
taxonomy, 375–376
temperature conditions, 386

tetraploid, 382–384
unpredictable gene expression in wheat background, 404
variation of chromosome length, 371
volunteering in, 387
water and, 385
wild perennial forms, 375
winter hardiness in, 365, 386
world production rankings, 367

## S

Salt tolerance
    in bread what hybrids, 76, 80, 82
    sorghum breeding for, 335
    in triticale, 409
Saponin production, in oat breeding, 223
Secondary gene pool
    barley, 243
    bread wheat, 63
    introgressions in oats, 216
    pearl millet, 283–284
Secondary trisomics
    in barley, 265
    in rice, 130
Seed fertility, in wheat-barley addition lines, 249
Seed parent development, in pearl millet, 295
Segmental triticale, 418
Shoot fly resistance, in sorghum, 343
Simple sequence repeat (SSR), in triticale, 413
Simply inherited traits, importance of emphasis on, 87
Single trait improvement, vs. productivity gains, 80
Soft opaques
    development in maize, 183
    genetic variation for kernel modification, 184
Soil acidity tolerance, in triticale, 409
Soil chemical toxicity tolerance, in sorghum, 334–336
Soluble arabinoxilans (SAX), 416
Soluble dietary fiber (SDF), in triticale, 416
Somaclonal variation
    in barley, 249
    in rice, 145enhancing alien introgression via, 145–146
Somatic cell hybridization, in rice, 146–147
Sorghum, vii, 14
    adaptation and productivity enhancement, 325–326
    African genetic resources, 315
    amino acid composition variability, 336–337
    aneuploid variation, 322
    for animal fodder, 311
    apomixis in, 322–323
    biochemical basis of resistance to salinity, 335–336
    biochemical variability, 319
    biotechnology improvements and, 340–344
    breeding behavior, 313
    breeding concepts, 324–325
    breeding for increased nutritional quality, 336–338
    breeding for insect resistance, 328–329
    breeding for soil salinity tolerance, 335
    breeding materials, 325
    china landrace collection, 315
    cold tolerance, 341

conversion programs, 323–324
core collection, 316
crop improvement, 309–311, 323–329
cytogenetic manipulation, 16–17, 309–311, 322–323
cytoplasmic-nuclear male sterility, 321
disease resistance, 329–330, 342–343
DNA marker technology for, 340–344
DNA variability, 319
domestication history, 312–313
downy mildew resistance in, 343
drought tolerance, 332–334, 341
environmental response characteristics, 323
euploid variation, 322
evaluation, characterization, documentation of,
    316–317
factors affecting global production, 311
farmers' participatory approach to crop improvement,
    339–340
genetic diversity efforts, 323–324
genetic male sterility, 320, 325
genetic resources, 309–311, 313–318
genetic transformation technology for, 344
genetic variability, 318–319
genetics and cytogenetics, 319–322
genetics of protein content and digestibility, 337
grain mold resistance, 342–343
grain yield component traits, 341
high protein digestibility and lysine content, 337–338
ICRISAT projects, 313–314, 328
increased micronutrient density in, 338–339
inheritance and breeding for soil acidity tolerance,
    334–335
inheritance and breeding for soil salinity tolerance, 335
insect resistance traits, 343–344
karyotype, 322
linkage map construction, 340
maintenance of genetic resources, 315–316
male sterility, 320–321
midge resistance, 343–344
morphological/phenotypic variability, 318–319
morphological trait genetics, 319–320
need for resource conservation, 313–314
new tools for improvement, 339–344
origin, 312
pedigree breeding program, 324
photoperiod response and, 342
pollination control, 313
protein content genetic variability, 336–337
races, 312
ranking in grain production, 310, 345
resistant traits, 319–320
resource utilization, 317–318
restorer lines, 324
selected trait range of variation, 317
shoot fly resistance, 343
soil chemical toxicity tolerance, 334–336
spotted stem borer resistance in, 344
stay-green resistance, 332–333
striga breeding strategies, 331–332, 340, 341–342
striga resistance in, 330–331
striga screening techniques, 331

tannin content genetic manipulation, 338
taxonomy, 311–312
trait-based breeding, 326–339
transgenics and conventional breeding integrated
    technology, 344
USDA collection project, 315
use in potable alcohol, 311
Species distribution, rice, 119
Spontaneous Robertsonian translocation, 80
Spot blotch resistance
    in bread wheat hybrids, 82, 83
    in rye, 386
Spotted stem borer resistance, 345
    in sorghum, 328, 344
Sprouting tolerance, 83
Stay-green resistance, in sorghum, 332–333
Stem rust resistance, in oats, 219
Stored grain pest resistance, in maize, 179–180
Stress resistance, 80
    in bread wheat hybrids, 76
    maize breeding for, 159–161, 177–182
Stress screening, synthetic hexaploid wheats, 73
Striga, 330–331, 340, 345
    screening techniques in sorghum, 331
    sorghum breeding strategies for resistance to, 331–332,
        341–342
Stripe rust resistance, in bread wheat hybrids, 81
Subarm aneuploids, development in oats, 221–222
Sugarcane aphid resistance, in sorghum, 328
Syngenta, 171
Syntenous rice, molecular and morphological maps, 267,
    268, 269, 270, 271, 272, 273
Synteny, 14–15, 103
    in crop cereals, 19
Synthetic hexaploid wheats, 70
    production of, 72–73
    resistances in, 83

**T**

Tall fescue, 8
Tan spot resistance, 83
Tannins, genetic manipulation in sorghum, 338
Target region amplification polymorphism (TRAP)
    technique, 101
Target traits, useful genetic variability in rice, 132–133
Telotrisomics
    in rice, 130
    in rye, 365–370
Tertiary gene pool
    barley, 243
    bread wheat, 63
    introgressions in oats, 216–219, 219
    pearl millet, 284
Tertiary trisomics, in barley, 265
Tetraploid breeding
    barley, 245
    in rye, 382–384
    triticale, 402–404
Tetraploid species
    barley, 241–242

oats, 204–205
rice, 149
Tifleaf 3, 293
Trait-based breeding
  disease resistance, 329–330
  forage and feed, 326–328
  grain yield and adaptation, 326
  increased micronutrient density, 338–339
  increased nutritional quality, 336–338
  insect resistance, 328–329
  soil chemical toxicity tolerance, 334–336
  striga resistance, 320–322
Transgenic technology, 18
  integration with conventional breeding in sorghum, 344
  rice examples with agronomically important genes, 151
  transgenic wheat production, 107
Triploid hybrids, oats, 208–209
Triploids, rice, 130
Trisomics
  in rice, 130–132
  in rye, 369–370
Triticale, vii, 17, 395–397
  abiotic stress, 409–410
  animal fodder use, 416
  barley yellow dwarf virus resistance in, 411–412
  biotechnology in, 413–415
  biotic stresses, 410–412
  bread-making quality of octoploid, 399
  breeding programs, 406–407
  brown rust resistance in, 411
  crosses with rye, 418–419
  crosses with wheat, 418
  cytogenetic architecture, 400–404
  cytogenetic manipulation, 16–17
  cytological instabilities, 417
  DANKO breeding program, 406
  detrimental heterozygosity in rye genome for, 205
  fermentation and storage problems, 415
  food use, 415–416
  forage use, 416–417
  fungal disease resistance in, 411
  Fusarium resistance in, 410
  gene bank, 398
  genetics, 404–406
  gluten content comparison with wheat, 415
  grain shriveling in, 407
  haploidy in, 414–415
  as healthiest cereal, 410
  hexaploid stability, 400
  historical origins, 398–400
  hybrid breeding, 412–413
  international testing program, 399
  lodging of, 408
  as man-made crop, 396
  meiotic behavior of rye chromosomes in F1 hybrids, 402
  molecular markers/mapping in, 413–414
  powdery mildew resistance in, 411
  preharvest sprouting in, 408–409
  quality, 415
  Rockefeller Foundation support for breeding, 399
  segmental, 418
  sources of genetic variation, 417–418
  tetraploid, 402–404
  and wheat-rye galaxy, 397–398
  winter hardiness, 407
  yield, 407
Tungro disease resistance
  gene mapping in rice, 141
  in rice, 135–136

V

Value-added traits
  genetic transformation in maize, 170–171
  golden rice example, 18
  in oats, 223–224
Viscosity, wheat vs. triticale, 416

W

Waterlogging tolerance, in maize, 181–182
Wheat. See also Allopolyploid wheats; Bread wheat;
    Durum wheat
  alien integration into, 12
  cytogenetic manipulation of, 11–12
  genomic makeup, 99–100, 102–107
  homology with rice, 105
  molecular markers in, 18, 99–100, 100–102
  pest-disease resistance in, 18
  Ph1 regulation of chromosome pairing in, 11
  rye genome transfer into, 387–388
  specificity of chromosome pairing, 12
  subarm deletion lines in, 221
  vernalization response, 106
  wheat transformation and application in breeding, 107–109
Wheat-barley addition lines, 246–249
Wheat ESTs, 102–105
Wheat genomics, 102
  map-based cloning, 106–107
  wheat ESTs and comparative mapping, 102–105
Wheat-rye hybrids, 62, 397–398
Wheat transformation, 107
  and abiotic stress tolerance, 108–109
  engineering insect pests and disease resistance, 108
  and grain quality improvement, 108
  transgenic wheat production, 107
Whole grains, preventive effect in cardiovascular disease, vii
Wide cross germplasm, 87
  in bread wheat, 79–80, 85
  characterization in rice using GISH, 153
  need for long-term commitment to, 88
  in rice, 134
Wide hybridization
  in barley, 246
  wheat-barley addition lines, 246–249
Winter hardiness
  in rye, 366, 386
  triticale, 407

# Y

Yellow dwarf virus resistance, in bread wheat, 80
Yellow rust resistance, in rye, 386
Yellow sugarcane aphid resistance, in sorghum, 328
Yields
    improvement by genetic engineering, vii
    increases since 1930, 18

and intraspecific hybridization in pearl millet,
    289–297
maize breeding for, 159–161, 181
in pearl millet hybridization, 293, 294
sorghum breeding methods for maximizing, 324, 326,
    341
triticale, 407
wheat, rye, and triticale, 396

Printed and bound by CPI Group (UK) Ltd, Croydon, CR0 4YY

29/10/2024

01780550-0001